Power Systems

Electrical power has been the technological foundation of industrial societies for many years. Although the systems designed to provide and apply electrical energy have reached a high degree of maturity, unforeseen problems are constantly encountered, necessitating the design of more efficient and reliable systems based on novel technologies. The book series Power Systems is aimed at providing detailed, accurate and sound technical information about these new developments in electrical power engineering. It includes topics on power generation, storage and transmission as well as electrical machines. The monographs and advanced textbooks in this series address researchers, lecturers, industrial engineers and senior students in electrical engineering. ** Power Systems is indexed in Scopus**

More information about this series at http://www.springer.com/series/4622

Mohammad Kiani-Moghaddam
Mojtaba Shivaie • Philip D. Weinsier

Modern Music-Inspired Optimization Algorithms for Electric Power Systems

Modeling, Analysis and Practice

Springer

Mohammad Kiani-Moghaddam
Department of Electrical Engineering
Shahid Beheshti University
Tehran, Iran

Mojtaba Shivaie
Faculty of Electrical Engineering and Robotic
Shahrood University of Technology
Shahrood, Iran

Philip D. Weinsier
Department of Applied Electrical Engineering
Bowling Green State University Firelands
Huron, OH, USA

ISSN 1612-1287 ISSN 1860-4676 (electronic)
Power Systems
ISBN 978-3-030-12043-6 ISBN 978-3-030-12044-3 (eBook)
https://doi.org/10.1007/978-3-030-12044-3

Library of Congress Control Number: 2019931017

This Springer imprint is published by the registered company Springer Nature Switzerland AG
The registered company address is: Gewerbestrasse 11, 6330 Cham, Switzerland

To my wife for her continued support and whose love sustains me.
To my parents for their unconditional love that has nurtured me.
And to those trying to bring nations together through science.

Foreword

Power systems optimization has received a great deal of attention in recent years. Even from the 1960s, when the foundation of classical power systems optimization was laid, it has encompassed a variety of complex power system problems including cost minimization, economic load dispatch, and unit commitment. Initially, mostly convex problems were tackled using gradient and quadratic optimization techniques. In addition, graphing techniques and linear and dynamic programming were employed to solve unconventional problems like security-constrained unit commitment. In the earlier years, computing speed was the main obstacle significantly limiting the scope of the investigation and the complexity of the system model. The evolution of computers, however, has removed many of the previous obstacles, thereby enabling the development of more complex and comprehensive system models and expanding the solution domain. In particular, the evolution of metaheuristic optimization algorithms enables the investigation of real-world, large-scale, non-convex, multi-objective problems including mixed programming. This covers a very wide variety of areas from power systems planning and operations to power quality applications. Examples include, but are not limited to, generation, transmission and sub-transmission expansion design, smart grid and micro-grid design optimization, renewables and energy storage systems design, load forecasting, unit commitment, bidding mechanism, competitive electric energy markets, demand-side management and energy savings, deployment of intelligent protection devices, and optimum deployment of harmonic power filters.

Sophisticated methods and algorithms inspired by nature have flourished in recent years and have demonstrated both effectiveness and robustness in dealing with very complex power system problems. Among them are genetic algorithms, swarm intelligence, and music-inspired algorithms. Since 2001, when Geem introduced the harmony search algorithm, music-inspired methods have been used in many areas including industry, signal and image processing, and power systems. The latter encompasses a variety of optimization problems such as cost minimization and static/dynamic economic load dispatch.

This book is a timely and important work in the field of power systems optimization that can benefit both the practicing engineer and the graduate student endeavoring in the field. From the first step, it provides a comprehensive introduction to traditional music-inspired algorithms, followed by several innovative architectures with the aim of overcoming the shortcomings of the traditional algorithms. This book also elaborates on a number of common and uncommon power system problems, ranging from power systems operation problems such as a bilateral bidding mechanism and competitive electric markets; power systems planning problems such as generation expansion planning, transmission expansion planning, coordination of generation and transmission expansion planning; distribution expansion planning; and harmonic power filter planning problems such as passive, active, and hybrid harmonic power filter planning.

Innovative models accounting for the stochastic nature of power systems are introduced and include uncertainty parameters and incomplete information. This is significant for modern power systems operation and planning, particularly with the integration of distributed generation and renewable resources into the network. The book also gives appropriate significance to risk as an important supplement to planning. With this aim in mind, the information-gap decision theory and the efficient two-point estimate method are widely employed to handle severe uncertainties in the solution of power systems operation and planning problems and also harmonic power filter planning problems, respectively. The book provides a very good treatment of this topic. Finally, a variety of test systems are provided.

The discipline of power systems optimization has come a long way since the days of gradient methods, and the tools of modern engineers have expanded in both scope and breadth. The work that this book is based on represents an important milestone in the ever-expanding and increasingly demanding field of power systems optimization.

Department of Electrical and Computer Engineering, Southern Illinois University, Carbondale, IL, USA
November 2018

Constantine J. Hatziadoniu

Preface

In today's world, the optimization process has become a commonly used concept across different branches of science, such as engineering, economics, management, operations research, and so on. The main purpose of this process is to effectively decrease wasted time and resources, design mistakes, and unnecessary costs in such a way that the predetermined objectives and constraints pertaining to optimization are still met with the highest level of satisfaction. On the one hand, the lack of complexity in the scientific challenges reported in the form of optimization problems and, on the other hand, the lack of access to powerful computing techniques for solving these optimization problems have given rise to the fact that optimization problems had a relatively simple architecture at the beginning of their development. By improving the efficiency of computing techniques and the emergence of a new generation of techniques with extraordinary powers, however, this promise has been given to the researchers and specialists who can put aside simplifications performed in scientific challenges, due to the lack of powerful computing techniques. In addition, the advent of new concepts in different branches of the sciences has forced researchers and specialists to move toward the definition of scientific challenges with more realistic characteristics during the last decade. As a result, researchers and specialists have encountered a new generation of optimization problems with new complexities, such as mixed-integer decision-making variables, multiple conflicting and heterogeneous objective functions, and non-convex, non-smooth, and nonlinear relationships.

At the same time, and with the development of optimization problems with more realistic characteristics, researchers and specialists have addressed a wide range of optimization algorithms in order to deal with these optimization problems. Meta-heuristic optimization algorithms are a new class of optimization algorithms, widely employed as a well-established optimization technique across different branches of science and owing to their unique characteristics and high strengths. Meta-heuristic optimization algorithms operate independently of the optimization problems. That is to say that, unlike other optimization algorithms, these algorithms are not dependent on the structure of the optimization problems and, thus, are able to solve the optimization problems without changing their structure. Meta-heuristic optimization algorithms can, therefore, be utilized in order to deal with a wide range of optimization problems having different structures.

Technically speaking, meta-heuristic music-inspired optimization algorithms employ a different architecture compared with their counterparts that have a source of inspiration other than music. In the architecture of meta-heuristic music-inspired optimization algorithms, the process of generating new solutions depends on the entire space of the nonempty feasible decision-making variables. As a consequence, achieving sufficient knowledge and its promotion to a relatively high level concerning both the varied characteristics of the optimization problems, the solution process, and the optimization algorithms, particularly the meta-heuristic and meta-heuristic music-inspired optimization algorithms, can dramatically help researchers and specialists to meet their scientific challenges as much as possible with satisfactory results.

Electrical power engineering is one of the principal subfields of electrical engineering in which most of its challenges are described in the form of optimization problems. On the one hand, it is well-known that most power system optimization problems are complicated, real-world, large-scale, non-convex, non-smooth optimization problems having a nonlinear, mixed-integer nature with big data. This means that the number of the local optimal points is greatly increased by enlarging the size of the power grid under study, after which traditional optimization algorithms are encountered with many difficulties in dealing with such optimization problems and finding the global optimal point. On the other hand, with emerging, newfound issues and challenges in power systems, such as unbundling, deregulation, integration of renewable and nonrenewable energy resources, global environmental policies, etc., new uncertainties have been introduced and existing ones escalated. Under these circumstances, optimization problems of power systems can be much more complicated than before.

Strictly speaking, power system researchers and specialists take into account two different processes in dealing with the optimization problems of power systems. The first group of researchers and specialists considered only a simple model of power system optimization problems, such as being a single-objective model, disregarding the uncertainties, regarding the traditional structure of power systems, lack of consideration practical limits, etc. Accordingly, the theoretical aspects have merely been applied in these power system optimization problems, the outputs for which are not applicable. A second group of researchers and specialists, however, considered a more complex and realistic model of the power system optimization problems. In order to solve these complex optimization problems, some simplifications, such as linearization of the nonlinear models, have been assumed in the solution process. These simplifications, which bring about the obtained outputs of linearized models, are not based on the actual conditions. There is a need, then, to use a simple model of the power system optimization problems and/or to use a more complex and realistic model of the power system optimization problems but with multiple simplifications derived from the lack of powerful optimization techniques. As a general result, the development of such powerful optimization techniques for dealing with these power system optimization problems with a minimal simplification of the vital aspects is considered as a critical requirement.

The contents of this book are briefly summarized below:

Chapter 1: This chapter begins with a concise definition of the optimization problem and its parameters, along with a mathematical description of an optimization problem with continuous and discrete decision-making variables whose objective functions are employed in a standard form of an optimization problem along with equality and inequality constraints. Subsequently, the authors address the classifications of an optimization problem from different perspectives, which deserve attention and can achieve full knowledge regarding an optimization problem and its parameters. Also mentioned are categorizations of an optimization problem from the point of view of the number of objective functions, the constraints, the nature of employed equations, the objective function landscape, the kind of decision-making variables, the number of decision-making variables, the separability of the employed equations, and the uncertainty. In addition, a succinct overview pertaining to the optimization algorithms with a focus on meta-heuristic optimization algorithms is reported. With that in mind, a well-organized classification of meta-heuristic optimization algorithms is demonstrated on the basis of the inspirational source and then categorized into four classes: (1) swarm intelligence-based meta-heuristic optimization algorithms, (2) biologically inspired meta-heuristic optimization algorithms not based on swarm intelligence, (3) physics- and chemistry-based meta-heuristic optimization algorithms, and (4) human behaviors and society-inspired meta-heuristic optimization algorithms. This topic is indispensable, because of its highly influential nature in choosing a reasonable and efficient meta-heuristic algorithm to solve a typical optimization problem with respect to the breadth and variety of meta-heuristic optimization algorithms.

Chapter 2: In this chapter, the necessity of utilizing a multi-objective optimization process is rigorously elucidated. Afterward, the fundamental concepts of optimization in multi-objective optimization problems are thoroughly described in five sections: (1) mathematical description of a multi-objective optimization problem; (2) concepts related to efficiency, efficient frontier, and dominance; (3) concepts relevant to Pareto optimality; (4) concepts associated with the vector of ideal objective functions and the vector of nadir objective functions; and (5) concepts pertaining to Pareto optimality investigation. In this chapter, the authors also provide an exhaustive classification of the multi-objective optimization algorithms by concentrating on the role of the decision-maker in the solution process, which are then broken down into two approaches: (1) non-interactive and (2) interactive. Non-interactive approaches are further divided into four classes, including basic, no-preference, a priori, and a posteriori approaches. The fuzzy satisfying method is then extensively expressed in order to select the final optimal compromise solution from the Pareto-optimal solutions set. This method is considered as the preferred multi-objective decision-making technique, due to its simplicity and similarity to human reasoning. Thorough coverage of these topics in this chapter is requisite, owing to their great effectiveness in attaining the Pareto-optimal solutions set and choosing the final optimal solution for a typical multi-objective optimization problem.

Chapter 3: This chapter illustrates the definition of music with regard to its historical roots then denotes the different interpretations of music from the standpoint of well-known philosophers and scientists. A concise history of music is also presented through a review of archaeological evidence. These topics are especially advantageous in the solution process and, therefore, are discussed in detail throughout the book, because the authors will introduce and develop modern music-inspired meta-heuristic optimization algorithms by borrowing the phenomena and fundamental concepts of music. Besides these initial topics, Chap. 3 deals with the music-inspired meta-heuristic optimization algorithms from past to present: the single-stage computational single-dimensional harmony search algorithm (SS-HSA), the single-stage computational single-dimensional improved harmony search algorithm (SS-IHSA), and the continuous two-stage computational, multi-dimensional, single-homogeneous melody search algorithm (TMS-MSA). In doing so, interdependencies of phenomena and fundamental concepts of music and the optimization problem are expressed, and then, the basic principles of the original SS-HSA, SS-IHSA, and original continuous TMS-MSA, along with their performance-driven architectures, are precisely addressed. This chapter will also help readers to identify the enhancements applied on the original SS-HSA in the form of a structural classification, including (1) the enhanced versions of the original SS-HSA, based on parameter adjustments; (2) enhanced versions of the original SS-HSA, according to a combination of this algorithm with other meta-heuristic optimization algorithms; and (3) enhanced versions of the original SS-HSA, in accordance with architectural principles. The first category of this structural classification describes how the original SS-IHSA is considered to be a well-known enhanced version of the original SS-HSA by investigating the basic differences between the SS-IHSA and original SS-HSA. Finally, the chapter elaborates on reasonability and applicability of the music-inspired meta-heuristic optimization algorithms from past to present for solving complicated, real-world, large-scale, non-convex, non-smooth optimization problems having a nonlinear, mixed-integer nature with big data and, subsequently, outlines a valuable background for elucidating innovative versions of the music-inspired meta-heuristic optimization algorithms in Chap. 4.

Chapter 4: This chapter complements the preceding chapter by providing multiple innovative versions of the modern music-inspired optimization algorithms. First, the authors propose an innovative continuous/discrete TMS-MSA by borrowing the basic principles of the original continuous TMS-MSA in order to deal with the complicated, real-world, large-scale, non-convex, non-smooth optimization problems with a simultaneous combination of the continuous and discrete decision-making variables. Then, an innovative improved version of the proposed continuous/discrete TMS-MSA, called a two-stage computational multi-dimensional single-homogeneous enhanced melody search algorithm (TMS-EMSA), is developed in order to increase the efficiency and efficacy of the performance of this optimization algorithm. Also described, with regard to today's world, are modern engineering challenges widely developed in the form of the multilevel optimization problems; therefore, the original SS-HSA, SS-IHSA, original continuous TMS-MSA, proposed continuous/discrete TMS-MSA, and proposed TMS-EMSA may not be able to maintain the most favorable performance in the solution process of such optimization problems. This owes to the fact that a single-homogeneous structure is employed in the architecture of the aforementioned optimization algorithms. As a consequence, an innovative version of the architecture of the proposed TMS-EMSA—a multi-stage computational multi-dimensional multiple-homogeneous enhanced melody search algorithm (MMM-EMSA), multi-stage computational multi-dimensional single-inhomogeneous enhanced melody search algorithm (MMS-EMSA), or symphony orchestra search algorithm (SOSA)—is rigorously developed in order to appreciably enhance its performance, flexibility, robustness, and parallel capability. The newly developed SOSA has a multi-stage computational multi-dimensional and multiple-homogeneous or multi-stage computational multi-dimensional and single-inhomogeneous structure.

Many real-world optimization problems in the engineering sciences, particularly electrical engineering, have more than one objective function and are introduced in the form of a multi-objective optimization problem. The most reasonable strategy for solving an optimization problem having multiple conflicting, non-commensurable, and correlated objective functions is the use of a multi-objective optimization process. The architecture of the original SS-HSA, SS-IHSA, and original continuous TMS-MSA described in the Chap. 3 and the architecture of the proposed continuous/discrete TMS-MSA, TMS-EMSA, and SOSA in this chapter have, however, been developed in such a way that they are only suitable for solving single-objective optimization problems and cannot be employed for dealing with multi-objective optimization problems. With that in mind, the chapter continues with the presentation of new multi-objective strategies for remodeling the architecture of the meta-heuristic music-inspired optimization algorithms or, more comprehensively, the meta-heuristic music-inspired optimization algorithms with a single-stage computational and a single-dimensional structure, such as the original SS-HSA and SS-IHSA, in dealing with multi-objective optimization problems. Afterwards, new multi-objective strategies are represented for remodeling the architecture of the meta-heuristic music-inspired optimization algorithms with a two-stage computational single-dimensional and single-homogeneous structure, such as the original continuous TMS-MSA, continuous/discrete TMS-MSA, and proposed TMS-EMSA in dealing with multi-objective optimization problems. Eventually, a new multi-objective strategy is addressed for remodeling the architecture of the SOSA, which has a multi-stage computational multi-dimensional and multiple-homogeneous structure or a multi-stage computational multi-dimensional and single-inhomogeneous structure, in dealing with multi-objective optimization problems.

Chapter 5: The second part of the book starts with Chap. 5, which is devoted to an innovative, two-level computational-logical framework for a bilateral bidding mechanism within a competitive security-constrained electricity market. In this chapter, a comprehensive survey related to game theory is meticulously presented. Subsequently, the authors describe the formulation of a two-level computational-logical framework, including its mathematical model of first and second levels. In the first level, the generation and distribution companies maximize their profits. In the second level, however, the independent system operator clears the competitive security-constrained electricity market by considering additional objectives containing CO_2, SO_2, and NO_x emissions. Exhaustive coverage of this topic in this chapter is indispensable, due to its being considered as a central core and or as a short-term operational slave problem for the strategic multilevel long-term expansion planning frameworks of the power systems planning studies in Chap. 6. Further, the proposed two-level computational-logical framework is applied on the modified six-machine eight-bus test network. In order to show the effectiveness of the performance associated with the two-level computational-logical framework, two cases are taken into account in the implementation process: (1) considering only a unilateral bidding mechanism and (2) considering a bilateral bidding mechanism. The comparative analysis demonstrates the sufficiency and profitableness of the proposed two-level computational-logical framework by considering a bilateral bidding mechanism. This chapter also compares the performance of the newly proposed single-objective SOSA, single-objective TMS-EMSA, and single-objective continuous/discrete TMS-MSA with the original single-objective SS-IHSA, original single-objective SS-HSA, and non-dominated sorting genetic algorithm-II (NSGA-II), as state-of-the-art optimization algorithms. The obtained results prove the practicality and high performance of the newly proposed single-objective music-inspired optimization algorithms, especially the single-objective SOSA, when compared with other optimization algorithms.

Chapter 6: This chapter begins with a general, yet rigorous treatment of power systems planning studies with a focus on important characteristics, such as power system structure, planning horizon, uncertainties, and solving algorithms. The chapter also explains why power systems need expansion planning studies. Afterward, four innovative strategic multilevel computational-logical frameworks are developed for pseudo-dynamic generation expansion planning (PD-GEP), pseudo-dynamic transmission expansion planning (PD-TEP), coordination of pseudo-dynamic generation and transmission expansion-planning (PD-G&TEP), and pseudo-dynamic distribution expansion planning (PD-DEP) in order to optimally supply, transfer, and distribute electric energy. The proposed PD-GEP and PD-TEP are formulated by a strategic tri-level computational-logical framework; similarly, the PD-G&TEP is organized by a strategic quad-level computational-logical framework. These formulations contain a short-term operational slave problem and a long-term planning master problem. Moreover, the PD-DEP is described by a techno-economic framework in the presence of the distributed generation resources (DGRs). The PD-DEP formulation includes a short-term loadability-based optimal power flow (LBOPF) problem and a long-term planning problem. This chapter widely employs a well-founded information-gap decision theory (IGDT) under a twofold envelope-bound uncertainty model in order to handle risks of the PD-GEP, PD-TEP, PD-G&TEP, and PD-DEP problems stemming from severe twofold uncertainties of demand and market price. In the IGDT-based formulation, the proposed PD-GEP, PD-TEP, PD-G&TEP, and PD-DEP problems are then remodeled under three risk-neutral, risk-averse, and risk-taker decision-making policies. The proposed PD-GEP and PD-G&TEP problems are applied on the modified large-scale 46-bus south Brazilian system. The PD-TEP problem is also implemented on both the modified

IEEE 30-bus test system and the modified large-scale Iranian 400-kV transmission network. The PD-DEP problem is also run on the modified three-phase 13.8 kV 27-node open-loop distribution test network. In the implementation process of this chapter, the performances of the strategic multilevel computational-logical frameworks are analyzed both by considering the IGDT risk-averse decision-making policy, the first case, and considering the IGDT risk-taker decision-making policy, the second case. This chapter also compares the performances of the newly proposed PD-GEP, PD-TEP, PD-G&TEP, and PD-DEP problems under the IGDT risk-averse policy with the performances of those under the robust optimization technique. The comparative investigation exhibits profitableness of the proposed frameworks by the IGDT risk-averse policy. Finally, the efficiencies of the newly proposed multi-objective SOSA, multi-objective TMS-EMSA, multi-objective continuous/discrete TMS-MSA, multi-objective SS-HSA, and multi-objective SS-IHSA are scrutinized on the PD-GEP, PD-TEP, PD-G&TEP, and PD-DEP problems. The efficiencies of the NSGA-II, as a well-known and powerful state-of-the-art multi-objective optimization algorithm, then compared to each of these multi-objective problems. The results indicate applicability and high efficiency of the newly proposed multi-objective music-inspired optimization algorithms, especially the multi-objective SOSA, when compared with the NSGA-II.

Chapter 7: In this chapter, the authors provide a succinct overview of harmonic power filter planning studies, including causes and malicious effects of nonlinear loads and detailed descriptions of passive and active harmonic power filters. Next, different methodologies for solving harmonic power-flow problems are precisely classified using three points of view: (1) a modeling technique, (2) a distribution network condition, and (3) a solution approach. Besides these outlines, the chapter develops the formulation of an innovative techno-economic multi-objective framework for the hybrid harmonic power filter (HHPF) planning problem in distribution networks, with consideration of uncertainty in demand and harmonic currents injected by nonlinear loads. The proposed framework is also broken down into a harmonic power flow problem and the HHPF planning problem. The harmonic power flow problem acts as a central core of the HHPF planning problem and is solved via a probabilistic decoupled harmonic power flow (PDHPF) methodology. This chapter widely utilizes an efficient two-point estimate method (two-PEM) in order to handle uncertainty in demand and harmonic currents injected by nonlinear loads in the proposed framework. With that in mind, the proposed PDHPF methodology, according to the efficient two-PEM, is implemented by a deterministic decoupled harmonic power flow (DDHPF) methodology. A loadability-based Newton-Raphson power flow (LBNRPF) methodology is also applied to solve the power-flow problem at the principal frequency, because the DDHPF methodology for the distribution network assessment at harmonic frequencies requires the acquisition of information determined from the power-flow problem at the principal frequency. Furthermore, the proposed techno-economic multi-objective framework is applied on both the modified IEEE 18-bus distorted test network and the modified 34-bus distribution test network. In order to illustrate the effectiveness of the performance related to the techno-economic multi-objective framework, three cases are considered in the implementation process: (1) considering only passive harmonic power filters, (2) considering only active harmonic power filters, and (3) considering both passive and active harmonic power filters (i.e., hybrid harmonic power filters). The filtering efficiency of allocated passive harmonic power filters is also scrutinized through the harmonic attenuation coefficient. The comparative evaluation reveals the flexibility and desirability of the proposed techno-economic multi-objective framework by considering both passive and active harmonic power filters. Eventually, the authors investigate the performances of the newly proposed multi-objective SOSA, multi-objective TMS-EMSA, multi-objective continuous/discrete TMS-MSA, multi-objective SS-HSA, and multi-objective SS-IHSA on the techno-economic multi-objective framework, which are subsequently compared to the performance of the NSGA-II as a well-known and powerful state-of-the-art multi-objective optimization algorithm. The results represent feasibility, reasonability, and high performance of the newly developed multi-objective music-inspired optimization algorithms, especially the multi-objective SOSA, when compared with the NSGA-II.

It is presumed that readers have an adequate knowledge of the fundamental concepts associated with the optimization process, meta-heuristic optimization algorithms, and power systems studies, including power systems operation and planning as well as power quality planning.

In the end, the authors encourage and welcome input from readers of this book, who wish to point out deficiencies, provide suggestions, or inquire about any issues.

Tehran, Iran — Mohammad Kiani-Moghaddam
Shahrood, Iran — Mojtaba Shivaie
Firelands, OH — Philip David Weinsier
November 2018

Acknowledgments

The authors would like to thank so many of their colleagues, researchers, specialists, and institutions for their support and help. The authors are truly grateful to Shahrood University of Technology for providing an outstanding research environment. The authors are also deeply grateful to Prof. Constantine Hatziadoniu for his generous comments.

The authors would also like to thank Amanda Quinn, Michael McCabe, Faith Pilacik, Brian Halm, Zoe Kennedy, Abhishek Ravi Shanker and Anup Kumar at Springer Nature for their help and professionalism.

Last but not the least, the authors offer their deepest personal thanks to their families for their love, encouragement, and unconditional support during the writing of this book.

November 2018

Mohammad Kiani-Moghaddam
Shahid Beheshti University, Tehran, Iran
Mojtaba Shivaie
Shahrood University of Technology, Shahrood, Iran
Philip D. Weinsier
Bowling Green State University Firelands, Huron, OH, USA

Contents

About the Authors

Mohammad Kiani-Moghaddam received the B.Sc. degree with first-class honors in Electrical Engineering from the Islamic Azad University of Najafabad, Isfahan, Iran, and the M.Sc. degree with first-class honors in Electrical Engineering from Shahid Beheshti University, Tehran, Iran. His emphasis is on the research, design, and application of complex mathematical models for use in the analysis of power systems with a particular focus on risk assessment, worth-based reliability evaluation, economic strategies, as well as artificial intelligence and optimization theory. He has served as a peer reviewer for over four international journals.

Mojtaba Shivaie is currently an assistant professor in the Faculty of Electrical and Robotic Engineering at Shahrood University of Technology, Shahrood, Iran. He obtained the B.Sc. degree with first-class honors in Electrical Engineering from Semnan University, Semnan, Iran, in 2008. He also received the M.Sc. and Ph.D. degrees with first-class honors, both in Electrical Engineering, from Shahid Beheshti University, Tehran, Iran, in 2010 and 2015, respectively. He has worked extensively in the areas of power systems, smart distribution grids, stochastic simulation, and optimization techniques, and he (with Mr. Kiani-Moghaddam and Prof. Weinsier) is the inventor of a modern optimization technique known as "symphony orchestra search algorithm" and an innovative architecture for competitive electricity markets known as "Hypaethral market." He was awarded the Dr. Shahriari's scholarship by the office of honor students of Shahid Beheshti University and the Dr. Kazemi-Ashtiani's award by Iran's National Elites Foundation for outstanding educational and research achievements. He has served as an editorial board member of the *International Transaction of Electrical and Computer Engineers System* journal and the *Control and Systems Engineering* journal and also a peer reviewer for over 12 high-impact journals. He was a recipient of the outstanding reviewer award of the *Applied Soft Computing* in 2014, the *Energy Conversion and Management* in 2016, and the *Electric Power Systems Research* in 2017.

Philip D. Weinsier is currently professor and Electrical/Electronic Engineering Technology Program director at Bowling Green State University Firelands. He received his B.S. degrees in Physics/Mathematics and Industrial Education/Teaching from Berry College in 1978 and M.S. degree in Industrial Education and Ed.D. degree in Vocational/Technical Education from Clemson University in 1979 and 1990, respectively. He is currently senior editor of the *International Journal of Modern Engineering* and the *International Journal of Engineering Research and Innovation* and editor-in-chief of *The Technology Interface International Journal.* He is a Fulbright Scholar, a lifetime member of the International Fulbright Association, and a member of the European Association for Research on Learning and Instruction since 1989.

Part I
Fundamental Concepts of Optimization Problems and Theory of Meta-Heuristic Music-Inspired Optimization Algorithms

Chapter 1
Introduction to Meta-heuristic Optimization Algorithms

1.1 Introduction

From the early years of the formation and evolution of modern man-made societies, willingness to perform activities originating from modernization in industrial, commercial, medical, and agriculture sectors with the least difficulty and cost was one of the most prominent concerns of humans. In doing so, many efforts have been made to provide relatively high prosperity in different dimensions of human life. In the field of engineering sciences, one of the most fundamental of these efforts was to determine optimal solutions for overcoming engineering challenges. Obtaining such solutions is accomplished during a process called optimization. The main aim of this process is to effectively decrease wasted time and resources, engineering mistakes, and unnecessary costs in such a way that the process objectives are still satisfied. As a result, the optimization process can appropriately provide a bridge between theory and practice across the wide range of different sciences, especially the engineering sciences. Specialists in different branches of the engineering sciences (e.g., electrical, civil, computer, mechanical, and aerospace) often make use of mathematical models associated with their specialty branches to optimize their time and resources. In the optimization process, specialists make decisions according to these optimal solutions while achieving minimum or maximum values of one or more objective functions under the predetermined constraints.

In recent years, the application of the optimization process in solving engineering challenges has dramatically grown, in view of the need of the specialists to identify optimal solutions under varying parameters. These challenges are often expressed as complicated, large-scale, non-convex optimization problems of a nonlinear mixed-integer nature. Under these circumstances, many serious difficulties arise in reaching a final optimal solution and, in the face of new concepts such as deregulation, multidisciplinary analysis, and big data, already serious difficulties can escalate.

M. Kiani-Moghaddam et al., *Modern Music-Inspired Optimization Algorithms for Electric Power Systems*, Power Systems,
https://doi.org/10.1007/978-3-030-12044-3_1

In the literature, numerous algorithms have been developed to solve a wide range of optimization problems. Nevertheless, given the architectural type of such algorithms, each of these techniques is appropriate only for a narrow range of problems. In other words, there is no specific optimization algorithm to solve optimization problems with different characteristics. Selecting a suitable algorithm to solve optimization problems depends not only on a thorough understanding of the optimization problem and its parameters, but also on full knowledge of the architecture of the algorithms and their characteristics.

In view of the wide variety of optimization problems and algorithms, providing an appropriate classification is needed. Hence, the authors will focus on two targets:

- Target 1: Provide a brief introduction pertaining to the optimization problems and their parameters.
- Target 2: Present a brief overview related to the classification of optimization algorithms with a focus on meta-heuristic optimization algorithms.

The rest of this chapter is organized as follows: First, the mathematical description of an optimization problem and its parameters are explained in Sect. 1.2. In Sect. 1.3, the classification of optimization problems and algorithms is reviewed from different points of view. Finally, the chapter ends with a brief summary and some concluding remarks in Sect. 1.4.

1.2 An Optimization Problem and Its Parameters

In a very general sense, optimization refers to finding one or more solutions from a set of possible solutions through minimizing/maximizing one or more objective functions under the predetermined constraints, if any.

1.2.1 Mathematical Description of an Optimization Problem

In mathematical terms, the standard form of an optimization problem can generally be represented by Eqs. (1.1) and (1.2):

$$\begin{aligned} &\underset{\mathrm{x}\in\mathrm{X}}{\text{Minimize}} \quad \mathrm{F(x)} = [f_1(\mathrm{x}),\ldots,f_a(\mathrm{x}),\ldots,f_{\mathrm{A}}(\mathrm{x})]; \quad \forall\{\mathrm{A}>0\},\forall\{a\in\Psi^{\mathrm{A}}\} \\ &\quad \text{subject to :} \\ &\quad \mathrm{G(x)} = [g_1(\mathrm{x}),\ldots,g_b(\mathrm{x}),\ldots,g_{\mathrm{B}}(\mathrm{x})] = 0; \quad \forall\{\mathrm{B}\geq 0\},\forall\{b\in\Psi^{\mathrm{B}}\} \\ &\quad \mathrm{H(x)} = [h_1(\mathrm{x}),\ldots,h_e(\mathrm{x}),\ldots,h_{\mathrm{E}}(\mathrm{x})] \leq 0; \quad \forall\{\mathrm{E}\geq 0\},\forall\{e\in\Psi^{\mathrm{E}}\} \end{aligned} \tag{1.1}$$

$$\begin{aligned}\mathrm{x} = \; & [x_1, \ldots, x_v, \ldots, x_{\mathrm{NDV}}]; \quad \forall\{v \in \Psi^{\mathrm{NDV}}\}, \forall\{\Psi^{\mathrm{NDV}} = \Psi^{\mathrm{NCDV+NDDV}}\}, \forall\{\mathrm{x} \in \mathrm{X}\}, \\ & \forall\{x_v^{\min} \le x_v \le x_v^{\max} \mid v \in \Psi^{\mathrm{NCDV}}\}, \\ & \{x_v \in \{x_v(1), \ldots, x_v(w), \ldots, x_v(W_v)\} \mid v \in \Psi^{\mathrm{NDDV}}\}\end{aligned} \tag{1.2}$$

In Eq. (1.1), A, B, and C indicate the number of objective functions that must be simultaneously optimized, and the number of equality and inequality constraints that must be satisfied, respectively. In this equation, $\mathrm{F(x)} \in \mathfrak{R}^{\mathrm{A}} \to \mathfrak{R}$, $\mathrm{G(x)} \in \mathfrak{R}^{\mathrm{B}} \to \mathfrak{R}$, and $\mathrm{H(x)} \in \mathfrak{R}^{\mathrm{E}} \to \mathfrak{R}$ also represent the vector of objective functions, the vector of equality constraints, and the vector of inequality constraints, respectively. Additionally, $f_a(\mathrm{x}) : \mathfrak{R}^{\mathrm{NDV}} \to \mathfrak{R}$, $g_b(\mathrm{x}) : \mathfrak{R}^{\mathrm{NDV}} \to \mathfrak{R}$, and $h_e(\mathrm{x}) : \mathfrak{R}^{\mathrm{NDV}} \to \mathfrak{R}$ express objective function a or component a of the vector of objective functions, equality constraint b or component b of the vector of equality constraints, and inequality constraint e or component e of the vector of inequality constraints, respectively. In Eq. (1.1), x describes the vector of decision-making variables—vector of design—which belongs to the nonempty feasible decision-making space—nonempty possible design space—$\mathrm{X} \subset \mathfrak{R}^{\mathrm{NDV}}$. In Eq. (1.2), x_v represents decision-making variable v or component v of the vector of decision-making variable, which can be continuous or discrete. The sum of the number of continuous decision-making variables (NCDV) and the number of discrete decision-making variables (NDDV) is considered as the total number of decision-making variables (NDV). If the decision-making variable x_v has a continuous nature—belongs to the set of the NCDV—the set of candidate admissible values for this decision-making variable is confined by a lower bound, $x_v^{\min}$, and an upper bound, $x_v^{\max}$. For the same reason, if the decision-making variable x_v has a discrete nature—belongs to the set of the NDDV—the set of candidate permissible values for this decision-making variable is defined as the set $\{x_v(1), \ldots, x_v(w_v), \ldots, x_v(W_v)\}$. The point to be made here is that the nonempty feasible decision-making space, X, which names the possible design space, is addressed as the set $\{\mathrm{x} | \mathrm{G(x)} = 0, \mathrm{H(x)} \le 0\}$. The vector of objective functions addresses the illustration of the vector of decision-making variables and contains the values of the objective functions, as given by Eq. (1.3):

$$\mathrm{z} = \mathrm{F(x)} = [f_1(\mathrm{x}), \ldots, f_a(\mathrm{x}), \ldots, f_{\mathrm{A}}(\mathrm{x})]; \quad \forall\{\mathrm{A} \ge 2\}, \forall\{a \in \Psi^{\mathrm{A}}\} \tag{1.3}$$

The illustration of the nonempty feasible decision-making space in the objective space is known as feasible objective space $\mathrm{Z} = \mathrm{F(X)}$ and is described by using the set of $\{\mathrm{F(x)} | \mathrm{x} \in \mathrm{X}\}$. It is necessary to mention that the domain of the feasible objective space in the multi-objective optimization problems is considerably larger than that in single-objective problems. More precisely, the domain of the feasible objective space in an optimization problem depends on the number of its objective functions. A solution is considered to be a possible solution if it does not lead to any violation of equality and inequality constraints.

Technically speaking, the standard form of an optimization problem is generally defined as a minimization problem. Nonetheless, a maximization problem can be addressed in the negative form of its equivalent minimization problem. More precisely, the maximization of the objective function a, $f_a(\mathrm{x})$, is indicated as a minimization of the negative of the corresponding objective function, $-f_a(\mathrm{x})$. Moreover, the inequality constraint is generally described as $h_e(\mathrm{x}) \leq 0$, according to the standard formulation of an optimization problem from Eq. (1.1). For this reason, if there is an inequality constraint, e, such as $h_e(\mathrm{x}) \geq 0$, in the formulation of a specific optimization problem, this constraint must be implemented as $-h_e(\mathrm{x}) \leq 0$.

1.3 Classification of an Optimization Problem

Broadly speaking, the classification of an optimization problem is a challenging task. This is due to fact that there are different tastes, ideas, and experiences for this classification. Here, the authors' aim is to present a brief classification for the standard form of an optimization problem, as demonstrated by Eqs. (1.1) and (1.2). Readers interested in a comprehensive discussion on this topic are referred to the work by Rao [1].

1.3.1 Classification of Optimization Problems from the Perspective of a Number of Objective Functions

Optimization problems can be broken down into two types, from the perspective of a number of objective functions: single- and multi-objective. In Eqs. (1.1) and (1.2), if A is equal to 1, the corresponding problem is converted to a single-objective optimization problem. The single-objective optimization problem can then be addressed by Eqs. (1.4) and (1.5):

$$\begin{aligned} \underset{\mathrm{x}\in\mathrm{X}}{\text{Minimize}} \quad & \mathrm{F}(\mathrm{x}) = [f(\mathrm{x})] \\ & \text{subject to}: \\ & \mathrm{G}(\mathrm{x}) = [g_1(\mathrm{x}),\ldots,g_b(\mathrm{x}),\ldots,g_\mathrm{B}(\mathrm{x})] = 0; \quad \forall\{\mathrm{B}\geq 0\},\forall\{b\in\Psi^{\mathrm{B}}\} \\ & \mathrm{H}(\mathrm{x}) = [h_1(\mathrm{x}),\ldots,h_e(\mathrm{x}),\ldots,h_\mathrm{E}(\mathrm{x})] \leq 0; \quad \forall\{\mathrm{E}\geq 0\},\forall\{e\in\Psi^{\mathrm{E}}\} \end{aligned} \tag{1.4}$$

$$\begin{aligned} \mathrm{x} = \; & [x_1,\ldots,x_v,\ldots,x_{\mathrm{NDV}}];\forall\{v\in\Psi^{\mathrm{NDV}}\},\forall\{\Psi^{\mathrm{NDV}}{=}\Psi^{\mathrm{NCDV+NDDV}}\},\forall\{\mathrm{x}\in\mathrm{X}\}, \\ & \forall\{x_v^{\min}\leq x_v\leq x_v^{\max}|v\in\Psi^{\mathrm{NCDV}}\}, \\ & \{x_v\in\{x_v(1),\ldots,x_v(w),\ldots,x_v(W_v)\}|v\in\Psi^{\mathrm{NDDV}}\} \end{aligned} \tag{1.5}$$

From Eqs. (1.4) and (1.5), a single-objective optimization problem consists of an individual objective function. Solving a single-objective optimization problem necessitates finding an individual response/output called the optimal solution. In Eqs. (1.1) and (1.2), however, if A is larger than 1, the corresponding problem is turned into to a multi-objective optimization problem. In this optimization problem, several different objective functions are simultaneously considered with conflicting interests relative to each other. Solving a multi-objective optimization problem gives rise to an optimal response/output set referred to as a Pareto-optimal solution set or non-dominated solution set. As a consequence, a multi-objective optimization problem is much more complicated than a single-objective optimization problem. The solution process for a multi-objective optimization problem is described in Chap. 2.

1.3.2 Classification of Optimization Problems from the Perspective of Constraints

Optimization problems can be divided into four types, from the perspective of a number of constraints: unconstrained, equality constrained, inequality constrained, and hybrid equality or inequality constrained. In the standard formulation of an optimization problem presented in Eqs. (1.1) and (1.2), if B plus E is equal to zero, $\mathrm{B} + \mathrm{E} = 0$, the corresponding problem is converted to an unconstrained optimization problem. The unconstrained optimization problem can, then, be given by Eqs. (1.6) and (1.7):

$$\underset{\mathrm{x}\in\mathrm{X}}{\text{Minimize}} \quad \mathrm{F}(\mathrm{x}) = [f_1(\mathrm{x}), \ldots, f_a(\mathrm{x}), \ldots, f_\mathrm{A}(\mathrm{x})]; \quad \forall\{\mathrm{A} > 0\}, \forall\{a \in \Psi^\mathrm{A}\} \tag{1.6}$$

$$\begin{aligned} \mathrm{x} = \; & [x_1, \ldots, x_v, \ldots, x_\mathrm{NDV}]; \quad \forall\{v \in \Psi^\mathrm{NDV}\}, \forall\{\Psi^\mathrm{NDV} = \Psi^\mathrm{NCDV+NDDV}\}, \forall\{\mathrm{x} \in \mathrm{X}\}, \\ & \forall\{x_v^\mathrm{min} \le x_v \le x_v^\mathrm{max} \,|\, v \in \Psi^\mathrm{NCDV}\}, \\ & \{x_v \in \{x_v(1), \ldots, x_v(w), \ldots, x_v(W_v)\} \,|\, v \in \Psi^\mathrm{NDDV}\} \end{aligned} \tag{1.7}$$

Solving an unconstrained optimization problem is much easier than a constrained optimization problem. Also from Eqs. (1.1) and (1.2), if B is greater than zero, $\mathrm{B} > 0$, and E is equal to zero, $\mathrm{E} = 0$, the corresponding problem is transformed into an equality constrained optimization problem, as shown in Eqs. (1.8) and (1.9):

$$\begin{aligned} & \underset{\mathrm{x}\in\mathrm{X}}{\text{Minimize}} \quad \mathrm{F}(\mathrm{x}) = [f(\mathrm{x})] \\ & \qquad \text{subject to}: \\ & \qquad \mathrm{G}(\mathrm{x}) = [g_1(\mathrm{x}), \ldots, g_b(\mathrm{x}), \ldots, g_\mathrm{B}(\mathrm{x})] = 0; \quad \forall\{\mathrm{B} \ge 0\}, \forall\{b \in \Psi^\mathrm{B}\} \end{aligned} \tag{1.8}$$

$$\begin{aligned} \mathrm{x} = \; & [x_1, \ldots, x_v, \ldots, x_{\mathrm{NDV}}]; \quad \forall\{v \in \Psi^{\mathrm{NDV}}\}, \forall\{\Psi^{\mathrm{NDV}} = \Psi^{\mathrm{NCDV+NDDV}}\}, \forall\{\mathrm{x} \in \mathrm{X}\}, \\ & \forall\{x_v^{\min} \leq x_v \leq x_v^{\max} | v \in \Psi^{\mathrm{NCDV}}\}, \\ & \{x_v \in \{x_v(1), \ldots, x_v(w), \ldots, x_v(W_v)\} | v \in \Psi^{\mathrm{NDDV}}\} \end{aligned} \tag{1.9}$$

Similarly, if B is equal to zero, B = 0, and E is greater than zero, E > 0, the corresponding problem is also changed to an inequality constrained optimization problem, as illustrated in Eqs. (1.10) and (1.11):

$$\begin{aligned} \underset{\mathrm{x} \in \mathrm{X}}{\text{Minimize}} \quad & \mathrm{F}(\mathrm{x}) = [f(\mathrm{x})] \\ & \text{subject to :} \\ & \mathrm{H}(\mathrm{x}) = [h_1(\mathrm{x}), \ldots, h_e(\mathrm{x}), \ldots, h_{\mathrm{E}}(\mathrm{x})] \leq 0; \quad \forall\{\mathrm{E} \geq 0\}, \forall\{e \in \Psi^{\mathrm{E}}\} \end{aligned} \tag{1.10}$$

$$\begin{aligned} \mathrm{x} = \; & [x_1, \ldots, x_v, \ldots, x_{\mathrm{NDV}}]; \quad \forall\{v \in \Psi^{\mathrm{NDV}}\}, \forall\{\Psi^{\mathrm{NDV}} = \Psi^{\mathrm{NCDV+NDDV}}\}, \forall\{\mathrm{x} \in \mathrm{X}\}, \\ & \forall\{x_v^{\min} \leq x_v \leq x_v^{\max} | v \in \Psi^{\mathrm{NCDV}}\}, \\ & \{x_v \in \{x_v(1), \ldots, x_v(w), \ldots, x_v(W_v)\} | v \in \Psi^{\mathrm{NDDV}}\} \end{aligned} \tag{1.11}$$

By the same token, if B and E are simultaneously greater than zero, B > 0 and E > 0, the corresponding problem is turned into a hybrid equality or inequality constrained optimization problem.

1.3.3 Classification of Optimization Problems from the Perspective of the Nature of Employed Equations

Optimization problems can be classified into two different types, from the perspective of the nature of the employed equations in objective functions and constraints: linear and nonlinear. In Eqs. (1.1) and (1.2), the problem will be linear provided that all objective functions and equality and inequality constraints are defined as linear functions of the decision-making variables. And the problem will be nonlinear if the conditions of the objective functions and equality and inequality constraints are nonlinear functions of the decision-making variables. Solving nonlinear optimization problems is much more difficult than linear optimization problems. As previously mentioned, each optimization algorithm is able to solve a specific range of optimization problems; accordingly, this classification is very useful for choosing an efficient optimization algorithm.

1.3.4 Classification of Optimization Problems from the Perspective of an Objective Functions Landscape

Optimization problems can be broken down into two different types, from the perspective of an objective functions landscape: unimodal and multimodal. Consider a single-objective optimization problem in accordance with Eqs. (1.4) to (1.5). In this formulation, the corresponding problem is unimodal if its objective function has only one peak. In this case, the local optimal point is also the global optimum point. A convex optimization problem is considered a special type of unimodal optimization problem if the objective function has not only a certain convexity but also a guaranteed optimality. Furthermore, if the objective function has more than one peak then the corresponding problem is considered to be a multimodal optimization problem. Basically, multimodal optimization problems are much more complicated in comparison to unimodal optimization problems.

1.3.5 Classification of Optimization Problems from the Perspective of the Kind of Decision-Making Variables

Optimization problems can be divided into three different types, from the perspective of the kind of the decision-making variables: continuous, discrete, and hybrid continuous and discrete. In Eqs. (1.1) and (1.2), if the decision-making variables have continuous values—real values over certain intervals—the corresponding problem is continuous. This optimization problem can be presented according to Eqs. (1.12) and (1.13):

$$\begin{aligned} &\underset{\mathrm{x}\in\mathrm{X}}{\text{Minimize}} \quad \mathrm{F(x)} = [f_1(\mathrm{x}),\ldots,f_a(\mathrm{x}),\ldots,f_\mathrm{A}(\mathrm{x})]; \quad \forall\{\mathrm{A}>0\},\forall\{a\in\Psi^\mathrm{A}\} \\ &\text{subject to}: \\ &\mathrm{G(x)} = [g_1(\mathrm{x}),\ldots,g_b(\mathrm{x}),\ldots,g_\mathrm{B}(\mathrm{x})] = 0; \quad \forall\{\mathrm{B}\geq 0\},\forall\{b\in\Psi^\mathrm{B}\} \\ &\mathrm{H(x)} = [h_1(\mathrm{x}),\ldots,h_e(\mathrm{x}),\ldots,h_\mathrm{E}(\mathrm{x})] \leq 0; \quad \forall\{\mathrm{E}\geq 0\},\forall\{e\in\Psi^\mathrm{E}\} \end{aligned} \tag{1.12}$$

$$\begin{aligned} \mathrm{x} = \ & [x_1,\ldots,x_v,\ldots,x_\mathrm{NCDV}]; \quad \forall\{v\in\Psi^\mathrm{NCDV}\},\forall\{\mathrm{x}\in\mathrm{X}\}, \\ & \forall\{x_v^{\min}\leq x_v\leq x_v^{\max}\,|\,v\in\Psi^\mathrm{NCDV}\} \end{aligned} \tag{1.13}$$

Using similar reasoning for Eqs. (1.1) and (1.2), if the decision-making variables have discrete values over certain intervals, the corresponding problem is discrete. This optimization problem can be given using Eqs. (1.14) and (1.15):

$$\begin{aligned}&\underset{x\in X}{\text{Minimize}}\quad F(x)=[f_1(x),\ldots,f_a(x),\ldots,f_A(x)];\quad \forall\{A>0\},\forall\{a\in\Psi^A\}\\&\quad\text{subject to}:\\&\quad G(x)=[g_1(x),\ldots,g_b(x),\ldots,g_B(x)]=0;\quad \forall\{B\geq 0\},\forall\{b\in\Psi^B\}\\&\quad H(x)=[h_1(x),\ldots,h_e(x),\ldots,h_E(x)]\leq 0;\quad \forall\{E\geq 0\},\forall\{e\in\Psi^E\}\end{aligned} \tag{1.14}$$

$$\begin{aligned}&x=[x_1,\ldots,x_v,\ldots,x_{NDDV}];\quad \forall\{v\in\Psi^{NDDV}\},\forall\{x\in X\},\\&\{x_v\in\{x_v(1),\ldots,x_v(w),\ldots,x_v(W_v)\}|v\in\Psi^{NDDV}\}\end{aligned} \tag{1.15}$$

In a discrete optimization problem, if the discrete decision-making variables have only binary values, {0, 1}, then the problem is known as a pure-binary optimization problem. Many real-world optimization problems are found to have a combination of continuous and discrete decision-making variables, making them mixed-integer problems—see Eqs. (1.1) and (1.2). In this situation, the complexity of the optimization problem dramatically increases. In the mixed-integer optimization problem, if the discrete decision-making variables have only binary values then the problem is recognized as a mixed-binary integer problem.

1.3.6 Classification of Optimization Problems from the Perspective of the Number of Decision-Making Variables

Optimization problems can be classified into two different types, from the perspective of the NDV: single-variable and multivariable. In the standard formulation of an optimization problem described by Eqs. (1.1) and (1.2), if the NDV is equal to 1 then the corresponding problem becomes a single-variable problem. A single-variable optimization problem is thus represented on the basis of Eqs. (1.16) and (1.17):

$$\begin{aligned}&\underset{x\in X}{\text{Minimize}}\quad F(x)=[f_1(x),\ldots,f_a(x),\ldots,f_A(x)];\quad \forall\{A>0\},\forall\{a\in\Psi^A\}\\&\quad\text{subject to}:\\&\quad G(x)=[g_1(x),\ldots,g_b(x),\ldots,g_B(x)]=0;\quad \forall\{B\geq 0\},\forall\{b\in\Psi^B\}\\&\quad H(x)=[h_1(x),\ldots,h_e(x),\ldots,h_E(x)]\leq 0;\quad \forall\{E\geq 0\},\forall\{e\in\Psi^E\}\end{aligned} \tag{1.16}$$

$$\begin{aligned}x=\ &[x];\ \forall\{x\in X\},\forall\{x^{min}\leq x\leq x^{max}|x:\text{continuous decision-making variable}\},\\&\{x\in\{x(1),\ldots,x(w),\ldots,x(W)\}|x:\text{discrete decision-making variable}\}\end{aligned} \tag{1.17}$$

Similarly, if the NDV is greater than 1 then the corresponding problem is converted into a multivariable problem. A multivariable optimization problem has a much higher computational burden compared to a single-variable optimization problem.

1.3.7 Classification of Optimization Problems from the Perspective of the Separability of the Employed Equations

Optimization problems can be categorized into two different types, from the perspective of the separability of the employed equations: separable and inseparable. In Eqs. (1.1) and (1.2), the corresponding problem is a separable optimization problem provided that all objective functions and equality and inequality constraints can be indicated as the sum of the NDV single-variable functions. The separable optimization problem is given according to Eqs. (1.18) and (1.19):

$$\begin{aligned}
&\underset{\mathrm{x}\in\mathrm{X}}{\text{Minimize}} \quad \mathrm{F(x)} = \left[\sum_{v\in\Psi^{\mathrm{NDV}}} f_{1,v}(x_v),\ldots,\sum_{v\in\Psi^{\mathrm{NDV}}} f_{a,v}(x_v),\ldots,\sum_{v\in\Psi^{\mathrm{NDV}}} f_{\mathrm{A},v}(x_v)\right];\\
&\quad \forall\{\mathrm{A}>0\},\forall\{a\in\Psi^{\mathrm{A}}\}\\
&\quad \text{subject to :}\\
&\quad \mathrm{G(x)} = \left[\sum_{v\in\Psi^{\mathrm{NDV}}} g_{1,v}(x_v),\ldots,\sum_{v\in\Psi^{\mathrm{NDV}}} g_{b,v}(x_v),\ldots,\sum_{v\in\Psi^{\mathrm{NDV}}} g_{\mathrm{B},v}(x_v)\right] = 0;\\
&\quad \forall\{\mathrm{B}\geq 0\},\forall\{b\in\Psi^{\mathrm{B}}\}\\
&\quad \mathrm{H(x)} = \left[\sum_{v\in\Psi^{\mathrm{NDV}}} h_{1,v}(x_v),\ldots,\sum_{v\in\Psi^{\mathrm{NDV}}} h_{e,v}(x_v),\ldots,\sum_{v\in\Psi^{\mathrm{NDV}}} h_{\mathrm{E},v}(x_v)\right] \leq 0;\\
&\quad \forall\{\mathrm{E}\geq 0\},\forall\{e\in\Psi^{\mathrm{E}}\}
\end{aligned} \tag{1.18}$$

$$\begin{aligned}
\mathrm{x} = \ & [x_1,\ldots,x_v,\ldots,x_{\mathrm{NDV}}]; \quad \forall\{v\in\Psi^{\mathrm{NDV}}\},\forall\{\Psi^{\mathrm{NDV}}=\Psi^{\mathrm{NCDV+NDDV}}\},\forall\{\mathrm{x}\in\mathrm{X}\},\\
& \forall\{x_v^{\min}\leq x_v\leq x_v^{\max}\,|\,v\in\Psi^{\mathrm{NCDV}}\},\\
& \{x_v\in\{x_v(1),\ldots,x_v(w),\ldots,x_v(W_v)\}\,|\,v\in\Psi^{\mathrm{NDDV}}\}
\end{aligned} \tag{1.19}$$

Moreover, if each of the objective functions and equality and inequality constraints cannot be expressed as the sum of the NDV single-variable functions, the corresponding problem is an inseparable problem.

1.3.8 Classification of Optimization Problems from the Perspective of Uncertainty

Optimization problems can be broken down into two different types, from the perspective of uncertainty: deterministic and nondeterministic. In Eqs. (1.1) and (1.2), if the values of the objective functions and equality and inequality constraints are precisely specified for each set of decision-making variables and input parameters, the corresponding problem is deterministic. Conversely, the problem is nondeterministic if some or all of the decision-making variables and input parameters have a nondeterministic or stochastic nature. Put simply, in the nondeterministic optimization problem, the values of the objective functions and equality and inequality constraints are not exactly determined and have uncertainty. Uncertainly in optimization problems introduces additional complexities and increases the effects of existing ones. The need to provide new techniques is, therefore, felt more than ever to accurately analyze and mitigate the effects of the different uncertainties and complexities arising from them.

1.4 Optimization Algorithms and Their Characteristics

Different optimization problems, with various structures and parameters, have given rise to the emergence of distinct optimization algorithms. Each of the optimization algorithms is only suitable for solving a specific range of optimization problems. In the literature, there is no unique, well-recognized approach for classifying the optimization algorithms. This challenging task, to develop diverse indices, has fallen on specialists and researchers in the field. In a broad sense, optimization algorithms can be classified into two main categories: deterministic and nondeterministic/stochastic. Deterministic algorithms are based on a very precise process such that the decision-making variables and employed functions are repeatable. Most classical and traditional optimization algorithms fall into the category of deterministic optimization algorithms. The Newton-Raphson (NR) algorithm is one of the well-known deterministic optimization algorithms, which requires the derivatives of the employed functions in order to solve an optimization problem. The NR algorithm is perfectly appropriate for solving unimodal optimization problems but, by contrast, does not have favorable performance in multimodal optimization problems or optimization problems with discontinuity in objective functions. Readers interested in a comprehensive discussion on this topic are referred to the work by Rao [1].

In addition, nondeterministic/stochastic optimization algorithms always have a stochastic behavior. These algorithms can be broken down into two different types: heuristic and meta-heuristic. Heuristic means finding a solution by trial and error. Heuristic optimization algorithms have a simple structure and using them, particularly on complicated and large-scale optimization problems, can yield relatively good solutions over a reasonable period of time. However, the main disadvantage of this category of the optimization algorithms is that there is no guarantee that an optimal solution or a set of optimal solutions can be found. Local search algorithms are the most well-known heuristic optimization algorithms. These algorithms move through limited variations from one solution to another in the search space so as to find a desirable solution. In general, the local search optimization algorithms are approximate or incomplete algorithms because, even if the best solution found by these algorithms is not optimal, the search process may be stopped. Stopping occurs even if the optimization algorithm is terminated due to its failure to improve the solution. In this condition, the optimal solution can be far from the proximity and neighborhood of the solutions that the algorithm has passed through. Readers interested in a comprehensive discussion on heuristic optimization algorithms are referred to the work by Rothlauf [2].

Developments to overcome the drawbacks of the heuristic optimization algorithms are known as meta-heuristic optimization algorithms. In the meta-heuristic phrase, the "meta" part means beyond or at a higher level. Hence, and as expected, meta-heuristic optimization algorithms have better performance compared to their heuristic counterparts. In the next section, a brief introduction to meta-heuristic optimization algorithms and their classification is presented.

1.5 Meta-heuristic Optimization Algorithms

In a very general sense, meta-heuristic algorithms are optimization techniques operating independently of the optimization problems. In other words, unlike other optimization algorithms, these algorithms are not dependent on the architecture of the optimization problems and can be used to solve a wide range of optimization problems with different structures. The main purpose of meta-heuristic optimization algorithms is not only their efficient work in the search space, but also their use of strategies to avoid getting stuck in a local optimum point and instead finding the semi-optimal or optimal solutions. Meta-heuristic optimization algorithms employ experience and data obtained during the search process in the form of a memory, with the aim of guiding this process into a global optimum point in the search space. These algorithms also rely on random search principles in all stages of the search process for the optimal solution or a set of optimal solutions.

Most meta-heuristic optimization algorithms use a somewhat similar mechanism in order to search for the optimal solution or a set of optimal solutions. The search process of these algorithms begins with the creation of one or more random responses—initial solutions—within the permissible range of the decision variables. Afterwards, one or more new random responses are generated by employing the predefined operators of the optimization algorithm. The new responses are compared to available responses in the memory of the optimization algorithm; the memory is then updated according to the best responses. This process continues until the stopping criterion of the optimization algorithm is reached. The most important part of the architecture of meta-heuristic optimization algorithms is the operators employed in order to generate one or more new responses. From a technical point of view, this part plays the role of a central core for the meta-heuristic optimization algorithms and expresses its intelligence capacity and performance capability.

Given the breadth and variety of meta-heuristic optimization algorithms, providing a well-organized classification for recognizing their characteristics is certainly a necessity. This category can be highly influential in choosing a meta-heuristic algorithm to solve a typical optimization problem. In view of the fact that meta-heuristic optimization algorithms can be based on different criteria, it is a complicated and challenging task to provide a structural classification system for them. In this chapter, a classification of meta-heuristic optimization algorithms is presented with a focus on the inspirational source. For a comprehensive discussion regarding different classifications of these optimization algorithms on the basis of other criteria, interested readers are directed to the work by Rao [1], Ng et al. [3], and Boussaïd et al. [4].

1.5.1 Classification of Meta-heuristic Optimization Algorithms with a Focus on Inspirational Sources

Nature is a huge and powerful source of inspiration and a highly diverse, dynamic, and strong exhibit of attractive phenomena for solving difficult and complicated problems across different sciences. The design of a wide range of meta-heuristic optimization algorithms has been inspired by natural phenomena. As a consequence, the inspirational source can be employed as a very worthy criterion for classifying meta-heuristic optimization algorithms. However, even with a focus on natural phenomena, depending on how much detail is considered, the classification of meta-heuristic optimization algorithms may have many subclassifications. By concentrating on the highest level of a given inspirational source, the algorithms can be broken down into four different categories [5]: (1) swarm intelligence-based meta-heuristic optimization algorithms (SI-MHOAs); (2) biologically inspired meta-heuristic optimization algorithms not based on swarm intelligence (BI-MHOAs-NSI); (3) physics- and chemistry-based meta-heuristic optimization algorithms (P&C-MHOAs); and (4) human behavior- and society-inspired meta-heuristic optimization algorithms (H&S-MHOAs). In the following sections, each of the aforementioned categories is briefly addressed.

1.5.1.1 Swarm Intelligence-Based Meta-heuristic Optimization Algorithms

This category of meta-heuristic optimization algorithms was inspired by existing swarm intelligence-oriented structures in nature. SI-MHOAs are organized by using a population of agents (e.g., flocks of birds, fish, ants, termites, bees, fireflies, bats) that interact locally with each other and globally with the environment. These multi-agent populations follow some simple rules. Although there is no centralized control structure that determines the behaviors of the agents, local interactions and partly random interactions among the agents give rise to the emergence of intelligent global behavior, which is impossible for an individual agent. Put another way, any available agent in the swarm group may be considered as unintelligible; however, the entire swarm group may demonstrate a self-organized behavior. Accordingly, a swarm group consisting of multiple agents will exhibit intelligent behaviors. Moreover, the structure of SI-MHOAs is such that data is shared among the different agents. This data sharing during various iterations of the optimization algorithm results in an increase in the efficiency of the related meta-heuristic optimization algorithms. On the other hand, the available agents in the swarm group can be effortlessly paralleled. The parallelism characteristic of SI-MHOAs can cause real-world, large-scale optimization to be more feasible from an implementation standpoint. As a result, meta-heuristic optimization algorithms are highly popular and widely employed for solving optimization problems with different structures across various sciences.

Examples of the most well-known SI-MHOAs include (1) particle swarm optimization algorithm [6]; (2) ant colony optimization algorithm [7]; (3) firefly search optimization algorithm [8]; (4) bat search optimization algorithm [9]; (5) bacteria foraging optimization algorithm [10]; (6) bee colony optimization algorithm [11]; (7) wolf search optimization algorithm [12]; and, (8) cuckoo search optimization algorithm [13]. Interested readers are directed to the work by Parpinelli and Hopes [14] for a comprehensive discussion of SI-MHOAs.

1.5.1.2 Biologically Inspired Meta-heuristic Optimization Algorithms Not Based on Swarm Intelligence

SI-MHOAs represent a subset of a larger set known as biologically inspired meta-heuristic optimization algorithms. Many of the biologically inspired algorithms do not directly apply the original characteristic of the SI-MHOAs—swarm/collective behavior—and are, therefore, known as a subset of the BI-MHOAs-NSI. As a result, the integration of the subset of the SI-MHOAs and the subset of the BI-MHOAs-NSI is organized as a set of biologically inspired meta-heuristic optimization algorithms.

Examples of the most well-known BI-MHOAs-NSI include (1) genetic optimization algorithm [15]; (2) brain storm optimization algorithm [16]; (3) dolphin echolocation optimization algorithm [17]; (4) shuffled frog leaping optimization algorithm [18]; and, (5) flower pollination optimization algorithm [19]. Readers interested in a comprehensive discussion on this topic are referred to the work by Kar [20].

1.5.1.3 Physics- and Chemistry-Based Meta-heuristic Optimization Algorithms

Biologically inspired meta-heuristic optimization algorithms belong to a larger set known as the nature-inspired meta-heuristic optimization algorithms. However, some of these algorithms cannot be included in the set of biologically inspired meta-heuristic optimization algorithms, due to some specific characteristics. These algorithms have been inspired by using physics and chemistry phenomena and, therefore, are identified as P&C-MHOAs. Combining the set of P&C-MHOAs and the set of biologically inspired meta-heuristic optimization algorithms forms a perfect set of nature-inspired meta-heuristic optimization algorithms.

Examples of the most well-known P&C-MHOAs include (1) harmony search optimization algorithm [21]; (2) gravitational search optimization algorithm [22]; (3) simulated annealing optimization algorithm [23]; (4) electromagnetism optimization algorithm [24]; and, (5) big bang-big crunch optimization algorithm [25]. Interested readers are directed to the work by Biswas et al. [26] and the work by Siddique and Adeli [27] for a comprehensive discussion in the context of P&C-MHOAs.

1.5.1.4 Human Behavior- and Society-Inspired Meta-heuristic Optimization Algorithms

Some of the meta-heuristic optimization algorithms have an inspirational source other than nature. It is, therefore, not possible to put these optimization algorithms in the categories mentioned thus far. These meta-heuristic optimization algorithms, which are generally inspired by human behavior and modern human societies, fall into the fourth category associated with the classification of the meta-heuristic optimization algorithms.

Examples of the most well-known H&S-MHOAs include (1) imperialist competitive optimization algorithm [28]; (2) anarchic society optimization algorithm [29]; (3) social emotional optimization algorithm [30]; (4) league championship optimization algorithm [31]; and, (5) artificial cooperative search optimization algorithm [32].

1.5.1.5 Some Hints Concerning the Architecture of Meta-heuristic Optimization Algorithms

As indicated earlier, nature is an extremely broad inspirational source for the design of meta-heuristic optimization algorithms. Nevertheless, specialists and researchers should note that many of the available inspirational sources in nature can resemble one another. Hence, considering each of these inspirational sources separately in the design of meta-heuristic optimization algorithms may decrease the novelty and originality characteristics for the meta-heuristic optimization algorithms. To illustrate, with about 33,600 different species, fish have one of the widest variety of species among wildlife species in nature. This does not mean that specialists and researchers should invent 33,600 meta-heuristic optimization algorithms. Nor does it mean that by inventing a meta-heuristic optimization algorithm from a single species of fish research and development should be neglected on other species. It is simply important to note that existing similarities can be found in inspirational sources, which ultimately leads to the homogeneity of the meta-heuristic optimization algorithms, which are identified and taken into account as much as possible in the architecture of the algorithms. In addition, it is suggested that instead of establishing new algorithms using inspirational sources similar to those for other algorithms, specialists and researchers should concentrate on developing the architecture, performance, and efficiency of available meta-heuristic optimization algorithms.

1.6 Conclusions

In this chapter, the authors presented a brief introduction to the optimization concepts and meta-heuristic optimization algorithms. First, the standard formulation of an optimization problem was explained, followed by a thorough classification for optimization problems with a focus on eight different points of view: (1) the number of objective functions; (2) the constraints; (3) the nature of the employed equations; (4) the objective functions landscape; (5) the kind of decision-making variables; (6) the number of decision-making variables; (7) the separability of the employed equations; and, (8) the uncertainty.

In addition, a thorough categorization was reported for the meta-heuristic optimization algorithms by concentrating on the inspirational source. This categorization was developed based on the highest level of the inspirational sources and was broken down into four classes: (1) swarm intelligence-based meta-heuristic optimization algorithms; (2) biologically inspired meta-heuristic optimization algorithms not based on swarm intelligence; (3) physics- and chemistry-based meta-heuristic optimization algorithms; and, (4) human behavior- and society-inspired meta-heuristic optimization algorithms.

Finally, based on hints concerning architecture of the meta-heuristic optimization algorithms, it was suggested that efficient meta-heuristic optimization algorithms should be addressed and invented by specialists and researchers so that real-world, large-size, non-convex optimization having a nonlinear, mixed-integer nature become more practical from an implementation perspective. As a result, the main focus of the authors is not only to investigate the advantages and disadvantages of meta-heuristic optimization algorithms, with the aim of achieving better insight into more beneficial algorithms, but also to develop algorithms to tackle the difficulties in solving complicated optimization problems.

Appendix 1: List of Abbreviations and Acronyms

NCDV	Number of continuous decision-making variables
NDDV	Number of discrete decision-making variables
NDV	Number of decision-making variables including continuous and discrete decision-making variables
NR	Newton-Raphson
SI-MHOAs	Swarm intelligence-based meta-heuristic optimization algorithms

(continued)

BI-MHOAs-NSI	Biologically inspired meta-heuristic optimization algorithms not based on swarm intelligence
P&C-MHOAs	Physics- and chemistry-based meta-heuristic optimization algorithms
H&S-MHOAs	Human behavior- and society-inspired meta-heuristic optimization algorithms

Appendix 2: List of Mathematical Symbols

Index:	
a	Index for objective functions running from 1 to A
b	Index for equality constraints running from 1 to B
e	Index for inequality constraints running from 1 to E
v	Index for decision-making variables, including the continuous and discrete decision-making variables, running from 1 to the NDV and an index for continuous decision-making variables running from 1 to the NCDV and also an index for discrete decision-making variables running from 1 to the NDDV
Set:	
Ψ^{A}	Set of indices of objective functions
Ψ^{B}	Set of indices of equality constraints
Ψ^{E}	Set of indices of inequality constraints
Ψ^{NCDV}	Set of indices of continuous decision-making variables
Ψ^{NDDV}	Set of indices of discrete decision-making variables
Ψ^{NDV}	Set of indices of decision-making variables, including the continuous and discrete decision-making variables
W_v	Set of indices of candidate permissible values of discrete decision-making variable v
$\mathfrak{R}$	Set of real numbers
$\mathfrak{R}^{B}$	B-dimensional set of real numbers
$\mathfrak{R}^{E}$	E-dimensional set of real numbers
$\mathfrak{R}^{NDV}$	NDV-dimensional set of real numbers
Parameters:	
x_v^{max}	Upper bound on the continuous decision-making variable v
x_v^{min}	Lower bound on the continuous decision-making variable v
X	Nonempty feasible decision-making space, including feasible continuous and discrete decision-making spaces
Z	Feasible objective space
Variables:	
$f_a(\mathrm{x})$	Objective function a of the optimization problem or component a of the vector of objective functions
F(x)	Vector of objective functions of the optimization problem
$g_b(\mathrm{x})$	Equality constraint b of the optimization problem or component b of the vector of equality constraints
G(x)	Vector of equality constraints of the optimization problem

(continued)

$h_e(x)$	Inequality constraint e of the optimization problem or component e of the vector of inequality constraints
H(x)	Vector of inequality constraints of the optimization problem
x_v	Continuous or discrete decision-making variable v
$x_v(w_v)$	Candidate permissible value w of discrete decision-making variable v
x	Vector of decision-making variables
z	Vector of objective functions

References

1. S.S. Rao, *Engineering Optimization, Theory and Practice*, 4th edn. (Wiley, New York, 2009)
2. F. Rothlauf, *Design of Modern Heuristics: Principles and Application* (Springer, New York, 2011)
3. K.K.H. Ng, C.K.M. Lee, F.T.S. Chan, Y. Lv, Review on meta-heuristics approaches for airside operation research. Appl. Soft Comput. **66**, 104–133 (2018)
4. I. Boussaïd, J. Lepagnot, P. Siarry, A survey on optimization metaheuristics. Inf. Sci. **237**, 82–117 (2013)
5. I. Fister Jr., X.S. Yang, I. Fister, J. Brest, D. Fister, A brief review of nature-inspired algorithms for optimization. Elektrotehniški Vestnik **80**(3), 1–7 (2013)
6. J. Kennedy, R. Eberhart, Particle swarm optimization. IEEE Int. Conf. Neural Netw. **4**, 1942–1948 (1995)
7. M. Dorigo, Optimization, learning and natural algorithms. Ph.D. Dissertation, Politecnico di Milano, Italy, 1992
8. X.S. Yang, Firefly algorithm, stochastic test functions and design optimization. Int. J. Bio-Inspired Comput. **2**(2), 78–84 (2010)
9. X.S. Yang, A new metaheuristic bat-inspired algorithm, in *Nature Inspired Cooperative Strategies for Optimization*, pp. 65–74, 2010
10. K.M. Passino, Biomimicry of bacterial foraging for distributed optimization and control. IEEE Control. Syst. **22**(3), 52–67 (2002)
11. D. Teodorovic, M. Dell'orco, Bee colony optimization–a cooperative learning approach to complex transportation problems, in *16th Mini-EURO Conference on Advanced OR and AI Methods in Transportation*, pp. 51–60, 2005
12. R. Tang, S. Fong, X.S. Yang, S. Deb, Wolf search algorithm with ephemeral memory, in *7th International Conference on Digital Information Management*, pp. 165–172, 2012
13. X.S. Yang, S. Deb, Cuckoo search via Lévy flights, in *World Congress on Nature & Biologically Inspired Computing*, pp. 210–214, 2009
14. R.S. Parpinelli, H.S. Lopes, New inspirations in swarm intelligence: a survey. Int. J. Bio-Inspired Comput. **3**(1), 1–16 (2011)
15. J.H. Holland, *Genetic Algorithms and Adaptation*, in *Adaptive Control of Ill-Defined Systems, NATO Conference Series (II Systems Science)*, vol. 16, (Springer, Boston, MA, 1984), pp. 317–333
16. Y. Shi, An optimization algorithm based on brainstorming process. Int. J. Swarm Intell. Res. **2** (4), 35–62 (2011)
17. A. Kaveh, N. Farhoudi, A new optimization method: dolphin echolocation. Adv. Eng. Softw. **59**, 53–70 (2013)
18. M.M. Eusuff, K.E. Lansey, Optimization of water distribution network design using the shuffled frog leaping algorithm. J. Water Resour. Plan. Manag. **129**(3), 210–225 (2003)

19. X.S. Yang, Flower pollination algorithm for global optimization, in *International Conference on Unconventional Computing and Natural Computation*, pp. 240–249, 2012
20. A.K. Kar, Bio inspired computing–a review of algorithms and scope of applications. Expert Syst. Appl. **59**, 20–32 (2016)
21. Z.W. Geem, J.H. Kim, G.V. Loganathan, A new heuristic optimization algorithm: harmony search. Simulation **76**(2), 60–68 (2001)
22. E. Rashedi, H. Nezamabadi-Pour, S. Saryazdi, GSA: a gravitational search algorithm. Inf. Sci. **179**(13), 2232–2248 (2009)
23. S. Kirkpatrick, C.D. Gelatt Jr., M.P. Vecchi, Optimization by simulated annealing. Science **220** (4598), 671–680 (1983)
24. E. Cuevas, D. Oliva, D. Zaldivar, M. Perez-Cisneros, H. Sossa, Circle detection using electromagnetism optimization. Inf. Sci. **182**(1), 40–55 (2012)
25. Z. Zandi, E. Afjei, M. Sedighizadeh, Reactive power dispatch using big bang-big crunch optimization algorithm for voltage stability enhancement, in *International Conference on Power and Energy*, pp. 239–244, 2012
26. A. Biswas, K.K. Mishra, S. Tiwari, A.K. Misra, Physics-inspired optimization algorithms: a survey. J. Opt. **2013**, 1–15 (2013)
27. N. Siddique, H. Adeli, Nature-inspired chemical reaction optimisation algorithms. Cogn. Comput. **9**(4), 411–422 (2017)
28. E. Atashpaz-Gargari, C. Lucas, Imperialist competitive algorithm: an algorithm for optimization inspired by imperialistic competition, in *IEEE Congress on Evolutionary Computation*, pp. 4661–4667, 2007
29. H. Shayeghi, J. Dadashpour, Anarchic society optimization based PID control of an automatic voltage regulator (AVR) system. Electr. Electron. Eng. **2**(4), 199–207 (2012)
30. Y. Xu, Z. Cui, J. Zeng, Social emotional optimization algorithm for nonlinear constrained optimization problems, in *International Conference on Swarm, Evolutionary, and Memetic Computing*, pp. 583–590, 2010
31. A. Husseinzadeh Kashan, League championship algorithm: a new algorithm for numerical function optimization, in *International Conference of Soft Computing and Pattern Recognition*, pp. 43–48, 2009
32. P. Civicioglu, Artificial cooperative search algorithm for numerical optimization problems. Inf. Sci. **229**, 58–76 (2013)

Chapter 2
Introduction to Multi-objective Optimization and Decision-Making Analysis

2.1 Introduction

In today's world, overcoming complicated real-world engineering challenges with big data has become an indispensable need for specialists and researchers in order to make better use of their time and resources. These engineering challenges are mostly addressed as a large-scale, non-convex optimization problem having a nonlinear, mixed-integer nature. As a result, the application of the optimization is significantly increased for solving real-world engineering challenges and achieving an optimal solution. Initially, optimization problems were organized as simple single-objective mathematical models in which one given objective function needed to be minimized or maximized. More precisely, a single-objective optimization problem consists of an individual objective function subject to some specified constraints in such a way that solving this particular optimization problem leads to finding an individual optimal solution. That is to say that the main goal of minimization or maximization of a single-objective optimization problem is to obtain the minimum or maximum value of the corresponding objective function, provided that this value does not violate any specified constraints—an optimal solution.

Conversely, in multi-objective optimization problems (MOOPs), a set of objective functions that are often conflicting should be simultaneously minimized or maximized. In this circumstance, solving a MOOP results in finding a set of different compromise solutions called a Pareto-optimal solution set or non-dominated optimal solution set. With that in mind, only one solution should be chosen from the Pareto-optimal solution set. Unlike a single-objective optimization problem, solving a MOOP is, therefore, composed of three important and completely different steps: (1) formation of a mathematical model; (2) optimization; and (3) decision-making [1]. In the optimization step, the Pareto-optimal solution set is determined; however, in the decision-making step, the most satisfactory solution is chosen from the Pareto-optimal solution set based on the preferences of the decision maker.

M. Kiani-Moghaddam et al., *Modern Music-Inspired Optimization Algorithms for Electric Power Systems*, Power Systems,
https://doi.org/10.1007/978-3-030-12044-3_2

To illustrate, consider a two-objective optimization problem for operation of a typical engineering system. Simultaneous minimization of operational costs and maximization of system reliability are two conflicting objectives considered in this optimization problem. This means that minimization of operational costs brings about a reduction in system reliability, and maximization of system reliability gives rise to an increase in operational costs. As a consequence, there is no individual solution that can simultaneously optimize these two conflicting objectives. In this condition, a Pareto-optimal solution set is achieved by solving this two-objective optimization problem. As an instance, take into account two solutions in the Pareto-optimal solution set. If the first solution, in terms of the operational costs, can overcome the second solution—given that the first solution has a lower cost than the second one—this solution with respect to system reliability cannot overcome the second solution—given that the first solution definitely has less reliability than the second one. Put another way, compared to the second solution, the first solution is a non-dominated response/output in terms of operational costs—the lowest operational costs—when it is a dominated response/output with respect to system reliability—the most unfavorable performance. After determination of the Pareto-optimal solution set, or non-dominated solution set, the decision maker, by considering its preferences, should choose the final solution so that a trade-off is made up between operational costs and system performance. Given the preferences of the decision maker, it is also possible that system performance would have a higher degree of significance compared to the operational costs and vice versa. Hence, the preferences of the decision maker dramatically affect the choice of the final solution from the Pareto-optimal solution set.

In the MOOPs, the complexities and difficulties of the solution process are dramatically increased in view of introducing and integrating new concepts compared to single-objective optimization problems. In addition, multi-objective optimization algorithms (MOOAs) are needed to solve the MOOPs. In related literature, many MOOAs have been developed to deal with a wide range of multi-objective optimization problems. However, each MOOA is appropriate for solving only a specific range of MOOPs. The choice of a well-suited MOOA depends first on a full understanding of the MOOP and its characteristics and second on having full knowledge of the architecture and features of the different MOOAs. Due to the different concepts of the optimization in the MOOPs and the diversity and variety of MOOAs, it is thoroughly indispensable to clarify the fundamental concepts of multi-objective optimization and provide a suitable classification for the MOOAs. In this chapter, then, the authors will concentrate on the following targets.

- Target 1: Providing a brief introduction associated with fundamental concepts of optimization in the MOOPs.
- Target 2: Presenting a brief overview pertaining to the classification of the MOOAs.

In this chapter, the authors do not present all of the details related to optimization concepts in the MOOPs, as it is assumed that the reader is already familiar with the elementary concepts of optimization. The main focus, then, will be on the fundamental concepts of optimization in the MOOPs, particularly the fundamental concepts discussed in the MOOPs that will be widely employed later in this book. Where appropriate, though, the reader will be referred to related studies that cover more details of concepts of optimization in the MOOPs.

The remainder of this chapter is arranged as follows: First, the necessity of using the multi-objective optimization process is reviewed in Sect. 2.2. Then, the fundamental concepts of optimization in the MOOPs are described in Sect. 2.3. In Sect. 2.4, the classification of the MOOAs is addressed from different points of view. A fuzzy satisfying method is also expounded upon in Sect. 2.5. Finally, the chapter ends with a brief summary and some concluding remarks in Sect. 2.6.

2.2 Necessity of Using Multi-objective Optimization

In a very general sense, many objective functions can be employed in real-world engineering problems. These objective functions usually have a conflicting, noncommensurable, and correlated nature with each other. In this way, the integration of objective functions of a MOOP, with the aim of forming a single-objective optimization problem and then employing the developed single-objective solvers, is a common misconception.

The conversion of a MOOP into a single-objective optimization problem causes the decision-making step to be transferred before the optimization step. It is, therefore, very difficult to specify the preferences of the decision maker before the optimization, and it may not match the obtained solution of the single-objective optimization problem with a determined solution of the MOOP that is selected from the Pareto-optimal solution set by the decision maker. Nevertheless, the implementation of the optimization process in a MOOP, without turning it into a single-objective problem, can force the decision-making step to be placed after the optimization step or these two steps to be transformed into a hybrid process. This structure helps the decision maker to better understand the MOOP and be able to make a more knowledgeable choice through the Pareto-optimal solution set with regard to its preferences. Achieving the Pareto-optimal solution set also enables the decision maker to perform a thorough analysis regarding the interdependencies among decision-making variables, objective functions, and constraints. Acquiring knowledge about these interactions can be employed in order to reconsider the mathematical model of the optimization problem with the aim of increasing the chances of determining a solution that not only aligns better with reality but also better matches the preferences of the decision maker. As a result, if an optimization problem consists of multiple conflicting, noncommensurable, and correlated objective functions, the most reasonable strategy is to take advantage of the multi-objective optimization process in order to solve the problem.

2.3 Fundamental Concepts of Optimization in the MOOPs

The concepts of optimization in the MOOPs are different from those in single-objective problems. In this section, the fundamental concepts of optimization in the MOOPs are briefly addressed. As previously mentioned, the optimization process in the MOOPs includes three general steps: (1) formation of a mathematical model; (2) optimization; and, (3) decision-making [1]. The mathematical description of an optimization problem—the formulation of an optimization problem by defining its decision-making variables, objective functions, and constraints—is considered as the first step in the optimization process. The next two steps in the optimization process depend on the structure and characteristics of the problem. Many studies carried out in the context of the optimization process implicitly suppose that the MOOP has been correctly determined. In practice, however, this assumption is not necessarily valid on all occasions. As a consequence, providing a rigorous mathematical model by considering the structure and characteristics of a MOOP can be practically helpful in the optimization process.

2.3.1 *Mathematical Description of a MOOP*

Technically speaking, a MOOP consists of multiple objective functions in such a way that these functions ordinarily have a conflicting, noncommensurable, and correlated nature with each other. The mathematical description of a MOOP can generally be expressed according to Eqs. (2.1) and (2.2) [1]:

$$\begin{aligned}
\underset{\mathrm{x}\in\mathrm{X}}{\text{Minimize}}\quad & \mathrm{F(x)} = [f_1(\mathrm{x}),\ldots,f_a(\mathrm{x}),\ldots,f_{\mathrm{A}}(\mathrm{x})];\quad \forall\{\mathrm{A}\geq 2\},\forall\{a\in\Psi^{\mathrm{A}}\}\\
& \text{subject to}:\\
& \mathrm{G(x)} = [g_1(\mathrm{x}),\ldots,g_b(\mathrm{x}),\ldots,g_{\mathrm{B}}(\mathrm{x})] = 0;\quad \forall\{\mathrm{B}\geq 0\},\forall\{b\in\Psi^{\mathrm{B}}\}\\
& \mathrm{H(x)} = [h_1(\mathrm{x}),\ldots,h_e(\mathrm{x}),\ldots,h_{\mathrm{E}}(\mathrm{x})] \leq 0;\quad \forall\{\mathrm{E}\geq 0\},\forall\{e\in\Psi^{\mathrm{E}}\}
\end{aligned} \tag{2.1}$$

$$\begin{aligned}
\mathrm{x} = \quad & [x_1,\ldots,x_v,\ldots,x_{\mathrm{NDV}}];\quad \forall\{v\in\Psi^{\mathrm{NDV}},\Psi^{\mathrm{NDV}}=\Psi^{\mathrm{NCDV+NDDV}},\mathrm{x}\in\mathrm{X}\},\\
& \forall\{x_v^{\min}\leq x_v\leq x_v^{\max}|v\in\Psi^{\mathrm{NCDV}}\},\\
& \forall\{x_v\in\{x_v(1),\ldots,x_v(w),\ldots,x_v(W_v)\}|v\in\Psi^{\mathrm{NDDV}}\}
\end{aligned} \tag{2.2}$$

The explanations associated with the parameters and variables from Eqs. (2.1) and (2.2) were previously defined in Sect. 1.2.1 of Chap. 1. The vector of objective functions addresses the illustration of the vector of decision-making variables and contains the values of the objective functions, as given by Eq. (2.3):

$$z = F(x) = [f_1(x), \ldots, f_a(x), \ldots, f_A(x)]; \quad \forall\{A \geq 2\}, \forall\{a \in \Psi^A\} \tag{2.3}$$

The explanations related to the parameters and variables from Eq. (2.3) were formerly described in Sect. 1.2.1 of Chap. 1.

2.3.2 Concepts Associated with Efficiency, Efficient frontier, and Dominance

In this section, efficiency, efficient frontier, and dominance, as main concepts of the MOOP, are thoroughly demonstrated.

Efficiency definition: A vector of decision-making variables $x^{**} \in X$ is efficient in the MOOP, given in Eqs. (2.1) and (2.2), if there is no another vector of decision-making variables like $x^{*} \in X$ so that $F(x^{*}) \leq F(x^{**})$ with at least one $f_a(x^{*}) < f_a(x^{**})$. Otherwise, the vector of decision-making variables $x^{**} \in X$ is inefficient [2].

Efficient frontier definition: The complete set of efficient vectors of decision-making variables is known as the efficient frontier [2].

Dominance definition: A vector of objective functions $F(x^{**}) \in Z$ is non-dominated in the MOOP, given in Eqs. (2.1) and (2.2), if there is no another vector of objective functions like $F(x^{*}) \in Z$ in such a way that $F(x^{*}) \leq F(x^{**})$ with at least one $f_a(x^{*}) < f_a(x^{**})$. Otherwise, the vector of objective functions $F(x^{**}) \in Z$ is dominated or has failed [2]. Put another way, the vector of objective functions $F(x^{*}) \in Z$ overcomes the vector of objective functions $F(x^{**}) \in Z$ in the MOOP, if the following two conditions are simultaneously met:

Condition 1: All components or elements of the vector of objective functions $F(x^{*}) \in Z$ are not worse than the corresponding components or elements of the vector of objective functions $F(x^{**}) \in Z$, as given by Eq. (2.4):

$$F(x^{*}) \leq F(x^{**}) \vee f_a(x^{*}) \leq f_a(x^{**}); \text{for all objective functions}, \forall\{a \in \Psi^A\} \tag{2.4}$$

Note that the symbol "∨" in Eq. (2.4) represents the operator "or."

Condition 2: At least one of the components or elements of the vector of objective functions $F(x^{*}) \in Z$ is better than components or elements of the vector of objective functions $F(x^{**}) \in Z$, as presented by Eq. (2.5):

$$f_k(x^{*}) < f_k(x^{**}); \text{for at least one objective function}, \forall\{k \in \Psi^A\} \tag{2.5}$$

Given what has been described, it can be found that the definitions of efficiency and dominance are analogous for all practical aims. However, it must be noted that the concepts associated with efficiency and dominance usually refer to the vector of decision-making variables in a feasible decision-making space and vector of objective functions in a feasible objective space, respectively.

2.3.3 Concepts Pertaining to Pareto Optimality

Unlike a single-objective optimization problem, there is no a single solution that can simultaneously optimize all objective functions in a MOOP. The objective functions of a MOOP are often in conflict with each other, and parameters related to some objective functions may not lead to optimality for other objective functions—even though these parameters can sometimes give rise to worse amounts for other objective functions. As a consequence, in contrast to solving a single-objective optimization problem that yields a single solution, solving a MOOP results in finding a set of solutions that represents a trade-off among the different objective functions. These solutions are also known collectively as a Pareto-optimal solution set or a non-dominated optimal solution set. With that in mind, Pareto optimally, weakly, and appropriately Pareto optimal are other fundamental concepts of the MOOPs which are briefly reviewed in this section.

Definition of a Pareto-optimal solution: A vector of decision-making variables $x^{**} \in X$ is a Pareto-optimal solution in the MOOP, given in Eqs. (2.1) and (2.2), if there is no another vector of decision-making variables like $x^{*} \in X$ so that $F(x^{*}) \leq F(x^{**})$ and $f_a(x^{*}) < f_a(x^{**})$ for at least one objective function. Otherwise, the vector of decision-making variables $x^{**} \in X$ is not a Pareto-optimal solution [1–3]. In other words, the vector of decision-making variables $x^{**} \in X$ is a Pareto-optimal solution if there is no another vector of decision-making variables like $x^{*} \in X$ that can simultaneously satisfy the conditions presented in Eqs. (2.4) and (2.5). That is to say that the vector of decision-making variables $x^{**} \in X$ is a Pareto-optimal solution if there does not exist another vector of decision-making variables like $x^{*} \in X$ that can improve at least one of the objective functions of the MOOP without worsening other objective functions. The set of Pareto-optimal vectors of decision-making variables is taken into account as P(X). Mutually, a vector of objective functions is a Pareto-optimal solution if the corresponding vector of decision-making variables is a Pareto-optimal solution. In this way, the set of Pareto-optimal vectors of objective functions is considered as P(Z). Many algorithms used for solving multi-objective optimization problems provide solutions that are not Pareto-optimal. These solutions can, however, meet other criteria. One of the most important of these criteria that can be very useful and effective in real-world MOOPs and provide useful information for the decision maker is the weak Pareto-optimal solution, which can be explained as follows:

Definition of a weak Pareto-optimal solution: A vector of decision-making variables $x^{**} \in X$ is a weak Pareto-optimal solution in the MOOP, given in Eqs. (2.1) and (2.2), if there is no another vector of decision-making variables like $x^{*} \in X$ in such a way that $F(x^{*}) < F(x^{**})$. Otherwise, the vector of decision-making variables $x^{**} \in X$ is not a weak Pareto-optimal solution [1–3]. In simple terms, the vector of decision-making variables $x^{**} \in X$ is a weak Pareto-optimal solution if there does not exist another vector of decision-making variables like $x^{*} \in X$ so that the response/output obtained by this vector of decision-making variables in all objective functions of the MOOP is better than the response/output calculated by the vector of decision-making variables $x^{**} \in X$ in the corresponding objective functions. Or, the vector of decision-making variables $x^{**} \in X$ is a weak Pareto-optimal solution if there is no another vector of decision-making variables like $x^{*} \in X$ that can simultaneously improve all objective functions of the MOOP. The set of weak Pareto-optimal vectors of decision-making variables is considered as WP(X). Correspondingly, a vector of objective functions represents a weak Pareto-optimal solution if the corresponding vector of decision variables is a weak Pareto-optimal solution. In this case, the set of weak Pareto-optimal vectors of objective functions is taken into account as WP(Z). As a result, the set of Pareto-optimal vectors of decision-making variables belongs to a larger set called the set of weak Pareto-optimal vectors of decision-making variables. Accordingly, if the vector of decision-making variables $x^{**} \in X$ is the Pareto-optimal solution, then this vector is the weak Pareto-optimal solution. However, if the vector of decision-making variables $x^{**} \in X$ is the weak Pareto-optimal solution, then this vector is not necessarily the Pareto-optimal solution. Each of the available responses or outputs in the Pareto-optimal solution set can be classified as either an appropriate or an inappropriate Pareto-optimal response or output. In related literature, there are different definitions for the appropriate Pareto-optimal concept, which are not equivalent. Here, the definition employed for an appropriate or inappropriate Pareto-optimal concept is derived according to Geoffrion [4].

Definition of an appropriate Pareto-optimal solution: A vector of decision-making variables $x^{**} \in X$ is an appropriate Pareto-optimal solution in the MOOP, given in Eqs. (2.1) and (2.2), if this vector is the Pareto-optimal solution and if there is some real number $J > 0$ not only for each objective function a and for each of the other vectors of decision-making variables like $x^{*} \in X$ satisfying $f_a(x^{*}) < f_a(x^{**})$, but also that there is at least one objective function k in the MOOP such that $f_k(x^{**}) < f_k(x^{*})$ and $\{f_a(x^{**}) - f_a(x^{*})\}/\{f_k(x^{*}) - f_k(x^{**})\} \leq J$. The quotient of the fraction $\{f_a(x^{**}) - f_a(x^{*})\}/\{f_k(x^{*}) - f_k(x^{**})\}$ refers to a compromise in the MOOP; that is, it indicates an increase in the objective function k originating from a decrease in the objective function a. Put simply, a Pareto-optimal solution is an appropriate Pareto-optimal solution if there exists at least one pair of objective functions such that a confined decrease in one objective function is possible only with an increase in the other objective function. The set of appropriate Pareto-optimal vectors of decision-making variables is AP(X). Mutually, a vector of objective functions is an appropriate Pareto-optimal solution if the corresponding vector of decision-making variables is an appropriate Pareto-optimal solution. In this manner, the set of appropriate Pareto-optimal vectors of objective functions is considered as AP(Z).

As a result, the set of appropriate Pareto-optimal vectors of decision-making variables belongs to a larger set called the set of Pareto-optimal vectors of decision-making variables. Therefore, if the vector of decision-making variables $x^{**} \in X$ is the appropriate Pareto-optimal solution, then this vector is the Pareto-optimal solution. However, if the vector of decision-making variables $x^{**} \in X$ is the Pareto-optimal solution, then this vector is not necessarily the appropriate Pareto-optimal solution.

Concepts pertaining to Pareto optimality and relationships among these concepts are demonstrated in Fig. 2.1. In Fig. 2.1, the set of appropriate Pareto-optimal

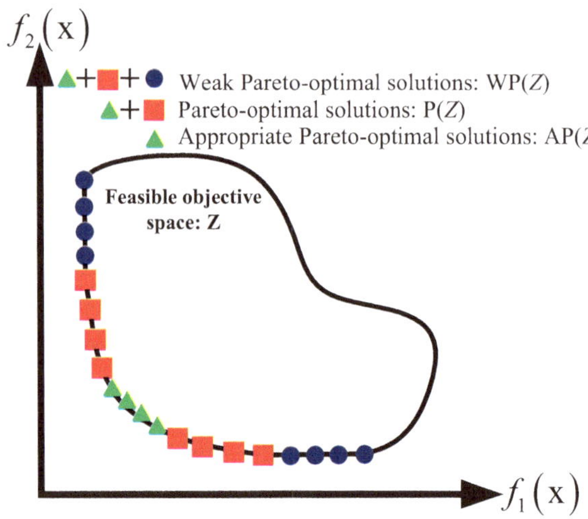

Fig. 2.1 Concepts pertaining to Pareto optimality

responses and outputs is shown with solid green triangles. The set of Pareto-optimal responses and outputs is illustrated as a sum of the solid green triangles and solid red squares. The set of weak Pareto-optimal responses and outputs is depicted as a sum of solid green triangles, solid red squares, and solid blue circles. Also from Fig. 2.1, it can be seen that the set of appropriate Pareto-optimal responses and outputs belongs to a larger set called the set of Pareto-optimal responses and outputs, as previously mentioned. Moreover, it can be seen that the set of Pareto-optimal responses/outputs belongs to a larger set called the set of weak Pareto-optimal responses/outputs, as stated earlier.

2.3.4 Concepts Related to the Vector of Ideal Objective Functions and the Vector of Nadir Objective Functions

Suppose that in the MOOP, given in Eq. (2.1) and (2.2), the objective functions are bounded on the feasible objectives space. In this circumstance, the upper and lower bounds associated with the set of Pareto-optimal responses and outputs in the feasible objectives space can provide very useful information about the MOOP. For this MOOP, the lower bounds related to the set of Pareto-optimal responses and outputs are available in the vector of ideal objective functions—$\mathrm{z}^{\mathrm{ideal}} \in \Re^{\mathrm{A}}$ [1–3]. The vector of ideal objective functions is defined using Eq. (2.6):

$$\mathrm{z}^{\mathrm{ideal}} = \left[z_1^{\mathrm{ideal}}, \ldots, z_a^{\mathrm{ideal}}, \ldots, z_{\mathrm{A}}^{\mathrm{ideal}}\right]; \quad \forall \left\{a \in \Psi^{\mathrm{A}}\right\} \tag{2.6}$$

The component or element a relevant to the vector of ideal objective functions z_a^{ideal} can be obtained by minimizing the objective function a of the MOOP as a single-objective optimization problem bounded by equality and inequality constraints, as given in Eqs. (2.7) and (2.8):

$$\begin{aligned}\underset{\mathrm{x}\in \mathrm{X}}{\text{Minimize}} \quad & \mathrm{F}(\mathrm{x}) = [f_a(\mathrm{x})]; \quad \forall\{a \in \Psi^{\mathrm{A}}\} \\ & \text{subject to :} \\ & \mathrm{G}(\mathrm{x}) = [g_1(\mathrm{x}), \ldots, g_b(\mathrm{x}), \ldots, g_{\mathrm{B}}(\mathrm{x})] = 0; \quad \forall\{\mathrm{B} \geq 0\}, \forall\{b \in \Psi^{\mathrm{B}}\} \\ & \mathrm{H}(\mathrm{x}) = [h_1(\mathrm{x}), \ldots, h_e(\mathrm{x}), \ldots, h_{\mathrm{E}}(\mathrm{x})] \leq 0; \quad \forall\{\mathrm{E} \geq 0\}, \forall\{e \in \Psi^{\mathrm{E}}\}\end{aligned} \tag{2.7}$$

$$\begin{aligned}\mathrm{x} = \; & [x_1, \ldots, x_v, \ldots, x_{\mathrm{NDV}}]; \quad \forall\{v \in \Psi^{\mathrm{NDV}}, \Psi^{\mathrm{NDV}} = \Psi^{\mathrm{NCDV+NDDV}}, \mathrm{x} \in \mathrm{X}\}, \\ & \forall\{x_v^{\min} \leq x_v \leq x_v^{\max} | v \in \Psi^{\mathrm{NCDV}}\}, \\ & \forall\{x_v \in \{x_v(1), \ldots, x_v(w), \ldots, x_v(W_v)\} | v \in \Psi^{\mathrm{NDDV}}\}\end{aligned} \tag{2.8}$$

A vector of objective functions strictly dominated by the vector of ideal objective functions is known as the vector of utopian objective functions—$\mathrm{z}^{\text{utopian}}$ [1–3]. The vector of the utopian objective functions is defined by Eq. (2.9):

$$\mathrm{z}^{\text{utopian}} = \left[z_1^{\text{utopian}}, \ldots, z_a^{\text{utopian}}, \ldots, z_{\mathrm{A}}^{\text{utopian}}\right]; \quad \forall\{a \in \Psi^{\mathrm{A}}\} \tag{2.9}$$

The relationship between the component or element a related to the vector of ideal objective functions and the component or element a relevant to the vector of utopian objective functions is defined using Eq. (2.10):

$$z_a^{\text{utopian}} = z_a^{\text{ideal}} - \varepsilon; \quad \forall\{a \in \Psi^{\mathrm{A}}\} \tag{2.10}$$

In Eq. (2.10), ε is a positive scalar number. For this same MOOP, the upper bounds associated with the set of Pareto-optimal responses and outputs are available in the vector of nadir objective functions—$\mathrm{z}^{\text{nadir}}$ [1–3]. The vector of nadir objective functions is described using Eq. (2.11):

$$\mathrm{z}^{\text{nadir}} = \left[z_1^{\text{nadir}}, \ldots, z_a^{\text{nadir}}, \ldots, z_{\mathrm{A}}^{\text{nadir}}\right]; \quad \forall\{a \in \Psi^{\mathrm{A}}\} \tag{2.11}$$

In MOOPs having a nonlinear nature, there is usually no useful well-recognized process to accurately calculate the vector of nadir objective functions. It is, therefore, generally difficult to precisely capture the components or elements relevant to the vector of nadir objective functions. These components or elements can be approximately estimated by using some decision-making analysis tools, such as the payoff table; however, the estimate resulting from these approaches may not be trustworthy [1].

2.3.5 Concepts Relevant to the Investigation of Pareto Optimality

In related literature, there are several methods generally used to investigate Pareto optimality of the vector of decision-making variables. One of the most well-known methods is to examine the Pareto optimality of the vector of decision-making variables, with the idea of forming an optimization problem [5]. In this regard, Pareto optimality of the vector of decision-making variables $x^{**} \in X$ can be investigated by solving the optimization problem given in Eq. (2.12):

$$\begin{aligned} \underset{x \in X, \gamma}{\text{Maximize}} \quad & \sum_{a \in \Psi^A} \gamma_a; \quad \forall\{a \in \Psi^A\} \\ & \text{subject to}: \\ & f_a(x) + \gamma_a = f_a(x^{**}); \quad \forall\{a \in \Psi^A\} \\ & \gamma_a \geq 0; \quad \forall\{a \in \Psi^A\} \\ & G(x) = [g_1(x), \ldots, g_b(x), \ldots, g_B(x)] = 0; \quad \forall\{B \geq 0\}, \forall\{b \in \Psi^B\} \\ & H(x) = [h_1(x), \ldots, h_e(x), \ldots, h_E(x)] \leq 0; \quad \forall\{E \geq 0\}, \forall\{e \in \Psi^E\} \end{aligned} \tag{2.12}$$

In Eq. (2.12), both $x \in \mathfrak{R}^{NDV}$ and $\gamma \in \mathfrak{R}^{A}_{+}$ are variables. The coefficients vector γ can be also indicated by using Eq. (2.13):

$$\gamma = [\gamma_1, \ldots, \gamma_a, \ldots, \gamma_A]; \quad \forall\{a \in \Psi^A\} \tag{2.13}$$

If the value of the objective function of the optimization problem given in Eq. (2.12) is equal to zero, the vector of decision-making variables $x^{**} \in X$ is then Pareto optimal. In other words, the vector of decision-making variables $x^{**} \in X$ is Pareto optimal, provided that all of the components or elements of the coefficients vector γ given in Eq. (2.13) are equal to zero. This strategy can also be employed to generate initial Pareto-optimal solutions for interactive MOOAs. Readers interested in a comprehensive discussion on this strategy are referred to the work by Benson [5].

2.4 Multi-objective Optimization Algorithms

Basically, the process of solving a MOOP in order to find the Pareto-optimal solution set, and then select a final optimal solution from this set, requires information related to the preferences of the decision maker. More precisely, the process of solving a MOOP should be established with regard to the preferences of the decision maker. In this way, the solution process can give rise to finding solutions that have more compatible with the preferences of the decision maker. The decision maker generally has sufficient insight into the MOOP. In addition, the decision maker can provide information pertaining to the preferences of different objective functions or different solutions in various structures.

In related literature, many MOOAs have been developed for finding the Pareto-optimal solution set and selecting the final optimal solution. Because the classification of these optimization algorithms can be carried out with a view to different criteria, it is a challenging task to provide a well-organized classification for these optimization algorithms. In related literature, different classifications have been reported on the basis of various criteria. MOOAs can be broken down into two types of approaches, according to the role of the decision maker in the solution process [1]: noninteractive and interactive.

2.4.1 Noninteractive Approaches

In a general classification, noninteractive approaches (NIAs) can be divided into four classes: (1) basic; (2) no preference; (3) a priori; and, (4) a posteriori.

2.4.1.1 Basic Approaches

Basic approaches are one of the most well-known and most used approaches for solving MOOPs. In order to employ solutions developed for single-objective optimization, these approaches transform a MOOP into a single-objective problem. Therefore, these approaches cannot actually be taken into account as a MOOA. The weighting coefficient approach and the ε-constraint approach are the most common basic approaches. Because of the widespread use and applicability of these approaches in solving MOOPs, an overview of these approaches is provided next.

Weighting coefficient approach: In the literature, the weighting coefficient approach is one of the simplest and most popular basic approaches for solving a MOOP. In this approach, the objective functions of the MOOP are transformed into a scalar objective function by using weighting coefficients [3, 6]. More precisely, in this approach, the MOOP, given in Eqs. (2.1) and (2.2), is turned into a single-objective optimization problem in accordance with Eqs. (2.14) and (2.15) through the weighting coefficients:

$$\begin{aligned}
&\underset{\mathrm{x}\in\mathrm{X}}{\text{Minimize}} \quad \mathrm{F(x)} = \left\{\sum_{a\in\Psi^{\mathrm{A}}} \omega_a . f_a(\mathrm{x})\right\}; \quad \forall\{\mathrm{A}\geq 2\}, \forall\{a\in\Psi^{\mathrm{A}}, \omega_a\in[0,1]\}\\
&\text{subject to}:\\
&\mathrm{G(x)} = [g_1(\mathrm{x}),\ldots,g_b(\mathrm{x}),\ldots,g_{\mathrm{B}}(\mathrm{x})] = 0; \quad \forall\{\mathrm{B}\geq 0\}, \forall\{b\in\Psi^{\mathrm{B}}\}\\
&\mathrm{H(x)} = [h_1(\mathrm{x}),\ldots,h_e(\mathrm{x}),\ldots,h_{\mathrm{E}}(\mathrm{x})] \leq 0; \quad \forall\{\mathrm{E}\geq 0\}, \forall\{e\in\Psi^{\mathrm{E}}\}
\end{aligned} \tag{2.14}$$

$$\begin{aligned} \mathrm{x} = \quad & [x_1, \ldots, x_v, \ldots, x_{\mathrm{NDV}}]; \quad \forall\{v \in \Psi^{\mathrm{NDV}}, \Psi^{\mathrm{NDV}} {=} \Psi^{\mathrm{NCDV+NDDV}}, \mathrm{x} \in \mathrm{X}\}, \\ & \forall\{x_v^{\min} \le x_v \le x_v^{\max} | v \in \Psi^{\mathrm{NCDV}}\}, \\ & \forall\{x_v \in \{x_v(1), \ldots, x_v(w), \ldots, x_v(W_v)\} | v \in \Psi^{\mathrm{NDDV}}\} \end{aligned} \tag{2.15}$$

In these equations, ω_a describes the weighting coefficient corresponding to objective function a of the MOOP, which is usually followed by Eqs. (2.16) and (2.17):

$$0 \le \omega_a \le 1; \quad \forall\{\mathrm{A} \ge 2\}, \forall\{a \in \Psi^{\mathrm{A}}\} \tag{2.16}$$

$$\sum_{a \in \Psi^{\mathrm{A}}} \omega_a = 1; \quad \forall\{\mathrm{A} \ge 2\}, \forall\{a \in \Psi^{\mathrm{A}}\} \tag{2.17}$$

In the weighting coefficient approach, the decision maker, by systematically changing the weighting coefficients, solves the single-objective optimization problem organized in Eqs. (2.14) and (2.15). Solving the single-objective optimization problem formed in Eqs. (2.14) and (2.15) for different weighting coefficients results in the estimation of the Pareto-optimal solutions. The solution specified by solving the optimization problem given in Eqs. (2.14) and (2.15) is a weak Pareto-optimal solution if the condition provided in Eq. (2.18) is satisfied:

$$\omega_a > 0; \quad \forall\{\mathrm{A} \ge 2\}, \forall\{a \in \Psi^{\mathrm{A}}\} \tag{2.18}$$

This solution is also the Pareto-optimal solution if it is unique [3]. The weighting coefficient approach is appropriate for a MOOP in which all objective functions are of the same type and have a common scale (e.g., all objective functions are of a cost type with a dollar scale). If objective functions of optimization problem are not of same type and not have a common scale, the use of the weighting coefficient approach is not efficient. In this situation, the recommended strategy for employing the weighting coefficient approach is to normalize the objective functions. Objective function a of this same MOOP is normalized through Eq. (2.19):

$$\tilde{f}_a(\mathrm{x}) = \frac{f_a^{\max}(\mathrm{x}) - f_a(\mathrm{x})}{f_a^{\max}(\mathrm{x}) - f_a^{\min}(\mathrm{x})}; \quad \forall\{a \in \Psi^{\mathrm{A}}\} \tag{2.19}$$

This equation refers to a situation in which the minimization of the objective function a of the MOOP is taken into account. Similarly, if maximization of the objective function a of the MOOP is regarded, Eq. (2.19) should be rewritten according to Eq. (2.20):

$$\tilde{f}_a(\mathrm{x}) = 1 - \frac{f_a^{\max}(\mathrm{x}) - f_a(\mathrm{x})}{f_a^{\max}(\mathrm{x}) - f_a^{\min}(\mathrm{x})}; \quad \forall\{a \in \Psi^{\mathrm{A}}\} \tag{2.20}$$

In Eqs. (2.19) and (2.20), the lower and upper bounds of objective function a of the MOOP are calculated by using a single-objective optimization. More precisely, the upper bound of objective function a, $f_a^{\max}(\mathrm{x})$, is achieved by single-objective maximization of the corresponding objective function. In the same way, the lower bound of objective function a, $f_a^{\min}(\mathrm{x})$, is determined by single-objective minimization of the corresponding objective function. After normalization of the objective functions of the MOOP given in Eqs. (2.1) and (2.2), this optimization problem can be rewritten based on Eqs. (2.21) and (2.22):

$$\begin{array}{ll}\underset{\mathrm{x}\in\mathrm{X}}{\text{Minimize}} & \mathrm{F}(\mathrm{x})=\left\{\sum_{a\in\Psi^{\mathrm{A}}}\omega_a\cdot\tilde{f}_a(\mathrm{x})\right\};\quad\forall\{\mathrm{A}\geq 2\},\forall\{a\in\Psi^{\mathrm{A}},\omega_a\in[0,1]\}\\ & \text{subject to}:\\ & \mathrm{G}(\mathrm{x})=[g_1(\mathrm{x}),\ldots,g_b(\mathrm{x}),\ldots,g_{\mathrm{B}}(\mathrm{x})]=0;\quad\forall\{\mathrm{B}\geq 0\},\forall\{b\in\Psi^{\mathrm{B}}\}\\ & \mathrm{H}(\mathrm{x})=[h_1(\mathrm{x}),\ldots,h_e(\mathrm{x}),\ldots,h_{\mathrm{E}}(\mathrm{x})]\leq 0;\quad\forall\{\mathrm{E}\geq 0\},\forall\{e\in\Psi^{\mathrm{E}}\}\end{array} \tag{2.21}$$

$$\begin{array}{rl}\mathrm{x}= & [x_1,\ldots,x_v,\ldots,x_{\mathrm{NDV}}];\quad\forall\{v\in\Psi^{\mathrm{NDV}}\},\forall\{\Psi^{\mathrm{NDV}}=\Psi^{\mathrm{NCDV+NDDV}}\},\forall\{\mathrm{x}\in\mathrm{X}\},\\ & \forall\{x_v^{\min}\leq x_v\leq x_v^{\max}|v\in\Psi^{\mathrm{NCDV}}\},\\ & \forall\{x_v\in\{x_v(1),\ldots,x_v(w),\ldots,x_v(W_v)\}|v\in\Psi^{\mathrm{NDDV}}\}\end{array} \tag{2.22}$$

One of the most important strengths of the weighting coefficients approach, making it widely utilized for solving a wide range of MOOPs, is the simplicity of its use. In this approach, one solution can be found through the Pareto-optimal solution set by changing the weighting coefficients. It has been proven, however, that this characteristic is reliable only in convex optimization problems. That is, in non-convex optimization problems, regardless of how the weighing coefficients are chosen, some Pareto-optimal solutions cannot be found. Furthermore, if some objective functions correlate with each other in the MOOPs, changing the weighting coefficients may not lead to finding Pareto-optimal solutions. As a result, the weighting coefficient approach does not have an appropriate performance for these MOOPs. It should be pointed out that the decision maker can employ the weighting coefficient approach either as an a priori approach or as an a posteriori approach.

ε-Constraint approach: In related literature, the ε-constraint approach is one of the most applicable basic approaches for solving MOOPs [7, 8]. At each step in this approach, one of the objective functions of the MOOP is chosen for optimization, while the remaining objective functions are considered as constraints. This process is repeated for all objective functions of the MOOP. By using the ε-constraint approach, the MOOP, again from Eqs. (2.1) and (2.2), is transformed into a single-objective optimization problem in accordance with Eqs. (2.23) and (2.24):

$$\begin{aligned} &\underset{x \in X}{\text{Minimize}} \quad \mathrm{F}(\mathrm{x}) = \{f_a(\mathrm{x})\}; \quad \forall\{a \in \Psi^{\mathrm{A}}\} \\ &\qquad \text{subject to :} \\ &\qquad \mathrm{G}(\mathrm{x}) = [g_1(\mathrm{x}), \ldots, g_b(\mathrm{x}), \ldots, g_{\mathrm{B}}(\mathrm{x})] = 0; \quad \forall\{\mathrm{B} \geq 0\}, \forall\{b \in \Psi^{\mathrm{B}}\} \\ &\qquad \mathrm{H}(\mathrm{x}) = [h_1(\mathrm{x}), \ldots, h_e(\mathrm{x}), \ldots, h_{\mathrm{E}}(\mathrm{x})] \leq 0; \quad \forall\{\mathrm{E} \geq 0\}, \forall\{e \in \Psi^{\mathrm{E}}\} \\ &\qquad f_k(\mathrm{x}) \leq \varepsilon_k^{\max}; \quad \forall\{k \in \Psi^{\mathrm{A}}\}, \forall\{k \neq a\} \end{aligned} \tag{2.23}$$

$$\begin{aligned} \mathrm{x} = \quad & [x_1, \ldots, x_v, \ldots, x_{\mathrm{NDV}}]; \quad \forall\{v \in \Psi^{\mathrm{NDV}}, \Psi^{\mathrm{NDV}} = \Psi^{\mathrm{NCDV+NDDV}}, \mathrm{x} \in \mathrm{X}\}, \\ & \forall\forall\{x_v^{\min} \leq x_v \leq x_v^{\max} | v \in \Psi^{\mathrm{NCDV}}\}, \\ & \forall\{x_v \in \{x_v(1), \ldots, x_v(w), \ldots, x_v(W_v)\} | v \in \Psi^{\mathrm{NDDV}}\} \end{aligned} \tag{2.24}$$

In Eq. (2.23), $\varepsilon_k^{\max}$ represents the upper bound for objective function k of the MOOP. The vector of decision-making variables $\mathrm{x}^{**} \in \mathrm{X}$ is Pareto optimal if and only if this vector solves the optimization problem organized in Eqs. (2.23) and (2.24) for each objective function of the MOOP, $f_a(\mathrm{x}^{**})$; $\forall$ $\{a \in \Psi^{\mathrm{A}}\}$, while satisfying $\varepsilon_k^{\max} = f_k(\mathrm{x}^{**}); \forall\{k \in \Psi^{\mathrm{A}}\}, \forall\{k \neq a\}$ [7, 8]. More precisely, to ensure that Pareto optimality corresponds to the vector of decision-making variables $\mathrm{x}^{**} \in \mathrm{X}$—finding one solution from the Pareto-optimal solution set—either the single-objective optimization problem formed in Eqs. (2.23) and (2.24) must be solved by the number of objective functions of the MOOP or one unique solution of the single-objective optimization problem formed in Eqs. (2.23) and (2.24) must be achieved. Nevertheless, in a MOOP, if the weak Pareto-optimal solution is satisfactory, from the perspective of the decision maker, solving the single-objective optimization problem organized in Eqs. (2.23) and (2.24) is sufficient for an objective function to find one solution from the weak Pareto-optimal solution set.

In contrast to the weighting coefficient approach, finding the Pareto-optimal solution set by using the ε-constraint approach does not depend on the convexity or non-convexity of the optimization problem. In other words, the ε-constraint approach has a desirable performance in dealing with convex or non-convex optimization problems.

In practice, the selection of the upper bounds associated with different objective functions of the MOOP has many complexities. These complexities are dramatically expanded by increasing the number of objective functions of the MOOP. The selection of the upper bounds must, therefore, be made meticulously. In this manner, the upper bounds selected for different objective functions of the MOOP must be within the feasible space; otherwise, the single-objective optimization problem formed in Eqs. (2.23) and (2.24) will not have a solution. If maximization of this MOOP is taken into account, then the MOOP is turned into a single-objective optimization problem based on Eqs. (2.25) and (2.26) by using the ε-constraint approach:

$$
\begin{aligned}
&\underset{x \in X}{\text{Maximize}} \quad F(x) = \{f_a(x)\}; \quad \forall\{a \in \Psi^{A}\} \\
&\qquad \text{subject to :} \\
&\qquad G(x) = [g_1(x), \ldots, g_b(x), \ldots, g_B(x)] = 0; \quad \forall\{B \geq 0\}, \forall\{b \in \Psi^{B}\} \\
&\qquad H(x) = [h_1(x), \ldots, h_e(x), \ldots, h_E(x)] \leq 0; \quad \forall\{E \geq 0\}, \forall\{e \in \Psi^{E}\} \\
&\qquad f_k(x) \geq \varepsilon_k^{\min}; \quad \forall\{k \in \Psi^{A}\}, \forall\{k \neq a\}
\end{aligned}
\tag{2.25}
$$

$$
\begin{aligned}
x = \quad & [x_1, \ldots, x_v, \ldots, x_{NDV}]; \quad \forall\{v \in \Psi^{NDV}, \Psi^{NDV} = \Psi^{NCDV+NDDV}, x \in X\}, \\
& \forall\{x_v^{\min} \leq x_v \leq x_v^{\max} | v \in \Psi^{NCDV}\}, \\
& \forall\{x_v \in \{x_v(1), \ldots, x_v(w), \ldots, x_v(W_v)\} | v \in \Psi^{NDDV}\}
\end{aligned}
\tag{2.26}
$$

In Eq. (2.25), $\varepsilon_k^{\min}$ describes the lower bound for objective function k of the MOOP. Similar to the weighting coefficient approach, the ε-constraint approach can be utilized by the decision maker either as an a priori approach or as an a posteriori approach.

2.4.1.2 No-Preference Approaches

In no-preference approaches, known as neutral-preference approaches, the preferences of the decision maker are not considered in the process of solving the MOOP. In these approaches, the MOOP is solved by using some relatively simple approaches, at which point the solution is taken at the disposal of the decision maker. The decision maker is also able to accept or reject the specified solution. Non-preference approaches are suitable for situations in which information related to the preferences of the decision maker is not available, or the decision maker does not consider particular preferences. The global criterion approach and neutral-compromise solution approach are the best-known no-preference approaches. Readers interested in a thorough discussion on these approaches are directed to the work by Yu [9] and Wierzbicki [10], respectively.

2.4.1.3 A Priori Approaches

In a priori approaches, the decision maker first determines the information related to his/her preferences, and then solves the MOOP by trying to find a Pareto-optimal solution that can, as much as possible, satisfy his/her preferences. Simply put, in a priori approaches, information related to the preferences of the decision maker is determined before the process of solving the MOOP begins.

A major disadvantage in a priori approaches is that the decision maker is not necessarily aware of the possibilities and restrictions of the MOOP in advance. As a result, it is possible that information about the preferences of the decision maker is overly optimistic or pessimistic. That is to say that the decision maker does not necessarily know in advance what the response/output is likely to be from the MOOP or how realistic his/her preferences are. In this situation, it is possible that the solution cannot satisfy the decision maker and encourage the decision maker to rectify his/her preferences. The most well-known a priori approaches can be referred to as value function approaches, lexicographic ordering approaches, and goal programming approaches. Readers interested in a comprehensive discussion on these approaches are referred to the work by Keeney and Raiffa [11], Fishburn [12], and Charnes and Cooper [13], respectively.

2.4.1.4 A Posteriori Approaches

The main idea of a posteriori approaches is established on the basis of finding the Pareto-optimal solution set and presenting it to the decision maker with the aim of choosing the final solution through the aforementioned set. More precisely, in a posteriori approaches, the process of solving the MOOP first tries to find the Pareto-optimal solution set. After determination of the Pareto-optimal solution set, this set is taken at the disposal of the decision maker. Finally, the decision maker chooses the most satisfactory solution from the set as the final optimal solution.

One of the strengths of a posteriori approaches, compared to a priori approaches, is that in a posteriori approaches, the Pareto-optimal solution set is completed before being presented to the decision maker. In this way, the decision maker has a complete overview of all solutions, making it easier and more realistic to choose the most satisfactory solution. Nonetheless, the major weakness of a posteriori approaches is their high computational burden. Additionally, the decision maker encounters a very large amount of information in optimization problems with more than two objective functions, which makes analysis of the information a difficult task.

The best-known a posteriori approaches can be referred to as weighted metrics approaches, achievement scalarizing function approaches, approximation approaches, and meta-heuristic MOOAs. Readers interested in a thorough discussion on these approaches are referred to the work by Miettinen [3], Wierzbicki [14], and Ruzika and Wiecek [15], respectively. It is important to be noted that detailed descriptions of some developed meta-heuristic MOOAs by the authors are provided in Chap. 4.

2.4.2 Interactive Approaches

Interactive approaches (IAs) are established on the basis of creating an iterative solution procedure or pattern that consists of different steps. In this approach to finding the most satisfactory solution, different steps of this iterative procedure are repeated and the decision maker progressively determines preference information during the solution process. In other words, after completion of each step of the iterative procedure, the information is taken at the disposal of the decision maker, at which point the decision maker assesses the information and then may specify additional details. This interactive process is repeated until the stopping criterion is satisfied—and as long as the most satisfactory solution has been specified by the decision maker. In this structure, the decision maker can modify and update his/her preference information. As a consequence, the decision maker straightforwardly directs the process in the IAs.

The main steps of an iterative procedure can be briefly expressed as follows:

- Step one: Initialization (i.e., determine the ideal vector of objective functions and nadir vector of objective vector and present these values to the decision maker).
- Step two: Produce a Pareto-optimal starting point (i.e., some neutral-compromise solution or solution specified by the decision maker that can be taken into account as the starting point).
- Step three: Specify the preference information by the decision maker (i.e., the number of new solutions to be produced).
- Step four: Produce one or more Pareto-optimal solutions by taking into account the preferences specified by the decision maker in the previous step and then showing this Pareto-optimal solution or solutions along with information associated with the MOOP to the decision maker.
- Step five: Select the most satisfactory solution by the decision maker through Pareto-optimal solutions achieved thus far, if multiple Pareto-optimal solutions have been produced in the fourth step. If a Pareto-optimal solution has been produced in the fourth step, this solution is considered as the most satisfactory solution by the decision maker in this step.
- Step six: Stop, if the consent of the decision maker is satisfied by the solution chosen in the fifth step; otherwise, go to the third step.

One of the strengths of the IAs is that the decision maker is able to update his/her preference information in each iteration of the process. Accordingly, by informing the decision maker about interdependencies between the iterative solution procedure and its preferences, the probability of achieving a satisfactory solution that meets the preferences of the decision maker is increased. In other words, because of the establishment of the IAs, based on an iterative procedure that allows the decision maker to specify or update preference information during the process, Pareto-optimal solutions are produced that can satisfy the decision maker. As a result, the structure of the IAs can give rise to a significant reduction in computational burden.

In recent years, a wide range of IAs have been developed for solving MOOPs. Basically, there is no unique IA that has a more preferred performance for solving the MOOPs with different features and structures as well as multiple decision makers compared to other approaches. This means that each approach is generally developed for a specific range of MOOPs and decision makers. In a wide classification, IAs can be broken down into three general classes: (1) compromise-driven or trade-off-based approaches; (2) reference point approaches; and, (3) classification-based approaches. Readers interested in a thorough discussion of these approaches are referred to the work by Branke et al. [1].

2.5 Selection of the Final Solution by Using a Fuzzy Satisfying Method

After solving our MOOP and determining the Pareto-optimal solution set, the next step is to select a flexible and realistic solution from the entire set of candidate solutions that represent a compromise among different objective functions of the MOOP. In related literature, there are several multi-objective decision-making tools for selecting the most satisfactory solution from the Pareto-optimal solution set; keep in mind, though, that a fuzzy satisfying method (FSM) is highly regarded in this situation, owing to its simplicity and similarity to human reasoning [16–19]. In this method, then, a fuzzy membership function, $\Phi_{f_a^n(\mathrm{x})}(\bar{x}_n)$, is defined for each given objective function,$f_a^n(\mathrm{x})$, in any available solution in the Pareto-optimal solution set, $\bar{x}_n$. The value of this membership function can vary from 0 to 1. The fuzzy membership function demonstrates a numerical description for the satisfaction level of the decision maker regarding the value of objective function a in the available solution n in the Pareto-optimal solution set. The fuzzy membership function with a value of 0, $\Phi_{f_a^n(\mathrm{x})}(\bar{x}_n) = 0$, represents a complete dissatisfaction of the decision maker. On the other hand, the fuzzy membership function with a value of 1, $\Phi_{f_a^n(\mathrm{x})}(\bar{x}_n) = 1$, represents full satisfaction of the decision maker. As a result, higher values of this membership function refer to higher levels of satisfaction of the decision maker regarding the value of objective function a in the available solution n in the Pareto-optimal solution set. Different types of fuzzy membership functions can generally be used by the decision maker, such as linear, convex exponential, concave exponential, piecewise linear, and hyperbolic types. Considering different types of fuzzy membership functions for different objective functions of the MOOP can affect the choice of the final solution through the Pareto-optimal solution set. As an example, suppose that the fuzzy membership function considered by the decision maker for objective function a of our MOOP is convex exponential and the fuzzy membership function regarded by the decision maker for other objective functions is linear. These conditions provide a priority for minimization of objective function a relative to other objective functions. This is due to the fact that a smaller fuzzy membership function in the neighborhood of the upper bound of the objective function a, $f_a^{\max}(\mathrm{x})$, has been assigned by the convex exponential membership function compared with the linear membership function.

Here, the fuzzy membership function considered for all of the existing objective functions in our MOOP is assumed to be a linear membership function. To clarify, the linear membership function corresponds to objective function a of the MOOP that is depicted in Fig. 2.2.

If the minimization of the objective functions of the MOOP is considered, the linear membership function related to objective function a is represented as a descending uniform function (see Fig. 2.2). The mathematical description of the linear membership function shown in Fig. 2.2 can also be set out using Eq. (2.27):

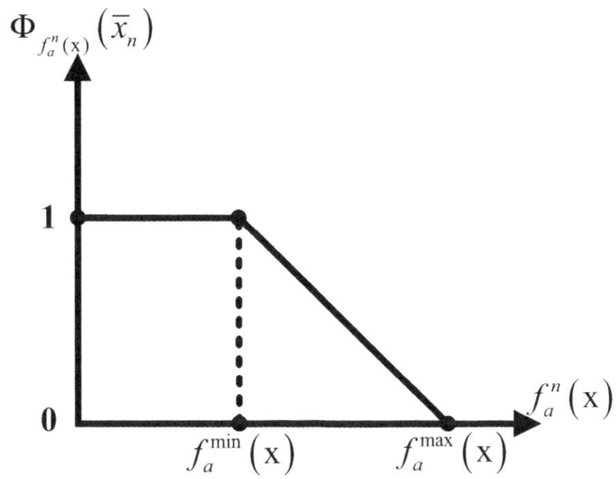

Fig. 2.2 The linear membership function corresponds to the minimization of objective function *a* of the MOOP

$$\Phi_{f_a^n(\mathrm{x})}(\bar{x}_n) = \begin{cases} 0 & ;\forall\{f_a^n(\mathrm{x}) > f_a^{\max}(\mathrm{x})\} \\ \dfrac{f_a^{\max}(\mathrm{x}) - f_a^n(\mathrm{x})}{f_a^{\max}(\mathrm{x}) - f_a^{\min}(\mathrm{x})} & ;\forall\{f_a^{\min}(\mathrm{x}) \le f_a^n(\mathrm{x}) \le f_a^{\max}(\mathrm{x})\} \\ 1 & ;\forall\{f_a^n(\mathrm{x}) < f_a^{\min}(\mathrm{x})\} \end{cases}$$
$$;\forall\{\mathrm{A} \ge 2\}, \forall\{a \in \Psi^{\mathrm{A}}, n \in \Psi^{\mathrm{N}}\} \tag{2.27}$$

As shown in Fig. 2.2 and formulated in Eq. (2.27), this membership function has both a lower bound, $f_a^{\min}(\mathrm{x})$, and an upper bound, $f_a^{\max}(\mathrm{x})$. These bounds are achieved by using a single-objective optimization. That is to say that the lower and upper bounds of objective function *a* of the MOOP are calculated by minimizing and maximizing only the corresponding objective function as a single-objective optimization problem, respectively. Similarly, if the maximization of the objective functions of the same MOOP is taken into account, the linear membership function relevant to objective function *a* is addressed as an ascending uniform function (see Fig. 2.3). The mathematical description of the linear membership function depicted in Fig. 2.3 can also be set out using Eq. (2.28):

$$\Phi_{f_a^n(\mathrm{x})}(\bar{x}_n) = \begin{cases} 1 & ;\forall\{f_a^n(\mathrm{x}) > f_a^{\max}(\mathrm{x})\} \\ 1 - \dfrac{f_a^{\max}(\mathrm{x}) - f_a^n(\mathrm{x})}{f_a^{\max}(\mathrm{x}) - f_a^{\min}(\mathrm{x})} & ;\forall\{f_a^{\min}(\mathrm{x}) \le f_a^n(\mathrm{x}) \le f_a^{\max}(\mathrm{x})\} \\ 0 & ;\forall\{f_a^n(\mathrm{x}) < f_a^{\min}(\mathrm{x})\} \end{cases}$$
$$;\forall\{\mathrm{A} \ge 2\}, \forall\{a \in \Psi^{\mathrm{A}}, n \in \Psi^{\mathrm{N}}\} \tag{2.28}$$

After describing the membership functions for all objective functions for all available solutions in the Pareto-optimal solutions set, the decision maker must specify the level of desirability of achieving each objective function of the MOOP,

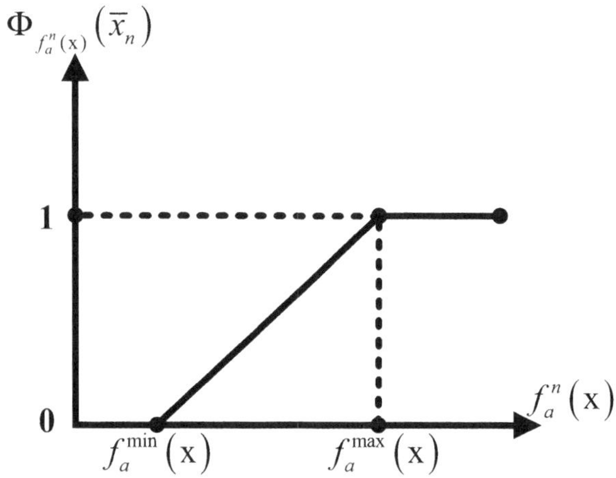

Fig. 2.3 The linear membership function relevant to the maximization of objective function a of the MOOP

$\tilde{\Phi}_{f_a(\mathrm{x})}$. The level of desirability of achieving objective function a of the MOOP, $f_a(\mathrm{x})$, is known as the reference level of achieving the corresponding objective function. After determination of the level of desirability of achieving each objective function of the MOOP, the decision maker should employ a well-suited decision-making analysis tool in order to choose the final optimal compromise solution from the Pareto-optimal solution set. To do this, there are generally many decision-making analysis tools developed using a variety of philosophies and from myriad perspectives. Here, the conservative and distance metric methodologies, as two applicable and well-known decision-making analysis tools, are reviewed and discussed.

2.5.1 Conservative Methodology

In the conservative methodology (CM)—the min-max formulation—conservative decision-making can be achieved by trying to find a solution whose minimum meets the maximum objective function. This means that the decision maker is willing to specify a solution that simultaneously achieves the highest level of satisfaction for all of the objective functions of the MOOP. In this methodology, the final optimal compromise solution is determined from all available solutions in the Pareto-optimal set by solving the optimization problem given in Eq. (2.29):

$$\min_{n \in \Psi^{\mathrm{N}}} \left\{ \max_{a \in \Psi^{\mathrm{A}}} \left\{ \left| \tilde{\Phi}_{f_a(\mathrm{x})} - \Phi_{f_a^n(\mathrm{x})}(\bar{x}_n) \right| \right\} \right\}; \quad \forall \{\mathrm{A} \geq 2\}, \forall \{a \in \Psi^{\mathrm{A}}, n \in \Psi^{\mathrm{N}}\} \tag{2.29}$$

If the decision maker is willing to achieve the highest level of satisfaction for all of the objective functions of the MOOP, $\tilde{\Phi}_{f_a(\mathrm{x})} = 1; \forall \{a \in \Psi^{\mathrm{A}}\}$, Eq. (2.29) must be rewritten as Eq. (2.30):

$$\min_{n \in \Psi^{\mathrm{N}}} \left\{ \max_{a \in \Psi^{\mathrm{A}}} \left\{ \left| 1 - \Phi_{f_a^n(\mathrm{x})}\left(\bar{x}_n\right) \right| \right\} \right\}; \quad \forall\{\mathrm{A} \geq 2\}, \forall\left\{a \in \Psi^{\mathrm{A}}, n \in \Psi^{\mathrm{N}}\right\} \tag{2.30}$$

In other words, Eq. (2.30) can also be expressed as Eq. (2.31):

$$\max_{n \in \Psi^{\mathrm{N}}} \left\{ \min_{a \in \Psi^{\mathrm{A}}} \left\{ \left| \Phi_{f_a^n(\mathrm{x})}\left(\bar{x}_n\right) \right| \right\} \right\}; \quad \forall\{\mathrm{A} \geq 2\}, \forall\left\{a \in \Psi^{\mathrm{A}}, n \in \Psi^{\mathrm{N}}\right\} \tag{2.31}$$

The CM ensures for the decision maker that all of the objective functions of the MOOP are well optimized. Interested readers are directed to the work by Sakawa and Yano [20] for a comprehensive discussion of the CM.

2.5.2 *Distance Metric Methodology*

In the distance metric methodology (DM), the final optimal compromise solution is obtained from all available solutions in the Pareto-optimal set by solving the optimization problem given in Eq. (2.32):

$$\min_{n \in \Psi^{\mathrm{N}}} \left\{ \sum_{a \in \Psi^{\mathrm{A}}} \left\{ \left| \tilde{\Phi}_{f_a(\mathrm{x})} - \Phi_{f_a^n(\mathrm{x})}\left(\bar{x}_n\right) \right|^u \right\} \right\}; \quad \forall\{\mathrm{A} \geq 2\}, \forall\left\{a \in \Psi^{\mathrm{A}}, n \in \Psi^{\mathrm{N}}, u \in [1, \infty)\right\} \tag{2.32}$$

It can be seen that Eq. (2.32) attempts to minimize the u-norm deviations from the values of the reference membership. The u quantity has a value between one and infinity, a value that has already been specified by the decision maker. Because the absolute difference of the level of desirability of achieving objective function a and its fuzzy membership function in available solution n in the Pareto-optimal set $\left|\tilde{\Phi}_{f_a(\mathrm{x})} - \Phi_{f_a^n(\mathrm{x})}\left(\bar{x}_n\right)\right|$ always has a value between zero and one, a larger value of u represents less sensitivity to reference levels and vice versa. It should be pointed out that if the decision maker is not satisfied by the solution, he/she is able to improve the corresponding solution by updating the levels of desirability of achieving different objective functions of the MOOP, $\tilde{\Phi}_{f_a(\mathrm{x})}$. Interested readers are directed to the work by Chen [21] for a comprehensive discussion of DM.

2.5.3 *Step-by-Step Process for Implementing the FSM*

As a general result, after solving the MOOP, as given in Eqs. (2.1) and (2.2), and specifying the Pareto-optimal solution set, the implementation of the FSM by the decision maker to select the final optimal compromise solution through the Pareto-optimal solution set can be expressed using the following step-by-step process:

- Step one: Set the number of objective functions of the MOOP equal to A.
- Step two: Set the counter of the objective function to $a = 1$.
- Step three: Determine the lower and the upper bounds of objective function a of the MOOP by minimizing and maximizing only the corresponding objective function as a single-objective optimization, respectively.
- Step four: Set the number of available solutions in the Pareto-optimal solution set specified by solving the MOOP equal to N.
- Step five: Set the counter of available solutions in the Pareto-optimal solution set to $n = 1$.
- Step six: Calculate the value of the linear membership function associated with objective function a in available solution n in the Pareto-optimal solution set by using Eq. (2.27).
- Step seven: If $n < \mathrm{N}$, set $n = n + 1$ and go to step six; otherwise, go to the next step.
- Step eight: If $a < \mathrm{A}$, set $a = a + 1$ and go to step three; otherwise, go to the next step.
- Step nine: Specify the level of desirability of achieving each objective function of the MOOP.
- Step ten: Determine the final optimal compromise solution from the Pareto-optimal solution set either by using the CM—min-max formulation given in Eq. (2.29)—or by using the DM—formulation given in Eq. (2.32).

2.6 Conclusions

In this chapter, the authors presented a brief introduction to the multi-objective optimization process. First, the necessity of employing the multi-objective optimization process instead of the single-objective optimization process was justified. Then, the fundamental concepts of optimization in the MOOPs were exhaustively addressed in the five sections: (1) mathematical description of a MOOP; (2) concepts associated with efficiency, efficient frontier, and dominance; (3) concepts pertaining to Pareto optimality; (4) concepts related to the vector of ideal objective functions and the vector of nadir objective functions; and, (5) concepts relevant to Pareto optimality investigation. In addition, a thorough classification was provided for the MOOAs with a focus on the role of the decision maker in the process of solving the MOOP. This classification was broken down into two approaches: NIAs and IAs. The NIAs were also classified into four different classes including basic, no preference, a priori, and a posteriori approaches. Finally, the FSM, as the most preferred multi-objective decision-making tool, was thoroughly described in order to select the final optimal compromise solution from the Pareto-optimal solution set.

Appendix 1: List of Abbreviations and Acronyms

CM	Conservative methodology
DM	Distance metric methodology
FSM	Fuzzy satisfying method
IAs	Interactive approaches
MOOAs	Multi-objective optimization algorithms
MOOPs	Multi-objective optimization problems
NIAs	Noninteractive approaches

Appendix 2: List of Mathematical Symbols

Index:	
a	Index for objective functions running from 1 to A
b	Index for equality constraints running from 1 to B
e	Index for inequality constraints running from 1 to E
k	Index for objective functions running from 1 to A
n	Index for available solutions in the Pareto-optimal solution set running from 1 to N
v	Index for decision-making variables, including the continuous and discrete decision-making variables, running from 1 to NDV, and an index for continuous decision-making variables running from 1 to NCDV and also an index for discrete decision-making variables running from 1 to NDDV
Set:	
Ψ^{A}	Set of indices of objective functions
Ψ^{B}	Set of indices of equality constraints
Ψ^{E}	Set of indices of inequality constraints
Ψ^{NCDV}	Set of indices of continuous decision-making variables
Ψ^{NDDV}	Set of indices of discrete decision-making variables
Ψ^{NDV}	Set of indices of decision-making variables, including the continuous and discrete decision-making variables
Ψ^{N}	Set of indices of available solutions in the Pareto-optimal solution set
AP(X)	Set of appropriate Pareto-optimal vectors of decision-making variables
AP(Z)	Set of appropriate Pareto-optimal vectors of objective functions
P(X)	Set of Pareto-optimal vectors of decision-making variables
P(Z)	Set of Pareto-optimal vectors of objective functions
$\Re^{A}$	A-dimensional set of real numbers
$\Re_{+}^{A}$	A-dimensional set of positive real numbers
$\Re^{E}$	E-dimensional set of real numbers
W_v	Set of indices of candidate permissible values of discrete decision-making variable v
WP(X)	Set of weak Pareto-optimal vectors of decision-making variables
WP(Z)	Set of weak Pareto-optimal vectors of objective functions

(continued)

Parameters:	
$f_a^{\max}(\mathrm{x})$	Upper bound of objective function *a*
$f_a^{\min}(\mathrm{x})$	Lower bound of objective function *a*
J	Some real number
u	Integer with a value between one and infinity
$x_v^{\max}$	Upper bound on continuous decision-making variable *v*
$x_v^{\min}$	Lower bound on continuous decision-making variable *v*
X	Nonempty feasible decision-making space
Z	Feasible objective space
ε	A positive scalar number
$\varepsilon_k^{\max}$	Upper bound of objective function *k*
$\varepsilon_k^{\min}$	Lower bound of objective function *k*
$\tilde{\Phi}_{f_a(\mathrm{x})}$	Level of desirability of achieving objective function *a*
Variables:	
$f_a(\mathrm{x})$	Objective function *a* of the optimization problem or component *a* of the vector of objective functions
$f_k(\mathrm{x})$	Objective function *k* of the optimization problem or component *k* of the vector of objective functions
$\tilde{f}_a(\mathrm{x})$	Normalized value of objective function *a* of the optimization problem or normalized value of component *a* of the vector of objective functions
$f_a^n(\mathrm{x})$	Given objective function *a* in available solution *n* in the Pareto-optimal solution set
F(x)	Vector of objective functions of the optimization problem
$g_b(\mathrm{x})$	Equality constraint *b* of the optimization problem or component *b* of the vector of equality constraints
G(x)	Vector of equality constraints of the optimization problem
$h_e(\mathrm{x})$	Inequality constraint *e* of the optimization problem or component *e* of the vector of inequality constraints
H(x)	Vector of inequality constraints of the optimization problem
$\bar{x}_n$	Solution *n* in the Pareto-optimal solution set
x_v	Continuous or discrete decision-making variable *v*
$x_v(w_v)$	Candidate permissible value *w* of discrete decision-making variable *v*
$\mathrm{x}, \mathrm{x}^*, \mathrm{x}^{**}$	Vector of decision-making variables
z	Vector of objective functions
$\mathrm{z}^{\mathrm{ideal}}$	Vector of ideal objective functions
z_a^{ideal}	Component or element *a* relevant to the vector of ideal objective functions
$\mathrm{z}^{\mathrm{utopian}}$	Vector of utopian objective functions
z_a^{utopian}	Component or element *a* relevant to the vector of utopian objective functions
$\mathrm{z}^{\mathrm{nadir}}$	Vector of the nadir objective functions
z_a^{nadir}	Component or element *a* relevant to the vector of nadir objective functions
$\mathrm{\gamma}$	Coefficient vector
γ_a	Component or element *a* relevant to the coefficient vector
ω_a	Weighting coefficient relevant to objective function *a*
$\Phi_{f_a^n(\mathrm{x})}(\bar{x}_n)$	Fuzzy membership function of objective function *a* in available solution *n* in the Pareto-optimal solution set

References

1. J. Branke, K. Deb, K. Miettinen, *Multiobjective Optimization: Interactive and Evolutionary Approaches* (Springer, New York, 2008)
2. R.T. Marler, J.S. Arora, Survey of multi-objective optimization methods for engineering. Struct. Multidiscip. Optim. **26**(6), 369–395 (2004)
3. K. Miettinen, *Nonlinear Multiobjective Optimization* (Kluwer Academic Publishers, Boston, MA, 1999)
4. A.M. Geoffrion, Proper efficiency and the theory of vector maximization. J. Math. Anal. Appl. **22**(3), 618–630 (1968)
5. H.P. Benson, Existence of efficient solutions for vector maximization problems. J. Optim. Theory Appl. **26**(4), 569–580 (1978)
6. L. Zadeh, Optimality and non-scalar-valued performance criteria. IEEE Trans. Autom. Control **8**(1), 59–60 (1963)
7. Y.Y. Haimes, L.S. Lasdon, D.A. Wismer, On a bicriterion formulation of the problems of integrated system identification and system optimization. IEEE Trans. Syst. Man Cybern. **1**(3), 296–297 (1971)
8. V. Chankong, Y.Y. Haimes, *Multiobjective Decision Making: Theory and Methodology* (Elsevier Science Publishing, New York, 1983)
9. P.L. Yu, A class of solutions for group decision problems. Manag. Sci. **19**(8), 936–946 (1973)
10. A.P. Wierzbicki, Reference point approaches, in *Multicriteria Decision Making, Chapter 9*, vol. 21, (Springer, New York, 1999), pp. 237–275
11. R.L. Keeney, H. Raiffa, *Decisions with Multiple Objectives: Preferences and Value Tradeoffs* (Wiley, Chichester, 1976)
12. P.C. Fishburn, Lexicographic orders, utilities and decision rules: a survey. Manag. Sci. **20**(11), 1442–1471 (1974)
13. A. Charnes, W.W. Cooper, Goal programming and multiple objective optimization: part 1. Eur. J. Oper. Res. **1**(1), 39–54 (1977)
14. A.P. Wierzbicki, A mathematical basis for satisficing decision making. Mathematical Modelling **3**(5), 391–405 (1982)
15. S. Ruzika, M.M. Wiecek, Approximation methods in multiobjective programming. J. Optim. Theory Appl. **126**(3), 473–501 (2005)
16. M. Sakawa, H. Yano, An interactive fuzzy satisfying method using augmented minimax problems and its application to environmental systems. IEEE Trans. Syst. Man Cybern. **17** (6), 720–729 (1985)
17. M. Sakawa, K. Yauchi, An interactive fuzzy satisficing method for multiobjective nonconvex programming problems through floating point genetic algorithms. Eur. J. Oper. Res. **117**(1), 113–124 (1999)
18. C. Kahraman, *Fuzzy Multi-Criteria Decision Making: Theory and Applications with Recent Developments* (Springer, New York, 2008)
19. M. Kiani-Moghaddam, M. Shivaie, A. Salemnia, M.T. Ameli, Probabilistic multi-objective framework for multiple active power filters planning. Electr. Power Compon. Syst. **45**(18), 2062–2077 (2017)
20. M. Sakawa, H. Yano, An interactive fuzzy satisfying method for multi-objective nonlinear programming problems with fuzzy parameters. Fuzzy Sets Syst. **30**(3), 221–238 (1989)
21. Y.L. Chen, An interactive fuzzy-norm satisfying method for multi-objective reactive power sources planning. IEEE Trans. Power Syst. **15**(3), 1154–1160 (2000)

Chapter 3
Music-Inspired Optimization Algorithms: From Past to Present

3.1 Introduction

Many engineering challenges are addressed in the form of an optimization problem. These engineering challenges must generally satisfy multiple conflicting and heterogeneous objectives, such as finding the lowest cost, the most profit, the shortest path, the maximum reliability, the best topology, etc. For such cases, achieving the most preferred response/output requires mathematical modeling of the corresponding challenge and solving it by using an optimization algorithm. In a general sense, the optimization problem refers to the process of finding the most satisfactory response/output under the specified conditions. Technically speaking, the optimization problem can also be defined as the process of finding the minimum or maximum value of one or more objective functions, provided that the equality and inequality constraints, if any, are not violated. In a broad sense, the optimization algorithms can be broken down into two main categories, deterministic and nondeterministic/stochastic, as outlined in Sect. 1.4 of Chap. 1.

Most traditional, or conventional, optimization algorithms (e.g., the Newton-Raphson algorithm) fall into the category of deterministic optimization algorithms. Basically, deterministic optimization algorithms need to have the derivatives of the objective functions in order to solve the optimization problems. Each of the deterministic optimization algorithms is only appropriate for solving a narrow range of optimization problems. More precisely, because most real-world optimization problems involve complexities—such as mixed-integer decision-making variables, multiple conflicting and heterogeneous objective functions, and non-convex, non-smooth, and nonlinear equations—there is no unique deterministic optimization algorithm that has the desirable performance to solve the real-world optimization problems with the aforementioned complexities. Moreover, the nondeterministic/stochastic algorithms always have a stochastic behavior and can be divided into two main categories—heuristic and meta-heuristic. One of the strengths of heuristic optimization algorithms is their uncomplicated architecture,

M. Kiani-Moghaddam et al., *Modern Music-Inspired Optimization Algorithms for Electric Power Systems*, Power Systems,
https://doi.org/10.1007/978-3-030-12044-3_3

compared to deterministic optimization algorithms. Accordingly, the implementation of heuristic optimization algorithms in different engineering optimization problems, particularly complicated large-scale optimization problems, can lead to finding relatively satisfactory solutions in a reasonable amount of time. Nevertheless, the main disadvantage of the heuristic optimization algorithms is that there is no guarantee that an optimal solution and/or a set of optimal solutions will be found. Developments designed to overcome the disadvantage of heuristic optimization algorithms are referred to as meta-heuristic optimization algorithms. The meta-heuristic optimization algorithms are the optimization techniques independent of the architecture of the optimization problems. That is to say that, unlike other optimization algorithms, the meta-heuristic optimization algorithms can extensively be employed to solve a wide range of optimization problems with different structures. A well-organized classification of the meta-heuristic optimization algorithms with a focus on inspirational source was exhaustively reported in Sect. 1.5.1 of Chap. 1.

From an implementation point of view, the existing meta-heuristic optimization algorithms bring about multiple undesirable difficulties, such as premature convergence, getting stuck in a local optimum point, low convergence rate, and extremely high dependency on accurate adjustments of initial values of algorithm parameters. Technically speaking, when the existing meta-heuristic optimization algorithms fall into a local optimum point, most of these algorithms do not have the ability to exit the local optimum point and to continue the search process for reaching a global optimal point; and, thus, premature convergence occurs. In most of the existing meta-heuristic optimization algorithms, the process of generating new solutions also depends on a confined decision-making space whose dependency can affect the favorable performance of these optimization techniques. Put another way, in each new generation, a solution vector is generated with respect to a finite set of solution vectors stored in the memory of the algorithm. For example, the genetic algorithm (GA) takes into account only two parent vectors stored in memory—mating pool—to generate a new solution vector. Consequently, most of the existing meta-heuristic optimization algorithms do not have a high chance of reaching a global optimum point in solving complicated, real-world, large-scale, non-convex, non-smooth optimization problems that have a nonlinear, mixed-integer nature with big data, due to the poor performance of these optimization techniques during the search process, along with the other difficulties identified above.

In 2001, a new population-based meta-heuristic optimization algorithm, referred to as a harmony search algorithm (HSA), was developed by the inspiration of music phenomena. The original HSA had a somewhat different architecture compared to other existing meta-heuristic optimization algorithms. In the proposed architecture for this optimization algorithm, the process of generating new solutions depends on the entire space of the nonempty feasible decision-making. Put simply, in each new generation, or improvisation, the HSA generates a new solution vector after sweeping over all of the solution vectors stored in the memory of the algorithm; this characteristic can appreciably enhance the performance of the HSA in the search process. With that in mind, the favorable performance of the

HSA compared to its counterparts has given rise to its widespread utilization for solving complicated, real-world, large-scale, non-convex, non-smooth optimization problems in different branches of the engineering sciences (e.g., electrical, civil, computer, mechanical, and aerospace). In addition, many enhanced versions of the HSA have been developed with the aim of improving the efficiently and efficacy of the performance of this algorithm in solving the complicated, real-world, large-scale, non-convex, non-smooth optimization problems by specialists and researchers. However, by increasing an unbalanced number of dimensions of complicated, real-world, large-scale, non-convex, non-smooth optimization problems with big data, the performance of most of the existing meta-heuristic optimization algorithms, even the HSA and its enhanced versions, is highly influenced and cannot maintain its favorable performance in the face of such optimization problems. This is due to the tenuous and vulnerable characteristics employed in the architecture of the existing meta-heuristic optimization algorithms: having only a single-stage computational structure; using single-dimensional structures; etc. In 2011, for the first time, a new meta-heuristic optimization algorithm, referred to as a melody search algorithm (MSA), was proposed. It had a very different architecture compared to other meta-heuristic optimization algorithms. The MSA was inspired by the phenomena and concepts of music and developed as a new version of architecture of the HSA. It has a two-stage (or level) computational, multi-dimensional, and single-homogenous structure. Organizing the MSA brings about an innovative direction in the architecture of the meta-heuristic algorithms in order to solve complicated, real-world, large-scale, non-convex, non-smooth optimization problems having a nonlinear, mixed-integer nature with big data. With regard to the well-designed architecture of the music-inspired optimization algorithms and their favorable performance, there may well be appropriate optimization techniques for overcoming the difficulties in solving complicated, real-world, large-scale, non-convex, non-smooth optimization problems and finding the most satisfactory response/output with higher accuracy and convergence speed compared to other existing meta-heuristic optimization algorithms.

For the reasons identified above, the authors have focused on two targets in the context of the music-inspired optimization algorithms.

- Target 1: Providing an extensive introduction to the HSA.
- Target 2: Presenting an extensive introduction to the MSA.

The remainder of this chapter is arranged as follows. First, the interdependencies of phenomena and concepts of music and the optimization problem are reviewed briefly in Sect. 3.2. An overview of the HSA is presented in Sect. 3.3. In Sect. 3.4, a general classification of the enhanced versions of the HSA is reported, followed by a thorough description of the improved harmony search algorithm (IHSA) in Sect. 3.5. In Sect. 3.6, an overview of the MSA is presented. Finally, the chapter ends with a brief summary and some concluding remarks in Sect. 3.7.

3.2 A Brief Review of Music

In this section, the authors briefly address the definition of music, its history, and the interdependencies of phenomena and concepts of music and the optimization problem.

3.2.1 The Definition of Music

Generally speaking, music, as a social and communicative tool, is the art of incorporating vocal or instrumental sounds (or both) in order to reach pleasant and melodious forms of hearing. The phrase "music" originated from the ancient Greek word "Mousiké," which spoke to each of the skills and arts imparted by the nine Muses—daughters of Zeus and Mnemosyne, who were inspirational goddesses of science and art in Greek mythology. In ancient Iran, however, music was referred to as "Khóniya." The phrase "Khóniya" comes from the words "Khóniyak" and "Hónavac," which evolved in two parts: "Hó" meaning beauty/pleasure and "Navak" meaning tone/song. The phrase "Khóniya," therefore, represents a beautiful/pleasant tone/song. From ancient times to present, music has earned a lot of consideration in view of its desirable effect on emotions and performances of humans. With that in mind, different interpretations and definitions have been reported for music by well-known philosophers and scientists. Some of the most significant definitions for music are as follows:

- Greek philosopher Plato: "Music is a moral law. It gives soul to the universe, wings to the mind, flight to the imagination, and charm and gaiety to life and to everything."
- Greek philosopher Aristotle: "Music has the power of producing a certain effect on the moral character of the soul, and if it has the power to do this, it is clear that the young must be directed to music and must be educated in it."
- German philosopher Friedrich Nietzsche: "Without music, life would be an error. The German imagines even God singing songs."
- Persian philosopher Abu Nasr Al-Farabi: "Music is the science of identifying tones and includes two parts: theory of music and practice of music."
- Persian polymath Avicenna—Abu Ali Sina or Ibn Sina: "Music is a mathematical science in which the quality of the tones in terms of rhythm and harmony and how to set the time among tones are exhaustively discussed."

Although Avicenna, as the most distinguished Persian philosopher, physician, astronomer, thinker, and writer, referred to music in the mathematical section of the Book of Healing—Al-Shifa—music can be generally considered an art. This is due to the fact that, unlike principles of mathematical science, music is adjustable and changeable with respect to the tastes, ideas, and experiences of the player/instrumentalist/musician. As a result, music has recently been represented as a combination of mathematical science and art.

3.2.2 A Brief Review of Music History

Music history is not rigorously known with a view to historical studies concerned with music from its origins to the present. Archaeological evidence, however, demonstrates the effects of the phenomena associated with music in the process of human life in territories such as ancient Iran, Greece, Abyssinia, Japan, and Germany, several thousand years ago. An ancient, unique cylinder seal was discovered in the Choghamish district,[1] which dates back to 3400 BC (i.e., the fourth millennium BC), suggesting that the oldest world music orchestra was in Dezful county, Khuzestan province, Iran [1]. This cylinder seal is actually the earliest historical evidence indicating that music was artistically organized. Figure 3.1 shows a depiction of this seal, which is currently on display at the National Museum of Iran.[2] As can be seen, a scene of music performance with a feasting man, a servant, a vocalist, and multiple players is depicted in this cylinder seal. In addition, string, wind, and percussion instrument are exhibited in one inscription for the first time, which reveals the origin of the harmonious and symphonious tones/songs. As a result, ancient Iran was one of the first civilizations in the world in which full knowledge pertaining to the fundamental concepts of music was widely provided and developed for different purposes several thousand years ago.

3.2.3 The Interdependencies of Phenomena and Concepts of Music and the Optimization Problem

Since the advent of music, humans have sought to take advantage of music capabilities to overcome difficulties and obstacles in various sciences. Music therapy is one of the most popular applications of music in medical science for treating patients. Music therapy is generally a clinical use of music consisting of three major processes: (1) induction of relaxation; (2) acceleration of the process of curing diseases; and, (3) enhancing mental performance and bringing health. Nevertheless, music capabilities were neglected when it came to dealing with engineering challenges and alleviating their complexities. The point to be made here is that these challenges are often expressed as an optimization problem.

In 2001, for the first time, a new optimization algorithm, referred to as an HSA, was developed through inspiration of the fundamental concepts of music to solve different optimization problems and achieve the optimal solution [2]. The HSA is based on the music improvisation process in such a way that players play their musical instruments step by step in order to achieve more harmony and better

---

[1] Choghamish district is a district in Dezful county, Khuzestan province, Iran.

[2] The National Museum of Iran is located in Tehran province, Iran. It was established in two parts: The Museum of Ancient Iran and the Museum of the Islamic Era whose inaugurations were in 1937 and 1972, respectively.

Fig. 3.1 The oldest world music orchestra in 3400 BC in the Choghamish district of Dezful county, Khuzestan province, Iran [1]

Table 3.1 Interdependencies of phenomena and concepts of music and the optimization problem modeled by the HSA

No.	Comparison factor	Concept of the optimization problem modeled by the HSA	Music concept
1	Structural pattern	Decision-making variable	Player
2	Component	Value of the decision-making variable	Pitch of the musical instrument
3	Decision-making space	Value range of the decision-making variable	Pitch range of the musical instrument
4	General structural pattern	Solution vector	Musical harmony
5	Target	Objective function	Aesthetic standard of the audience
6	Process unit	Iteration	Time/practice
7	Memory	Solution vector matrix	Experience of the players
8	Best state	Global optimum point	Best harmony
9	Search process	Local and global optimum searches	Improvisation of the players

sound. This process is virtually the same as the optimization process in solving engineering challenges in which the optimal solution can be explored by the evaluation of the objective function. Table 3.1 gives the interdependencies of phenomena and concepts of music and the optimization problem modeled by means of the HSA.

3.3 Harmony Search Algorithm

The HSA is a population-based music-inspired meta-heuristic optimization algorithm that was developed in 2001 [2]. This algorithm was inspired by the improvisation process of jazz players seeking to find the best harmony and generate the most beautiful music possible. At each music performance, these players would try to enhance the sound of their musical instruments in order to produce more mature and beautiful music. As set out in Table 3.1, the concepts of music are equivalently expressed with the concept of an optimization problem modeled by the HSA. With that in mind, each player or music instrument, the pitch of the musical instrument at the disposal of each player, and the pitch range of the musical instrument at the disposal of each player are virtually the same as for each decision-making variable, the value of the decision-making variable corresponding to the relevant player, and the value range of the decision-making variable corresponding to the relevant player, respectively. By the same token, the musical harmony, aesthetic standard of the audience, and time/practice refer to the solution vector, objective function, and iteration, respectively. Additionally, the experience of the players, the best harmony and improvisation of the players are equivalent to the solution vector matrix, global optimum point, and local and global optimum searches, respectively. By enhancing the musical harmony by the players in each practice, compared to before practice from the viewpoint of the aesthetic standard of the audience, the solution vector related to the optimization problem is improved in each iteration, compared to before each iteration from the perspective of the proximity to the optimal global point. Form the standpoint of algorithm architecture, the HSA has two main characteristics: single-stage computational structure and single-dimensional structure. The HSA is, therefore, referred to as a single-stage computational, single-dimensional harmony search algorithm (SS-HSA).

The prerequisite for comprehending these characteristics is that you scrutinize features employed in the architecture of the MSA and symphony orchestra search algorithm (SOSA), which are thoroughly discussed in Sect. 3.6 of this chapter and Sect. 4.4 of Chap. 4, respectively. After a detailed investigation of the architecture associated with the MSA and SOSA, you will discover the reasons for the characteristics expressed for the SS-HSA.

The performance-driven architecture of the SS-HSA is generally broken down into four stages [2–4], as follows:

- Stage 1—Definition stage: Definition of the optimization problem and its parameters.
- Stage 2—Initialization stage.
  - Sub-stage 2.1: Initialization of the parameters of the SS-HSA.
  - Sub-stage 2.2: Initialization of the harmony memory (HM).
- Stage 3—Computational stage.
  - Sub-stage 3.1: Improvisation of a new harmony vector.
  - Sub-stage 3.2: Update of the HM.

 - Sub-stage 3.3: Check of the stopping criterion of the SS-HSA.
- Stage 4—Selection stage: Selection of the final optimal solution—the best harmony.

3.3.1 Stage 1: Definition Stage—Definition of the Optimization Problem and its Parameters

In order to solve an optimization problem using the SS-HSA, stage 1 is used to meticulously define the optimization problem and its parameters. In mathematical terms, the standard form of an optimization problem can be generally indicated based on Eqs. (1.1) and (1.2), which were given in Sect. 1.2.1 of Chap. 1. However, because the original version of the SS-HSA was developed to solve the single-objective optimization problems, now the standard form of an optimization problem must be rewritten according to Eqs. (3.1) and (3.2):

$$\begin{aligned} \underset{\mathrm{x} \in \mathrm{X}}{\text{Minimize}} \quad & \mathrm{F}(\mathrm{x}) = [f(\mathrm{x})] \\ & \text{subject to :} \\ & \mathrm{G}(\mathrm{x}) = [g_1(\mathrm{x}), \ldots, g_b(\mathrm{x}), \ldots, g_{\mathrm{B}}(\mathrm{x})] = 0; \quad \forall\{\mathrm{B} \geq 0\}, \quad \forall\{b \in \Psi^{\mathrm{B}}\} \\ & \mathrm{H}(\mathrm{x}) = [h_1(\mathrm{x}), \ldots, h_e(\mathrm{x}), \ldots, h_{\mathrm{E}}(\mathrm{x})] \leq 0; \quad \forall\{\mathrm{E} \geq 0\}, \quad \forall\{e \in \Psi^{\mathrm{E}}\} \end{aligned} \tag{3.1}$$

$$\begin{aligned} \mathrm{x} = \quad & [x_1, \ldots, x_v, \ldots, x_{\mathrm{NDV}}]; \quad \forall\{v \in \Psi^{\mathrm{NDV}}, \Psi^{\mathrm{NDV}} = \Psi^{\mathrm{NCDV+NDDV}}, \mathrm{x} \in \mathrm{X}\}, \\ & \forall\{x_v^{\min} \leq x_v \leq x_v^{\max} | v \in \Psi^{\mathrm{NCDV}}\}, \\ & \{x_v \in \{x_v(1), \ldots, x_v(w), \ldots, x_v(W_v)\} | v \in \Psi^{\mathrm{NDDV}}\} \end{aligned} \tag{3.2}$$

The explanations related to the parameters and variables from Eqs. (3.1) and (3.2) were previously defined in Sect. 1.2.1 of Chap. 1. The vector of the objective function elucidates the illustration of the vector of decision-making variables and contains the value of the objective function, as given by Eq. (3.3):

$$\mathrm{z} = \mathrm{f}(\mathrm{x}) \tag{3.3}$$

It should be pointed out that the illustration of the nonempty feasible decision-making space is recognized as a feasible objective space in objective space Z = f (X) and is explained by the set {f(x)|x ∈ X}. If a solution does not result in any violation in equality and inequality constraints, it is also considered as a feasible solution.

The SS-HSA explores the entire space of the nonempty feasible decision-making in order to find the vector of optimal decision-making variables, or solution vector. The optimal vector has the lowest possible value for the objective function given in

Eq. (3.1). Basically, the SS-HSA merely considers the objective function given in Eq. (3.1). Nonetheless, if the solution vector obtained by the SS-HSA gives rise to any violation in equality and/or inequality constraints given in Eq. (3.1), the algorithm can employ one of the two following processes from the perspective of the decision-maker in dealing with this solution vector:

- First process: The SS-HSA ignores the obtained solution vector.
- Second process: The SS-HSA takes into account the obtained solution vector by applying a specified penalty coefficient to the objective function of the optimization problem.

3.3.2 Stage 2: Initialization Stage

After completion of stage 1 and a thorough mathematical description of the optimization problem, stage 2 is employed. This stage is formed by two sub-stages: initialization of the parameters of the SS-HSA and initialization of the HM, which is discussed in detail below.

3.3.2.1 Sub-stage 2.1: Initialization of the Parameters of the SS-HSA

In sub-stage 2.1, the parameter adjustments of the SS-HSA should be initialized with specific values. Table 3.2 provides a detailed description of the parameter adjustments of the SS-HSA. In the SS-HSA, the HM is a place for storing the solution, or harmony vectors. The HM in the SS-HSA is virtually the same as the mating pool in the GA. The harmony memory size (HMS) represents the number of solution vectors stored in the HM.

The HMS is equivalent to the population size in the GA. In the improvisation process of a new harmony vector, the harmony memory considering rate (HMCR)

Table 3.2 Adjustment parameters of the SS-HSA

No.	SS-HSA parameter	Abbreviation	Parameter range
1	Harmony memory	HM	–
2	Harmony memory size	HMS	HMS ≥ 1
3	Harmony memory considering rate	HMCR	$0 \leq$ HMCR ≤ 1
4	Pitch adjusting rate	PAR	$0 \leq$ PAR ≤ 2
5	Bandwidth	BW	$0 \leq$ BW $< +\infty$
6	Number of continuous of decision-making variables	NCDV	NCDV ≥ 1
7	Number of discrete decision-making variables	NDDV	NDDV ≥ 1
8	Number of decision-making variables	NDV	NDV ≥ 2
9	Maximum number of improvisations/iterations	MNI	MNI ≥ 1

is employed in order to determine whether the value of a decision-making variable related to a new harmony vector is derived from the HM or from the entire space of the nonempty feasible decision-making. Put another way, the HMCR expresses the rate at which the value of a decision-making variable from a new harmony vector is randomly selected with respect to the player's memory, or more comprehensively from the HM. In this regards, 1-HMCR indicates the rate at which the value of a decision-making variable from a new harmony vector is haphazardly chosen in terms of the entire space of the nonempty feasible decision-making. By the same token, in the improvisation process of a new harmony vector, the pitch adjusting rate (PAR) is utilized to specify whether the value of a decision-making variable selected from the HM needs an update to its neighbor value or not. More precisely, the PAR describes the rate at which the value of a decision-making variable selected with the HMCR rate from the player's memory, or more comprehensively from the HM, is altered. With that in mind, 1-PAR clarifies the rate at which the value of a decision-making variable, chosen with the HMCR rate from the player's memory or more comprehensively from the HM, is not changed. The bandwidth (BW)—fret width—is considered to be an optional length and is exclusively defined for continuous decision-making variables. In music literature, the fret width is a significant element on the neck of a string musical instrument (e.g., a bass guitar) in such a way that the neck of a string musical instrument is broken up into fixed-length segments at intervals pertaining to the musical framework. In the string musical instruments (e.g., the guitar family), each fret illustrates a semitone, and 12 semitones make up an octave in the standard Western style. In the SS-HSA, however, the frets represent arbitrary points that divide the entire space of the nonempty feasible continuous decision-making into fixed parts. The fret width—BW—is defined as the distance between two neighbor frets. The number of decision-making variables (NDV), which is dependent on the optimization problem given in Eqs. (3.1) and (3.2), consists of the sum of the number of continuous decision-making variables (NCDV) and the number of discrete decision-making variables (NDDV). The NDV characterizes the dimensions of the harmony vector in the SS-HSA. The maximum number of improvisations/iterations (MNI) addresses the number of times that the computational stage is repeated in the SS-HSA. The point to be made here is that the SS-HSA improvises a harmony vector in each improvisation/iteration. The MNI is usually employed as a stopping criterion in the SS-HSA.

3.3.2.2 Sub-stage 2.2: Initialization of the HM

After finalization of sub-stage 2.1 and parameter adjustments of the SS-HSA, the HM must be initialized in sub-stage 2.2. In this sub-stage, the HM matrix, which has a dimension equal to $\{\text{HMS}\} \cdot \{\text{NDV} + 1\}$, is filled with a large number of solution vectors generated randomly according to Eqs. (3.4) through (3.6):

$$\text{HM} = \begin{bmatrix} \mathrm{x}^1 \\ \vdots \\ \mathrm{x}^s \\ \vdots \\ \mathrm{x}^{\text{HMS}} \end{bmatrix} = \left[\begin{array}{ccccc|c} x_1^1 & \cdots & x_v^1 & \cdots & x_{\text{NDV}}^1 & f(\mathrm{x}^1) \\ \vdots & & \vdots & & \vdots & \vdots \\ x_1^s & \cdots & x_v^s & \cdots & x_{\text{NDV}}^s & f(\mathrm{x}^s) \\ \vdots & & \vdots & & \vdots & \vdots \\ x_1^{\text{HMS}} & \cdots & x_v^{\text{HMS}} & \cdots & x_{\text{NDV}}^{\text{HMS}} & f(\mathrm{x}^{\text{HMS}}) \end{array}\right]; \quad (3.4)$$

$$\forall\left\{v \in \Psi^{\text{NDV}}, s \in \Psi^{\text{HMS}}, \Psi^{\text{NDV}} = \Psi^{\text{NCDV+NDDV}}\right\}$$

$$x_v^s = x_v^{\min} + \text{U}(0,1) \cdot \left(x_v^{\max} - x_v^{\min}\right); \quad \forall\left\{v \in \Psi^{\text{NCDV}}, s \in \Psi^{\text{HMS}}\right\} \quad (3.5)$$

$$x_v^s = x_v(y); \quad \forall\left\{v \in \Psi^{\text{NDDV}}, s \in \Psi^{\text{HMS}}, y \sim \text{U}\{x_v(1), \ldots, x_v(w_v), \ldots, x_v(W_v)\}\right\} \quad (3.6)$$

Equation (3.4) represents the HM. Equations (3.5) and (3.6) are also considered for continuous and discrete decision-making variables, respectively. In Eq. (3.5), U (0, 1) indicates a random number with a uniform distribution between 0 and 1. In addition, Eq. (3.5) expresses how the value of the continuous decision-making variable v from the harmony vector s stored in the HM is randomly determined using the set of candidate admissible values for this decision-making variable, which is confined by lower bound $x_v^{\min}$ and upper bound $x_v^{\max}$. In Eq. (3.6), the y index describes a random integer with a uniform distribution through the set $\{x_v(1), \ldots, x_v(w_v), \ldots, x_v(W_v)\}$—$y \sim \text{U}\{x_v(1), \ldots, x_v(w_v), \ldots, x_v(W_v)\}$. Equation (3.6) describes how the value of the discrete decision-making variable v from the harmony vector s stored in the HM is randomly specified using the set of candidate allowable values for this decision-making variable, which is demonstrated by the set $\{x_v(1), \ldots, x_v(w_v), \ldots, x_v(W_v)\}$. Table 3.3 gives the pseudocode associated with initialization of the HM in the SS-HSA. After filling the HM with random solution vectors, the solution vectors stored in the HM must be sorted from the lowest value to the highest value—in an ascending order—with regard to the value of the objective function of the optimization problem. Table 3.4 presents the pseudocode related to sorting the solution vectors stored in the HM under the SS-HSA.

3.3.3 Stage 3: Computational Stage

After completion of stage 2 and initialization of the parameters of the SS-HSA and the HM, this computational stage must be performed. This stage consists of three sub-stages: (1) improvisation of a new harmony vector; (2) update of the HM; and (3) check of the stopping criterion of the SS-HSA. The mathematical equations expressed at this stage must depend on the improvisation/iteration index—index m—because of the repeatability of the computational stage in the SS-HSA.

Table 3.3 Pseudocode associated with initialization of the HM in the SS-HSA

Algorithm 1: Pseudocode for initialization of the HM in the SS-HSA

Input: HMS, NCDV, NDDV, NDV, $x_v^{\min}$, $x_v^{\max}$, $\{x_v(1), \ldots, x_v(w_v), \ldots, x_v(W_v)\}$
Output: HM

start main body

begin
construct the HM matrix with dimension {HMS} · {NDV + 1} and zero initial value
for *harmony vector s* $[s \in \Psi^{\text{HMS}}]$ **do**
  construct the harmony vector x^s with dimension {1} · {NDV + 1} and zero initial value
  for *decision-making variable v* $[v \in \Psi^{\text{NDV}}]$ **do**
    $x_v^s = x_v^{\min} + \text{U}(0,1) \cdot \left(x_v^{\max} - x_v^{\min}\right)$; for CDVs
    $x_v^s = x_v(y); \quad \forall y \sim \text{U}\{x_v(1), \ldots, x_v(w_v), \ldots, x_v(W_v)\}$; for DDVs
    allocate x_v^s to element $(1, v)$ of the harmony vector x^s
  end for
  calculate the value of objective function, fitness function, derived from the harmony vector x^s as $f(\text{x}^s)$
  allocate $f(\text{x}^s)$ to element $(1, \text{NDV} + 1)$ of the harmony vector x^s
  add harmony vector x^s to the row s of the HM matrix
end for
terminate

end main body

Note: Continuous decision-making variable (CDVs), discrete decision-making variable (DDVs)

Table 3.4 Pseudocode related to sorting the solution vectors stored in the HM under the SS-HSA

Algorithm 2: Pseudocode for sorting the solution vectors stored in the HM under the SS-HSA

Input: Unsorted HM
Output: Sorted HM

start main body

begin
$\text{F}^{\text{sort}} = \text{sort}(\text{HM}(1{:}\ \text{HMS}, \text{NDV} + 1), \text{'ascend'})$
for *harmony vector s* $[s \in \Psi^{\text{HMS}}]$ **do**
  for *harmony vector* s^* $[s^* \in \Psi^{\text{HMS}}]$ **do**
    if $\text{F}^{\text{sort}}(s) == \text{HM}(s^*, \text{NDV} + 1)$ **then**
      $\text{HM}^{\text{sort}}(s, 1{:}\ \text{NDV} + 1) = \text{HM}(s^*, 1{:}\ \text{NDV} + 1)$;
    end if
  end for
end for
$\text{HM} = \text{HM}^{\text{sort}}$
terminate

end main body

3.3.3.1 Sub-stage 3.1: Improvisation of a New Harmony Vector

In the jazz improvisation process, a musical note can generally be played by a player based on one of the three different styles: (1) selection of a musical note from the corresponding player's memory; (2) creation of a slight alteration in the selected musical note from the corresponding player's memory; and, (3) random selection of a musical note from the entire playable range. In the SS-HSA, however, improvisation process refers to the process of producing a harmony vector. Similarly, in the SS-HSA, selection of the value of a decision-making variable corresponding to a player can be accomplished according to one of the three different methods: (1) selection of the value of a decision-making variable from the HM_m; (2) creation of a slight alteration in the value of the selected decision-making variable from the HM_m; and, (3) selection of the value of a decision-making variable from the entire space of the nonempty feasible decision-making. In an exhaustive definition, the improvisation process of a new harmony vector—$\mathrm{x}_m^{\text{new}} = \left(x_{m,1}^{\text{new}}, \ldots, x_{m,v}^{\text{new}}, \ldots, x_{\mathrm{m,NDV}}^{\text{new}}\right)$—in the SS-HSA can be expressed by three rules: (1) harmony memory consideration; (2) pitch adjustment; and, (3) random selection.

Rule 1: In the harmony memory consideration rule, the values of a new harmony vector are randomly selected from the available harmony vectors in the HM_m with the probability of the HMCR. More precisely, the value of the first decision-making variable from a new harmony vector, $x_{m,1}^{\text{new}}$, is randomly chosen from the available corresponding decision-making variable in the harmony vectors stored in the HM_m, $\left(x_{m,1}^{1}, \ldots, x_{m,1}^{s}, \ldots, x_{m,1}^{\text{HMS}}\right)$, with the probability of the HMCR. The values for other decision-making variables are also selected in the same way. Applying the harmony memory consideration rule to determine the value of the decision-making variable v from a new harmony vector, $x_{m,v}^{\text{new}}$, is performed using Eq. (3.7):

$$x_{m,v}^{\text{new}} = x_{m,v}^{r}; \quad \forall\left\{m \in \Psi^{\text{MNI}}, v \in \Psi^{\text{NDV}}, r \sim \text{U}\{1,2,\ldots,\text{HMS}\}, \Psi^{\text{NDV}} = \Psi^{\text{NCDV+NDDV}}\right\} \tag{3.7}$$

Equation (3.7) is employed for continuous and discrete decision-making variables. It is also important to point out that index r is a random integer with a uniform distribution through the set $\{1, 2, \ldots, \text{HMS}\}$—$r \sim \text{U}\{1, 2, \ldots, \text{HMS}\}$. In other words, in Eq. (3.7), the value of index r is randomly determined through the set of allowable values illustrated by the set $\{1, 2, \ldots, \text{HMS}\}$. Determination of this index is represented in accordance with Eq. (3.8):

$$r = \text{int}(\text{U}(0, 1) \cdot \text{HMS}) + 1 \tag{3.8}$$

It should be pointed out that other distributions can be utilized for index r, such as $(\text{U}(0, 1))^2$. The use of this distribution gives rise to the selection of lower values for index r.

Rule 2: In the pitch adjustment rule, the values of a new harmony vector, which are randomly selected through the existing harmony vectors in the HM_m with the probability of the HMCR, are updated with the probability of the PAR to the available values in the neighborhood of the current values. Put another way, after the value of the first decision-making variable from a new harmony vector, $x_{m,1}^{\text{new}}$, is randomly selected from the available corresponding decision-making variable in the harmony vectors stored in the HM_m with the probability of the HMCR, this decision-making variable is updated with the probability of the PAR to one of the available values in the neighborhood of its current value. The update process to one of the available values in the neighborhood for this decision-making variable is done by adding a specific value to its current value. The values for other decision-making variables are also selected in the same way. Applying the pitch adjustment rule to specify the value of the decision-making variable v from a new harmony vector, $x_{m,v}^{\text{new}}$, is carried out by using Eqs. (3.9) and (3.10):

$$x_{m,v}^{\text{new}} = x_{m,v}^{\text{new}} \pm \mathrm{U}(0,1) \cdot \mathrm{BW}; \quad \forall \{m \in \Psi^{\text{MNI}}, v \in \Psi^{\text{NCDV}}\} \tag{3.9}$$

$$\begin{aligned} x_{m,v}^{\text{new}} = \; & x_{m,v}^{\text{new}}(y+t); \\ & \forall \{m \in \Psi^{\text{MNI}}, v \in \Psi^{\text{NDDV}}, y \sim \mathrm{U}\{x_v(1), \ldots, x_v(w_v), \ldots, \\ & x_v(W_v)\}, t \sim \mathrm{U}\{-1,+1\}\} \end{aligned} \tag{3.10}$$

Equations (3.9) and (3.10) are used for the continuous and discrete decision-making variables, respectively. In Eq. (3.10), t represents the neighborhood index. The neighborhood index t is a random integer with a uniform distribution through the set $\{-1,+1\}$—$t \sim \mathrm{U}\{-1,+1\}$. In other words, in Eq. (3.10), the value of index t is randomly determined through the set of allowable values illustrated by the set $\{-1,+1\}$.

Rule 3: In the random selection rule, the values of a new harmony vector are randomly chosen from the entire space of the nonempty feasible decision-making with the probability of the 1-HMCR. In simple terms, the value of the first decision-making variable from a new harmony vector, $x_{m,1}^{\text{new}}$, is randomly selected from the entire space of the nonempty feasible decision-making with the probability of the 1-HMCR. The values for other decision-making variables are also chosen in the same way. The point to be made here is that the random selection rule was already utilized in sub-stage 2.2 for initialization of the HM. Applying the random selection rule to characterize the value of the decision-making variable v from a new harmony vector, $x_{m,v}^{\text{new}}$, is done using Eqs. (3.11) and (3.12):

$$x_{m,v}^{\text{new}} = x_v^{\min} + \mathrm{U}(0,1) \cdot \left(x_v^{\max} - x_v^{\min}\right); \quad \forall \{m \in \Psi^{\text{MNI}}, v \in \Psi^{\text{NCDV}}\} \tag{3.11}$$

$$x_{m,v}^{\text{new}} = x_v(y); \quad \forall \{m \in \Psi^{\text{MNI}}, v \in \Psi^{\text{NDDV}}, y \sim \mathrm{U}\{x_v(1), \ldots, x_v(w_v), \ldots, x_v(W_v)\}\} \tag{3.12}$$

Equations (3.11) and (3.12) are used for the continuous and discrete decision-making variables, respectively. As further elucidation, assume that the parameter adjustments for the HMCR and PAR are considered to be 0.75 and 0.65, respectively. First, a random number with a uniform distribution between 0 and 1, U (0, 1), is generated. If the generated random number has a lower value than the value of the HMCR parameter (i.e., 0.75), the value of the first decision-making variable from a new harmony vector, $x_{m,1}^{\text{new}}$, is randomly selected from the available corresponding decision-making variable among the harmony vectors stored in the HM_m, $\left(x_{m,1}^1, \ldots, x_{m,1}^s, \ldots, x_{m,1}^{\text{HMS}}\right)$, with the probability of 0.75. Correspondingly, the value of the first decision-making variable from a new harmony vector, $x_{m,1}^{\text{new}}$, is randomly chosen from the entire space of the nonempty feasible decision-making with the probability of (1 – 0.75), provided that the random number generated has a value higher than the value of the HMCR parameter (0.75).

After the value of the first decision-making variable from a new harmony vector, $x_{m,1}^{\text{new}}$, has been randomly selected from the available corresponding decision-making variable in the harmony vectors stored in the HM_m with the probability of 0.75, one more random uniform number between 0 and 1, U(0, 1), is generated. If this random number has a value lower than the value of the PAR parameter (0.65), the value of the first decision-making variable from a new harmony vector, $x_{m,1}^{\text{new}}$, is updated to one of the available values in the neighborhood of its current value chosen from the HM_m with the probability of 0.65. Mutually, the value of the first decision-making variable from a new harmony vector, $x_{m,1}^{\text{new}}$, which was haphazardly selected from available corresponding decision-making variable in the harmony vectors stored in the HM_m, $\left(x_{m,1}^1, \ldots, x_{m,1}^s, \ldots, x_{m,1}^{\text{HMS}}\right)$, with the probability of 0.75, is not changed if the generated random number has a value higher than the value of the PAR parameter (0.65).

As a general result, the probability that the value of the first decision-making variable from a new harmony vector, $x_{m,1}^{\text{new}}$, can be determined by applying the harmony memory consideration, pitch adjustment, and random selection rules is equal to HMCR × (1 – PAR), HMCR × PAR, and 1 – HMCR, respectively. The values for other decision-making variables are also chosen in the same way. Table 3.5 presents the pseudocode pertaining to improvisation of a new harmony vector in the SS-HSA.

3.3.3.2 Sub-stage 3.2: Update of the HM

After finalization of sub-stage 3.1 and improvisation of a new harmony vector, the update process of the HM_m must be carried out in sub-stage 3.2. In this sub-stage, a new harmony vector is evaluated and compared with the worst available harmony vector in the HM_m—the harmony vector stored in the HMS row of the HM_m—from the perspective of the objective function. If a new harmony vector has a better value than the worst available harmony vector in the HM_m, from the perspective of the objective function, this new harmony vector replaces the worst harmony vector available in the HM_m; the worst available harmony vector is then eliminated from the HM_m. Table 3.6 shows the pseudocode related to the update of the HM_m in the SS-HSA.

Table 3.5 Pseudocode pertaining to improvisation of a new harmony vector in the SS-HSA

Algorithm 3: Pseudocode for improvisation of a new harmony vector in the SS-HSA

Input: BW, HMCR, HMS, NCDV, NDDV, NDV, PAR, $x_v^{\min}, x_v^{\max}, \{x_v(1), \ldots, x_v(w_v), \ldots x_v(W_v)\}$

Output: $\mathrm{x}_m^{\mathrm{new}}$

start main body

begin

construct the new harmony vector $\mathrm{x}_m^{\mathrm{new}}$ with dimension $\{1\} \cdot \{\mathrm{NDV} + 1\}$ and zero initial value

for *decision-making variable v* $[v \in \Psi^{\mathrm{NDV}}]$ **do**

  if $\mathrm{U}(0, 1) \leq \mathrm{HMCR}$ **then**

    Rule 1: The harmony memory consideration with the probability of the HMCR

   $x_{m,v}^{\mathrm{new}} = x_{m,v}^{r}; \quad \forall r \sim \mathrm{U}\{1, 2, \ldots, \mathrm{HMS}\}$; for CDVs and DDVs

   if $\mathrm{U}(0, 1) \leq \mathrm{PAR}$ **then**

    Rule 2: The pitch adjustment with the probability of the HMCR · PAR

    $x_{m,v}^{\mathrm{new}} = x_{m,v}^{\mathrm{new}} \pm \mathrm{U}(0, 1) \cdot \mathrm{BW}$; for CDVs

    $x_{m,v}^{\mathrm{new}} = x_{m,v}^{\mathrm{new}}(y + t); y \sim \mathrm{U}\{x_v(1), \ldots, x_v(w_v), \ldots, x_v(W_v)\}, \ \forall t \sim \mathrm{U}\{-1, +1\}$; for DDVs

   end if

  else if

   Rule 3: The random selection with the probability of the 1 − HMCR

   $x_{m,v}^{\mathrm{new}} = x_v^{\min} + \mathrm{U}(0, 1) \cdot \left(x_v^{\max} - x_v^{\min}\right)$; for CDVs

   $x_{m,v}^{\mathrm{new}} = x_v(y); \forall y \sim \mathrm{U}\{x_v(1), \ldots, x_v(w_v), \ldots, x_v(W_v)\}$; for DDVs

   end if

end for

calculate the value of objective function, fitness function, derived from the harmony vector $\mathrm{x}_m^{\mathrm{new}}$ as $f\left(\mathrm{x}_m^{\mathrm{new}}\right)$

allocate $f\left(\mathrm{x}_m^{\mathrm{new}}\right)$ to element (1, NDV + 1) of the new harmony vector $\mathrm{x}_m^{\mathrm{new}}$

terminate

end main body

Note: Continuous decision-making variable (CDVs), discrete decision-making variable (DDVs)

Table 3.6 Pseudocode related to update of the HM_m in the SS-HSA

Algorithm 4: Pseudocode for the update of the HM_m in the SS-HSA

Input: Not update HM_m, $\mathrm{x}_m^{\mathrm{new}}$

Output: Updated HM_m

start main body

begin

set $\mathrm{x}^{\mathrm{worst}} = \mathrm{x}_m^{\mathrm{HMS}}$

set $f(\mathrm{x}^{\mathrm{worst}}) = f\left(\mathrm{x}_m^{\mathrm{HMS}}\right)$

if $f\left(\mathrm{x}_m^{\mathrm{new}}\right) < f(\mathrm{x}^{\mathrm{worst}})$ **then**

  $\mathrm{x}_m^{\mathrm{new}} \in HM_m$

  $\mathrm{x}^{\mathrm{worst}} \notin HM_m$

end if

terminate

end main body

Table 3.7 Pseudocode relevant to sorting the solution vectors stored in the HM_m under the SS-HSA

Algorithm 5: Pseudocode for sorting the solution vectors stored in the HM_m under the SS-HSA

Input: Unsorted HM_m
Output: Sorted HM_m

start main body

begin
$F_m^{\text{sort}} = \text{sort}\left(HM_m(1{:}\ \text{HMS}, \text{NDV} + 1), '\text{ascend}'\right)$
for *harmony vector* s $[s \in \Psi^{\text{HMS}}]$ **do**
  for *harmony vector* s^* $[s^* \in \Psi^{\text{HMS}}]$ **do**
    if $F_m^{\text{sort}}(s) == HM_m(s^*, \text{NDV} + 1)$ **then**
      $HM_m^{\text{sort}}(s, 1{:}\ \text{NDV} + 1) = HM_m(s^*, 1{:}\ \text{NDV} + 1);$
    end if
  end for
end for
$HM_m = HM_m^{\text{sort}}$
terminate

end main body

It should be pointed out that the update process of the HM_m is not accomplished if the new harmony vector is not notably better than the worst available harmony vector in the HM_m, from the standpoint of the objective function. After completion of this process, harmony vectors stored in the HM_m must be re-sorted based on the value of objective function—fitness function—in an ascending order. The pseudocode related to sorting the solution vectors stored in the HM was already provided in Table 3.4. Given the dependence of the HM to the improvisation/iteration index of the computational stage—index m—the aforementioned pseudocode must be rewritten according to Table 3.7.

3.3.3.3 Sub-stage 3.3: Check of the Stopping Criterion of the SS-HSA

After completion of sub-stage 3.2 and an update of the HM, the check process of the stopping criterion of the SS-HSA must be done in sub-stage 3.3. In this sub-stage, the computational efforts of the SS-HSA are terminated when its stopping criterion—the MNI—is satisfied. Otherwise, sub-stages 3.1 and 3.2 are repeated.

3.3.4 Stage 4: Selection Stage—Selection of the Final Optimal Solution—The Best Harmony

After finalization of stage 3, or accomplishment of the computational stage, the selection of the final optimal solution—the best harmony—must be performed in stage 4. In this stage, the best harmony vector stored in the HM, x^1, is taken as the

Table 3.8 Pseudocode relevant to the selection of the final optimal solution in the SS-HAS

Algorithm 6: Pseudocode for the selection of the final optimal solution in the SS-HSA
Input: HM
Output: $\mathrm{x}^{\mathrm{best}}$
start main body
begin
set $\mathrm{x}^{\mathrm{best}} = \mathrm{x}^{1}$
terminate
end main body

Table 3.9 Pseudocode pertaining to performance-driven architecture of the SS-HSA

Algorithm 7: Pseudocode for performance-driven architecture of the SS-HSA
Input: BW, HMCR, HMS, MNI, NCDV, NDDV, NDV, PAR, $x_v^{\min}$, $x_v^{\max}$, $\{x_v(1), \ldots, x_v(w_v), \ldots, x_v(W_v)\}$
Output: $\mathrm{x}^{\mathrm{best}}$
start main body
begin
Stage 1—Definition stage: Definition of the optimization problem and its parameters
Stage 2—Initialization stage
Sub-stage 2.1: Initialization of the parameters of the SS-HSA
Sub-stage 2.2: Initialization of the of the HM
Algorithm 1: Pseudocode for initialization of the HM in the SS-HSA
Algorithm 2: Pseudocode for sorting the solution vectors stored in the HM under the SS-HSA
Stage 3—Computational stage
set *improvisation/iteration* $m = 1$
set $HM_m = \mathrm{HM}$
while $m \leq \mathrm{MNI}$ **do**
Sub-stage 3.1: Improvisation of a new harmony vector
Algorithm 3: Pseudocode for improvisation of a new harmony vector in the SS-HSA
Sub-stage 3.2: Update of the HM
Algorithm 4: Pseudocode for the update of the HM_m in the SS-HSA
Algorithm 5: Pseudocode for sorting the solution vectors stored in the HM_m under the SS-HSA
set *improvisation/iteration* $m = m + 1$
end while
Stage 4—Selection stage: Selection of the final optimal solution—The best harmony
Algorithm 6: Pseudocode for the selection of the final optimal solution in the SS-HSA
terminate
end main body

final optimal solution. Table 3.8 gives the pseudocode relevant to the selection of the final optimal solution in the SS-HSA. The designed pseudocode in different stages and sub-stages of the SS-HSA is located in a regular sequence and forms the performance-driven architecture of this algorithm. Table 3.9 presents the pseudocode pertaining to the performance-driven architecture of the SS-HSA. In here, sub-stage 3.3—the check process of the stopping criterion of the SS-HSA—is defined by the WHILE loop in the pseudocode pertaining to the performance-driven architecture of the SS-HSA (see Table 3.9).

3.4 Enhanced Versions of the Single-Stage Computational, Single-Dimensional Harmony Search Algorithm

As previously mentioned, the original SS-HSA was introduced in 2001. Readers interested in a comprehensive discussion on different applications of the SS-HSA are referred to the work by Manjarres et al. [5]. From 2001 to the present, many enhanced versions of the original SS-HSA have been developed to solve a wide range of optimization problems in the engineering sciences (e.g., electrical, civil, computer, mechanical, and aerospace). In related literature, different classifications for the enhanced versions of the SS-HSA have been presented. Providing a structural classification for the enhanced versions of the SS-HSA can dramatically help interested readers to reasonably understand how to enhance the SS-HSA. By investigating all enhanced versions of the SS-HSA, it can be concluded that all enhancements, from the perspective of implementation, can be broken down into three general categories, as follows:

- Category 1: Enhancements applied on the SS-HSA from the perspective of the parameter adjustments. The most well-known existing enhanced version of this category is the IHSA.
- Category 2: Enhancements accomplished on the SS-HSA from the standpoint of a combination of this algorithm with other meta-heuristic optimization algorithms. Enhanced versions of this category are divided into two subcategories.
  - Subcategory 2.1: Enhancements performed on the SS-HSA from the viewpoint of integration of some components associated with other meta-heuristic optimization algorithms in the architecture of the SS-HSA. The best known existing enhanced version of this subcategory is the global-best harmony search algorithm.
  - Subcategory 2.2: Enhancements carried out by the SS-HSA from the perspective of integration of some components pertaining to the SS-HSA in the architecture of other meta-heuristic optimization algorithms. The most well-known existing enhanced version of this subcategory is the adaptive GA using the SS-HSA.
- Category 3: Enhancements implemented on the SS-HSA from the standpoint of architectural principles. The first existing enhanced version of this category is the MSA.

More detailed descriptions regarding the enhanced versions of the SS-HSA are beyond the scope of this chapter, but the interested reader may look to the work by Moh'd-Alia and Mandava [6] for a thorough discussion of these enhanced versions.

As the IHSA and MSA are widely employed in the second part of this book for comparison purposes, these two existing optimization techniques will be discussed extensively in Sects. 3.5 and 3.6 of this chapter, respectively.

3.5 Improved Harmony Search Algorithm

As previously mentioned, the IHSA, as the most well-known existing enhanced version of the SS-HSA, was developed by dynamically changing the parameter adjustments in each improvisation/iteration. The architecture of the IHSA is, therefore, quite similar to the architecture of the SS-HSA. In more detail, the IHSA has two main characteristics: a single-stage computational structure and a single-dimensional structure. With that in mind, the IHSA is referred to as the single-stage computational, single-dimensional improved harmony search algorithm (SS-IHSA). In the architecture of the SS-HSA, the PAR and BW parameters play a pivotal role in adjusting the convergence rate of the algorithm to achieve the final optimal solution. Accordingly, desirable performance of the SS-HSA is highly dependent on precise and proper adjustment of these parameters. In view of the fact that the BW parameter can have any value in the range of zero to positive infinity, fine-tuning this parameter is more difficult than the PAR parameter. In the SS-HSA, the values of the PAR and BW parameters are adjusted in stage 2.1 and cannot be changed during subsequent improvisations/iterations. Simply put, the SS-HSA employs invariant values for the PAR and BW parameters in all improvisations/iterations. The main disadvantage of these parameter adjustments appears in the number of iterations required by the SS-HSA to find the final optimal solution. Considering small values for the PAR parameter with large values for the BW parameter can generally bring about a poor performance for the SS-HSA and a significant increase in the number of iterations needed to reach the final optimal solution. Although considering smaller values for the BW parameter in the terminative improvisations/iterations strengthens the probability of more precise adjustment of solution vectors, taking into account large values for the BW parameter in initial improvisations/iterations is certainly a necessity for increasing diversity in the solution vectors of the SS-HSA. Similarly, considering large values for the PAR parameter with small values for the BW parameter can usually improve the solutions in the terminative improvisations/iterations in such a way that the SS-HSA converges towards the optimal solution vector.

In 2007, to overcome the difficulties associated with the invariant values of the BW and PAR parameters, the SS-IHSA was introduced and variant values were employed for the PAR and BW parameters [7]. Given the fact that the different stages in the SS-IHSA are virtually the same as the different stages in the SS-HSA, only the differences caused by the use of variant values for the PAR and BW parameters are referred to here. The major differences between the SS-IHSA and the SS-HSA appear only in sub-stage 2.1 (initialization of the parameters of the algorithm) and in sub-stage 3.1 (improvisation of a new harmony vector). In sub-stage 2.1, the parameter adjustments of the offered SS-HSA are characterized according to Table 3.1, which is presented in Sect. 3.3.2.1 of this chapter. As is clear from Table 3.1, the SS-HSA considers invariant values for the PAR and BW parameters. In this sub-stage, however, the SS-IHSA replaces the PAR and BW parameters with parameters minimum pitch adjusting rate ($\mathrm{PAR}^{\min}$) and maximum pitch adjusting rate ($\mathrm{PAR}^{\max}$) and minimum bandwidth ($\mathrm{BW}^{\min}$) and maximum bandwidth ($\mathrm{BW}^{\max}$), respectively. Other parameters presented in Table 4.1 remain unchanged for the SS-IHSA. As a result, the detailed descriptions relevant to the adjustment parameters of the SS-IHSA are thoroughly represented in Table 3.10.

In sub-stage 3.1, unlike the SS-HSA, which uses invariant values for the PAR and BW parameters in the improvisation process of a new harmony vector, the SS-IHSA utilizes the updated values for the PAR and BW parameters in the improvisation process of a new harmony vector. In this sub-stage, the values associated with the PAR and BW parameters are dynamically changed and updated in each improvisation/iteration of the SS-IHSA by using Eqs. (3.13) and (3.14), respectively:

$$BW_m = \mathrm{BW}^{\max} \cdot \exp\left(\frac{\ln\left(\mathrm{BW}^{\max}/\mathrm{BW}^{\min}\right)}{\mathrm{MNI}} \cdot m\right); \quad \forall\left\{m \in \Psi^{\mathrm{MNI}}\right\} \tag{3.13}$$

$$PAR_m = \mathrm{PAR}^{\min} + \left(\frac{\mathrm{PAR}^{\max} - \mathrm{PAR}^{\min}}{\mathrm{MNI}}\right) \cdot m; \quad \forall\left\{m \in \Psi^{\mathrm{MNI}}\right\} \tag{3.14}$$

In Eq. (3.13), the value of the BW_m parameter is represented as an exponential function of the improvisation/iteration index—index m. In this equation, the value of the BW_m parameter is exponentially decreased by increasing the value of the improvisation/iteration index.

Table 3.10 Adjustment parameters of the SS-IHSA

No.	SS-IHSA parameter	Abbreviation	Parameter range
1	Harmony memory	HM	–
2	Harmony memory size	HMS	$\mathrm{HMS} \geq 1$
3	Harmony memory considering rate	HMCR	$0 \leq \mathrm{HMCR} \leq 1$
4	Minimum pitch adjusting rate	$\mathrm{PAR}^{\min}$	$0 \leq \mathrm{PAR}^{\min} \leq 2$
5	Maximum pitch adjusting rate	$\mathrm{PAR}^{\max}$	$0 \leq \mathrm{PAR}^{\max} \leq 2$
6	Minimum bandwidth	$\mathrm{BW}^{\min}$	$0 \leq \mathrm{BW}^{\min} <+\infty$
7	Maximum bandwidth	$\mathrm{BW}^{\max}$	$0 \leq \mathrm{BW}^{\max} <+\infty$
8	Number of continuous decision-making variables	NCDV	$\mathrm{NCDV} \geq 1$
9	Number of discrete decision-making variables	NDDV	$\mathrm{NDDV} \geq 1$
10	Number of decision-making variables	NDV	$\mathrm{NDV} \geq 2$
11	Maximum number of improvisations/iterations	MNI	$\mathrm{MNI} \geq 1$

That is to say that the value of the BW_m parameter, by altering the improvisation/iteration index from zero to the MNI, $m \in \{0 \rightarrow \text{MNI}\}$, is exponentially changed from the value of the $\text{BW}^{\max}$ parameter to the value of the $\text{BW}^{\min}$ parameter, $BW_m \in \{\text{BW}^{\max} \rightarrow \text{BW}^{\min}\}$. In Eq. (3.14), the value of the PAR_m parameter is expressed as a linear function of the improvisation/iteration index—index m. In this equation, the value of the PAR_m parameter is linearly grown by increasing the value of the improvisation/iteration index. Put simply, the value of the PAR_m parameter, by changing the improvisation/iteration index from zero to the MNI, $m \in \{0 \rightarrow \text{MNI}\}$, is linearly altered from the value of the $\text{PAR}^{\min}$ parameter to the value of the $\text{PAR}^{\max}$ parameter, $PAR_m \in \{\text{PAR}^{\min} \rightarrow \text{PAR}^{\max}\}$. Table 3.11 shows the rectified

Table 3.11 Pseudocode associated with improvisation of a new harmony vector in the SS-IHSA

Algorithm 8: Pseudocode for improvisation of a new harmony vector in the SS-IHSA

Input: $\text{BW}^{\max}$, $\text{BW}^{\min}$, HMCR, HMS, MNI, NCDV, NDDV, NDV, $\text{PAR}^{\max}$, $\text{PAR}^{\min}$, $x_v^{\min}$, $x_v^{\max}$, $\{x_v(1), \ldots, x_v(w_v), \ldots, x_v(W_v)\}$

Output: x_m^{new}

start main body

begin

$BW_m = \text{BW}^{\max} \cdot \exp\left[(\ln(\text{BW}^{\max}/\text{BW}^{\min})/\text{MNI}) \cdot m\right]$

$PAR_m = \text{PAR}^{\min} - \left[((\text{PAR}^{\max} - \text{PAR}^{\min})/\text{MNI}) \cdot m\right]$

construct the new harmony vector x_m^{new} with dimension $\{1\} \cdot \{\text{NDV} + 1\}$ and zero initial value

for *decision-making variable v* $[v \in \Psi^{\text{NDV}}]$ **do**

  if $\text{U}(0, 1) \leq \text{HMCR}$ **then**

      Rule 1: The harmony memory consideration with the probability of the HMCR

    $x_{m,v}^{\text{new}} = x_{m,v}^{r}; \forall r \sim \text{U}\{1, 2, \ldots, \text{HMS}\}$; for CDVs and DDVs

    if $\text{U}(0, 1) \leq PAR_m$ **then**

      Rule 2: The pitch adjustment with the probability of the $\text{HMCR} \cdot PAR_m$

      $x_{m,v}^{\text{new}} = x_{m,v}^{\text{new}} \pm \text{U}(0, 1) \cdot BW_m$; for CDVs

      $x_{m,v}^{\text{new}} = x_{m,v}^{\text{new}}(y + t); y \sim \text{U}\{x_v(1), \ldots, x_v(w_v), \ldots, x_v(W_v)\}, \quad \forall t \sim \text{U}\{-1, +1\}$; for DDVs

    end if

  else if

      Rule 3: The random selection with the probability of the $1 - \text{HMCR}$

  $x_{m,v}^{\text{new}} = x_v^{\min} + \text{U}(0, 1) \cdot \left(x_v^{\max} - x_v^{\min}\right)$; for CDVs

  $x_{m,v}^{\text{new}} = x_v(y); \forall y \sim \text{U}\{x_v(1), \ldots, x_v(w_v), \ldots, x_v(W_v)\}$; for DDVs

  end if

end for

calculate the value of objective function, fitness function, derived from the harmony vector x_m^{new} as $f\left(\text{x}_m^{\text{new}}\right)$

allocate $f\left(\text{x}_m^{\text{new}}\right)$ to element $(1, \text{NDV} + 1)$ of the new harmony vector x_m^{new}

terminate

end main body

Note: Continuous decision-making variable (CDVs), discrete decision-making variable (DDVs)

Table 3.12 Pseudocode related to the performance-driven architecture of the SS-IHSA

Algorithm 9: Pseudocode for performance-driven architecture of the SS-IHSA
Input: BW^{max}, BW^{min}, HMCR, HMS, MNI, NCDV, NDDV, NDV, PAR^{max}, PAR^{min}, x_v^{min}, x_v^{max}, $\{x_v(1), \ldots, x_v(w_v), \ldots, x_v(W_v)\}$
Output: x^{best}
start main body
begin
Stage 1—Definition stage: Definition of the optimization problem and its parameters
Stage 2—Initialization stage
Sub-stage 2.1: Initialization of the parameters of the SS-IHSA
Sub-stage 2.2: Initialization of the of the HM
Algorithm 1: Pseudocode for initialization of the HM in the SS-IHSA
Algorithm 2: Pseudocode for sorting the solution vectors stored in the HM under the SS-IHSA
Stage 3—Computational stage
set *improvisation/iteration* $m = 1$
set HM_m = HM
while $m \leq$ MNI **do**
Sub-stage 3.1: Improvisation of a new harmony vector
Algorithm 8: Pseudocode for improvisation of a new harmony vector in the SS-IHSA
Sub-stage 3.2: Update of the HM
Algorithm 4: Pseudocode for the update of the HM_m in the SS-IHSA
Algorithm 5: Pseudocode for sorting the solution vectors stored in the HM_m under the SS-IHSA
set *improvisation/iteration* $m = m + 1$
end while
Stage 4—Selection stage: Selection of the final optimal solution—The best harmony
Algorithm 6: Pseudocode for the selection of the final optimal solution in the SS-IHSA
terminate
end main body

pseudocode associated with improvisation of a new harmony vector in the SS-IHSA. Table 3.12 gives the pseudocode related to the performance-driven architecture of the SS-IHSA.

3.6 Melody Search Algorithm

In a general sense, playing more than one musical note at a time is referred to as a harmony. The difference in the pitch between the two musical notes is called their interval. Given this definition, consider a few simple scenarios: two-note harmonies have one interval; three-note harmonies have three intervals; and four-note harmonies have six intervals. The impressiveness and diversity of each harmony increase geometrically with the addition of each musical note. More precisely, if the number of musical notes played at a given time increases, the richness and variety of harmony increase, owing to the fact that a combination of musical notes is utilized in order to create a beautiful and pleasant tone or complete song. Harmonies with three or more musical notes are called chords. Chords generally make a harmonic structure or a background mode for a piece of music. In these harmonies, intervals are considered structural blocks of the chords.

That is, in music, harmony is the use of simultaneous pitches or chords. In this circumstance, the investigation of harmony involves chords, their construction, and progressions; connection principles are accomplished to form a melody. It is important to note that harmony refers to vertical aspects of the music space, due to simultaneous playing of available music notes in harmony. In contrast to harmony, a melody is a linear sequence of musical notes that can be recognized as a single song by a listener. More precisely, a melody consists of a linear sequence of individual pitches or musical notes, one followed by another one in a certain order. A hybrid ordering of musical notes, then, makes up a song. The point to be made here is that melody refers to the horizontal aspect of the music space, because the available musical notes are played in a linear sequence and read mostly horizontally from left to right. Figure 3.2 depicts the major difference between the structures related to harmony and melody. Harmony is able to convey different types of emotions, impulses, and coloring to the melody. Harmony therefore causes deepening and richness of the melody. Stated another way, if a melody is a boat, harmony is a river along which the boat floats. Where the river is deeper and without stones and obstacles, the boat can move, or flow, more easily—more fluently and beautifully. More detailed descriptions of the entire concepts pertaining to harmony and melody in music are out of the scope of this chapter, but the interested reader may look to the work by Martineau [8] and Sturman [9] for an exhaustive discussion regarding these concepts.

According to what has been described in Sect. 3.4, the SS-HSA and its enhanced versions—categories 1 and 2—have a single-stage computational and one-dimensional structure. These characteristics cause the performance of the SS-HSA and its enhanced versions to be greatly influenced by the process of solving complicated, real-world, large-scale, non-convex, non-smooth optimization

Fig. 3.2 The major difference between the structure of harmony and melody

problems having a nonlinear, mixed-integer nature with big data in such a way that these algorithms cannot maintain their affordable performance. In order to tackle the disadvantages of the SS-HSA and its enhanced versions, a new meta-heuristic optimization algorithm, referred to as an MSA, was introduced in 2011 [10]. Subsequently, the completed version of this algorithm was presented in 2013 [11]. The MSA is an innovative, population-oriented, meta-heuristic optimization algorithm, which was inspired by borrowing the phenomena and concepts of music as well as principles employed in the SS-HSA.

This newly developed optimization technique basically has a different architecture compared to other meta-heuristic optimization algorithms, because it imitates the process of music performance and interactive relationships among members of a musical group, while each player is looking for the best set of pitches within a melody line. In such a musical group, the presence of multiple players with different tastes, ideas, styles, and experiences under interactive relationships among players can effectively result in attaining the most desirable sequence of pitches more quickly. This process is virtually the same as the optimization process in engineering sciences in which the optimal solution can be explored by evaluating the objective function. Table 3.13 shows the interdependencies of phenomena and concepts of music and the optimization problem modeled by the MSA. As set out in Table 3.13, the concepts of music are equivalently indicated with the concept of the optimization

Table 3.13 Interdependencies of phenomena and concepts of music and the optimization problem modeled by the MSA

No.	Comparison factor	Concept of the optimization problem modeled by the MSA	Music concept
1	Structural pattern	Decision-making variable	Pitch in a particular melody played by a particular player in the musical group
2	Component	Value of decision-making variable	Value of each pitch in a particular melody played by a particular player in the musical group
3	Decision-making space	Value range of decision-making variable	Range of each pitch in a particular melody played by a particular player in the musical group
4	General structural pattern	Solution vector	Musical melody played by each existing player in the musical group
5	Target	Objective function	Aesthetic standard of the audience
6	Process unit	Iteration	Time/practice invested by all existing players in the musical group
7	Memory	Solution vectors matrix	Experience of all existing players in the musical group
8	Best state	Global optimum point	Best melody selected from among all melodies played by all existing players in the musical group
9	Search process	Local and global optimum searches	Improvisation of all existing players in the musical group

problem modeled by the MSA. With that in mind, each pitch in a particular melody played by a particular player in the musical group, the value of each pitch in a particular melody played by a particular player in the musical group, and the range of each pitch in a particular melody played by a particular player in the musical group are virtually the same as each decision-making variable, value of each decision-making variable, and value range of each decision-making variable, respectively. In the same way, the musical melody played by each existing player in the musical group, aesthetic standard of the audience, and time and practice invested by all existing players in the musical group refer to the solution vector, objective function, and iteration, respectively. Moreover, the experience of all existing players in the musical group, the best melody selected from among all melodies played by all existing players in the musical group, and improvisation of all existing players in the musical group are equivalent to the solution vectors matrix, global optimum point, and local and global optimum searches, respectively. By improving the musical melodies played by all existing players in the group at each practice, compared to before practice from the perspective of the aesthetic standard of audience, the solution vector pertaining to the optimization problem is also enhanced in each iteration, compared to the situation prior to each iteration from the standpoint of proximity to the optimal global point. Although the MSA was designed by employing the phenomena and concepts of music and the principles of the SS-HSA, its architecture is entirely different from the SS-HSA.

Unlike the SS-HSA, which employs a single-stage computational structure, the MSA utilizes a two-stage computational structure in order to achieve the optimal solution: (1) a single computational stage or single improvisation stage (SIS) and (2) a pseudo-group computational stage or pseudo-group improvisation stage (PGIS). In the SIS, each musician, or player, improvises the melody individually, without the influence of other players in the group. In the PGIS, however, the MSA has a pseudo-group performance. More precisely, in this stage, each player improvises the melody interactively with the influence of other players in the group. Different melodies available in the group can direct the players to select better, albeit random, pitches and strengthen the probability of playing a better melody in the next improvisation/iteration. Furthermore, and in contrast to the SS-HSA, which uses a single HM, the MSA employs multiple player memories (PMs). Multiple PMs form a melody memory (MM). As a result, the SS-HSA is referred to as a single-stage computational (or single-level computational stage), one-dimensional optimization technique, because it has a single improvisation stage and a single or an individual memory. Conversely, the MSA is called as a two-stage (or two-level) computational, multi-dimensional, single-homogeneous MSA (TMS-MSA) owing to the fact that it has two improvisation stages and multiple memories. The point to be made here is that the prerequisite of understanding of the characteristics, such as single-homogeneous and pseudo-group computational stage in the TMS-MSA, is that you first scrutinize the features employed in the architecture of the proposed two-stage (or two-level) computational, multi-dimensional, single-homogeneous enhanced melody search algorithm (TMS-EMSA) and the proposed SOSA, which is described extensively in Sects. 4.3 and 4.4 of Chap. 4, respectively. It is also necessary to state that, unlike the SS-HSA in which the feasible range of each continuous decision-making variable is not changed during different improvisations/iterations, the feasible range of each continuous decision-making variable in any improvisation/iteration associated with the PGIS of the TMS-MSA is updated only for random selection.

The performance-driven architecture of the TMS-MSA is generally broken down into five stages [11], as follows:

- Stage 1—Definition stage: Definition of the optimization problem and its parameters.
- Stage 2—Initialization stage.
  - Sub-stage 2.1: Initialization of the parameters of the TMS-MSA.
  - Sub-stage 2.2: Initialization of the MM.
- Stage 3—Single computational stage or SIS.
  - Sub-stage 3.1: Improvisation of a new melody vector by each player.
  - Sub-stage 3.2: Update of each PM.
  - Sub-stage 3.3: Check of the stopping criterion of the SIS.
- Stage 4—Pseudo-group computational stage or PGIS.
  - Sub-stage 4.1: Improvisation of a new melody vector by each player taking into account the feasible ranges of the updated pitches.
  - Sub-stage 4.2: Update of each PM.
  - Sub-stage 4.3: Update of the feasible ranges of pitches—continuous decision-making variables—for the next improvisation—only for random selection.
  - Sub-stage 4.4: Check of the stopping criterion of the PGIS.
- Stage 5—Selection stage: Selection of the final optimal solution—the best melody.

3.6.1 Stage 1: Definition Stage—Definition of the Optimization Problem and its Parameters

In order to solve an optimization problem using the TMS-MSA, stage 1 is needed to precisely describe the optimization problem and its parameters. In mathematical terms, the standard form of an optimization problem can generally be expressed using Eqs. (1.1) and (1.2), which were presented in Sect. 1.2.1 of Chap. 1. However, because the original version of the TMS-MSA was developed to solve only single-objective optimization problems with continuous decision-making variables, the standard form of an optimization problem must be rewritten, as shown in Eqs. (3.13) and (3.14):

$$\begin{aligned}
&\underset{x \in X}{\text{Minimize}} \quad F(x) = [f(x)] \\
&\qquad \text{subject to :} \\
&\qquad G(x) = [g_1(x), \ldots, g_b(x), \ldots, g_B(x)] = 0; \quad \forall\{B \geq 0\}, \quad \forall\{b \in \Psi^B\} \\
&\qquad H(x) = [h_1(x), \ldots, h_e(x), \ldots, h_E(x)] \leq 0; \quad \forall\{E \geq 0\}, \quad \forall\{e \in \Psi^E\}
\end{aligned} \tag{3.13}$$

$$\begin{aligned}
x = \quad & [x_1, \ldots, x_v, \ldots, x_{NCDV}]; \quad \forall\{v \in \Psi^{NCDV}, x \in X\}, \\
& \forall\{x_v^{min} \leq x_v \leq x_v^{max} | v \in \Psi^{NCDV}\}
\end{aligned} \tag{3.14}$$

The explanations associated with the parameters and variables from Eqs. (3.13) and (3.14) were also previously represented in Sect. 1.2.1 of Chap. 1. In the SIS and PGIS of the TMS-MSA, each player—without and with the influence of other players in the musical group, respectively—explores the entire space of the nonempty feasible decision-making in order to find the optimal decision-making (solution) vector. The optimal decision-making vector has the lowest possible value for the objective function given in Eq. (3.13). Basically, each player in the group merely takes into account the objective function given in Eq. (3.13) in order to solve the optimization problem presented in Eqs. (3.13) and (3.14). Nevertheless, if the solution vector determined by the corresponding player results in any violation in any equality or inequality constraints provided in Eq. (3.13), the player would have to utilize one of the following two processes with respect to the standpoint of the decision maker in dealing with this solution vector:

- First process: The corresponding player ignores the solution vector.
- Second process: The corresponding player considers the solution vector by applying a specified penalty coefficient to the objective function of the optimization problem.

3.6.2 *Stage 2: Initialization Stage*

After finalization of stage 1 and a thorough mathematical description of the optimization problem, stage 2 must be processed. This stage is organized into two sub-stages: initialization of the parameters of the TMS-MSA and initialization of the MM, which is described at length below.

3.6.2.1 Sub-stage 2.1: Initialization of the Parameters of the TMS-MSA

In sub-stage 2.1, the parameter adjustments of the TMS-MSA should be initialized with specific values. Table 3.14 gives a detailed description of the parameter adjustments related to the TMS-MSA. In the TMS-MSA, the MM is a place for storing the solution vectors for all existing players in the musical group. The number

Table 3.14 Adjustment parameters of the TMS-MSA

No.	TMS-MSA parameter	Abbreviation	Parameter range
1	Melody memory	MM	–
2	Player number	PN	$\text{PN} \geq 1$
3	Player memory of player p	PM_p	–
4	Player memory size	PMS	$\text{PMS} \geq 1$
5	Player memory considering rate	PMCR	$0 \leq \text{PMCR} \leq 1$
6	Minimum pitch adjusting rate	$\text{PAR}^{\min}$	$0 \leq \text{PAR}^{\min} \leq 2$
7	Maximum pitch adjusting rate	$\text{PAR}^{\max}$	$0 \leq \text{PAR}^{\max} \leq 2$
8	Minimum bandwidth	$\text{BW}^{\min}$	$0 \leq \text{BW}^{\min} \leq +\infty$
9	Maximum bandwidth	$\text{BW}^{\max}$	$0 \leq \text{BW} < +\infty$
10	Number of continuous decision-making variables	NCDV	$\text{NCDV} \geq 1$
11	Maximum number of improvisations/iterations of the SIS	MNI-SIS	$\text{MNI - SIS} \geq 1$
12	Maximum number of improvisations/iterations of the PGIS	MNI-PGIS	$\text{MNI - PGIS} \geq 1$

of player (PN) parameter represents the number of existing players in the group. Each player in the group has a memory defined by a PM parameter. The memory of the player p in the group is a place for storing the corresponding player's solution vectors. And, multiple PMs form the MM. The player memory size (PMS) describes the number of solution vectors stored in a player's memory. In the improvisation process of a new melody vector by a particular player under sub-stages 3.1 and 4.1, the player memory considering rate (PMCR) is used to specify whether the value of a decision-making variable relevant to a new melody vector played by the corresponding player is derived from the player's PM or from the entire space of the nonempty feasible decision-making. In other words, the PMCR indicates the rate at which the value of a decision-making variable from a new melody vector played by a particular player is randomly chosen according to its PM. Conversely, 1-PMCR expresses the rate at which the value of a decision-making variable from a new melody vector played by a particular player is randomly selected in accordance with the entire space of the nonempty feasible decision-making. The $\text{PAR}^{\min}$ and the $\text{PAR}^{\max}$ parameters are used to calculate the PAR parameter in the iteration m of the SIS and PGIS—PAR_m.

This parameter is dynamically changed and updated in each improvisation/ iteration of the SIS and PGIS. For the same reason, in the improvisation process of a new melody vector by a particular player under sub-stages 3.1 and 4.1, the PAR_m is employed to determine whether the value of a decision-making variable chosen from the corresponding PM needs an update to its neighbor's value or not. Put simply, the PAR_m clarifies the rate at which the value of a decision-making variable selected with the PMCR rate from the corresponding PM is changed. Therefore, $1 - PAR_m$ addresses the rate at which the value of a decision-making variable chosen with the PMCR rate from the corresponding PM is not altered. The $\text{BW}^{\min}$ and the $\text{BW}^{\max}$ parameters are employed to determine the BW parameter in the iteration m of the SIS and PGIS—BW_m. This parameter is dynamically changed and updated in each improvisation/iteration of the SIS and PGIS. The BW_m is taken to be an optional length and is merely defined for continuous decision-making variables. Detailed descriptions of the BW_m were provided in sub-stage 2.1 of the SS-HSA (see Sect. 3.3.2.1 of this chapter). The NCDV depends on the optimization problem given in Eqs. (3.13) and (3.14). This parameter specifies the dimension of the melody vector in the TMS-MSA. The maximum number of improvisations/iterations of the SIS (MNI-SIS) denotes the number of times a single computational stage is repeated in the TMS-MSA. Similarly, the maximum number of improvisations/iterations of the PGIS (MNI-PGIS) signifies the number of times a pseudo-group computational stage is repeated in the TMS-MSA. It should be pointed out that each player in the musical group improvises one melody individually without the influence of any other players in each improvisation/iteration of the SIS. The corresponding player also improvises one melody interactively with the influence of other players in each improvisation/iteration of the PGIS. The sum of the MNI-SIS and the MNI-PGIS is employed as a stopping criterion in the TMS-MSA.

3.6.2.2 Sub-stage 2.2: Initialization of the MM

After completion of sub-stage 2.1 and parameter adjustments of the TMS-MSA, the MM must be initialized in sub-stage 2.2. As previously mentioned, the MM is composed of multiple PMs. With that in mind, Fig. 3.3 shows the architecture of the MM in the TMS-MSA. Given the above descriptions, the MM matrix with the dimensions of {PMS} · {(NCDV + 1) · PN} consists of multiple PM submatrices with the dimensions of {PMS} · {NCDV + 1}.

In the TMS-MSA, the number of PMs forming the MM is specified by the PN parameter. The MM matrix and PM submatrices are filled with a large number of solution vectors generated randomly and based on Eqs. (3.15) through (3.17):

$$\mathrm{MM} = \left[PM_1 \;\cdots\; PM_p \;\cdots\; PM_{\mathrm{PN}}\right]; \quad \forall\left\{p \in \Psi^{\mathrm{PN}}\right\} \tag{3.15}$$

$$PM_p = \begin{bmatrix} \mathrm{x}_p^1 \\ \vdots \\ \mathrm{x}_p^s \\ \vdots \\ \mathrm{x}_p^{\mathrm{PMS}} \end{bmatrix} = \begin{bmatrix} x_{p,1}^1 & \cdots & x_{p,v}^1 & \cdots & x_{p,\mathrm{NCDV}}^1 & | & f\left(\mathrm{x}_p^1\right) \\ \vdots & & \vdots & & \vdots & & \vdots \\ x_{p,1}^s & \cdots & x_{p,v}^s & \cdots & x_{p,\mathrm{NCDV}}^s & | & f\left(\mathrm{x}_p^s\right) \\ \vdots & & \vdots & & \vdots & & \vdots \\ x_{p,1}^{\mathrm{PMS}} & \cdots & x_{p,v}^{\mathrm{PMS}} & \cdots & x_{p,\mathrm{NCDV}}^{\mathrm{PMS}} & | & f\left(\mathrm{x}_p^{\mathrm{PMS}}\right) \end{bmatrix};$$

$$\forall\left\{p \in \Psi^{\mathrm{PN}}, v \in \Psi^{\mathrm{NCDV}}, s \in \Psi^{\mathrm{PMS}}\right\} \tag{3.16}$$

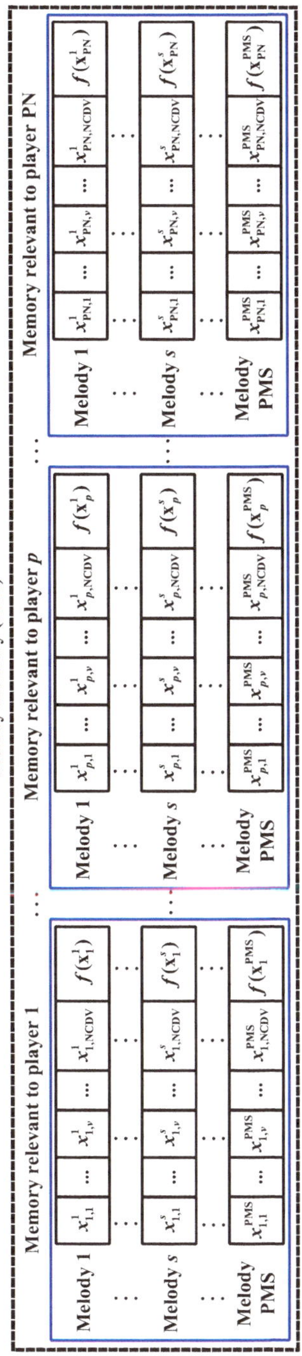

Fig. 3.3 The architecture of the MM in the TMS-MSA

$$x_{p,v}^{s} = x_{v}^{\min} + \mathrm{U}(0,1) \cdot \left(x_{v}^{\max} - x_{v}^{\min}\right); \quad \forall\left\{p \in \Psi^{\mathrm{PN}}, v \in \Psi^{\mathrm{NCDV}}, s \in \Psi^{\mathrm{PMS}}\right\} \tag{3.17}$$

Equation (3.15) denotes the MM. Equation (3.16) represents the memory relevant to existing player p (PM_p) in the musical group. In Eq. (3.17), U(0, 1) indicates a random number with a uniform distribution between 0 and 1. Furthermore, Eq. (3.17) tells us that the value of the continuous decision-making variable v from melody vector s stored in the memory related to player p ($x_{p,v}^{s}$) is randomly specified by the set of candidate-admissible values for this decision-making variable, limited by lower bound $x_{v}^{\min}$ and upper bound $x_{v}^{\max}$. Table 3.15 gives the pseudocode relevant to initialization of the entire set of PMs or MM in the TMS-MSA. After filling all of the PMs or the MM with random solution vectors, the solution vectors stored in each PM must be sorted from the lowest value to the highest value—in an ascending order—with respect to the value of the objective function of the optimization problem. Table 3.16 illustrates the pseudocode pertaining to sorting the solution vectors stored in the PMs or MM in the TMS-MSA.

Table 3.15 Pseudocode relevant to initialization of the entire set of PMs or MM in the TMS-MSA

Algorithm 10: Pseudocode for initialization of the entire set of PMs or MM in the TMS-MSA

Input: NCDV, PMS, PN, $x_{v}^{\min}$, $x_{v}^{\max}$
Output: MM

start main body

begin
construct the matrix MM with dimension {PMS} · {(NCDV + 1) · PN} and zero initial value
for *music player p* [$p \in \Psi^{\mathrm{PN}}$] **do**
  construct the submatrix PM_p with dimension {PMS} · {NCDV + 1} and zero initial value
  for *melody vector s* [$s \in \Psi^{\mathrm{PMS}}$] **do**
    construct melody vector s of music player p, x_{p}^{s}, with dimension {1} · {NCDV + 1} and zero initial value
    for *continuous decision-making variable v* [$v \in \Psi^{\mathrm{NCDV}}$] **do**
      $x_{p,v}^{s} = x_{v}^{\min} + \mathrm{U}(0,1) \cdot \left(x_{v}^{\max} - x_{v}^{\min}\right)$
      allocate $x_{p,v}^{s}$ to element $(1, v)$ of melody vector x_{p}^{s}
    end for
    calculate the value of the objective function, fitness function, derived from melody vector x_{p}^{s} as $f\left(\mathrm{x}_{p}^{s}\right)$
    allocate $f\left(\mathrm{x}_{p}^{s}\right)$ to element (1, NCDV + 1) of melody vector x_{p}^{s}
    add melody vector x_{p}^{s} to the row s of the submatrix PM_p
  end for
  add submatrix PM_p to the rows 1 to PMS and columns $1 + [(p-1) \cdot (\mathrm{NCDV}+1)]$ to $[p \cdot (\mathrm{NCDV}+1)]$ of the matrix MM
end for
terminate

end main body

Table 3.16 Pseudocode pertaining to sorting the solution vectors stored in the PMs or MM in the TMS-MSA

Algorithm 11: Pseudocode for sorting the solution vectors stored in the PMs or MM in the TMS-MSA

Input: Unsorted MM
Output: Sorted MM

start main body

begin
for *music player p* $[p \in \Psi^{\text{PN}}]$ **do**
  set $PM_p = \text{MM}(1 : \text{PMS}, [(p-1) \cdot (\text{NCDV}+1)] : [p \cdot (\text{NCDV}+1)])$
  $F_p^{\text{sort}} = \text{sort}\left(PM_p(1 : \text{PMS}, (\text{NCDV}+1)), '\text{ascend}'\right)$
  for *melody vector s* $[s \in \Psi^{\text{PMS}}]$ **do**
    for *melody vector* s^* $[s^* \in \Psi^{\text{PMS}}]$ **do**
      if $F_p^{\text{sort}}(s) == PM_p(s^*, (\text{NCDV}+1))$ **then**
        $PM_p^{\text{sort}}(s, 1 : (\text{NCDV}+1)) = PM_p(s^*, 1 : (\text{NCDV}+1))$
      end if
    end for
  end for
  $\text{MM}^{\text{sort}}(1 : \text{PMS}, 1 + [(p-1) \cdot (\text{NCDV}+1)] : [p \cdot (\text{NCDV}+1)]) = PM_p^{\text{sort}}$
end for
$\text{MM} = \text{MM}^{\text{sort}}$
terminate

end main body

3.6.3 Stage 3: Single Computational Stage or SIS

After finalization of stage 2 and initialization of the parameters of the TMS-MSA and the MM, the single computational stage, or SIS, must be completed. This stage contains three sub-stages: (1) improvisation of a new melody vector by each player; (2) update of each PM; and, (3) check of the stopping criterion of the SIS, which is described below.

The mathematical equations expressed at this stage must depend on the improvisation/iteration index—index m—due to the repeatability of the SIS in the TMS-MSA.

3.6.3.1 Sub-stage 3.1: Improvisation of a New Melody Vector by Each Player

In sub-stage 3.1, the improvisation process of a new melody vector by each player in the musical group must be carried out. In this sub-stage, each player improvises a new melody vector individually, without the influence of other players. In the TMS-MSA, a new melody vector or a new melody line played by player

p—$\mathrm{x}_{m,p}^{\text{new}} = \left(x_{m,p,1}^{\text{new}}, \ldots, x_{m,p,v}^{\text{new}}, \ldots, x_{m,p,\text{NCDV}}^{\text{new}}\right)$—is generated through a new alternative improvisation procedure (AIP) established according to the main concepts of improvisation of a harmony vector in the SS-HSA. The AIP will be explained in Sect. 3.6.6. The improvisation process of a new melody vector is carried out by other players in the same way.

3.6.3.2 Sub-stage 3.2: Update of Each PM

After completion of sub-stage 3.1 and improvisation of a new melody vector by each player in the musical group, the update process of the PMs or MM must be done in sub-stage 3.2. To illustrate, consider the memory relevant to player p ($PM_{m,\,p}$). In this sub-stage, a new melody vector played by player p— $\mathrm{x}_{m,p}^{\text{new}} = \left(x_{m,p,1}^{\text{new}}, \ldots, x_{m,p,v}^{\text{new}}, \ldots, x_{m,p,\text{NCDV}}^{\text{new}}\right)$ —is evaluated and compared with the worst available melody vector in the $PM_{m,p}$—the melody vector stored in the PMS row of the $PM_{m,p}$—from the perspective of the objective function. If the new melody vector played by player p has a better value than the worst available melody vector in the $PM_{m,p}$, from the standpoint of the objective function, this new melody vector replaces the worst available melody vector in the $PM_{m,p}$; the worst available melody vector is then eliminated from the $PM_{m,p}$. This process is also performed for other players in the group. Table 3.17 gives the pseudocode associated with the update of the memory of all existing players in the musical group or the update of the MM_m.

Table 3.17 Pseudocode associated with the update of the memory of all existing players in the musical group or the update of the MM_m in the TMS-MSA

Algorithm 12: Pseudocode for the update of the memory of all existing players in the musical group or the update of the MM_m in the TMS-MSA

Input: Not updated MM_m, $\mathrm{x}_{m,p}^{\text{new}}$
Output: Updated MM_m

Start main body

begin
for *music player p* [$p \in \Psi^{\text{PN}}$] **do**
  set $\mathrm{x}_{p}^{\text{worst}} = \mathrm{x}_{m,p}^{\text{PMS}}$
  set $f\left(\mathrm{x}_{p}^{\text{worst}}\right) = f\left(\mathrm{x}_{m,p}^{\text{PMS}}\right)$
  if $f\left(\mathrm{x}_{m,p}^{\text{new}}\right) < f\left(\mathrm{x}_{p}^{\text{worst}}\right)$ **then**
    $\mathrm{x}_{m,p}^{\text{new}} \in PM_{m,p}$
    $\mathrm{x}_{p}^{\text{worst}} \notin PM_{m,p}$
  end if
end for
terminate

end main body

Table 3.18 Pseudocode pertaining to sorting the solution vectors stored in the PMs or MM_m in the TMS-MSA

Algorithm 13: Pseudocode for sorting the solution vectors stored in the PMs or MM_m in the TMS-MSA

Input: Unsorted MM_m
Output: Sorted MM_m

start main body

begin
for *music player p* $[p \in \Psi^{\mathrm{PN}}]$ **do**
  set $PM_{m,\,p} = MM_m(1{:}\ \mathrm{PMS}, [(p-1)\cdot(\mathrm{NCDV}+1)] : [p\cdot(\mathrm{NCDV}+1)])$
  $F^{\mathrm{sort}}_{m,p} = \mathrm{sort}\left(PM_{m,p}(1{:}\ \mathrm{PMS}, (\mathrm{NCDV}+1)), '\mathrm{ascend}'\right)$
  for *melody vector s* $[s \in \Psi^{\mathrm{PMS}}]$ **do**
    for *melody vector* s^* $[s^* \in \Psi^{\mathrm{PMS}}]$ **do**
      if $F^{\mathrm{sort}}_{m,p}(s) == PM_{m,\,p}(s^*, (\mathrm{NCDV}+1))$ **then**
        $PM^{\mathrm{sort}}_{m,p}(s, 1{:}\ (\mathrm{NCDV}+1)) = PM_{m,p}(s^*, 1 : (\mathrm{NCDV}+1))$
      end if
    end for
  end for
  $MM^{\mathrm{sort}}_m(1{:}\ \mathrm{PMS}, 1 + [(p-1)\cdot(\mathrm{NCDV}+1)]{:}\ [p\cdot(\mathrm{NCDV}+1)]) = PM^{\mathrm{sort}}_{m,p}$
end for
$MM_m = MM^{\mathrm{sort}}_m$
terminate

end main body

The update process of the $PM_{m,p}$ is not performed if the new melody vector played by player p in the musical group is not notably better than the worst available melody vector in its memory, from the standpoint of the objective function. After completion of this process, melody vectors stored in the memory of all existing players in the musical group or the MM_m must be re-sorted based on the value of objective function—fitness function—in an ascending order.

The pseudocode pertaining to sorting the solution vectors stored in the memory of all existing players in the musical group or the MM was formerly presented in Table 3.16. Given the dependence of each PM or more comprehensively the MM to the improvisation/iteration index of the SIS—index m—this pseudocode must be rewritten according to Table 3.18.

3.6.3.3 Sub-stage 3.3: Check of the Stopping Criterion of the SIS

After completion of sub-stage 3.2 and an update of all PMs, the process of checking the stopping criterion of the single computational stage must be accomplished. If the stopping criterion of the SIS—the MNI-SIS—is satisfied, its computational efforts are terminated. Otherwise, sub-stages 3.1 and 3.2 are repeated.

3.6.4 Stage 4: Pseudo-Group Computational Stage or PGIS

After finalization of stage 3, or accomplishment of the SIS, the pseudo-group computational stage or the PGIS must be performed. This stage consists of four sub-stages: (1) improvisation of a new melody vector by each player taking into account the feasible ranges of the updated pitches; (2) update of each PM; (3) update of the feasible ranges of pitches—continuous decision-making variables—for the next improvisation—only for random selection; and, (4) check of the stopping criterion of the PGIS. The mathematical equations expressed at this stage must depend on the improvisation/iteration index—index m—due to the repeatability of the PGIS in the TMS-MSA.

3.6.4.1 Sub-stage 4.1: Improvisation of a New Melody Vector by Each Player Taking into Account the Feasible Ranges of the Updated Pitches

In sub-stage 4.1, the improvisation process of a new melody vector by each player in the group must be performed. In this sub-stage, each player improvises a new melody vector interactively with the influence of other players. In other words, in this sub-stage, player p improvises a new melody vector—$\mathrm{x}_{m,p}^{\mathrm{new}} = \left(x_{m,p,1}^{\mathrm{new}}, \ldots, x_{m,p,v}^{\mathrm{new}}, \ldots, x_{m,p,\mathrm{NCDV}}^{\mathrm{new}}\right)$—by the AIP taking into account the feasible range of the pitches—continuous decision-making variables—which are updated in different improvisations/iterations of the PGIS. The improvisation process of a new melody vector is carried out by other players in the same way.

3.6.4.2 Sub-stage 4.2: Update of Each PM

After completion of sub-stage 4.1 and improvisation of a new melody vector by each player in the group, the update process of the PMs or MM must be performed. This process is similar to sub-stage 3.2 of the SIS, which was explained in Sect. 3.6.3.2.

3.6.4.3 Sub-stage 4.3: Update of the Feasible Ranges of Pitches—Continuous Decision-Making Variables—for the Next Improvisation—Only for Random Selection

This sub-stage is a major part of the architecture of the TMS-MSA, which can give rise to a remarkable difference between this optimization technique and the SS-HSA. In the SS-HSA, the feasible ranges of continuous decision-making variables in the harmony vector are not changed during different improvisations/iterations. In the

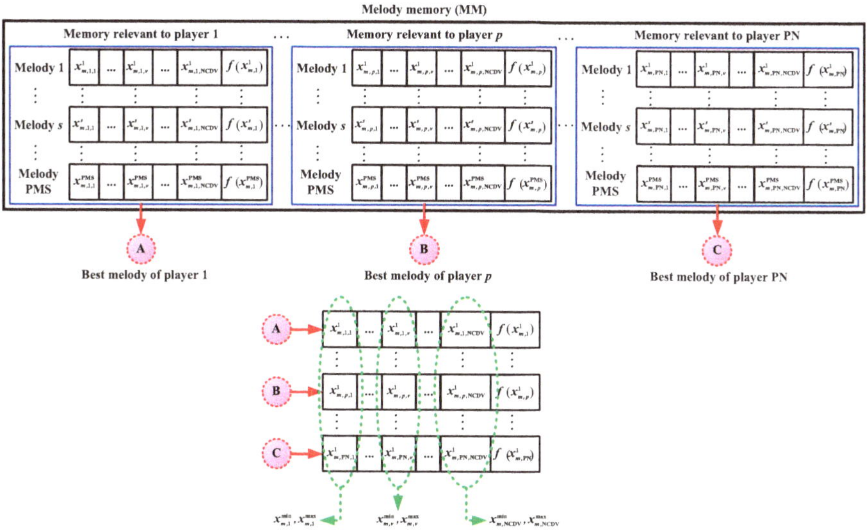

Fig. 3.4 Update of the feasible ranges of the continuous decision-making variables in the TMS-MSA

TMS-MSA, however, the feasible ranges of continuous decision-making variables in the melody vector are altered and updated during each improvisation/iteration of the PGIS, but only for random selection. This means that the lower bound of the continuous decision-making variable v ($x_v^{\min}$) and the upper bound of the continuous decision-making variable v ($x_v^{\max}$) in the PGIS depend on the improvisation/iteration index of the PGIS and change in the form of $x_{m,v}^{\min}$ and $x_{m,v}^{\max}$, respectively. Figure 3.4 displays the process of updating the feasible ranges of continuous decision-making variables in the TMS-MSA. Table 3.19 provides the pseudocode relevant to the update of the feasible ranges of the continuous decision-making variables in the TMS-MSA.

3.6.4.4 Sub-stage 4.4: Check of the Stopping Criterion of the PGIS

After finalization of sub-stage 4.3 and update of the feasible ranges of the continuous decision-making variables for the next improvisation of the PGIS, the checking process of the stopping criterion of this computational stage must be accomplished. In this sub-stage, the computational efforts of the PGIS are terminated in case if its stopping criterion—the MNI-PGIS—is satisfied. Otherwise, sub-stages 4.1, 4.2, and 4.3 are repeated.

Table 3.19 Pseudocode relevant to the update of the feasible ranges of the continuous decision-making variables in the TMS-MSA

Algorithm 14: Pseudocode for the update of the feasible ranges of the continuous decision-making variables in the TMS-MSA

Input: $\mathrm{x}^{1}_{m,p}$
Output: $x^{\min}_{m,v}, x^{\max}_{m,v}$

start main body

begin
for *music player p* $[p \in \Psi^{\mathrm{PN}}]$ **do**
  set $\mathrm{x}^{\mathrm{best}}_{p} = \mathrm{x}^{1}_{m,p}$
end for
for *continuous decision-making variable v* $[v \in \Psi^{\mathrm{NCDV}}]$ **do**
  $x^{\min}_{m,v} = \min\left(x^{\mathrm{best}}_{p,v}\right); \forall p \in \Psi^{\mathrm{PN}}, \forall\{m \in \Psi^{\text{MNI-PGIS}}\}$
  $x^{\max}_{m,v} = \max\left(x^{\mathrm{best}}_{p,v}\right); \forall p \in \Psi^{\mathrm{PN}}, \forall\{m \in \Psi^{\text{MNI-PGIS}}\}$
end for
terminate

end main body

Table 3.20 Pseudocode related to the selection of the final optimal solution in the TMS-MSA

Algorithm 15: Pseudocode for the selection of the final optimal solution in the TMS-MSA

Input: MM
Output: $\mathrm{x}^{\mathrm{best}}$

start main body

begin
for *music player p* $[p \in \Psi^{\mathrm{PN}}]$ **do**
  set $\mathrm{x}^{\mathrm{best}}_{p} = \mathrm{x}^{1}_{p}$
end for
$\mathrm{x}^{\mathrm{best}} = \min\left(\mathrm{x}^{\mathrm{best}}_{p}\right); \forall p \in \Psi^{\mathrm{PN}}$
terminate

end main body

3.6.5 Stage 5: Selection Stage—Selection of the Final Optimal Solution—The Best Melody

After completion of stage 4, or accomplishment of the PGIS, the selection of the final optimal solution must be made in stage 5. In this stage, the best melody vector stored in the memory of each existing player in the musical group is determined. Then, the best melody vector is selected from among these melody vectors as the final optimal solution. Table 3.20 shows the pseudocode related to the selection of the final optimal solution in the TMS-MSA.

3.6.6 Alternative Improvisation Procedure

As indicated earlier, in the TMS-MSA, player p in the group improvises a new melody vector—$\mathrm{x}_{m,p}^{\text{new}} = \left(x_{m,p,1}^{\text{new}}, \ldots, x_{m,p,v}^{\text{new}}, \ldots, x_{m,p,\text{NCDV}}^{\text{new}}\right)$—using the AIP. This procedure was developed with regard to the fundamental concepts of improvisation of a harmony vector in the SS-HSA. Implementing the AIP by player p is carried out according to three rules: (1) player memory consideration; (2) pitch adjustment; and, (3) random selection.

Rule 1: In consideration of a player's memory, the values of the new melody vector for player p are randomly selected from the melody vectors stored in the $PM_{m,p}$ with the probability of the PMCR. In this rule, two principles are alternately employed, with each principle consisting of a linear combination of a decision-making variable chosen from the $PM_{m,p}$ and a ratio of the BW_m. If the first principle is activated, the value of the first decision-making variable from the new melody vector played by player p, $x_{m,p,1}^{\text{new}}$, is randomly selected from the available corresponding continuous decision-making variable in the melody vectors stored in the $PM_{m,p}$—$\left(x_{m,p,1}^{1}, \ldots, x_{m,p,1}^{s}, \ldots, x_{m,p,1}^{\text{PMS}}\right)$—with the probability of the PMCR and updated by the BW_m parameter. Conversely, the value of the first decision-making variable from the new melody vector played by player p, $x_{m,p,1}^{\text{new}}$, is randomly chosen from the entire set of available continuous decision-making variables stored in the $PM_{m,\ p}$—$\left\{\left(x_{m,p,1}^{1}, \ldots, x_{m,p,1}^{s}, \ldots, x_{m,p,1}^{\text{PMS}}\right), \ldots, \left(x_{m,p,v}^{1}, \ldots, x_{m,p,v}^{s}, \ldots, x_{m,p,v}^{\text{PMS}}\right), \ldots, \left(x_{m,p,\text{NCDV}}^{1} \cdots x_{m,p,\text{NCDV}}^{s}, \cdots x_{m,p,\text{NCDV}}^{\text{PMS}}\right)\right\}$—with the probability of the PMCR and updated by the BW_m parameter, provided that the first principle is not activated or the second principle is activated. The values for other continuous decision-making variables are also selected in the same way.

Implementing the player memory consideration rule to specify the value of the continuous decision-making variable v from a new melody vector played by player p, $x_{m,p,v}^{\text{new}}$, is done using Eqs. (3.18) and (3.19):

$$x_{m,p,v}^{\text{new}} = x_{m,p,v}^{r} \pm \mathrm{U}(0,1) \cdot BW_m;$$
$$\forall\left\{m \in \Psi^{(\text{MNI-SIS})+(\text{MNI-PGIS})}, p \in \Psi^{\text{PN}}, v \in \Psi^{\text{NCDV}}, r \sim \mathrm{U}\{1,2,\ldots,\text{PMS}\}\right\} \tag{3.18}$$

$$\begin{aligned} x_{m,p,v}^{\text{new}} = \;& x_{m,p,k}^{r} \pm \mathrm{U}(0,1) \cdot BW_m; \\ & \forall\Big\{m \in \Psi^{(\text{MNI-SIS})+(\text{MNI-PGIS})}, p \in \Psi^{\text{PN}}, v \in \Psi^{\text{NCDV}}, \\ & r \sim \mathrm{U}\{1,2,\ldots,\text{PMS}\}, k \sim \mathrm{U}\{1,2,\ldots,\text{NCDV}\}\Big\} \end{aligned} \tag{3.19}$$

Equations (3.18) and (3.19) are used for the first and second principles of the player memory consideration rule, respectively. Here, index r is a random integer with a uniform distribution through the set $\{1, 2, \ldots, \text{PMS}\}$—$r \sim \mathrm{U}\{1, 2, \ldots, \text{PMS}\}$—and index k is a random integer with a uniform distribution through the set

$\{1, 2, \ldots, \text{NCDV}\}$—$k \sim \text{U}\{1, 2, \ldots, \text{NCDV}\}$. Put another way, in Eq. (3.18), the value of index r is randomly specified through the set of admissible values demonstrated by the set $\{1, 2, \ldots, \text{PMS}\}$. Determination of this index is elucidated on the basis of Eq. (3.20):

$$r = \text{int}(\text{U}(0, 1) \cdot \text{PMS}) + 1 \tag{3.20}$$

In Eq. (3.19), the value of index k is also randomly characterized through the set of permissible values displayed by the set $\{1, 2, \ldots, \text{NCDV}\}$. Determination of this index is described based on Eq. (3.21):

$$k = \text{int}(\text{U}(0, 1) \cdot \text{NCDV}) + 1 \tag{3.21}$$

The point to be made here is that other distributions can be employed for indexes r and k, such as $(\text{U}(0, 1))^2$. The utilization of this distribution results in the selection of lower values for these indexes. In the player memory consideration rule, the first and second principles can effectively give rise to a more desirable convergence and a more substantial increase in the diversity of the generated solutions for the TMS-MSA. Applying the player memory consideration rule is also accomplished for other players in the same way.

Rule 2: In the pitch adjustment rule, the values of a new melody vector played by player p, haphazardly selected from among the existing melody vectors in the $PM_{m,\ p}$ with the probability of the PMCR, are updated with the probability of the PAR_m. More precisely, after the value of the first continuous decision-making variable from a new melody vector by player p, $x_{m,p,1}^{\text{new}}$, is haphazardly chosen from the melody vectors stored in the $PM_{m,p}$ with the probability of the PMCR, this continuous decision-making variable is updated with the probability of the PAR_m. The update process for this continuous decision-making variable is performed by replacing it with the value of the first continuous decision-making variable from the best melody vector available in the $PM_{m,p}$, $x_{m,p,1}^{\text{best}}$. The values for other continuous decision-making variables are also updated in the same way. Implementing the pitch adjustment rule to determine the value of the continuous decision-making variable v from a new melody vector played by player p, $x_{m,p,v}^{\text{new}}$, is done by using Eq. (3.22):

$$x_{m,p,v}^{\text{new}} = x_{m,p,v}^{\text{best}}; \quad \forall \left\{ m \in \Psi^{(\text{MNI-SIS})+(\text{MNI-PGIS})}, p \in \Psi^{\text{PN}}, v \in \Psi^{\text{NCDV}} \right\} \tag{3.22}$$

Applying the pitch adjustment rule is also carried out for other players in the same way.

Rule 3: In the random selection rule, the values of a new melody vector played by player p are haphazardly selected from the entire space of the nonempty feasible decision-making with the probability of the 1-PMCR. In here, the random selection rule is organized in accordance with two different principles. The first and second principles are activated in the SIS and PGIS, respectively. If the first principle of the random selection rule is activated, the value of the first continuous decision-making variable from the new melody vector played by the player p, $x_{m,p,1}^{\text{new}}$, is haphazardly selected from the entire space of the nonempty feasible decision-making related to this decision-making variable with the probability of the 1-PMCR. In here, the entire space of the nonempty feasible decision-making relevant to the corresponding decision-making variable is characterized by an invariable lower bound, $x_1^{\min}$, and an invariable upper bound, $x_1^{\max}$, which are defined in the first stage of the TMS-MSA—definition of the optimization problem and its parameters—and unchanged in all improvisations/iterations of the SIS. The values for other continuous decision-making variables are also chosen in the same way. This principle of the random selection rule was previously used in sub-stage 2.2 for initialization of the MM.

Implementing the first principle of random selection rule to specify the value of the continuous decision-making variable v from a new melody vector played by player p, $x_{m,p,v}^{\text{new}}$, is performed by using Eq. (3.23):

$$x_{m,p,v}^{\text{new}} = x_v^{\min} + \mathrm{U}(0,1) \cdot \left(x_v^{\max} - x_v^{\min}\right); \quad \forall\left\{m \in \Psi^{(\text{MNI-SIS})}, p \in \Psi^{\text{PN}}, v \in \Psi^{\text{NCDV}}\right\} \tag{3.23}$$

Equation (3.23) tells us that in the first principle of the random selection rule, the player p in order to determine the value of the continuous decision-making variable v from the new vector can utilize the entire space of the nonempty feasible decision-making relevant to this decision-making variable which is specified in the first stage of the TMS-MSA and unchanged in all improvisations/iterations of the SIS.

Conversely, the value of the first continuous decision-making variable from the new melody vector played by player p, $x_{m,p,1}^{\text{new}}$, is randomly chosen from the entire space of the nonempty feasible decision-making pertaining to this continuous decision-making variable with the probability of the 1-PMCR provided that the first principle is not activated or the second principle is activated. In here, the entire space of the nonempty feasible decision-making associated with the corresponding decision-making variable is determined by a variable lower bound, $x_{m,1}^{\min}$, and a variable upper bound, $x_{m,1}^{\max}$, which are dynamically changed and updated in each improvisation/iteration of the PGIS. The values for other continuous decision-making variables are also chosen in the same way.

Implementing the second principle of random selection rule to specify the value of the continuous decision-making variable v from a new melody vector played by player p, $x_{m,p,v}^{\text{new}}$, is performed by using Eq. (3.24):

$$x_{m,p,v}^{\text{new}} = x_{m,v}^{\min} + \mathrm{U}(0,1) \cdot \left(x_{m,v}^{\max} - x_{m,v}^{\min}\right); \quad \forall\left\{m \in \Psi^{(\text{MNI-PGIS})}, p \in \Psi^{\text{PN}}, v \in \Psi^{\text{NCDV}}\right\} \tag{3.24}$$

Equation (3.24) tells us that in the second principle of the random selection rule, the player p in order to specify the value of the continuous decision-making variable v from the new melody vector can merely use the space of the nonempty feasible decision-making related to this decision-making variable which is updated in all improvisation/iteration of the PGIS. Applying the random selection rule is also accomplished for other players in the same way.

As a general consequence, the probability that the value of the continuous decision-making variable v from a new melody vector played by player p, $x_{m,p,v}^{\text{new}}$, can be obtained by applying the player memory consideration, pitch adjustment, and random selection rules equals to PMCR × (1 − PAR_m), PMCR × PAR_m, and 1 − PMCR, respectively. In order to provide a more favorable convergence, as well as a more significant increase in the diversity of solution vectors for the TMS-MSA, each player in the musical group employs the updated values of the PAR_m and BW_m parameters in the improvisation process of its melody vector. The PAR_m and BW_m parameters are updated in each improvisation/iteration of the SIS and the PGIS by using Eqs. (3.25) and (3.26), respectively:

$$BW_m = \text{BW}^{\max} \cdot \exp\left(\frac{\ln\left(\text{BW}^{\max}/\text{BW}^{\min}\right)}{(\text{MNI-SIS}) + (\text{MNI-PGIS})} \cdot m\right); \quad \forall\left\{m \in \Psi^{(\text{MNI-SIS})+(\text{MNI-PGIS})}\right\} \tag{3.25}$$

$$PAR_m = \text{PAR}^{\min} + \left(\frac{\text{PAR}^{\max} - \text{PAR}^{\min}}{(\text{MNI-SIS}) + (\text{MNI-PGIS})}\right) \cdot m; \quad \forall\left\{m \in \Psi^{(\text{MNI-SIS})+(\text{MNI-PGIS})}\right\} \tag{3.26}$$

The update process of the PAR_m and BW_m parameters by Eq. (3.25) and (3.26) is virtually the same as the update process used in the SS-IHSA. Hence, further explanations are related to the update process of these parameters available in Sect. 3.5 of this chapter.

Table 3.21 gives the pseudocode pertaining to the improvisation of a new melody vector by each player in the musical group of the TMS-MSA. The designed pseudocode in different stages and sub-stages of the TMS-MSA is located in a regular sequence and forms the performance-driven architecture of this algorithm.

Table 3.22 also gives the pseudocode associated with the performance-driven architecture of the TMS-MSA. In here, sub-stages 3.3 and 4.4—the check process of the stopping criterion of the SIS and PGIS—are defined by the first and second WHILE loops in the pseudocode pertaining to the performance-driven architecture of the TMS-MSA (see Table 3.22).

Table 3.21 Pseudocode pertaining to improvisation of a new melody vector by each player in the musical group of the TMS-MSA

Algorithm 16: Pseudocode for improvisation of a new melody vector by each player in the musical group of the TMS-MSA

Input: $\text{BW}^{\max}$, $\text{BW}^{\min}$, MNI-SIS, MNI-PGIS, NCDV, $\text{PAR}^{\max}$, $\text{PAR}^{\min}$, PMCR, PMS, PN, $x_v^{\min}$, $x_v^{\max}$

Output: $\text{x}_{m,p}^{\text{new}}$

start main body

begin
$BW_m = \text{BW}^{\max} \cdot \exp[(\ln(\text{BW}^{\max}/\text{BW}^{\min})/((\text{MNI-SIS}) + (\text{MNI-PGIS}))) \cdot m]$
$PAR_m = \text{PAR}^{\min} - [((\text{PAR}^{\max} - \text{PAR}^{\min})/((\text{MNI-SIS}) + (\text{MNI-PGIS}))) \cdot m]$
for *music player p* $[p \in \Psi^{\text{PN}}]$ **do**
  construct the new melody vector for music player p, $\text{x}_{m,p}^{\text{new}}$, with dimension $\{1\} \cdot \{\text{NCDV} + 1\}$ and zero initial value
  for *continuous decision-making variable v* $[v \in \Psi^{\text{NCDV}}]$ **do**
    if $\text{U}(0,1) \leq \text{PMCR}$ **then**
      Rule 1: Harmony memory consideration with probability PMCR
      if *improvisation/iteration m* $[m \in \Psi^{(\text{MNI-SIS}) + (\text{MNI-PGIS})}]$ is odd **then**
        Principle 1: First combination
        $x_{m,p,v}^{\text{new}} = x_{m,p,v}^{r} \pm \text{U}(0,1) \cdot BW_m; \forall r \sim \text{U}\{1,2,\ldots,\text{PMS}\}$
      else
        Principle 2: Second combination
        $x_{m,p,v}^{\text{new}} = x_{m,p,k}^{r} \pm \text{U}(0,1) \cdot BW_m; \forall r \sim \text{U}\{1,2,\ldots,\text{PMS}\}, \forall k \sim \text{U}\{1,2,\ldots,\text{NCDV}\}$
      end if
      if $\text{U}(0,1) \leq PAR_m$ **then**
        Rule 2: Pitch adjustment with probability HMCR · PAR_m
        $x_{m,p,v}^{\text{new}} = x_{m,p,v}^{\text{best}}$
      end if
    else if
      Rule 3: Random selection with probability 1-HMCR
    switch 1
    case *improvisation/iteration m* $[m \in \Psi^{(\text{MNI-SIS}) + (\text{MNI-PGIS})}] \leq (\text{MNI-SIS})$ **then**
    $x_{m,p,v}^{\text{new}} = x_v^{\min} + \text{U}(0,1) \cdot \left(x_v^{\max} - x_v^{\min}\right)$
    case *improvisation/iteration m* $[m \in \Psi^{(\text{MNI-SIS}) + (\text{MNI-PGIS})}] > (\text{MNI-SIS})$ and *improvisation/iteration m* $[m \in \Psi^{(\text{MNI-SIS}) + (\text{MNI-PGIS})}] \leq (\text{MNI-SIS}) + (\text{MNI-PGIS})$ **then**
    $x_{m,p,v}^{\text{new}} = x_{m,v}^{\min} + \text{U}(0,1) \cdot \left(x_{m,v}^{\max} - x_{m,v}^{\min}\right)$
    end switch
    end if
  end for
  calculate the value of objective function, fitness function, derived from melody vector $\text{x}_{m,p}^{\text{new}}$ as $f\left(\text{x}_{m,p}^{\text{new}}\right)$
  allocate $f\left(\text{x}_{m,p}^{\text{new}}\right)$ to element $(1, \text{NCDV} + 1)$ of the new melody vector $\text{x}_{m,p}^{\text{new}}$
end for
terminate

end main body

Table 3.22 Pseudocode associated with the performance-driven architecture of the TMS-MSA

Algorithm 17: Pseudocode for performance-driven architecture of the TMS-MSA

Input: $\mathrm{BW}^{\max}$, $\mathrm{BW}^{\min}$, MNI-SIS, MNI-PGIS, NCDV, $\mathrm{PAR}^{\max}$, $\mathrm{PAR}^{\min}$, PMCR, PMS, PN, $x_v^{\min}$, $x_v^{\max}$

Output: $\mathrm{x}^{\mathrm{best}}$

start main body

begin
Stage 1—Definition stage: Definition of the optimization problem and its parameters
Stage 2—Initialization stage
  Sub-stage 2.1: Initialization of the parameters of the TMS-MSA
  Sub-stage 2.2: Initialization of the of the MM
    Algorithm 10: Pseudocode for initialization of the entire set of PMs or MM in the TMS-MSA
    Algorithm 11: Pseudocode for sorting the solution vectors stored in the PMs or MM in the TMS-MSA
Stage 3—Single computational stage or SIS
  set *improvisation/iteration* $m = 1$
  set MM_m = MM
  while $m \leq$ (MNI-SIS) **do**
    Sub-stage 3.1: Improvisation of a new melody vector by each player
      Algorithm 16: Pseudocode for improvisation a new melody vector by each player in the musical group of the TMS-MSA
    Sub-stage 3.2: Update of each PM
      Algorithm 12: Pseudocode for the update of the memory of all existing players in the musical group or the update of the MM_m in the TMS-MSA
      Algorithm 13: Pseudocode for sorting the solution vectors stored in the PMs or MM_m in the TMS-MSA
    set *improvisation/iteration* $m = m + 1$
  end while
Stage 4—Pseudo-group computational stage or PGIS
  while $m >$ (MNI-SIS) and $m \leq$ (MNI-PGIS) **do**
    Sub-stage 4.1: Improvisation of a new melody vector by each player with taking into account the feasible ranges of the updated pitches
      Algorithm 16: Pseudocode for improvisation of a new melody vector by each player in the musical group of the TMS-MSA
    Sub-stage 4.2: Update of each PM
      Algorithm 12: Pseudocode for the update of the memory of all existing players in the musical group or the update of the MM_m in the TMS-MSA
      Algorithm 13: Pseudocode for sorting the solution vectors stored in the PMs or MM_m in the TMS-MSA
    Sub-stage 4.3: Update of the feasible ranges of pitches—Continuous decision-making variables—For the next improvisation—Only for random selection
      Algorithm 14: Pseudocode for the update of the feasible ranges of the continuous decision-making variables in the TMS-MSA
    set *improvisation/iteration* $m = m + 1$
  end while
Stage 5—Selection stage: Selection of the final optimal solution—The best melody
  Algorithm 15: Pseudocode for the selection of the final optimal solution in the TMS-MSA
terminate

end main body

3.7 Conclusions

In this chapter, the music-inspired meta-heuristic optimization algorithms were reviewed from past to present, with a focus on the SS-HSA, SS-IHSA, and TMS-MSA. First, a brief review of the definition of music, its history, and the interdependencies of phenomena and concepts of music and the optimization problem was addressed. Second, the fundamental principles of the SS-HSA and its performance-driven architecture were rigorously described. In addition, a structural classification for the enhanced versions of the SS-HSA was provided. In this regard, the basic differences between the SS-IHSA, as a well-known enhanced version of the SS-HSA, and the SS-HSA were carefully examined in detail. Third, the fundamental principles of the TMS-MSA and its performance-driven architecture were meticulously expressed. Given related literature, and after presentation of the different versions of the music-inspired meta-heuristic optimization algorithms and their implementation on optimization problems in different branches of the engineering sciences (e.g., electrical, civil, computer, mechanical, and aerospace), it was observed that these optimization techniques may represent a reasonable and applicable method for solving complicated, real-world, large-scale, non-convex, non-smooth optimization problems having a nonlinear, mixed-integer nature with big data. This is due to the fact that the music-inspired meta-heuristic optimization algorithms have a distinctive and flexible architecture for facing optimization problems compared with other optimization techniques. It can be seen, then, that the willingness of specialists and researchers in different branches of the engineering sciences to employ music-inspired meta-heuristic optimization algorithms with the aim of overcoming difficulties in solving complicated, real-world, large-scale, non-convex, non-smooth optimization problems over recent years has been appreciably increasing.

As a result, this chapter can serve as an aid to using the music-inspired meta-heuristic optimization algorithms. Moreover, this chapter can effectively provide a precious background for explaining innovative versions of music-inspired meta-heuristic optimization algorithms, which will be discussed in the next chapter.

Appendix 1: List of Abbreviations and Acronyms

AIP	Alternative improvisation procedure
BW	Bandwidth
CDVs	Continuous decision-making variables
DDVs	Discrete decision-making variables
GA	Genetic algorithm
HM	Harmony memory
HMCR	Harmony memory considering rate
HMS	Harmony memory size

(continued)

HSA	Harmony search algorithm
IHSA	Improved harmony search algorithm
MM	Melody memory
MNI	Maximum number of improvisations/iterations
MNI-PGIS	Maximum number of improvisations/iterations of the pseudo-group improvisation stage
MNI-SIS	Maximum number of improvisations/iterations of the single improvisation stage
MSA	Melody search algorithm
NCDV	Number of continuous decision-making variables
NDDV	Number of discrete decision-making variables
NDV	Number of decision-making variables including continuous and discrete decision-making variable
PAR	Pitch adjusting rate
PGIS	Pseudo-group improvisation stage
PMCR	Player memory considering rate
PMs	Player memories
PMS	Player memory size
PN	Player number
SIS	Single improvisation stage
SOSA	Symphony orchestra search algorithm
SS-HSA	Single-stage computational, single-dimensional harmony search algorithm
SS-IHSA	Single-stage computational, single-dimensional improved harmony search algorithm
TMS-EMSA	Two-stage computational, multidi-mensional, single-homogeneous enhanced melody search algorithm
TMS-MSA	Two-stage computational, multidi-mensional, single-homogeneous melody search algorithm

Appendix 2: List of Mathematical Symbols

Index:	
b	Index for equality constraints running from 1 to B
e	Index for inequality constraints running from 1 to E
m	Index for improvisations/iterations running from 1 to MNI in the SS-HSA and also running from 1 to (MNI-SIS) + (MNI-PGIS) in the TMS-MSA
p	Index for existing players in a music group running from 1 to PN
s, s^*	Index for harmony vectors stored in the HM running from 1 to HMS in the SS-HSA and also an index for melody vectors stored in each PM running from 1 to PMS in the TMS-MSA
v	Index for decision-making variables, including the continuous and discrete decision-making variables, running from 1 to NDV in the SS-HSA and also an index for continuous decision-making variables running from 1 to NCDV in the TMS-MSA
w_v	Index for candidate permissible values of discrete decision-making variable v running from 1 to W_v in the SS-HSA

(continued)

Set:	
Ψ^{B}	Set of indices of equality constraints
Ψ^{E}	Set of indices of inequality constraints
Ψ^{HMS}	Set of indices of harmony vectors stored in the HM
Ψ^{MNI}	Set of indices of improvisations/iterations in the SS-HSA
$\Psi^{(MNI\text{-}SIS)+(MNI\text{-}PGIS)}$	Set of indices of improvisations /iterations in the TMS-MSA
Ψ^{NDV}	Set of indices of decision-making variables, including the continuous and discrete decision-making variables
Ψ^{NCDV}	Set of indices of continuous decision-making variables
Ψ^{NDDV}	Set of indices of discrete decision-making variables
Ψ^{PMS}	Set of indices of melody vectors stored in each PM
Ψ^{PN}	Set of indices of existing players in a music group
W_v	Set of indices of candidate permissible values of discrete decision-making variable *v*
Parameters:	
BW	Bandwidth
BW^{max}	Maximum bandwidth
BW^{min}	Minimum bandwidth
HMCR	Harmony memory considering rate
HMS	Harmony memory size
MNI	Maximum number of improvisations/iterations in the SS-HSA
MNI-SIS	Maximum number of iterations of the SIS in the TMS-MSA
MNI-PGIS	Maximum number of iterations of the PGIS in the TMS-MSA
PAR	Pitch adjusting rate
PAR^{max}	Maximum pitch adjusting rate
PAR^{min}	Minimum pitch adjusting rate
PMCR	Player memory considering rate
PMS	Player memory size
x_v^{max}	Upper bound on the decision-making variable *v*
x_v^{min}	Lower bound on the decision-making variable *v*
X	Nonempty feasible decision-making space
Z	Feasible objective space
Variables:	
BW_m	Bandwidth in improvisation/iteration *m* of the SS-HSA or bandwidth in improvisation/iteration *m* of the TMS-MSA
$f(\mathrm{x})$	Objective function of the optimization problem
$f(\mathrm{x}^s)$	Value of the objective function—Fitness function—Derived from the harmony vector *s* stored in the HM matrix
$f\left(\mathrm{x}_p^s\right)$	Value of the objective function—Fitness function—Derived from the melody vector *s* stored in memory submatrix relevant to existing player *p* in the musical group
$f\left(\mathrm{x}_m^{new}\right)$	Value of the objective function—Fitness function—Derived from the new harmony vector in improvisation/iteration *m* of the SS-HSA

(continued)

$f\left(\mathrm{x}_{m,p}^{\mathrm{new}}\right)$	Value of the objective function—Fitness function—Derived from the new melody vector played by existing player p in the musical group in improvisation/iteration m of the TMS-MSA
F(x)	Vector of objective function of the optimization problem
$g_b(\mathrm{x})$	Equality constraint b of the optimization problem or component b of the vector of equality constraints
G(x)	Vector of equality constraints of the optimization problem
$h_e(\mathrm{x})$	Inequality constraint e of the optimization problem or component e of the vector of inequality constraints
H(x)	Vector of inequality constraints of the optimization problem
HM	Harmony memory matrix
k	Random integer with a uniform distribution through the set $\{1, 2, \ldots, \mathrm{NCDV}\}$ in the TMS-MSA
MM	Melody memory matrix
PAR_m	Pitch adjusting rate in improvisation/iteration m of the SS-HSA or pitch adjusting rate in improvisation/iteration m of the TMS-MSA
PM_p	Memory submatrix relevant to existing player p in the musical group
r	Random integer with a uniform distribution through the set $\{1, 2, \ldots, \mathrm{HMS}\}$ in the SS-HSA and random integer with a uniform distribution through the set $\{1, 2, \ldots, \mathrm{PMS}\}$ in the TMS-MSA
t	Random integer with a uniform distribution through the set $\{-1, +1\}$
U(0, 1)	Random number with a uniform distribution between 0 and 1
x_v	Decision-making variable v or component v of the vector of decision-making variable
$x_{m,v}^{\mathrm{new}}$	Element v of the new harmony vector in improvisation/iteration m of the SS-HSA
$x_{m,p,v}^{\mathrm{best}}$	Element v of the best melody vector stored in the memory submatrix relevant to existing player p in the musical group in improvisation/iteration m of the TMS-MSA
$x_{m,p,v}^{\mathrm{new}}$	Element v of the new melody vector played by existing player p in the musical group in improvisation/iteration m of the TMS-MSA
x_v^s	Element v of the harmony vector s stored in the HM matrix
$x_{p,v}^s$	Element v of the melody vector s stored in the memory submatrix relevant to existing player p in the musical group
$x_v(w_v)$	Candidate permissible value w of discrete decision-making variable v
x	Vector of decision-making variables
$\mathrm{x}_m^{\mathrm{new}}$	New harmony vector in improvisation/iteration m of the SS-HSA
$\mathrm{x}_{m,p}^{\mathrm{new}}$	New melody vector played by existing player p in the musical group in improvisation/iteration m of the TMS-MSA
$\mathrm{x}^{\mathrm{best}}$	Best harmony vector stored in the HM matrix in the SS-HSA and also best melody vector stored in the MM matrix in the TMS-MSA
$\mathrm{x}_p^{\mathrm{best}}$	Best melody vector stored in the memory submatrix relevant to existing player p in the musical group
x^s	Harmony vector s stored in the HM matrix
x_p^s	Melody vector s stored in the memory submatrix relevant to existing player p in the musical group

(continued)

x^{worst}	Worst harmony vector stored in the HM matrix
x_p^{worst}	Worst melody vector stored in the memory submatrix relevant to existing player p in the musical group
y	Random integer with a uniform distribution through the set $\{x_v(1), \ldots, x_v(w_v), \ldots, x_v(W_v)\}$
z	Vector of the objective function

References

1. B. Badiozamani, G. Badiozamani, *Iran and America: Rekindling a Love Lost* (East-West Understanding Press, San Diego, CA, 2005)
2. Z.W. Geem, G.H. Kim, G.V. Loganathan, A new heuristic optimization algorithm: harmony search. SIMULATION **76**(2), 60–68 (2001)
3. Z.W. Geem, *Music-Inspired Harmony Search Algorithm: Theory and Applications* (Springer, Berlin, 2009)
4. Z.W. Geem, State-of-the-art in the structure of harmony search algorithm, Chapter 1, in *Recent Advances in Harmony Search Algorithm* (Springer, Berlin, 2010), pp. 1–10
5. D. Manjarres, I. Landa-Torres, S. Gil-Lopez, J. Del Ser, M.N. Bilbao, S. Salcedo-Sanz, Z.W. Geem, A survey on applications of the harmony search algorithm. Eng. Appl. Artif. Intell. **26**(8), 1818–1831 (2013)
6. O. Moh'd-Alia, R. Mandava, The variants of the harmony search algorithm: an overview. Artif. Intell. Rev. **36**(1), 49–68 (2011)
7. M. Mahdavi, M. Fesanghary, E. Damangir, An improved harmony search algorithm for solving optimization problems. Appl. Math. Comput. **188**(2), 1567–1579 (2007)
8. J. Martineau, *The Elements of Music: Melody, Rhythm and Harmony* (Walker & Company, USA, Santa Monia, CA, 2008)
9. P. Sturman, *Harmony, Melody and Composition* (Cambridge University Press, Cambridge, 1983)
10. S.M. Ashrafi, A.B. Dariane, A novel and effective algorithm for numerical optimization: melody search (MS), in *11th International Conference on Hybrid Intelligence Systems (HIS)*, 2011
11. S.M. Ashrafi, A.B. Dariane, Performance evaluation of an improved harmony search algorithm for numerical optimization: Melody search (MS). Eng. Appl. Artif. Intell. **26**(4), 1301–1321 (2013)

Chapter 4
Advances in Music-Inspired Optimization Algorithms

4.1 Introduction

As mentioned in the preceding chapter, most of the existing meta-heuristic optimization algorithms have given rise to many difficulties in solving complicated, real-world, large-scale, non-convex, non-smooth optimization problems such as slow convergence rate, premature convergence, getting stuck in a local optimum point, and extremely high dependency on precise adjustments of initial values of algorithm parameters. In addition, the process of generating new responses or outputs by these algorithms depends on a limited space for nonempty feasible decision-making. More precisely, in each new generation, a solution vector is produced with regard to a narrow set of solution vectors stored in the memory of the algorithm. This dependency can, therefore, dramatically decrease the desirable performance of the meta-heuristic optimization algorithms in solving these complicated optimization problems having a nonlinear, mixed-integer nature with big data. According to what has been described, most of the existing meta-heuristic optimization algorithms have neither appropriate efficiency nor sufficiency to achieve a global optimum point in solving the optimization problems identified above.

In the previous chapter, a single-stage computational single-dimensional harmony search algorithm (SS-HSA), as a new population-based meta-heuristic optimization algorithm, was developed with inspiration from music phenomena for the first time in 2001 [1]. The architecture of the original SS-HSA is rather different from other existing meta-heuristic optimization algorithms, in that this optimization technique has been conceptualized from the improvisation process of jazz players. In this improvisation process, each jazz player seeks to find the best harmony and generate the most beautiful music possible. In this case, the jazz players try to improve the sound of their musical instruments at each music performance with the aim of providing more mature and beautiful music. In the architecture of the original SS-HSA, the process of generating new responses/outputs depends on the

M. Kiani-Moghaddam et al., *Modern Music-Inspired Optimization Algorithms for Electric Power Systems*, Power Systems,
https://doi.org/10.1007/978-3-030-12044-3_4

entire space of nonempty feasible decision-making. That is to say that this optimization technique generates a new solution vector in each new improvisation after sweeping the entire solution vectors stored in the memory of the algorithm. This characteristic can considerably strengthen the performance of the SS-HSA in solving complicated real-world optimization problems having a nonlinear, mixed-integer nature with big data. Detailed descriptions pertaining to the SS-HSA were exhaustively represented in Sect. 3.3 of Chap. 3.

In recent years, the use of the SS-HSA has grown remarkably for solving the optimization problems noted earlier in different branches of the engineering sciences (e.g., electrical, civil, computer, mechanical and aerospace, and engineering), because of its desirable performance compared to other meta-heuristic optimization algorithms. Many attempts have been made by specialists and researchers to provide more efficient and effective versions of the SS-HSA. Readers interested in achieving a thorough systematic classification of enhanced versions of the SS-HSA are referred to Sect. 3.4 of Chap. 3. Nevertheless, the performance of most existing meta-heuristic optimization algorithms, even the SS-HSA and many of its enhanced versions, is strongly affected by an unbalanced increase in dimensions of the big-data optimization problems. For such cases, these optimization techniques cannot satisfactorily maintain their desirable performance. The poor performance of most existing meta-heuristic optimization algorithms can be attributed to tenuous and vulnerable features utilized in their architecture, such as having only a single-stage computational structure, using a single-dimensional structure, etc.

In order to enhance the performance of the SS-HSA and remove its tenuous and vulnerable features, the two-stage computational multi-dimensional single-homogeneous melody search algorithm (TMS-MSA), as a new architectural version of the SS-HSA, was reported for the first time in 2011 [2]. Although this optimization technique was inspired by the phenomena and concepts of music and the fundamental principles of the SS-HSA, it has a markedly different architecture compared to the SS-HSA and other existing meta-heuristic optimization algorithms. The original TMS-MSA had a two-stage computational, multi-dimensional, and single-homogenous structure. Establishing the TMS-MSA with relatively strengthened and robust features brings about an innovative direction in the architecture of meta-heuristic algorithms in order to solve complicated, real-world, large-scale, non-convex, non-smooth optimization problems having a nonlinear, mixed-integer nature with big data. The detailed descriptions associated with the original TMS-MSA were thoroughly described in Sect. 3.6 of Chap. 3.

The original TMS-MSA was merely developed for optimization problems with continuous decision-making variables. The original TMS-MSA is, therefore, referred to as the continuous TMS-MSA. Consequently, the continuous TMS-MSA does not have the ability to solve complicated, real-world, large-scale, non-convex, non-smooth optimization problems with a concurrent combination of continuous and discrete decision-making variables. With that in mind, specialists and researchers in different branches of the engineering sciences are enthusiastic about employing innovative alternatives in order to tackle the complexities of our complicated real-world optimization problems.

In this chapter, then, the authors begin by proposing a new continuous/discrete TMS-MSA in order to deal with the complicated, real-world, large-scale, non-convex, non-smooth optimization problems with a simultaneous combination of the continuous and discrete decision-making variables. In addition, the authors develop an innovative improved version of the continuous/discrete TMS-MSA, or two-stage computational multi-dimensional single-homogeneous enhanced melody search algorithm (TMS-EMSA), with the aim of heightening efficiency and efficacy of the performance of this optimization technique.

In today's world, however, modern engineering challenges with multilevel dimensions in different branches of the engineering sciences, particularly electrical engineering, have been widely encountered; therefore, these challenges cannot be represented in the form of single-level optimization problems, or as traditional optimization problems. Instead, modern engineering challenges with multilevel dimensions must be developed in the form of new nontraditional optimization problems—multilevel optimization problems—because of their specific characteristics. In this case, many of the optimization algorithms, even the original TMS-MSA, the proposed continuous/discrete TMS-MSA, and the TMS-EMSA, may not be able to maintain the most desirable performance in solving such multilevel optimization problems. This is due to the fact that a single-homogeneous structure has been used in the architecture of the aforementioned optimization algorithms. Consequently, developing an innovative optimization algorithm is certainly a necessity for maintaining its favorable performance for solving multilevel optimization problems. In this chapter, then, the authors develop an innovative architectural version of the proposed TMS-EMSA, which is referred to as either a multi-stage computational multi-dimensional multiple-homogeneous enhanced melody search algorithm (MMM-EMSA), or a multi-stage computational multi-dimensional single-inhomogeneous enhanced melody search algorithm (MMS-EMSA), or a symphony orchestra search algorithm (SOSA), in order to appreciably enhance its performance, flexibility, robustness, and parallel capability. The SOSA has a multi-stage computational multi-dimensional multiple-homogeneous—or multi-stage computational multi-dimensional single-inhomogeneous—structure.

It should be noted that the SS-HSA, the single-stage computational single-dimensional improved harmony search algorithm (SS-IHSA), the continuous TMS-MSA, the continuous/discrete TMS-MSA, the proposed TMS-EMSA, and the proposed SOSA can be solely employed for solving single-objective optimization problems. If an optimization problem consists of multiple conflicting, noncommensurable, and correlated objective functions, the most reasonable strategy is to take advantage of the multi-objective optimization process in order to deal with such optimization problems, as described in Sect. 2.2 of Chap. 2. As a result, to overcome this acute weakness in the architecture of the aforementioned optimization algorithms and to provide appropriate strategies for solving multi-objective optimization problems, the authors propose the multi-objective versions of these optimization algorithms.

For the reasons explained above, the authors have concentrated on five targets in the area of advances in music-inspired optimization algorithms.

- Target 1: Present a new hybrid version of the original TMS-MSA, referred to as the continuous/discrete TMS-MSA.
- Target 2: Propose a new enhanced version of the proposed continuous/discrete TMS-MSA, referred to as the TMS-EMSA.
- Target 3: Develop an innovative version of architecture of the proposed TMS-EMSA, referred to as the MMM-EMSA, MMS-EMSA, or SOSA.
- Target 4: Provide a new multi-objective strategy for remodeling the architecture of the single-stage computational single-dimensional meta-heuristic music-inspired optimization algorithms (i.e., the SS-HSA and SS-IHSA).
- Target 5: Provide a new multi-objective strategy for remodeling the architecture of the two-stage computational multi-dimensional single-homogeneous meta-heuristic music-inspired optimization algorithms (i.e., the original TMS-MSA, proposed continuous/discrete TMS-MSA, and TMS-EMSA).
- Target 6: Provide an innovative multi-objective strategy for remodeling the architecture of the multi-stage computational multi-dimensional multiple-homogeneous meta-heuristic music-inspired optimization algorithm (i.e., the proposed SOSA).

The rest of this chapter is arranged as follows: First, the proposed continuous/discrete TMS-MSA is thoroughly addressed in Sect. 4.2. Then, the proposed TMS-EMSA is reviewed in Sect. 4.3. In Sect. 4.4, the newly developed SOSA is represented more in depth. The implementation strategies of multi-objective versions pertaining to the SS-HSA, SS-IHSA, offered continuous/discrete TMS-MSA, proposed TMS-EMSA, and proposed SOSA are meticulously explained in Sect. 4.5. Finally, the chapter ends with a brief summary and some concluding remarks in Sect. 4.6.

4.2 Continuous/Discrete TMS-MSA

As previously mentioned, the original TMS-MSA—the continuous TMS-MSA—was addressed only for solving optimization problems with continuous decision-making variables. Most real-world optimization problems in different branches of the engineering sciences (e.g., electrical, civil, computer, mechanical, and aerospace) are involved with a concurrent combination of the continuous and discrete decision-making variables. Therefore, it is not feasible to employ this optimization technique in order to deal with this range of optimization problems. In this section, then, the authors propose a continuous/discrete TMS-MSA in order to deal with a wide range of optimization problems. The performance-driven architecture of the proposed TMS-MSA is generally broken down into five stages, as follows:

- Stage 1—Definition stage: Definition of the optimization problem and its parameters
- Stage 2—Initialization stage

  - Sub-stage 2.1: Initialization of the parameters of the continuous/discrete TMS-MSA
  - Sub-stage 2.2: Initialization of the melody memory (MM)

- Stage 3—Single computational stage or single improvisation stage (SIS)
  - Sub-stage 3.1: Improvisation of a new melody vector by each player
  - Sub-stage 3.2: Update each player's memory (PM)
  - Sub-stage 3.3: Check of the stopping criterion of the SIS

- Stage 4—Pseudo-group computational stage or pseudo-group improvisation stage (PGIS)
  - Sub-stage 4.1: Improvisation of a new melody vector by each player with taking into account the feasible ranges of the updated pitches
  - Sub-stage 4.2: Update of each PM
  - Sub-stage 4.3: Update of the feasible ranges of pitches—continuous decision-making variables—for the next improvisation—only for random selection
  - Sub-stage 4.4: Check of the stopping criterion of the PGIS

- Stage 5—Selection stage: Selection of the final optimal solution—the best melody

It is clear that the performance-driven architecture of the proposed continuous/discrete TMS-MSA is quite similar to the performance-driven architectures of the continuous TMS-MSA, which were expressed in Sect. 3.6 of Chap. 3. The proposed continuous/discrete TMS-MSA, however, has fundamental differences in some stages and sub-stages of the performance-driven architecture when compared with its continuous version, due to the concurrent presence of continuous and discrete decision-making variables. In here, given that this optimization technique is taken into account as a foundation for the subsequent proposed TMS-EMSA, its different stages and sub-stages are entirely represented.

4.2.1 Stage 1: Definition Stage—Definition of the Optimization Problem and Its Parameters

In order to solve an optimization problem using the proposed continuous/discrete TMS-MSA, stage 1 must precisely describe the optimization problem and its parameters. In mathematical terms, the standard form of an optimization problem can generally be defined using Eqs. (1.1) and (1.2), which were presented in Sect. 1.2.1 of Chap. 1. Because the proposed continuous/discrete TMS-MSA was developed for solving single-objective optimization problems, the standard form of an optimization problem must be rewritten according to Eqs. (4.1) and (4.2):

$$
\begin{aligned}
\underset{\mathrm{x}\in\mathrm{X}}{\text{Minimize}}\quad & \mathrm{F}(\mathrm{x})=[f(\mathrm{x})] \\
& \text{and subject to :} \\
& \mathrm{G}(\mathrm{x})=[g_1(\mathrm{x}),\ldots,g_b(\mathrm{x}),\ldots,g_{\mathrm{B}}(\mathrm{x})]=0;\quad \forall\{\mathrm{B}\geq 0\},\ \forall\{b\in\Psi^{\mathrm{B}}\} \\
& \mathrm{H}(\mathrm{x})=[h_1(\mathrm{x}),\ldots,h_e(\mathrm{x}),\ldots,h_{\mathrm{E}}(\mathrm{x})]\leq 0;\quad \forall\{\mathrm{E}\geq 0\},\ \forall\{e\in\Psi^{\mathrm{E}}\}
\end{aligned}
\tag{4.1}
$$

$$
\begin{aligned}
&\mathrm{x}=[x_1,\ldots,x_v,\ldots,x_{\mathrm{NDV}}];\quad \forall\{v\in\Psi^{\mathrm{NDV}},\Psi^{\mathrm{NDV}}=\Psi^{\mathrm{NCDV+NDDV}},\mathrm{x}\in\mathrm{X}\},\\
&\forall\{x_v^{\min}\leq x_v\leq x_v^{\max}|v\in\Psi^{\mathrm{NCDV}}\},\{x_v\in\{x_v(1),\ldots,x_v(w),\ldots,x_v(W_v)\}|v\in\Psi^{\mathrm{NDDV}}\}
\end{aligned}
\tag{4.2}
$$

Detailed descriptions associated with variables and parameters of Eqs. (4.1) and (4.2) were previously explained in Sect. 1.2.1 of Chap. 1.

4.2.2 Stage 2: Initialization Stage

After completion of stage 1 and an exhaustive mathematical description of the optimization problem, stage 2 is started. For the proposed continuous/discrete TMS-MSA, this stage is organized into two sub-stages: initialization of the parameters of the proposed continuous/discrete TMS-MSA and initialization of the MM.

4.2.2.1 Sub-stage 2.1: Initialization of the Parameters of the Proposed Continuous/Discrete TMS-MSA

In sub-stage 2.1, the parameter adjustments of the proposed continuous/discrete TMS-MSA should be initialized with specific values. Detailed descriptions pertaining to parameter adjustments of the proposed continuous/discrete TMS-MSA are given in Table 4.1, where the sum of the number of continuous decision-making variables (NCDV) and the number of discrete decision-making variables (NDDV) is considered as the total number of decision-making variables (NDV). Unlike the continuous TMS-MSA in which the dimension of the melody vector, or solution vector, is determined by the NCDV, in the proposed continuous/discrete TMS-MSA, it is characterized by the number of the NDV. Detailed descriptions relevant to other parameters presented in Table 4.1 were formerly clarified in Sect. 3.6.2.1 of Chap. 3.

4.2.2.2 Sub-stage 2.2: Initialization of the MM

After finalization of sub-stage 2.1 and parameter adjustments of the proposed continuous/discrete TMS-MSA, the MM must be initialized in sub-stage 2.2. Similar to the continuous TMS-MSA, multiple PMs are placed next to each other to form the

Table 4.1 Adjustment parameters of the proposed continuous/discrete TMS-MSA

No.	The proposed continuous/discrete TMS-MSA parameter	Abbreviation	Parameter range
1	Melody memory	MM	–
2	Player number	PN	$\text{PN} \geq 1$
3	Player memory of player *p*	PM_p	–
4	Player memory size	PMS	$\text{PMS} \geq 1$
5	Player memory considering rate	PMCR	$0 \leq \text{PMCR} \leq 1$
6	Minimum pitch adjusting rate	$\text{PAR}^{\min}$	$0 \leq \text{PAR}^{\min} \leq 2$
7	Maximum pitch adjusting rate	$\text{PAR}^{\max}$	$0 \leq \text{PAR}^{\max} \leq 2$
8	Minimum bandwidth	$\text{BW}^{\min}$	$0 \leq \text{BW}^{\min} < +\infty$
9	Maximum bandwidth	$\text{BW}^{\max}$	$0 \leq \text{BW}^{\max} < +\infty$
10	Number of continuous decision-making variables	NCDV	$\text{NCDV} \geq 1$
11	Number of discrete decision-making variables	NDDV	$\text{NDDV} \geq 1$
12	Number of decision-making variables	NDV	$\text{NDV} \geq 2$
13	Maximum number of improvisations/iterations of the SIS	MNI-SIS	$\text{MNI-SIS} \geq 1$
14	Maximum number of improvisations/iterations of the PGIS	MNI-PGIS	$\text{MNI-PGIS} \geq 1$

MM in the proposed continuous/discrete TMS-MSA. In a more precise expression, the MM matrix, which has a dimension equal to $\{\text{PMS}\} \cdot \{(\text{NDV} + 1) \cdot \text{PN}\}$, consists of multiple PM submatrices so that each PM has a dimension equal to $\{\text{PMS}\} \cdot \{\text{NDV} + 1\}$.

The MM matrix and PM submatrices are filled with a large number of solution vectors generated randomly and based on Eqs. (4.3) (4.6):

$$\text{MM} = \left[PM_1 \ \cdots \ PM_p \ \cdots \ PM_{\text{PN}}\right]; \quad \forall\left\{p \in \Psi^{\text{PN}}\right\} \tag{4.3}$$

$$PM_p = \begin{bmatrix} \mathrm{x}_p^1 \\ \vdots \\ \mathrm{x}_p^s \\ \vdots \\ \mathrm{x}_p^{\text{PMS}} \end{bmatrix} = \left[\begin{array}{ccccc|c} x_{p,1}^1 & \cdots & x_{p,v}^1 & \cdots & x_{p,\text{NDV}}^1 & f\left(\mathrm{x}_p^1\right) \\ \vdots & & \vdots & & \vdots & \vdots \\ x_{p,1}^s & \cdots & x_{p,v}^s & \cdots & x_{p,\text{NDV}}^s & f\left(\mathrm{x}_p^s\right) \\ \vdots & & \vdots & & \vdots & \vdots \\ x_{p,1}^{\text{PMS}} & \cdots & x_{p,v}^{\text{PMS}} & \cdots & x_{p,\text{NDV}}^{\text{PMS}} & f\left(\mathrm{x}_p^{\text{PMS}}\right) \end{array}\right]; \tag{4.4}$$
$$\forall\left\{p \in \Psi^{\text{PN}}, v \in \Psi^{\text{NDV}}, s \in \Psi^{\text{PMS}}, \Psi^{\text{NDV}} = \Psi^{\text{NCDV+NDDV}}\right\}$$

$$x_{p,v}^s = x_v^{\min} + \text{U}(0,1) \cdot \left(x_v^{\max} - x_v^{\min}\right); \quad \forall\left\{p \in \Psi^{\text{PN}}, v \in \Psi^{\text{NCDV}}, s \in \Psi^{\text{PMS}}\right\} \tag{4.5}$$

$$x_{p,v}^{s} = x_v(y);\ \forall\{p \in \Psi^{\mathrm{PN}}, v \in \Psi^{\mathrm{NDDV}}, s \in \Psi^{\mathrm{PMS}}, y \sim \mathrm{U}\{x_v(1), \ldots, x_v(w_v), \ldots, x_v(W_v)\}\} \tag{4.6}$$

Detailed descriptions related to Eqs. (4.3)–(4.5) were given in Sect. 3.6.2.2 of Chap. 3. In Eq. (4.6), index y is a random integer with a uniform distribution through the set of $\{x_v(1), \ldots, x_v(w_v), \ldots, x_v(W_v)\}$—$y \sim \mathrm{U}\{x_v(1), \ldots, x_v(w_v), \ldots, x_v(W_v)\}$. Equation (4.6) tells us that the value of the discrete decision-making variable v from the melody vector s stored in the memory associated with player p ($x_{p,v}^{s}$) is randomly determined by the set of candidate permissible values for this decision, which is shown by the set $\{x_v(1), \ldots, x_v(w_v), \ldots, x_v(W_v)\}$. Table 4.2 presents the pseudocode related to initialization of the entire set of PMs or MM in the proposed continuous/discrete TMS-MSA. After filling all of the PMs or MM with random solution vectors, the solution vectors stored in each PM must be sorted from the lowest value to the highest value—in an ascending order—from the perspective of the value of the objective function of the optimization problem. Table 4.3 illustrates the pseudocode for sorting the solution vectors stored in the PMs or MM in the proposed continuous/discrete TMS-MSA.

4.2.3 *Stage 3: Single Computational Stage or SIS*

After finalization of stage 2 and initialization of the parameters of the proposed continuous/discrete TMS-MSA and the MM, the single computational stage, or SIS, must be started. This stage is formed by three sub-stages: (1) improvisation of a new melody vector by each player; (2) update of each PM; and, (3) check of the stopping criterion of the SIS, all of which are described below.

The mathematical equations defined at this stage must depend on the improvisation/iteration index—index m—due to the repeatability of the SIS in the proposed continuous/discrete TMS-MSA.

4.2.3.1 Sub-stage 3.1: Improvisation of a New Melody Vector by Each Player

In sub-stage 3.1, a new melody vector is individually improvised by each player in the musical group without the influence of other players. In the continuous TMS-MSA, generating a new melody vector or a new melody line played by existing player p in the musical group—$\mathrm{x}_{m,p}^{\mathrm{new}} = \left(x_{m,p,1}^{\mathrm{new}}, \ldots, x_{m,p,v}^{\mathrm{new}}, \ldots, x_{m,p,\mathrm{NDV}}^{\mathrm{new}}\right)$—is accomplished by using an alternative improvisation procedure (AIP). The AIP employed in the continuous TMS-MSA was developed for optimization problems with continuous decision-making variables. A detailed description related to the AIP was presented in Sect. 3.6.5 of Chap. 3.

Table 4.2 Pseudocode related to initialization of the entire set of PMs or MM in the proposed continuous/discrete TMS-MSA

Algorithm 1: Pseudocode for initialization of the entire set of PMs or MM in the proposed continuous/discrete TMS-MSA

Input: PN, PMS, NCDV, NDDV, NDV, $x_v^{\min}$, $x_v^{\max}$, $\{x_v(1), \ldots, x_v(w_v), \ldots, x_v(W_v)\}$
Output: MM

start main body

begin
construct the matrix MM with dimension {PMS} · {(NDV + 1) · PN} and zero initial value
for *music player p* $[p \in \Psi^{\text{PN}}]$ **do**
  construct the submatrix PM_p with dimension {PMS} · {NDV + 1} and zero initial value
  for *melody vector s* $[s \in \Psi^{\text{PMS}}]$ **do**
    construct melody vector s of music player p, x_p^s, with dimension {1} · {NDV + 1} and zero initial value
    for *decision-making variable v* $[v \in \Psi^{\text{NDV}}]$ **do**
      $x_{p,v}^s = x_v^{\min} + \text{U}(0,1) \cdot \left(x_v^{\max} - x_v^{\min}\right)$; for CDVs
      $x_{p,v}^s = x_v(y)$; $\forall y \sim \text{U}\{x_v(1), \ldots, x_v(w_v), \ldots, x_v(W_v)\}$; for DDVs
      allocate $x_{p,v}^s$ to element $(1, v)$ of melody vector x_p^s
    end for
    calculate the value of the objective function, fitness function, derived from melody vector x_p^s as $f\left(\mathrm{x}_p^s\right)$
    allocate $f\left(\mathrm{x}_p^s\right)$ to element (1, NDV + 1) of melody vector x_p^s
    add melody vector x_p^s to the row s of the submatrix PM_p
  end for
  add submatrix PM_p to the rows 1 to PMS and columns $1 + [(p-1) \cdot (\text{NDV}+1)]$ to $[p \cdot (\text{NDV}+1)]$ of the matrix MM
end for
terminate

end main body

Note: Continuous decision-making variable (CDVs), discrete decision-making variables (DDVs)

The continuous/discrete TMS-MSA was developed in order to solve optimization problems with a concurrent combination of the continuous and discrete decision-making variables; and, accordingly, the AIP is not functional for these optimization problems and, therefore, cannot be employed by each player in the musical group for improvisation of a new melody vector in this optimization technique. Thus, the authors propose a new continuous/discrete AIP. In the continuous/discrete TMS-MSA, a new melody vector or a new melody line played by player p in the musical group—$\mathrm{x}_{m,p}^{\text{new}} = \left(x_{m,p,1}^{\text{new}}, \ldots, x_{m,p,v}^{\text{new}}, \ldots, x_{m,p,\text{NDV}}^{\text{new}}\right)$—is generated by using the continuous/discrete AIP, which will be thoroughly explained in Sect. 4.2.6 of

Table 4.3 Pseudocode relevant to the sorting of the solution vectors stored in the PMs or MM in the proposed continuous/discrete TMS-MSA

Algorithm 2: Pseudocode for sorting the solution vectors stored in the PMs or MM in the proposed continuous/discrete TMS-MSA

Input: Unsorted MM
Output: Sorted MM

start main body

begin
for *music player p* $[p \in \Psi^{\mathrm{PN}}]$ **do**
  set $PM_p = \mathrm{MM}(1:\mathrm{PMS}, [(p-1)\cdot(\mathrm{NDV}+1)] : [p\cdot(\mathrm{NDV}+1)])$
  $F_p^{\mathrm{sort}} = \mathrm{sort}\left(PM_p(1:\mathrm{PMS}, (\mathrm{NDV}+1)), '\mathrm{ascend}'\right)$
  for *melody vector s* $[s \in \Psi^{\mathrm{PMS}}]$ **do**
    for *melody vector* s^* $[s^* \in \Psi^{\mathrm{PMS}}]$ **do**
      if $F_p^{\mathrm{sort}}(s) == PM_p(s^*, (\mathrm{NDV}+1))$ **then**
        $PM_p^{\mathrm{sort}}(s, 1:(\mathrm{NDV}+1)) = PM_p(s^*, 1:(\mathrm{NDV}+1))$
      end if
    end for
  end for
  $\mathrm{MM}^{\mathrm{sort}}(1:\mathrm{PMS}, 1+[(p-1)\cdot(\mathrm{NDV}+1)] : [p\cdot(\mathrm{NDV}+1)]) = PM_p^{\mathrm{sort}}$
end for
$\mathrm{MM} = \mathrm{MM}^{\mathrm{sort}}$
terminate

end main body

this chapter. The improvisation process of a new melody vector is carried out by other players in the same way.

4.2.3.2 Sub-stage 3.2: Update of Each PM

After completion of sub-stage 3.1 and improvisation of a new melody vector by each player in the musical group, the update process of the PMs or MM must be performed in sub-stage 3.2. To clarify, consider the memory associated with player p ($PM_{m,p}$). In this sub-stage, a new melody vector played by player p is evaluated and compared with the worst available melody vector in the $PM_{m,p}$—the melody vector stored in the PMS row of the $PM_{m,p}$—from the perspective of the objective function. If the new melody vector played by player p has a better value than the worst available melody vector in the $PM_{m,p}$, from the standpoint of the objective function, this new melody vector replaces the worst available melody vector in the $PM_{m,p}$; the worst available melody vector is then discarded from the $PM_{m,p}$. This process is also done for other players in the group. Table 4.4 presents the pseudocode pertaining to the update of the memory of all existing players in the musical group or the update of the MM_m.

The update process of the $PM_{m,p}$ is not accomplished if the new melody vector played by player p in the musical group is not notably better than the worst available melody vector in its memory, from the perspective of the objective

Table 4.4 Pseudocode pertaining to the update of the memory of all existing players in the musical group or the update of the MM_m in the proposed continuous/discrete TMS-MSA

Algorithm 3: Pseudocode for the update of the memory of all existing players in the musical group or the update of the MM_m in the proposed continuous/discrete TMS-MSA

Input: Not updated MM_m, $\mathrm{x}_{m,p}^{\mathrm{new}}$
Output: Updated MM_m

Start main body

begin
for *music player p* $[p \in \Psi^{\mathrm{PN}}]$ **do**
  set $\mathrm{x}_p^{\mathrm{worst}} = \mathrm{x}_{m,p}^{\mathrm{PMS}}$
  set $f\left(\mathrm{x}_p^{\mathrm{worst}}\right) = f\left(\mathrm{x}_{m,p}^{\mathrm{PMS}}\right)$
  if $f\left(\mathrm{x}_{m,p}^{\mathrm{new}}\right) < f\left(\mathrm{x}_p^{\mathrm{worst}}\right)$ **then**
    $\mathrm{x}_{m,p}^{\mathrm{new}} \in PM_{m,p}$
    $\mathrm{x}_p^{\mathrm{worst}} \notin PM_{m,p}$
  end if
end for
terminate

end main body

function. After completion of this process, melody vectors stored in the memory of all existing players in the musical group or the MM_m must be re-sorted based on the value of objective function—fitness function—in an ascending order. The pseudocode pertaining to sorting the solution vectors stored in the memory of all existing players in the musical group or the MM was formerly presented in Table 4.3. Given the dependence of each PM or more comprehensively the MM to the improvisation/iteration index of the SIS—index m—this pseudocode must be rewritten according to Table 4.5.

4.2.3.3 Sub-stage 3.3: Check of the Stopping Criterion of the SIS

After completion of sub-stage 3.2 and an update of all PMs, the check process of the stopping criterion of the single computational stage must be accomplished. The computational efforts of the SIS are terminated if its stopping criterion—the MNI-SIS—is satisfied. Otherwise, sub-stages 3.1 and 3.2 are repeated.

4.2.4 Stage 4: Pseudo-Group Computational Stage or PGIS

After finalization of stage 3, or accomplishment of the SIS, the pseudo-group computational stage or the PGIS must be performed. This stage is organized into four sub-stages: (1) improvisation of a new melody vector by each player, taking into

Table 4.5 Pseudocode relevant to the sorting of the solution vectors stored in the PMs or MM_m in the proposed continuous/discrete TMS-MSA

Algorithm 4: Pseudocode for sorting the solution vectors stored in the PMs or MM_m in the proposed continuous/discrete TMS-MSA

Input: Unsorted MM_m
Output: Sorted MM_m

start main body

begin
for *music player p* $[p \in \Psi^{\text{PN}}]$ **do**
  set $PM_{m,p} = MM_m\,(1\text{: PMS}, [(p-1)\cdot(\text{NDV}+1)]\text{: }[p\cdot(\text{NDV}+1)])$
  $F^{\text{sort}}_{m,p} = \text{sort}\big(PM_{m,p}(1:\text{PMS}, (\text{NCDV}+1)), '\text{ascend}'\big)$
  for *melody vector s* $[s \in \Psi^{\text{PMS}}]$ **do**
    for *melody vector* s^* $[s^* \in \Psi^{\text{PMS}}]$ **do**
      if $F^{\text{sort}}_{m,p}(s) == PM_{m,p}(s^*, (\text{NCDV}+1))$ **then**
        $PM^{\text{sort}}_{m,p}(s, 1:(\text{NDV}+1)) = PM_{m,p}(s^*, 1:(\text{NDV}+1))$
      end if
    end for
  end for
  $MM^{\text{sort}}_m(1:\text{PMS}, 1+[(p-1)\cdot(\text{NDV}+1)]:[p\cdot(\text{NDV}+1)]) = PM^{\text{sort}}_{m,p}$
end for
$MM_m = MM^{\text{sort}}_m$
terminate

end main body

account the feasible ranges of the updated pitches; (2) update of each PM; (3) update of the feasible ranges of pitches—continuous decision-making variables—for the next improvisation—only for random selections; and, (4) check of the stopping criterion of the PGIS. The mathematical equations expressed at this stage must depend on the improvisation/iteration index—index m—due to the repeatability of the PGIS in the proposed continuous/discrete TMS-MSA.

4.2.4.1 Sub-stage 4.1: Improvisation of a New Melody Vector by Each Player

In sub-stage 4.1, the improvisation process of a new melody vector by each player in the group must be performed. In this sub-stage, each player improvises a new melody under the influence of other players. In other words, the improvisation process of a new melody vector by player p—$\mathrm{x}^{\text{new}}_{m,p} = \left(x^{\text{new}}_{m,p,1}, \ldots, x^{\text{new}}_{m,p,v}, \ldots, x^{\text{new}}_{m,p,\text{NDV}}\right)$—is accomplished by using the proposed continuous/discrete AIP, taking into account the feasible range of the pitches, which are updated in different improvisations/iterations of the PGIS. The improvisation process of a new melody vector is done by other players in the same way.

4.2.4.2 Sub-stage 4.2: Update of Memory of Each Player

After completion of sub-stage 4.1 and improvisation of a new melody vector by each player in the group, the update process of the PMs or MM must be done. This process is virtually the same as for sub-stage 3.2 of the SIS, which was discussed in Sect. 4.2.3.2.

4.2.4.3 Sub-stage 4.3: Update of the Feasible Ranges of Pitches—Continuous Decision-Making Variables for the Next Improvisation—Only for Random Selection

After completion of sub-stage 4.2 and an update of the PMs, the update process of the feasible ranges of pitches—continuous decision-making variables for the next improvisation—only for random selection must be performed. Similar to the continuous TMS-MSA, in the continuous/discrete TMS-MSA, the feasible ranges of continuous decision-making variables in the melody vector are changed and updated during each improvisation/iteration of the PGIS, but only for random selection. This means that the lower bound of the continuous decision-making variable v ($x_v^{\min}$) and the upper bound of the continuous decision-making variable v ($x_v^{\max}$) in the PGIS depend on the improvisation/iteration index of the PGIS and change in the form of $x_{m,v}^{\min}$ and $x_{m,v}^{\max}$, respectively. Table 4.6 gives the pseudocode associated with the update of the feasible ranges of continuous decision-making variables in the proposed continuous/discrete TMS-MSA.

4.2.4.4 Sub-stage 4.4: Check of the Stopping Criterion of the Pseudo-Group Improvisation Stage

After finalization of sub-stage 4.3 and the update of the feasible ranges of the continuous decision-making variables for the next improvisation of the PGIS, the check process of the stopping criterion of this computational stage must be carried out. If the stopping criterion of the PGIS—the MNI-PGIS—is satisfied, its computational efforts are then terminated. Otherwise, sub-stages 4.1, 4.2, and 4.3 are repeated.

4.2.5 Stage 5: Selection Stage—Selection of the Final Optimal Solution—The Most Favorable Melody

After completion of stage 4, or accomplishment of the PGIS, the selection of the final optimal solution must be done in stage 5. In this stage, the best melody vector stored

Table 4.6 The pseudocode associated with the update of the feasible ranges of the continuous decision-making variables in the proposed continuous/discrete TMS-MSA

Algorithm 5: Pseudocode for the update of the feasible ranges of the continuous decision-making variables in the proposed continuous/discrete TMS-MSA

Input: $\mathrm{x}_{m,p}^{1}$
Output: $x_{m,v}^{\min}, x_{m,v}^{\max}$

start main body

begin
for *music player p* $[p \in \Psi^{\mathrm{PN}}]$ **do**
  set $\mathrm{x}_{p}^{\mathrm{best}} = \mathrm{x}_{m,p}^{1}$
end for
for *continuous decision-making variable v* $[v \in \Psi^{\mathrm{NCDV}}]$ **do**
  $x_{m,v}^{\min} = \min\left(x_{p,v}^{\mathrm{best}}\right); \forall p \in \Psi^{\mathrm{PN}}, \forall\left\{m \in \Psi^{\mathrm{MNI\text{-}PGIS}}\right\}$
  $x_{m,v}^{\max} = \max\left(x_{p,v}^{\mathrm{best}}\right); \forall p \in \Psi^{\mathrm{PN}}, \forall\left\{m \in \Psi^{\mathrm{MNI\text{-}PGIS}}\right\}$
end for
terminate

end main body

Table 4.7 Pseudocode relevant to the selection of the final optimal solution in the proposed continuous/discrete TMS-MSA

Algorithm 6: Pseudocode for the selection of the final optimal solution in the proposed continuous/discrete TMS-MSA

Input: MM
Output: $\mathrm{x}^{\mathrm{best}}$

start main body

begin
for *music player p* $[p \in \Psi^{\mathrm{PN}}]$ **do**
  set $\mathrm{x}_{p}^{\mathrm{best}} = \mathrm{x}_{p}^{1}$
end for
  $\mathrm{x}^{\mathrm{best}} = \min\left(\mathrm{x}_{p}^{\mathrm{best}}\right); \forall p \in \Psi^{\mathrm{PN}}$
terminate

end main body

in the memory of each existing player in the musical group is characterized; then, the best melody vector is chosen from among these melody vectors as the final optimal solution. Table 4.7 provides the pseudocode relevant to the selection of the final optimal solution in the proposed continuous/discrete TMS-MSA.

4.2.6 Continuous/Discrete Alternative Improvisation Procedure

As previously mentioned, existing player p in the musical group improvises a new melody vector—$x_{m,p}^{\text{new}} = \left(x_{m,p,1}^{\text{new}}, \ldots, x_{m,p,v}^{\text{new}}, \ldots, x_{m,p,\text{NDV}}^{\text{new}}\right)$—by using the proposed continuous/discrete AIP in the continuous/discrete TMS-MSA. Employing the continuous/discrete AIP by existing player p in the musical group is accomplished on the basis of three rules: (1) player memory consideration; (2) pitch adjustment; and, (3) random selection.

Rule 1: In the player memory consideration rule, the values of the new melody vector for player p are randomly selected from the melody vectors stored in the $PM_{m,p}$ with the probability of the PMCR. In this rule, two principles are alternately employed.

Applying of the first principle of the player memory consideration rule to determine the value of the continuous or discrete decision-making variable v from a new melody vector played by player p in the music group, $x_{m,p,v}^{\text{new}}$, is carried out using Eqs. (4.7) and (4.8), respectively:

$$x_{m,p,v}^{\text{new}} = x_{m,p,v}^{r} \pm \text{U}(0,1) \cdot BW_m;$$
$$\forall \left\{ m \in \Psi^{(\text{MNI-SIS})+(\text{MNI-PGIS})}, p \in \Psi^{\text{PN}}, v \in \Psi^{\text{NCDV}}, r \sim \text{U}\{1,2,\ldots,\text{PMS}\} \right\} \tag{4.7}$$

$$x_{m,p,v}^{\text{new}} = x_{m,p,v}^{r}; \quad \forall \left\{ m \in \Psi^{(\text{MNI-SIS})+(\text{MNI-PGIS})}, p \in \Psi^{\text{PN}}, v \in \Psi^{\text{NDDV}} \right\},$$
$$\forall \{ r \sim \text{U}\{1,2,\ldots,\text{PMS}\}\} \tag{4.8}$$

Equation (4.7) tells us that the value of the continuous decision-making variable v from the new melody vector played by player p, $x_{m,p,v}^{\text{new}}$, is randomly selected from the available corresponding continuous decision-making variable in the melody vectors stored in the $PM_{m,p}$—$\left(x_{m,p,v}^{1}, \ldots, x_{m,p,v}^{s}, \ldots, x_{m,p,v}^{\text{PMS}}\right)$—with the probability of the PMCR and updated by the BW_m parameter. Equation (4.8) also tells us that the value of the discrete decision-making variable v from the new melody vector played by player p, $x_{m,p,v}^{\text{new}}$, is haphazardly chosen from the available corresponding discrete decision-making variable in the melody vectors stored in the $PM_{m,p}$—$\left(x_{m,p,v}^{1}, \ldots, x_{m,p,v}^{s}, \ldots, x_{m,p,v}^{\text{PMS}}\right)$—with the probability of the PMCR. Implementing of the second principle of the player memory consideration rule to specify the value of the continuous or discrete decision-making variable v from a new melody vector played by the existing player p in the music group, $x_{m,p,v}^{\text{new}}$, is done using Eqs. (4.9) and (4.10), respectively:

$$x_{m,p,v}^{\text{new}} = x_{m,p,k}^{r} \pm \text{U}(0,1) \cdot BW_m;$$
$$\forall \Big\{ m \in \Psi^{(\text{MNI-SIS})+(\text{MNI-PGIS})}, p \in \Psi^{\text{PN}}, v \in \Psi^{\text{NCDV}}, r \sim \text{U}\{1,2,\ldots,\text{PMS}\},$$
$$k \sim \text{U}\{1,2,\ldots,\text{NCDV}\}\Big\} \tag{4.9}$$

$$x_{m,p,v}^{\text{new}} = x_{m,p,l}^{r};$$
$$\forall \Big\{ m \in \Psi^{(\text{MNI-SIS})+(\text{MNI-PGIS})}, p \in \Psi^{\text{PN}}, v \in \Psi^{\text{NDDV}}, r \sim \text{U}\{1,2,\ldots,\text{PMS}\},$$
$$l \sim \text{U}\{1,2,\ldots,\text{NDDV}\}\Big\} \tag{4.10}$$

Equation (4.9) tells us that the value of the continuous decision-making variable v from the new melody vector played by player p is randomly chosen from the entire set of available continuous decision-making variables stored in the $PM_{m,p}$— $\Big\{\Big(x_{m,p,1}^{1},\ldots,x_{m,p,1}^{s},\ldots,x_{m,p,1}^{\text{PMS}}\Big),\ldots,\Big(x_{m,p,v}^{1},\ldots,x_{m,p,v}^{s},\ldots,x_{m,p,v}^{\text{PMS}}\Big),\ldots,$ $\Big(x_{m,p,\text{NCDV}}^{1},\ldots,x_{m,p,\text{NCDV}}^{s},\ldots,x_{m,p,\text{NCDV}}^{\text{PMS}}\Big)\Big\}$ —with the probability of the PMCR and updated by the BW_m parameter. Also, Eq. (4.10) tells us that the value of the discrete decision-making variable v from the new melody vector played by player p is randomly chosen from the entire set of available discrete decision-making variables stored in the $PM_{m,p}$—$\Big\{\Big(x_{m,p,1}^{1},\ldots,x_{m,p,1}^{s},\ldots,x_{m,p,1}^{\text{PMS}}\Big),\ldots,\Big(x_{m,p,v}^{1},\ldots,x_{m,p,v}^{s},\ldots,x_{m,p,v}^{\text{PMS}}\Big),\ldots,$ $\Big(x_{m,p,\text{NDDV}}^{1},\ldots,x_{m,p,\text{NDDV}}^{s},\ldots,x_{m,p,\text{NDDV}}^{\text{PMS}}\Big)\Big\}$—with the probability of the PMCR.

Index r in Eqs. (4.7)–(4.10) is a random integer with a uniform distribution through the set $\{1,2,\ \ldots\ ,\text{PMS}\}$—$r \sim \text{U}\{1,2,\ \ldots\ ,\text{PMS}\}$. Put simply, the value of index r in Eqs. (4.7)–(4.10) is randomly specified through the set of permissible values demonstrated by the set $\{1,2,\ \ldots\ ,\text{PMS}\}$. Determination of this index is performed according to Eq. (4.11):

$$r = \text{int}(\text{U}(0,1) \cdot \text{PMS}) + 1 \tag{4.11}$$

Index k in Eq. (4.9) is also a random integer with a uniform distribution through the set $\{1,2,\ \ldots\ ,\text{NCDV}\}$—$k \sim \text{U}\{1,2,\ \ldots\ ,\text{NCDV}\}$. Put simply, the value of index k in Eq. (4.9) is randomly characterized through the set of permissible values displayed by the set $\{1,2,\ \ldots\ ,\text{NCDV}\}$. Determination of this index is done in accordance with Eq. (4.12):

$$k = \text{int}(\text{U}(0,1) \cdot \text{NCDV}) + 1 \tag{4.12}$$

Moreover, indices l in Eq. (4.10) represent a random integer with a uniform distribution through the set $\{1,2,\ \ldots\ ,\text{NDDV}\}$—$l \sim \text{U}\{1,2,\ \ldots\ ,\text{NDDV}\}$. Put simply, the value of index l in Eq. (4.10) is randomly characterized through the set of permissible values displayed by the set $\{1,2,\ \ldots,\text{NDDV}\}$. Determination of this index is performed according to Eq. (4.13):

$$l = \text{int}(\text{U}(0,1) \cdot \text{NDDV}) + 1 \tag{4.13}$$

Applying the player memory consideration rule is also performed for other players in the same way.

Rule 2: In the pitch adjustment rule, the values of a new melody vector played by player p, randomly selected from among the existing melody vectors in the $PM_{m,p}$ with the probability of the PMCR, are updated with the probability of the PAR_m. Applying of the pitch adjustment rule to determine the value of the continuous or discrete decision-making variable v from a new melody vector played by player p, $x^{\text{new}}_{m,p,v}$, is carried out by using Eqs. (4.14) and (4.15), respectively:

$$x^{\text{new}}_{m,p,v} = x^{\text{best}}_{m,p,v}; \quad \forall \left\{ m \in \Psi^{(\text{MNI-SIS})+(\text{MNI-PGIS})}, p \in \Psi^{\text{PN}}, v \in \Psi^{\text{NCDV}} \right\} \tag{4.14}$$

$$x^{\text{new}}_{m,p,v} = x^{\text{best}}_{m,p,v}; \quad \forall \left\{ m \in \Psi^{(\text{MNI-SIS})+(\text{MNI-PGIS})}, p \in \Psi^{\text{PN}}, v \in \Psi^{\text{NDDV}} \right\} \tag{4.15}$$

Equations (4.14) and (4.15) tell us that after the value of the continuous or discrete decision-making variable v from a new melody vector by player p was haphazardly chosen from the melody vectors stored in the $PM_{m,p}$ with the probability of the PMCR, this continuous or discrete decision-making variable is updated with the probability of the PAR_m. The update process for the continuous or discrete decision-making variable v is done by replacing it with the value of the corresponding continuous or discrete decision-making variable from the best melody vector available in the $PM_{m,p}$, $x^{\text{best}}_{m,p,1}$. Applying the pitch adjustment rule is also done for other players in the same way.

Rule 3: In the random selection rule, the values of a new melody vector played by player p are randomly selected from the entire space of the nonempty feasible decision-making with the probability of the 1-PMCR. In this rule, two different principles are used. The first and second principles are activated in the SIS and PGIS, respectively. Applying of the first and second principles of the random selection rule to determine the value of the continuous decision-making variable v from a new melody vector played by player p in the music group, $x^{\text{new}}_{m,p,v}$, is accomplished by using Eqs. (4.16) and (4.17), respectively:

$$x^{\text{new}}_{m,p,v} = x^{\min}_{v} + \text{U}(0,1) \cdot \left(x^{\max}_{v} - x^{\min}_{v}\right); \quad \forall \left\{ m \in \Psi^{(\text{MNI-SIS})}, p \in \Psi^{\text{PN}}, v \in \Psi^{\text{NCDV}} \right\} \tag{4.16}$$

$$x^{\text{new}}_{m,p,v} = x^{\min}_{m,v} + \text{U}(0,1) \cdot \left(x^{\max}_{m,v} - x^{\min}_{m,v}\right); \quad \forall \left\{ m \in \Psi^{(\text{MNI-PGIS})}, p \in \Psi^{\text{PN}}, v \in \Psi^{\text{NCDV}} \right\} \tag{4.17}$$

Equation (4.16) tells us that the value of the continuous decision-making variable v from the new melody vector played by player p, $x^{\text{new}}_{m,p,1}$, is randomly chosen from the entire space of the nonempty feasible decision-making relevant to this decision-making variable, which is characterized by an invariable lower bound, $x_v^{\min}$, and an invariable upper bound, $x_v^{\max}$, with the probability of the 1-PMCR. Equation (4.17) also tells us that the value of the continuous decision-making variable v from the new melody vector played by player p, $x_{m,p,1}^{\text{new}}$, is haphazardly selected from the entire space of the nonempty feasible decision-making pertaining to this decision-making variable, which is specified by a variable lower bound, $x_{m,v}^{\min}$, and a variable upper bound, $x_{m,v}^{\max}$, with the probability of the 1-PMCR.

The first and second principles of the random-selection rule associated with discrete decision-making variables are quite similar. Applying of the first and second principles of the random-selection rule in order to determine the value of the discrete decision-making variable v from a new melody vector played by the existing player p in the music group, $x_{m,p,v}^{\text{new}}$, is performed by using Eq. (4.18):

$$x_{m,p,v}^{\text{new}} = x_v(y);$$
$$\forall\left\{m \in \Psi^{(\text{MNI-SIS})+(\text{MNI-PGIS})}, p \in \Psi^{\text{PN}}, v \in \Psi^{\text{NDDV}}, y \sim \text{U}\{x_v(1), \ldots, x_v(w_v), \ldots, x_v(W_v)\}\right\} \tag{4.18}$$

Equation (4.18) tells us that the value of the discrete decision-making variable v from a new melody vector played by player $p, x_{m,p,v}^{\text{new}}$, is randomly specified through the set of candidate permissible values for corresponding discrete decision-making variable shown by the set of $\{x_v(1), \ldots, x_v(w_v), \ldots, x_v(W_v)\}$. In Eq. (4.18), index y is a random integer with a uniform distribution through the set of $\{x_v(1), \ldots, x_v(w_v), \ldots, x_v(W_v)\}$—$y \sim \text{U}\{x_v(1), \ldots, x_v(w_v), \ldots, x_v(W_v)\}$. Applying of the random selection rule is also accomplished for other players in the same way. Similar to the continuous TMS-MSA, in the proposed continuous/discrete TMS-MSA, each player in the musical group employs the updated values of the BW_m and PAR_m parameters in the improvisation process of its melody vector. The BW_m and PAR_m parameters are updated in each improvisation/iteration of the SIS and the PGIS by using Eqs. (4.19) and (4.20), respectively:

$$BW_m = \text{BW}^{\max} \cdot \exp\left(\frac{\ln\left(\text{BW}^{\max}/\text{BW}^{\min}\right)}{(\text{MNI-SIS}) + (\text{MNI-PGIS})} \cdot m\right); \quad \forall\left\{m \in \Psi^{(\text{MNI-SIS})+(\text{MNI-PGIS})}\right\} \tag{4.19}$$

$$PAR_m = \text{PAR}^{\min} + \left(\frac{\text{PAR}^{\max} - \text{PAR}^{\min}}{(\text{MNI-SIS}) + (\text{MNI-PGIS})}\right) \cdot m; \quad \forall\left\{m \in \Psi^{(\text{MNI-SIS})+(\text{MNI-PGIS})}\right\} \tag{4.20}$$

Table 4.8 presents the pseudocode relevant to the improvisation of a new melody vector by each player in the musical group of the proposed continuous/discrete

Table 4.8 Pseudocode relevant to improvisation of a new melody vector by each player in the musical group of the proposed continuous/discrete TMS-MSA

Algorithm 7: Pseudocode for improvisation of a new melody vector by each player in the musical group of the proposed continuous/discrete TMS-MSA

Input: $\text{BW}^{\max}$, $\text{BW}^{\min}$, MNI-SIS, MNI-PGIS, NCDV, NDDV, NDV, $\text{PAR}^{\max}$, $\text{PAR}^{\min}$, PMCR, PMS, PN, $x_v^{\min}$, $x_v^{\max}$, $\{x_v(1), \ldots, x_v(w_v), \ldots, x_v(W_v)\}$

Output: $\text{x}_{m,p}^{\text{new}}$

start main body

begin

$BW_m = \text{BW}^{\max} \cdot \exp\left[(\ln(\text{BW}^{\max}/\text{BW}^{\min})/((\text{MNI-SIS}) + (\text{MNI-PGIS}))) \cdot m\right]$

$PAR_m = \text{PAR}^{\min} - \left[((\text{PAR}^{\max} - \text{PAR}^{\min})/((\text{MNI-SIS}) + (\text{MNI-PGIS}))) \cdot m\right]$

for *music player* p $[p \in \Psi^{\text{PN}}]$ **do**

construct the new melody vector for music player p, $\text{x}_{m,p}^{\text{new}}$, with dimension $\{1\} \cdot \{\text{NDV} + 1\}$ and zero initial value

for *decision-making variable* v $[v \in \Psi^{\text{NDV}}]$ **do**

if $\text{U}(0,1) \leq \text{PMCR}$ **then**

Rule 1: harmony memory consideration with probability PMCR

if *improvisation/iteration* $m[m \in \Psi^{(\text{MNI-SIS}) + (\text{MNI-PGIS})}]$ is odd **then**

Principle 1: first combination

$x_{m,p,v}^{\text{new}} = x_{m,p,v}^{r} \pm \text{U}(0,1) \cdot BW_m;\ \forall r \sim \text{U}\{1, 2, \ldots, \text{PMS}\}$; for CDVs

$x_{m,p,v}^{\text{new}} = x_{m,p,v}^{r};\ \forall r \sim \text{U}\{1, 2, \ldots, \text{PMS}\}$; for DDVs

else

Principle 2: second combination

$x_{m,p,v}^{\text{new}} = x_{m,p,k}^{r} \pm \text{U}(0,1) \cdot BW_m;\ \forall r \sim \text{U}\{1, 2, \ldots, \text{PMS}\},\ \forall k \sim \text{U}\{1, 2, \ldots, \text{NCDV}\}$; for CDVs

$x_{m,p,v}^{\text{new}} = x_{m,p,l}^{r};\ \forall r \sim \text{U}\{1, 2, \ldots, \text{PMS}\},\ \forall l \sim \text{U}\{1, 2, \ldots, \text{NDDV}\}$; for DDVs

end if

if $\text{U}(0,1) \leq PAR_m$ **then**

Rule 2: pitch adjustment with probability HMCR $\cdot$ PAR_m

$x_{m,p,v}^{\text{new}} = x_{m,p,v}^{\text{best}}$; for CDVs and DDVs

end if

else if

Rule 3: random selection with probability 1-HMCR

switch 1

case *improvisation/iteration* $m[m \in \Psi^{(\text{MNI-SIS}) + (\text{MNI-PGIS})}] \leq (\text{MNI-SIS})$ **then**

$x_{m,p,v}^{\text{new}} = x_v^{\min} + \text{U}(0,1) \cdot \left(x_v^{\max} - x_v^{\min}\right)$; for CDVs

$x_{m,p,v}^{\text{new}} = x_v(y);\ \forall y \sim \text{U}\{x_v(1), \ldots, x_v(w_v), \ldots, x_v(W_v)\}$; for DDVs

case *improvisation/iteration* m $[m \in \Psi^{(\text{MNI-SIS}) + (\text{MNI-PGIS})}] > (\text{MNI-SIS})$ and *improvisation/iteration* m $[m \in \Psi^{(\text{MNI-SIS}) + (\text{MNI-PGIS})}] \leq (\text{MNI-SIS}) + (\text{MNI-PGIS})$ **then**

$x_{m,p,v}^{\text{new}} = x_{m,v}^{\min} + \text{U}(0,1) \cdot \left(x_{m,v}^{\max} - x_{m,v}^{\min}\right)$; for CDVs

$x_{m,p,v}^{\text{new}} = x_v(y);\ \forall y \sim \text{U}\{x_v(1), \ldots, x_v(w_v), \ldots, x_v(W_v)\}$; for DDVs

end switch

end if

end for

calculate the value of objective function, fitness function, derived from melody vector $\text{x}_{m,p}^{\text{new}}$ as $f\left(\text{x}_{m,p}^{\text{new}}\right)$

allocate $f\left(\text{x}_{m,p}^{\text{new}}\right)$ to element $(1, \text{NDV} + 1)$ of the new melody vector $\text{x}_{m,p}^{\text{new}}$

end for

terminate

end main body

Note: Continuous decision-making variable (CDVs), discrete decision-making variable (DDVs)

TMS-MSA. The designed pseudocodes in various stages and sub-stages of the continuous/discrete TMS-MSA are organized in a regular sequence and form the performance-driven architecture of this algorithm. Table 4.9 presents the pseudocode associated with the performance-driven architecture of the proposed continuous/discrete TMS-MSA.

4.3 Enhanced Version of the Proposed Continuous/Discrete TMS-MSA

In the preceding section, the continuous/discrete TMS-MSA was developed in order to deal with complicated, real-world, large-scale, non-convex, non-smooth optimization problems involved with a concurrent combination of continuous and discrete decision-making variables. In order to appreciably improve the performance, flexibility, and robustness of the continuous/discrete TMS-MSA, an enhanced version of the continuous/discrete TMS-MSA—the TMS-EMSA—will now be presented.

The performance-driven architecture of the TMS-EMSA is generally broken down into five stages, as follows:

- Stage 1—Definition stage: Definition of the optimization problem and its parameters
- Stage 2—Initialization stage
  - Sub-stage 2.1: Initialization of the parameters of the TMS-EMSA
  - Sub-stage 2.2: Initialization of the MM
- Stage 3—Single computational stage or SIS
  - Sub-stage 3.1: Improvisation of a new melody vector by each player
  - Sub-stage 3.2: Update of each PM
  - Sub-stage 3.3: Check of the stopping criterion of the SIS
- Stage 4—Group computational stage or group improvisation stage (GIS)
  - Sub-stage 4.1: Improvisation of a new melody vector by each player taking into account the feasible ranges of the updated pitches
  - Sub-stage 4.2: Update of each PM
  - Sub-stage 4.3: Update of the feasible ranges of pitches—continuous decision-making variables for the next improvisation—only for random selection
  - Sub-stage 4.4: Check of the stopping criterion of the GIS
- Stage 5—Selection stage: Selection of the final optimal solution—the best melody

It should be obvious that the performance-driven architecture of the TMS-EMSA is quite similar to the performance-driven architectures of the continuous/discrete TMS-MSA, which were reported in Sect. 4.2 of this chapter. Nevertheless, the proposed TMS-EMSA brings about fundamental differences in some stages and

Table 4.9 Pseudocode associated with the performance-driven architecture of the proposed continuous/discrete TMS-MSA

Algorithm 8: Pseudocode for performance-driven architecture of the proposed continuous/discrete TMS-MSA

Input: $\mathrm{BW}^{\max}$, $\mathrm{BW}^{\min}$, MNI-SIS, MNI-PGIS, NCDV, $\mathrm{PAR}^{\max}$, $\mathrm{PAR}^{\min}$, PMCR, PMS, PN, $x_v^{\min}$, $x_v^{\max}$

Output: $\mathrm{x}^{\mathrm{best}}$

start main body

begin

Stage 1—Definition stage: Definition of the optimization problem and its parameters

Stage 2—Initialization stage

  Sub-stage 2.1: Initialization of the parameters of the continuous/discrete TMS-MSA

  Sub-stage 2.2: Initialization of the of the MM

    Algorithm 1: Pseudocode for initialization of the entire set of PMs or MM in the proposed continuous/discrete TMS-MSA

    Algorithm 2: Pseudocode for sorting the solution vectors stored in the PMs or MM in the proposed continuous/discrete TMS-MSA

Stage 3—Single computational stage or SIS

  set *improvisation/iteration* $m = 1$

  set MM_m = MM

  while $m \leq$ (MNI-SIS) **do**

   Sub-stage 3.1: Improvisation of a new melody vector by each player

    Algorithm 7: Pseudocode for improvisation of a new melody vector by each player in the musical group of the proposed continuous/discrete TMS-MSA

   Sub-stage 3.2: Update of each PM

    Algorithm 3: Pseudocode for the update of the memory of all existing players in the musical group or the update of the MM_m in the proposed continuous/discrete TMS-MSA

    Algorithm 4: Pseudocode for sorting the solution vectors stored in the PMs or MM_m in the proposed continuous/discrete TMS-MSA

   set *improvisation/iteration* $m = m + 1$

  end while

Stage 4—Pseudo-group computational stage or PGIS

  while $m >$ (MNI-SIS) and $m \leq$ (MNI-SIS) + (MNI-PGIS) **do**

   Sub-stage 4.1: Improvisation of a new melody vector by each player taking into account the feasible ranges of the updated pitches

    Algorithm 7: Pseudocode for improvisation of a new melody vector by each player in the musical group of the proposed continuous/discrete TMS-MSA

   Sub-stage 4.2: Update of each PM

    Algorithm 3: Pseudocode for the update of the memory of all existing players in the musical group or the update of the MM_m in the proposed continuous/discrete TMS-MSA

    Algorithm 4: Pseudocode for sorting the solution vectors stored in the PMs or MM_m in the proposed continuous/discrete TMS-MSA

(continued)

Table 4.9 (continued)

Sub-stage 4.3: Update of the feasible ranges of pitches—continuous decision-making variables—for the next improvisation—only for random selection
Algorithm 5: Pseudocode for the update of the feasible ranges of the continuous decision-making variables in the proposed continuous/discrete TMS-MSA
set *improvisation/iteration* $m = m + 1$
end while
Stage 5—Selection stage: Selection of the final optimal solution—the best melody
Algorithm 6: Pseudocode for the selection of the final optimal solution in the proposed continuous/discrete TMS-MSA
terminate
end main body

sub-stages of the performance-driven architecture over the continuous/discrete TMS-MSA, because of its enhancements. Here, only the different stages and sub-stages altered in the TMS-EMSA, in comparison with the continuous/discrete TMS-MSA, are redefined to prevent repeating the previous descriptions in Sect. 4.2 of this chapter.

Stage 1 is related to the definition of the optimization problem and its parameters. This stage of the proposed TMS-EMSA is virtually the same as the corresponding stage of the continuous/discrete TMS-MSA, previously described in Sect. 4.2.1 of this chapter.

Sub-stage 2.1 is related to the initialization of the parameters of the optimization algorithm. In this sub-stage, the parameter adjustments of the continuous/discrete TMS-MSA are characterized in Table 4.1, which was presented in Sect. 4.2.2.1 of this chapter. As for Table 4.1, the PMS parameter represents the number of melody vectors—solution vectors—stored in the memory of each existing player in the musical group. In the music literature, this parameter refers to the number of melodies that an existing player in the musical group is capable of storing in its own mind. Table 4.1 shows that in the continuous/discrete TMS-MSA, the PMS parameter is considered to be invariant for all of the existing players in the musical group. That is to say that the number of melody vectors stored in the memory of all existing players in the musical group is equal. In reality, however, the number of melodies that a player can store in its own mind is a unique parameter. Stated another way, the number of melodies that existing players in the musical group can store in their minds is not equal. In this regard, the PMS parameter in the TMS-EMSA can be different for each player in the musical group. For this purpose, the PMS parameter must be dependent on the index of the player, p, or consequently the PMS parameter in the TMS-EMSA is substituted with the PMS of player p—PMS_p.

In Table 4.1, the PMCR parameter is employed in the improvisation process of a new melody vector by an existing player in the musical group under sub-stages 3.1

and 4.1 with the aim of determining whether the value of a decision-making variable from a new melody vector played by the corresponding player is derived from its PM or from the entire space of the nonempty feasible decision-making related to the corresponding variable. In the music literature, the activation of this parameter for an existing player in the musical group is equivalent to the player randomly choosing notes of its new melody vector through the notes of the melody vectors stored in its own mind. Conversely, the inactivation of the PMCR parameter for an existing player in the musical group brings about haphazardly selected notes for this player from all of the notes available to be played. Table 4.1 also shows that the PMCR parameter is taken into account to be invariant for all existing players in the musical group in the continuous/discrete TMS-MSA. In reality, however, determining whether the value of a note from a new melody vector played by an existing player in the musical group should be randomly chosen from the notes of melody vectors stored in the mind of the corresponding player or from all of the available notes is a unique parameter. That is, the PMCR parameter for existing players in the musical group is not equal. In the proposed TMS-EMSA, the PMCR parameter can be different for each player in the musical group. The PMCR parameter must, then, be dependent on the index of player—p. As a result, the PMCR parameter in the TMS-EMSA is replaced by the PMCR of player p—$PMCR_p$.

Also from Table 4.1, the PAR^{min} and PAR^{max} parameters are generally employed in order to determine the value of the PAR parameter in improvisation/iteration m of the SIS and PGIS—PAR_m. The update process of the PAR_m parameter was previously described by using Eq. (4.20) in Sect. 4.2.6 of this chapter. In the improvisation process of a new melody vector by an existing player in the musical group under sub-stages 3.1 and 4.1, the PAR_m parameter is exploited with the aim of specifying whether the value of a decision-making variable, randomly selected from the memory relevant to the corresponding player with the probability of the PMCR, needs a change/update or not. In the music literature, the activation of this parameter for an existing player in the musical group is equivalent to the player making a change/update in notes of its new melody vector, which is haphazardly selected from the notes of the melody vectors stored in its own mind. In fact, the inactivation of the PAR_m parameter for an existing player gives rise to the player not creating any changes or updates in notes of its new melody vector, which is randomly chosen from the notes of the melody vectors stored in its own mind. As set out in Table 4.1, in the continuous/discrete TMS-MSA, the PAR^{min} and PAR^{max} parameters are considered to be invariant for all players in the musical group. That is to say that the PAR_m parameter, as a pitch adjusting rate parameter in improvisation/iteration m of the SIS and PGIS, is the same for all players in the musical group. In fact, the rate that specifies whether or not the value of a decision-making variable from a new melody vector played by an existing player haphazardly selected from the memory relevant to the corresponding player with the probability of the PMCR needs a change or update is a unique parameter. Put another way, in reality, the PAR_m parameter for existing players in the musical group is not equal. In the TMS-EMSA, this parameter can, therefore, be different for each player. In doing so, the PAR^{min} and PAR^{max} parameters must be dependent on the index of player—p. This dependency causes the PAR_m parameter, which is calculated by using the PAR^{min} and PAR^{max} parameters, which also depends on the index of player—p. As a consequence, the PAR^{min}, PAR^{max}, and PAR_m parameters in the TMS-EMSA are substituted with a minimum pitch adjusting rate of player p (PAR_p^{min}), a maximum pitch adjusting rate of player p (PAR_p^{max}), and a pitch adjusting rate of player p in improvisation/iteration m of the SIS and GIS ($PAR_{m,p}$), respectively.

Again from Table 4.1, the BW^{min} and BW^{max} parameters are generally used in order to specify the value of the BW parameter in improvisation/iteration m of the SIS and PGIS—BW_m. For this parameter, the update process was previously expressed using Eq. (4.19) in Sect. 4.2.6 of this chapter. The BW_m parameter is exploited in order to change/update the value of continuous decision-making variable chosen from the memory relevant to an existing player in the musical group with the probability of the PMCR, in the improvisation process of a new melody vector by the corresponding player under sub-stages 3.1 and 4.1. The point to be made here is that the BW_m parameter is of an optional length and is solely defined for continuous decision-making variables. In the music literature, after activation of the PMCR for an existing player, the player must make a change or an update in notes of its new melody vector, haphazardly selected from the notes of the melody vectors stored in its own mind. The corresponding player performs this change/update by using the BW_m parameter. Table 4.1 further shows that, in the continuous/discrete TMS-MSA, the BW^{min} and BW^{max} parameters are considered to be invariant for all players. This means that the BW_m parameter, as BW parameter in improvisation/iteration m of the SIS and PGIS, is the same for all players. In reality, however, the BW_m parameter for changing/updating the value of a decision-making variable from a new melody vector played by an existing player, randomly chosen from the memory relevant to the corresponding player with the probability of the PMCR, is a unique parameter. Put simply, the BW_m parameter for existing players in the musical group is not equal. In this regard, the BW_m parameter can be different for each player in the TMS-EMSA. To do this, the BW^{min} and BW^{max} parameters must be dependent on the index of player—p. This dependency gives rise to the BW_m parameter, which is obtained by using the BW^{min} and BW^{max} parameters, and depends on the index of player—p. As a result, the BW^{min}, BW^{max}, and BW_m parameters in the TMS-EMSA are replaced by the minimum bandwidth of player p (BW_p^{min}), the maximum bandwidth of player p (BW_p^{max}), and the bandwidth of player p in improvisation/iteration m of the SIS and GIS ($BW_{m,p}$), respectively.

Another adjustment parameter of the continuous/discrete TMS-MSA presented in Table 4.1 is the maximum number of improvisations/iterations of the pseudo-group improvisation stage—MNI-PGIS. Given the architecture of the TMS-EMSA, the PGIS becomes the GIS. The main reason for the transformation of the PGIS into the GIS originates from an innovative, well-organized improvisation procedure developed by the authors for the proposed TMS-EMSA to appreciably improve its performance, flexibility, and robustness. Accordingly, the MNI-PGIS is substituted with the maximum number of improvisations/iterations of the group improvisation stage—MNI-GIS—in the TMS-EMSA. Other parameters presented in Table 4.1

Table 4.10 Adjustment parameters of the proposed TMS-EMSA

No.	Proposed TMS-EMSA parameter	Abbreviation	Parameter range
1	Melody memory	MM	–
2	Player number	PN	PN ≥ 1
3	Player memory	PM_p	–
4	Memory size of player *p*	PMS_p	$PMS_p \geq 1$
5	Memory considering rate of player *p*	$PMCR_p$	$0 \leq PMCR_p \leq 1$
6	Minimum pitch adjusting rate of player *p*	$PAR_p^{\min}$	$0 \leq PAR_p^{\min} \leq 2$
7	Maximum pitch adjusting rate of player *p*	$PAR_p^{\max}$	$0 \leq PAR_p^{\max} \leq 2$
8	Minimum bandwidth of player *p*	$BW_p^{\min}$	$0 \leq BW_p^{\min} < +\infty$
9	Maximum bandwidth of player *p*	$BW_p^{\max}$	$0 \leq BW_p^{\max} < +\infty$
10	Number of continuous decision-making variables	NCDV	NCDV ≥ 1
11	Number of discrete decision-making variables	NDDV	NDDV ≥ 1
12	Number of decision-making variables	NDV	NDV ≥ 2
13	Maximum number of improvisations/iterations of the SIS	MNI-SIS	MNI-SIS ≥ 1
14	Maximum number of improvisations/iterations of the GIS	MNI-GIS	MNI-GIS ≥ 1

remain unchanged for the TMS-EMSA. Detailed descriptions for the adjustment parameters of the TMS-EMSA are presented in Table 4.10.

Sub-stage 2.2 is related to initialization of the MM. In this sub-stage of the continuous/discrete TMS-MSA, initialization of the MM is carried out according to Eqs. (4.3)–(4.6), given in Sect. 4.2.2.2 of this chapter. In the TMS-EMSA, however, in view of the changes applied in the parameters under sub-stage 2.1, the initialization process of the MM must be rewritten using Eqs. (4.21)–(4.24), as follows:

$$\mathrm{MM} = \left[PM_1 \quad \cdots \quad PM_p \quad \cdots \quad PM_{\mathrm{PN}} \right]; \quad \forall \left\{ p \in \Psi^{\mathrm{PN}} \right\} \tag{4.21}$$

$$PM_p = \begin{bmatrix} \mathrm{x}_p^1 \\ \vdots \\ \mathrm{x}_p^s \\ \vdots \\ \mathrm{x}_p^{PMS_p} \end{bmatrix} = \begin{bmatrix} x_{p,1}^1 & \cdots & x_{p,v}^1 & \cdots & x_{p,\mathrm{NDV}}^1 & | & f\left(\mathrm{x}_p^1\right) \\ \vdots & & \vdots & & \vdots & & \vdots \\ x_{p,1}^s & \cdots & x_{p,v}^s & \cdots & x_{p,\mathrm{NDV}}^s & | & f\left(\mathrm{x}_p^s\right) \\ \vdots & & \vdots & & \vdots & & \vdots \\ x_{p,1}^{PMS_p} & \cdots & x_{p,v}^{PMS_p} & \cdots & x_{p,\mathrm{NDV}}^{PMS_p} & | & f\left(\mathrm{x}_p^{PMS_p}\right) \end{bmatrix};$$

$$\forall \left\{ p \in \Psi^{\mathrm{PN}}, v \in \Psi^{\mathrm{NDV}}, s \in \Psi^{PMS_p}, \Psi^{\mathrm{NDV}} = \Psi^{\mathrm{NCDV+NDDV}} \right\} \tag{4.22}$$

Table 4.11 Pseudocode related to initialization of the entire set of PMs or MM in the proposed TMS-EMSA

Algorithm 9: Pseudocode for initialization of the entire set of PMs or MM in the proposed TMS-EMSA

Input: NCDV, NDDV, NDV, PN, PMS_p, $x_v^{\min}$, $x_v^{\max}$, $\{x_v(1), \ldots, x_v(w_v), \ldots, x_v(W_v)\}$
Output: MM

start main body

begin
set $\text{PMS}^{\max} = \max(PMS_p); \forall p \in \Psi^{\text{PN}}$
construct the matrix MM with dimension $\{\text{PMS}^{\max}\} \cdot \{(\text{NDV} + 1) \cdot \text{PN}\}$ and zero initial value
for *music player p* $[p \in \Psi^{\text{PN}}]$ **do**
  construct the submatrix PM_p with dimension $\{PMS_p\} \cdot \{\text{NDV} + 1\}$ and zero initial value
  for *melody vector s* $\left[s \in \Psi^{PMS_p}\right]$ **do**
    construct the melody vector *s* of music player *p*, x_p^s, with dimension $\{1\} \cdot \{\text{NDV} + 1\}$ and zero initial value
    for *decision-making variable v* $[v \in \Psi^{\text{NDV}}]$ **do**
      $x_{p,v}^s = x_v^{\min} + \text{U}(0,1) \cdot \left(x_v^{\max} - x_v^{\min}\right)$; for CDVs
      $x_{p,v}^s = x_v(y); \forall y \sim \text{U}\{x_v(1), \ldots, x_v(w_v), \ldots, x_v(W_v)\}$; for DDVs
      allocate $x_{p,v}^s$ to element $(1, v)$ of the melody vector x_p^s
    end for
    calculate the value of the objective function, fitness function, derived from the melody vector x_p^s as $f\left(\text{x}_p^s\right)$
    allocate $f\left(\text{x}_p^s\right)$ to element $(1, \text{NDV} + 1)$ of the melody vector x_p^s
    add melody vector x_p^s to the row *s* of the submatrix PM_p
  end for
  add submatrix PM_p to the rows 1 to PMS_p and columns $1 + (p - 1) \cdot (\text{NDV} + 1)$ to $p \cdot (\text{NDV} + 1)$ of the matrix MM
end for
terminate

end main body

Note: Continuous decision-making variable (CDVs), discrete decision-making variable (DDVs)

$$x_{p,v}^s = x_v^{\min} + \text{U}(0,1) \cdot \left(x_v^{\max} - x_v^{\min}\right); \quad \forall\left\{p \in \Psi^{\text{PN}}, v \in \Psi^{\text{NCDV}}, s \in \Psi^{PMS_p}\right\} \tag{4.23}$$

$$x_{p,v}^s = x_v(y); \quad \forall\left\{p \in \Psi^{\text{PN}}, v \in \Psi^{\text{NDDV}}, s \in \Psi^{PMS_p}, y \sim \text{U}\{x_v(1), \ldots, x_v(w_v), \ldots, x_v(W_v)\}\right\} \tag{4.24}$$

Table 4.12 Pseudocode relevant to the sorting of the solution vectors stored in the PMs or MM in the proposed TMS-EMSA

Algorithm 10: Pseudocode for sorting of the solution vectors stored in the PMs or MM in the proposed TMS-EMSA

Input: Unsorted MM
Output: Sorted MM

start main body

begin
for *music player p* $[p \in \Psi^{\mathrm{PN}}]$ **do**
  set $PM_p = \mathrm{MM}(1 : PMS_p, [(p-1) \cdot (\mathrm{NDV}+1)] : [p \cdot (\mathrm{NDV}+1)])$
  $F_p^{\mathrm{sort}} = \mathrm{sort}\left(PM_p\left(1 : PMS_p, (\mathrm{NDV}+1)\right), '\mathrm{ascend}'\right)$
  for *melody vector s* $\left[s \in \Psi^{PMS_p}\right]$ **do**
    for *melody vector* s^* $\left[s^* \in \Psi^{PMS_p}\right]$ **do**
      if $F_p^{\mathrm{sort}}(s) == PM_p(s^*, (\mathrm{NDV}+1))$ **then**
        $PM_p^{\mathrm{sort}}(s, 1 : (\mathrm{NDV}+1)) = PM_p(s^*, 1 : (\mathrm{NDV}+1))$
      end if
    end for
  end for
  $\mathrm{MM}^{\mathrm{sort}}\left(1 : PMS_p, 1 + [(p-1) \cdot (\mathrm{NDV}+1)] : [p \cdot (\mathrm{NDV}+1)]\right) = PM_p^{\mathrm{sort}}$
end for
$\mathrm{MM} = \mathrm{MM}^{\mathrm{sort}}$
terminate

end main body

Detailed descriptions associated with Eqs. (4.21)–(4.24) were presented in Sect. 3.6.2.2 of Chap. 3 and Sect. 4.2.2.2 of this chapter. Table 4.11 presents the pseudocode related to initialization of the entire set of PMs or MM in the proposed TMS-EMSA. After filling all of the PMs or MM with random solution vectors, the solution vectors stored in each PM must be sorted from the lowest value to the highest value—in an ascending order—from the standpoint of the value of the objective function of the optimization problem. In the continuous/discrete TMS-MSA, sorting the stored solution vectors for all PMs or MM is performed according to the pseudocode presented in Table 4.3. In the TMS-EMSA, however, due to the changes made in the parameters under sub-stage 2.1, the sorting process for all PMs or MM must be restructured. As a result, Table 4.12 gives the pseudocode relevant to the sorting of the solution vectors stored in the PMs or MM in the proposed TMS-EMSA. Sub-stages 3.1 and 4.1 are related to the improvisation of a new melody vector by each player in the musical group without and with the influence of other players in the musical group, respectively.

Simply put, in sub-stages 3.1 and 4.1, each player improvises a new melody vector both without and with taking into account interactive relationships among existing players in the musical group, respectively. In fact, the ultimate purpose of each player is to achieve the most desirable sequence of pitches in a melody that represents a truly fantastic melody. For example, consider player p in the musical group. First, the SIS is run by player p in order to achieve a superior melody. In the SIS, the corresponding player improvises its melodies individually without the influence of other players in the musical group. After the corresponding player has achieved an appropriate sequence of pitches in its melodies, the GIS is run. In the GIS, the corresponding player improves its sequence of pitches of melodies played in the SIS interactively with the influence of other players. More precisely, in the SIS, player p merely seeks to achieve a superior melody with an individual exercise, while in the GIS the corresponding player has the ability to learn and imitate the best existing player in the group with the aim of attaining the superior melody. In sub-stages 3.1 and 4.1 of the continuous/discrete TMS-MSA, player p improvises a new melody—$\mathrm{x}_{m,p}^{\text{new}} = \left(x_{m,p,1}^{\text{new}}, \ldots, x_{m,p,v}^{\text{new}}, \ldots, x_{m,p,\text{NDV}}^{\text{new}}\right)$—by using the continuous/discrete AIP. Detailed descriptions pertaining to the continuous/discrete AIP were presented in Sect. 4.2.6 of this chapter. In the continuous/discrete TMS-MSA, the sole interactive relationship among existing players in the musical group is the update process of the feasible ranges of continuous decision-making variables. In this regard, the upper and lower bounds of continuous decision-making variables are updated in each improvisation/iteration of the PGIS. Detailed descriptions relevant to this update process in the continuous/discrete TMS-MSA were described in Sect. 4.2.4.3 of this chapter. In this optimization technique, the capability of learning and imitating each player in the musical group is also ignored from the best player in the group. The pseudo-group performance is, therefore, widely employed to improvise a new melody vector in stage 4 of the continuous/discrete TMS-MSA. That is, stage 4 of the continuous/discrete TMS-MSA is referred to as the PGIS, due to a lack of thorough modeling of the interactive relationships among the different players in the group. In the proposed TMS-EMSA, however, the update process of the feasible ranges of continuous decision-making variables is concurrently considered with the possibility of learning and imitating each player from the best existing player through interactive relationships among the different players. For this reason, stage 4 of the TMS-EMSA is referred to as the GIS. In sub-stages 3.1 and 4.1, an enhanced alternative improvisation procedure (EAIP) is developed to improvise a new melody— $\mathrm{x}_{m,p}^{\text{new}} = \left(x_{m,p,1}^{\text{new}}, \ldots, x_{m,p,v}^{\text{new}}, \ldots, x_{m,p,\text{NDV}}^{\text{new}}\right)$ —from player p, which is extensively described later in this chapter.

Sub-stages 3.2 and 4.2 relate to the update process of all PMs or MM. In sub-stages 3.2 and 4.2 of the continuous/discrete TMS-MSA, the update process of all PMs or MM is done on the basis of the pseudocode presented in Table 4.4. Due to the changes applied to the parameters of the TMS-EMSA in sub-stage 2.1, the update process of all PMs or MM must be restructured. As a result, Table 4.13 illustrates the pseudocode associated with the update of the memory of all existing players in the music group or the update of the MM_m in the proposed TMS-EMSA. In sub-stages 3.2 and 4.2, the update process of the memory of an existing player

Table 4.13 Pseudocode associated with the update of the memory of all existing players in the music group or the update of the MM_m in the proposed TMS-EMSA

Algorithm 11: Pseudocode for the update of the memory of all existing players in the music group or the update of the MM_m in the proposed TMS-EMSA
Input: Not updated MM_m, $\mathrm{x}_{m,p}^{\mathrm{new}}$
Output: Updated MM_m
Start main body
begin
for *music player p* $[p \in \Psi^{\mathrm{PN}}]$ **do**
set $\mathrm{x}_p^{\mathrm{worst}} = \mathrm{x}_{m,p}^{PMS_p}$
set $f\left(\mathrm{x}_p^{\mathrm{worst}}\right) = f\left(\mathrm{x}_{m,p}^{PMS_p}\right)$
if $f\left(\mathrm{x}_{m,p}^{\mathrm{new}}\right) < f\left(\mathrm{x}_p^{\mathrm{worst}}\right)$ **then**
$\mathrm{x}_{m,p}^{\mathrm{new}} \in PM_{m,p}$
$\mathrm{x}_p^{\mathrm{worst}} \notin PM_{m,p}$
end if
end for
terminate
end main body

in the group is not accomplished if the new melody vector played by a corresponding player is not notably better than the worst available melody vector in its memory, from the perspective of the objective function.

After completion of this process, melody vectors stored in the memory of all existing players in the musical group or the MM_m must be re-sorted based on the value of objective function—fitness function—in an ascending order. In the continuous/discrete TMS-MSA, the pseudocode pertaining to sorting the solution vectors stored in the memory of all existing players in the musical group or the MM_m was formerly presented in Table 4.5. Due to the changes applied to the parameters of the TMS-EMSA in sub-stage 2.1, this pseudocode must be restructured according to Table 4.14. Sub-stages 3.3 and 4.4 are associated with the check process of the stopping criterion of the SIS and GIS. These sub-stages of the TMS-EMSA are virtually the same as the corresponding sub-stages of the continuous/discrete TMS-MSA, which are previously described in Sects. 4.2.3.3 and 4.2.4.4 of this chapter, respectively.

Only in sub-stage 4.4 must the MNI-PGIS parameter be replaced with the MNI-GIS parameter.

Sub-stage 4.3 is related to the update of the feasible range of pitches—continuous decision-making variables for the next improvisation—only for random selection. This sub-stage of the TMS-EMSA is similar to the corresponding sub-stage of the continuous/discrete TMS-MSA, previously described in Sect. 4.2.4.3 of this chapter. Only in sub-stage 4.3 must the MNI-PGIS parameter be replaced with the MNI-GIS

Table 4.14 Pseudocode related to the sorting of the solution vectors stored in the PMs or MM_m in the proposed TMS-EMSA

Algorithm 12: Pseudocode for sorting the solution vectors stored in the PMs or MM_m in the proposed TMS-EMSA

Input: Unsorted MM_m
Output: Sorted MM_m

start main body

begin
for *music player p* $[p \in \Psi^{\text{PN}}]$ **do**
  set $PM_{m,p} = MM_m(1 : PMS_p, [(p-1) \cdot (\text{NDV}+1)] : [p \cdot (\text{NDV}+1)])$
  $F^{\text{sort}}_{m,p} = \text{sort}\left(PM_{m,p}\left(1 : PMS_p, (\text{NDV}+1)\right), '\text{ascend}'\right)$
  for *melody vector s* $\left[s \in \Psi^{PMS_p}\right]$ **do**
    for *melody vector* s^* $\left[s^* \in \Psi^{PMS_p}\right]$ **do**
      if $F^{\text{sort}}_{m,p}(s) == PM_{m,p}(s^*, (\text{NCDV}+1))$ **then**
        $PM^{\text{sort}}_{m,p}(s, 1 : (\text{NDV}+1)) = PM_{m,p}(s^*, 1 : (\text{NDV}+1))$
      end if
    end for
  end for
  $MM^{\text{sort}}_m\left(1 : PMS_p, 1 + [(p-1) \cdot (\text{NDV}+1)] : [p \cdot (\text{NDV}+1)]\right) = PM^{\text{sort}}_{m,p}$
end for
$MM_m = MM^{\text{sort}}_m$
terminate

end main body

parameter. Stage 5 is related to the selection of the final optimal solution, or the best melody. This stage of the TMS-EMSA is similar to the corresponding stage of the continuous/discrete TMS-MSA, previously described in Sect. 4.2.5 of this chapter.

As previously mentioned in sub-stages 3.1 and 4.1 of the TMS-EMSA, player p in the musical group improvises a new melody vector— $\mathrm{x}^{\text{new}}_{m,p} = \left(x^{\text{new}}_{m,p,1}, \ldots, x^{\text{new}}_{m,p,v}, \ldots, x^{\text{new}}_{m,p,\text{NDV}}\right)$—by using the proposed EAIP.

Employing the proposed EAIP by player p is performed on the basis of three rules: (1) player memory consideration; (2) pitch adjustment; and, (3) random selection.

Rule 1: As previously mentioned, in the player memory consideration rule of the continuous/discrete AIP, the values of a new melody vector played by player p are randomly selected from melody vectors stored in the $PM_{m,p}$ with the probability of the PMCR. In this rule of the continuous/discrete AIP, two principles are alternately employed. Applying of the first and second principles of the player memory consideration rule in the continuous/discrete AIP in order to determine the value of the continuous or discrete decision-making variable v from a new melody vector played by player p, $x_{m,p,v}^{\text{new}}$, is carried out using Eqs. (4.7) and (4.8) and Eqs. (4.9) and (4.10) from Sect. 4.2.6 of this chapter, respectively.

In the player memory consideration rule of the proposed EAIP, however, the values of a new melody vector played by player p are randomly selected from melody vectors stored in the $PM_{m,p}$ with the probability of the $PMCR_p$. In this rule, two principles are alternately used. In the presence of the continuous decision-making variables, each principle consists of a linear combination of a continuous decision-making variable discrete from the $PM_{m,p}$ and a ratio of the $BW_{m,p}$. Conversely, each principle consists of a discrete decision-making variable chosen from the $PM_{m,p}$ in the presence of the discrete decision-making variables. As a consequence, to specify the value of the continuous or discrete decision-making variable v from a new melody vector played by player p, $x_{m,p,v}^{\text{new}}$, the first principle of the player memory consideration rule in the proposed EAIP is rewritten and accomplished in accordance with Eqs. (4.25) and (4.26), because of the changes applied to the parameters of the proposed TMS-EMSA in sub-stage 2.1:

$$x_{m,p,v}^{\text{new}} = x_{m,p,v}^{r_p} \pm \text{U}(0,1) \cdot BW_{m,p};$$
$$\forall\left\{m \in \Psi^{(\text{MNI-SIS})+(\text{MNI-GIS})}, p \in \Psi^{\text{PN}}, v \in \Psi^{\text{NCDV}}, r_p \sim \text{U}\{1,2,\ldots,PMS_p\}\right\} \tag{4.25}$$

$$x_{p,v}^{\text{new}} = x_{m,p,v}^{r_p}; \quad \forall\left\{m \in \Psi^{(\text{MNI-SIS})+(\text{MNI-GIS})}, p \in \Psi^{\text{PN}}, v \in \Psi^{\text{NDDV}}, r_p \sim \text{U}\{1,2,\ldots,PMS_p\}\right\} \tag{4.26}$$

Detailed descriptions related to Eqs. (4.25) and (4.26) are virtually the same as the detailed explanations pertaining to Eqs. (4.7) and (4.8), previously reported in Sect. 4.2.6 of this chapter. By the same token, to determine the value of the continuous or discrete decision-making variable v from a new melody vector played by player p, $x_{m,p,v}^{\text{new}}$, the second principle of the player memory consideration rule in the proposed EAIP is rewritten and performed according to Eqs. (4.27) and (4.28):

$$x_{m,p,v}^{\text{new}} = x_{m,p,k}^{r_p} \pm \text{U}(0,1) \cdot BW_{m,p}; \quad \forall\left\{m \in \Psi^{(\text{MNI-SIS})+(\text{MNI-GIS})}, p \in \Psi^{\text{PN}}, v \in \Psi^{\text{NCDV}}\right\},$$
$$\forall\{r_p \sim \text{U}\{1,2,\ldots,PMS_p\}, k \sim \text{U}\{1,2,\ldots,\text{NCDV}\}\} \tag{4.27}$$

$$x_{m,p,v}^{\text{new}} = x_{m,p,l}^{r_p}; \quad \forall\left\{m \in \Psi^{(\text{MNI-SIS})+(\text{MNI-GIS})}, p \in \Psi^{\text{PN}}, v \in \Psi^{\text{NDDV}}\right\},$$
$$\forall\{r_p \sim \text{U}\{1,2,\ldots,PMS_p\}, l \sim \text{U}\{1,2,\ldots,\text{NDDV}\}\} \tag{4.28}$$

Detailed descriptions associated with Eqs. (4.27) and (4.28) are virtually the same as those corresponding to Eqs. (4.9) and (4.10), represented in Sect. 4.2.6 of this chapter.

In Eqs. (4.25)–(4.28), index r related to player p in the musical group—r_p—is a random integer with uniform distribution through the set of $\{1, 2, \ldots, PMS_p\}$—$r_p \sim \mathrm{U}\{1, 2, \ldots, PMS_p\}$. Stated another way, index r is randomly characterized through the set of the permissible values given by using the set of $\{1, 2, \ldots, PMS_p\}$. Determination of this index is described according to Eq. (4.29):

$$r_p = \mathrm{int}\big(\mathrm{U}(0, 1) \cdot PMS_p\big) + 1; \quad \forall\{p \in \Psi^{\mathrm{PN}}\} \tag{4.29}$$

In addition, detailed descriptions of the indices k and l were already presented in Sect. 4.2.6 of this chapter. Determination of these indices was previously indicated according to Eqs. (4.12) and (4.13), given in Sect. 4.2.6 of this chapter, respectively. In the proposed EAIP, applying of the player memory consideration rule is also done for other players in the same way.

Rule 2: One of the fundamental differences between the continuous/discrete AIP and the proposed EAIP appears in the pitch adjustment rule. In the pitch adjustment rule of the continuous/discrete AIP, the values of a new melody vector played by player p, haphazardly chosen from the available melody vectors in the $PM_{m,p}$ with the probability of the PMCR, are updated with the probability of the PAR_m. The pitch adjustment rule is organized based on a single principle in the continuous/discrete AIP. Implementing of the pitch adjustment rule to specify the value of the continuous or discrete decision-making variable v from a new melody vector played by player p, $x^{\mathrm{new}}_{m,p,v}$, is carried out using Eqs. (4.14) and (4.15), given in Sect. 4.2.6 of this chapter, respectively. These equations indicate that, by the activation of the pitch adjustment rule in the continuous/discrete AIP, the value of the continuous or discrete decision-making variable v from a new melody vector played by player p, $x^{\mathrm{new}}_{m,p,v}$, randomly selected from the melody vectors stored in the $PM_{m,p}$ with the probability of the PMCR, is updated with the probability of the PAR_m. The update process for this continuous or discrete decision-making variable is accomplished by substituting it with the value of the continuous or discrete decision-making variable v from the best available melody vector in the $PM_{m,p}$—$x^{\mathrm{best}}_{m,p,v}$. Simply put, in the pitch adjustment rule of the continuous/discrete AIP, player p tries to emulate the best melody stored in its own memory.

In the pitch adjustment rule of the proposed EAIP, however, the values of a new melody vector played by player p, haphazardly chosen from the available melody vectors in the $PM_{m,p}$ with the probability of the $PMCR_p$, are updated with the probability of the $PAR_{m,p}$. The pitch adjustment rule is established according to two principles in the EAIP. The first and second principles of the pitch adjustment rule in the EAIP are activated in the SIS and GIS, respectively. If the first principle is activated, the value of the continuous or discrete decision-making variable v from a new melody vector played by player p, $x^{\mathrm{new}}_{m,p,v}$, randomly selected from the melody vectors stored in the $PM_{m,p}$ with the probability of the $PMCR_p$, is updated with the probability of the $PAR_{m,p}$. The update process for this continuous or discrete decision-making variable is performed by replacing it with the value of the continuous or discrete decision-making variable v from the best available melody vector in the $PM_{m,p}$—$x^{\text{best}}_{m,p,v}$. As a result, applying of the first principle of the pitch adjustment rule in the EAIP to determine the value of the continuous or discrete decision-making variable v from a new melody vector played by player p, $x^{\text{new}}_{m,p,v}$, is done by using Eqs. (4.30) and (4.31), respectively:

$$x^{\text{new}}_{m,p,v} = x^{\text{best}}_{m,p,v}; \quad \forall \left\{ m \in \Psi^{(\text{MNI-SIS})+(\text{MNI-GIS})}, p \in \Psi^{\text{PN}}, v \in \Psi^{\text{NCDV}} \right\} \tag{4.30}$$

$$x^{\text{new}}_{m,p,v} = x^{\text{best}}_{m,p,v}; \quad \forall \left\{ m \in \Psi^{(\text{MNI-SIS})+(\text{MNI-GIS})}, p \in \Psi^{\text{PN}}, v \in \Psi^{\text{NDDV}} \right\} \tag{4.31}$$

Detailed descriptions related to Eqs. (4.30) and (4.31) are virtually the same as the detailed explanations relevant to Eqs. (4.14) and (4.15), previously reported in Sect. 4.2.6 of this chapter. Equations (4.30) and (4.31) state that in the first principle of the pitch adjustment rule in the EAIP, player p tries to emulate the best melody stored in its own memory. If the first principle of the pitch adjustment rule in the EAIP is deactivated, or in other words the second principle is activated, the continuous or discrete decision-making variable v from a new melody vector played by player p, $x^{\text{new}}_{m,p,v}$, haphazardly chosen from the melody vectors stored in the $PM_{m,p}$ with the probability of the $PMCR_p$, is updated with the probability of the $PAR_{m,p}$. The update process for this continuous or discrete decision-making variable is carried out by substituting it with the value of the continuous or discrete decision-making variable v from the best available melody vector in the MM_m—$x^{\text{best}}_{m,\text{best},v}$. As a consequence, implementing of the second principle of the pitch adjustment rule in the EAIP to characterize the value of the continuous or discrete decision-making variable v from a new melody vector played by player p, $x^{\text{new}}_{m,p,v}$, is accomplished by using Eqs. (4.32) and (4.33), respectively:

$$x^{\text{new}}_{m,p,v} = x^{\text{best}}_{m,\text{best},v}; \quad \forall \left\{ m \in \Psi^{(\text{MNI-SIS})+(\text{MNI-GIS})}, p \in \Psi^{\text{PN}}, v \in \Psi^{\text{NCDV}} \right\} \tag{4.32}$$

$$x^{\text{new}}_{m,p,v} = x^{\text{best}}_{m,\text{best},v}; \quad \forall \left\{ m \in \Psi^{(\text{MNI-SIS})+(\text{MNI-GIS})}, p \in \Psi^{\text{PN}}, v \in \Psi^{\text{NDDV}} \right\} \tag{4.33}$$

Equations (4.32) and (4.33) state that in the second principle of the pitch adjustment rule in the EAIP, player p tries to emulate the best melody stored in the memory of the best existing player in the musical group. In the proposed EAIP, applying the pitch adjustment rule is also performed for other players in the same way.

Rule 3: In the random selection rule of the continuous/discrete AIP, the values of a new melody vector played by player p are randomly selected from the entire nonempty feasible decision-making space with the probability of the 1-PMCR. In this rule, two principles are employed. The first and second principles are activated in the SIS and PGIS, respectively. Applying of the first and second principles of the random selection rule in the continuous/discrete AIP to determine the value of the continuous decision-making variable v from a new melody vector played by player p, $x^{\text{new}}_{m,p,v}$, is accomplished by using Eqs. (4.16) and (4.17), given in Sect. 4.2.6 of this chapter, respectively. Also, implementing of the first and second principles of the random selection rule in the continuous/discrete AIP to determine the value of the discrete decision-making variable v from a new melody vector played by player p, $x^{\text{new}}_{m,p,v}$, is performed by using Eq. (4.18), given in Sect. 4.2.6 of this chapter. In the random selection rule of the proposed EAIP, however, the values of a new melody vector played by player p in the musical group are haphazardly chosen from the entire nonempty feasible decision-making space with the probability of the $1 - PMCR_p$. In this rule, two principles are used. The first and second principles are activated in the SIS and GIS, respectively. The application of the first and second principles of the random selection rule in the proposed EAIP to specify the value of the continuous decision-making variable v from a new melody vector played by player p, $x^{\text{new}}_{m,p,v}$, is rewritten and accomplished in accordance with Eqs. (4.34) and (4.35), respectively, because of the changes applied on the parameters of the proposed TMS-EMSA in sub-stage 2.1:

$$x^{\text{new}}_{m,p,v} = x^{\min}_{v} + \mathrm{U}(0,1) \cdot \left(x^{\max}_{v} - x^{\min}_{v}\right); \quad \forall\left\{m \in \Psi^{\text{MNI-SIS}}, p \in \Psi^{\text{PN}}, v \in \Psi^{\text{NCDV}}\right\} \tag{4.34}$$

$$x^{\text{new}}_{m,p,v} = x^{\min}_{m,v} + \mathrm{U}(0,1) \cdot \left(x^{\max}_{m,v} - x^{\min}_{m,v}\right); \quad \forall\left\{m \in \Psi^{\text{MNI-GIS}}, p \in \Psi^{\text{PN}}, v \in \Psi^{\text{NCDV}}\right\} \tag{4.35}$$

Detailed descriptions pertaining to Eqs. (4.34) and (4.35) are virtually the same as those corresponding to Eqs. (4.16) and (4.17), represented in Sect. 4.2.6 of this chapter, respectively. By the same token, to determine the value of the discrete decision-making variable v from a new melody vector played by player p, $x^{\text{new}}_{m,p,v}$, the first and second principles of the random selection rule in the proposed EAIP are rewritten and performed according to Eq. (4.36):

$$x^{\text{new}}_{m,p,v} = x_v(y); \quad \forall\left\{m \in \Psi^{(\text{MNI-SIS})+(\text{MNI-GIS})}, p \in \Psi^{\text{PN}}, v \in \Psi^{\text{NDDV}}, y \sim \mathrm{U}\{x_v(1), \ldots, x_v(w_v), \ldots, x_v(W_v)\}\right\} \tag{4.36}$$

Detailed descriptions pertaining to Eq. (4.36) are virtually the same as those corresponding to Eq. (4.18), represented in Sect. 4.2.6 of this chapter. In the proposed EAIP, implementing of the random selection rule is also done for other players in the same way.

In the proposed continuous/discrete TMS-MSA, each player in the musical group employs the updated values of the BW_m and PAR_m parameters in the improvisation process of its melody vector. The update process of the BW_m and PAR_m parameters in each improvisation/iteration of the SIS and PGIS of the continuous/discrete TMS-MSA is defined according to Eqs. (4.19) and (4.20), given in Sect. 4.2.6 of this chapter, respectively. In the TMS-EMSA, however, in view of the changes applied in the parameters under sub-stage 2.1, the update process of the $BW_{m,p}$ and $PAR_{m,p}$ parameters in each improvisation/iteration of the SIS and GIS must be rewritten using Eqs. (4.37) and (4.38), as follows:

$$BW_{m,p} = BW_p^{\max} \cdot \exp\left(\frac{\ln\left(BW_p^{\max}/BW_p^{\min}\right)}{(\text{MNI-SIS}) + (\text{MNI-GIS})} \cdot m\right); \quad \forall\left\{m \in \Psi^{(\text{MNI-SIS})+(\text{MNI-GIS})}\right\} \tag{4.37}$$

$$PAR_{m,p} = PAR_p^{\min} + \left(\frac{PAR_p^{\max} - PAR_p^{\min}}{(\text{MNI-SIS}) + (\text{MNI-GIS})}\right) \cdot m; \quad \forall\left\{m \in \Psi^{(\text{MNI-SIS})+(\text{MNI-GIS})}\right\} \tag{4.38}$$

Equations (4.37) and (4.38) tell us that in the TMS-EMSA each player in the musical group employs the updated values of its $BW_{m,p}$ and $PAR_{m,p}$ parameters in the improvisation process of its melody vector. Stated another way, in the continuous/discrete TMS-MSA, all existing players in the music group employ the same updated BW_m and PAR_m parameters in the improvisation process of their melody vectors, while in the TMS-EMSA each player in the musical group employs its own unique updated $BW_{m,p}$ and $PAR_{m,p}$ parameters in the improvisation process of its melody vector. Table 4.15 presents the pseudocode associated with the improvisation of a new melody vector by each player in the musical group of the proposed TMS-EMSA. The designed pseudocode in different stages and sub-stages of the TMS-EMSA is located in a regular sequence and forms the performance-driven architecture of this algorithm.

Table 4.16 illustrates the pseudocode relevant to the performance-driven architecture of the proposed TMS-EMSA. As a general result, the changes applied in different stages and sub-stages of the TMS-EMSA give rise to more realistically and appreciably improved performance, flexibility, and robustness, in comparison to the continuous/discrete TMS-MSA. That is to say that the changes applied in the TMS-EMSA and, consequently, its new architecture can bring about well-suited characteristics for dealing with the complicated, real-world, large-scale, non-convex, non-smooth optimization problems: (1) increasing the diversity of melody vectors stored in the memory of each existing player in the musical group; (2) increasing the probability of achieving a more appropriate sequence of pitches in the melody vectors played by each existing player in the musical group; (3) increasing the convergence rate of algorithms to attain the final optimal solution; etc.

Table 4.15 Pseudocode associated with the improvisation of a new melody vector by each player in the musical group of the proposed TMS-EMSA

Algorithm 13: Pseudocode for improvisation of a new melody vector by each player in the musical group of the proposed TMS-EMSA

Input: $BW_p^{\max}$, $BW_p^{\min}$, MNI-SIS, MNI-GIS, NCDV, NDDV, NDV, $PAR_p^{\max}$, $PAR_p^{\min}$, $PMCR_p$, PMS_p, PN, $x_v^{\min}$, $x_v^{\max}$, $\{x_v(1), \ldots, x_v(w_v), \ldots, x_v(W_v)\}$

Output: $\mathrm{x}_{m,p}^{\text{new}}$

start main body

begin

$BW_{m,p} = BW_p^{\max} \cdot \exp\left[\left(\ln\left(BW_p^{\max}/BW_p^{\min}\right)/((\text{MNI-SIS}) + (\text{MNI-GIS}))\right) \cdot m\right]$

$PAR_{m,p} = PAR_p^{\min} - \left[\left(\left(PAR_p^{\max} - PAR_p^{\min}\right)/((\text{MNI-SIS}) + (\text{MNI-GIS}))\right) \cdot m\right]$

for *music player p* $[p \in \Psi^{\text{PN}}]$ **do**

construct the new melody vector for music player p, $\mathrm{x}_{m,p}^{\text{new}}$, with dimension $\{1\} \cdot \{\text{NDV} + 1\}$ and zero initial value

for *decision-making variable v* $[v \in \Psi^{\text{NDV}}]$ **do**

if $\mathrm{U}(0,1) \leq PMCR_p$ **then**

Rule 1: player memory consideration with probability $PMCR_p$

if *improvisation/iteration m* $[m \in \Psi^{(\text{MNI-SIS}) + (\text{MNI-GIS})}]$ is odd **then**

Principle 1 of Rule 1: first combination

$x_{m,p,v}^{\text{new}} = x_{m,p,v}^{r_p} \pm \mathrm{U}(0,1) \cdot BW_{m,p}; \forall r_p \sim \mathrm{U}\{1, 2, \ldots, PMS_p\}$; for CDVs

$x_{m,p,v}^{\text{new}} = x_{m,p,v}^{r_p}; \forall r_p \sim \mathrm{U}\{1, 2, \ldots, PMS_p\}$; for DDVs

else

Principle 2 of Rule 1: second combination

$x_{m,p,v}^{\text{new}} = x_{m,p,k}^{r_p} \pm \mathrm{U}(0,1) \cdot BW_{m,p}; \forall r_p \sim \mathrm{U}\{1, 2, \ldots, PMS_p\}, \forall k \sim \mathrm{U}\{1, 2, \ldots, \text{NCDV}\}$; for CDVs

$x_{m,p,v}^{\text{new}} = x_{m,p,l}^{r_p}; \forall r_p \sim \mathrm{U}\{1, 2, \ldots, PMS_p\}, \forall l \sim \mathrm{U}\{1, 2, \ldots, \text{NDDV}\}$; for DDVs

end if

if $\mathrm{U}(0,1) \leq PAR_{m,p}$ **then**

Rule 2: pitch adjustment with probability $PMCR_p \cdot PAR_{m,p}$

switch 1

case *improvisation/iteration m* $[m \in \Psi^{(\text{MNI-SIS}) + (\text{MNI-GIS})}] \leq$ (MNI-SIS) **then**

Principle 1 of Rule 2: first choice

$x_{m,p,v}^{\text{new}} = x_{m,p,v}^{\text{best}}$; for CDVs and DDVs

case *improvisation/iteration m* $[m \in \Psi^{(\text{MNI-SIS}) + (\text{MNI-GIS})}] >$ (MNI-SIS) and *improvisation/iteration m* $[m \in \Psi^{(\text{MNI-SIS}) + (\text{MNI-GIS})}] \leq$ (MNI-SIS) + (MNI-GIS) **then**

Principle 2 of Rule 2: second choice

$x_{m,p,v}^{\text{new}} = x_{m,\text{best},v}^{\text{best}}$; for CDVs and DDVs

end switch

end if

else if

Rule 3: random selection with probability $1 - PMCR_p$

switch 1

case *improvisation/iteration m* $[m \in \Psi^{(\text{MNI-SIS}) + (\text{MNI-GIS})}] \leq$ (MNI-SIS) **then**

(continued)

Table 4.15 (continued)

Principle 1 of Rule 2: first choice

$x_{m,p,v}^{\text{new}} = x_v^{\min} + \text{U}(0,1) \cdot \left(x_v^{\max} - x_v^{\min}\right)$; for CDVs

$x_{m,p,v}^{\text{new}} = x_v(y); \forall y \sim \text{U}\{x_v(1), \ldots, x_v(w_v), \ldots, x_v(W_v)\}$; for DDVs

case *improvisation/iteration m* $[m \in \Psi^{(\text{MNI-SIS}) + (\text{MNI-GIS})}] >$ (MNI-SIS) and *improvisation/iteration m* $[m \in \Psi^{(\text{MNI-SIS}) + (\text{MNI-GIS})}] \leq$ (MNI-SIS) + (MNI-GIS) **then**

Principle 2 of Rule 2: second choice

$x_{m,p,v}^{\text{new}} = x_{m,v}^{\min} + \text{U}(0,1) \cdot \left(x_{m,v}^{\max} - x_{m,v}^{\min}\right)$; for CDVs

$x_{m,p,v}^{\text{new}} = x_v(y); \forall y \sim \text{U}\{x_v(1), \ldots, x_v(w_v), \ldots, x_v(W_v)\}$; for DDVs

end switch

end if

end for

calculate the value of objective function, fitness function, derived from melody vector $\text{x}_{m,p}^{\text{new}}$ as $f\left(\text{x}_{m,p}^{\text{new}}\right)$

allocate $f\left(\text{x}_{m,p}^{\text{new}}\right)$ to element (1, NDV + 1) of the new melody vector $\text{x}_{m,p}^{\text{new}}$

end for

terminate

end main body

Table 4.16 Pseudocode relevant to the performance-driven architecture of the proposed TMS-EMSA

Algorithm 14: Pseudocode for performance-driven architecture of the proposed TMS-EMSA

Input: $BW_p^{\max}$, $BW_p^{\min}$, MNI-SIS, MNI-GIS, NCDV, NDDV, NDV, $PAR_p^{\max}$, $PAR_p^{\min}$, $PMCR_p$, PMS_p, PN, $x_v^{\min}$, $x_v^{\max}$, $\{x_v(1), \ldots, x_v(w_v), \ldots, x_v(W_v)\}$

Output: x^{best}

start main body

begin

Stage 1—Definition stage: Definition of the optimization problem and its parameters

Stage 2—Initialization stage

Sub-stage 2.1: Initialization of the parameters of the TMS-EMSA

Sub-stage 2.2: Initialization of the of the MM

Algorithm 9: Pseudocode for initialization of the entire set of PMs or MM in the proposed TMS-EMSA

Algorithm 10: Pseudocode for sorting of the solution vectors stored in the PMs or MM in the proposed TMS-EMSA

Stage 3—Single computational stage or SIS

set *improvisation/iteration* $m = 1$

set MM_m = MM

while $m \leq$ (MNI-SIS) **do**

Sub-stage 3.1: Improvisation of a new melody vector by each player

(continued)

Table 4.16 (continued)

Algorithm 13: Pseudocode for improvisation of a new melody vector by each player in the musical group of the proposed TMS-EMSA
Sub-stage 3.2: Update each PM
Algorithm 11: Pseudocode for the update of the memory of all existing players in the music group or the update of the MM_m in the proposed TMS-EMSA
Algorithm 12: Pseudocode for sorting of the solution vectors stored in the PMs or MM_m in the proposed TMS-EMSA
set *improvisation/iteration* $m = m + 1$
end while
Stage 4—Group computational stage or GIS
while $m >$ (MNI-SIS) and $m \leq$ (MNI-SIS) + (MNI-GIS) **do**
Sub-stage 4.1: Improvisation of a new melody vector by each player taking into account the feasible ranges of the updated pitches
Algorithm 13: Pseudocode for improvisation of a new melody vector by each player in the musical group of the proposed TMS-EMSA
Sub-stage 4.2: Update of each PM
Algorithm 11: Pseudocode for the update of the memory of all existing players in the music group or the update of the MM_m in the proposed TMS-EMSA
Algorithm 12: Pseudocode for sorting the solution vectors stored in the PMs or MM_m in the proposed TMS-EMSA
Sub-stage 4.3: Update of the feasible ranges of pitches—continuous decision-making variables—for the next improvisation—only for random selection
Algorithm 5: Pseudocode for the update of the feasible ranges of the continuous decision-making variables in the proposed TMS-EMSA; while the MNI-PGIS parameter is replaced with the MNI-GIS
set *improvisation/iteration* $m = m + 1$
end while
Stage 5—Selection stage: Selection of the final optimal solution—the best melody
Algorithm 6: Pseudocode for the selection of the final optimal solution in the proposed TMS-EMSA
terminate
end main body

4.4 Multi-stage Computational Multi-dimensional Multiple-Homogeneous Enhanced Melody Search Algorithm: Symphony Orchestra Search Algorithm

Technically speaking, the original TMS-MSA, the proposed continuous/discrete TMS-MSA, and the TMS-EMSA all have a two-stage computational, multi-dimensional and single-homogeneous structure, as mentioned in the preceding sections. This structure causes these optimization techniques to maintain their desirable performance to an acceptable degree in solving complicated, real-world, large-scale, non-convex, non-smooth optimization problems having a nonlinear, mixed-integer nature with big data.

As previously mentioned, though, modern engineering challenges with multilevel dimensions in different branches of the engineering sciences, particularly electrical engineering, have been extensively developed. With that in mind, these modern challenges cannot be addressed in the form of conventional optimization problems—single-level optimization problems. Modern engineering challenges with multilevel dimensions, due to their specific characteristics, must be raised in the form of new, unconventional optimization problems—multilevel optimization problems. Multilevel optimization problems, unlike their single-level counterparts, are much more complex. Basically, the different levels of multilevel optimization problems may be involved with the multiple objective functions and constraints, and may be solved from the different perspectives of the specialists and researches, and also may be aligned together and follow a single goal.

The structure of multilevel optimization problems is in such a way that the information flow among the different levels is frequently repeated until a convergence is achieved in the results. More precisely, in multilevel optimization problems, the output information arising from the implementation of the optimization process in one or more levels is considered to be input information for the implementation of the optimization process in one or more other levels, and vice versa. As a result, unlike the single-level optimization problems, the amount of information involved in the optimization process of multilevel optimization problems is dramatically increased.

The point to be made here is that an instance of the two-level optimization problems will be thoroughly presented in Chap. 5 and entitled a computational-logical framework for a bilateral bidding mechanism within a competitive electricity market. Two examples of three-level optimization problems will also be extensively addressed in Sects. 6.3 and 6.4 of Chap. 6 and entitled a strategic tri-level computational-logical framework for the pseudo-dynamic generation expansion planning and a strategic tri-level computational-logical framework for the pseudo-dynamic transmission expansion planning, respectively. In addition, an instance of four-level optimization problems will be exhaustively described in Sect. 6.5 of Chap. 6 and entitled a strategic quad-level computational-logical framework for the pseudo-dynamic coordinated generation and transmission expansion planning.

Remarkable increases in the information involved in the optimization process in the different levels of the multilevel optimization problem and the interdependency of these different levels can give rise to the performance of the optimization algorithms; even the original TMS-MSA, the proposed continuous/discrete TMS-MSA, and the TMS-EMSA are influenced in the optimization process; and consequently, these algorithms cannot maintain their favorable performance to an acceptable degree.

In this section, then, the authors develop an innovative architectural version of the proposed TMS-EMSA, which will be referred to as either MMM-EMSA, MMS-EMSA, or SOSA, to more efficiently and effectively overcome the disadvantages of the optimization algorithms in the optimization process of the multilevel optimization problems. The newly developed SOSA was inspired by the phenomena and concepts of music, available principles in a symphony orchestra structure, and fundamental concepts of the TMS-EMSA. Nonetheless, the proposed SOSA differs from the perspective of architecture with other meta-heuristic optimization algorithms, because of its multi-stage computational multi-dimensional and multi-homogeneous structure.

The word "symphony," initially derived from the ancient Greek word "Συμφωνία," referred to the concord of the sound. This word originally evolved as two parts: (1) syn- (συν), meaning with/together, and, (2) phone (φωνή), meaning sound/sounding. In ancient Greece, the symphony was employed in order to represent the general concept of concord, not only between consecutive sounds but also in the unison of concurrent sounds. Moreover, the word orchestra, initially derived from the ancient Greek word όρχήστρα, referred to an area in the front of an ancient Greek stage reserved for the Greek vocalists. In the music literature, the orchestra refers to a great musical ensemble in which there are different musical instruments from various families and tries to perform specific musical pieces. The musical instruments employed in an orchestra can be generally classified into four main families with regard to the technique of sound production by the instruments: (1) string instruments; (2) woodwind instruments; (3) brass instruments; and, (4) percussion instruments. The string instruments are the largest family of the musical instruments in the orchestra and include two groups. The first group of the string instruments consists of four sizes. From smallest to largest they are (1) violin; (2) viola; (3) cello; and, (4) double-bass—contrabass. The second group, which has structural differences with the first group, includes only the harp. The woodwind instruments are another family of musical instruments in the orchestra and include four groups. The first group of the woodwind instruments includes (1) flute and (2) piccolo. The oboe and English horn make up the second group of woodwind instruments. The third group of woodwind instruments contains (1) clarinet; (2) e-flat clarinet; and, (3) bass clarinet. Eventually, the bassoon and contrabassoon were organized into a fourth group of woodwind instruments. Another family of musical instruments in the orchestra is the brass instruments. These musical instruments form four groups in such a way that each group has only one instrument. The musical instruments of groups one to four are respectively, (1) trumpet; (2) French horn; (3) trombone; and, (4) tuba. The percussion instruments represent yet another family of instruments and contain two groups. The first group of percussion instruments consists only of a piano. The second group, however, includes other musical instruments: (1) timpani; (2) xylophone; (3) cymbals; (4) triangle; (5) snare drum; (6) bass drum; (7) tambourine; (8) maracas; (9) gong; (10) chimes; (11) castanets; and, (12) celesta.

Technically speaking, each orchestra is named depending on the type and number of musical instruments in it. A full-size orchestra—the largest type—which includes all of the musical instrument families and almost all of the Western classical instruments, is recognized as the symphony orchestra. The number of players employed in a symphony orchestra to play a specific musical piece, depending on the size of the venue and the characteristics of the musical piece to be played, may vary from 70 to more than 100 players. The arrangement of different musical instruments in a symphony orchestra is illustrated in Fig. 4.1.

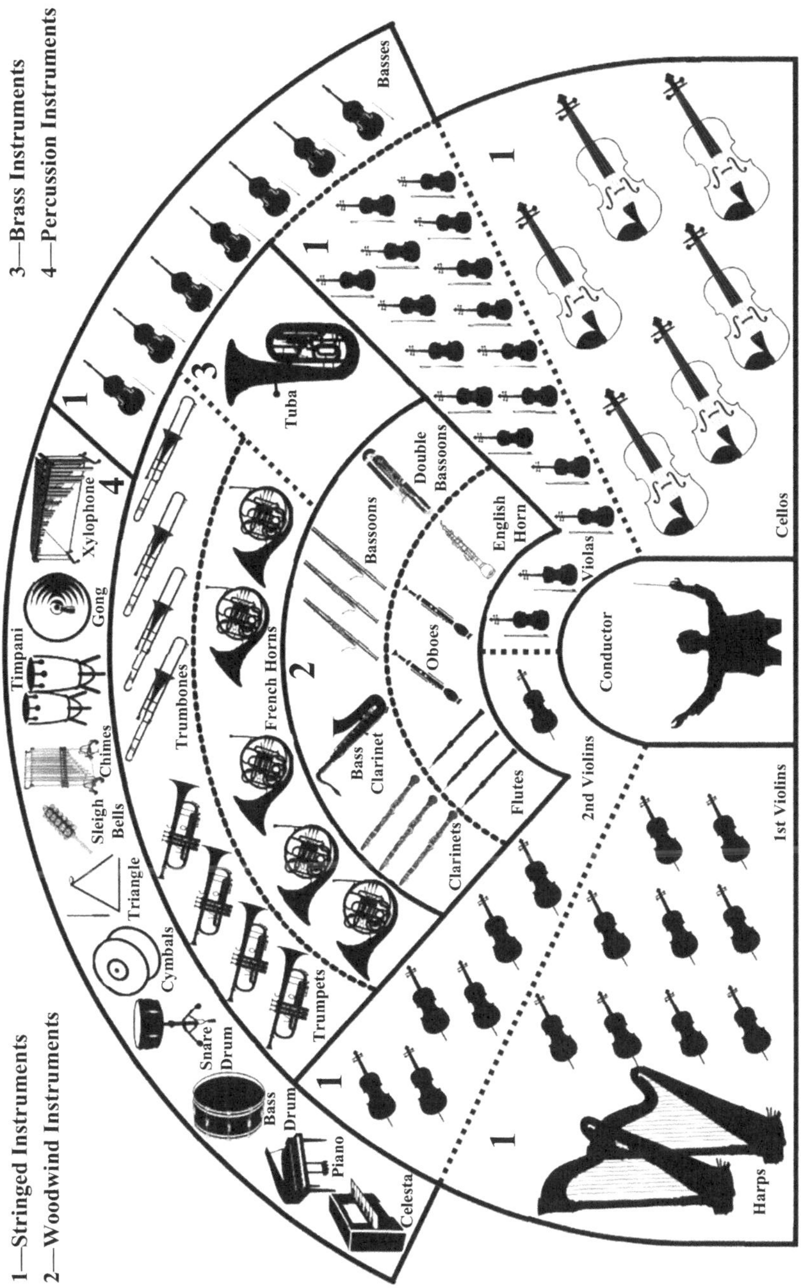

Fig. 4.1 The arrangement of different musical instruments in a symphony orchestra

The number of players in an orchestra may be about 50 players or fewer. In this circumstance, this small-size orchestra is known as a chamber orchestra. It is, therefore, obvious that the chamber orchestra is smaller than the symphony orchestra in terms of the number of players and musical instruments.

In some cases, other names are utilized for the orchestras, which do not refer to the type of the musical instruments or the role of these instruments in these orchestras (e.g., philharmonic orchestra). The structure of a philharmonic orchestra is similar to a symphony orchestra in terms of the musical instruments. This structure is used to identify different orchestras in a region or city (e.g., the London symphony orchestra or the London philharmonic orchestra). More detailed descriptions of the definitions and concepts related to the symphony orchestra are beyond the scope of this chapter, but interested readers may look to the work by Rimsky-Korsakov [3] for a thorough discussion regarding these definitions and concepts.

As previously described, the original TMS-MSA, the proposed continuous/discrete TMS-MSA, and the TMS-EMSA imitate not only the processes of music performance, but also the interactive relationships among members of a musical group, while each player in this musical group is looking for the best set of the pitches within a melody line. The architecture of these optimization algorithms has been developed in such a way that all players in this musical group must have the same musical instruments. Stated another way, the interactive relationships among members of a homogenous musical group—a musical group with similar instruments (e.g., a group of violinists)—are widely employed by these optimization algorithms. In a homogenous musical group, there are different players having the same musical instrument; however, at the same time, their different tastes, ideas, styles, and experiences can give rise to the achievement of the most desirable sequence of fastest pitches. On the other hand, a symphony orchestra refers to a great musical ensemble that includes multiple groups of players with different musical instruments from diverse families and endeavors to carry out specific musical pieces, as formerly depicted in Fig. 4.1. With that in mind, a symphony orchestra alludes to an inhomogeneous musical ensemble organized by multiple homogeneous groups of players with different musical instruments (e.g., a group of violinists, a group of harp players, a group of clarinet players). More precisely, it can be stated that a symphony orchestra—an inhomogeneous musical ensemble—consists of several homogeneous musical groups. Each homogeneous musical group includes a number of players with similar musical instruments. The proposed SOSA mimics the processes of the music performance and interactive relationships among existing players in the symphony orchestra, while each player is looking for the best set of the pitches within a melody line. This is due to the fact that the architecture of this optimization algorithm is established in accordance with the definitions and concepts related to the symphony orchestra structure.

In addition, the interactive relationships among existing players in the symphony orchestra consist of two types: (1) the interactive relationships among existing players in each existing homogeneous musical group in the symphony orchestra and (2) the interactive relationships among existing players in different homogeneous musical groups in the symphony orchestra or the inhomogeneous musical

Table 4.17 Interdependencies of phenomena and concepts of the music and the optimization problem modeled by the proposed SOSA

No.	Comparison factor	Concept of the optimization problem modeled by the SOSA	Music concept
1	Structural pattern	Decision-making variable	Pitch in a specific melody played by a particular existing player in a special homogeneous musical group of the symphony orchestra
2	Component	Value of the decision-making variable	Value of each pitch in a specific melody played by a particular existing player in a special homogeneous musical group of the symphony orchestra
3	Decision-making space	Value range of the decision-making variable	Range of each pitch in a specific melody played by a particular existing player in a special homogeneous musical group of the symphony orchestra
4	General structural pattern	Solution vector	Musical melody played by a particular existing player in a special homogeneous musical group of the symphony orchestra
5	Target	Objective function	Aesthetic standard of the audience
6	Process unit	Iteration	Time/practice invested by all existing players in the symphony orchestra
7	Memory	Solution vector matrix	Experience of all existing players in the symphony orchestra
8	Best state	Global optimum point	Best melody selected from among all melodies played by all existing players in the symphony orchestra
9	Search process	Local and global optimum searches	Improvisation of all existing players in the symphony orchestra

ensemble. In the symphony orchestra, there are different players having diverse musical instruments from various families. These different players with diverse tastes, ideas, styles, and experiences can lead to the attainment of the best sequence of fastest pitches. The interdependencies of phenomena and concepts of the music and the optimization problem modeled by the proposed SOSA are demonstrated in Table 4.17. As set out in Table 4.17, each pitch in a specific melody played by a particular existing player in a special homogeneous musical group of the symphony orchestra, the value of each pitch in a specific melody played by a particular existing player in a special homogeneous musical group of the symphony orchestra, and the range of each pitch in a specific melody played by a particular existing player in a special homogeneous musical group of the symphony orchestra are virtually the same as each decision-making variable, the value of each decision-making variable, and the range of each decision-making variable, respectively.

By the same token, the musical melody played by a particular existing player in a special homogeneous musical group of the symphony orchestra, aesthetic standard of the audience, and time and practice invested by all existing players in the symphony orchestra refer to the solution vector, objective function, and iteration, respectively. Moreover, the experience of all existing players in the symphony orchestra, the best melody selected from among all melodies played by all existing players in the symphony orchestra, and the improvisation of all existing players in the symphony orchestra are equivalent to the solution vectors matrix, global optimum point, and local and global optimum searches, respectively. By improving the musical melody played by existing players in the symphony orchestra in each practice in comparison to before practice from the perspective of the aesthetic standard of audience, the solution vector associated with the optimization problem is also enhanced in each iteration in comparison to the before iteration from the standpoint of the proximity to the optimal global point.

Although the main concepts employed in the TMS-EMSA along with the phenomena and concepts of the symphony orchestra are the original inspirational sources for the SOSA, the architecture of the proposed SOSA has fundamental differences with the architecture of the TMS-EMSA. Unlike the TMS-EMSA, which consists of the SIS and GIS, the proposed SOSA utilizes three computational stages in order to achieve an optimal response/output: (1) SIS; (2) group computational stage for each homogeneous musical group or group improvisation stage for each homogeneous musical group (GISHMG); and, (3) group computational stage for inhomogeneous musical ensemble or group improvisation stage for inhomogeneous musical ensemble (GISIME). In the SIS, each player in the symphony orchestra improvises its melody individually, without the influence of other players in the homogeneous musical group to which it belongs, and without the influence of other players in other homogeneous musical groups to which it does not belong.

In the GISHMG, the proposed SOSA has a homogeneous group performance. Stated another way, in this computational stage, each player in the symphony orchestra improvises its melody interactively only under the influence of other players in the homogeneous musical group to which it belongs. In the GISHMG, the different melodies in the memory of different players in a homogeneous musical group of the symphony orchestra not only can direct the players of corresponding homogeneous musical group to better choices of random pitches, but also can increase the probability of playing a better melody by the players of corresponding homogeneous musical groups in the next improvisation/iteration of this stage. In the GISIME, however, the proposed SOSA has an inhomogeneous ensemble performance. Put another way, in this computational stage, each player in the symphony orchestra improvises its melody interactively, both with the influence of other players in the homogeneous musical group to which it belongs and under the influence of other players in other homogeneous musical groups to which it does not belong. In the GISIME, the different melodies in the memory of different players in all of the existing homogeneous musical groups in the symphony orchestra can not only effectively handle the players of all of the homogeneous musical groups to better choose random pitches, but also efficiently increase the probability of playing a better melody by the players of all of the homogeneous musical groups in the next improvisation/iteration of this stage.

On the other hand, unlike the TMS-EMSA, which utilizes a single MM, the proposed SOSA employs multiple MMs. A symphony orchestra memory (SOM) is then organized by incorporating the many MMs. The point to be made here is that these different MMs interact with each other, similar to the performance of different homogeneous groups in a symphony orchestra. With that in mind, in the proposed SOSA, an MM indicates the memory related to a homogeneous musical group in a symphony orchestra. Similar to the TMS-EMSA, in the proposed SOSA, each MM consists of numerous PMs. As a consequence, the TMS-EMSA is called a two-stage computational multi-dimensional and single-homogeneous optimization algorithm, due to its two improvisation stages, multiple single PMs, and a homogeneous group memory. This definition can be expressed for the original TMS-MSA and the proposed continuous/discrete TMS-MSA in the same way. Similarly, in view of the fact that the proposed SOSA has three improvisation stages, multiple single PMs, and multiple homogeneous group memories or an inhomogeneous ensemble memory, it is referred to as the MMM-EMSA or the MMS-EMSA.

In the TMS-EMSA, the feasible ranges of each pitch—each decision-making variable—are merely updated for the random selection in each improvisation/iteration of the GIS. In the proposed SOSA, however, the feasible ranges of each pitch are solely updated for the random selection both in each improvisation/iteration of the GISHMG and in each improvisation/iteration of the GISIME.

The performance-driven architecture of the proposed SOSA is generally broken down into six stages, as follows:

- Stage 1—Definition stage: Definition of the optimization problem and its parameters
- Stage 2—Initialization stage
  - Sub-stage 2.1: Initialization of the parameters of the SOSA
  - Sub-stage 2.2: Initialization of the SOM
- Stage 3—Single computational stage or SIS
  - Sub-stage 3.1: Improvisation of a new melody vector by each player in the symphony orchestra
  - Sub-stage 3.2: Update of each available PM in the symphony orchestra
  - Sub-stage 3.3: Check of the stopping criterion of the SIS
- Stage 4—Group computational stage for each homogeneous musical group or GISHMG
  - Sub-stage 4.1: Improvisation of a new melody vector by each existing player in the symphony orchestra taking into account the feasible ranges of the updated pitches for each homogeneous musical group
  - Sub-stage 4.2: Update of each available PM in the symphony orchestra
  - Sub-stage 4.3: Update of the feasible ranges of the pitches—continuous decision-making variables—for each homogeneous musical group in the next improvisation—only for random selection
  - Sub-stage 4.4: Check of the stopping criterion of the GISHMG

- Stage 5—Group computational stage for the inhomogeneous musical ensemble or GISIME
  - Sub-stage 5.1: Improvisation of a new melody vector by each player in the symphony orchestra taking into account the feasible ranges of the updated pitches for the inhomogeneous musical ensemble
  - Sub-stage 5.2: Update of each available PM in the symphony orchestra
  - Sub-stage 5.3: Update of the feasible ranges of the pitches—continuous decision-making variables—for the inhomogeneous musical ensemble in the next improvisation—only for random selection
  - Sub-stage 5.4: Check of the stopping criterion of the GISIME
- Stage 6—Selection stage: Selection of the final optimal solution—the best melody

4.4.1 Stage 1: Definition Stage—Definition of the Optimization Problem and Its Parameters

In order to solve an optimization problem using the proposed SOSA, stage 1 must rigorously describe the optimization problem and its parameters. As a mathematical expression, the standard form of an optimization problem can generally be expressed in accordance with Eqs. (4.39) and (4.40):

$$\begin{aligned} &\underset{\mathrm{x}\in\mathrm{X}}{\text{Minimize}} \quad f(\mathrm{x}) \\ &\qquad \text{subject to :} \\ &\qquad \mathrm{G(x)} = [g_1(\mathrm{x}), \ldots, g_b(\mathrm{x}), \ldots, g_{\mathbf{B}}(\mathrm{x})] = 0; \quad \forall\{\mathrm{B} \geq 0\}, \forall\{b \in \Psi^{\mathrm{B}}\} \\ &\qquad \mathrm{H(x)} = [h_1(\mathrm{x}), \ldots, h_e(\mathrm{x}), \ldots, h_{\mathbf{E}}(\mathrm{x})] \leq 0; \quad \forall\{\mathrm{E} \geq 0\}, \forall\{e \in \Psi^{\mathrm{E}}\} \end{aligned} \tag{4.39}$$

$$\begin{aligned} &\mathrm{x} = [x_1, \ldots, x_v, \ldots, x_{\mathrm{NDV}}]; \quad \forall\{v \in \Psi^{\mathrm{NDV}}\}, \forall\{\Psi^{\mathrm{NDV}} = \Psi^{\mathrm{NCDV+NDDV}}\}, \forall\{\mathrm{x} \in \mathrm{X}\}, \\ &\forall\{x_v^{\min} \leq x_v \leq x_v^{\max} | v \in \Psi^{\mathrm{NCDV}}\}, \{x_v \in \{x_v(1), \ldots, x_v(w), \ldots, x_v(W_v)\} | v \in \Psi^{\mathrm{NDDV}}\} \end{aligned} \tag{4.40}$$

The detailed descriptions relevant to the variables and parameters of Eqs. (4.39) and (4.40) were previously represented in Sect. 1.2.1 of Chap. 1.

In the proposed SOSA, each player in the symphony orchestra explores the entire space of the nonempty feasible decision-making variable in order to find the vector of optimal decision-making—solution vector. In the SIS, implementing this search process is carried out by each player without taking into account the interactive relationships with other players in the homogeneous musical group to which it belongs, and without taking into account the interactive relationships with other existing players in other homogeneous musical groups to which it does not belong. In the GISHMG and GISIME, however, applying of this search process is accomplished by each player, while taking into account the interactive relationships with other players in the homogeneous musical group to which it belongs, and both with and without taking into account the interactive relationships with other players in other homogeneous musical groups to which it does not belong, respectively.

The optimal vector is a vector that gives the lowest possible value for the objective function presented in Eq. (4.39). Besides, in order to solve the optimization problem presented in Eqs. (4.39) and (4.40), each player in the symphony orchestra exclusively considers the objective function given in Eq. (4.39). If the solution vector determined by the corresponding player gives rise to a violation in each of the equality and/or inequality constraints provided in Eq. (4.39), this player can apply one of the following two processes, depending on the standpoint of the decision maker in dealing with this solution vector:

- First process: The corresponding player ignores the determined solution vector.
- Second process: The corresponding player considers the determined solution vector by applying a specific penalty coefficient to the objective function of the optimization problem.

4.4.2 Stage 2: Initialization Stage

After finalization of stage 1 and a complete mathematical description of the optimization problem, stage 2 must be carried out. This stage is formed by two sub-stages: (1) initialization of the parameters of the proposed SOSA and (2) initialization of the SOM.

4.4.2.1 Sub-stage 2.1: Initialization of the Parameters of the SOSA

The parameter adjustments of the proposed SOSA should be initialized with specific values in sub-stage 2.1. Table 4.18 presents a detailed description of the parameter adjustments pertaining to the proposed SOSA. The SOM is a place for storing the solution vectors of all players in the symphony orchestra in the proposed SOSA. The number of homogeneous musical groups (NHMG) describes the number of the available homogeneous musical groups in the symphony orchestra.

To illustrate, each player in the violinists group, harp players group, clarinet players group, and other existing groups in the symphony orchestra is solely considered as an instance of a homogeneous musical group in the symphony orchestra—a group of players with similar instruments.

These homogeneous musical groups also form a symphony orchestra—an inhomogeneous musical ensemble. The melody memory of the homogeneous musical group g in the symphony orchestra (MM_g) is a place for storing the solution vectors of all players in this available homogeneous musical group in the symphony

Table 4.18 Adjustment parameters of the proposed SOSA

No.	The proposed SOSA parameters	Abbreviation	Parameter range
1	Symphony orchestra memory	SOM	–
2	Number of homogeneous musical groups	NHMG	NHMG ≥ 1
3	Melody memory of homogeneous musical group g	MM_g	–
4	Existing number of players in the homogeneous musical group g	PN_g	$PN_g \geq 1$
5	Player memory of existing player p in the homogeneous musical group g	$PM_{g,\,p}$	–
6	Player memory size of player p in the homogeneous musical group g	$PMS_{g,\,p}$	$PMS_{g,\,p} \geq 1$
7	Player memory considering rate of player p in the homogeneous musical group g	$PMCR_{g,\,p}$	$0 \leq PMCR_{g,\,p} \leq 1$
8	Minimum pitch adjusting rate of player p in the homogeneous musical group g	$PAR_{g,p}^{\min}$	$0 \leq PAR_{g,p}^{\min} \leq 2$
9	Maximum pitch adjusting rate of player p in the homogeneous musical group g	$PAR_{g,p}^{\max}$	$0 \leq PAR_{g,p}^{\max} \leq 2$
10	Minimum bandwidth of player p in the homogeneous musical group g	$BW_{g,p}^{\min}$	$0 \leq BW_{g,p}^{\min} < +\infty$
11	Maximum bandwidth of player p in the homogeneous musical group g	$BW_{g,p}^{\max}$	$0 \leq BW_{g,p}^{\max} < +\infty$
12	Number of continuous decision-making variables	NCDV	NCDV ≥ 1
13	Number of discrete decision-making variables	NDDV	NDDV ≥ 1
14	Number of decision-making variables	NDV	NDV ≥ 2
15	Maximum number of improvisations/iterations of the SIS	MNI-SIS	MNI-SIS ≥ 1
16	Maximum number of improvisations/iterations of the GISHMG	MNI-GISHMG	MNI-GISHMG ≥ 1
17	Maximum number of improvisations/iterations of the GISIME	MNI-GISIME	MNI-GISIME ≥ 1

orchestra. The SOM is organized by multiple melody memories related to all homogeneous musical groups in the symphony orchestra. The number of existing players in the homogeneous musical group g in the symphony orchestra (PN_g) indicates the number of players available in the homogeneous musical group g in the symphony orchestra. Player p in the available homogeneous musical group g in the symphony orchestra has a memory defined by the player memory of player p in the homogeneous musical group g ($PM_{g,\,p}$). The player memory size of player p in the homogeneous musical group g ($PMS_{g,p}$) also represents the number of solution vectors stored in the player memory of player p in the available homogeneous musical group g in the symphony orchestra. In the improvisation process of a new melody vector by player p in sub-stages 3.1, 4.1, and 5.1 of the proposed SOSA, the player memory considering the rate of player p in the available homogeneous musical group g in the symphony orchestra ($PMCR_{g,p}$) is employed in order to determine whether the value of a decision-making variable associated with a new melody vector played by the corresponding player is derived from its PM or from the entire space of nonempty feasible decision-making. More precisely, the $PMCR_{g,p}$ parameter expresses the rate at which the value of a decision-making variable from a new melody vector played by player p in the available homogeneous musical group g in the symphony orchestra is randomly selected, based on its PM. In this regard, $1 - PMCR_{g,p}$ parameter addresses the rate at which the value of a decision-making variable from a new melody vector played by player p in the available homogeneous musical group g in the symphony orchestra is haphazardly chosen on the basis of the entire space of the nonempty feasible decision-making. The minimum pitch adjusting rate of player p in the homogeneous musical group g in the symphony orchestra ($PAR_{g,p}^{\min}$) and the maximum pitch adjusting rate of player p in the homogeneous musical group g in the symphony orchestra ($PAR_{g,p}^{\max}$) are utilized in order to specify the value of the pitch adjusting rate parameter relevant to player p in the available homogeneous musical group g in the symphony orchestra under improvisation/iteration m of the SIS, GISHMG, and GISIME ($PAR_{m,g,p}$). In the improvisation process of a new melody vector by player p in the available homogeneous musical group g in the in the symphony orchestra under sub-stages 3.1, 4.1, and 5.1 of the proposed SOSA, the $PAR_{m,g,p}$ parameter is used to characterize whether or not the value of a decision-making variable chosen from the memory of the corresponding player needs a change or update. Simply put, the $PAR_{m,g,p}$ parameter represents the rate at which the value of a decision-making variable selected with the probability of the $PMCR_{g,p}$ from the memory pertaining to the corresponding player is changed or updated. With that in mind, the $1 - PAR_{m,g,p}$ parameter describes the rate at which the value of a decision-making variable chosen with the probability of the $PMCR_{g,p}$ from memory pertaining to the corresponding player remains unchanged. The minimum bandwidth of player p in the homogeneous musical group g in the symphony orchestra ($BW_{g,p}^{\min}$) and maximum bandwidth of player p in the homogeneous musical group g in the symphony orchestra ($BW_{g,p}^{\max}$) are employed in order to determine the value of the bandwidth parameter corresponding to player p in the available homogeneous musical group g in the symphony orchestra under improvisation/iteration m of the SIS, GISHMG, and GISIME ($BW_{m,g,p}$). In the improvisation process of a new melody vector by player p in the available homogeneous musical group g in the in the symphony orchestra under sub-stages 3.1, 4.1, and 5.1 of the proposed SOSA, the $BW_{m,g,p}$ parameter is utilized in order to change or update the value of a continuous decision-making variable selected from the memory associated with the corresponding player. It is simply important to note that the $BW_{m,g,p}$ parameter is considered to be an optional length and is solely defined for continuous decision-making variables.

The NCDV and NDDV parameters depend on the optimization problem provided in Eqs. (4.39) and (4.40). The sum of the NCDV and NDDV parameters is taken into account as the NDV parameter. The dimension of the melody vector in the proposed SOSA depends directly on these parameters. The MNI-SIS and maximum number of improvisation/iteration of the GISHMG (MNI-GISHMG) and maximum number of improvisation/iteration of the GISIME (MNI-GISIME) express the number of times

Fig. 4.2 The architecture of the SOM in the proposed SOSA

the SIS, GISHMG, and GISIME are repeated in the proposed SOSA, respectively. The sum of the MNI-SIS, MNI-GISHMG, and MNI-GISIME parameters is regarded as a stopping criterion for the proposed SOSA.

4.4.2.2 Sub-stage 2.2: Initialization of the Symphony Orchestra Memory

After completion of sub-stage 2.1 and parameter adjustments of the proposed SOSA, the SOM must be initialized in sub-stage 2.2. As previously indicated, the SOM is organized into multiple MMs, while each MM is also composed of multiple PMs. The architecture of the SOM in the proposed SOSA is rigorously demonstrated in Fig. 4.2. Given the above descriptions, the SOM matrix, with the dimensions of $\left\{\sum_{g\in\Psi^{\mathrm{NHMG}}} PMS_g^{\max}\right\}\cdot\{(\mathrm{NDV}+1)\cdot \mathrm{PN}^{\max}\}$, consists of multiple MM submatrices with the dimensions of $\left\{PMS_g^{\max}\right\}\cdot\{(\mathrm{NDV}+1)\cdot PN_g\}$.

In the proposed SOSA, the number of the MMs forming the SOM is determined by the number of the NHMG parameter. In addition, each MM_g submatrix is organized into multiple $PM_{g,p}$ submatrices with the dimensions $\{PMS_{g,p}\}\cdot\{\mathrm{NDV}+1\}$.

The number of the $PM_{g,p}$ forming the MM_g is also characterized by the number of the PN_g parameter. The SOM matrix, MM_g, and $PM_{g,p}$ submatrices are filled with a large number of solution vectors generated haphazardly according to Eqs. (4.41)–(4.45), as follows:

$$\mathrm{SOM} = \left[MM_1 \quad \cdots \quad MM_g \quad \cdots \quad MM_{\mathrm{NHMG}}\right]^{\mathrm{T}}; \quad \forall\left\{g \in \Psi^{\mathrm{NHMG}}\right\} \tag{4.41}$$

$$MM_g = \left[PM_{g,1} \quad \cdots \quad PM_{g,p} \quad \cdots \quad PM_{g,PN_g}\right]; \quad \forall\left\{g \in \Psi^{\mathrm{NHMG}}, p \in \Psi^{PN_g}\right\} \tag{4.42}$$

$$PM_{g,p} = \begin{bmatrix} X^1_{g,p} \\ \vdots \\ X^s_{g,p} \\ \vdots \\ X^{PMS_{g,p}}_{g,p} \end{bmatrix} = \begin{bmatrix} x^1_{g,p,1} & \cdots & x^1_{g,p,v} & \cdots & x^1_{g,p,\mathrm{NDV}} & | & f\left(\mathrm{x}^1_{g,p}\right) \\ \vdots & & \vdots & & \vdots & & \vdots \\ x^s_{g,p,1} & \cdots & x^s_{g,p,v} & \cdots & x^s_{g,p,\mathrm{NDV}} & | & f\left(\mathrm{x}^s_{g,p}\right) \\ \vdots & & \vdots & & \vdots & & \vdots \\ x^{PMS_{g,p}}_{g,p,1} & \cdots & x^{PMS_{g,p}}_{g,p,v} & \cdots & x^{PMS_{g,p}}_{g,p,\mathrm{NDV}} & | & f\left(\mathrm{x}^{PMS_{g,p}}_{g,p}\right) \end{bmatrix};$$

$$\forall\left\{g \in \Psi^{\mathrm{NHMG}}, p \in \Psi^{PN_g}, v \in \Psi^{\mathrm{NDV}}, s \in \Psi^{PMS_{g,p}}\right\} \tag{4.43}$$

$$x^s_{g,p,v} = x^{\min}_v + \mathrm{U}(0,1) \cdot \left(x^{\max}_v - x^{\min}_v\right); \quad \forall\left\{g \in \Psi^{\mathrm{NHMG}}, p \in \Psi^{PN_g}, v \in \Psi^{\mathrm{NCDV}}, s \in \Psi^{PMS_{g,p}}\right\} \tag{4.44}$$

$$x^s_{g,p,v} = x_v(y); \quad \forall\left\{g \in \Psi^{\mathrm{NHMG}}, p \in \Psi^{PN_g}, v \in \Psi^{\mathrm{NDDV}}, s \in \Psi^{PMS_{g,p}}, y \sim \mathrm{U}\{x_v(1), \ldots, x_v(w_v), \ldots, x_v(W_v)\}\right\} \tag{4.45}$$

Equation (4.41) indicates the SOM. Equation (4.42) also denotes the melody memory pertaining to the available homogeneous musical group g in the symphony orchestra—MM_g. Equation (4.43) describes the memory associated with player p in the available homogeneous musical group g in the symphony orchestra—$PM_{g,p}$. In Eq. (4.44), $\mathrm{U}(0,1)$ expresses a random number with a uniform distribution between 0 and 1. Equation (4.44) also indicates that the value of the continuous decision-making variable v from the melody vector s stored in the memory relevant to player p in the available homogeneous musical group g in the symphony orchestra ($x^s_{g,p,v}$) is randomly determined by the set of candidate permissible values for the corresponding decision-making variable, restricted by lower bound $x^{\min}_v$ and upper bound $x^{\max}_v$. In Eq. (4.45), index y represents a random integer with uniform distribution through the set $\{x_v(1), \ldots, x_v(w_v), \ldots, x_v(W_v)\}$—$y \sim \mathrm{U}\{x_v(1), \ldots, x_v(w_v), \ldots, x_v(W_v)\}$. Equation (4.45) indicates that the value of the discrete decision-making variable v from melody vector s stored in the memory related to player p in the available homogeneous musical group g in the symphony orchestra ($x^{s}_{g,p,v}$) is haphazardly obtained through the set of candidate permissive values for this decision-making variable, given by the set $\{x_v(1), \ldots, x_v(w_v), \ldots, x_v(W_v)\}$. Table 4.19 illustrates the pseudocode relevant to the initialization of the entire set of PMs, or entire set of MMs, or SOM in the proposed SOSA. After filling all of the PMs, MMs, or SOM with random solution vectors, the solution vectors stored in each PM must be sorted from the lowest value to the highest value—in an ascending order—with respect to the value of the objective function of the optimization problem. Table 4.20 gives the pseudocode related to sorting the solution vectors stored in the PMs, MMs, or SOM in the proposed SOSA.

4.4.3 Stage 3: Single Computational Stage or SIS

After finalization of stage 2 and initialization of the parameters of the proposed SOSA and the SOM, the single computational stage, or SIS, must be carried out. This stage is formed by three sub-stages: (1) improvisation of a new melody vector by each player in the symphony orchestra; (2) update of each available PM in the symphony orchestra; and, (3) check of the stopping criterion of the SIS. The mathematical equations expressed at this stage must depend on the improvisation/iteration index—index m—due to the repeatability of the SIS in the proposed SOSA.

4.4.3.1 Sub-stage 3.1: Improvisation of a New Melody Vector by Each Player in the Symphony Orchestra

In this sub-stage, each player in the symphony orchestra improvises a new melody vector individually without the influence of other existing players in the homogeneous musical group to which it belongs, and without the influence of other players in other homogeneous musical groups to which it does not belong. That is to say that, in sub-stage 3.1, player p in the available homogeneous musical group g in the symphony orchestra is looking to achieve a fantastic melody only by practicing individually and without learning from the other players in the homogeneous musical group to which it belongs and the players in homogeneous musical groups to which it does not belong.

The point to be made here is that, in the proposed SOSA, a new melody vector or a new melody line played by player p in the homogeneous musical group g in the symphony orchestra—$\mathrm{x}^{\mathrm{new}}_{m,g,p} = \left(x^{\mathrm{new}}_{m,g,p,1}, \ldots, x^{\mathrm{new}}_{m,g,p,v}, \ldots, x^{\mathrm{new}}_{m,g,p,\mathrm{NDV}}\right)$—is generated by using a novel improvisation procedure (NIP), which is established based on fundamental concepts of the improvisation of both a harmony vector and a melody vector in the SS-HSA and TMS-EMSA, respectively.

Table 4.19 Pseudocode relevant to initialization of the entire set of PMs, or entire set of MMs, or SOM in the proposed SOSA

Algorithm 15: Pseudocode for initialization of the entire set of PMs, or entire set of MMs, or SOM in the proposed SOSA

Input: NHMG, PN_g, $PMS_{g,p}$, NCDV, NDDV, NDV, $x_v^{\min}$, $x_v^{\max}$, $\{x_v(1), \ldots, x_v(w_v), \ldots, x_v(W_v)\}$

Output: SOM

start main body

begin
set $PMS_0^{\max} = 0$
set $PMS_g^{\max} = \max(PMS_{g,p}); \forall p \in \Psi^{PN_g}$
set $\text{PN}^{\max} = \max(PN_g); \forall g \in \Psi^{\text{NHMG}}$
construct the matrix SOM with dimension $\left\{\sum_{g \in \Psi^{\text{NHMG}}} PMS_g^{\max}\right\} \cdot \{(\text{NDV}+1) \cdot \text{PN}^{\max}\}$ and zero initial value
for *homogeneous musical group g* $[g \in \Psi^{\text{NHMG}}]$ **do**
    construct the submatrix MM_g with dimension $\left\{PMS_g^{\max}\right\} \cdot \{(\text{NDV}+1) \cdot PN_g\}$ and zero initial value
    for *music player p of the homogeneous musical group g* $\left[p \in \Psi^{PN_g}\right]$ **do**
        construct the submatrix $PM_{g,p}$ with dimension $\{PMS_{g,p}\} \cdot \{\text{NDV}+1\}$ and zero initial value
        for *melody vector s* $\left[s \in \Psi^{PMS_{g,p}}\right]$ **do**
            construct the melody vector s of music player p of the homogeneous musical group g, $\mathrm{x}_{g,p}^s$, with dimension $\{1\} \cdot \{\text{NDV}+1\}$ and zero initial value
            for *decision-making variable v* $[v \in \Psi^{\text{NDV}}]$ **do**
                $x_{g,p,v}^s = x_v^{\min} + \text{U}(0,1) \cdot \left(x_v^{\max} - x_v^{\min}\right)$; for CDVs
                $x_{g,p,v}^s = x_v(y); \forall y \sim \text{U}\{x_v(1), \ldots, x_v(w_v), \ldots, x_v(W_v)\}$; for DDVs
                allocate $x_{g,p,v}^s$ to element $(1, v)$ of the melody vector $\mathrm{x}_{g,p}^s$
            end for
            calculate the value of the objective function, fitness function, derived from the melody vector $\mathrm{x}_{g,p}^s$ as $f\left(\mathrm{x}_{g,p}^s\right)$
            allocate $f\left(\mathrm{x}_{g,p}^s\right)$ to element $(1, \text{NDV}+1)$ of the melody vector $\mathrm{x}_{g,p}^s$
            add melody vector $\mathrm{x}_{g,p}^s$ to the row s of the submatrix $PM_{g,p}$
        end for
        add submatrix $PM_{g,p}$ to the rows 1 to $PMS_{g,p}$ and columns $1 + (p-1) \cdot (\text{NDV}+1)$ to $p \cdot (\text{NDV}+1)$ of the submatrix MM_g
    end for
    add submatrix MM_g to the rows $1 + \sum_{u \in \Psi^{g-1}} PMS_u^{\max}$ to $\sum_{z \in \Psi^g} PMS_z^{\max}$ and columns 1 to $(\text{NDV}+1) \cdot PN_g$ of the matrix SOM
end for
terminate

end main body

Note: Continuous decision-making variables (CDVs), discrete decision-making variable (DDVs)

Table 4.20 Pseudocode related to the sorting of the solution vectors stored in the PMs, MMs, or SOM in the proposed SOSA

Algorithm 16: Pseudocode for sorting of the solution vectors stored in the PMs, MMs, or SOM in the proposed SOSA

Input: Unsorted SOM
Output: Sorted SOM

start main body

begin
set $PMS_0^{\max} = 0$
set $PMS_g^{\max} = \max(PMS_{g,p})$; $\forall p \in \Psi^{PN_g}$
set $\text{PN}^{\max} = \max(PN_g)$; $\forall g \in \Psi^{\text{NHMG}}$
for *homogeneous musical group g* $[g \in \Psi^{\text{NHMG}}]$ **do**
    set $MM_g = \text{SOM}\left(1 + \sum_{u \in \Psi^{g-1}} PMS_u^{\max} : \sum_{z \in \Psi^g} PMS_z^{\max}, 1 : [(\text{NDV} + 1) \cdot PN_g]\right)$
    for *music player p of the homogeneous musical group g* $[p \in \Psi^{PN_g}]$ **do**
        set $PM_{g,p} = MM_g(1 : PMS_p, 1 + [(p - 1) \cdot (\text{NDV} + 1)] : [p \cdot (\text{NDV} + 1)])$
        $\text{F}_{g,p}^{\text{sort}} = \text{sort}\left(PM_{g,p}(1 : PMS_p, (\text{NDV} + 1)), '\text{ascend}'\right)$
        for *melody vector s* $[s \in \Psi^{PMS_p}]$ **do**
            for *melody vector* s^* $[s^* \in \Psi^{PMS_p}]$ **do**
                if $\text{F}_{g,p}^{\text{sort}}(s) == PM_{g,p}(s^*, (\text{NDV} + 1))$ **then**
                    $PM_{g,p}^{\text{sort}}(s, 1 : (\text{NDV} + 1)) = PM_{g,p}(s^*, 1 : (\text{NDV} + 1))$
                end if
            end for
        end for
        $MM_g^{\text{sort}}(1 : PMS_p, 1 + [(p - 1) \cdot (\text{NDV} + 1)] : [p \cdot (\text{NDV} + 1)]) = PM_{g,p}^{\text{sort}}$
    end for
    $\text{SOM}^{\text{sort}}\left(1 + \sum_{u \in \Psi^{g-1}} PMS_u^{\max} : \sum_{z \in \Psi^g} PMS_z^{\max}, 1 : [(\text{NDV} + 1) \cdot PN_g]\right) = MM_g^{\text{sort}}$
end for
$\text{SOM} = \text{SOM}^{\text{sort}}$
terminate

end main body

The NIP will be presented in Sect. 4.4.6 of this chapter. Here, the improvisation process of a new melody vector is performed by players in available homogeneous musical group g and by other players in other available homogeneous musical groups in the symphony orchestra in the same way.

4.4.3.2 Sub-stage 3.2: Update of Each Available PM in the Symphony Orchestra

After completion of sub-stage 3.1 and improvisation of a new melody vector by each player in the symphony orchestra, the update process of each available PM in the symphony orchestra must be accomplished in sub-stage 3.2. Consider player p in the homogeneous musical group g in the symphony orchestra. In this sub-stage, a new melody vector played by player p in the homogeneous musical group g in the symphony orchestra—$\mathrm{x}_{m,g,p}^{\mathrm{new}} = \left(x_{m,g,p,1}^{\mathrm{new}}, \ldots, x_{m,g,p,v}^{\mathrm{new}}, \ldots, x_{m,g,p,\mathrm{NDV}}^{\mathrm{new}}\right)$—is assessed and compared with the worst available melody vector in its PM—the melody vector stored in the $PMS_{g,p}$ row of the $PM_{m,g,p}$—from the perspective of the objective function. If the new melody vector played by the corresponding player has a better value than the worst available melody vector in the $PM_{m,g,p}$, from the viewpoint of the objective function, this new melody vector substitutes the worst available melody vector in the $PM_{m,g,p}$; the worst melody vector is thus eliminated from the $PM_{m,g,p}$. This process is also done for other players in homogeneous musical group g and for all existing players in other available homogeneous musical groups in the symphony orchestra in the same way. Table 4.21 gives the pseudocode related to the update of the memory of all existing players in the symphony orchestra or the update of the SOM_m in the proposed SOSA. It should be pointed out that the update process of the $PM_{m,g,p}$ is not performed if the new melody vector played by player p in the homogeneous musical group g in the symphony orchestra is not notably better than the worst available melody vector in its memory, from the standpoint of the objective function. After completion of this process, solution vectors stored in the memory of all existing players in the symphony orchestra or the SOM_m must be re-sorted, based on the value of objective function—fitness function—in an ascending order.

The pseudocode related to the sorting of the solution vectors stored in the PMs, MMs, or SOM in the proposed SOSA was formerly presented in Table 4.20. Given the dependence of each PM, MM, or, more comprehensively, SOM on the improvisation/iteration index of the SIS—index m—this pseudocode must be rewritten according to Table 4.22.

4.4.3.3 Sub-stage 3.3: Check of the Stopping Criterion of the SIS

After completion of sub-stage 3.2 and an update of each available PM in the symphony orchestra, the check process of the stopping criterion of the SIS must be done. In this sub-stage, if the stopping criterion of the SIS—the MNI-SIS—is satisfied, its computational efforts are terminated. Otherwise, sub-stages 3.1 and 3.2 are repeated.

4.4.4 Stage 4: Group Computational Stage for Each Homogeneous Musical Group or GISHMG

After finalization of stage 3, or implementation of the SIS, the group computational stage for each homogeneous musical group or GISHMG must be carried out. This stage is organized into four sub-stages: (1) improvisation of a new melody vector by

Table 4.21 Pseudocode related to the update of the memory of all existing players in the symphony orchestra or the update of the SOM_m in the proposed SOSA

Algorithm 17: Pseudocode for the update of the memory of all existing players in the symphony orchestra or the update of the SOM_m in the proposed SOSA

Input: Not updated SOM_m, $\mathrm{x}_{m,g,p}^{\mathrm{new}}$
Output: Updated SOM_m

start main body

begin
for *homogeneous musical group g* $[g \in \Psi^{\mathrm{NHMG}}]$ **do**
    for *music player p of the homogeneous musical group g* $[p \in \Psi^{PN_g}]$ **do**
        set $\mathrm{x}_{g,p}^{\mathrm{worst}} = \mathrm{x}_{m,g,p}^{PMS_{g,p}}$
        set $f\left(\mathrm{x}_{g,p}^{\mathrm{worst}}\right) = f\left(\mathrm{x}_{m,g,p}^{PMS_{g,p}}\right)$
        if $f\left(\mathrm{x}_{m,g,p}^{\mathrm{new}}\right) < f\left(\mathrm{x}_{g,p}^{\mathrm{worst}}\right)$ **then**
            $\mathrm{x}_{m,g,p}^{\mathrm{new}} \in PM_{m,g,p}$
            $\mathrm{x}_{g,p}^{\mathrm{worst}} \notin PM_{m,g,p}$
        end if
    end for
end for
terminate

end main body

each player in the symphony orchestra, taking into account the feasible ranges of the updated pitches for each homogeneous musical group; (2) update of each available PM in the symphony orchestra; (3) update of the feasible ranges of the pitches—continuous decision-making variables—for each homogeneous musical group in the next improvisation—only for random selection; and, (4) check of the stopping criterion of the GISHMG. The mathematical equations expressed at this stage must depend on the improvisation/iteration index—index m—due to the repeatability of the GISHMG in the proposed SOSA.

4.4.4.1 Sub-stage 4.1: Improvisation of a New Melody Vector by Each Player in the Symphony Orchestra Taking into Account the Feasible Ranges of the Updated Pitches for Each Homogeneous Musical Group

In sub-stage 4.1, the improvisation process of a new melody vector by each player in the symphony orchestra must be accomplished. That is, each player improvises a new melody vector interactively under the influence of other players in the homogeneous musical group to which it belongs, and individually without the influence of other players in other homogeneous musical groups to which it does not belong.

Table 4.22 Pseudocode related to the sorting of the solution vectors stored in the PMs, MMs, or SOM_m in the proposed SOSA

Algorithm 18: Pseudocode for sorting of the solution vectors stored in the PMs, MMs, or SOM_m in the proposed SOSA

Input: Unsorted SOM_m
Output: Sorted SOM_m

start main body

begin
set $PMS_0^{\max} = 0$
set $PMS_g^{\max} = \max(PMS_{g,p}); \forall p \in \Psi^{PN_g}$
set $\mathrm{PN}^{\max} = \max(PN_g); \forall g \in \Psi^{\mathrm{NHMG}}$
for *homogeneous musical group g* $[g \in \Psi^{\mathrm{NHMG}}]$ **do**
  set $MM_{m,g} = SOM_m\left(1 + \sum_{u \in \Psi^{g-1}} PMS_u^{\max} : \sum_{z \in \Psi^g} PMS_z^{\max}, 1 : \left[(\mathrm{NDV}+1) \cdot PN_g\right]\right)$
  for *music player p of the homogeneous musical group g* $\left[p \in \Psi^{PN_g}\right]$ **do**
    set $PM_{m,g,p} = MM_{m,g}(1 : PMS_p, 1 + [(p-1) \cdot (\mathrm{NDV}+1)] : [p \cdot (\mathrm{NDV}+1)])$
    $\mathrm{F}_{m,g,p}^{\mathrm{sort}} = \mathrm{sort}\left(PM_{m,g,p}(1 : PMS_p, (\mathrm{NDV}+1)), \text{'ascend'}\right)$
    for *melody vector s* $\left[s \in \Psi^{PMS_p}\right]$ **do**
      for *melody vector* s^* $\left[s^* \in \Psi^{PMS_p}\right]$ **do**
        if $\mathrm{F}_{m,g,p}^{\mathrm{sort}}(s) == PM_{m,g,p}(s^*, (\mathrm{NDV}+1))$ **then**
          $PM_{m,g,p}^{\mathrm{sort}}(s, 1 : (\mathrm{NDV}+1)) = PM_{m,g,p}(s^*, 1 : (\mathrm{NDV}+1))$
        end if
      end for
    end for
    $MM_{m,g}^{\mathrm{sort}}\left(1 : PMS_p, 1 + [(p-1) \cdot (\mathrm{NDV}+1)] : [p \cdot (\mathrm{NDV}+1)]\right) = PM_{m,g,p}^{\mathrm{sort}}$
  end for
  $SOM_m^{\mathrm{sort}}\left(1 + \sum_{u \in \Psi^{g-1}} PMS_u^{\max} : \sum_{z \in \Psi^g} PMS_z^{\max}, 1 : \left[(\mathrm{NDV}+1) \cdot PN_g\right]\right) = MM_{m,g}^{\mathrm{sort}}$
end for
$SOM_m = SOM_m^{\mathrm{sort}}$
terminate

end main body

Put another way, player p in the available homogeneous musical group g is looking to attain a fantastic melody only by practicing, learning, and imitating the best player in the homogeneous musical group to which it belongs. The update process of feasible ranges of pitches—continuous decision-making variables—for homogeneous musical group g along with the ability to learn and imitate player p in homogeneous musical group g from the best player in the homogeneous musical group to which it belongs is considered as the interactive relationship among existing players in homogeneous musical group g. In sub-stage 4.1, player p in the homogeneous musical group g improvises a new melody—$\mathrm{x}_{m,g,p}^{\mathrm{new}} = \left(x_{m,g,p,1}^{\mathrm{new}}, \ldots, x_{m,g,p,v}^{\mathrm{new}}, \ldots, x_{m,g,p,\mathrm{NDV}}^{\mathrm{new}}\right)$—by using the proposed NIP and taking into account the interactive relationships among different players in homogeneous musical group g. Here, the improvisation process of a new melody vector is accomplished by players in homogeneous musical group g and by other players in other homogeneous musical groups in the symphony orchestra in the same way.

4.4.4.2 Sub-stage 4.2: Update of Each Available PM in the Symphony Orchestra

After completion of sub-stage 4.1 and improvisation of a new melody vector by each player in the symphony orchestra, taking into account the feasible ranges of the updated pitches for each homogeneous musical group, the update process of each available PM in the symphony orchestra must be accomplished in sub-stage 4.2. The update process for each available PM in the symphony orchestra is virtually the same as sub-stage 3.2 of the SIS, which is extensively reported in Sect. 4.4.3.2 of this chapter.

4.4.4.3 Sub-stage 4.3: Update of the Feasible Ranges of the Pitches—Continuous Decision-Making Variables—for Each Homogeneous Musical Group in the Next Improvisation—Only for Random Selection

After finalization of sub-stage 4.2 and the update of each available PM in the symphony orchestra, the update process of the feasible ranges of pitches—continuous decision-making variables—for each homogeneous musical group in the next improvisation—only for random selection—must be done. In the proposed SOSA, the feasible ranges of the continuous decision-making variables in the melody vector for each available homogeneous musical group in the symphony orchestra are changed and updated during different improvisations/iterations of the GISHMG, but only for random selection. Figure 4.3 demonstrates the update process of the feasible ranges of the continuous decision-making variables for homogeneous musical group g in the symphony orchestra in the proposed SOSA. The update process of the feasible ranges of the continuous decision-making variables for other homogeneous musical groups in the symphony orchestra is similar to Fig. 4.3. Table 4.23 presents the pseudocode related to the update of the feasible ranges of the continuous decision-making variables for each homogeneous musical group in the symphony orchestra in the proposed SOSA.

4.4.4.4 Sub-stage 4.4: Check of the Stopping Criterion of the GISHMG

After completion of sub-stage 4.3 and the update of the feasible ranges of the continuous decision-making variables for each homogeneous musical group in the next improvisation—only for random selection—the check process of the stopping

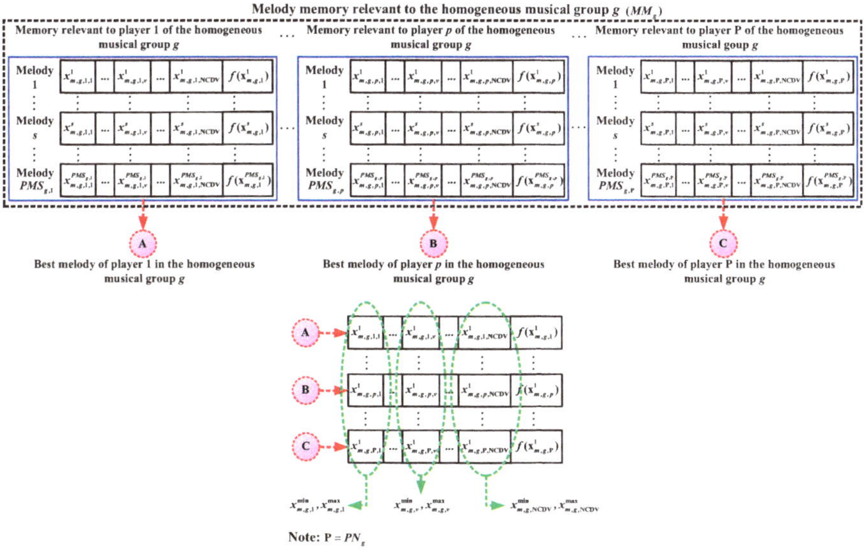

Fig. 4.3 Update process of the feasible ranges of the continuous decision-making variables for homogeneous musical group g in the symphony orchestra in the proposed SOSA

criterion of the GISHMG must be performed. In this sub-stage, the computational efforts of the GISHMG are terminated, if its stopping criterion—the MNI-GISHMG—is satisfied. Otherwise, sub-stages 4.1, 4.2, and 4.3 are repeated.

4.4.5 Stage 5: Group Computational Stage for the Inhomogeneous Musical Ensemble or GISIME

After finalization of stage 4, or implementation of the GISHMG, the group computational stage for the inhomogeneous musical ensemble or the GISIME must be accomplished. This stage is made up of four sub-stages: (1) improvisation of a new melody vector by each player in the symphony orchestra, taking into account the feasible ranges of the updated pitches for the inhomogeneous musical ensemble; (2) update of each available PM in the symphony orchestra; (3) update of the feasible ranges of the pitches—continuous decision-making variables—for the inhomogeneous musical ensemble in the next improvisation—only for random selection; and, (4) check of the stopping criterion of the GISIME. The mathematical equations expressed at this stage must depend on the improvisation/iteration index—index m—due to the repeatability of the GISIME in the proposed SOSA.

Table 4.23 The pseudocode related to the update of the feasible ranges of the continuous decision-making variables for each homogeneous musical group in the symphony orchestra in the proposed SOSA

Algorithm 19: Pseudocode for update of the feasible ranges of the continuous decision-making variables for each homogeneous musical group in the symphony orchestra in the proposed SOSA

Input: $\mathrm{x}_{m,g,p}^{1}$
Output: $x_{m,g,v}^{\min}$, $x_{m,g,v}^{\max}$

start main body

begin
for *homogeneous musical group g* $[g \in \Psi^{\mathrm{NHMG}}]$ **do**
  for *music player p of the homogeneous musical group g* $\left[p \in \Psi^{PN_g}\right]$ **do**
    set $\mathrm{x}_{g,p}^{\mathrm{best}} = \mathrm{x}_{m,g,p}^{1}$
  end for
end for
for *homogeneous musical group g* $[g \in \Psi^{\mathrm{NHMG}}]$ **do**
  for *continuous decision-making variable v* $[v \in \Psi^{\mathrm{NCDV}}]$ **do**
    $x_{m,g,v}^{\min} = \min\left(x_{g,p,v}^{\mathrm{best}}\right); \forall p \in \Psi^{PN_g}$
    $x_{m,g,v}^{\max} = \max\left(x_{g,p,v}^{\mathrm{best}}\right); \forall p \in \Psi^{PN_g}$
  end for
end for
terminate

end main body

4.4.5.1 Sub-stage 5.1: Improvisation of a New Melody Vector by Each Player in the Symphony Orchestra Taking into Account the Feasible Ranges of the Updated Pitches for the Inhomogeneous Musical Ensemble

In sub-stage 5.1, the improvisation process of a new melody vector by each player in the symphony orchestra must be carried out. In this sub-stage, each player improvises a new melody vector interactively both under the influence of other players in the homogeneous musical group to which it belongs and under the influence of other players in other homogeneous musical groups to which it does not belong. That is to say that player p in homogeneous musical group g is looking to obtain a fantastic melody both by learning and imitating from the best existing player in the inhomogeneous musical ensemble—the best player in the symphony orchestra. The update process of feasible ranges of pitches—continuous decision-making variables—for the inhomogeneous musical ensemble along with the ability to learn and imitate player p in the homogeneous musical group g from the best player in the inhomogeneous musical ensemble—the best existing player in the symphony orchestra—is regarded as the interactive relationship among different players in the inhomogeneous musical ensemble. In sub-stage 5.1, each player p in the homogeneous musical group g improvises a new melody—$\mathrm{x}_{m,g,p}^{\text{new}} = \left(x_{m,g,p,1}^{\text{new}}, \ldots, x_{m,g,p,v}^{\text{new}}, \ldots, x_{m,g,p,\text{NDV}}^{\text{new}}\right)$ —by using the proposed NIP and taking into account the interactive relationships among different players in the inhomogeneous musical ensemble. In this sub-stage, the improvisation process of a new melody vector is done by players in available homogeneous musical group g and by other players in other available homogeneous musical groups in the symphony orchestra in the same way.

4.4.5.2 Sub-stage 5.2: Update of Each Available PM in the Symphony Orchestra

After completion of sub-stage 5.1 and improvisation of a new melody vector by each player in the symphony orchestra by considering the feasible ranges of the updated pitches for the inhomogeneous musical group, the update process of each available PM in the symphony orchestra must be accomplished in sub-stage 5.2. The update process of each available PM is virtually the same as sub-stage 3.2 of the SIS and sub-stage 4.2 of the GISHMG, which were addressed in Sect. 4.4.3.2 of this chapter.

4.4.5.3 Sub-stage 5.3: Update of the Feasible Ranges of the Pitches—Continuous Decision-Making Variables—for the Inhomogeneous Musical Group in the Next Improvisation—Only for Random Selection

After finalization of sub-stage 5.2 and the update of each available PM in the symphony orchestra, the update process of the feasible ranges of pitches—continuous decision-making variables—for the inhomogeneous musical ensemble in the next improvisation—only for random selection—must be performed. In the proposed SOSA, the feasible ranges of the continuous decision-making variables in the melody vector for the inhomogeneous musical ensemble are changed and updated during different improvisations/iterations of the GISIME, but only for random selection. The update process of the feasible ranges of the continuous decision-making variables for the inhomogeneous musical ensemble in the proposed SOSA is depicted in Fig. 4.4. Table 4.24 presents the pseudocode associated with the update of the feasible ranges of the continuous decision-making variables for the inhomogeneous musical ensemble in the proposed SOSA.

4.4.5.4 Sub-stage 5.4: Check of the Stopping Criterion of the GISIME

After completion of sub-stage 5.3 and the update of the feasible ranges of the pitches for the inhomogeneous musical ensemble in the next improvisation—only for random selection—the check process of the stopping criterion of the GISIME must be accomplished. In this sub-stage, the computational efforts of the GISIME

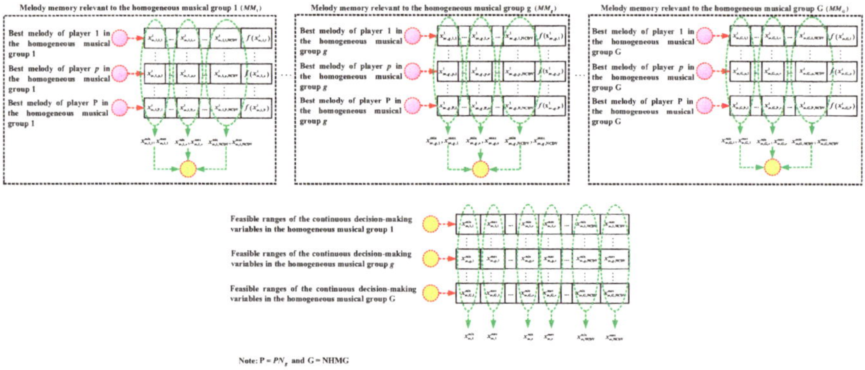

Fig. 4.4 Update process of the feasible ranges of the continuous decision-making variables for the inhomogeneous musical ensemble in the proposed SOSA

Table 4.24 The pseudocode associated with the update of the feasible ranges of the continuous decision-making variables for the inhomogeneous musical ensemble in the proposed SOSA

Algorithm 20: Pseudocode for the update of the feasible ranges of the continuous decision-making variables for the inhomogeneous musical ensemble in the proposed SOSA

Input: $\mathrm{x}^{1}_{m,g,p}$
Output: $x^{\min}_{m,v}$, $x^{\max}_{m,v}$

start main body

begin
for *homogeneous musical group g* $[g \in \Psi^{\mathrm{NHMG}}]$ **do**
  for *music player p of the homogeneous musical group g* $\left[p \in \Psi^{PN_g}\right]$ **do**
    set $\mathrm{x}^{\mathrm{best}}_{g,p} = \mathrm{x}^{1}_{m,g,p}$
  end for
end for
for *homogeneous musical group g* $[g \in \Psi^{\mathrm{NHMG}}]$ **do**
  for *continuous decision-making variable v* $[v \in \Psi^{\mathrm{NCDV}}]$ **do**
    $x^{\min}_{m,g,v} = \min\left(x^{\mathrm{best}}_{g,p,v}\right); \forall p \in \Psi^{PN_g}$
    $x^{\max}_{m,g,v} = \max\left(x^{\mathrm{best}}_{g,p,v}\right); \forall p \in \Psi^{PN_g}$
  end for
end for
for *continuous decision-making variable v* $[v \in \Psi^{\mathrm{NCDV}}]$ **do**
  $x^{\min}_{m,v} = \min\left(x^{\min}_{g,v}\right); \forall g \in \Psi^{\mathrm{NHMG}}$
  $x^{\max}_{m,v} = \max\left(x^{\max}_{g,v}\right); \forall g \in \Psi^{\mathrm{NHMG}}$
end for
terminate

end main body

Table 4.25 Pseudocode associated with the selection of the final optimal solution in the proposed SOSA

Algorithm 21: Pseudocode for the selection of the final optimal solution in the proposed SOSA

Input: SOM
Output: x^{best}

start main body

begin
for *homogeneous musical group g* $[g \in \Psi^{\text{NHMG}}]$ **do**
  for *music player p of the homogeneous musical group g* $\left[p \in \Psi^{PN_g}\right]$ **do**
    set $x_{g,p}^{\text{best}} = x_{g,p}^{1}$
  end for
end for
for *each homogeneous musical group g* $[g \in \Psi^{\text{NHMG}}]$ **do**
  $x_{g}^{\text{best}} = \min\left(x_{g,p}^{\text{best}}\right); \forall p \in \Psi^{PN_g}$
end for
$x^{\text{best}} = \min\left(x_{g}^{\text{best}}\right); \forall g \in \Psi^{\text{NHMG}}$
terminate

end main body

are terminated when its stopping criterion—the MNI-GISIME—is satisfied. Otherwise, sub-stages 5.1, 5.2, and 5.3 are repeated.

4.4.6 Stage 6: Selection Stage—Selection of the Final Optimal Solution—the Best Melody

After completion of stage 5, or accomplishment of the GISIME, the selection of the final optimal solution—the best melody—must be carried out in stage 6. In this stage, the best melody vector stored in the memory of each player in each existing homogeneous musical group in the symphony orchestra is specified. Then, the best melody vector is selected from among the best melody vectors of the existing music players in each homogeneous musical group as the best melody vector of the corresponding homogeneous musical group. Eventually, the best melody vector is chosen from among the best available melody vectors in all homogeneous musical groups as the final optimal solution.

Table 4.25 gives the pseudocode associated with the selection of the final optimal solution in the proposed SOSA.

4.4.7 Novel Improvisation Procedure

As previously described, in the proposed SOSA, player p in the homogeneous musical group g improvises a new melody vector—$\mathrm{x}_{m,g,p}^{\text{new}} = \left(x_{m,g,p,1}^{\text{new}}, \ldots, x_{m,g,p,v}^{\text{new}}, \ldots, x_{m,g,p,\text{NDV}}^{\text{new}}\right)$—by using the NIP.

This improvisation procedure was established on the basis of fundamental concepts of the improvisation of both a harmony vector and a melody vector in the SS-HSA and TMS-EMSA, respectively, previously reported in Sect. 3.6.5 of Chap. 3 and Sect. 4.3 of this chapter. The proposed NIP is organized according to three rules: (1) player memory consideration; (2) pitch adjustment; and, (3) random selection.

Rule 1: In the player memory consideration rule, the values of the new melody vector played by player p are randomly chosen from the melody vectors stored in the $PM_{m,g,p}$ with the probability of the $PMCR_{g,p}$. In this rule, two principles are alternately used. In the presence of the continuous decision-making variables, each principle consists of a linear combination of a decision-making variable chosen from the $PM_{m,g,p}$ and a ratio of the $BW_{m,g,p}$. In the presence of the discrete decision-making variables, however, each principle consists of a discrete decision-making variable selected from the $PM_{m,g,p}$. If the first principle is activated, the value of the continuous decision-making variable v from the new melody vector played by player p ($x_{m,g,p,v}^{\text{new}}$) is haphazardly chosen from the available corresponding continuous decision-making variable in the melody vectors stored in the $PM_{m,g,p}$—$\left(x_{m,g,p,v}^{1}, \ldots, x_{m,g,p,v}^{s}, \ldots, x_{m,g,p,v}^{PMS_{g,p}}\right)$—with the probability of the $PMCR_{g,p}$ and updated by the $BW_{m,g,p}$ parameter. By the same token, the value of the discrete decision-making variable v from the new melody vector played by player p is randomly selected from the available corresponding discrete decision-making variables in the melody vectors stored in the $PM_{m,g,p}$—$\left(x_{m,g,p,v}^{1}, \ldots, x_{m,g,p,v}^{s}, \ldots, x_{m,g,p,v}^{PMS_{g,p}}\right)$—with the probability of the $PMCR_{g,p}$ in case the first principle is activated. Conversely, if the first principle is deactivated—or, in other words, the second principle is activated—the value of the continuous decision-making variable v from the new melody vector played by player p is haphazardly chosen through the entire set of available continuous decision-making variables in the melody vectors stored in the $PM_{m,g,p}$—$\left\{\left(x_{m,g,p,1}^{1}, \ldots, x_{m,g,p,1}^{s}, \ldots, x_{m,g,p,1}^{PMS_{g,p}}\right), \ldots, \left(x_{m,g,p,v}^{1}, \ldots, x_{m,g,p,v}^{s}, \ldots, x_{m,g,p,v}^{PMS_{g,p}}\right), \ldots, \left(x_{m,g,p,\text{NCDV}}^{1}, \ldots, x_{m,g,p,\text{NCDV}}^{s}, \ldots, x_{m,g,p,\text{NCDV}}^{PMS_{g,p}}\right)\right\}$—with the probability of $PMCR_{g,p}$ and updated by the $BW_{m,g,p}$ parameter. For the same reason, the value of the discrete decision-making variable v from the new melody vector played by player p is randomly selected through the entire set of available decision-making variables in the melody vectors stored in the $PM_{m,g,p}$—$\left\{\left(x_{m,g,p,1}^{1}, \ldots, x_{m,g,p,1}^{s}, \ldots, x_{m,g,p,1}^{PMS_{g,p}}\right), \ldots, \left(x_{m,g,p,v}^{1}, \ldots, x_{m,g,p,v}^{s}, \ldots, x_{m,g,p,v}^{PMS_{g,p}}\right), \ldots, \left(x_{m,g,p,\text{NDDV}}^{1}, \ldots, x_{m,g,p,\text{NDDV}}^{s}, \ldots, x_{m,g,p,\text{NDDV}}^{PMS_{g,p}}\right)\right\}$—with the probability of $PMCR_{g,p}$ provided that the first principle is not activated—or, in other words, the second principle is activated. As a result, in the proposed NIP, implementing the first principle of the player memory consideration rule to determine the value of the continuous or discrete decision-making variable

v from a new melody vector played by player p is performed according to Eqs. (4.46) and (4.47), respectively:

$$x_{m,g,p,v}^{\text{new}} = x_{m,g,p,v}^{r_{g,p}} \pm \text{U}(0,1) \cdot BW_{m,g,p}; \\ \forall \left\{ g \in \Psi^{\text{NHMG}}, p \in \Psi^{PN_g}, v \in \Psi^{\text{NCDV}}, m \in \Psi^{(\text{MNI-SIS})+(\text{MNI-GISHMG})+(\text{MNI-GISIME})} \right\}, \\ \forall \left\{ r_{g,p} \sim \text{U}\{1,2,\ldots,PMS_{g,p}\} \right\} \tag{4.46}$$

$$x_{m,g,p,v}^{\text{new}} = x_{m,g,p,v}^{r_{g,p}}; \quad \forall \left\{ g \in \Psi^{\text{NHMG}}, p \in \Psi^{PN_g}, v \in \Psi^{\text{NDDV}} \right\}, \\ \forall \left\{ r_{g,p} \sim \text{U}\{1,2,\ldots,PMS_{g,p}\} \right\} \tag{4.47}$$

Similarly, applying of the second principle of the player memory consideration rule to characterize the value of the continuous or discrete decision-making variable v from a new melody vector played by player p is accomplished in accordance with Eqs. (4.48) and (4.49), respectively:

$$x_{m,g,p,v}^{\text{new}} = x_{m,g,p,k}^{r_{g,p}} \pm \text{U}(0,1) \cdot BW_{m,g,p}; \\ \forall \left\{ g \in \Psi^{\text{NHMG}}, p \in \Psi^{PN_g}, v \in \Psi^{\text{NCDV}}, m \in \Psi^{(\text{MNI-SIS})+(\text{MNI-GISHMG})+(\text{MNI-GISIME})} \right\}, \\ \forall \left\{ r_{g,p} \sim \text{U}\{1,2,\ldots,PMS_{g,p}\} \right\}, \forall \{ k \sim \text{U}\{1,2,\ldots,\text{NCDV}\} \} \tag{4.48}$$

$$x_{m,g,p,v}^{\text{new}} = x_{m,g,p,l}^{r_{g,p}}; \quad \forall \left\{ g \in \Psi^{\text{NHMG}}, p \in \Psi^{PN_g}, v \in \Psi^{\text{NDDV}} \right\}, \\ \forall \left\{ r_{g,p} \sim \text{U}\{1,2,\ldots,PMS_{g,p}\} \right\}, \forall \{ l \sim \text{U}\{1,2,\ldots,\text{NDDV}\} \} \tag{4.49}$$

In Eqs. (4.46)–(4.49), index r associated with player p in the homogeneous musical group g in the symphony orchestra ($r_{g,p}$) is a random integer with uniform distribution through the set of $\{1, 2, \ldots, PMS_{g,p}\}$—$r_{g,p} \sim \text{U}\{1, 2, \ldots, PMS_{g,p}\}$. More precisely, the value of index r is randomly obtained through the set of allowable values which is shown by the set of $\{1, 2, \ldots, PMS_{g,p}\}$. Determination of this index is expressed based on Eq. (4.50):

$$r_{g,p} = \text{int}\big(\text{U}(0,1) \cdot PMS_{g,p}\big) + 1; \quad \forall \left\{ g \in \Psi^{\text{NHMG}}, p \in \Psi^{PN_g} \right\} \tag{4.50}$$

Index k is also a random integer with uniform distribution through the set of $\{1, 2, \ldots, \text{NCDV}\}$—$k \sim \text{U}\{1, 2, \ldots, \text{NCDV}\}$. In other words, the value of index k is haphazardly achieved through the set of permissive values which is illustrated by the set of $\{1, 2, \ldots, \text{NCDV}\}$. Determination of this index is addressed according to Eq. (4.51):

$$k = \text{int}(\text{U}(0, 1) \cdot \text{NCDV}) + 1 \tag{4.51}$$

In addition to this, index l is a random integer with uniform distribution through the set of $\{1, 2, \ldots, \text{NDDV}\}$—$l \sim \text{U}\{1, 2, \ldots, \text{NDDV}\}$. Stated another way, the value of index l is randomly specified through the set of permissible values which is displayed by the set of $\{1, 2, \ldots, \text{NDDV}\}$. Determination of this index is addressed in accordance with Eq. (4.52):

$$l = \text{int}(\text{U}(0, 1) \cdot \text{NDDV}) + 1 \tag{4.52}$$

It is important to note that other distributions can be employed for indices r, k, and l such as $(\text{U}(0, 1))^2$. The employment of this distribution gives rise to selection of lower values for these indices. In the proposed NIP, the first and second principles of the player memory consideration rule can efficiently lead to a more favorable convergence and a more significant increase in the diversity of the generated solutions for the newly developed SOSA. Implementing the player memory consideration rule in the proposed NIP is also carried out for other players in homogeneous musical group g and for other players in other homogeneous musical groups in the symphony orchestra in the same way.

Rule 2: In the pitch adjustment rule, the values of a new melody vector played by player p, haphazardly chosen through the existing melody vectors in the $PM_{m,g,p}$ with the probability of the $PMCR_{g,p}$, are updated with the probability of the $PAR_{m,g,p}$. In the proposed NIP, the pitch adjustment rule is established based on three principles. The first, second, and third principles of the pitch adjustment rule in the proposed NIP are activated in the SIS, GISEMG, and GISIME, respectively. If the first principle of the pitch adjustment rule in the proposed NIP is activated, the update process for the continuous or discrete decision-making variable v is accomplished by substituting it with the value of the continuous or discrete decision-making variable v from the best available melody vector in the $PM_{m,g,p}$—$x^{\text{best}}_{m,g,p,v}$. In the proposed NIP, therefore, implementing the first principle of the pitch adjustment rule to obtain the value of the continuous or discrete decision-making variable v from a new melody vector played by player p is done by using Eqs. (4.53) and (4.54), respectively:

$$x^{\text{new}}_{m,g,p,v} = x^{\text{best}}_{m,g,p,v}; \quad \forall\left\{g \in \Psi^{\text{NHMG}}, p \in \Psi^{PN_g}, v \in \Psi^{\text{NCDV}}\right\} \tag{4.53}$$

$$x^{\text{new}}_{m,g,p,v} = x^{\text{best}}_{m,g,p,v}; \quad \forall\left\{g \in \Psi^{\text{NHMG}}, p \in \Psi^{PN_g}, v \in \Psi^{\text{NDDV}}\right\} \tag{4.54}$$

Equations (4.53) and (4.54) tell us that, in the first principle of the pitch adjustment rule under the proposed NIP, player p tries to imitate the best melody stored in its own memory. If the second principle of the pitch adjustment rule in the proposed NIP is activated, the update process for the continuous or discrete decision-making variable v is carried out by replacing it with the value of the continuous or discrete decision-making variable v from the best available melody vector in the $MM_{m,g}$—$x^{\text{best}}_{m,g,\text{best},v}$. Applying of the second principle of the pitch adjustment rule to specify the value of the continuous or discrete decision-making variable v from a new melody vector played by player p is performed by using Eqs. (4.55) and (4.56), respectively:

$$x_{m,g,p,v}^{\text{new}} = x_{m,g,\text{best},v}^{\text{best}}; \quad \forall\left\{g \in \Psi^{\text{NHMG}}, p \in \Psi^{PN_g}, v \in \Psi^{\text{NCDV}}\right\} \tag{4.55}$$

$$x_{m,g,p,v}^{\text{new}} = x_{m,g,\text{best},v}^{\text{best}}; \quad \forall\left\{g \in \Psi^{\text{NHMG}}, p \in \Psi^{PN_g}, v \in \Psi^{\text{NDDV}}\right\} \tag{4.56}$$

Equations (4.55) and (4.56) tell us that, in the second principle of the pitch adjustment rule under the proposed NIP, player p tries to mimic the best melody stored in the memory of the best player in the homogeneous musical group g in the symphony orchestra. If the third principle of the pitch adjustment rule in the proposed NIP is activated, the update process for the continuous or discrete decision-making variable v is accomplished by substituting it with the value of the continuous or discrete decision-making variable v from the best available melody vector in the SOM_m—$x_{m,\text{best},\text{best},v}^{\text{best}}$. Subsequently, implementing of the third principle of the pitch adjustment rule to characterize the value of the continuous or discrete decision-making variable v from a new melody vector played by player p is done by using Eqs. (4.57) and (4.58), respectively:

$$x_{m,g,p,v}^{\text{new}} = x_{m,\text{best},\text{best},v}^{\text{best}}; \quad \forall\left\{g \in \Psi^{\text{NHMG}}, p \in \Psi^{PN_g}, v \in \Psi^{\text{NCDV}}\right\} \tag{4.57}$$

$$x_{m,g,p,v}^{\text{new}} = x_{m,\text{best},\text{best},v}^{\text{best}}; \quad \forall\left\{g \in \Psi^{\text{NHMG}}, p \in \Psi^{PN_g}, v \in \Psi^{\text{NDDV}}\right\} \tag{4.58}$$

Equations (4.57) and (4.58) tell us that, in the third principle of the pitch adjustment rule under the proposed NIP, player p tries to emulate the best melody stored in the memory of the best player in the best homogeneous musical group or the best melody stored in the memory of the best player in the inhomogeneous musical ensemble. In the proposed NIP, the first, second, and third principles of the pitch adjustment rule, by modeling the interactive relationships both among existing players in each homogeneous musical group and among existing players in the inhomogeneous musical ensemble, can purposefully enhance the desirable convergence, diversity of the generated solutions, and ability to achieve a fantastic melody in the entire space of the nonempty feasible decision-making for the proposed SOSA. Applying the pitch adjustment rule in the proposed NIP is also performed for players in homogeneous musical group g and for other players in other available homogeneous musical groups in the symphony orchestra in the same way.

Rule 3: In the random selection rule of the proposed NIP, the values of a new melody vector played by player p are randomly chosen from the entire space of the nonempty feasible decision-making variables with the probability of the $1 - PMCR_{g,p}$. Here, the random selection rule related to continuous decision-making variables is established according to three principles. The first, second, and third principles of the random selection rule are activated in the SIS, GISHMG, and GISIME, respectively. If the first principle of the random selection rule for continuous decision-making variables is activated, the value of the continuous decision-making variable v from a new melody vector played by player p is haphazardly selected from the entire space of the nonempty feasible decision-making relevant to this continuous decision-making variable, limited by an invariable lower bound ($x_v^{\min}$) and an invariable upper bound ($x_v^{\max}$). These invariable lower and upper bounds for the continuous decision-making variable v, specified in stage 1 of the proposed SOSA—definition of the optimization problem and its parameters—are unchanged in all improvisations/iterations of the SIS. Accordingly, implementing the first principle of the random selection rule for continuous decision-making variables to obtain the value of the continuous decision-making variable v from a new melody vector played by player p is accomplished by using Eq. (4.59):

$$x_{m,g,p,v}^{\text{new}} = x_v^{\min} + \text{U}(0,1) \cdot \left(x_v^{\max} - x_v^{\min}\right); \quad \forall\left\{g \in \Psi^{\text{NHMG}}, p \in \Psi^{PN_g}, v \in \Psi^{\text{NCDV}}\right\} \tag{4.59}$$

Equation (4.59) states that, in the first principle of the random selection rule for continuous decision-making variables, player p, characterized by the value of the continuous decision-making variable v from a new melody vector, has the strength to seek the entire space of the nonempty feasible decision-making associated with this continuous decision-making variable, defined in stage 1 of the proposed SOSA. If the second principle of the random selection rule for continuous decision-making variables is activated, the value of the continuous decision-making variable v from a new melody vector played by player p is randomly chosen from the entire space of the nonempty feasible decision-making pertaining to this continuous decision-making variable, restricted by a variable lower bound ($x_{m,g,v}^{\min}$) and a variable upper bound ($x_{m,g,v}^{\max}$). These variable lower and upper bounds for the continuous decision-making variable v are updated in each improvisation/iteration of the GISHMG and for the available homogeneous musical group g in the symphony orchestra. As a result, applying of the second principle of random selection rule for continuous decision-making variables to determine the value of the continuous decision-making variable v from a new melody vector played by player p is done by using Eq. (4.60):

$$x_{m,g,p,v}^{\text{new}} = x_{m,g,v}^{\min} + \text{U}(0,1) \cdot \left(x_{m,g,v}^{\max} - x_{m,g,v}^{\min}\right); \quad \forall\left\{g \in \Psi^{\text{NHMG}}, p \in \Psi^{PN_g}, v \in \Psi^{\text{NCDV}}, m \in \Psi^{(\text{MNI-GISHMG})}\right\} \tag{4.60}$$

Equation (4.60) addresses the second principle of the random selection rule for continuous decision-making variables, where player p specifies that the value of the continuous decision-making variable v from a new melody vector merely has the ability to seek the space of the nonempty feasible decision-making related to this continuous decision-making variable, which is updated in each improvisation/

iteration of the GISHMG and for the available homogeneous musical group g in the symphony orchestra. The update process of the feasible ranges of the continuous decision-making variable v in each improvisation/iteration of the GISHMG and for the available homogeneous musical group g in the symphony orchestra was discussed in Sect. 4.4.4.3 of this chapter. Similarly, the value of the continuous decision-making variable v from a new melody vector played by player p is haphazardly selected from the entire space of the nonempty feasible decision-making relevant to this continuous decision-making variable bounded by a variable lower bound ($x_{m,v}^{\min}$) and a variable upper bound ($x_{m,v}^{\max}$), provided that the third principle of the random selection rule for continuous decision-making variables is activated. These variable lower and upper bounds for the continuous decision-making variable v are updated in each improvisation/iteration of the GISIME and for the inhomogeneous musical group or symphony orchestra. Consequently, implementing of the third principle of random selection rule for continuous decision-making variables to obtain the value of the continuous decision-making variable v from a new melody vector played by player p is carried out by using Eq. (4.61):

$$x_{m,g,p,v}^{\text{new}} = x_{m,v}^{\min} + \text{U}(0,1) \cdot \left(x_{m,v}^{\max} - x_{m,v}^{\min}\right);$$
$$\forall\left\{g \in \Psi^{\text{NHMG}}, p \in \Psi^{PN_g}, v \in \Psi^{\text{NCDV}}, m \in \Psi^{(\text{MNI-GISIME})}\right\} \tag{4.61}$$

Equation (4.61) describes the third principle of the random selection rule for continuous decision-making variables, where player p characterizes the value of the continuous decision-making variable v from a new melody vector solely that has the strength to seek the space of the nonempty feasible decision-making associated with this continuous decision-making variable, updated in each improvisation/iteration of the GISIME and for the inhomogeneous musical group or symphony orchestra. The update process of the feasible ranges of the continuous decision making variable v in each improvisation/iteration of the GISIME and for the inhomogeneous musical group was previously reported in Sect. 4.4.5.3 of this chapter. Similar to the three principles of the pitch adjustment rule, the first, second, and third principles of the random selection rule for continuous decision-making variable by modeling the interactive relationships both among existing players in each homogeneous musical group and among existing players in the inhomogeneous musical ensemble can appreciably promote favorable convergence, diversity of the generated solutions, and ability to attain a fantastic melody in the entire space of the nonempty feasible decision-making for the proposed SOSA. The point to be made here is that the first, second, and third principles of the random selection rule for discrete decision-making variables are virtually the same. Applying of the first, second, and third principles of the random selection rule for discrete decision-making variables to determine the value of the discrete decision-making variable v from a new melody vector played by player p is performed by using Eq. (4.62):

$$x_{m,g,p,v}^{\text{new}} = x_v(y); \quad \forall\{g \in \Psi^{\text{NHMG}}, p \in \Psi^{PN_g}, v \in \Psi^{\text{NDDV}}, \\ y \sim \text{U}\{x_v(1), \ldots, x_v(w_v), \ldots, x_v(W_v)\}\} \tag{4.62}$$

In Eq. (4.62), index y is a random integer with a uniform distribution through the set of $\{x_v(1), \ldots, x_v(w_v), \ldots, x_v(W_v)\}$—$y \sim \text{U}\{x_v(1), \ldots, x_v(w_v), \ldots, x_v(W_v)\}$. Equation (4.62) also indicates that the value of the discrete decision-making variable v from the new melody vector played by player p is randomly characterized through the set of candidate allowable values for this decision-making variable demonstrated by the set $\{x_v(1), \ldots, x_v(w_v), \ldots, x_v(W_v)\}$. Implementing of the random selection rule in the proposed NIP for continuous and discrete decision-making variables is accomplished for other players in homogeneous musical group g and for other players in other homogeneous musical groups in the symphony orchestra in the same way.

In the newly developed SOSA, player p utilizes the updated values of the $BW_{m,g,p}$ and $PAR_{m,g,p}$ parameters in the improvisation process of its melody vector. These parameters in each improvisation/iteration of the SIS, GISHMG, and GISIME of the proposed SOSA are updated according to Eqs. (4.63) and (4.64), respectively:

$$BW_{m,g,p} = BW_{g,p}^{\max} \cdot \exp\left(\frac{\ln\left(BW_{g,p}^{\max}/BW_{g,p}^{\min}\right)}{(\text{MNI-SIS}) + (\text{MNI-GISHMG}) + (\text{MNI-GISIME})} \cdot m\right); \\ \forall\left\{g \in \Psi^{\text{NHME}}, p \in \Psi^{PN_g}, m \in \Psi^{(\text{MNI-SIS})+(\text{MNI-GISHMG})+(\text{MNI-GISIME})}\right\} \tag{4.63}$$

$$PAR_{m,g,p} = PAR_{g,p}^{\min} + \left(\frac{PAR_{g,p}^{\max} - PAR_{g,p}^{\min}}{(\text{MNI-SIS}) + (\text{MNI-GISHMG}) + (\text{MNI-GISIME})}\right) \cdot m; \\ \forall\left\{g \in \Psi^{\text{NHME}}, p \in \Psi^{PN_g}, m \in \Psi^{(\text{MNI-SIS})+(\text{MNI-GISHMG})+(\text{MNI-GISIME})}\right\} \tag{4.64}$$

Equation (4.63) tells us that the value of the bandwidth parameter relevant to player p ($BW_{m,g,p}$) is addressed as an exponential function of the improvisation/iteration index—index m—in such a way that the value of the $BW_{m,g,p}$ parameter is exponentially decreased by increasing the value of this index. Stated another way, by altering the improvisation/iteration index from zero to the maximum number of the improvisation/iteration—$m \in \{0 \rightarrow (\text{MNI-SIS}) + (\text{MNI-GISHMG}) + (\text{MNI-GISIME})\}$—the value of the $BW_{m,g,p}$ parameter is exponentially changed/updated from the $BW_{g,p}^{\max}$ parameter to the $BW_{g,p}^{\min}$ parameter— $BW_{m,g,p} \in \left\{BW_{g,p}^{\max} \rightarrow BW_{g,p}^{\min}\right\}$. Equation (4.64) also tells us that the value of the pitch adjusting rate parameter associated with player p ($PAR_{m,g,p}$) is described as a linear function of the improvisation/iteration index—index m—so that the value of the $PAR_{m,g,p}$ parameter is linearly increased by increasing the value of this index. In other words, by varying the improvisation/iteration index from zero to the maximum number of the improvisation/iteration—$m \in \{0 \rightarrow \text{(MNI-SIS)} + \text{(MNI-GISHMG)} + \text{(MNI-GISIME)}\}$—the value of the $PAR_{m,g,p}$ parameter is linearly changed/updated from the $PAR_{g,p}^{\min}$ parameter to the $PAR_{g,p}^{\max}$ parameter— $PAR_{m,g,p} \in \left\{PAR_{g,p}^{\max} \rightarrow PAR_{g,p}^{\min}\right\}$. Table 4.26 gives the pseudocode pertaining to improvisation of a new melody vector by each player in the symphony orchestra of the proposed SOSA. The designed pseudocode in different stages and sub-stages of the proposed SOSA is located in a regular sequence and forms the performance-driven architecture of this algorithm. Table 4.27 presents the pseudocode related to the performance-driven architecture of the proposed SOSA. Here, sub-stages 3.3, 4.4, and 5.4—the check process of the stopping criterion of the SIS, GISHMG, and GISIME—are defined by the first, second, and third WHILE loops in the pseudocode pertaining to the performance-driven architecture of the proposed SOSA (see Table 4.27).

4.4.8 Some Hints Regarding the Architecture of the Proposed SOSA

One of the substantial points regarding the newly developed SOSA is its computational burden. To clarify, consider three homogeneous musical groups in the symphony orchestra: (1) a homogeneous group of four violinists; (2) a homogeneous group of three clarinet players; and, (3) a homogeneous group of two celesta players. In these special circumstances, the homogeneous group of four violinists has an equal computational burden compared to the TMS-EMSA with four players. By the same token, the homogeneous group of three clarinet players and the homogeneous group of two celesta players have the same computational burden in comparison with the TMS-EMSA with three and two players, respectively. In addition, each of the existing players in each of the aforementioned homogeneous musical groups has an equal computational burden compared to the SS-IHSA. As a result, the proposed SOSA has three and nine times more computational burden than the TMS-EMSA and SS-IHSA. Stated another way, the proposed SOSA has as much computational burden as NHMG—the number of homogeneous musical groups in the symphony orchestra—times greater than the TMS-EMSA and as much as $\sum_{g \in \Psi^{\text{NHMG}}} PN_g$—the number of the existing players in the symphony orchestra—times greater than the SS-IHSA. Given this fact, the principal question is whether employing the proposed SOSA with this high computational burden is rational and cost effective? In answer to this question, two perspectives must be considered, which depend on the time needed to solve the optimization problem. The first perspective is related to real-time optimization problems with small data, which must be solved in real time or near real time (e.g., every few minutes or every few hours) and the responses/outputs must be made available to the specialist/researcher/planner in order to decide the next time step. Charging and

Table 4.26 The pseudocode pertaining to improvisation of a new melody vector by each existing player in the symphony orchestra of the proposed SOSA

Algorithm 22: Pseudocode for improvisation of a new melody vector by each existing player in the symphony orchestra of the proposed SOSA

Input: $BW_{g,p}^{\max}$, $BW_{g,p}^{\min}$, MNI-GISHME, MNI-GISIME, MNI-SIS, NCDV, NDDV, NDV, $PAR_{g,p}^{\max}$, $PAR_{g,p}^{\min}$, $PMCR_{g,p}$, $PMS_{g,p}$, PN_g, $x_v^{\min}$, $x_v^{\max}$, $\{x_v(1), \ldots, x_v(w_v), \ldots, x_v(W_v)\}$

Output: $\mathrm{x}_{m,g,p}^{\mathrm{new}}$

start main body

begin

$BW_{m,g,p} = BW_{g,p}^{\max} \cdot \exp\left[\left(\ln\left(BW_{g,p}^{\max}/BW_{g,p}^{\min}\right)/((\text{MNI-SIS}) + (\text{MNI-GISHMG}) + (\text{MNI-GISIME}))\right) \cdot m\right]$

$PAR_{m,g,p} = PAR_{g,p}^{\min} - \left[\left(\left(PAR_{g,p}^{\max} - PAR_{g,p}^{\min}\right)/((\text{MNI-SIS}) + (\text{MNI-GISHMG}) + (\text{MNI-GISIME}))\right) \cdot m\right]$

for *homogeneous musical group g* $[g \in \Psi^{\text{NHMG}}]$ **do**

  for *music player p of the homogeneous musical group g* $\left[p \in \Psi^{PN_g}\right]$ **do**

    construct the new melody vector for the music player p of the homogeneous music group g, $\mathrm{x}_{m,g,p}^{\mathrm{new}}$,with dimension $\{1\} \cdot \{\text{NDV} + 1\}$ and zero initial value

    for *decision-making variable v* $[v \in \Psi^{\text{NDV}}]$ **do**

      if $\mathrm{U}(0,1) \le PMCR_{g,p}$ **then**

        Rule 1: harmony memory consideration with probability $PMCR_{g,\,p}$

      if *improvisation/iteration m* $[m \in \Psi^{(\text{MNI-SIS}) + (\text{MNI-GISHMG}) + (\text{MNI-GISIME})}]$ is odd **then**

        Principle 1 of Rule 1: first combination

        $x_{m,g,p,v}^{\text{new}} = x_{m,g,p,v}^{r_{g,p}} \pm \mathrm{U}(0,1) \cdot BW_{m,g,p}; \forall r_{g,p} \sim \mathrm{U}\{1, 2, \ldots, PMS_{g,p}\}$; for CDVs

        $x_{m,g,p,v}^{\text{new}} = x_{m,g,p,v}^{r_{g,p}}; \forall r_{g,p} \sim \mathrm{U}\{1, 2, \ldots, PMS_{g,\,p}\}$; for DDVs

      else

        Principle 2 of Rule 1: second combination

        $x_{m,g,p,v}^{\text{new}} = x_{m,g,p,k}^{r_{g,p}} \pm \mathrm{U}(0,1) \cdot BW_{m,g,p}; \forall r_{g,p} \sim \mathrm{U}\{1, 2, \ldots, PMS_p\}, \forall k \sim \mathrm{U}\{1, 2, \ldots,$ NCDV$\}$; for CDVs

        $x_{m,g,p,v}^{\text{new}} = x_{m,g,p,l}^{r_{g,p}}; \forall r_{g,p} \sim \mathrm{U}\{1, 2, \ldots, PMS_{g,p}\}, \forall l \sim \mathrm{U}\{1, 2, \ldots, \text{NDDV}\}$; for DDVs

      end if

      if $\mathrm{U}(0,1) \le PAR_{m,g,p}$ **then**

        Rule 2: pitch adjustment with probability $PMCR_{g,p} \cdot PAR_{g,p}$

        switch 1

        case *improvisation/iteration m* $[m \in \Psi^{(\text{MNI-SIS}) + (\text{MNI-GISHMG}) + (\text{MNI-GISIME})}]$ $\le$ (MNI-SIS) **then**

          Principle 1 of Rule 2: first choice

          $x_{m,g,p,v}^{\text{new}} = x_{m,g,p,v}^{\text{best}}$; for CDVs and DDVs

        case *improvisation/iteration m* $[m \in \Psi^{(\text{MNI-SIS}) + (\text{MNI-GISHMG}) + (\text{MNI-GISIME})}]$ $>$ (MNI-SIS) and *improvisation/iteration m* $[m \in \Psi^{(\text{MNI-SIS}) + (\text{MNI-GISHMG}) + (\text{MNI-GISIME})}]$ $\le$ (MNI-SIS) + (MNI-GISHMG) **then**

          Principle 2 of Rule 2: second choice

          $x_{m,g,p,v}^{\text{new}} = x_{m,g,\text{best},v}^{\text{best}}$; for CDVs and DDVs

        case *improvisation/iteration m* $[m \in \Psi^{(\text{MNI-SIS}) + (\text{MNI-GISHMG}) + (\text{MNI-GISIME})}]$ $>$ (MNI-SIS) + (MNI-GISHMG) and *improvisation/iteration m* $[m \in \Psi^{(\text{MNI-SIS}) + (\text{MNI-GISHMG}) + (\text{MNI-GISIME})}]$ $\le$ (MNI-SIS) + (MNI-GISHMG) + (MNI-GISIME) **then**

          Principle 3 of Rule 2: third choice

          $x_{m,g,p,v}^{\text{new}} = x_{m,\text{best},\text{best},v}^{\text{best}}$; for CDVs and DDVs

        end switch

      end if

    else if

      Rule 3: random selection with probability $1 - PMCR_{g,p}$

    switch 1

    case *improvisation/iteration m* $[m \in \Psi^{(\text{MNI-SIS}) + (\text{MNI-GISHMG}) + (\text{MNI-GISIME})}]$ $\le$ (MNI-SIS) **then**

      Principle 1 of Rule 3: first choice

      $x_{m,g,p,v}^{\text{new}} = x_v^{\min} + \mathrm{U}(0,1) \cdot \left(x_v^{\max} - x_v^{\min}\right)$; for CDVs

      $x_{m,g,p,v}^{\text{new}} = x_v(y); \forall y \sim \mathrm{U}\{x_v(1), \ldots, x_v(w_v), \ldots, x_v(W_v)\}$; for DDVs

(continued)

Table 4.26 (continued)

case *improvisation/iteration m* $[m \in \Psi^{\text{(MNI-SIS) + (MNI-GISHMG) + (MNI-GISIME)}}] >$ (MNI-SIS) and *improvisation/iteration m* $[m \in \Psi^{\text{(MNI-SIS) + (MNI-GISHMG) + (MNI-GISIME)}}] \leq$ (MNI-SIS) + (MNI-GISHMG) **then**
Principle 2 of Rule 3: second choice
$x^{\text{new}}_{m,g,p,v} = x^{\text{min}}_{m,g,v} + \text{U}(0,1) \cdot \left(x^{\text{max}}_{m,g,v} - x^{\text{min}}_{m,g,v}\right)$; for CDVs
$x^{\text{new}}_{m,g,p,v} = x_v(y); \forall y \sim \text{U}\{x_v(1), \ldots, x_v(w_v), \ldots, x_v(W_v)\}$; for DDVs
case *improvisation/iteration m* $[m \in \Psi^{\text{(MNI-SIS) + (MNI-GISHMG) + (MNI-GISIME)}}] >$ (MNI-SIS) + (MNI-GISHMG) and *improvisation/iteration m* $[m \in \Psi^{\text{(MNI-SIS) + (MNI-GISHMG) + (MNI-GISIME)}}] \leq$ (MNI-SIS) + (MNI-GISHMG) + (MNI-GISIME) **then**
Principle 3 of Rule 3: third choice
$x^{\text{new}}_{m,g,p,v} = x^{\text{min}}_{m,v} + \text{U}(0,1) \cdot \left(x^{\text{max}}_{m,v} - x^{\text{min}}_{m,v}\right)$; for CDVs
$x^{\text{new}}_{m,g,p,v} = x_v(y); \forall y \sim \text{U}\{x_v(1), \ldots, x_v(w_v), \ldots, x_v(W_v)\}$; for DDVs
end switch
end if
end for
calculate the value of objective function, fitness function, derived from melody vector $\text{x}^{\text{new}}_{m,g,p}$ as $f\left(\text{x}^{\text{new}}_{m,g,p}\right)$
allocate $f\left(\text{x}^{\text{new}}_{m,g,p}\right)$ to element $(1, \text{NDV} + 1)$ of the new melody vector $\text{x}^{\text{new}}_{m,g,p}$
end for
end for
terminate
end main body

Note: Continuous decision-making variable (CDVs), discrete decision-making variable (DDVs)

discharging management problems of plug-in hybrid electric vehicles in order to improve technical criteria of electrical distribution networks is an instance of real-time optimization problems in the field of electrical power engineering. In the solution process of such optimization problems, the well-adjusted computational burden of the optimization algorithm utilized is crucial and decisive.

Technically speaking, the architecture of the newly developed SOSA has been developed in such a way that it has a very high flexibility and robustness with parallel functionality. These distinctive characteristics effectively help to easily convert the computational burden of the proposed SOSA into an adjustable and controllable parameter. In doing so, the number of available homogeneous musical groups in the symphony orchestra and the number of existing players in each available homogeneous musical group in the symphony orchestra must also be fine-tuned in the architecture of the SOSA. As a consequence, by employing these parameters, the specialist/researcher/planner can purposefully adapt the computational burden of the proposed SOSA with time steps in which real-time or near-real-time optimization problems should be solved. With that in mind, the proposed SOSA may be a reasonable and applicable optimization algorithm for dealing with real-time or near-real-time optimization problems, due to its favorable flexibility in tuning the computational burden and high robustness compared to other optimization algorithms.

The second perspective is relevant to non-real-time optimization problems with big data, which do not need to be solved in real time or near real time.

Table 4.27 Pseudocode related to the performance-driven architecture of the proposed SOSA

Algorithm 23: Pseudocode for performance-driven architecture of the proposed SOSA

Input: $BW_{g,p}^{\max}$, $BW_{g,p}^{\min}$, MNI-GISHME, MNI-GISIME, MNI-SIS, NCDV, NDDV, NDV, $PAR_{g,p}^{\max}$, $PAR_{g,p}^{\min}$, $PMCR_{g,p}$, $PMS_{g,p}$, PN_g, $x_v^{\min}$, $x_v^{\max}$, $\{x_v(1), \ldots, x_v(w_v), \ldots, x_v(W_v)\}$

Output: x^{best}

start main body

begin
Stage 1—Definition stage: Definition of the optimization problem and its parameters
Stage 2—Initialization stage
    Sub-stage 2.1: Initialization of the parameters of the SOSA
    Sub-stage 2.2: Initialization of the of the SOM
        Algorithm 15: Pseudocode for initialization of the entire set of PMs or entire set of MMs or SOM in the proposed SOSA
        Algorithm 16: Pseudocode for sorting of the solution vectors stored in the PMs or MMs or SOM in the proposed SOSA
Stage 3—Single computational stage or SIS
    set *improvisation/iteration* $m = 1$
    set SOM_m = SOM
    while $m \leq$ (MNI-SIS) **do**
        Sub-stage 3.1: Improvisation of a new melody vector by each existing player in the symphony orchestra
            Algorithm 22: Pseudocode for improvisation of a new melody vector by each existing player in the symphony orchestra of the proposed SOSA
        Sub-stage 3.2: Update of each available PM in the symphony orchestra
            Algorithm 17: Pseudocode for the update of the memory of all existing players in the symphony orchestra or the update of the SOM_m in the proposed SOSA
            Algorithm 18: Pseudocode for sorting of the solution vectors stored in the PMs or MMs or SOM_m in the proposed SOSA
        set *improvisation/iteration* $m = m + 1$
    end while
Stage 4—Group computational stage for each homogeneous musical group or GISHMG
    while $m >$ (MNI-SIS) and $m \leq$ (MNI-SIS) + (MNI-GISHMG) **do**
        Sub-stage 4.1: Improvisation of a new melody vector by each existing player in the symphony orchestra taking into account the feasible ranges of the updated pitches for each homogeneous musical group
            Algorithm 22: Pseudocode for improvisation of a new melody vector by each existing player in the symphony orchestra of the proposed SOSA
        Sub-stage 4.2: Update of each available PM in the symphony orchestra
            Algorithm 17: Pseudocode for the update of the memory of all existing players in the symphony orchestra or the update of the SOM_m in the proposed SOSA
            Algorithm 18: Pseudocode for sorting of the solution vectors stored in the PMs or MMs or SOM_m in the proposed SOSA

(continued)

Table 4.27 (continued)

Sub-stage 4.3: Update of the feasible ranges of the pitches—continuous decision-making variables—for each homogeneous musical group in the next improvisation—only for random selection

Algorithm 19: Pseudocode for update of the feasible ranges of the continuous decision-making variables for each homogeneous musical group in the symphony orchestra in the proposed SOSA

set *improvisation/iteration* $m = m + 1$

end while

Stage 5—Group computational stage for inhomogeneous musical ensemble or GISIME

while $m >$ (MNI-SIS) + (MNI-GISHMG) and $m \leq$ (MNI-SIS) + (MNI-GISHMG) + (MNI-GISIME) **do**

Sub-stage 5.1: Improvisation of a new melody vector by each existing player in the symphony orchestra taking into account the feasible ranges of the updated pitches for the inhomogeneous musical ensemble

Algorithm 22: Pseudocode for improvisation of a new melody vector by each existing player in the symphony orchestra of the proposed SOSA

Sub-stage 5.2: Update of each available PM in the symphony orchestra

Algorithm 17: Pseudocode for the update of the memory of all existing players in the symphony orchestra or the update of the SOM_m in the proposed SOSA

Algorithm 18: Pseudocode for sorting of the solution vectors stored in the PMs or MMs or SOM_m in the proposed SOSA

Sub-stage 5.3: Update of the feasible ranges of the pitches—continuous decision-making variables—for the inhomogeneous musical ensemble in the next improvisation—only for random selection

Algorithm 20: Pseudocode for the update of the feasible ranges of the continuous decision-making variables for the inhomogeneous musical ensemble in the proposed SOSA

set *improvisation/iteration* $m = m + 1$

end while

Stage 6—Selection stage: Selection of the final optimal solution—the best melody

Algorithm 21: Pseudocode for the selection of the final optimal solution in the proposed SOSA

terminate

end main body

Pseudo-dynamic generation, transmission, and distribution expansion planning problems are instances of non-real-time optimization problems with big data, which will be extensively represented in Chap. 6. Although, to solve the non-real-time optimization problems with big data, employing the proposed SOSA can bring about high computational burden compared to other optimization algorithms, achieving the optimal value of objective functions is more noteworthy than the computational burden in such optimization problems. To illustrate, consider the pseudo-dynamic transmission expansion planning problem. One of the significant and widely used objective functions of this problem is the amount of investment costs. The scale of the investment costs in the pseudo-dynamic transmission expansion planning problem is generally over a few hundred million dollars or even a few billion dollars. In this regard, if the use of the proposed SOSA instead of other optimization algorithms in the solution process of this optimization problem gives rise to a reduction in the investment costs by 10%, a few tens of millions of dollars or a few hundred million dollars will be saved. For the same reason, this amount of savings may mostly be observed in other objective functions of the pseudo-dynamic transmission expansion planning problem. It can, therefore, be found that increasing computational burden in the solution process of non-real-time optimization problems with big data is less important than the amount of savings in objective functions, and then finding a most favorable response/output. Similar to real-time optimization problems with small data, in non-real-time optimization problems with big data, the computational burden of the proposed SOSA can be handled by adjusting the number of available homogeneous musical groups in the symphony orchestra and the number of existing players in each available homogeneous musical group in the symphony orchestra by the specialist/researcher/planner. As a result, desirable flexibility in adjusting the computational burden and high robustness of the proposed SOSA compared to its other counterparts may make this optimization algorithm an advisable and applicable option for dealing with non-real-time optimization problems with big data.

Another thing to highlight about the newly developed SOSA is that this optimization algorithm is an innovative architectural version of the TMS-EMSA. The TMS-EMSA is itself a novel structural version of the SS-HSA. Stated another way, the authors invented and developed the SOSA as a more efficient and effective optimization algorithm for tackling the difficulties in solving optimization problems by restructuring the architecture of the TMS-EMSA and the SS-HSA. Accordingly, many of the improvements employed in the SS-HSA, from the perspective of the parameter adjustments, can be readily implemented on the SOSA, due to its unique architecture. As a consequence, the performance of the SOSA can be further enhanced by using the improvements inspired by the SS-HSA and matching these improvements to the architecture of the SOSA.

Moreover, given the distinctive architecture of the SOSA, the calculations associated with different available homogeneous musical groups in the symphony orchestra and/or the computations related to the different existing players in each available homogeneous musical group can be performed using parallel processing/parallel computing. As a result, this characteristic of the SOSA can result in multiple advantages of parallel functionality in the solution process of complicated, real-world, large-scale, non-convex, non-smooth optimization problems having a nonlinear, mixed-integer nature with big data.

4.5 Multi-objective Strategies for the Music-Inspired Optimization Algorithms

Many real-world engineering optimization problems, particularly optimization problems pertaining to electrical engineering, have more than one objective function; these are referred to as multi-objective optimization problems (MOOPs). The objective functions of the MOOPs mostly have a conflicting, noncommensurable, and correlated nature. As discussed earlier in Sect. 2.2 of Chap. 2, the most reasonable strategy is to take advantage of the multi-objective optimization process in dealing with an optimization problem with multiple conflicting, noncommensurable, and correlated objective functions. An exhaustive introduction associated with the fundamental concepts of optimization in the MOOPs, along with a thorough categorization related to multi-objective optimization algorithms (MOOAs) with a focus on the role of the decision maker in the optimization process, is described in Sects. 2.3 and 2.4 of Chap. 2, respectively. The meta-heuristic MOOAs are considered to be one of the methods for solving MOOPs, which have been increasingly employed by specialists and researchers in various engineering sciences in recent years. The meta-heuristic MOOAs are a subset of a posteriori approaches, while a posteriori approaches are a subset of the noninteractive approaches, as previously indicated in Sect. 2.4.1 of Chap. 2.

In recent years, many meta-heuristic MOOAs have been developed by modifying the architecture of their single-objective versions. In regard to these multi-objective strategies, each of the meta-heuristic MOOAs has different strengths and weaknesses. The most well-known meta-heuristic MOOA is the non-dominated sorting genetic algorithm II (NSGA-II), which was first introduced in 2002 [4]. Although the NSGA-II was introduced many years ago, it is recognized as one of the most suitable and most powerful meta-heuristic MOOAs, due to its well-organized architecture in dealing with the MOOPs. Moreover, the architectures of the meta-heuristic music-inspired optimization algorithms described in Chap. 3 and this chapter, including the SS-HSA, the SSI-IHSA, the continuous TMS-MSA, the proposed continuous/discrete TMS-MSA, the proposed TMS-EMSA, and the proposed SOSA, have been developed in such a way that they are only suitable for solving single-objective optimization problems (SOOPs) and cannot be utilized for solving MOOPs.

In this section, then, the authors propose various multi-objective strategies for restructuring the architecture of these meta-heuristic music-inspired optimization algorithms in such a way that they are capable of solving MOOPs.

4.5.1 Multi-objective Strategies for the Meta-heuristic Music-Inspired Optimization Algorithms with Single-Stage Computational and Single-Dimensional Structure

In this section, the authors outline a new multi-objective strategy for modifying the architecture of the meta-heuristic music-inspired optimization algorithms with the single-stage computational and single-dimensional structure (i.e., the SS-HSA and SS-IHSA).

4.5.1.1 Multi-objective Strategy for the SS-HSA

The performance-driven architecture of the original SS-HSA—the single-objective SS-HSA—was previously discussed in Sect. 3.3 of Chap. 3. Due to the difference in the process of solving MOOPs compared to SOOPs, the performance-driven architecture of the SS-HSA must be restructured. Therefore, by implementing the proposed multi-objective strategy on the architecture of the single-objective SS-HSA, a multi-objective SS-HSA was developed in order to solve the MOOPs. The performance-driven architecture of the proposed multi-objective SS-HSA is generally broken down into four stages, as follows:

- Stage 1—Definition stage: Definition of the MOOP and its parameters
- Stage 2—Initialization stage
  - Sub-stage 2.1: Initialization of the parameters of the multi-objective SS-HSA
  - Sub-stage 2.2: Initialization of the input harmony memory (IHM)
- Stage 3—External computational stage
  - Sub-stage 3.1: Internal computational sub-stage

    Sub-stage 3.2.1: Improvisation of a new harmony vector
    Sub-stage 3.2.2: Update of the IHM
    Sub-stage 3.2.3: Check of the stopping criterion of the internal computational sub-stage
  - Sub-stage 3.2: Integration and separation procedures of harmony vectors
  - Sub-stage 3.3: Check of the stopping criterion of the external computational sub-stage
- Stage 4—Selection stage: Selection of the final optimal solution—the best harmony

Stage 1 is related to the definition of the MOOP and its parameters and is equivalent to stage 1 of the single-objective SS-HSA. In stage 1 of the single-objective SS-HSA, definition of the SOOP and its parameters is performed according to Eqs. (3.1) and (3.2), which were presented in Sect. 3.3.1 of Chap. 3. In the proposed multi-objective SS-HSA, however, this stage must be redefined. In stage 1 of the multi-objective SS-HSA, the standard form of a MOOP can be generally defined according to Eqs. (4.65) and (4.66):

$$\begin{aligned}
\underset{x\in X}{\text{Minimize}}\quad & \mathrm{F(x)} = [f_1(\mathrm{x}),\ldots,f_a(\mathrm{x}),\ldots,f_{\mathrm{A}}(\mathrm{x})];\quad \forall\{\mathrm{A}>0\},\forall\{a\in\Psi^{\mathrm{A}}\}\\
& \text{subject to :}\\
& \mathrm{G(x)} = [g_1(\mathrm{x}),\ldots,g_b(\mathrm{x}),\ldots,g_{\mathrm{B}}(\mathrm{x})] = 0;\quad \forall\{\mathrm{B}\geq 0\},\forall\{b\in\Psi^{\mathrm{B}}\}\\
& \mathrm{H(x)} = [h_1(\mathrm{x}),\ldots,h_e(\mathrm{x}),\ldots,h_{\mathrm{E}}(\mathrm{x})] \leq 0;\quad \forall\{\mathrm{E}\geq 0\},\forall\{e\in\Psi^{\mathrm{E}}\}
\end{aligned} \tag{4.65}$$

$$\mathrm{x} = [x_1, \ldots, x_v, \ldots, x_{\mathrm{NDV}}]; \quad \forall\{v \in \Psi^{\mathrm{NDV}}, \Psi^{\mathrm{NDV}} = \Psi^{\mathrm{NCDV+NDDV}}, \mathrm{x} \in \mathrm{X}\},$$
$$\forall\{x_v^{\min} \le x_v \le x_v^{\max} | v \in \Psi^{\mathrm{NCDV}}\}, \{x_v \in \{x_v(1), \ldots, x_v(w), \ldots, x_v(W_v)\} | v \in \Psi^{\mathrm{NDDV}}\} \tag{4.66}$$

The explanations associated with the parameters and variables from Eqs. (4.65) and (4.66) were previously described in Sect. 1.2.1 of Chap. 1. The vector of objective functions addresses the illustration of the vector of decision-making variables and contains the values of the objective functions, as illustrated in Eq. (4.67):

$$\mathrm{z} = \mathrm{F}(\mathrm{x}) = [f_1(\mathrm{x}), \ldots, f_a(\mathrm{x}), \ldots, f_{\mathrm{A}}(\mathrm{x})]; \quad \forall\{\mathrm{A} \ge 2\}, \forall\{a \in \Psi^{\mathrm{A}}\} \tag{4.67}$$

The illustration of the nonempty feasible decision-making space in the objective space is known as feasible objective space Z = F(X) and is defined by using the set {F(x)|x ∈ X}. Unlike the single-objective SS-HSA that explores the entire space of the nonempty feasible decision-making in order to find the vector of optimal decision-making variables or solution vector, the proposed multi-objective SS-HSA performs this task in order to achieve the set of Pareto-optimal vectors of decision-making variables or the Pareto-optimal solution set or non-dominated optimal solution set. A detailed description relevant to the Pareto-optimal solution set was previously provided in Chap. 2.

Basically, the multi-objective SS-HSA simultaneously considers the objective functions given in Eq. (4.65). Nonetheless, if the solution vector obtained by the multi-objective SS-HSA gives rise to any violation in equality and/or inequality constraints given in Eq. (4.65), the algorithm can employ the penalty function strategy. In this strategy, the multi-objective SS-HSA takes into account the obtained solution vector by applying a specified penalty function to each objective function of the MOOP. The penalty function added to each objective function of the MOOP is compatible with the type and nature of the corresponding objective function.

Sub-stage 2.1 is related to the initialization of the parameters of the multi-objective SS-HSA. This sub-stage is equivalent to sub-stage 2.1 of the single-objective SS-HSA. In this sub-stage, the parameter adjustments of the single-objective SS-HSA are characterized according to Table 3.2, which was presented in Sect. 3.3.2.1 of Chap. 3. As for Table 3.2, the harmony memory (HM) is a place for storing the solution, or harmony vectors. Unlike the single-objective SS-HSA that has a unique HM, the performance-driven architecture of the multi-objective SS-HSA has been developed in such a way that it requires the use of three HMs: (1) the IHM; (2) the output harmony memory (OHM); and, (3) the hybrid harmony memory (HHM). The IHM is considered to be the input for the internal computational sub-stage (sub-stage 3.1) and is updated in each iteration of the external computational stage (stage 3). The OHM is also extracted after completion of the internal computational sub-stage (sub-stage 3.1). Put simply, the OHM is obtained by applying the internal computational sub-step (sub-stage 3.1) on the IHM.

In addition, the HHM is constructed from the integration of the IHM and OHM and used to create the IHM in the next iteration of the external computational stage or the new IHM. Further explanations pertaining to these parameters are reported in the related sub-stages. The single-objective SS-HSA has a computational stage in which this stage is repeated to the maximum number of improvisations/iterations (MNI). In contrast, the multi-objective SS-HSA has an external computational stage and an internal computational sub-stage. The external computational stage is repeated to the maximum number of iteration of the external computational stage (MNI-E). In each iteration of the external computational stage, the internal computational sub-stage is also repeated to the maximum number of improvisation/iteration of the internal computational stage (MNI-I). As a consequence, the multi-objective SS-HSA employs the MNI-E and MNI-I parameters instead of the MNI parameter. Other parameters presented in Table 3.2 remain unchanged for the multi-objective SS-HSA. The detailed descriptions associated with the adjustment parameters of the multi-objective SS-HSA are provided in Table 4.28. Sub-stage 2.2 is related to the initialization of the IHM. This sub-stage is equivalent to sub-stage 2.2 of the single-objective SS-HSA. In this sub-stage of the single-objective SS-HSA, the initialization of the HM is performed according to Eqs. (3.4)–(3.6) given in Sect. 3.3.2.2 of Chap. 3.

In the multi-objective SS-HSA, due to the replacement of the SOOP with the MOOP in stage 1, the initialization of the IHM will be faced with changes. In the multi-objective SS-HSA, the IHM matrix, which has a dimension equal to $\{HMS\} \cdot \{NDV + A\}$, is filled with a large number of solution vectors generated randomly according to Eqs. (4.68)–(4.70). It is clear that Eq. (4.68) is different with its counterpart in the single-objective SS-HSA; however, Eqs. (4.69) and (4.70) remain unchanged. The pseudocode associated with initialization of the

Table 4.28 Adjustment parameters of the proposed multi-objective SS-HSA

No.	The proposed multi-objective SS-HSA parameter	Abbreviation	Parameter range
1	Input harmony memory	IHM	–
2	Output harmony memory	OHM	–
3	Hybrid harmony memory	HHM	–
4	Harmony memory size	HMS	$HMS \geq 1$
5	Harmony memory considering rate	HMCR	$0 \leq HMCR \leq 1$
6	Pitch adjusting rate	PAR	$0 \leq PAR \leq 2$
7	Bandwidth	BW	$0 \leq BW < +\infty$
8	Number of continuous decision-making variables	NCDV	$NCDV \geq 1$
9	Number of discrete decision-making variables	NDDV	$NDDV \geq 1$
10	Number of decision-making variables	NDV	$NDV \geq 2$
11	Maximum number of iterations of the external computational stage	MNI-E	$MNI\text{-}E \geq 1$
12	Maximum number of improvisations/iterations of the internal computational sub-stage	MNI-I	$MNI\text{-}I \geq 1$

HM in the single-objective SS-HSA is presented in Table 3.3. In the multi-objective SS-HSA, however, due to the replacement of the SOOP with the MOOP in stage 1, this pseudocode must be restructured. As a result, Table 4.29 gives the pseudocode related to initialization of the IHM in the multi-objective SS-HSA.

$$\mathrm{IHM}=\begin{bmatrix} \mathrm{x}^1 \\ \vdots \\ \mathrm{x}^s \\ \vdots \\ \mathrm{x}^{\mathrm{HMS}} \end{bmatrix}=\left[\begin{array}{ccccc|ccccc} x_1^1 & \cdots & x_v^1 & \cdots & x_{\mathrm{NDV}}^1 & f(\mathrm{x}_1^1) & \cdots & f(\mathrm{x}_a^1) & \cdots & f(\mathrm{x}_{\mathrm{A}}^1) \\ \vdots & & \vdots & & \vdots & \vdots & & \vdots & & \vdots \\ x_1^s & \cdots & x_v^s & \cdots & x_{\mathrm{NDV}}^s & f(\mathrm{x}_1^s) & \cdots & f(\mathrm{x}_a^s) & \cdots & f(\mathrm{x}_{\mathrm{A}}^s) \\ \vdots & & \vdots & & \vdots & \vdots & & \vdots & & \vdots \\ x_1^{\mathrm{HMS}} & \cdots & x_v^{\mathrm{HMS}} & \cdots & x_{\mathrm{NDV}}^{\mathrm{HMS}} & f(\mathrm{x}_1^{\mathrm{HMS}}) & \cdots & f(\mathrm{x}_a^{\mathrm{HMS}}) & \cdots & f(\mathrm{x}_{\mathrm{A}}^{\mathrm{HMS}}) \end{array}\right];$$
$$\forall\left\{v\in\Psi^{\mathrm{NDV}}, s\in\Psi^{\mathrm{HMS}}, a\in\Psi^{\mathrm{A}}, \Psi^{\mathrm{NDV}}=\Psi^{\mathrm{NCDV+NDDV}}\right\} \tag{4.68}$$

Table 4.29 Pseudocode related to initialization of the IHM in the proposed multi-objective SS-HSA

Algorithm 24: Pseudocode for initialization of the IHM in the proposed multi-objective SS-HSA

Input: A, HMS, NCDV, NDDV, NDV, $x_v^{\min}$, $x_v^{\max}$, $\{x_v(1), \ldots, x_v(w_v), \ldots, x_v(W_v)\}$
Output: IHM

start main body

begin
construct the IHM matrix with dimension {HMS} · {NDV + A} and zero initial value
for *harmony vector s* $[s \in \Psi^{\mathrm{HMS}}]$ **do**
  construct the harmony vector x^s with dimension {1} · {NDV + A} and zero initial value
  for *decision-making variable v* $[v \in \Psi^{\mathrm{NDV}}]$ **do**
    $x_v^s = x_v^{\min} + \mathrm{U}(0,1)\cdot\left(x_v^{\max} - x_v^{\min}\right)$; for CDVs
    $x_v^s = x_v(y); \forall y \sim \mathrm{U}\{x_v(1), \ldots, x_v(w_v), \ldots, x_v(W_v)\}$; for DDVs
    allocate x_v^s to element $(1, v)$ of the harmony vector x^s
  end for
  for *objective function a* $[a \in \Psi^{\mathrm{A}}]$ **do**
    calculate the value of objective function *a*, fitness function, derived from the harmony vector x^s as $f(\mathrm{x}_a^s)$
    allocate $f(\mathrm{x}_a^s)$ to element $(1, \mathrm{NDV} + a)$ of the harmony vector x^s
  end for
  add harmony vector x^s to the row *s* of the IHM matrix
end for
terminate

end main body

Note: Continuous decision-making variable (CDVs), discrete decision-making variable (DDVs)

$$x_v^s = x_v^{\min} + \mathrm{U}(0,1) \cdot \left(x_v^{\max} - x_v^{\min}\right); \quad \forall\left\{v \in \Psi^{\mathrm{NCDV}}, s \in \Psi^{\mathrm{HMS}}\right\} \tag{4.69}$$

$$x_v^s = x_v(y); \quad \forall\left\{v \in \Psi^{\mathrm{NDDV}}, s \in \Psi^{\mathrm{HMS}}, y \sim \mathrm{U}\{x_v(1), \ldots, x_v(w_v), \ldots, x_v(W_v)\}\right\} \tag{4.70}$$

In the single-objective SS-HSA, after filling the HM with random solution vectors, the solution vectors stored in the HM must be sorted from the lowest value to the highest value—in an ascending order—with regard to the value of the objective function of the SOOP. The pseudocode related to sorting the solution vectors stored in the HM under the single-objective SS-HSA is presented in Table 3.4 in Sect. 3.3.2.2 of Chap. 3. As can be seen from Table 3.4, the sorting process of solution vectors stored in the HM under the single-objective SS-HSA is a simple and straightforward process. However, this pseudocode is efficient only for single-objective SS-HSA and is practically not possible to use in the multi-objective SS-HSA.

In the multi-objective optimization literature, there are various strategies to sort the solution vectors stored in the memory of a meta-heuristic MOOA. Each of these strategies has several advantages and disadvantages in terms of how it is implemented. One of the most appropriate and efficient strategies for sorting the solution vectors stored in the memory of a meta-heuristic MOOA is the strategy employed in the NSGA-II [4]. The strategy utilized in the NSGA-II is based on the application of two approaches: (1) the fast non-dominated sorting approach and (2) the crowded-comparison approach. By matching the original fast non-dominated sorting approach and the original crowded-comparison approach with the concepts of the multi-objective SS-HSA, the authors propose two modified versions of these approaches as a modified strategy in order to sort the harmony vectors—solution vectors—stored in the IHM under the multi-objective SS-HSA.

The modified fast non-dominated sorting approach (MFNDSA) is employed in order to determine the non-dominated front—rank—pertaining to each harmony vector stored in the IHM. In the MFNDSA, first two entities are calculated for each harmony vector stored in the IHM. To illustrate, consider the harmony vector s stored in the IHM. Two entities that must be calculated for this harmony vector are (1) the number of harmony vectors that dominate harmony vector s—domination count d_s—and (2) a set of harmony vectors where harmony vector s dominates—set S_s. Before continuing the description, it should be pointed out that the concepts related to dominance were previously presented in Sect. 2.3.2 of Chap. 2. Next, all harmony vectors with zero domination count will be placed in the first non-dominated front. For each harmony vector with zero domination count or more comprehensively for each harmony vector available on the first non-dominated front—like harmony vector s—each member or harmony vector—like harmony vector s^*—of its set S_s is recalled and one of its domination count is reduced. If the domination count becomes zero for any member or harmony vector s^*, this harmony vector will be placed in a separate collection (collection Q). All members or harmony vectors available in collection Q belong to the second non-dominated front. The above process will be then applied to the harmony vectors available on the second non-dominated front in order to identify the third non-dominated front. This process will continue until all non-dominated fronts are identified. Once all non-dominated fronts have been identified, the rank associated with each harmony vector stored in the IHM can be determined. The rank of each harmony vector stored in the IHM is the non-dominated front number on which the corresponding harmony vector is located on that front. Table 4.30 gives the pseudocode related to the MFNDSA in order to determine the non-dominated front—rank—of each harmony vector stored in the IHM under the multi-objective SS-HSA.

The modified crowded-comparison approach (MCCA) is employed in order to specify the crowding distance of each harmony vector stored in each identified non-dominated front or, more comprehensively, in the IHM. To explain the MCCA, a brief definition of the density-estimation metric will first be presented. In order to get an estimate of the density of harmony vectors surrounding a particular harmony vector available in each identified non-dominated front, the average distance of two harmony vectors on either side of this particular harmony vector along each of the objective functions of the MOOP is computed.

This quantity—the crowding distance—serves as an estimate of the perimeter of the cuboid formed by using the nearest neighbors as the vertices. Consider harmony vector s in the identified non-dominated front r of a double-objective optimization problem (see Fig. 4.5).

The crowding distance of this harmony vector, shown with a hollow square, in its non-dominated front, illustrated as a sum of the solid squares, is the average side length of the cuboid (depicted with a dotted box). In order to perform the crowding distance computation process of all harmony vectors available in each identified non-dominated front, it is necessary that these harmony vectors be sorted from the lowest value to the highest value in an ascending order with regard to the value of each objective function of this double-objective optimization problem. Then, for each objective function of this double-objective optimization problem, a crowding distance with an infinite amount is dedicated to the boundary harmony vectors—harmony vectors with the lowest and highest values of the objective function. A crowding distance with a value equal to the absolute normalized difference in the values of the objective function associated with two adjacent harmony vectors is also allocated to all other intermediate harmony vectors. For another objective function of this double-objective optimization problem, the above calculation process is repeated. The overall crowding distance value of each harmony vector available in each non-dominated front is computed as the sum of individual crowding distance values of this harmony vector and corresponding to each objective function of the double-objective optimization problem. It should be noted that each objective function of the double-objective optimization problem must be normalized before calculating the crowding distance. After assigning a crowding distance value to each harmony vector available in each non-dominated front, comparison of two harmony vectors for their extent of proximity to other harmony vectors can be performed. A harmony vector with a smaller value of this distance measure is, in some sense, more crowded by other harmony vectors. Although Fig. 4.5 demonstrates the crowding

Table 4.30 Pseudocode related to the MFNDSA used in order to determine the non-dominated front—rank—of each harmony vector stored in the IHM under the proposed multi-objective SS-HSA

Algorithm 25: Pseudocode for the MFNDSA used in order to determine the non-dominated front—rank—of each harmony vector stored in the IHM under the proposed multi-objective SS-HSA

Input: IHM
Output: Rank of each harmony vector stored in the IHM

start main body

begin
for *harmony vector s* $[s \in \Psi^{\text{HMS}}]$ **do**
  set $d_{x^s} = 0$ {the number of harmony vectors which dominate the harmony vector s}
  set $S_{x^s} = \varnothing$ {the set of harmony vectors that the harmony vector s dominates}
  for *harmony vector* s^* $[s^* \in \Psi^{\text{HMS}}, s^* \neq s]$ **do**
    if $s \prec s^*$ {harmony vector s dominates harmony vector s^*} **then**
      $S_{x^s} \cup \{x^{s^*}\}$ {add harmony vector s^* to the set of harmony vectors dominated by harmony vector s}
    else if $s^* \prec s$ {harmony vector s^* dominates harmony vector s}
      $d_{x^s} = d_{x^s} + 1$ {increment the domination count of harmony vector s}
    end if
  end for
  if $d_{x^s} = 0$ **then** {harmony vector s belongs to the first non-dominated front}
    $\text{rank}_{x^s} = 1$
    $\mathcal{F}_1 = \mathcal{F}_1 \cup \{x^s\}$
  end if
end for
set *non-dominated front* $r = 1$
while $\mathcal{F}_r \neq \varnothing$
  set $Q = \varnothing$ {employed to store the harmony vectors of the next non-dominated front}
  for *harmony vector s* $[s \in \mathcal{F}_r]$ **do**
    for *harmony vector* s^* $[s^* \in S_{x^s}]$ **do**
      $d_{x^{s^*}} = d_{x^{s^*}} - 1$ {decrement the domination count of harmony vector s^*}
      if $d_{x^{s^*}} = 0$ **then** {harmony vector s^* belongs to the non-dominated front $r + 1$}
        $\text{rank}_{x^{s^*}} = r + 1$
        $Q = Q \cup \{x^{s^*}\}$
      end if
    end for
  end for
  $r = r + 1$ {increment the non-dominated front r}
  $\mathcal{F}_r = Q$
end while
for *harmony vector s* $[s \in \Psi^{\text{HMS}}]$ **do**
  for *non-dominated front r* $[r \in \Psi^{\text{R}}]$ **do**
    if $s \in \mathcal{F}_r$ **do**
      $\text{rank}_{x^s} = r$ {assigning rank r to the harmony vector s}
    end if
  end for
end for
terminate

end main body

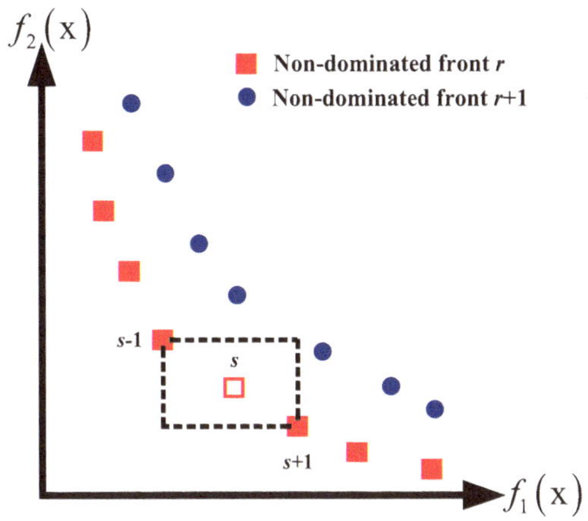

Fig. 4.5 Crowding distance of harmony vector s

distance computation process for a double-objective optimization problem, this process can be easily applied to an optimization problem with objective functions greater than 2. Table 4.31 presents the pseudocode related to the MCCA used in order to characterize the crowding distance of each harmony vector stored in the IHM under the multi-objective SS-HSA. The convergence to the Pareto-optimal solution set and maintenance of diversity in solutions of the Pareto-optimal set are introduced by using the MFNDSA and MCCA, respectively. After performing the MFNDSA and MCCA, rank and crowding distance of all harmony vectors stored in the IHM are determined. Then, by using these two parameters, the harmony vectors in the IHM will be sorted. The sorting process of harmony vectors stored in the IHM is accomplished in such a way that between two harmony vectors or more comprehensively two non-dominated fronts with different non-domination ranks, the harmony vector or non-dominated front with the lower—better—rank will be preferred. In addition, between two harmony vectors with same non-domination rank, the harmony vector with the larger—better—crowding distance will be preferred. Table 4.32 illustrates the pseudocode associated with sorting the harmony vectors stored in the IHM under the multi-objective SS-HSA. Unlike the single-objective SS-HSA, which employs a single-stage computational structure to find the optimal solution, the multi-objective SS-HSA utilizes an interconnected computational structure to achieve the Pareto-optimal solution set. This interconnected computational structure has an external computational stage and an internal computational sub-stage.

The internal computational sub-stage acts as a central core for the external computational stage. Stated another way, in each iteration of the external computational stage, the internal computational sub-stage is repeated to the MNI-I.

The stage 3 is related to the external computational stage. This stage forms the main body of the multi-objective SS-HSA and consists of three sub-stages: (1) the internal computational sub-stage; (2) the integration and separation procedures of harmony vectors; and, (3) a check of the stopping criterion of the external

Table 4.31 Pseudocode related to the MCCA used in order to determine the crowding distance of each harmony vector stored in the IHM under the proposed multi-objective SS-HSA

Algorithm 26: Pseudocode for the MCCA used in order to determine the crowding distance of each harmony vector stored in the IHM under the proposed multi-objective SS-HSA

Input: IHM
Output: Crowding distance of each harmony vector stored in the IHM

start main body

begin
for *non-dominated front r* $[r \in \Psi^{R}]$ **do**
  set $h_r = |\mathcal{F}_r|$ {the number of harmony vectors available in the non-dominated front r}
  for *harmony vector s* $[s \in \mathcal{F}_r]$ **do**
    set $\text{distance}_{x^s} = 0$ {initialize crowding distance of the harmony vectors s available in the non-dominated front r}
  end for
  for *objective function a* $[a \in \Psi^{A}]$ **do**
    $F^{\text{sort}}_{r,a} = \text{sort}(\mathcal{F}_r(1 : h_r, \text{NDV} + a), '\text{ascend}')$
    for *harmony vector s* $[s \in \mathcal{F}_r]$ **do** {sort the harmony vectors available in the non-dominated front r from the perspective of the value of objective function a}
      for *harmony vector* s^* $[s^* \in \mathcal{F}_r]$ **do**
        if $F^{\text{sort}}_{r,a}(s) == \mathcal{F}_r(s^*, \text{NDV} + a)$ **then**
          $\mathcal{F}^{\text{sort}}_{r,a}(s, 1 : \text{NDV} + \text{A}) = \mathcal{F}_r(s^*, 1 : \text{NDV} + \text{A});$
        end if
      end for
    end for
    set $\text{distance}_{x^1,a} = \text{distance}_{x^{h_r},a} = \infty$ {the first and the last harmony vectors stored in the sorted non-dominated front r from the perspective of the value of objective function a are assigned the crowding distance equal to infinity}
    for *harmony vector s* $[s \in \mathcal{F}_r, s \neq 1, s \neq h_r]$ **do**
      $\text{distance}_{x^s,a} = \left(f_a^{\mathcal{F}^{\text{sort}}_{r,a,s+1}} - f_a^{\mathcal{F}^{\text{sort}}_{r,a,s-1}}\right) / \left(f_a^{\max} - f_a^{\min}\right)$
    end for
    for *harmony vector s* $[s \in \mathcal{F}_r]$ **do**
      for *harmony vector* s^* $[s^* \in \mathcal{F}_r]$ **do**
        if $\mathcal{F}_r(s, 1 : \text{NDV}) == \mathcal{F}^{\text{sort}}_{r,a}(s^*, 1 : \text{NDV})$ **then**
          $\text{distance}_{x^s} = \text{distance}_{x^s} + \text{distance}_{x^{s^*},a}$ {compute the crowding distance of the harmony vectors s}
        end if
      end for
    end for
  end for
end for
terminate

end main body

Table 4.32 Pseudocode associated with sorting the harmony vectors stored in the IHM under the proposed multi-objective SS-HSA

Algorithm 27: Pseudocode for sorting the harmony vectors stored in the IHM under the proposed multi-objective SS-HSA

Input: Unsorted IHM
Output: Sorted IHM

start main body

begin

set *non-dominated front* $r = 1$

for *non-dominated front* r $[r \in \Psi^{R}]$ **do**

  set $\mathcal{F}_r = \varnothing$ {the non-dominated front r}

  for *harmony vector* s $[s \in \Psi^{HMS}]$ **do**

    if $\text{rank}_{x^s} == r$ **then**

      $\mathcal{F}_r \cup \{x^s\}$ {add harmony vector s to the non-dominated front r}

    end if

  end for

  set $h_r = |\mathcal{F}_r|$ {the number of harmony vectors available in the non-dominated front r}

  for *harmony vector* s $[s \in \mathcal{F}_r]$ **do**

    set *superiority* $q = 0$

    for *harmony vector* s^* $[s^* \in \mathcal{F}_r, s^* \neq s]$ **do**

      if $\text{distance}_{x^s} > \text{distance}_{x^{s^*}}$ **then**

        $q = q + 1$ {increment the superiority q}

      end if

    end for

    $Q(s) = q$

  end for

  for *harmony vector* s $[s \in \mathcal{F}_r]$ **do**

    for *harmony vector* s^* $[s^* \in \mathcal{F}_r]$ **do**

      if $Q(s^*) == h_r - s$ **then**

        $\mathcal{F}_r^{\text{sort}}(s, 1 \cdot \text{NDV} + \text{A})$ $\mathcal{F}_r(s^*)$ {sort the harmony vectors available in non-dominated front r}

      end if

    end for

  end for

  $\text{IHM}^{\text{sort}}\left(1 + \sum_{z \in \Psi^{r-1}} h_z : \sum_{r \in \Psi^{R}} h_r, 1 : \text{NDV} + \text{A}\right) = \mathcal{F}_r^{\text{sort}}$ {add the sorted non-dominated front r to the IHM}

end for

$\text{IHM} = \text{IHM}^{\text{sort}}$

terminate

end main body

computational stage. The mathematical equations expressed at the external computational stage must depend on the iteration index of this stage—index n—because of the repeatability of the external computational stage in the multi-objective SS-HSA.

It is important to highlight that in the first iteration of the external computational stage (stage 3), the IHM, which was initialized in sub-stage 2.2, is considered to be the input for the internal computational sub-stage (sub-stage 3.1). In the second iteration of the external computational stage (stage 3), however, the IHM obtained after the completion of sub-stage 3.2 in the first iteration of the external computational stage (stage 3) is considered to be the input for the internal computational sub-stage (sub-stage 3.1), and so on.

Simply put, except for the first iteration of the external computational stage (stage 3), the IHM initialized in sub-stage 2.1 is regarded as the input for the internal computational sub-stage; the IHM obtained after the completion of sub-stage 3.2 in iteration $n - 1$ of the external computational stage (stage 3) is considered to be the input for the internal computational sub-stage (sub-stage 3.1) in iteration n of the external computational stage (stage 3). Sub-stage 3.1 is associated with the internal computational stage. This sub-stage is equivalent to stage 3 of the single-objective SS-HSA. In the multi-objective SS-HSA, the internal computational sub-stage (sub-stage 3.1) is the first sub-stage of the external computational stage, which itself consists of three sub-stages: (1) improvisation of a new harmony vector; (2) update of the IHM; and, (3) check of the stopping criterion of the internal computational sub-stage. The mathematical equations expressed at sub-stage 3.1, in addition to the dependence on the iteration index of the external computational stage (index n), must also depend on the improvisation/iteration index of the internal computational sub-stage (index m), because of the repeatability of the internal computational sub-stage in the multi-objective SS-HSA.

Sub-stage 3.1.1 is relevant to the improvisation of a new harmony vector. This sub-stage of the multi-objective SS-HSA is virtually the same as sub-stage 3.1 of the single-objective SS-HSA, previously described in Sect. 3.3.3.1 of Chap. 3. Hence, in order to avoid repetition, the description related to this sub-stage is ignored. However, due to the addition of the external computational stage in the multi-objective SS-HSA, the iteration index of the external computational stage (index n) must be added to all mathematical equations expressed in Sect. 3.3.3.1 of Chap. 3, which have the improvisation/iteration index of the internal computational step (index m). The pseudocode pertaining to improvisation of a new harmony vector in the single-objective SS-HSA is presented in Table 3.5 of Sect. 3.3.3.1 of Chap. 3. In the multi-objective SS-HSA, however, due to the replacement of the SOOP with the MOOP in stage 1, this pseudocode must be restructured. As a result, Table 4.33 gives the pseudocode pertaining to improvisation of a new harmony vector in the multi-objective SS-HSA.

Sub-stage 3.1.2 is associated with the update process of the IHM. This sub-stage is equivalent to sub-stage 3.2 of the single-objective SS-HSA. A detailed description related to the update process of the HM_m in the single-objective SS-HSA was previously reported in Sect. 3.3.3.2 of Chap. 3. The pseudocode relevant to the update of the HM_m in the single-objective SS-HSA is also presented in Table 3.6 of Sect. 3.3.3.2 of Chap. 3. In the multi-objective SS-HSA, due to the replacement of the SOOP with the MOOP in stage 1, the update process of the IHM will be faced with changes. In this sub-stage of the multi-objective SS-HSA, a new harmony vector is evaluated and compared with the

Table 4.33 Pseudocode pertaining to improvisation of a new harmony vector in the proposed multi-objective SS-HSA

Algorithm 28: Pseudocode for improvisation of a new harmony vector in the proposed multi-objective SS-HSA

Input: A, BW, HMCR, HMS, NCDV, NDDV, NDV, PAR, $x_v^{\min}$, $x_v^{\max}$, $\{x_v(1), \ldots, x_v(w_v), \ldots, x_v(W_v)\}$

Output: $\mathrm{x}_{n,m}^{\mathrm{new}}$

start main body

begin
construct the new harmony vector $\mathrm{x}_{n,m}^{\mathrm{new}}$ with dimension $\{1\} \cdot \{\mathrm{NDV} + \mathrm{A}\}$ and zero initial value
for *decision-making variable v* $[v \in \Psi^{\mathrm{NDV}}]$ **do**
  if $\mathrm{U}(0,1) \leq \mathrm{HMCR}$ **then**
   Rule 1: The harmony memory consideration with the probability of the HMCR
   $x_{n,m,v}^{\mathrm{new}} = x_{n,m,v}^{r}; \forall r \sim \mathrm{U}\{1, 2, \ldots, \mathrm{HMS}\}$; for CDVs and DDVs
   if $\mathrm{U}(0,1) \leq \mathrm{PAR}$ **then**
    Rule 2: The pitch adjustment with the probability of the HMCR. PAR
    $x_{n,m,v}^{\mathrm{new}} = x_{n,m,v}^{\mathrm{new}} \pm \mathrm{U}(0,1) \cdot \mathrm{BW}$; for CDVs
    $x_{n,m,v}^{\mathrm{new}} = x_{n,m,v}^{\mathrm{new}}(y+t); y \sim \mathrm{U}\{x_v(1), \ldots, x_v(w_v), \ldots, x_v(W_v)\}, \quad \forall t \sim \mathrm{U}\{-1, +1\}$; for DDVs
   end if
  else if
    Rule 3: The random selection with the probability of the 1 − HMCR
  $x_{n,m,v}^{\mathrm{new}} = x_v^{\min} + \mathrm{U}(0,1).\left(x_v^{\max} - x_v^{\min}\right)$; for CDVs
  $x_{n,m,v}^{\mathrm{new}} = x_v(y); \forall y \sim \mathrm{U}\{x_v(1), \ldots, x_v(w_v), \ldots, x_v(W_v)\}$; for DDVs
  end if
end for
for *objective function a* $[a \in \Psi^{\mathrm{A}}]$ **do**
  calculate the value of objective function *a*, fitness function, derived from the harmony vector $\mathrm{x}_{n,m}^{\mathrm{new}}$ as $f\left(\mathrm{x}_{n,m,a}^{\mathrm{new}}\right)$
  allocate $f\left(\mathrm{x}_{n,m,a}^{\mathrm{new}}\right)$ to element $(1, \mathrm{NDV} + a)$ of the new harmony vector $\mathrm{x}_{n,m}^{\mathrm{new}}$
end for
terminate

end main body

Note: Continuous decision-making variable (CDVs), discrete decision-making variable (DDVs)

worst available harmony vector in the $IHM_{n,m}$—the harmony vector with the biggest rank and the smallest crowding distance located in the HMS row of the $IHM_{n,m}$—from the perspective of the objective functions. If a new harmony vector dominates the worst available harmony vector in the $IHM_{n,m}$, from the perspective of the objective functions, this new harmony vector replaces the worst harmony vector available in the $IHM_{n,m}$; the worst available harmony vector is then eliminated from the $IHM_{n,m}$. Table 4.34 shows the pseudocode related to the update of the $IHM_{n,m}$ in the multi-objective SS-HSA.

Table 4.34 Pseudocode related to update of the $IHM_{n,m}$ in the proposed multi-objective SS-HSA

Algorithm 29: Pseudocode for the update of the $IHM_{n,m}$ in the proposed multi-objective SS-HSA

Input: Not update $IHM_{n,m}$, $x_{n,m}^{new}$
Output: Updated $IHM_{n,m}$

start main body

1:	**begin**
2:	set $x^{worst} = x_{n,m}^{HMS}$
3:	**if** $f(x_{n,m}^{new}) \prec f(x^{worst})$ **then** {new harmony vector dominates worst harmony vector}
4:	$x_{n,m}^{new} \in IHM_{n,m}$
5:	$x^{worst} \notin IHM_{n,m}$
6:	**end if**
7:	**terminate**

end main body

The point to be made here is that the update process of the $IHM_{n,m}$ is not performed if the new harmony vector does not dominate the worst available harmony vector in the $IHM_{n,m}$, from the standpoint of the objective functions. After completion of this process, harmony vectors stored in the $IHM_{n,m}$ must be resorted. The pseudocode related to sorting the harmony vectors stored in the IHM was already provided in Tables 4.30, 4.31, and 4.32. Given the dependence of the IHM on the iteration index of the external computational stage (index n) and improvisation/iteration index of the internal computational stage (index m), the aforementioned pseudocode must be redefined by adding the indices n and m to all mathematical equations used in these pseudocodes.

Sub-stage 3.1.3 is relevant to the check of the stopping criterion of the internal computational sub-stage. This sub-stage of the multi-objective SS-HSA is virtually the same as sub-stage 3.3 of the single-objective SS-HSA, previously represented in Sect. 3.3.3.3 of Chap. 3. Only in the multi-objective SS-HSA does the MNI parameter have to be replaced with the MNI-I parameter. In sub-stage 3.1.3 of the multi-objective SS-HSA, the computational efforts of the internal computational sub-stage are terminated when its stopping criterion (the MNI-I) is satisfied. Otherwise, sub-stages 3.1.1 and 3.1.2 are repeated.

Sub-stage 3.2 is related to the integration and separation procedures of harmony vectors. The main goal of the implementation of this sub-stage is to construct the IHM for the next iteration of the external computational stage or the new IHM. The construction of the new IHM requires the IHM in the current iteration of the external computational stage and two other memories called the OHM and HHM. Implementation of this sub-stage in iteration n of the external computational stage begins by determining the IHM_n and OHM_n. First, the IHM for the internal computational sub-stage (sub-stage 3.1) in iteration n of the external computational stage is considered as the IHM_n. The IHM_n obtained after the completion of the internal computational sub-stage (sub-stage 3.1) is also regarded as the OHM_n. Then, the HHM_n must be created by applying the integration procedure on the IHM_n and OHM_n. Since all harmony vectors in the IHM_n and OHM_n are transferred to the

HHM_n, elitism is ensured. The size of the HHM_n is also twice the size of the IHM_n and the OHM_n. This means that unlike the IHM_n and OHM_n that have a size equal to {HMS} · {NDV + A}, the HHM_n has a size of {2 · HMS} · {NDV + A}. Then, all harmony vectors available in the HHM_n must be sorted according to their rank and crowding distance by employing the pseudocode provided in Tables 4.30, 4.31, and 4.32. However, due to the fact that HHM_n in sub-stage 3.1.2 depends on the iteration index of the external computational stage (index n), the aforementioned pseudocode must be redefined by adding index n to all mathematical equations used in these pseudocodes. Next, the new IHM_n (IHM_{n+1}) must be created by applying the separation procedure on the HHM_n. Since the number of harmony vectors available in the HHM_n is twice the size of the IHM_{n+1}, the main question is which of the harmony vectors sorted in the HHM_n are used to fill the IHM_{n+1}. Basically, harmony vectors with lower—better—rank and higher—better—crowding distance are utilized to fill the IHM_{n+1}. Put another way, the first non-dominated front ($\mathcal{F}_1$) of the HHM_n is selected. If the size of $\mathcal{F}_1$ is smaller than the size of the IHM_{n+1}—{HMS}· {NDV + A}—all harmony vectors stored in $\mathcal{F}_1$ are transferred to the IHM_{n+1}. Otherwise, the harmony vectors with higher crowding distance of $\mathcal{F}_1$ are chosen for transfer to the IHM_{n+1}. If $\mathcal{F}_1$ fails to fill the IHM_{n+1} completely, the subsequent non-dominated front ($\mathcal{F}_2$) of the HHM_n will be selected to fill the remaining harmony vectors of the IHM_{n+1}. In other words, the remaining harmony vectors of the IHM_{n+1} are chosen from subsequent non-dominated front $\mathcal{F}_2$ of the HHM_n. This process is repeated until the IHM_{n+1} is filled. The IHM_{n+1} is taken into account as the input for the internal computational sub-stage in the iteration n+1 of the external computational stage. Table 4.35 gives the pseudocode pertaining to the integration and separation procedures of harmony vectors in the multi-objective SS-HSA.

The integration and separation procedures of harmony vectors in the multi-objective SS-HSA are also depicted in Fig. 4.6.

Sub-stage 3.3 is related to check of the stopping criterion of the external computational stage. In this sub-stage of the multi-objective SS-HSA, the computational efforts of the external computational stage are terminated when its stopping criterion (MNI-E) is satisfied. Otherwise, sub-stages 3.1 and 3.2 are repeated.

Stage 4 is associated with the selection of the final solution from the identified Pareto-optimal solution set. This stage is equivalent to stage 4 of the single-objective SS-HSA.

In the single-objective SS-HSA, the selection of the final optimal solution is a simple and straightforward process such that the best harmony vector stored in the HM or the harmony vector stored in the first row of the HM is taken as the final optimal solution. In the multi-objective SS-HSA, however, the selection of the final optimal solution from the identified Pareto-optimal solution set is a complicated process. Here, the fuzzy satisfying method (FSM) is employed for selecting the most satisfactory solution from the identified Pareto-optimal solution set. A detailed description relevant to the FSM is presented in Sect. 2.5 of Chap. 2.

The designed pseudocode in different stages and sub-stages of the multi-objective SS-HSA is located in a regular sequence and forms the performance-driven

Table 4.35 Pseudocode pertaining to the integration and separation procedures of harmony vectors in the proposed multi-objective SS-HSA

Algorithm 30: Pseudocode for the integration and separation procedures of harmony vectors in the proposed multi-objective SS-HSA

Input: $IHM_{n,1}$, $IHM_{n,\text{ MNI-I}}$
Output: IHM_{n+1}

start main body

1:	**begin**
2:	set $IHM_n = IHM_{n,1}$
3:	set $OHM_n = IHM_{n,\text{ MNI-I}}$
4:	construct the HHM_n with dimension $\{2 \cdot \text{HMS}\} \cdot \{\text{NDV} + \text{A}\}$ and zero initial value
5:	set $HHM_n = IHM_n \cup OHM_n$
6:	sort the HHM_n
7:	Algorithm 25: Pseudocode for the MFNDSA used in order to determine the non-dominated front—rank—of each harmony vector stored in the HHM_n under the proposed multi-objective SS-HSA, while the n index is added to all equations in this pseudocode
8:	Algorithm 26: Pseudocode for the MCCA used in order to determine the crowding distance of each harmony vector stored in the HHM_n under the proposed multi-objective SS-HSA, while the n index is added to all equations in this pseudocode
9:	Algorithm 27: Pseudocode for sorting the harmony vectors stored in the HHM_n under the proposed multi-objective SS-HSA, while the n index is added to all equations in this pseudocode
10:	construct the IHM_{n+1} with dimension $\text{HMS} \times (\text{NDV} + \text{A})$ and zero initial value
11:	$IHM_{n+1} = HHM_n(1 : \text{HMS}, 1 : \text{NDV} + \text{A})$
12:	**terminate**

end main body

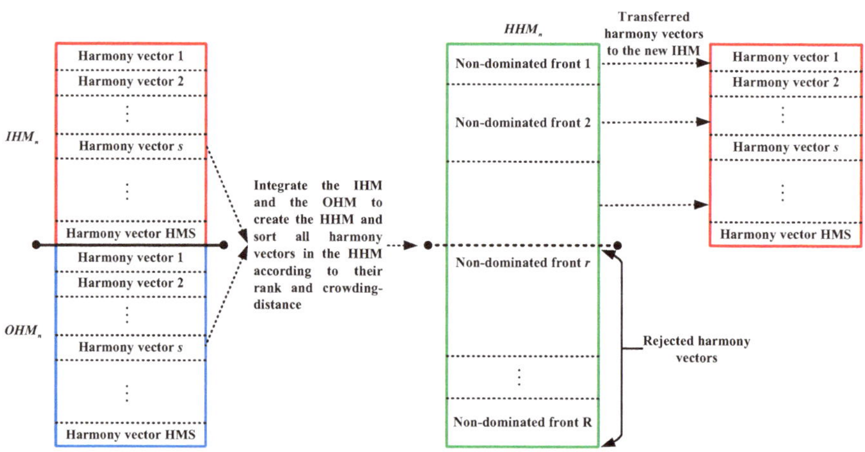

Fig. 4.6 Integration and separation procedures of harmony vectors in the proposed multi-objective SS-HSA

architecture of this algorithm. Table 4.36 illustrates the pseudocode related to the performance-driven architecture of the multi-objective SS-HSA. Here, sub-stages 3.3 and 3.1.3—the check process of the stopping criterion of the external computational stage and the internal computational sub-stage—are defined by the first and second WHILE loops in the pseudocode pertaining to the performance-driven architecture of the multi-objective SS-HSA (see Table 4.36).

4.5.1.2 Multi-objective Strategy for the SS-IHSA

The single-objective SS-HSA employs invariant values for the PAR and BW parameters in all improvisations/iterations of the computational stage. The disadvantages associated with using the invariant values of these parameters in the single-objective SS-HSA were previously described in Sect. 3.5 of Chap. 3. Since the multi-objective SS-HSA uses invariant values for the BW and PAR parameters like its single-objective version, these disadvantages also appear in the multi-objective SS-HSA. To cope the disadvantages related to using the invariant BW and PAR parameters in the single-objective SS-HSA, the original SS-IHSA—the single-objective SS-IHSA—was first introduced in 2007 [5]. The performance-driven architecture of the single-objective SS-HISA was previously reported in Sect. 3.5 of Chap. 3.

Technically speaking, the single-objective SS-IHSA was developed for solving only the SOOPs and cannot be used in dealing with MOOPs. Given the fact that the single-objective SS-IHSA similar to the single-objective SS-HSA has a single-stage computational and single-dimensional structure, the multi-objective strategy for the single-objective SS-HSA can be easily implemented for the single-objective SS-IHSA. Therefore, by implementing the multi-objective strategy on the structure of the single-objective SS-IHSA, a multi-objective SS-IHSA is demonstrated.

The performance-driven architecture of the multi-objective SS-IHSA is virtually the same as the performance-driven architecture of the multi-objective SS-HSA that was described in the previous section.

Hence, only the stages and sub-stages caused by the use of variant values for the PAR and BW parameters are referred to here. The major differences between the proposed multi-objective SS-IHSA and the multi-objective SS-HSA appear only in sub-stage 2.1 (initialization of the parameters of the algorithm) and in sub-stage 3.1.1 (improvisation of a new harmony vector). Other stages and sub-stages remain unchanged. As shown in Table 4.28, in sub-stage 2.1, the multi-objective SS-HSA employs invariant values for the PAR and BW parameters. In this sub-stage, however, the multi-objective SS-IHSA replaces the PAR and BW parameters with minimum pitch adjusting rate (PAR^{min}) and maximum pitch adjusting rate (PAR^{max}) and minimum bandwidth (BW^{min}) and maximum bandwidth (BW^{max}) parameters, respectively. Other parameters presented in Table 4.24 remain unchanged for the multi-objective SS-IHSA. Detailed descriptions relevant to the

Table 4.36 Pseudocode related to the performance-driven architecture of the proposed multi-objective SS-HSA

Algorithm 31: Pseudocode for performance-driven architecture of the proposed multi-objective SS-HSA

Input: A, BW, HMCR, HMS, MNI-I, MNI-E, NCDV, NDDV, NDV, PAR, $x_v^{\min}$, $x_v^{\max}$, $\{x_v(1), \ldots, x_v(w_v), \ldots, x_v(W_v)\}$

Output: Pareto optimal solution set and also x^{best}

start main body

begin
Stage 1—Definition stage: Definition of the MOOP and its parameters
Stage 2—Initialization stage
  Sub-stage 2.1: Initialization of the parameters of the multi-objective SS-HSA
  Sub-stage 2.2: Initialization of the of the IHM
    Algorithm 24: Pseudocode for initialization of the IHM in the proposed multi-objective SS-HSA
    Algorithm 25: Pseudocode for the MFNDSA used in order to determine the non-dominated front—rank—of each harmony vector stored in the IHM under the proposed multi-objective SS-HSA
    Algorithm 26: Pseudocode for the MCCA used in order to determine the crowding distance of each harmony vector stored in the IHM under the proposed multi-objective SS-HSA
    Algorithm 27: Pseudocode for sorting the harmony vectors stored in the IHM under the proposed multi-objective SS-HSA
Stage 3—External computational stage
  set *iteration of the external computational stage* $n = 1$
  set IHM_n = IHM
  while $n \leq$ MNI-E **do**
    Sub-stage 3.1—Internal computational sub-stage
      set *improvisation/iteration of the internal computational sub-stage* $m = 1$
      set $IHM_{n,m} = IHM_n$
      while $m \leq$ MNI-I **do**
        Sub-stage 3.1.1: Improvisation of a new harmony vector
          Algorithm 28: Pseudocode for improvisation of a new harmony vector in the proposed multi-objective SS-HSA
        Sub-stage 3.1.2: Update of the IHM
          Algorithm 29: Pseudocode for the update of the $IHM_{n,m}$ in the proposed multi-objective SS-HSA
          Algorithm 25: Pseudocode for the MFNDSA used in order to determine the non-dominated front—rank—of each harmony vector stored in the $IHM_{n,m}$ under the proposed multi-objective SS-HSA; while the n and m indices are added to all equations in this pseudocode
          Algorithm 26: Pseudocode for the MCCA used in order to determine the crowding distance of each harmony vector stored in the $IHM_{n,m}$ under the proposed multi-objective

(continued)

Table 4.36 (continued)

	SS-HSA, while the n and m indices are added to all equations in this pseudocode
24:	Algorithm 27: Pseudocode for sorting the harmony vectors stored in the $IHM_{n,m}$ under the proposed multi-objective SS-HSA, while the n and m indices are added to all equations in this pseudocode
25:	**set** *improvisation/iteration of the internal computational sub-stage* $m = m + 1$
26:	**end while**
27:	Sub-stage 3.2—Integration and separation procedures of harmony vectors
28:	Algorithm 30: Pseudocode for the integration and separation procedures of harmony vectors in the proposed multi-objective SS-HSA
29:	**set** *iteration of the external computational stage* $n = n + 1$
30:	**end while**
31:	Stage 4—Selection stage: Selection of the final optimal solution—the best harmony
32:	Step-by-step process of the FSM presented in Sect. 2.5.3 of Chap. 2
33:	**terminate**
	end main body

adjustment parameters of the multi-objective SS-IHSA are represented in Table 4.37.

In sub-stage 3.1.1, unlike the multi-objective SS-HSA, which utilizes invariant values for the PAR and BW parameters in the improvisation process of a new harmony vector, the multi-objective SS-IHSA uses the updated values for the PAR and BW parameters in the improvisation process of a new harmony vector. In this sub-stage, the values related to the PAR and BW parameters are dynamically changed and updated in each improvisation/iteration of the internal computational sub-stage by using Eqs. (4.71) and (4.72), respectively:

$$BW_{n,m} = \mathrm{BW}^{\max} \cdot \exp\left(\frac{\ln\left(\mathrm{BW}^{\max}/\mathrm{BW}^{\min}\right)}{\text{MNI-I}} \cdot m\right); \quad \forall\left\{m \in \Psi^{\text{MNI-I}}\right\} \tag{4.71}$$

$$PAR_{n,m} = \mathrm{PAR}^{\min} + \left(\frac{\mathrm{PAR}^{\max} - \mathrm{PAR}^{\min}}{\text{MNI-I}}\right) \cdot m; \quad \forall\left\{m \in \Psi^{\text{MNI-I}}\right\} \tag{4.72}$$

Detailed descriptions related to Eqs. (4.71) and (4.72) are virtually the same as the detailed explanations pertaining to Eqs. (3.13) and (3.14), previously reported in Sect. 3.5 of Chap. 3. Table 4.38 gives the rectified pseudocode related to improvisation of a new harmony vector in the multi-objective SS-IHSA. Table 4.39 gives the pseudocode related to the performance-driven architecture of the multi-objective SS-IHSA.

Table 4.37 Adjustment parameters of the proposed multi-objective SS-IHSA

No.	The proposed multi-objective SS-IHSA parameter	Abbreviation	Parameter range
1	Input harmony memory	IHM	–
2	Output harmony memory	OHM	–
3	Hybrid harmony memory	HHM	–
4	Harmony memory size	HMS	HMS ≥ 1
5	Harmony memory considering rate	HMCR	$0 \leq$ HMCR ≤ 1
6	Minimum pitch adjusting rate	PAR^{min}	$0 \leq \text{PAR}^{\text{min}} \leq 2$
7	Maximum pitch adjusting rate	PAR^{max}	$0 \leq \text{PAR}^{\text{max}} \leq 2$
8	Minimum bandwidth	BW^{min}	$0 \leq \text{BW}^{\text{min}} < +\infty$
9	Maximum bandwidth	BW^{max}	$0 \leq \text{BW}^{\text{max}} < +\infty$
10	Number of continuous decision-making variables	NCDV	NCDV ≥ 1
11	Number of discrete decision-making variables	NDDV	NDDV ≥ 1
12	Number of decision-making variables	NDV	NDV ≥ 2
13	Maximum number of iteration of the external computational stage	MNI-E	MNI-E ≥ 1
14	Maximum number of improvisation/iteration of the internal computational sub-stage	MNI-I	MNI-I ≥ 1

4.5.2 Multi-objective Strategies for the Meta-heuristic Music-Inspired Optimization Algorithms with Two-Stage Computational Multi-dimensional and Single-Homogeneous Structure

In this section, the authors propose a new multi-objective strategy to modify the architecture of the meta-heuristic music-inspired optimization algorithms with a two-stage computational multi-dimensional and single-homogeneous structure (i.e., the continuous TMS-MSA, continuous/discrete TMS-MSA, and proposed TMS-EMSA). The continuous TMS-MSA is achieved when the discrete decision-making variables are neglected in the continuous/discrete TMS-MSA. The continuous TMS-MSA is then considered as a specific version of the continuous/discrete TMS-MSA and, subsequently, its descriptions are ignored in order to prevent repetition.

4.5.2.1 Multi-objective Strategy for the Proposed Continuous/Discrete TMS-MSA

The performance-driven architecture of the single-objective continuous/discrete TMS-MSA was previously discussed in Sect. 4.2 of this chapter. This single-objective optimization algorithm is only suitable for solving the SOOPs. Given the fact that there are fundamental differences in the process of solving SOOPs and MOOPs, it is not possible to employ this single-objective optimization algorithm in

Table 4.38 Pseudocode pertaining to improvisation of a new harmony vector in the proposed multi-objective SS-IHSA

Algorithm 32: Pseudocode for improvisation of a new harmony vector in the proposed multi-objective SS-IHSA

Input: A, BW^{max}, BW^{min}, HMCR, HMS, MNI, NCDV, NDDV, NDV, PAR^{max}, PAR^{min}, x_v^{min}, x_v^{max}, $\{x_v(1), \ldots, x_v(w_v), \ldots, x_v(W_v)\}$
Output: $x_{n,m}^{new}$

start main body

begin
$BW_{n,m} = BW^{max} \cdot \exp[(\ln(BW^{max}/BW^{min})/\text{MNI-I}) \cdot m]$
$PAR_{n,m} = PAR^{min} - [((PAR^{max} - PAR^{min})/\text{MNI-I}) \cdot m]$
construct the new harmony vector $x_{n,m}^{new}$ with dimension $\{1\} \cdot \{\text{NDV} + \text{A}\}$ and zero initial value
for *decision-making variable* v $[v \in \Psi^{NDV}]$ **do**
  if $U(0,1) \leq \text{HMCR}$ **then**
   Rule 1: The harmony memory consideration with the probability of the HMCR
   $x_{n,m,v}^{new} = x_{n,m,v}^{r}; \forall r \sim U\{1, 2, \ldots, \text{HMS}\}$; for CDVs and DDVs
   if $U(0,1) \leq PAR_{n,m}$ **then**
    Rule 2: The pitch adjustment with the probability of the HMCR $\cdot$ $PAR_{n,m}$
    $x_{n,m,v}^{new} = x_{n,m,v}^{new} \pm U(0,1) \cdot BW_{n,m}$; for CDVs
    $x_{n,m,v}^{new} = x_{n,m,v}^{new}(y+t); y \sim U\{x_v(1), \ldots, x_v(w_v), \ldots, x_v(W_v)\}, \quad \forall t \sim U\{-1, +1\}$; for DDVs
   end if
  else if
    Rule 3: The random selection with the probability of the $1 - \text{HMCR}$
  $x_{n,m,v}^{new} = x_v^{min} + U(0,1) \cdot (x_v^{max} - x_v^{min})$; for CDVs
  $x_{n,m,v}^{new} = x_v(y); \forall y \sim U\{x_v(1), \ldots, x_v(w_v), \ldots, x_v(W_v)\}$; for DDVs
  end if
end for
for *objective function* a $[a \in \Psi^{A}]$ **do**
  calculate the value of objective function a, fitness function, derived from the harmony vector $x_{n,m}^{new}$ as $f(x_{n,m,a}^{new})$
  allocate $f(x_{n,m,a}^{new})$ to element $(1, \text{NDV} + a)$ of the new harmony vector $x_{n,m}^{new}$
end for
terminate

end main body

Note: Continuous decision-making variable (CDVs), discrete decision-making variable (DDVs)

its current structure to solve the MOOPs. To cope with this inadequacy, the performance-driven architecture of the single-objective continuous/discrete TMS-MSA must be restructured. Hence, by implementing the proposed multi-objective strategy on the structure of the single-objective continuous/discrete TMS-MSA, a new multi-objective continuous/discrete TMS-MSA is represented.

Table 4.39 Pseudocode related to the performance-driven architecture of the proposed multi-objective SS-IHSA

Algorithm 33: Pseudocode for performance-driven architecture of the proposed multi-objective SS-IHSA

Input: A, BW^{max}, BW^{min}, HMCR, HMS, MNI, NCDV, NDDV, NDV, PAR^{max}, PAR^{min}, x_v^{min}, x_v^{max}, $\{x_v(1), \ldots, x_v(w_v), \ldots, x_v(W_v)\}$

Output: Pareto optimal solutions set and also x^{best}

start main body

begin

Stage 1—Definition stage: Definition of the MOOP and its parameters

Stage 2—Initialization stage

  Sub-stage 2.1: Initialization of the parameters of the multi-objective SS-IHSA

  Sub-stage 2.2: Initialization of the of the IHM

    Algorithm 24: Pseudocode for initialization of the IHM in the proposed multi-objective SS-IHSA

    Algorithm 25: Pseudocode for the MFNDSA used in order to determine the non-dominated front—rank—of each harmony vector stored in the IHM under the proposed multi-objective SS-IHSA

    Algorithm 26: Pseudocode for the MCCA used in order to determine the crowding distance of each harmony vector stored in the IHM under the proposed multi-objective SS-IHSA

    Algorithm 27: Pseudocode for sorting the harmony vectors stored in the IHM under the proposed multi-objective SS-IHSA

Stage 3—External computational stage

  set *iteration of the external computational stage* $n = 1$

  set IHM_n = IHM

  while $n \leq$ MNI-E **do**

    Sub-stage 3.1—Internal computational sub-stage

      set *improvisation/iteration of the internal computational sub-stage* $m = 1$

      set $IHM_{n,m} = IHM_n$

      while $m \leq$ MNI-I **do**

        Sub-stage 3.1.1: Improvisation of a new harmony vector

          Algorithm 32: Pseudocode for improvisation of a new harmony vector in the proposed multi-objective SS-IHSA

        Sub-stage 3.1.2: Update of the IHM

          Algorithm 29: Pseudocode for the update of the $IHM_{n,m}$ in the proposed multi-objective SS-IHSA

          Algorithm 25: Pseudocode for the MFNDSA used in order to determine the non-dominated front—rank—of each harmony vector stored in the $IHM_{n,m}$ under the proposed multi-objective SS-IHSA, while the n and m indices are added to all equations in this pseudocode

          Algorithm 26: Pseudocode for the MCCA used in order to determine the crowding-distance of each harmony vector stored in the $IHM_{n,m}$ under the proposed multi-objective SS-IHSA, while the n and m indices are added to all equations in this pseudocode

(continued)

Table 4.39 (continued)

Algorithm 27: Pseudocode for sorting the harmony vectors stored in the $IHM_{n,m}$ under the proposed multi-objective SS-IHSA, while the n and m indices are added to all equations in this pseudocode
set *improvisation/iteration of the internal computational stage* $m = m + 1$
end while
Sub-stage 3.2—Integration and separation procedure of harmony vectors
Algorithm 30: Pseudocode for the integration and separation procedures of harmony vectors in the proposed multi-objective SS-IHSA
set *iteration of the external computational stage* $n = n + 1$
end while
Stage 4—Selection stage: Selection of the final optimal solution—the best harmony
Step-by-step process of the FSM presented in Sect. 2.5.3 of Chap. 2
terminate
end main body

The performance-driven architecture of the multi-objective continuous/discrete TMS-MSA is generally broken down into four stages, as follows:

- Stage 1—Definition stage: Definition of the MOOP and its parameters
- Stage 2—Initialization stage
  - Sub-stage 2.1: Initialization of the parameters of the multi-objective continuous/discrete TMS-MSA
  - Sub-stage 2.2: Initialization of the input melody memory (IMM)
- Stage 3—External computational stage
  - Sub-stage 3.1: Internal computational sub stage

    Sub-stage 3.1.1: Single computational sub-stage or single improvisation sub-stage (SISS)

      Sub-stage 3.1.1.1: Improvisation of a new melody vector by each player
      Sub-stage 3.1.1.2: Update of each input player memory (IPM)
      Sub-stage 3.1.1.3: Check of the stopping criterion of the SISS

    Sub-stage 3.1.2: Pseudo-group computational sub-stage or pseudo-group improvisation sub-stage (PGISS)

      Sub-stage 3.1.2.1: Improvisation of a new melody vector by each player taking into account the feasible ranges of the updated pitches
      Sub-stage 3.1.2.2: Update of each IPM
      Sub-stage 3.1.2.3: Update of the feasible ranges of pitches—continuous decision-making variables—for the next improvisation—only for random selection
      Sub-stage 3.1.2.4: Check of the stopping criterion of the PGISS

- Sub-stage 3.2: Integration and separation procedures of melody vectors
- Sub-stage 3.3: Check of the stopping criterion of the external computational sub-stage

- Stage 4—Selection stage: Selection of the final optimal solution—the best melody

Stage 1 is related to the definition of the MOOP and its parameters. This stage is equivalent to stage 1 of the single-objective continuous/discrete TMS-MSA. In this stage of the single-objective continuous/discrete TMS-MSA, definition of the SOOP and its parameters is carried out in accordance with Eqs. (4.1) and (4.2), described in Sect. 4.2.1 of this chapter. In the multi-objective continuous/discrete TMS-MSA, however, this stage must be redefined. In this stage of the multi-objective continuous/discrete TMS-MSA, the standard form of a MOOP can be generally described according to Eqs. (4.65) and (4.66), which were presented in Sect. 4.5.1 of this chapter.

Sub-stage 2.1 is relevant to the initialization of the parameters of the multi-objective continuous/discrete TMS-MSA. This sub-stage is equivalent to sub-stage 2.1 of the single-objective continuous/discrete TMS-MSA. In this sub-stage, the parameter adjustments of the single-objective continuous/discrete TMS-MSA are identified according to Table 4.1, which were presented in Sect. 4.2.2.1 of this chapter. As for Table 4.1, the MM is a place for storing the solution, or melody vectors for all existing players in the musical group. The single-objective continuous/discrete TMS-MSA has only one unique MM, which consists of multiple PMs—equal to the PN parameters.

The memory of player p in the group is also a place for storing the corresponding player's solution vectors. Nevertheless, one of the requirements for designing the performance-driven architecture of the multi-objective continuous/discrete TMS-MSA is the use of three MMs that include (1) the IMM that includes multiple IPMs—equal to the PN parameters; (2) the output melody memory (OMM) that includes multiple OPMs—equal to the PN parameters; and, (3) the hybrid melody memory (HMM) that includes multiple HPMs—equal to the PN parameters. The IMM is taken into account as the input for the internal computational sub-stage (sub-stage 3.1) and is updated in each iteration of the external computational stage (stage 3). Also, the OMM is derived after finalization of the internal computational sub-stage (sub-stage 3.1). That is to say that the OMM is determined by applying the internal computational sub-stage (sub-stage 3.1) on the IMM. In addition, the IMM and OMM are integrated and form the HMM. The newly developed HMM is then employed to create the IMM in the next iteration of the external computational stage or the new IMM. More details about the IMM, OMM, and HMM will be provided in the relevant sub-stages.

The single-objective continuous/discrete TMS-MSA has two computational stages, including the SIS and the PGIS, which are repeated for the MNI-SIS and MNI-PGIS, respectively. The multi-objective continuous/discrete TMS-MSA, however, consists of an external computational stage and an internal computational sub-stage. The external computational stage is repeated for the MNI-E. The internal computational sub-stage is also composed of two computational sub-stages, identified as the SISS and the PGISS. Put another way, the SIS and the PGIS of the single-objective continuous/discrete TMS-MSA are regarded as the computational sub-stages of the internal computational sub-stage in the multi-objective continuous/discrete TMS-MSA by renaming the SISS and PGISS, respectively. Subsequently, the MNI-SIS and MNI-PGIS parameters are renamed as the maximum number of improvisation/iteration of the single improvisation sub-stage (MNI-SISS) and the maximum number of improvisation/iteration of the pseudo-group improvisation sub-stage (MNI-PGISS), respectively. In each iteration of the external computational stage, the SISS and PGISS of the internal computational sub-stage are repeated for the MNI-SISS and MNI-PGISS, respectively. The sum of the MNI-SISS and MNI-PGISS parameters is taken into account as the MNI-I parameter. Other parameters presented in Table 4.1 remain unchanged for the multi-objective continuous/discrete TMS-MSA. The detailed descriptions associated with the adjustment parameters of the multi-objective continuous/discrete TMS-MSA are described in Table 4.40.

Sub-stage 2.2 pertains to the initialization of the IMM. This sub-stage is equivalent to sub-stage 2.2 of the single-objective continuous/discrete TMS-MSA. In this sub-stage of the single-objective continuous/discrete TMS-MSA, the initialization of the MM is carried out based on Eqs. (4.3)–(4.6), given in Sect. 4.2.2.2 of this chapter.

In the multi-objective continuous/discrete TMS-MSA, due to the replacement of the SOOP with the MOOP in stage 1, the initialization of the IMM must be redefined.

In the multi-objective continuous/discrete TMS-MSA, the IMM matrix, which has a dimension equal to {PMS} · {(NDV + A) · PN}, is composed of multiple IPM submatrices. Each IPM has also a dimension equal to {PMS} · {NDV + A}. The IMM matrix and IPM submatrices are filled with a large number of solution vectors generated randomly according to Eqs. (4.73)–(4.76):

$$\text{IMM} = \left[IPM_1 \quad \cdots \quad IPM_p \quad \cdots \quad IPM_{\text{PN}} \right]; \quad \forall \left\{ p \in \Psi^{\text{PN}} \right\} \tag{4.73}$$

$$IPM_p = \begin{bmatrix} \mathrm{x}_p^1 \\ \vdots \\ \mathrm{x}_p^s \\ \vdots \\ \mathrm{x}_p^{\text{PMS}} \end{bmatrix} = \left[\begin{array}{ccc|ccc} x_{p,1}^1 \cdots & x_{p,v}^1 \cdots & x_{p,\text{NDV}}^1 & f\left(\mathrm{x}_{p,1}^1\right) \cdots & f\left(\mathrm{x}_{p,a}^1\right) \cdots & f\left(\mathrm{x}_{p,\text{A}}^1\right) \\ \vdots & \vdots & \vdots & \vdots & \vdots & \vdots \\ x_{p,1}^s \cdots & x_{p,v}^s \cdots & x_{p,\text{NDV}}^s & f\left(\mathrm{x}_{p,1}^s\right) \cdots & f\left(\mathrm{x}_{p,a}^s\right) \cdots & f\left(\mathrm{x}_{p,\text{A}}^s\right) \\ \vdots & \vdots & \vdots & \vdots & \vdots & \vdots \\ x_{p,1}^{\text{PMS}} \cdots & x_{p,v}^{\text{PMS}} \cdots & x_{p,\text{NDV}}^{\text{PMS}} & f\left(\mathrm{x}_{p,1}^{\text{PMS}}\right) \cdots & f\left(\mathrm{x}_{p,a}^{\text{PMS}}\right) \cdots & f\left(\mathrm{x}_{p,\text{A}}^{\text{PMS}}\right) \end{array} \right];$$

$$\forall \left\{ p \in \Psi^{\text{PN}}, v \in \Psi^{\text{NDV}}, s \in \Psi^{\text{PMS}}, \Psi^{\text{NDV}} = \Psi^{\text{NCDV+NDDV}} \right\} \tag{4.74}$$

$$x_{p,v}^s = x_v^{\min} + \text{U}(0,1) \cdot \left(x_v^{\max} - x_v^{\min} \right); \quad \forall \left\{ p \in \Psi^{\text{PN}}, v \in \Psi^{\text{NCDV}}, s \in \Psi^{\text{PMS}} \right\} \tag{4.75}$$

Table 4.40 Adjustment parameters of the proposed multi-objective continuous/discrete TMS-MSA

No.	The proposed multi-objective continuous/discrete TMS-MSA parameters	Abbreviation	Parameter range
1	Input melody memory	IMM	–
2	Output melody memory	OMM	–
3	Hybrid melody memory	HMM	–
4	Player number	PN	PN $\geq$ 1
5	Input player memory of player *p*	IPM_p	–
6	Output player memory of player *p*	OPM_p	–
7	Hybrid player memory of player *p*	HPM_p	–
8	Player memory size	PMS	PMS $\geq$ 1
9	Player memory considering rate	PMCR	$0 \leq \text{PMCR} \leq 1$
10	Minimum pitch adjusting rate	PAR^{min}	$0 \leq \text{PAR}^{\text{min}} \leq 2$
11	Maximum pitch adjusting rate	PAR^{max}	$0 \leq \text{PAR}^{\text{max}} \leq 2$
12	Minimum bandwidth	BW^{min}	$0 \leq \text{BW}^{\text{min}} < +\infty$
13	Maximum bandwidth	BW^{max}	$0 \leq \text{BW}^{\text{max}} < +\infty$
14	Number of continuous decision-making variables	NCDV	NCDV $\geq$ 1
15	Number of discrete decision-making variables	NDDV	NDDV $\geq$ 1
16	Number of decision-making variables	NDV	NDV $\geq$ 2
17	Maximum number of iterations of the external computational stage	MNI-E	MNI-E $\geq$ 1
18	Maximum number of improvisations/iterations of the SISS	MNI-SISS	MNI-SISS $\geq$ 1
19	Maximum number of improvisations/iterations of the PGISS	MNI-PGISS	MNI-PGISS $\geq$ 1
20	Maximum number of improvisation/iterations of the internal computational stage	MNI-I	MNI-I $\geq$ 2

$$x^{s}_{p,v} = x_v(y);\ \ \forall\{p \in \Psi^{\text{PN}}, v \in \Psi^{\text{NDDV}}, s \in \Psi^{\text{PMS}}, y \sim \text{U}\{x_v(1), \ldots, x_v(w_v), \ldots, x_v(W_v)\}\} \tag{4.76}$$

It is clear that Eqs. (4.73) and (4.74) have been indirectly and directly changed, compared to their counterparts in the single-objective continuous/discrete TMS-MSA, respectively. However, Eqs. (4.75) and (4.76) remain unchanged.

The pseudocode related to initialization of all PMs, or the MM in the single-objective continuous/discrete TMS-MSA, is illustrated in Table 4.2, presented in Sect. 4.2.2.2 of this chapter. In the multi-objective continuous/discrete TMS-MSA, however, due to the replacement of the SOOP with the MOOP in stage 1, this pseudocode needs to be restructured. Consequently, Table 4.41 demonstrates the pseudocode pertaining to initialization of the entire set of IPMs, or IMM in the multi-objective continuous/discrete TMS-MSA. In the single-objective continuous/discrete TMS-MSA, after filling all of the PMs or MM with random solution vectors, the solution vectors stored in each PM must be sorted from the lowest value to the

Table 4.41 Pseudocode pertaining to initialization of the entire set of IPMs or IMM in the proposed multi-objective continuous/discrete TMS-MSA

Algorithm 34: Pseudocode for initialization of the entire set of IPMs or IMM in the proposed multi-objective continuous/discrete TMS-MSA

Input: A, NCDV, NDDV, NDV, PMS, PN, $x_v^{\min}$, $x_v^{\max}$, $\{x_v(1), \ldots, x_v(w_v), \ldots, x_v(W_v)\}$
Output: IMM

start main body

begin
construct the matrix IMM with dimension {PMS} · {(NDV + A) · PN} and zero initial value
for *music player p* $[p \in \Psi^{\mathrm{PN}}]$ **do**
  construct the submatrix IPM_p with dimension {PMS} · {NDV + A} and zero initial value
  for *melody vector s* $[s \in \Psi^{\mathrm{PMS}}]$ **do**
    construct melody vector s of music player p, x_p^s, with dimension {1} · {NDV + A} and zero initial value
    for *decision-making variable v* $[v \in \Psi^{\mathrm{NDV}}]$ **do**
      $x_{p,v}^s = x_v^{\min} + \mathrm{U}(0,1) \cdot \left(x_v^{\max} - x_v^{\min}\right)$; for CDVs
      $x_{p,v}^s = x_v(y); \forall y \sim \mathrm{U}\{x_v(1), \ldots, x_v(w_v), \ldots, x_v(W_v)\}$; for DDVs
      allocate $x_{p,v}^s$ to element $(1, v)$ of melody vector x_p^s
    end for
    for *objective function a* $[a \in \Psi^{\mathrm{A}}]$ **do**
      calculate the value of the objective function a, fitness function, derived from melody vector x_p^s as $f\left(\mathrm{x}_{p,a}^s\right)$
      allocate $f\left(\mathrm{x}_{p,a}^s\right)$ to element $(1, \mathrm{NDV} + a)$ of melody vector x_p^s
    end for
    add melody vector x_p^s to the row s of the submatrix IPM_p
  end for
  add submatrix IPM_p to the rows 1 to PMS and columns $1 + [(p-1) \cdot (\mathrm{NDV} + \mathrm{A})]$ to $[p \cdot (\mathrm{NDV} + \mathrm{A})]$ of the matrix IMM
end for
terminate

end main body

Note: Continuous decision-making variable (CDVs), discrete decision-making variable (DDVs)

highest value—in an ascending order—from the standpoint of the value of the objective function of the SOOP. The pseudocode relevant to the sorting the solution vectors stored in the PMs or the MM in the single-objective continuous/discrete TMS-MSA is presented in Table 4.3 in Sect. 4.2.2.2 of this chapter.

This pseudocode, as a simple and straightforward process, can be used only for single-objective optimization algorithms with the two-stage computational multi-dimensional and single-homogeneous structure (i.e., the proposed single-objective continuous/discrete TMS-MSA). More precisely, this pseudocode lacks the required efficiency to sort the solution vectors stored in the memory of a meta-heuristic MOOA with a two-stage computational multi-dimensional and single-homogeneous structure (i.e., the proposed multi-objective continuous/discrete TMS-MSA). In Sect. 4.5.1.1 of this chapter, the authors present the MFNDSA and MCCA in order to determine the non-dominated front—rank—and crowding distance of each harmony vector stored in the IHM under the multi-objective SS-HSA and multi-objective SS-IHSA. By identifying the rank and the crowding distance of all harmony vectors stored in the IHM, it is possible to sort them in the multi-objective SS-HSA and multi-objective SS-IHSA. However, the proposed MFNDSA and MCCA are compatible with meta-heuristic MOOAs with the single-stage computational and single-dimensional structure (i.e., the proposed multi-objective SS-HSA and the proposed multi-objective SS-IHSA). So, by matching the MFNDSA and MCCA with the structure of the multi-objective continuous/discrete TMS-MSA, they are employed in order to determine the rank and crowding distance of all melody vectors stored in all of the IPMs or IMM. The authors consider the same previous descriptions associated with the fundamental concepts of the MFNDSA and MCCA, as outlined in Sect. 4.5.1.1 of this chapter, for avoiding repetition in here. Table 4.42 represents the pseudocode related to the MFNDSA utilized in order to specify the non-dominated front—rank—of each melody vector stored in the IPMs or IMM under the multi-objective continuous/discrete TMS-MSA. Table 4.43 also gives the pseudocode associated with the MCCA used in order to characterize the crowding distance of each melody vector stored in the IPMs or IMM under the multi-objective continuous/discrete TMS-MSA. Table 4.44 illustrates the pseudocode relevant to the sorting of the melody vectors stored in the IPMs or IMM under the multi-objective continuous/discrete TMS-MSA.

The single-objective continuous/discrete TMS-MSA employs a two-stage computational structure in order to find the optimal solution: the SIS and the PGIS. The SIS and PGIS are accomplished in a timed sequence. Put another way, the SIS is carried out first and then the PGIS. The multi-objective continuous/discrete TMS-MSA, however, uses an interconnected computational structure in order to obtain the Pareto-optimal solution set. This interconnected computational structure is composed of an external computational stage and an internal computational sub-stage. The internal computational sub-stage is considered as the central core for the external computational stage. This sub-stage, in order to perform the assigned task as the central core, must be fully implemented in each iteration of the external computational stage.

Stage 3 is associated with the external computational stage. This stage is the main body of the multi-objective continuous/discrete TMS-MSA consisting of three sub-stages: (1) sub-stage 3.1 or the internal computational sub-stage; (2) sub-stage 3.2 or the integration and separation procedures of melody vectors; and, (3) sub-stage 3.3 or the check of the stopping criterion of the external computational stage. The mathematical equations expressed at the external computational stage depend on the iteration index of this stage (index n) because of the repeatability of the external computational stage in the multi-objective continuous/discrete TMS-MSA.

The IMM initialized in sub-stage 2.2 is considered as the input for the internal computational sub-stage (sub-stage 3.1), if the external computational stage is in the first iteration; otherwise, the IMM obtained after the completion of sub-stage 3.2 in

Table 4.42 Pseudocode pertaining to the MFNDSA used in order to determine the non-dominated front—rank—of each melody vector stored in the IPMs or IMM under the proposed multi-objective continuous/discrete TMS-MSA

Algorithm 35: Pseudocode for the MFNDSA used in order to determine the non-dominated front—rank—of each melody vector stored in the IPMs or IMM under the proposed multi-objective continuous/discrete TMS-MSA

Input: IMM
Output: Rank of each melody vector stored in the IPMs or IMM

start main body

begin
  for *music player p* $[p \in \Psi^{\text{PN}}]$ **do**
    for *melody vector s* $[s \in \Psi^{\text{PMS}}]$ **do**
      set $d_{\mathrm{x}_p^s} = 0$ {the number of melody vectors available in the IPM_p which dominate the melody vector s of the music player p}
      set $S_{\mathrm{x}_p^s} = \varnothing$ {the set of melody vectors available in the IPM_p that the melody vector s of the music player p dominates}
      for *melody vector* s^* $[s^* \in \Psi^{\text{PMS}}, s^* \neq s]$ **do**
        if $s \prec s^*$ **then** {melody vector s of the music player p dominates melody vector s^* of the music player p}
          $S_{\mathrm{x}_p^s} \cup \left\{\mathrm{x}_p^{s^*}\right\}$ {add melody vector s^* of the music player p to the set of melody vectors of the music player p dominated by melody vector s of the music player p }
        else if $s^* \prec s$ {melody vector s^* of the music player p dominates melody vector s of the music player p}
          $d_{\mathrm{x}_p^s} = d_{\mathrm{x}_p^s} + 1$ {increment the domination count of melody vector s of the music player p}
        end if
      end for
      if $d_{\mathrm{x}_p^s} = 0$ **then** {melody vector s of the music player p belongs to the first non-dominated front related to the music player p}
        $\text{rank}_{\mathrm{x}_p^s} = 1$
        $\mathcal{F}_{p,1} = \mathcal{F}_{p,1} \cup \left\{\mathrm{x}_p^s\right\}$
      end if
    end for
    set *non-dominated front r related to the music player p* $r_p = 1$
    while $\mathcal{F}_{p,r_p} \neq \varnothing$
      set $\mathrm{Q}_p = \varnothing$ {employed to store the melody vectors of the next non-dominated front related to the music player p}
      for *melody vector s* $\left[s \in \mathcal{F}_{p,r_p}\right]$ **do**
        for *melody vector* s^* $\left[s^* \in S_{\mathrm{x}_p^s}\right]$ **do**
          $d_{\mathrm{x}_p^{s^*}} = d_{\mathrm{x}_p^{s^*}} - 1$ {decrement the domination count of melody vector s^* of the music player p}
          if $d_{\mathrm{x}_p^{s^*}} = 0$ **then** {melody vector s^* of the music player p belongs to the non-dominated front $r + 1$ related to the music player p $r_p + 1$}

(continued)

Table 4.42 (continued)

$\text{rank}_{x_p^{s^*}} = r_p + 1$

$Q_p = Q_p \cup \left\{x_p^{s^*}\right\}$

end if

end for

end for

$r_p = r_p + 1$ {increment the non-dominated front r related to the music player p r_p}

$\mathcal{F}_{p,r_p} = Q_p$

end while

end for

for *music player p* $[p \in \Psi^{\text{PN}}]$ **do**

for *melody vector s* $[s \in \Psi^{\text{PMS}}]$ **do**

for *non-dominated front r related to the music player p* r_p $\left[r_p \in \Psi_p^{\text{R}}\right]$ **do**

if $s \in \mathcal{F}_{p,r_p}$ **do**

$\text{rank}_{x_p^s} = r_p$ {assigning rank r_p to the melody vector s of the music player p}

end if

end for

end for

end for

terminate

end main body

Table 4.43 Pseudocode associated with the MCCA used in order to determine the crowding distance of each melody vector stored in the IPMs or IMM under the proposed multi-objective continuous/discrete TMS-MSA

Algorithm 36: Pseudocode for the MCCA used in order to determine the crowding distance of each melody vector stored in the IPMs or IMM under the proposed multi-objective continuous/discrete TMS-MSA

Input: IMM
Output: Crowding distance of each melody vector stored in the IPMs or IMM

start main body

begin

for *music player p* $[p \in \Psi^{\text{PN}}]$ **do**

for *non-dominated front r related to the music player p* r_p $\left[r_p \in \Psi_p^{\text{R}}\right]$ **do**

set $h_{p,r_p} = \left|\mathcal{F}_{p,r_p}\right|$ {the number of melody vectors available in the non-dominated front r related to the music player p}

for *melody vector s* $\left[s \in \mathcal{F}_{p,r_p}\right]$ **do**

set $\text{distance}_{x_p^s} = 0$ {initialize crowding distance of the melody vectors s available in the non-dominated front r related to the music player p}

end for

(continued)

Table 4.43 (continued)

for *objective function a* $[a \in \Psi^{A}]$ **do**
$F^{\text{sort}}_{p,r_p,a} = \text{sort}\left(\mathcal{F}_{p,r_p}\left(1 : h_{p,r_p}, \text{NDV} + a\right), '\text{ascend}'\right)$
for *melody vector s* $\left[s \in \mathcal{F}_{p,r_p}\right]$ **do** {sort the melody vectors available in the non-dominated front *r* related to the music player *p* from the perspective of the value of objective function a}
for *melody vector* s^* $\left[s^* \in \mathcal{F}_{p,r_p}\right]$ **do**
if $F^{\text{sort}}_{p,r_p,a}(s) == \mathcal{F}_{p,r_p}(s^*, \text{NDV} + a)$ **then**
$\mathcal{F}^{\text{sort}}_{p,r_p,a}(s, 1 : \text{NDV} + \text{A}) = \mathcal{F}_{p,r_p}(s^*, 1 : \text{NDV} + \text{A});$
end if
end for
end for
set $\text{distance}_{\mathrm{x}^{1}_{p},a} = \text{distance}_{\mathrm{x}^{h_{p,r_p}}_{p},a} = \infty$ {the first and the last melody vectors stored in the sorted non-dominated front *r* related to the music player *p* from the perspective of the value of objective function a are assigned the crowding distance equal to infinity}
for *melody vector s* $\left[s \in \mathcal{F}_{p,r_p}, s \neq 1, s \neq h_{p,r_p}\right]$ **do**
$\text{distance}_{\mathrm{x}^{s}_{p},a} = \left(f_a^{\mathcal{F}^{\text{sort}}_{p,r_p,a,s+1}} - f_a^{\mathcal{F}^{\text{sort}}_{p,r_p,a,s-1}}\right) / \left(f_a^{\max} - f_a^{\min}\right)$
end for
for *melody vector s* $\left[s \in \mathcal{F}_{p,r_p}\right]$ **do**
for *melody vector* s^* $\left[s^* \in \mathcal{F}_{p,r_p}\right]$ **do**
if $\mathcal{F}_{p,r_p}(s, 1 : \text{NDV}) == \mathcal{F}^{\text{sort}}_{p,r_p,a}(s^*, 1 : \text{NDV})$ **then**
$\text{distance}_{\mathrm{x}^{s}_{p}} = \text{distance}_{\mathrm{x}^{s}_{p}} + \text{distance}_{\mathrm{x}^{s^*}_{p},a}$ {compute the crowding distance of the melody vectors s of the music player p}
end if
end for
end for
end for
end for
end for
terminate
end main body

iteration $n - 1$ of the external computational stage is regarded as the input for the internal computational sub-stage (sub-stage 3.1) in iteration n of the external computational stage.

Sub-stage 3.1 is related to the internal computational sub-stage. This sub-stage is equivalent to the sum of stages 3 and 4 of the single-objective continuous/discrete TMS-MSA. In simple terms, stage 3 (the SIS) and stage 4 (the PGIS) of the single-objective continuous/discrete TMS-MSA by changing the names to SISS and PGISS are taken into account as the sub-stages for the internal computational sub-stage in

Table 4.44 Pseudocode related to the sorting of the melody vectors stored in the IPMs or IMM under the proposed multi-objective continuous/discrete TMS-MSA

Algorithm 37: Pseudocode for sorting the melody vectors stored in the IPMs or IMM under the proposed multi-objective continuous/discrete TMS-MSA

Input: Unsorted IMM
Output: Sorted IMM

start main body

begin
for *music player p* $[p \in \Psi^{\mathrm{PN}}]$ **do**
set *non-dominated front r related to the music player p* $r_p = 1$
  for *non-dominated front r related to the music player p* $\left[r_p \in \Psi_p^{\mathrm{R}}\right]$ **do**
    set $\mathcal{F}_{p,r_p} = \varnothing$ {the non-dominated front *r* related to the music player *p*}
    for *melody vector s* $[s \in \Psi^{\mathrm{PMS}}]$ **do**
      if $\mathrm{rank}_{\mathrm{x}_p^s} == r_p$ **then**
        $\mathcal{F}_{p,r_p} \cup \left\{\mathrm{x}_p^s\right\}$ {add melody vector *s* of the music player *p* to the non-dominated front *r* related to the music player *p*}
      end if
    end for
    set $h_{p,r_p} = \left|\mathcal{F}_{p,r_p}\right|$ {the number of melody vectors available in the non-dominated front *r* related to the music player *p*}
    for *melody vector s* $\left[s \in \mathcal{F}_{p,r_p}\right]$ **do**
      set *superiority index q of the music player p* $q_p = 0$
      for *melody vector* s^* $\left[s^* \in \mathcal{F}_{p,r_p}, s^* \neq s\right]$ **do**
        if $\mathrm{distance}_{\mathrm{x}_p^s} > \mathrm{distance}_{\mathrm{x}_p^{s^*}}$ **then**
          $q_p = q_p + 1$ {increment the superiority index *q* of the music player *p*}
        end if
      end for
      $\mathrm{Q}_p(s) = q_p$
    end for
    for *melody vector s* $\left[s \in \mathcal{F}_{p,r_p}\right]$ **do**
      for *melody vector* s^* $\left[s^* \in \mathcal{F}_{p,r_p}\right]$ **do**
        if $\mathrm{Q}_p(s^*) == h_{p,r_p} - s$ **then**
          $\mathcal{F}_{p,r_p}^{\mathrm{sort}}(s, 1 : \mathrm{NDV} + \mathrm{A}) = \mathcal{F}_{p,r_p}(s^*)$ {sort the melody vectors available in non-dominated front *r* related to the music player *p*}
        end if
      end for
    end for
    $IPM_p^{\mathrm{sort}}\left(1 + \sum_{z \in \Psi^{r_p - 1}} h_{p,z} : \sum_{r_p \in \Psi_p^{\mathrm{R}}} h_{p,r_p}, 1 : \mathrm{NDV} + \mathrm{A}\right) = \mathcal{F}_{p,r_p}^{\mathrm{sort}}$ {add the sorted non-dominated front *r* related to the music player *p* to the IPM_p^{sort}}
  end for
  $IPM_p = IPM_p^{\mathrm{sort}}$
  $\mathrm{IMM}^{\mathrm{sort}}(1 : \mathrm{PMS}, 1 + [(p - 1) \cdot (\mathrm{NDV} + \mathrm{A})] : [p \cdot (\mathrm{NDV} + \mathrm{A})]) = IPM_p$
end for
$\mathrm{IMM} = \mathrm{IMM}^{\mathrm{sort}}$
terminate

end main body

the multi-objective continuous/discrete TMS-MSA. In the multi-objective continuous/discrete TMS-MSA, therefore, the internal computational sub-stage (sub-stage 3.1) consists of two sub-stages: (1) sub-stage 3.1.1 or the SISS and, (2) sub-stage 3.1.2 or the PGISS. The mathematical equations represented at sub-stage 3.1, in addition to the dependence on the iteration index of the external computational stage (index n), must also depend on the improvisation/iteration index of the internal computational sub-stage (index m) because of the repeatability of the internal computational sub-stage in the multi-objective continuous/discrete TMS-MSA. Sub-stage 3.1.1 or the SISS is composed of three sub-stages: (1) sub-stage 3.1.1.1 or the improvisation of a new melody vector by each player; (2) sub-stage 3.1.1.2 or the update of each IPM; and, (3) sub-stage 3.1.1.3 or the check of the stopping criterion of the SISS. Sub-stage 3.1.2 or the PGISS is also composed of four sub-stages: (1) sub-stage 3.1.2.1 or the improvisation of a new melody vector by each player taking into account the feasible ranges of the updated pitches; (2) sub-stage 3.1.2.2 or the update of each IPM; (3) sub-stage 3.1.2.3 or the update of the feasible ranges of pitches—continuous decision-making variables—for the next improvisation—only for random selection; and, (4) sub-stage 3.1.2.4 or the check of the stopping criterion of the PGISS.

Sub-stages 3.1.1.1 and 3.1.2.1 are relevant to the improvisation of a new melody vector by each existing player in the musical group without and with the influence of other existing players in the musical group, respectively. These sub-stages of the multi-objective continuous/discrete TMS-MSA are conceptually similar to sub-stages 3.1 and 4.1 of the single-objective continuous/discrete TMS-MSA, previously described in Sect. 4.2.6 of this chapter. In order to avoid repetition, a detailed description regarding these sub-stages is neglected. However, due to the existence of the external computational stage in the multi-objective continuous/discrete TMS-MSA, the iteration index of the external computational stage (index n) must be added to all of the mathematical equations indicated in Sect. 4.2.6 of this chapter that have the improvisation/iteration index of the internal computational step (index m).

The pseudocode relevant to the improvisation of a new melody vector by each player in the musical group in the single-objective continuous/discrete TMS-MSA is presented in Table 4.8 in Sect. 4.2.6 of this chapter. In the multi-objective continuous/discrete TMS-MSA, however, due to the replacement of the SOOP with the MOOP in stage 1, this pseudocode must be restructured. As a consequence, Table 4.45 demonstrates the pseudocode related to improvisation of a new melody vector by each player in the musical group in the multi-objective continuous/discrete TMS-MSA.

Sub-stages 3.1.1.2 and 3.1.2.2 are relevant to the process of updating all of the IPMs or the IMM. These sub-stages are equivalent to sub-stages 3.2 and 4.2 of the single-objective continuous/discrete TMS-MSA. A detailed description concerning the process of updating the memory of all existing players in the musical group or the MM_m in the single-objective continuous/discrete TMS-MSA was provided in Sect. 4.2.3.2 of this chapter. The pseudocode pertaining to the updating of the memory of all existing players in the musical group or the MM_m is also presented in Table 4.4 in Sect. 4.2.3.2 of this chapter. In the multi-objective continuous/

Table 4.45 Pseudocode related to improvisation of a new melody vector by each player in the musical group in the proposed multi-objective continuous/discrete TMS-MSA

Algorithm 38: Pseudocode for improvisation of a new melody vector by each player in the musical group in the proposed multi-objective continuous/discrete TMS-MSA

Input: A, $\text{BW}^{\max}$, $\text{BW}^{\min}$, MNI-SISS, MNI-PGISS, NCDV, NDDV, NDV, $\text{PAR}^{\max}$, $\text{PAR}^{\min}$, PMCR, PMS, PN, $x_v^{\min}$, $x_v^{\max}$, $\{x_v(1), \ldots, x_v(w_v), \ldots, x_v(W_v)\}$

Output: $\text{x}_{n,m,p}^{\text{new}}$

start main body

begin

$BW_{n,m} = \text{BW}^{\max} \cdot \exp\left[\left(\ln\left(\text{BW}^{\max}/\text{BW}^{\min}\right)/\left((\text{MNI-SISS}) + (\text{MNI-PGISS})\right)\right) \cdot m\right]$

$PAR_{n,m} = \text{PAR}^{\min} - \left[\left(\left(\text{PAR}^{\max} - \text{PAR}^{\min}\right)/\left((\text{MNI-SISS}) + (\text{MNI-PGISS})\right)\right) \cdot m\right]$

for *music player p* $[p \in \Psi^{\text{PN}}]$ **do**

  construct the new melody vector for music player p, $\text{x}_{n,m,p}^{\text{new}}$, with dimension $\{1\} \cdot \{\text{NDV} + \text{A}\}$ and zero initial value

  for *decision-making variable v* $[v \in \Psi^{\text{NDV}}]$ **do**

    if $\text{U}(0,1) \leq \text{PMCR}$ **then**

      Rule 1: harmony memory consideration with probability PMCR

      if *improvisation/iteration of the internal computational sub-stage m* $[m \in \Psi^{(\text{MNI-SISS}) + (\text{MNI-PGISS})}]$ is odd **then**

        Principle 1: first combination

        $x_{n,m,p,v}^{\text{new}} = x_{n,m,p,v}^{r} \pm \text{U}(0,1) \cdot BW_{n,m}; \forall r \sim \text{U}\{1,2,\ldots,\text{PMS}\}$; for CDVs

        $x_{n,m,p,v}^{\text{new}} = x_{n,m,p,v}^{r}; \forall r \sim \text{U}\{1,2,\ldots,\text{PMS}\}$; for DDVs

      else

        Principle 2: second combination

        $x_{n,m,p,v}^{\text{new}} = x_{n,m,p,k}^{r} \pm \text{U}(0,1) \cdot BW_{n,m}; \forall r \sim \text{U}\{1,2,\ldots,\text{PMS}\}, \forall k \sim \text{U}\{1,2,\ldots,\text{NCDV}\}$; for CDVs

        $x_{n,m,p,v}^{\text{new}} = x_{n,m,p,l}^{r}; \forall r \sim \text{U}\{1,2,\ldots,\text{PMS}\}, \forall l \sim \text{U}\{1,2,\ldots,\text{NDDV}\}$; for DDVs

      end if

      if $\text{U}(0,1) \leq PAR_{n,m}$ **then**

        Rule 2: pitch adjustment with probability PMCR $\cdot$ $PAR_{n,m}$

        $x_{n,m,p,v}^{\text{new}} = x_{n,m,p,v}^{\text{best}}$; for CDVs and DDVs

      end if

    else if

      Rule 3: random selection with probability 1-PMCR

    switch 1

    case *improvisation/iteration of the internal computational sub-stage m* $[m \in \Psi^{(\text{MNI-SISS}) + (\text{MNI-PGISS})}] \leq (\text{MNI-SISS})$ **then**

    $x_{n,m,p,v}^{\text{new}} = x_v^{\min} + \text{U}(0,1) \cdot \left(x_v^{\max} - x_v^{\min}\right)$; for CDVs

    $x_{n,m,p,v}^{\text{new}} = x_v(y); \forall y \sim \text{U}\{x_v(1), \ldots, x_v(w_v), \ldots, x_v(W_v)\}$; for DDVs

    case *improvisation/iteration of the internal computational sub-stage m* $[m \in \Psi^{(\text{MNI-SISS}) + (\text{MNI-PGISS})}] > (\text{MNI-SISS})$ and *improvisation/iteration of the internal computational sub-stage m* $[m \in \Psi^{(\text{MNI-SISS}) + (\text{MNI-PGISS})}] \leq (\text{MNI-SISS}) + (\text{MNI-PGISS})$ **then**

    $x_{n,m,p,v}^{\text{new}} = x_{n,m,v}^{\min} + \text{U}(0,1) \cdot \left(x_{n,m,v}^{\max} - x_{n,m,v}^{\min}\right)$; for CDVs

(continued)

Table 4.45 (continued)

$x_{n,m,p,v}^{\text{new}} = x_v(y); \forall y \sim \text{U}\{x_v(1), \ldots, x_v(w_v), \ldots, x_v(W_v)\}$; for DDVs
end switch
end if
end for
for *objective function a* $[a \in \Psi^{\text{A}}]$ **do**
calculate the value of objective function a, fitness function, derived from melody vector $\text{x}_{n,m,p}^{\text{new}}$ as $f\left(\text{x}_{n,m,p,a}^{\text{new}}\right)$
allocate $f\left(\text{x}_{n,m,p,a}^{\text{new}}\right)$ to element $(1, \text{NDV} + a)$ of the new melody vector $\text{x}_{n,m,p}^{\text{new}}$
end for
end for
terminate
end main body

Note: Continuous decision-making variable (CDVs), discrete decision-making variable (DDVs)

discrete TMS-MSA, due to the replacement of the SOOP with the MOOP in stage 1, the process of updating all of the IPMs or the IMM must be redefined. To illustrate these sub-stages of the multi-objective continuous/discrete TMS-MSA, consider the memory relevant to player p ($IPM_{n,m,p}$).

A new melody vector played by player p—$\text{x}_{n,m,p}^{\text{new}} = \left(x_{n,m,p,1}^{\text{new}}, \ldots, x_{n,m,p,v}^{\text{new}}, \ldots, x_{n,m,p,\text{NDV}}^{\text{new}}\right)$—is evaluated and compared with the worst available melody vector in the $IPM_{n,m,p}$—the melody vector with the biggest rank and the smallest crowding distance, which is placed in the PMS row of the $IPM_{n,m,p}$—from the perspective of the objective functions. If the new melody vector played by player p dominates the worst available melody vector in the $IPM_{n,m,p}$, from the standpoint of the objective functions, this new melody vector replaces the worst melody vector available in the $IPM_{n,m,p}$; the worst available melody vector is then eliminated from the $IPM_{n,m,p}$. This process is performed for other players in the group in a similar manner.

Table 4.46 illustrates the pseudocode relevant to the update of the memory of all existing players in the musical group or the $IMM_{n,m}$ in the multi-objective continuous/discrete TMS-MSA. The update process of the $IPM_{n,m,p}$ is not performed if the new melody vector played by player p in the musical group does not dominate the worst available melody vector in the $IPM_{n,m,p}$, from the viewpoint of the objective functions.

After completion of this process, melody vectors stored in the memory of all existing players in the musical group or the $IMM_{n,m}$ must be re-sorted. The pseudocode related to sorting of the melody vectors stored in the IMM was presented in Tables 4.42, 4.43, and 4.44. However, given the fact that the IMM in sub-stages 3.1.1.2 and 3.1.2.2 depends on the iteration index of the external computational stage (index n) and improvisation/iteration index of the internal computational stage

Table 4.46 Pseudocode related to the update of the memory of all existing players in the musical group or the $IMM_{n,m}$ in the proposed multi-objective continuous/discrete TMS-MSA

Algorithm 39: Pseudocode for the update of the memory of all existing players in the musical group or the $IMM_{n,m}$ in the proposed multi-objective continuous/discrete TMS-MSA

Input: Not updated $IMM_{n,m}$, $\mathrm{x}_{n,m,p}^{\mathrm{new}}$
Output: Updated $IMM_{n,m}$

start main body

1:	**begin**
2:	**for** *music player p* $[p \in \Psi^{\mathrm{PN}}]$ **do**
3:	set $\mathrm{x}_p^{\mathrm{worst}} = \mathrm{x}_{n,m,p}^{\mathrm{PMS}}$
4:	set $f\left(\mathrm{x}_p^{\mathrm{worst}}\right) = f\left(\mathrm{x}_{n,m,p}^{\mathrm{PMS}}\right)$
5:	**if** $f\left(\mathrm{x}_{n,m,p}^{\mathrm{new}}\right) \prec f\left(\mathrm{x}_p^{\mathrm{worst}}\right)$ **then** {new melody vector played by player p dominates worst melody vector in the memory of this player}
6:	$\mathrm{x}_{n,m,p}^{\mathrm{new}} \in IPM_{n,m,p}$
7:	$\mathrm{x}_p^{\mathrm{worst}} \notin IPM_{n,m,p}$
8:	**end if**
9:	**end for**
10:	**terminate**

end main body

(index m), the aforementioned pseudocode must be redefined by adding the indices n and m to all the mathematical equations used in these pseudocodes. Sub-stage 3.1.2.3 is associated with the update of the feasible ranges of pitches, continuous decision-making variables, for the next improvisation (only for random selection). This sub-stage is equivalent to sub-stage 4.3 of the single-objective continuous/discrete TMS-MSA, previously presented in Sect. 4.2.4.3 of this chapter. In sub-stage 4.3 of the single-objective continuous/discrete TMS-MSA, the process of updating the feasible ranges of pitches, continuous decision-making variables, is carried out according to the pseudocode presented in Table 4.6. Sub-stage 3.1.2.3 of the multi-objective continuous/discrete TMS-MSA is conceptually similar to sub-stage 4.3 of the single-objective continuous/discrete TMS-MSA. In the multi-objective continuous/discrete TMS-MSA, however, due to the existence of the external computational stage, this pseudocode must be restructured. As a result, Table 4.47 gives the pseudocode related to the update of the feasible ranges of the continuous decision-making variables in the multi-objective continuous/discrete TMS-MSA. Sub-stages 3.1.1.3 and 3.1.2.4 are relevant to the checking of the stopping criterion of the SISS and PGISS, respectively. These sub-stages of the multi-objective continuous/discrete TMS-MSA are virtually the same as sub-stages 3.3 and 4.4 of the single-objective continuous/discrete TMS-MSA, respectively, previously described in Sects. 4.2.3.3 and 4.2.4.4 of this chapter, respectively.

Only in sub-stages 3.1.1.3 and 3.1.2.4 of the multi-objective continuous/discrete TMS-MSA the MNI-SIS and the MNI-PGIS parameters must be replaced with the

Table 4.47 The pseudocode related to the update of the feasible ranges of the continuous decision-making variables in the proposed multi-objective continuous/discrete TMS-MSA

Algorithm 40: Pseudocode for the update of the feasible ranges of the continuous decision-making variables in the proposed multi-objective continuous/discrete TMS-MSA

Input: $\mathrm{x}^{1}_{n,m,p}$
Output: $x^{\min}_{n,m,v}$, $x^{\max}_{n,m,v}$

	start main body
1:	**begin**
2:	**for** *music player p* $[p \in \Psi^{\mathrm{PN}}]$ **do**
3:	set $\mathrm{x}^{\mathrm{best}}_{p} = \mathrm{x}^{1}_{n,m,p}$
4:	**end for**
5:	**for** *continuous decision-making variable v* $[v \in \Psi^{\mathrm{NCDV}}]$ **do**
6:	$x^{\min}_{n,m,v} = \min\left(x^{\mathrm{best}}_{p,v}\right); \forall p \in \Psi^{\mathrm{PN}}, \forall \{m \in \Psi^{\mathrm{MNI\text{-}PGISS}}\}$
7:	$x^{\max}_{n,m,v} = \max\left(x^{\mathrm{best}}_{p,v}\right); \forall p \in \Psi^{\mathrm{PN}}, \forall \{m \in \Psi^{\mathrm{MNI\text{-}PGISS}}\}$
8:	**end for**
9:	**terminate**
	end main body

MNI-SISS and MNI-PGISS parameters, respectively. Accordingly, in sub-stage 3.1.1.3 of the multi-objective continuous/discrete TMS-MSA, the computational efforts of the SISS are terminated when its stopping criterion—the MNI-SISS—is satisfied. Otherwise, sub-stages 3.1.1.1 and 3.1.1.2 are repeated. In sub-stage 3.1.2.4 of the multi-objective continuous/discrete TMS-MSA, the computational efforts of the PGISS are also terminated when its stopping criterion—the MNI-PGISS—is satisfied. Otherwise, sub-stages 3.1.2.1, 3.1.2.2, and 3.1.2.3 are repeated.

Sub-stage 3.2 of the multi-objective continuous/discrete TMS-MSA pertains to the integration and separation procedures of the melody vectors. The main goal of the implementation of this sub-stage is to construct the IMM for the next iteration of the external computational stage or the new IMM. The construction of the new IMM requires the IMM in the current iteration of the external computational stage and two other memories called the OMM and HMM. Similar to the IMM, which consists of multiple IPMs, the OMM and HMM are composed of multiple OPMs and HPMs, respectively. Implementation of this sub-stage in iteration n of the external computational stage begins by determining the IMM_n and OMM_n. First, the IMM for the internal computational sub-stage (sub-stage 3.1) in iteration n of the external computational stage is considered as the IMM_n. The IMM_n obtained after the completion of the internal computational sub-stage (sub-stage 3.1) is also regarded as the OMM_n. Then, the HMM_n must be created by applying the integration procedure on the IMM_n and OMM_n. The point to be made here is that all existing players in the musical group are involved in this procedure. To clarify, consider player p in the musical group. This player integrates the $IPM_{n,p}$ and $OPM_{n,p}$ in order to form the $HPM_{n,p}$. Given the fact that all of the melody vectors available in the $IPM_{n,\,p}$ and $OPM_{n,p}$ are transferred to the

$HPM_{n,p}$, elitism for this player of the musical group is ensured. This transfer also makes the size of the $HPM_{n,p}$ twice as large as the $IPM_{n,p}$ and $OPM_{n,p}$. This means that, unlike the $IPM_{n,p}$ and $OPM_{n,p}$ that have a size equal to {PMS} · {NDV + A}, the $HPM_{n,p}$ has a size of {2 · PMS} · {NDV + A}. This procedure is carried out by other players in the same way. After implementing the integration procedure by all players, all obtained values of the $HPM_{n,p}$ are located next to each other and form the HMM_n. It is obvious that the size of the HMM_n will be twice the size of the IMM_n and OMM_n. Simply put, unlike the IMM_n and OMM_n, which have a size equal to {PMS} · {(NDV + A) · PN}, the HMM_n has a size of {2 · PMS} · {(NDV + A) · PN}. Afterwards, all melody vectors available in the HMM_n must be sorted according to their rank and crowding distance by employing the pseudocodes presented in Tables 4.42, 4.43, and 4.44. However, due to the fact that HMM_n in sub-stage 3.1.2 depends on the iteration index of the external computational stage (index n), the aforementioned pseudocode must be redefined by adding index n to all of the mathematical equations used in these pseudocodes. Next, the new $IMM_n(IMM_{n+1})$ must be created by applying the separation procedure on the HMM_n. Likewise, all existing players in the musical group are involved in this procedure. To elucidate, consider player p in the musical group. This player employs the melody vectors stored in the $HPM_{n,p}$ to create the new $IPM_p(IPM_{n+1,p})$. But considering that the size of the $HPM_{n,p}$ is twice the size of the $IPM_{n+1,p}$, only half of the melody vectors available in the $HPM_{n,p}$ can be transferred to the $IPM_{n+1,p}$. The melody vectors with lower—better—rank and higher—better—crowding distance available in the $HPM_{n,p}$ have a higher priority for transferring to the $IPM_{n+1,p}$. Put another way, the first non-dominated front ($\mathcal{F}_{p,1}$) of the $HPM_{n,p}$ is selected. If the size of the $\mathcal{F}_{p,1}$ is smaller than the size of the $IPM_{n+1,p}$—{HMS} · {NDV + A}—all melody vectors stored in the $\mathcal{F}_{p,1}$ are transferred to the $IPM_{n+1,p}$. Otherwise, the melody vectors with a higher crowding distance of the $\mathcal{F}_{p,1}$ are chosen for transferring to the $IPM_{n+1,p}$. If the $\mathcal{F}_{p,1}$ fails to fill the $IPM_{n+1,p}$ completely, the subsequent non-dominated front ($\mathcal{F}_{p,2}$) of the $HPM_{n,p}$ will be selected to fill the remaining melody vectors of the $IPM_{n+1,p}$. In simple terms, the remaining melody vectors of the $IPM_{n+1,p}$ are chosen from the subsequent non-dominated front ($\mathcal{F}_{p,2}$) of the $HPM_{n,p}$. This process is repeated until the $IPM_{n+1,p}$ is completely filled. This procedure is performed by other players in the same way. After implementing the separation procedure by all players, all obtained $IPM_{n+1,p}$ are located next to each other and form the IMM_{n+1}. The IMM_{n+1} is taken into account as the IMM for the internal computational sub-stage in the iteration $n + 1$ of the external computational stage. Table 4.48 represents the pseudocode related to the integration and separation procedures of melody vectors in the multi-objective continuous/discrete TMS-MSA.

Sub-stage 3.3 is associated with the check of the stopping criterion of the external computational stage. In this sub-stage of the multi-objective continuous/discrete TMS-MSA, if the stopping criterion of the external computational stage (the MNI-E) is satisfied, its computational efforts are terminated. Otherwise, sub-stages 3.1 and 3.2 are repeated.

Stage 4 is related to selection of the final solution from the identified Pareto-optimal solution set. This stage is equivalent to stage 4 of the single-objective

Table 4.48 Pseudocode related to the integration and separation procedures of melody vectors in the proposed multi-objective continuous/discrete TMS-MSA

Algorithm 41: Pseudocode for the integration and separation procedures of melody vectors in the proposed multi-objective continuous/discrete TMS-MSA

Input: $IMM_{n,1}$, $IMM_{n,\text{MNI-I}}$
Output: IMM_{n+1}

start main body

begin
set $IMM_n = IMM_{n,1}$
set $OMM_n = IMM_{n,\text{MNI-I}}$
construct the HMM_n with dimension $\{2 \times \text{PMS}\} \cdot \{(\text{NDV} + \text{A}) \cdot \text{PN}\}$ and zero initial value
for *music player p* $[p \in \Psi^{\text{PN}}]$ **do**
    $IPM_{n,\,p} = IMM_n(1 : \text{PMS},\ 1 + [(p - 1) \cdot (\text{NDV} + \text{A})]:[p \cdot (\text{NDV} + \text{A})])$
    $OPM_{n,\,p} = OMM_n(1 : \text{PMS},\ 1 + [(p - 1) \cdot (\text{NDV} + \text{A})]:[p \cdot (\text{NDV} + \text{A})])$
    $HPM_{n,\,p} = IPM_{n,\,p} \cup OPM_{n,\,p}$
    $HMM_n(1 : 2 \cdot \text{PMS},\ 1 + [(p - 1) \cdot (\text{NDV} + \text{A})]:[p \cdot (\text{NDV} + \text{A})])) = HPM_{n,p}$
end for
sort the HMM_n
    Algorithm 35: Pseudocode for the MFNDSA used in order to determine the non-dominated front—rank—of each melody vector stored in the HMM_n under the proposed multi-objective continuous/discrete TMS-MSA, while the n index is added to all equations in this pseudocode
    Algorithm 36: Pseudocode for the MCCA used in order to determine the crowding distance of each melody vector stored in the HMM_n under the proposed multi-objective continuous/discrete TMS-MSA, while the n index is added to all equations in this pseudocode
    Algorithm 37: Pseudocode for sorting the melody vectors stored in the HMM_n under the proposed multi-objective continuous/discrete TMS-MSA, while the n index is added to all equations in this pseudocode
    construct the IMM_{n+1} with dimension $\{\text{PMS}\} \cdot \{(\text{NDV} + \text{A}) \cdot \text{PN}\}$ and zero initial value
for *music player p* $[p \in \Psi^{\text{PN}}]$ **do**
    $IPM_{n+1,p} = HMM_n(1 : \text{PMS}, 1 + [(p - 1) \cdot (\text{NDV} + \text{A})] : [p \cdot (\text{NDV} + \text{A})])$
    $IMM_{n+1}(1 : \text{PMS}, 1 + [(p - 1) \cdot (\text{NDV} + \text{A})] : [p \cdot (\text{NDV} + \text{A})]) = IPM_{n+1,p}$
end for
terminate

end main body

continuous/discrete TMS-MSA. In the single-objective continuous/discrete TMS-MSA, the final optimal solution will be chosen using a simple and straightforward process. In this algorithm, first, the best melody vector stored in the memory of each existing player in the musical group is determined. Then, the best melody vector is chosen from among these melody vectors as the final optimal solution.

In the multi-objective continuous/discrete TMS-MSA, however, the selection of the final optimal solution from the identified Pareto-optimal solution set is a complicated process. Here, the FSM is employed in order to select the most satisfactory solution from the identified Pareto-optimal solution set. Please see Sect. 2.5 of Chap. 2 for a modern introduction to the FSM. The performance-driven architecture of the multi-objective continuous/discrete TMS-MSA is formed by placing the designed pseudocode in different stages and sub-stages of this algorithm in a regular sequence. Table 4.49 gives the pseudocode related to the performance-driven architecture of the multi-objective continuous/discrete TMS-MSA. Here, sub-stages 3.3, 3.1.1.3, and 3.1.2.4 (the check process of the stopping criterion of the external computational stage, the SISS and the PGISS) are defined by the first, second, and third WHILE loops in the pseudocode related to the performance-driven architecture of the multi-objective continuous/discrete TMS-MSA (see Table 4.49).

4.5.2.2 Multi-objective Strategy for the Proposed TMS-EMSA

As previously mentioned in Sect. 4.2 of this chapter, the single-objective continuous/discrete TMS-MSA is addressed in order to solve the SOOPs. Moreover, to appreciably improve the performance, flexibility, and robustness of the single-objective continuous/discrete TMS-MSA, an enhanced version of this single-objective optimization algorithm—the TMS-EMSA—is developed in Sect. 4.3 of this chapter. It is obvious that these single-objective optimization algorithms cannot be employed in dealing with the MOOPs. The multi-objective continuous/discrete TMS-MSA is accordingly elaborated to solve the MOOPs in Sect. 4.5.2.1 of this chapter. The authors also propose a new multi-objective TMS-EMSA by applying the improvements on the multi-objective continuous/discrete TMS-MSA. These improvements were formerly implemented on the single-objective continuous/discrete TMS-MSA with the aim of forming the proposed single-objective TMS-EMSA.

The performance-driven architecture of the proposed multi-objective TMS-EMSA is generally broken down into four stages, as follows:

- Stage 1—Definition stage: Definition of the MOOP and its parameters
- Stage 2—Initialization stage
  - Sub-stage 2.1: Initialization of the parameters of the multi-objective TMS-EMSA
  - Sub-stage 2.2: Initialization of the IMM
- Stage 3—External computational stage
  - Sub-stage 3.1: Internal computational sub-stage

    Sub-stage 3.1.1: Single computational sub-stage or the SISS

      Sub-stage 3.1.1.1: Improvisation of a new melody vector by each player
      Sub-stage 3.1.1.2: Update each IPM
      Sub-stage 3.1.1.3: Check of the stopping criterion of the SISS

Table 4.49 Pseudocode related to the performance-driven architecture of the proposed multi-objective continuous/discrete TMS-MSA

Algorithm 42: Pseudocode for the performance-driven architecture of the proposed multi-objective continuous/discrete TMS-MSA

Input: A, BW^{max}, BW^{min}, MNI-E, MNI-I, MNI-SISS, MNI-PGISS, NCDV, NDDV, NDV, PAR^{max}, PAR^{min}, PMCR, PMS, PN, x_v^{min}, x_v^{max}, $\{x_v(1), \ldots, x_v(w_v), \ldots, x_v(W_v)\}$

Output: Pareto optimal solutions set and also x^{best}

start main body

begin
Stage 1—Definition stage: Definition of the MOOP and its parameters
Stage 2—Initialization stage
    Sub-stage 2.1: Initialization of the parameters of the proposed multi-objective continuous/discrete TMS-MSA
    Sub-stage 2.2: Initialization of the of the IMM
        Algorithm 34: Pseudocode for initialization of the entire set of IPMs or IMM in the proposed multi-objective continuous/discrete TMS-MSA
        Algorithm 35: Pseudocode for the MFNDSA used in order to determine the non-dominated front—rank—of each melody vector stored in the IPMs or IMM under the proposed multi-objective continuous/discrete TMS-MSA
        Algorithm 36: Pseudocode for the MCCA used in order to determine the crowding distance of each melody vector stored in the IPMs or IMM under the proposed multi-objective continuous/discrete TMS-MSA
        Algorithm 37: Pseudocode for sorting the melody vectors stored in the IPMs or IMM under the proposed multi-objective continuous/discrete TMS-MSA
Stage 3—External computational stage
    set *iteration of the external computational stage* $n = 1$
    set IMM_n = IMM
    while $n \leq$ MNI-E **do**
        Sub-stage 3.1—Internal computational sub-stage
            set *improvisation/iteration of the internal computational sub-stage* $m = 1$
            set $IMM_{n,m} = IMM_n$
            Sub-stage 3.1.1—single computational sub-stage or the SISS
            while $m \leq$ MNI-SISS **do**
                Sub-stage 3.1.1.1: Improvisation of a new melody vector by each player
                    Algorithm 38: Pseudocode for improvisation of a new melody vector by each player in the musical group in the proposed multi-objective continuous/discrete TMS-MSA
                Sub-stage 3.1.1.2: Update each IPM
                    Algorithm 39: Pseudocode for the update of the memory of all existing players in the musical group or the update of the $IMM_{n,m}$ in the proposed multi-objective continuous/discrete TMS-MSA

(continued)

Table 4.49 (continued)

Algorithm 35: Pseudocode for the MFNDSA used in order to determine the non-dominated front—rank—of each melody vector stored in the $IMM_{n,m}$ under the proposed multi-objective continuous/discrete TMS-MSA, while the n and m indices are added to all equations in this pseudocode
Algorithm 36: Pseudocode for the MCCA used in order to determine the crowding distance of each melody vector stored in the $IMM_{n,m}$ under the proposed multi-objective continuous/discrete TMS-MSA, while the n and m indices are added to all equations in this pseudocode
Algorithm 37: Pseudocode for sorting the melody vectors stored in the $IMM_{n,m}$ under the proposed multi-objective continuous/discrete TMS-MSA, while the n and m indices are added to all equations in this pseudocode
set *improvisation/iteration of the internal computational sub-stage* $m = m + 1$
end while
Sub-stage 3.1.2—pseudo-group computational sub-stage or the PGISS
while $m >$ (MNI-SISS) & $m \leq$ (MNI-SISS) + (MNI-PGISS) **do**
Sub-stage 3.1.2.1: Improvisation of a new melody vector by each player
Algorithm 38: Pseudocode for improvisation of a new melody vector by each player in the musical group in the proposed multi-objective continuous/discrete TMS-MSA
Sub-stage 3.1.2.2: Update each IPM
Algorithm 39: Pseudocode for the update of the memory of all existing players in the musical group or the update of the $IMM_{n,m}$ in the proposed multi-objective continuous/discrete TMS-MSA
Algorithm 35: Pseudocode for the MFNDSA used in order to determine the non-dominated front—rank—of each melody vector stored in the $IMM_{n,m}$ under the proposed multi-objective continuous/discrete TMS-MSA, while the n and m indices are added to all equations in this pseudocode
Algorithm 36: Pseudocode for the MCCA used in order to determine the crowding distance of each melody vector stored in the $IMM_{n,m}$ under the proposed multi-objective continuous/discrete TMS-MSA, while the n and m indices are added to all equations in this pseudocode
Algorithm 37: Pseudocode for sorting the melody vectors stored in the $IMM_{n,m}$ under the proposed multi-objective continuous/discrete TMS-MSA, while the n and m indices are added to all equations in this pseudocode
Sub-stage 3.1.2.3: Update of the feasible ranges of pitches—continuous decision-making variables—for the next improvisation—only for random selection
Algorithm 40: Pseudocode for the update of the feasible ranges of the continuous decision-making variables in the proposed multi-objective continuous/discrete TMS-MSA

(continued)

Table 4.49 (continued)

set *improvisation/iteration of the internal computational sub-stage* $m = m + 1$
end while
Sub-stage 3.2—Integration and separation procedures of melody vectors
Algorithm 41: Pseudocode for the integration and separation procedures of melody vectors in the proposed multi-objective continuous/discrete TMS-MSA
set *iteration of the external computational stage* $n = n + 1$
end while
Stage 4—Selection stage: Selection of the final optimal solution—the best melody
Step-by-step process of the FSM presented in Sect. 2.5.3 of Chap. 2
terminate
end main body

Sub-stage 3.1.2: Group computational sub-stage or group improvisation sub-stage (GISS)

Sub-stage 3.1.2.1: Improvisation of a new melody vector by each player taking into account the feasible ranges of the updated pitches
Sub-stage 3.1.2.2: Update of each IPM
Sub-stage 3.1.2.3: Update of the feasible ranges of pitches—continuous decision-making variables—for the next improvisation—only for random selection
Sub-stage 3.1.2.4: Check of the stopping criterion of the GISS

- Sub-stage 3.2: Integration and separation procedures of melody vectors
- Sub-stage 3.3: Check of the stopping criterion of the external computational sub-stage

• Stage 4—Selection stage: Selection of the final optimal solution—the best melody

It is clear that the performance-driven architecture of the multi-objective TMS-EMSA is virtually the same as the performance-driven architecture of the multi-objective continuous/discrete TMS-MSA, which was described in Sect. 4.5.2.1 of this chapter. However, the multi-objective TMS-EMSA gives rise to substantial differences in some stages and sub-stages of the performance-driven architecture over the multi-objective continuous/discrete TMS-MSA, owing to its enhancements. Only the different stages and sub-stages changed in the multi-objective TMS-EMSA, in comparison with the multi-objective continuous/discrete TMS-MSA, and are redefined to prevent repeating the previous descriptions in Sect. 4.5.2.1 of this chapter.

Sub-stage 2.1 is related to the initialization of the parameters of the multi-objective TMS-EMSA. In this sub-stage, the parameter adjustments of the multi-objective continuous/discrete TMS-MSA are defined in accordance with Table 4.40, which was presented in Sect. 4.5.2.1 of this chapter. In the multi-objective TMS-EMSA, the

PMS and PMCR parameters are substituted with the PMS_p and $PMCR_p$ parameters, respectively. The $\text{PAR}^{\min}$ and $\text{PAR}^{\max}$ parameters are also substituted for the $PAR_p^{\min}$ and $PAR_p^{\max}$ parameters, respectively. In the multi-objective continuous/discrete TMS-MSA, the $\text{PAR}^{\min}$ and $\text{PAR}^{\max}$ parameters are generally used to determine the value of the PAR parameter in iteration n of the external computational stage and in the improvisation/iteration m of the SISS and PGISS ($PAR_{n,m}$). Replacing the $\text{PAR}^{\min}$ and $\text{PAR}^{\max}$ parameters with the $PAR_p^{\min}$ and $PAR_p^{\max}$ parameters in the multi-objective TMS-EMSA leads to a change in the $PAR_{n,m}$ parameter to the pitch adjusting rate of player p in iteration n of the external computational stage and in the improvisation/iteration m of the SISS and GISS ($PAR_{n,m,p}$). In addition, the $\text{BW}^{\min}$ and $\text{BW}^{\max}$ parameters are replaced by the $BW_p^{\min}$ and $BW_p^{\max}$ parameters, respectively. Likewise, in the multi-objective continuous/discrete TMS-MSA, the $\text{BW}^{\min}$ and $\text{BW}^{\max}$ parameters are generally used to specify the value of the BW parameter in iteration n of the external computational stage and in the improvisation/ iteration m of the SISS and PGISS ($BW_{n,m}$). Replacing the $\text{BW}^{\min}$ and $\text{BW}^{\max}$ parameters with the $BW_p^{\min}$ and $BW_p^{\max}$ parameters in the multi-objective TMS-EMSA results in a change in the $BW_{n,m}$ parameter to the bandwidth of player p in iteration n of the external computational stage and also in the improvisation/iteration m of the SISS and GISS—$BW_{n,m,p}$. Besides, the MNI-PGISS is substituted with the maximum number of improvisation/iteration of the group improvisation sub-stage (MNI-GISS). Other parameters presented in Table 4.40 remain unchanged for the multi-objective TMS-EMSA. The reasons for the above changes were previously explained in detail in Sect. 4.3 of this chapter. The detailed descriptions relevant to the adjustment parameters of the multi-objective TMS-EMSA are provided in Table 4.50.

Sub-stage 2.2 is concerned with the initialization of the IMM. In this sub-stage of the multi-objective continuous/discrete TMS-MSA, the initialization of the IMM is performed in accordance with Eqs. (4.73)–(4.76) given in Sect. 4.5.2.1 of this chapter. In the multi-objective TMS-EMSA, however, in view of the changes applied in the parameters under sub-stage 2.1, the initialization process of the IMM must be rewritten according to Eqs. (4.77)–(4.80):

$$\text{IMM} = [IPM_1 \quad \cdots \quad IPM_p \quad \cdots \quad IPM_{\text{PN}}]; \quad \forall\{p \in \Psi^{\text{PN}}\} \tag{4.77}$$

$$IPM_p = \begin{bmatrix} \mathrm{x}_p^1 \\ \vdots \\ \mathrm{x}_p^s \\ \vdots \\ \mathrm{x}_p^{PMS_p} \end{bmatrix} = \left[\begin{array}{ccccc|ccccc} x_{p,1}^1 & \cdots & x_{p,v}^1 & \cdots & x_{p,\text{NDV}}^1 & f\left(\mathrm{x}_{p,1}^1\right) & \cdots & f\left(\mathrm{x}_{p,a}^1\right) & \cdots & f\left(\mathrm{x}_{p,\text{A}}^1\right) \\ \vdots & & \vdots & & \vdots & \vdots & & \vdots & & \vdots \\ x_{p,1}^s & \cdots & x_{p,v}^s & \cdots & x_{p,\text{NDV}}^s & f\left(\mathrm{x}_{p,1}^s\right) & \cdots & f\left(\mathrm{x}_{p,a}^s\right) & \cdots & f\left(\mathrm{x}_{p,\text{A}}^s\right) \\ \vdots & & \vdots & & \vdots & \vdots & & \vdots & & \vdots \\ x_{p,1}^{PMS_p} & \cdots & x_{p,v}^{PMS_p} & \cdots & x_{p,\text{NDV}}^{PMS_p} & f\left(\mathrm{x}_{p,1}^{PMS_p}\right) & \cdots & f\left(\mathrm{x}_{p,a}^{PMS_p}\right) & \cdots & f\left(\mathrm{x}_{p,\text{A}}^{PMS_p}\right) \end{array}\right];$$

$$\forall\{p \in \Psi^{\text{PN}}, v \in \Psi^{\text{NDV}}, s \in \Psi^{PMS_p}, \Psi^{\text{NDV}} = \Psi^{\text{NCDV+NDDV}}\} \tag{4.78}$$

Table 4.50 Adjustment parameters of the proposed multi-objective TMS-EMSA

No.	The proposed multi-objective TMS-EMSA parameter	Abbreviation	Parameter range
1	Input melody memory	IMM	–
2	Output melody memory	OMM	–
3	Hybrid melody memory	HMM	–
4	Player number	PN	PN ≥ 1
5	Input player memory of player p	IPM_p	–
6	Output player memory of player p	OPM_p	–
7	Hybrid player memory of player p	HPM_p	–
8	Player memory size of player p	PMS_p	$PMS_p \geq 1$
9	Player memory considering rate of player p	$PMCR_p$	$0 \leq PMCR_p \leq 1$
10	Minimum pitch adjusting rate of player p	$PAR_p^{\min}$	$0 \leq PAR_p^{\min} \leq 2$
11	Maximum pitch adjusting rate of player p	$PAR_p^{\max}$	$0 \leq PAR_p^{\max} \leq 2$
12	Minimum bandwidth of player p	$BW_p^{\min}$	$0 \leq BW_p^{\min} < +\infty$
13	Maximum bandwidth of player p	$BW_p^{\max}$	$0 \leq BW_p^{\max} < +\infty$
14	Number of continuous decision-making variables	NCDV	NCDV ≥ 1
15	Number of discrete decision-making variables	NDDV	NDDV ≥ 1
16	Number of decision-making variables	NDV	NDV ≥ 2
17	Maximum number of iterations of the external computational stage	MNI-E	MNI-E ≥ 1
18	Maximum number of improvisations/iterations of the SISS	MNI-SISS	MNI-SISS ≥ 1
19	Maximum number of improvisations/iterations of the GISS	MNI-GISS	MNI-GISS ≥ 1
20	Maximum number of improvisations/iterations of the internal computational stage	MNI-I	MNI-I ≥ 2

$$x_{p,v}^{s} = x_v^{\min} + \mathrm{U}(0,1) \cdot \left(x_v^{\max} - x_v^{\min}\right); \quad \forall \left\{p \in \Psi^{\mathrm{PN}}, v \in \Psi^{\mathrm{NCDV}}, s \in \Psi^{PMS_p}\right\} \tag{4.79}$$

$$x_{p,v}^{s} = x_v(y); \; \forall \left\{p \in \Psi^{\mathrm{PN}}, v \in \Psi^{\mathrm{NDDV}}, s \in \Psi^{PMS_p}, y \sim \mathrm{U}\{x_v(1), \ldots, x_v(w_v), \ldots, x_v(W_v)\}\right\} \tag{4.80}$$

The pseudocode pertaining to initialization of the entire set of IPMs or IMM in the multi-objective continuous/discrete TMS-MSA is presented in Table 4.41. In the multi-objective TMS-EMSA, however, due to the changes made in the parameters under sub-stage 2.1, this pseudocode needs to be restructured. As a result, Table 4.51 presents the pseudocode related to the initialization of the entire set of IPMs or IMM in the multi-objective TMS-EMSA.

After filling all of the IPMs or the IMM with random solution vectors, the solution vectors stored in each IPM must be sorted. In the multi-objective continuous/discrete TMS-MSA, sorting of the stored solution vectors in all of the PMs or the MM is

Table 4.51 Pseudocode related to the initialization of the entire set of IPMs or IMM in the proposed multi-objective TMS-EMSA

Algorithm 43: Pseudocode for the initialization of the entire set of IPMs or IMM in the proposed multi-objective TMS-EMSA

Input: A, NCDV, NDDV, NDV, PMS_p, PN, $x_v^{\min}$, $x_v^{\max}$, $\{x_v(1), \ldots, x_v(w_v), \ldots, x_v(W_v)\}$

Output: IMM

start main body

begin
set $\mathrm{PMS}^{\max} = \max\,(PMS_p); \forall p \in \Psi^{\mathrm{PN}}$
construct the matrix IMM with dimension $\{\mathrm{PMS}^{\max}\} \cdot \{(\mathrm{NDV} + \mathrm{A}) \cdot \mathrm{PN}\}$ and zero initial value
for *music player p* $[p \in \Psi^{\mathrm{PN}}]$ **do**
  construct the submatrix IPM_p with dimension $\{PMS_p\} \cdot \{\mathrm{NDV} + \mathrm{A}\}$ and zero initial value
  for *melody vector s* $\left[s \in \Psi^{PMS_p}\right]$ **do**
    construct melody vector s of music player p, x_p^s, with dimension $\{1\} \cdot \{\mathrm{NDV} + \mathrm{A}\}$ and zero initial value
    for *decision-making variable v* $[v \in \Psi^{\mathrm{NDV}}]$ **do**
      $x_{p,v}^s = x_v^{\min} + \mathrm{U}(0,1) \cdot \left(x_v^{\max} - x_v^{\min}\right)$; for CDVs
      $x_{p,v}^s = x_v(y); \forall y \sim \mathrm{U}\{x_v(1), \ldots, x_v(w_v), \ldots, x_v(W_v)\}$; for DDVs
      allocate $x_{p,v}^s$ to element $(1, v)$ of melody vector x_p^s
    end for
    for *objective function a* $[a \in \Psi^{\mathrm{A}}]$ **do**
      calculate the value of the objective function a, fitness function, derived from melody vector x_p^s as $f\left(\mathrm{x}_{p,a}^s\right)$
      allocate $f\left(\mathrm{x}_{p,a}^s\right)$ to element $(1, \mathrm{NDV} + a)$ of melody vector x_p^s
    end for
    add melody vector x_p^s to the row s of the submatrix IPM_p
  end for
  add submatrix IPM_p to the rows 1 to PMS_p and columns $1 + [(p - 1) \cdot (\mathrm{NDV} + \mathrm{A})]$ to $[p \cdot (\mathrm{NDV} + \mathrm{A})]$ of the matrix IMM
end for
terminate

end main body

Note: Continuous decision-making variable (CDVs), discrete decision-making variable (DDVs)

performed according to the pseudocodes presented in Tables 4.42, 4.43, and 4.44. In the multi-objective TMS-EMSA, however, due to the changes made in the parameters under sub-stage 2.1, these pseudocodes must be restructured.

The pseudocodes presented in Tables 4.42, 4.43, and 4.44 can be consequently employed in sorting the stored solution vectors in the IPMs or IMM in the TMS-EMSA, if the PMS parameter is replaced by the PMS_p parameter.

Sub-stages 3.1.1.1 and 3.1.2.1 are concerned with the improvisation of a new melody vector by each existing player in the musical group without and with the influence of other existing players in the musical group, respectively. The improvisation process of a new melody vector by each existing player in the musical group under the proposed multi-objective continuous/discrete TMS-MSA is conceptually similar to the corresponding process under the single-objective continuous/discrete TMS-MSA (the continuous/discrete AIP), previously presented in Sect. 4.2.3.1 of this chapter.

The pseudocode related to improvisation of a new melody vector by each player in the musical group under the multi-objective continuous/discrete TMS-MSA is presented in Table 4.45 in Sect. 4.5.2.1 of this chapter. In the multi-objective TMS-EMSA, however, a new improvisation process is employed by each existing player to create a new melody vector. This new improvisation process is conceptually similar to the improvisation process of a new melody vector by each player in the musical group under the single-objective TMS-EMSA (the proposed EAIP), previously presented in Sect. 4.3 of this chapter. To avoid repetition, a detailed description about this improvisation process will be ignored here. However, due to the existence of the external computational stage in the multi-objective TMS-EMSA, the iteration index of the external computational stage (index n) must be added to all of the mathematical equations used in the EAIP that have the improvisation/iteration index of the SISS and the GISS (index m).

The pseudocode associated with the improvisation of a new melody vector by each player in the musical group in the single-objective TMS-EMSA is presented in Table 4.15 in Sect. 4.3 of this chapter. In the multi-objective TMS-EMSA, however, due to the replacement of the SOOP with the MOOP in stage 1 and addition of the external computational stage, this pseudocode must be restructured. Table 4.52 gives the pseudocode pertaining to the improvisation of a new melody vector by each player under the multi-objective TMS-EMSA.

Sub-stages 3.1.1.2 and 3.1.2.2 are relevant to the process of updating all of the IPMs or the IMM. A detailed description related to the update process of all IPMs or the IMM in the multi-objective TMS-EMS is conceptually similar to this description in the multi-objective continuous/discrete TMS-MSA, previously provided in Sect. 4.5.2.1 of this chapter. Only in this description the PMS parameter must be replaced with the PMS_p parameter.

The pseudocode related to the updating of the memory of all existing players in the musical group or the $IMM_{n,m}$ in the multi-objective continuous/discrete TMS-MSA is also presented in Table 4.46 in Sect. 4.5.2.1 of this chapter. In the multi-objective TMS-EMSA, however, due to the changes applied on the parameters in sub-stage 2.1, this pseudocode must be restructured. The pseudocode presented in Table 4.46 can be employed to update the memory of all existing players or the $IMM_{n,m}$ in the TMS-EMSA, if the PMS parameter is replaced by the PMS_p parameter. After completion of this process, melody vectors stored in the memory of all existing players or the $IMM_{n,m}$ must be re-sorted. In the multi-objective continuous/discrete TMS-MSA, the pseudocodes presented in Tables 4.42, 4.43, and 4.44 are employed to sort the melody vectors stored in the memory of the players in the

Table 4.52 Pseudocode pertaining to the improvisation of a new melody vector by each player in the musical group in the proposed multi-objective TMS-EMSA

Algorithm 44: Pseudocode for the improvisation of a new melody vector by each player in the musical group in the proposed multi-objective TMS-EMSA

Input: A, $BW_p^{\max}$, $BW_p^{\min}$, MNI-SISS, MNI-GISS, NCDV, NDDV, NDV, $PAR_p^{\max}$, $PAR_p^{\min}$, $PMCR_p$, PMS_p, PN, $x_v^{\min}$, $x_v^{\max}$, $\{x_v(1), \ldots, x_v(w_v), \ldots, x_v(W_v)\}$

Output: $\mathrm{x}_{n,m,p}^{\text{new}}$

start main body

begin

$BW_{n,m,p} = BW_p^{\max} \cdot \exp\left[\left(\ln\left(BW_p^{\max}/BW_p^{\min}\right)/((\text{MNI-SISS}) + (\text{MNI-GISS}))\right) \cdot m\right]$

$PAR_{n,m,p} = PAR_p^{\min} - \left[\left(\left(PAR_p^{\max} - PAR_p^{\min}\right)/((\text{MNI-SISS}) + (\text{MNI-GISS}))\right) \cdot m\right]$

for *music player p* $[p \in \Psi^{\text{PN}}]$ **do**

  construct the new melody vector for music player p, $\mathrm{x}_{n,m,p}^{\text{new}}$, with dimension $\{1\} \cdot \{\text{NDV} + 1\}$ and zero initial value

  for *decision-making variable v* $[v \in \Psi^{\text{NDV}}]$ **do**

    if $\mathrm{U}(0,1) \le PMCR_p$ **then**

      Rule 1: player memory consideration with probability $PMCR_p$

      if *improvisation/iteration of the internal computational sub-stage* $m\ [m \in \Psi^{(\text{MNI-SISS}) + (\text{MNI-GISS})}]$ is odd **then**

        Principle 1 of Rule 1: first combination

        $x_{n,m,p,v}^{\text{new}} = x_{n,m,p,v}^{r_p} \pm \mathrm{U}(0,1) \cdot BW_{n,m,p}; \forall r_p \sim \mathrm{U}\{1, 2, \ldots, PMS_p\}$; for CDVs

        $x_{n,m,p,v}^{\text{new}} = x_{n,m,p,v}^{r_p}; \forall r_p \sim \mathrm{U}\{1, 2, \ldots, PMS_p\}$; for DDVs

      else

        Principle 2 of Rule 1: second combination

        $x_{n,m,p,v}^{\text{new}} = x_{n,m,p,k}^{r_p} \pm \mathrm{U}(0,1) \cdot BW_{n,m,p}; \forall r_p \sim \mathrm{U}\{1, 2, \ldots, PMS_p\}, \forall k \sim \mathrm{U}\{1, 2, \ldots, \text{NCDV}\}$; for CDVs

        $x_{n,m,p,v}^{\text{new}} = x_{n,m,p,l}^{r_p}; \forall r_p \sim \mathrm{U}\{1, 2, \ldots, PMS_p\}, \forall l \sim \mathrm{U}\{1, 2, \ldots, \text{NDDV}\}$; for DDVs

      end if

      if $\mathrm{U}(0,1) \le PAR_{n,m,p}$ **then**

        Rule 2: pitch adjustment with probability $PMCR_p \cdot PAR_{n,m,p}$

        switch 1

        case *improvisation/iteration of the internal computational sub-stage* $m\ [m \in \Psi^{(\text{MNI-SISS}) + (\text{MNI-GISS})}] \le (\text{MNI-SISS})$ **then**

          Principle 1 of Rule 2: first choice

          $x_{n,m,p,v}^{\text{new}} = x_{n,m,p,v}^{\text{best}}$; for CDVs and DDVs

        case *improvisation/iteration of the internal computational sub-stage* $m\ [m \in \Psi^{(\text{MNI-SISS}) + (\text{MNI-GISS})}] > (\text{MNI-SISS})$ and *improvisation/iteration of the internal computational sub-stage* $m\ [m \in \Psi^{(\text{MNI-SISS}) + (\text{MNI-GISS})}] \le (\text{MNI-SISS}) + (\text{MNI-GISS})$ **then**

          Principle 2 of Rule 2: second choice

          $x_{n,m,p,v}^{\text{new}} = x_{n,m,\text{best},v}^{\text{best}}$; for CDVs and DDVs

        end switch

(continued)

Table 4.52 (continued)

end if
else if
Rule 3: random selection with probability $1 - PMCR_p$
switch 1
case *improvisation/iteration of the internal computational sub-stage* m $[m \in \Psi^{(\text{MNI-SISS})+(\text{MNI-GISS})}] \leq (\text{MNI-SISS})$ **then**
Principle 1 of Rule 2: first choice
$x^{\text{new}}_{n,m,p,v} = x^{\min}_{v} + \text{U}(0,1) \cdot \left(x^{\max}_{v} - x^{\min}_{v}\right)$; for CDVs
$x^{\text{new}}_{n,m,p,v} = x_v(y); \forall y \sim \text{U}\{x_v(1), \ldots, x_v(w_v), \ldots, x_v(W_v)\}$; for DDVs
case *improvisation/iteration of the internal computational sub-stage* m $[m \in \Psi^{(\text{MNI-SISS})+(\text{MNI-GISS})}] > (\text{MNI-SISS})$ and *improvisation/iteration of the internal computational sub-stage* m $[m \in \Psi^{(\text{MNI-SISS})+(\text{MNI-GISS})}] \leq (\text{MNI-SISS}) + (\text{MNI-GISS})$ **then**
Principle 2 of Rule 2: second choice
$x^{\text{new}}_{n,m,p,v} = x^{\min}_{n,m,v} + \text{U}(0,1) \cdot \left(x^{\max}_{n,m,v} - x^{\min}_{n,m,v}\right)$; for CDVs
$x^{\text{new}}_{n,m,p,v} = x_v(y); \forall y \sim \text{U}\{x_v(1), \ldots, x_v(w_v), \ldots, x_v(W_v)\}$; for DDVs
end switch
end if
end for
for *objective function* a $[a \in \Psi^{\text{A}}]$ **do**
calculate the value of objective function a, fitness function, derived from melody vector $\text{x}^{\text{new}}_{n,m,p}$ as $f\left(\text{x}^{\text{new}}_{n,m,p,a}\right)$
allocate $f\left(\text{x}^{\text{new}}_{n,m,p,a}\right)$ to element $(1, \text{NDV} + a)$ of the new melody vector $\text{x}^{\text{new}}_{n,m,p}$
end for
end for
terminate
end main body

musical group or the $IMM_{n,m}$, while the iteration index of the external computational stage (index n) and improvisation/iteration index of the internal computational stage (index m) are added to all of the mathematical equations used in these pseudocodes. In the multi-objective TMS-EMSA, however, the pseudocodes presented in Tables 4.42, 4.43, and 4.44 can be utilized to sort the melody vectors stored in the memory of all existing players in the musical group or the $IMM_{n,m}$, while the iteration index of the external computational stage (index n) and improvisation/iteration index of the internal computational stage (index m) are added to all of the mathematical equations in these pseudocodes and the PMS parameter is replaced with the PMS_p parameter in these pseudocodes.

Sub-stage 3.1.2.3 is associated with the update of the feasible ranges of pitches (continuous decision-making variables) for the next improvisation (only for random selection). This sub-stage of the multi-objective TMS-EMSA is conceptually similar to corresponding sub-stage of the multi-objective continuous/discrete TMS-MSA, described in Sect. 4.5.2.1 of this chapter. The pseudocode related to the update of the feasible ranges of continuous decision-making variables in the multi-objective TMS-EMSA is also virtually the same as this pseudocode in the multi-objective continuous/discrete TMS-MSA, which was presented in Table 4.47. Only in this pseudocode the MNI-PGISS parameter must be replaced with the MNI-GISS parameter.

Sub-stages 3.1.1.3 and 3.1.2.4 are relevant to the check of the stopping criterion of the SISS and GISS. These sub-stages of the multi-objective TMS-EMSA are similar to the corresponding sub-stages of the multi-objective continuous/discrete TMS-MSA, previously described in Sect. 4.5.2.1 of this chapter. Only in sub-stage 3.1.2.4 must the MNI-PGISS parameter be replaced with the MNI-GISS parameter.

Sub-stage 3.2 is concerned with the integration and separation procedures of melody vectors. A detail description relevant to the integration and separation procedures in the multi-objective TMS-EMS is similar to this description in the multi-objective continuous/discrete TMS-MSA, previously provided in Sect. 4.5.2.1 of this chapter. Only in this description the PMS parameter must be replaced with the PMS_p parameter. The pseudocode related to the integration and separation procedures of melody vectors in the multi-objective continuous/discrete TMS-MSA is also presented in Table 4.48. In the multi-objective TMS-EMSA, however, due to the changes applied to the parameters of sub-stage 2.1, this pseudocode must be restructured. Table 4.53 gives the pseudocode pertaining to the integration and separation procedure of melody vectors in the multi-objective TMS-EMSA. The pseudocode related to the performance-driven architecture of the multi-objective continuous/discrete TMS-MSA is provided in Table 4.49. However, this pseudocode must be restructured, owing to the incorporation of some improvements from stages and sub-stages in the multi-objective TMS-EMSA. Table 4.54 presents the pseudocode related to the performance-driven architecture of the multi-objective TMS-EMSA.

Sub-stages 3.3, 3.1.1.3, and 3.1.2.4 (the process of checking the stopping criterion of the external computational stage, the SISS, and the GISS) are defined by the first, second, and third WHILE loops in the pseudocode related to the performance-driven architecture of the multi-objective TMS-EMSA (see Table 4.54).

4.5.3 Multi-objective Strategy for the Meta-heuristic Music-Inspired Optimization Algorithms with Multi-stage Computational Multi-dimensional and Multiple-Homogeneous Structure

The authors presented earlier a new meta-heuristic music-inspired optimization algorithm referred to as the single-objective SOSA in Sect. 4.4 of this chapter. The performance-driven architecture of the single-objective SOSA was developed in

Table 4.53 Pseudocode pertaining to the integration and separation procedures of melody vectors in the proposed multi-objective TMS-EMSA

Algorithm 45: Pseudocode for the integration and separation procedures of melody vectors in the proposed multi-objective TMS-EMSA

Input: $IMM_{n,1}$, $IMM_{n,\,\text{MNI-I}}$
Output: IMM_{n+1}

start main body

begin
set $IMM_n = IMM_{n,1}$
set $OMM_n = IMM_{n,\text{MNI-I}}$
set $\text{PMS}^{\max} = \max(PMS_p);\ \forall p \in \Psi^{\text{PN}}$
construct the HMM_n with dimension $\{2 \times \text{PMS}^{\max}\} \cdot \{(\text{NDV} + \text{A}) \cdot \text{PN}\}$ and zero initial value
for *music player p* $[p \in \Psi^{\text{PN}}]$ **do**
    $IPM_{n,p} = IMM_n(1 : PMS_p,\ 1 + [(p-1) \cdot (\text{NDV} + \text{A})]:[p \cdot (\text{NDV} + \text{A})])$
    $OPM_{n,p} = OMM_n(1 : PMS_p,\ 1 + [(p-1) \cdot (\text{NDV} + \text{A})]:[p \cdot (\text{NDV} + \text{A})])$
    $HPM_{n,\,p} = IPM_{n,\,p} \cup OPM_{n,p}$
    $HMM_n(1 : 2 \times PMS_p,\ 1 + [(p-1) \cdot (\text{NDV} + \text{A})]:[p \cdot (\text{NDV} + \text{A})])) = HPM_{n,p}$
end for
sort the HMM_n
    Algorithm 35: Pseudocode for the MFNDSA used in order to determine the non-dominated front—rank—of each melody vector stored in the HMM_n under the proposed multi-objective continuous/discrete TMS-MSA, while the n index is added to all equations and also the PMS parameters are replaced with the PMS_p parameter in this pseudocode
    Algorithm 36: Pseudocode for the MCCA used in order to determine the crowding distance of each melody vector stored in the HMM_n under the proposed multi-objective continuous/discrete TMS-MSA, while the n index is added to all equations and also the PMS parameters are replaced with the PMS_p parameter in this pseudocode
    Algorithm 37: Pseudocode for sorting the melody vectors stored in the HMM_n under the proposed multi-objective continuous/discrete TMS-MSA, while the n index is added to all equations and also the PMS parameters are replaced with the PMS_p parameter in this pseudocode
    construct the IMM_{n+1} with dimension $\{\text{PMS}^{\max}\} \cdot \{(\text{NDV} + \text{A}) \cdot \text{PN}\}$ and zero initial value
for *music player p* $[p \in \Psi^{\text{PN}}]$ **do**
    $IPM_{n+1,p} = HMM_n(1 : PMS_p,\ 1 + [(p-1) \cdot (\text{NDV} + \text{A})] : [p \cdot (\text{NDV} + \text{A})])$
    $IMM_{n+1}(1 : PMS_p,\ 1 + [(p-1) \cdot (\text{NDV} + \text{A})] : [p \cdot (\text{NDV} + \text{A})]) = IPM_{n+1,p}$
end for
terminate

end main body

such a way that it was only suitable for solving the SOOPs and could not be used to solve the MOOPs. To cope with this deficiency in the architecture of the single-objective SOSA, it is necessary to address a new multi-objective strategy for altering

Table 4.54 Pseudocode related to the performance-driven architecture of the proposed multi-objective TMS-EMSA

Algorithm 46: Pseudocode for the performance-driven architecture of the proposed multi-objective continuous/discrete TMS-MSA

Input: A, $BW_p^{\max}$, $BW_p^{\min}$, MNI-E, MNI-I, MNI-SISS, MNI-GISS, NCDV, NDDV, NDV, $PAR_p^{\max}$, $PAR_p^{\min}$, $PMCR_p$, PMS_p, PN, $x_v^{\min}$, $x_v^{\max}$, $\{x_v(1), \ldots, x_v(w_v), \ldots, x_v(W_v)\}$

Output: Pareto optimal solution set and also $\mathrm{x}^{\mathrm{best}}$

start main body

begin

Stage 1—Definition stage: Definition of the MOOP and its parameters

Stage 2—Initialization stage

  Sub-stage 2.1: Initialization of the parameters of the multi-objective TMS-EMSA

  Sub-stage 2.2: Initialization of the of the IMM

    Algorithm 43: Pseudocode for the initialization of the entire set of IPMs or IMM in the proposed multi-objective TMS-EMSA

    Algorithm 35: Pseudocode for the MFNDSA used in order to determine the non-dominated front—rank—of each melody vector stored in the IPMs or IMM under the proposed multi-objective TMS-EMSA, while the PMS parameter is replaced by the PMS_p parameter

    Algorithm 36: Pseudocode for the MCCA used in order to determine the crowding distance of each melody vector stored in the IPMs or IMM under the proposed multi-objective TMS-EMSA, while the PMS parameter is replaced by the PMS_p parameter

    Algorithm 37: Pseudocode for sorting the melody vectors stored in the IPMs or IMM under the proposed multi-objective TMS-EMSA, while the PMS parameter is replaced by the PMS_p parameter

Stage 3—External computational stage

  set *iteration of the external computational stage* $n = 1$

  set IMM_n = IMM

  while $n \leq$ MNI-E **do**

    Sub-stage 3.1—Internal computational sub-stage

      set *improvisation/iteration of the internal computational sub-stage* $m = 1$

      set $IMM_{n,\,m} = IMM_n$

      Sub-stage 3.1.1—single computational sub-stage or the SISS

      while $m \leq$ (MNI-SISS) **do**

        Sub-stage 3.1.1.1: Improvisation of a new melody vector by each player

          Algorithm 44: Pseudocode for the improvisation of a new melody vector by each player in the musical group in the proposed multi-objective TMS-EMSA

        Sub-stage 3.1.1.2: Update of each IPM

          Algorithm 39: Pseudocode for the update of the memory of all existing players in the musical group or the update of the $IMM_{n,m}$ in the proposed multi-objective TMS-EMSA, while the PMS parameter is replaced by the PMS_p parameter

(continued)

Table 4.54 (continued)

Algorithm 35: Pseudocode for the MFNDSA used in order to determine the non-dominated front—rank—of each melody vector stored in the IPMs or IMM under the proposed multi-objective TMS-EMSA, while the n and m indices are added to all equations and the PMS parameter is replaced by the PMS_p parameter in this pseudocode

Algorithm 36: Pseudocode for the MCCA used in order to determine the crowding distance of each melody vector stored in the IPMs or IMM under the proposed multi-objective TMS-EMSA, while the n and m indices are added to all equations and the PMS parameter is replaced by the PMS_p parameter in this pseudocode

Algorithm 37: Pseudocode for sorting the melody vectors stored in the IPMs or IMM under the proposed multi-objective TMS-EMSA, while the n and m indices are added to all equations and the PMS parameter is replaced by the PMS_p parameter in this pseudocode

set *improvisation/iteration of the internal computational sub-stage* $m = m + 1$

end while

Sub-stage 3.1.2—Group computational sub-stage or the GISS

while $m >$ (MNI-SISS) & $m \leq$ (MNI-SISS) + (MNI-GISS) **do**

Sub-stage 3.1.2.1: Improvisation of a new melody vector by each player

Algorithm 44: Pseudocode for the improvisation of a new melody vector by each player in the musical group in the proposed multi-objective TMS-EMSA

Sub-stage 3.1.1.2: Update of each IPM

Algorithm 39: Pseudocode for the update of the memory of all existing players in the musical group or the update of the $IMM_{n,m}$ in the proposed multi-objective TMS-EMSA, while the PMS parameter is replaced by the PMS_p parameter

Algorithm 35: Pseudocode for the MFNDSA used in order to determine the non-dominated front—rank—of each melody vector stored in the IPMs or IMM under the proposed multi-objective TMS-EMSA, while the n and m indices are added to all equations and the PMS parameter is replaced by the PMS_p parameter in this pseudocode

Algorithm 36: Pseudocode for the MCCA used in order to determine the crowding distance of each melody vector stored in the IPMs or IMM under the proposed multi-objective TMS-EMSA, while the n and m indices are added to all equations and the PMS parameter is replaced by the PMS_p parameter in this pseudocode

Algorithm 37: Pseudocode for sorting the melody vectors stored in the IPMs or IMM under the proposed multi-objective TMS-EMSA, while the n and m indices are added to all equations and the PMS parameter is replaced by the PMS_p parameter in this pseudocode

(continued)

Table 4.54 (continued)

Sub-stage 3.1.1.3: Update of the feasible ranges of pitches—continuous decision-making variables—for the next improvisation—only for random selection
Algorithm 40: Pseudocode for the update of the feasible ranges of the continuous decision-making variables in the proposed multi-objective TMS-EMSA, while the MNI-PGISS parameter is replaced with the MNI-GISS parameter
set *improvisation/iteration of the internal computational sub-stage* $m = m + 1$
end while
Sub-stage 3.2—Integration and separation procedures of melody vectors
Algorithm 45: Pseudocode for the integration and separation procedures of melody vectors in the proposed multi-objective TMS-EMSA
set *iteration of the external computational stage* $n = n + 1$
end while
Stage 4—Selection stage: Selection of the final optimal solution—the best melody
Step-by-step process of the FSM presented in Sect. 2.5.3 of Chap. 2
terminate
end main body

the performance-driven architecture of this single-objective algorithm in dealing with the MOOPs.

With that in mind, in Sect. 4.5.1 of this chapter, the authors proposed new multi-objective strategies for restructuring the architecture of the meta-heuristic music-inspired optimization algorithms or, more comprehensively, the meta-heuristic optimization algorithms with a single-stage computational and single-dimensional structure. Furthermore, by reforming the multi-objective strategies, the authors described new multi-objective strategies for restructuring the architecture of the meta-heuristic music-inspired optimization algorithms with a two-stage computational multi-dimensional and single-homogeneous structure in Sect. 4.5.2 of this chapter.

However, the single-objective SOSA has a multi-stage computational multi-dimensional and multiple-homogeneous structure, or a multi-stage computational multi-dimensional and single-inhomogeneous structure. Because of the different structure of the SOSA compared to its counterparts, the multi-objective strategies in Sects. 4.5.1 and 4.5.2 of this chapter cannot be implemented on the structure of the single-objective SOSA. Hence, the authors developed a new multi-objective strategy for the single-objective SOSA by modifying the multi-objective strategies reported in Sects. 4.5.1 and 4.5.2 of this chapter. The performance-driven architecture of the multi-objective SOSA is generally broken down into four stages, as follows:

- Stage 1—Definition stage: Definition of the MOOP and its parameters
- Stage 2—Initialization stage
  - Sub-stage 2.1: Initialization of the parameters of the multi-objective SOSA
  - Sub-stage 2.2: Initialization of the input symphony orchestra memory (ISOM)
- Stage 3—External computational stage
  - Sub-stage 3.1: Internal computational sub-stage

    Sub-stage 3.1.1: Single computational sub-stage or the SISS

      Sub-stage 3.1.1.1: Improvisation of a new melody vector by each player in the symphony orchestra
      Sub-stage 3.1.1.2: Update of each available IPM in the symphony orchestra
      Sub-stage 3.1.1.3: Check of the stopping criterion of the SISS

    Sub-stage 3.1.2: Group computational sub-stage for each homogeneous musical group or group improvisation sub-stage for each homogeneous musical group (GISSHMG)

      Sub-stage 3.1.2.1: Improvisation of a new melody vector by each player in the symphony orchestra taking into account the feasible ranges of the updated pitches for each homogeneous musical group
      Sub-stage 3.1.2.2: Update of each available IPM in the symphony orchestra
      Sub-stage 3.1.2.3: Update of the feasible ranges of the pitches—continuous decision-making variables—for each homogeneous musical group in the next improvisation—only for random selection
      Sub-stage 3.1.2.4: Check of the stopping criterion of the GISSHMG

    Sub-stage 3.1.3: Group computational sub-stage for the inhomogeneous musical ensemble or group improvisation sub-stage for the inhomogeneous musical ensemble (GISSIME)

      Sub-stage 3.1.2.1: Improvisation of a new melody vector by each player in the symphony orchestra taking into account the feasible ranges of the updated pitches for the inhomogeneous musical ensemble
      Sub-stage 3.1.2.2: Update of each available IPM in the symphony orchestra
      Sub-stage 3.1.2.3: Update of the feasible ranges of the pitches—continuous decision-making variables—for the inhomogeneous musical ensemble in the next improvisation—only for random selection
      Sub-stage 3.1.2.4: Check of the stopping criterion of the GISSIME

  - Sub-stage 3.2: Integration and separation procedures of melody vectors
  - Sub-stage 3.3: Check of the stopping criterion of the external computational sub-stage
- Stage 4—Selection stage: Selection of the final optimal solution—the best melody

Stage 1 is related to the definition of the MOOP and its parameters. This stage is equivalent to stage 1 of the single-objective SOSA. In this stage of the single-objective SOSA, definitions of the SOOP and its parameters are performed according to Eqs. (4.39) and (4.40), which were described in Sect. 4.4.1 of this chapter. In the multi-objective SOSA, however, this stage must be redefined. In this stage of the multi-objective SOSA, the standard form of a MOOP can be generally described in accordance with Eqs. (4.65) and (4.66), presented in Sect. 4.5.1.1 of this chapter.

Sub-stage 2.1 is associated with the initialization of the parameters of the multi-objective SOSA. This sub-stage is equivalent to sub-stage 2.1 of the single-objective SOSA. In this sub-stage, the parameter adjustments of the single-objective SOSA are defined according to Table 4.18, which was presented in Sect. 4.4.2.1 of this chapter. Based on the explanations related to Table 4.18, the SOM is a place for storing the solution, or melody vectors, of the entire set of players in the symphony orchestra. The single-objective SOSA has only one unique SOM. This SOM consists of multiple MMs, equal to the NHMG parameters. The MM related to homogeneous musical group g consists of multiple PMs, equal to the PN_g parameter. The memory of homogeneous musical group g in the symphony orchestra is also a place for storing the solution, or melody vectors, of the entire set of players in corresponding homogeneous musical group. The memory of player p in homogeneous musical group g in the symphony orchestra is a place for storing the corresponding player's solution, or melody vectors.

In designing the performance-driven architecture of the multi-objective SOSA, one of the most important requirements is the use of three SOMs: (1) the ISOM; (2) the output symphony orchestra memory (OSOM); and, (3) the hybrid symphony orchestra memory (HSOM). The structure of the ISOM, OSOM, and HSOM in the multi-objective SOSA is similar to the structure of the SOM in the single-objective SOSA. This means that the ISOM, OSOM, and HSOM are organized by multiple IMMs, multiple OMMs, and multiple HMMs, respectively, equal to the NHMG parameter. The IMM, OMM, and the HMM related to homogeneous musical group g are also composed of multiple IPMs, multiple OPMs, and multiple HPMs, respectively, equal to the PN_g parameter. The ISOM is taken as the input for the internal computational sub-stage (sub-stage 3.1) and is updated in each iteration of the external computational stage (stage 3). The OSOM is extracted after the completion of the internal computational sub-stage (sub-stage 3.1). In other words, by implementing the internal computational sub-stage (sub-stage 3.1) on the ISOM, the OSOM is achieved. In addition, the HSOM comes from the integration of the ISOM and the OSOM and is used to create the ISOM in the next iteration of the external computational stage (stage 3) or the new ISOM. More detailed information about the ISOM, OSOM, and HSOM will be given in the relevant sub-stages.

The single-objective SOSA has three computational stages—the SIS, GISHMG, and GISIME—which are repeated for the MNI-SIS, MNI-GISHMG, and MNI-GISIME, respectively. The multi-objective SOSA, however, is composed of an external computational stage and an internal computational sub-stage.

The external computational stage is repeated for the MNI-E. The internal computational sub-stage is composed of three computational sub-stages specified as the SISS, GISSHMG, and GISSIME. More precisely, the SIS, GISHMG, and GISIME of the single-objective SOSA are regarded as the computational sub-stages of the internal computational sub-stage in the multi-objective SOSA by renaming them as the SISS, GISSHMG, and GISSIME, respectively. Subsequently, the MNI-SIS, MNI-GISHMG, and MNI-GISIME parameters are renamed as the MNI-SISS, the maximum number of improvisations/iterations of the group improvisation sub-stage for each homogeneous musical group (MNI-GISSHMG), and the maximum number of improvisations/iterations of the group improvisation sub-stage for inhomogeneous musical ensemble (MNI-GISSIME), respectively. In each iteration of the external computational stage, the SISS, GISSHMG, and GISSIME of the internal computational sub-stage are repeated for the MNI-SISS, MNI-GISSHMG, and MNI-GISSIME, respectively. The sum of the MNI-SISS, MNI-GISSHMG, and MNI-GISSIME parameters is taken into account as the MNI-I parameter. Other parameters presented in Table 4.18 remain unchanged for the multi-objective SOSA. Detailed descriptions pertaining to the adjustment parameters of the multi-objective SOSA are provided in Table 4.55.

Sub-stage 2.2 is relevant to the initialization of the ISOM. This sub-stage is equivalent to sub-stage 2.2 of the single-objective SOSA. In this sub-stage of the single-objective SOSA, the initialization of the SOM is accomplished based on Eqs. (4.41)–(4.45), given in Sect. 4.4.2.2 of this chapter.

In the multi-objective SOSA, and due to the replacement of the SOOP with the MOOP in stage 1, the initialization of the ISOM must be redefined.

In the multi-objective SOSA, the ISOM matrix, which has a dimension equal to $\left\{\sum_{g\in\Psi^{\mathrm{NHMG}}} PMS_g^{\max}\right\}\cdot\left\{(\mathrm{NDV}+\mathrm{A})\cdot \mathrm{PN}^{\max}\right\}$ is composed of multiple IMM submatrices in which each IMM has a dimension equal to $\left\{PMS_g^{\max}\right\}\cdot\left\{(\mathrm{NDV}+1)\cdot PN_g\right\}$. Each MM_g submatrix is also organized by multiple $PM_{g,p}$ submatrices with the dimension $\{PMS_{g,p}\}\cdot\{\mathrm{NDV}+\mathrm{A}\}$. The ISOM matrix, IMM submatrices, and IPM submatrices are filled with a large number of solution vectors generated haphazardly according to Eqs. (4.81)–(4.85):

$$\mathrm{ISOM}=\left[IMM_1 \quad \cdots \quad IMM_g \quad \cdots \quad IMM_{\mathrm{NHMG}}\right]^{\mathrm{T}};\quad \forall\left\{g\in\Psi^{\mathrm{NHMG}}\right\} \tag{4.81}$$

$$IMM_g=\left[IPM_{g,1} \quad \cdots \quad IPM_{g,p} \quad \cdots \quad IPM_{g,PN_g}\right];\quad \forall\left\{g\in\Psi^{\mathrm{NHMG}},p\in\Psi^{PN_g}\right\} \tag{4.82}$$

$$IPM_{g,p}=\begin{bmatrix} \mathrm{x}_{g,p}^{1} \\ \vdots \\ \mathrm{x}_{g,p}^{s} \\ \vdots \\ \mathrm{x}_{g,p}^{PMS_{g,p}} \end{bmatrix}=\left[\begin{array}{ccccc|ccccc} x_{g,p,1}^{1} & \cdots & x_{g,p,v}^{1} & \cdots & x_{g,p,\mathrm{NDV}}^{1} & f\left(\mathrm{x}_{g,p,1}^{1}\right) & \cdots & f\left(\mathrm{x}_{g,p,a}^{1}\right) & \cdots & f\left(\mathrm{x}_{g,p,\mathrm{A}}^{1}\right) \\ \vdots & & \vdots & & \vdots & \vdots & & \vdots & & \vdots \\ x_{g,p,1}^{s} & \cdots & x_{g,p,v}^{s} & \cdots & x_{g,p,\mathrm{NDV}}^{s} & f\left(\mathrm{x}_{g,p,1}^{s}\right) & \cdots & f\left(\mathrm{x}_{g,p,a}^{s}\right) & \cdots & f\left(\mathrm{x}_{g,p,\mathrm{A}}^{s}\right) \\ \vdots & & \vdots & & \vdots & \vdots & & \vdots & & \vdots \\ x_{g,p,1}^{PMS_{g,p}} & \cdots & x_{g,p,v}^{PMS_{g,p}} & \cdots & x_{g,p,\mathrm{NDV}}^{PMS_{g,p}} & f\left(\mathrm{x}_{g,p,1}^{PMS_{g,p}}\right) & \cdots & f\left(\mathrm{x}_{g,p,a}^{PMS_{g,p}}\right) & \cdots & f\left(\mathrm{x}_{g,p,\mathrm{A}}^{PMS_{g,p}}\right) \end{array}\right];$$

$$\forall\left\{g\in\Psi^{\mathrm{NHMG}},p\in\Psi^{PN_g},v\in\Psi^{\mathrm{NDV}},s\in\Psi^{PMS_{g,p}},\Psi^{\mathrm{NDV}}=\Psi^{\mathrm{NCDV+NDDV}}\right\} \tag{4.83}$$

Table 4.55 Adjustment parameters of the proposed multi-objective SOSA

No.	The proposed multi-objective SOSA parameters	Abbreviation	Parameter range
1	Input symphony orchestra memory	ISOM	–
2	Output symphony orchestra memory	OSOM	–
3	Hybrid symphony orchestra memory	HSOM	–
4	Number of homogeneous musical groups	NHMG	NHMG ≥ 1
5	Input melody memory of homogeneous musical group g	IMM_g	–
6	Output melody memory of homogeneous musical group g	OMM_g	–
7	Hybrid melody memory of homogeneous musical group g	HMM_g	–
8	Existing player number in homogeneous musical group g	PN_g	$PN_g \geq 1$
9	Input player memory of player p in homogeneous musical group g	$IPM_{g,p}$	–
10	Output player memory of player p in homogeneous musical group g	$OPM_{g,p}$	–
11	Hybrid player memory of player p in homogeneous musical group g	$HPM_{g,\ p}$	–
12	Player memory size of the player p in homogeneous musical group g	$PMS_{g,\ p}$	$PMS_{g,p} \geq 1$
13	Player memory considering rate of player p in homogeneous musical group g	$PMCR_{g,p}$	$0 \leq PMCR_{g,p} \leq 1$
14	Minimum pitch adjusting rate of player p in homogeneous musical group g	$PAR_{g,p}^{\min}$	$0 \leq PAR_{g,p}^{\min} \leq 2$
15	Maximum pitch adjusting rate of player p in homogeneous musical group g	$PAR_{g,p}^{\max}$	$0 \leq PAR_{g,p}^{\max} \leq 2$
16	Minimum band width of player p in homogeneous musical group g	$BW_{g,p}^{\min}$	$0 \leq BW_{g,p}^{\min} < +\infty$
17	Maximum band width of player p in homogeneous musical group g	$BW_{g,p}^{\max}$	$0 \leq BW_{g,p}^{\max} < +\infty$
18	Number of continuous decision-making variables	NCDV	NCDV ≥ 1
19	Number of discrete decision-making variables	NDDV	NDDV ≥ 1
20	Number of decision-making variables	NDV	NDV ≥ 2
21	Maximum number of iterations of the external computational stage	MNI-E	MNI-I ≥ 1
22	Maximum number of improvisations/iterations of the SISS	MNI-SISS	MNI-SISS ≥ 1
23	Maximum number of improvisations/iterations of the GISSHMG	MNI-GISSHMG	MNI-GISSHMG ≥ 1
24	Maximum number of improvisations/iterations of the GISSIME	MNI-GISSIME	MNI-GISDIME ≥ 1
25	Maximum number of improvisations/iterations of the internal computational sub-stage	MNI-I	MNI-I ≥ 3

$$x_{g,p,v}^{s} = x_{v}^{\min} + \mathrm{U}(0,1) \cdot \left(x_{v}^{\max} - x_{v}^{\min}\right); \quad \forall\left\{g \in \Psi^{\mathrm{NHMG}}, p \in \Psi^{PN_g}, v \in \Psi^{\mathrm{NCDV}}, s \in \Psi^{PMS_{g,p}}\right\} \tag{4.84}$$

$$x_{g,p,v}^{s} = x_{v}(y); \quad \forall\left\{g \in \Psi^{\mathrm{NHMG}}, p \in \Psi^{PN_g}, v \in \Psi^{\mathrm{NDDV}}, s \in \Psi^{PMS_{g,p}}, y \sim \mathrm{U}\{x_v(1), \ldots, x_v(w_v), \ldots, x_v(W_v)\}\right\} \tag{4.85}$$

It is obvious that Eqs. (4.81) and (4.82) have been indirectly changed compared to their counterparts in the single-objective SOSA. Equation (4.83) has also been directly changed compared to its counterpart in the single-objective SOSA. However, Eqs. (4.84) and (4.85) remain unchanged. The pseudocode relevant to initialization of the entire set of PMs or entire set of MMs or SOM in the single-objective SOSA is given in Table 4.19, which was presented in Sect. 4.4.2.2 of this chapter. In the multi-objective SOSA, however, due to the replacement of the SOOP with the MOOP in stage 1, this pseudocode needs to be restructured. As a result, Table 4.56 illustrates the pseudocode concerned with the initialization of the entire set of PMs or entire set of MMs or SOM in the multi-objective SOSA. In the single-objective SOSA, after filling all of the PMs or all of the MMs or the MM with random solution vectors, the solution vectors stored in each PM must be sorted from the lowest value to the highest value—in an ascending order—from the standpoint of the value of the objective function of the SOOP. The pseudocode related to the sorting of the solution vectors stored in the PMs or MMs or SOM in the single-objective SOSA is presented in Table 4.20 in Sect. 4.5.2.2 of this chapter. This pseudocode demonstrates a simple and straightforward process and, accordingly, it can only be applied to the proposed single-objective SOSA and employed to solve the SOOPs. More precisely, this pseudocode suffers from the inability to sort the solution vectors stored in the memory of the proposed multi-objective SOSA. Basically, only by specifying the non-dominated front—rank—and crowding distance of all solution vectors stored in the memory of a meta-heuristic MOOA can they be sorted. In Sect. 4.5.1.1 of this chapter, the authors presented the MFNDSA and MCCA in order to determine the non-dominated front—rank—and crowding distance of each solution vector stored in the memory of the meta-heuristic music-inspired MOOAs with a single-stage computational single-dimensional structure (i.e., the proposed multi-objective SS-HSA and SS-IHSA). Then, in Sect. 4.5.2.1 of this chapter, by modifying the MFNDSA and MCCA, the authors used them to sort the solution vectors stored in the memory of the meta-heuristic music-inspired MOOAs with a two-stage computational multi-dimensional single-homogeneous structure (i.e., the proposed multi-objective continuous/discrete TMS-MSA and the proposed multi-objective TMS-EMSA).

However, by taking into account the fact that the multi-objective SOSA has a multi-stage computational multi-dimensional multiple-homogeneous structure or a multi-stage computational multi-dimensional single-inhomogeneous structure, the proposed MFNDSA and MCCA in their current formats cannot be used to characterize the non-dominated front—rank—and crowding distance of each solution vector stored in the memory of this algorithm, respectively. Hence, by matching the proposed MFNDSA and MCCA with the structure of the proposed

Table 4.56 Pseudocode concerned with the initialization of the entire set of PMs or entire set of MMs or SOM in the proposed multi-objective SOSA

Algorithm 47: Pseudocode for initialization of the entire set of PMs or entire set of MMs or SOM in the proposed multi-objective SOSA

Input: A, NCDV, NDDV, NDV, NHMG, $PMS_{g,p}$, PN_g, $x_v^{\min}$, $x_v^{\max}$, $\{x_v(1), \ldots, x_v(w_v), \ldots, x_v(W_v)\}$

Output: SOM

start main body

begin

$PMS_0^{\max} = 0$

$PMS_g^{\max} = \max\left(PMS_{g,p}\right); \forall g \in \Psi^{\mathrm{NHMG}}, \forall p \in \Psi^{PN_g}$

$\mathrm{PN}^{\max} = \max\left(PN_g\right); \forall g \in \Psi^{\mathrm{NHMG}}$

construct the matrix ISOM with dimension $\left\{\sum_{g \in \Psi^{\mathrm{NHMG}}} PMS_g^{\max}\right\} \cdot \left\{(\mathrm{NDV} + \mathrm{A}) \cdot \mathrm{PN}^{\max}\right\}$ and zero initial value

for *homogeneous musical group g* $[g \in \Psi^{\mathrm{NHMG}}]$ **do**

construct the submatrix IMM_g with dimension $\left\{PMS_g^{\max}\right\} \cdot \left\{(\mathrm{NDV} + \mathrm{A}) \cdot PN_g\right\}$ and zero initial value

for *music player p of the homogeneous musical group g* $\left[p \in \Psi^{PN_g}\right]$ **do**

construct the submatrix $IPM_{g,p}$ with dimension $\{PMS_{g,p}\} \cdot \{\mathrm{NDV} + \mathrm{A}\}$ and zero initial value

for *melody vector s* $\left[s \in \Psi^{PMS_{g,p}}\right]$ **do**

construct the melody vector *s* of the music player *p* of the homogeneous musical group *g*, $\mathrm{x}_{g,p}^{s}$, with dimension $\{1\} \cdot \{\mathrm{NDV} + \mathrm{A}\}$ and zero initial value

for *decision-making variable v* $[v \in \Psi^{\mathrm{NDV}}]$ **do**

$x_{g,p,v}^{s} = x_v^{\min} + \mathrm{U}(0,1) \cdot \left(x_v^{\max} - x_v^{\min}\right)$; for CDVs

$x_{g,p,v}^{s} = x_v(y); \forall y \sim \mathrm{U}\{x_v(1), \ldots, x_v(w_v), \ldots, x_v(W_v)\}$; for DDVs

allocate $x_{g,p,v}^{s}$ to element $(1, v)$ of the melody vector $\mathrm{x}_{g,p}^{s}$

end for

for *objective function a* $[a \in \Psi^{\mathrm{A}}]$ **do**

calculate the value of the objective function *a*, fitness function, derived from the melody vector $\mathrm{x}_{g,p}^{s}$ as $f\left(\mathrm{x}_{g,p,a}^{s}\right)$

allocate $f\left(\mathrm{x}_{g,p,a}^{s}\right)$ to element $(1, \mathrm{NDV} + a)$ of the melody vector $\mathrm{x}_{g,p}^{s}$

end for

add melody vector $\mathrm{x}_{g,p}^{s}$ to the row *s* of the submatrix $IPM_{g,p}$

end for

add submatrix $IPM_{g,p}$ to the rows 1 to $PMS_{g,p}$ and columns $1 + (p - 1) \cdot (\mathrm{NDV} + \mathrm{A})$ to $p \cdot (\mathrm{NDV} + \mathrm{A})$ of the submatrix IMM_g

end for

add submatrix IMM_g to the rows $1 + \sum_{u \in \Psi^{g-1}} PMS_u^{\max}$ to $\sum_{z \in \Psi^{g}} PMS_z^{\max}$ and columns 1 to $(\mathrm{NDV} + \mathrm{A}) \cdot PN_g$ of the matrix ISOM

end for

terminate

end main body

Note: Continuous decision-making variables (CDVs), discrete decision-making variable (DDVs) multi-objective SOSA, they are employed to specify the rank and crowding distance of all melody vectors stored in the IPMs or IMMs or ISOM. The detailed descriptions associated with the fundamental concepts of the MFNDSA and MCCA were previously explained in Sect. 4.5.1.1 of this chapter.

Table 4.57 addresses the pseudocode pertaining to the MFNDSA used to determine the non-dominated front—rank—of each melody vector stored in the IPMs or IMMs or ISOM under the multi-objective SOSA. Table 4.58 also presents the pseudocode related to the MCCA employed to specify the crowding distance of each melody vector stored in the IPMs or IMMs or ISOM under the multi-objective SOSA. In addition, Table 4.59 gives the pseudocode associated with the sorting of the melody vectors stored in the IPMs or IMMs or ISOM under the multi-objective SOSA. The single-objective SOSA utilizes a multi-stage computational structure in order to find the optimal solution: (1) the SIS; (2) the GISHMG; and, (3) the GISIME. The computational structure of the single-objective SOSA begins with the implementation of the SIS, proceeding with the implementation of the GISHMG, and ending with the implementation of the GISIME. That is say to that the implementation of the different computational stages in the single-objective SOSA must be performed in a timed sequence. On the other hand, the multi-objective SOSA employs an interconnected computational structure in order to find the Pareto-optimal solution set. This interconnected computational structure is formed by an external computational stage and an internal computational sub-stage. The internal computational sub-stage is taken into account as the central core for the external computational stage. More precisely, the internal computational sub-stage must be completely implemented in each iteration of the external computational stage.

Stage 3 is related to the external computational stage. This stage, which acts as the main body of the multi-objective SOSA, is composed of three sub-stages: (1) sub-stage 3.1 or the internal computational sub-stage; (2) sub-stage 3.2 or the integration and separation procedures of melody vectors; and, (3) sub-stage 3.3 or the check of the stopping criterion of the external computational stage. The mathematical equations expressed at the external computational stage must depend on the iteration index of this stage (index n), because of the repeatability of the external computational stage in the multi-objective SOSA. The ISOM initialized in sub-stage 2.2 is considered as the input for the internal computational sub-stage (sub-stage 3.1), if the external computational stage is in the first iteration.

Otherwise, the ISOM obtained after the completion of sub-stage 3.2 in iteration $n - 1$ of the external computational stage is regarded as the input for the internal computational sub-stage (sub-stage 3.1) in iteration n of the external computational stage. Sub-stage 3.1 is related to the internal computational sub-stage. This sub-stage is equivalent to the sum of stages 3, 4, and 5 of the single-objective SOSA. Put another way, by changing the names of stage 3 (the SIS), stage 4 (the GISHMG), and stage 5 (the GISIME) of the single-objective SOSA to SISS, GISSHMG, and GISSIME are taken into account as the sub-stages for the internal computational sub-stage in the multi-objective SOSA. Therefore, in the multi-objective SOSA, the internal computational sub-stage (sub-stage 3.1) consists of three sub-stages: (1) sub-stage 3.1.1 or SISS; (2) sub-stage 3.1.2 or GISSHMG; and, (3) sub-stage 3.1.3 or GISSIME.

Table 4.57 Pseudocode pertaining to the MFNDSA used in order to determine the non-dominated front—rank—of each melody vector stored in the IPMs or IMMs or ISOM under the proposed multi-objective SOSA

Algorithm 48: Pseudocode for the MFNDSA used in order to determine the non-dominated front—rank—of each melody vector stored in the IPMs or IMMs or ISOM under the proposed multi-objective SOSA

Input: ISOM
Output: Rank of each melody vector stored in the IPMs or IMMs or ISOM

start main body

begin
for *homogeneous musical group g* $[g \in \Psi^{\mathrm{NHMG}}]$ **do**
  for *music player p of the homogeneous musical group g* $[p \in \Psi^{PN_g}]$ **do**
    for *melody vector s* $[s \in \Psi^{PMS_{g,p}}]$ **do**
      set $d_{\mathrm{x}^s_{g,p}} = 0$ {the number of melody vectors available in the $IPM_{g,p}$ which dominate the melody vector s of the music player p of the homogeneous musical group g}
      set $S_{\mathrm{x}^s_{g,p}} = \varnothing$ {the set of melody vectors available in the $IPM_{g,p}$ that the melody vector s of the music player p of the homogeneous musical group g dominates}
      for *melody vector* s^* $[s^* \in \Psi^{PMS_{g,p}}, s^* \neq s]$ **do**
        if $s \prec s^*$ **then** {melody vector s of the music player p dominates melody vector s^* of the music player p}
          $S_{\mathrm{x}^s_{g,p}} \cup \left\{\mathrm{x}^{s^*}_{g,p}\right\}$ {add melody vector s^* of the music player p of the homogeneous musical group g to the set of melody vectors of the music player p of the homogeneous musical group g dominated by melody vector s of the music player p of the homogeneous musical group g}
        else if $s^* \prec s$ {melody vector s^* of the music player p of the homogeneous musical group g dominates melody vector s of the music player p of the homogeneous musical group g}
          $d_{\mathrm{x}^s_{g,p}} = d_{\mathrm{x}^s_{g,p}} + 1$ {increment the domination count of melody vector s of the music player p of the homogeneous musical group g}
        end if
      end for
      if $d_{\mathrm{x}^s_{g,p}} = 0$ **then** {melody vector s of the music player p of the homogeneous musical group g belongs to the first non-dominated front related to the music player p of the homogeneous musical group g}
        $\mathrm{rank}_{\mathrm{x}^s_{g,p}} = 1$
        $\mathcal{F}_{g,p,1} = \mathcal{F}_{g,p,1} \cup \left\{\mathrm{x}^s_{g,p}\right\}$
      end if
    end for
    set *non-dominated front r related to the music player p of the homogeneous musical group g* $r_{g,p} = 1$
    while $\mathcal{F}_{g,p,r_{g,p}} \neq \varnothing$

(continued)

Table 4.57 (continued)

set $Q_{g,p} = \varnothing$ {employed to store the melody vectors of the next non-dominated front related to the music player p of the homogeneous musical group g}

for *melody vector s* $\left[s \in \mathcal{F}_{g,p,r_{g,p}}\right]$ **do**

for *melody vector* s^* $\left[s^* \in S_{\mathrm{x}^s_{g,p}}\right]$ **do**

$d_{\mathrm{x}^{s^*}_{g,p}} = d_{\mathrm{x}^{s^*}_{g,p}} - 1$ {decrement the domination count of melody vector s^* of the music player p of the homogeneous musical group g}

if $d_{\mathrm{x}^{s^*}_{g,p}} = 0$ **then** {melody vector s^* of the music player p of the homogeneous musical group g belongs to the non-dominated front $r + 1$ related to the music player p of the homogeneous musical group g $r_{g,p} + 1$}

$\mathrm{rank}_{\mathrm{x}^{s^*}_{g,p}} = r_{g,p} + 1$

$Q_{g,p} = Q_{g,p} \cup \left\{\mathrm{x}^{s^*}_{g,p}\right\}$

end if

end for

end for

$r_{g,p} = r_{g,p} + 1$ {increment the non-dominated front r related to the music player p of the homogeneous musical group g $r_{g,p}$}

$\mathcal{F}_{g,p,r_{g,p}} = Q_{g,p}$

end while

end for

end for

for *homogeneous musical group g* $[g \in \Psi^{\mathrm{NHMG}}]$ **do**

for *music player p of the homogeneous musical group g* $\left[p \in \Psi^{PN_g}\right]$ **do**

for *melody vector s* $\left[s \in \Psi^{PMS_{g,p}}\right]$ **do**

for *non-dominated front r related to the music player p of the homogeneous musical group g* $r_{g,p}$ $\left[r_{g,p} \in \Psi^{\mathrm{R}}_{g,p}\right]$ **do**

if $s \in \mathcal{F}_{g,p,r_{g,p}}$ **do**

$\mathrm{rank}_{\mathrm{x}^s_{g,p}} = r_{g,p}$ {assigning rank $r_{g,p}$ to the melody vector s of the music player p of the homogeneous musical group g}

end if

end for

end for

end for

end for

terminate

end main body

Table 4.58 Pseudocode related to the MCCA used in order to determine the crowding distance of each melody vector stored in the IPMs or IMMs or ISOM under the proposed multi-objective SOSA

Algorithm 49: Pseudocode for the MCCA used in order to determine the crowding distance of each melody vector stored in the IPMs or IMMs or ISOM under the proposed multi-objective SOSA

Input: ISOM
Output: Crowding distance of each melody vector stored in the IPMs or IMMs or ISOM

start main body

begin
for *homogeneous musical group g* $[g \in \Psi^{\text{NHMG}}]$ **do**
  for *music player p of the homogeneous musical group g* $\left[p \in \Psi^{PN_g}\right]$ **do**
    for *non-dominated front r related to the music player p of the homogeneous musical group g* $r_{g,p}$ $\left[r_{g,p} \in \Psi^{\text{R}}_{g,p}\right]$ **do**
      set $h_{g,p,r_{g,p}} = \left|\mathcal{F}_{g,p,r_{g,p}}\right|$ {the number of melody vectors available in the non-dominated front *r* related to the music player *p* of the homogeneous musical group *g*}
      for *melody vector s* $\left[s \in \mathcal{F}_{g,p,r_{g,p}}\right]$ **do**
        set $\text{distance}_{\text{x}^s_{g,p}} = 0$ {initialize crowding distance of the melody vectors *s* available in the non-dominated front *r* related to the music player *p* of the homogeneous musical group *g*}
      end for
      for *objective function a* $[a \in \Psi^{\text{A}}]$ **do**
        $F^{\text{sort}}_{g,p,r_{g,p},a} = \text{sort}\left(\mathcal{F}_{g,p,r_{g,p}}\left(1 : h_{g,p,r_{g,p}}, \text{NDV} + a\right), '\text{ascend}'\right)$
        for *melody vector s* $\left[s \in \mathcal{F}_{g,p,r_{g,p}}\right]$ **do** {sort the melody vectors available in the non-dominated front *r* related to the music player *p* of the homogeneous musical group *g* from the perspective of the value of objective function *a*}
          for *melody vector* s^* $\left[s^* \in \mathcal{F}_{g,p,r_{g,p}}\right]$ **do**
            if $F^{\text{sort}}_{g,p,r_{g,p},a}(s) == \mathcal{F}_{g,p,r_{g,p}}(s^*, \text{NDV} + a)$ **then**
              $\mathcal{F}^{\text{sort}}_{g,p,r_{g,p},a}(s, 1 : \text{NDV} + \text{A}) = \mathcal{F}_{g,p,r_{g,p}}(s^*, 1 : \text{NDV} + \text{A});$
            end if
          end for
        end for
        set $\text{distance}_{\text{x}^1_{g,p},a} = \text{distance}_{\text{x}^{h_{g,p,r_{g,p}}}_{g,p},a} = \infty$ {the first and the last melody vectors stored in the sorted non-dominated front *r* related to the music player *p* of the homogeneous musical group *g* from the perspective of the value of objective function *a* are assigned the crowding distance equal to infinity}
        for *melody vector s* $\left[s \in \mathcal{F}_{g,p,r_{g,p}}, s \neq 1, s \neq h_{g,p,r_{g,p}}\right]$ **do**
          $\text{distance}_{\text{x}^s_{g,p},a} = \left(f_a^{\mathcal{F}^{\text{sort}}_{g,p,r_{g,p},a,s+1}} - f_a^{\mathcal{F}^{\text{sort}}_{g,p,r_{g,p},a,s-1}}\right) / \left(f_a^{\max} - f_a^{\min}\right)$
        end for
        for *melody vector s* $\left[s \in \mathcal{F}_{g,p,r_{g,p}}\right]$ **do**
          for *melody vector* s^* $\left[s^* \in \mathcal{F}_{g,p,r_{g,p}}\right]$ **do**
            if $\mathcal{F}_{g,p,r_{g,p}}(s, 1 : \text{NDV}) == \mathcal{F}^{\text{sort}}_{g,p,r_{g,p},a}(s^*, 1 : \text{NDV})$ **then**
              $\text{distance}_{\text{x}^s_{g,p}} = \text{distance}_{\text{x}^s_{g,p}} + \text{distance}_{\text{x}^{s^*}_{g,p},a}$ {compute the crowding distance of the melody vectors *s* of the music player *p* of the homogeneous musical group *g*}
            end if

(continued)

Table 4.58 (continued)

end for
end for
end for
end for
end for
end for
terminate

end main body

Table 4.59 Pseudocode associated with the sorting of the melody vectors stored in the IPMs or IMMs or ISOM under the proposed multi-objective SOSA

Algorithm 50: Pseudocode for sorting the melody vectors stored in the IPMs or IMMs or ISOM under the proposed multi-objective SOSA

Input: Unsorted ISOM

Output: Sorted ISOM

start main body

begin
$PMS_0^{\max} = 0$
for *homogeneous musical group g* $[g \in \Psi^{\mathrm{NHMG}}]$ **do**
  for *music player p of the homogeneous musical group g* $\left[p \in \Psi^{PN_g}\right]$ **do**
    set *non-dominated front r related to the music player p of the homogeneous musical group g* $r_{g,p} = 1$
    for *non-dominated front r related to the music player p of the homogeneous musical group g* $\left[r_{g,p} \in \Psi^{\mathrm{R}}_{g,p}\right]$ **do**
      set $\mathcal{F}_{g,p,r_{g,p}} = \varnothing$ {the non-dominated front r related to the music player p of the homogeneous musical group g}
      for *melody vector s* $\left[s \in \Psi^{PMS_{g,p}}\right]$ **do**
        if $\mathrm{rank}_{\mathrm{x}^s_{g,p}} == r_{g,p}$ **then**
          $\mathcal{F}_{g,p,r_{g,p}} \cup \left\{\mathrm{x}^s_{g,p}\right\}$ {add melody vector s of the music player p of the homogeneous musical group g to the non-dominated front r related to the music player p of the homogeneous musical group g}
        end if
      end for
      set $h_{g,p,r_{g,p}} = \left|\mathcal{F}_{g,p,r_{g,p}}\right|$ {the number of melody vectors available in the non-dominated front r related to the music player p of the homogeneous musical group g}
      for *melody vector s* $\left[s \in \mathcal{F}_{g,p,r_{g,p}}\right]$ **do**
        set *superiority index q of the music player p* of the homogeneous musical group g $q_{g,\,p} = 0$
        for *melody vector* s^* $\left[s^* \in \mathcal{F}_{g,p,r_{g,p}}, s^* \neq s\right]$ **do**
          if $\mathrm{distance}_{\mathrm{x}^s_{g,p}} > \mathrm{distance}_{\mathrm{x}^{s^*}_{g,p}}$ **then**
            $q_{g,p} = q_{g,p} + 1$ {increment the superiority index q of the music player p of the homogeneous musical group g}
          end if

(continued)

Table 4.59 (continued)

end for
$Q_{g,\,p}(s) = q_{g,\,p}$
end for
for *melody vector s* $\left[s \in \mathcal{F}_{g,p,r_{g,p}}\right]$ **do**
for *melody vector* s^* $\left[s^* \in \mathcal{F}_{g,p,r_{g,p}}\right]$ **do**
if $Q_{g,p}(s^*) == h_{g,p,r_{g,p}} - s$ **then**
$\mathcal{F}^{\text{sort}}_{g,p,r_{g,p}}(s, 1 : \text{NDV} + \text{A}) = \mathcal{F}_{g,p,r_{g,p}}(s^*)$ {sort the melody vectors available in non-dominated front r related to the music player p of the homogeneous musical group g}
end if
end for
end for
$IPM^{\text{sort}}_{g,p}\left(1 + \sum_{z \in \Psi^{r_{g,p}-1}} h_{g,p,z} : \sum_{r_{g,p} \in \Psi^{\text{R}}_{g,p}} h_{g,p,r_{g,p}}, 1 : \text{NDV} + \text{A}\right) = \mathcal{F}^{\text{sort}}_{g,p,r_{g,p}}$ {add the sorted non-dominated front r related to the music player p of the homogeneous musical group g to the $IPM^{\text{sort}}_{g,p}$}
end for
$IPM_{g,p} = IPM^{\text{sort}}_{g,p}$
$IMM^{\text{sort}}_{g}(1 : PMS_{g,p}, 1 + [(p-1) \cdot (\text{NDV} + \text{A})] : [p \cdot (\text{NDV} + \text{A})]) = IPM_{g,p}$
end for
$IMM_g = IMM^{\text{sort}}_{g}$
$\text{ISOM}^{\text{sort}}\left(1 + \sum_{u \in \Psi^{g-1}} PMS^{\max}_{u} : \sum_{z \in \Psi^{g}} PMS^{\max}_{z}, 1 : (\text{NDV} + \text{A}) \cdot PN_g\right) = IMM_g$
end for
$\text{ISOM} = \text{ISOM}^{\text{sort}}$
terminate

end main body

The mathematical equations expressed in sub-stage 3.1, in addition to the dependence on the iteration index of the external computational stage (index n), must also depend on the improvisation/iteration index of the internal computational sub-stage (index m), because of the repeatability of the internal computational sub-stage in the multi-objective SOSA. Sub-stage 3.1.1 or SISS is composed of three sub-stages: (1) sub-stage 3.1.1.1 or the improvisation of a new melody vector by each existing player in the symphony orchestra; (2) sub-stage 3.1.1.2 or the update of each available IPM in the symphony orchestra; and, (3) sub-stage 3.1.1.3 or the check of the stopping criterion of the SISS. Sub-stage 3.1.2 or GISSHMG is also composed of four sub-stages: (1) sub-stage 3.1.2.1 or the improvisation of a new melody vector by each existing player in the symphony orchestra, and taking into account the feasible ranges of the updated pitches for each homogeneous musical group; (2) sub-stage 3.1.2.2 or the update of each available IPM in the symphony orchestra; (3) sub-stage 3.1.2.3 or the update of the feasible ranges of the pitches, continuous decision-making variables, for each homogeneous musical group in the next improvisation (only for random selection); and, (4) sub-stage 3.1.2.4 or the check of the stopping criterion of the

GISSHMG. In addition, sub-stage 3.1.3 or GISSIME is composed of four sub-stages: (1) sub-stage 3.1.3.1 or the improvisation of a new melody vector by each player, and taking into account the feasible ranges of the updated pitches for the inhomogeneous musical ensemble; (2) sub-stage 3.1.3.2 or the update each available IPM in the symphony orchestra; (3) sub-stage 3.1.3.3 or the update of the feasible ranges of the pitches, continuous decision-making variables, for the inhomogeneous musical ensemble in the next improvisation (only for random selection); and, (4) sub-stage 3.1.3.4 or the check of the stopping criterion of the GISSIME.

Sub-stages 3.1.1.1, 3.1.2.1, and 3.1.3.1 are related to the improvisation of a new melody vector by each player. In sub-stage 3.1.1.1, each player improvises a new melody vector individually without the influence of other players in the homogeneous musical group to which it belongs, and without the influence of other players in other homogeneous musical groups to which it does not belong. In sub-stage 3.1.2.1, each player also improvises a new melody vector interactively only under the influence of other players in the homogeneous musical group to which it belongs. In addition, in sub-stage 3.1.3.1, each player improvises a new melody vector interactively both with the influence of other players in the homogeneous musical group to which it belongs and with the influence of other players in other homogeneous musical groups to which it does not belong. The improvisation process of a new melody vector by each player under the multi-objective SOSA is conceptually similar to this process under the single-objective SOSA (the proposed NIP), which was previously presented in Sect. 4.4.7 of this chapter. However, due to the existence of the external computational stage in the multi-objective SOSA, the iteration index of the external computational stage (index n) must be added to all mathematical equations indicated in Sect. 4.4.7 of this chapter that have the improvisation/iteration index of the internal computational step (index m). The pseudocode pertaining to improvisation of a new melody vector by each player in the symphony orchestra of the single-objective SOSA is presented in Table 4.26 in Sect. 4.4.7 of this chapter. In the multi-objective SOSA, however, due to the replacement of the SOOP with the MOOP in stage 1, this pseudocode must be restructured. As a result, Table 4.60 presents the pseudocode related to improvisation of a new melody vector by each existing player in the symphony orchestra in the multi-objective SOSA. Sub-stages 3.1.1.2, 3.1.2.2, and 3.1.3.2 are relevant to the process of updating all of the IPMs or all of the IMMs or ISOM. These sub-stages are equivalent to sub-stages 3.2, 4.2, and 5.2 of the single-objective SOSA. A detailed description of the process of updating the SOM in the single-objective SOSA was presented in Sect. 4.4.3.2 of this chapter. The pseudocode related to the update of the memory of all existing players in the symphony orchestra or the update of the SOM_m in the single-objective SOSA is also presented in Table 4.21 in Sect. 4.4.3.2 of this chapter. In the multi-objective SOSA, due to the replacement of the SOOP with the MOOP in stage 1, the update of the memory of all players in the symphony orchestra or the update of the $ISOM_{n,m}$ must be redefined.

To illustrate these sub-stages of the multi-objective SOSA, consider the memory relevant to player p in homogeneous musical group g in the symphony orchestra ($IPM_{n,m,g,p}$). A new melody vector played by player p in homogeneous musical group

Table 4.60 The pseudocode related to improvisation of a new melody vector by each existing player in the symphony orchestra in the proposed multi-objective SOSA

Algorithm 51: Pseudocode for improvisation of a new melody vector by each existing player in the symphony orchestra in the proposed multi-objective SOSA

Input: A, $BW_{g,p}^{\max}$, $BW_{g,p}^{\min}$, MNI-SISS, MNI-GISSHMG, MNI-GISSIME, NCDV, NDV, NDDV, $PAR_{g,p}^{\max}$, $PAR_{g,p}^{\min}$, $PMCR_{g,p}$, $PMS_{g,p}$, PN_g, $x_v^{\min}$, $x_v^{\max}$, $\{x_v(1), \ldots, x_v(w_v), \ldots, x_v(W_v)\}$

Output: $\mathrm{x}_{n,m,g,p}^{\mathrm{new}}$

start main body

begin

$BW_{n,m,g,p} = BW_{g,p}^{\max} \cdot \exp\left[\left(\ln\left(BW_{g,p}^{\max}/BW_{g,p}^{\min}\right)/((\text{MNI-SIS}) + (\text{MNI-GISHMG}) + (\text{MNI-GISIME}))\right) \cdot m\right]$

$PAR_{m,g,p} = PAR_{g,p}^{\min} - \left[\left(\left(PAR_{g,p}^{\max} - PAR_{g,p}^{\min}\right)/((\text{MNI-SIS}) + (\text{MNI-GISHMG}) + (\text{MNI-GISIME}))\right) \cdot m\right]$

for *homogeneous musical group g* $[g \in \Psi^{\mathrm{NHMG}}]$ **do**

  for *music player p of the homogeneous musical group g* $\left[p \in \Psi^{PN_g}\right]$ **do**

    construct the new melody vector for the music player *p* of the homogeneous musical group *g*, $\mathrm{x}_{n,m,g,p}^{\mathrm{new}}$, with dimension {1} · {NDV + A} and zero initial value

    for *decision-making variable v* $[v \in \Psi^{\mathrm{NDV}}]$ **do**

      if $U(0,1) \leq PMCR_{g,p}$ **then**

        Rule 1: harmony memory consideration with probability $PMCR_{g,p}$

        if *improvisation/iteration of the internal computational sub-stage m* $[m \in \Psi^{(\text{MNI-SISS}) + (\text{MNI-GISSHMG}) + (\text{MNI-GISSIME})}]$ is odd **then**

          Principle 1 of Rule 1: first combination

          $x_{n,m,g,p,v}^{\mathrm{new}} = x_{n,m,g,p,v}^{r_{g,p}} \pm U(0,1) \cdot BW_{m,g,p};\ \forall r_{g,p} \sim U\{1, 2, \ldots, PMS_{g,p}\}$; for CDVs

          $x_{n,m,g,p,v}^{\mathrm{new}} = x_{n,m,g,p,v}^{r_{g,p}};\ \forall r_{g,p} \sim U\{1, 2, \ldots, PMS_{g,p}\}$; for DDVs

        else

          Principle 2 of Rule 1: second combination

          $x_{n,m,g,p,v}^{\mathrm{new}} = x_{n,m,g,p,k}^{r_{g,p}} \pm U(0,1) \cdot BW_{m,g,p};\ \forall r_{g,p} \sim U\{1, 2, \ldots, PMS_{g,p}\}, \forall k \sim U\{1, 2, \ldots, \text{NCDV}\}$; for CDVs

          $x_{n,m,g,p,v}^{\mathrm{new}} = x_{n,m,g,p,l}^{r_{g,p}};\ \forall r_{g,p} \sim U\{1, 2, \ldots, PMS_{g,p}\}, \forall l \sim U\{1, 2, \ldots, \text{NDDV}\}$; for DDVs

        end if

        if $U(0,1) \leq PAR_{m,g,p}$ **then**

          Rule 2: pitch adjustment with probability $PMCR_{g,p} \cdot PAR_{g,p}$

          switch 1

          case *improvisation/iteration m* $[m \in \Psi^{(\text{MNI-SISS}) + (\text{MNI-GISSHMG}) + (\text{MNI-GISSIME})}] \leq (\text{MNI-SISS})$ **then**

            Principle 1 of Rule 2: first choice

            $x_{n,m,g,p,v}^{\mathrm{new}} = x_{n,m,g,p,v}^{\mathrm{best}}$; for CDVs and DDVs

          case *improvisation/iteration of the internal computational sub-stage m* $[m \in \Psi^{(\text{MNI-SISS}) + (\text{MNI-GISSHMG}) + (\text{MNI-GISSIME})}] > (\text{MNI-SISS})$ and *improvisation/iteration of the internal computational sub-stage m* $[m \in \Psi^{(\text{MNI-SISS}) + (\text{MNI-GISSHMG}) + (\text{MNI-GISSIME})}] \leq (\text{MNI-SISS}) + (\text{MNI-GISSHMG})$ **then**

            Principle 2 of Rule 2: second choice

            $x_{n,m,g,p,v}^{\mathrm{new}} = x_{n,m,g,\mathrm{best},v}^{\mathrm{best}}$; for CDVs and DDVs

          case *improvisation/iteration of the internal computational sub-stage m* $[m \in \Psi^{(\text{MNI-SISS}) + (\text{MNI-GISSHMG}) + (\text{MNI-GISSIME})}] > (\text{MNI-SISS}) + (\text{MNI-GISSHMG})$ and *improvisation/iteration of the internal computational sub-stage m* $[m \in \Psi^{(\text{MNI-SISS}) + (\text{MNI-GISSHMG}) + (\text{MNI-GISSIME})}] \leq (\text{MNI-SISS}) + (\text{MNI-GISSHMG}) + (\text{MNI-GISSIME})$ **then**

            Principle 3 of Rule 2: third choice

            $x_{n,m,g,p,v}^{\mathrm{new}} = x_{n,m,\mathrm{best},\mathrm{best},v}^{\mathrm{best}}$; for CDVs and DDVs

          end switch

(continued)

Table 4.60 (continued)

end if
else if
Rule 3: random selection with probability $1 - PMCR_{g,p}$
switch 1
case *improvisation/iteration of the internal computational sub-stage* m $[m \in \Psi^{(\text{MNI-SISS}) + (\text{MNI-GISSHMG}) + (\text{MNI-GISSIME})}] \leq (\text{MNI-SISS})$ **then**
Principle 1 of Rule 3: first choice
$x^{\text{new}}_{n,m,g,p,v} = x^{\min}_{v} + \mathrm{U}(0,1) \cdot (x^{\max}_{v} - x^{\min}_{v})$; for CDVs
$x^{\text{new}}_{n,m,g,p,v} = x_v(y); \forall y \sim \mathrm{U}\{x_v(1), \ldots, x_v(w_v), \ldots, x_v(W_v)\}$; for DDVs
case *improvisation/iteration of the internal computational sub-stage m* $[m \in \Psi^{(\text{MNI-SISS}) + (\text{MNI-GISSHMG}) + (\text{MNI-GISSIME})}] > (\text{MNI-SISS})$ and *improvisation/iteration of the internal computational sub-stage* m $[m \in \Psi^{(\text{MNI-SISS}) + (\text{MNI-GISSHMG}) + (\text{MNI-GISSIME})}] \leq (\text{MNI-SISS}) + (\text{MNI-GISSHMG})$ **then**
Principle 2 of Rule 3: second choice
$x^{\text{new}}_{n,m,g,p,v} = x^{\min}_{n,m,g,v} + \mathrm{U}(0,1) \cdot \left(x^{\max}_{n,m,g,v} - x^{\min}_{n,m,g,v}\right)$; for CDVs
$x^{\text{new}}_{n,m,g,p,v} = x_v(y); \forall y \sim \mathrm{U}\{x_v(1), \ldots, x_v(w_v), \ldots, x_v(W_v)\}$; for DDVs
case *improvisation/iteration of the internal computational sub-stage* m $[m \in \Psi^{(\text{MNI-SISS}) + (\text{MNI-GISSHMG}) + (\text{MNI-GISSIME})}] > (\text{MNI-SISS}) + (\text{MNI-GISSHMG})$ and *improvisation/iteration of the internal computational sub-stage* m $[m \in \Psi^{(\text{MNI-SISS}) + (\text{MNI-GISSHMG}) + (\text{MNI-GISSIME})}] \leq (\text{MNI-SISS}) + (\text{MNI-GISSHMG}) + (\text{MNI-GISSIME})$ **then**
Principle 3 of Rule 3: third choice
$x^{\text{new}}_{n,m,g,p,v} = x^{\min}_{n,m,v} + \mathrm{U}(0,1) \cdot (x^{\max}_{n,m,v} - x^{\min}_{n,m,v})$; for CDVs
$x^{\text{new}}_{n,m,g,p,v} = x_v(y); \forall y \sim \mathrm{U}\{x_v(1), \ldots, x_v(w_v), \ldots, x_v(W_v)\}$; for DDVs
end switch
end if
end for
for *objective function* a $[a \in \Psi^{\mathrm{A}}]$ **do**
calculate the value of objective function a, fitness function, derived from melody vector $\mathrm{x}^{\text{new}}_{n,m,g,p}$ as
$f\left(\mathrm{x}^{\text{new}}_{n,m,g,p,a}\right)$
allocate $f\left(\mathrm{x}^{\text{new}}_{n,m,g,p,a}\right)$ to element $(1, \text{NDV} + a)$ of the new melody vector $\mathrm{x}^{\text{new}}_{n,m,g,p}$
end for
end for
end for
terminate

end main body

Note: Continuous decision-making variable (CDVs), discrete decision-making variable (DDVs)

g in the symphony orchestra—$\mathbf{x}^{\text{new}}_{n,m,g,p} = \left(x^{\text{new}}_{n,m,g,p,1}, \ldots, x^{\text{new}}_{n,m,g,p,v}, \ldots, x^{\text{new}}_{n,m,g,p,\text{NDV}}\right)$—is evaluated and compared with the worst available melody vector in the $IPM_{n,m,g,p}$—the melody vector with the biggest rank and the smallest crowding distance, which is placed in the $PMS_{g,p}$ row of the $IPM_{n,m,g,p}$—from the perspective of the objective functions. If the new melody vector played by player p in homogeneous musical

Table 4.61 Pseudocode related to the update of the memory of all existing players in the symphony orchestra or the update of the $ISOM_{n,m}$ in the proposed multi-objective SOSA

Algorithm 52: Pseudocode for the update of the memory of all existing players in the symphony orchestra or the update of the $ISOM_{n,m}$ in the proposed multi-objective SOSA

Input: Not updated $ISOM_{n,m}$, $\mathrm{x}_{n,m,g,p}^{\mathrm{new}}$

Output: Updated $ISOM_{n,m}$

start main body

begin

for *homogeneous musical group g* $[g \in \Psi^{\mathrm{NHMG}}]$ **do**

  for *music player p of the homogeneous musical group g* $\left[p \in \Psi^{PN_g}\right]$ **do**

    set $\mathrm{x}_{g,p}^{\mathrm{worst}} = \mathrm{x}_{n,m,g,p}^{PMS_{g,p}}$

    set $f\left(\mathrm{x}_{g,p}^{\mathrm{worst}}\right) = f\left(\mathrm{x}_{n,m,g,p}^{PMS_{g,p}}\right)$

    if $f\left(\mathrm{x}_{n,m,g,p}^{\mathrm{new}}\right) \prec f\left(\mathrm{x}_{g,p}^{\mathrm{worst}}\right)$ **then** {new melody vector played by player p of the homogeneous musical group g dominates worst melody vector in the memory of this player}

      $\mathrm{x}_{n,m,g,p}^{\mathrm{new}} \in IPM_{n,m,g,p}$

      $\mathrm{x}_{g,p}^{\mathrm{worst}} \notin IPM_{n,m,g,p}$

    end if

  end for

end for

terminate

end main body

group g dominates the worst available melody vector in the $IPM_{n,m,g,p}$, from the perspective of the objective functions, this new melody vector replaces the worst melody vector available in the $IPM_{n,m,g,p}$; the worst available melody vector is then eliminated from the $IPM_{n,m,g,p}$. This process is also accomplished for other players in homogeneous musical group g and for all other players in other available homogeneous musical groups in the symphony orchestra in the same way. Table 4.61 gives the pseudocode related to the update of the memory of all players in the orchestra or the update of the $ISOM_m$ in the multi-objective SOSA. The process of updating the $IPM_{n,m,g,p}$ is not performed if the new melody vector played by player p in homogeneous musical group g does not dominate as the worst available melody vector in the $IPM_{n,m,g,p}$, from the standpoint of the objective functions. After completion of this process, melody vectors stored in the memory of all players or the $ISOM_{n,m}$ must be re-sorted. The pseudocode related to sorting the melody vectors stored in the ISOM was presented in Tables 4.57, 4.58, and 4.59. However, due to the fact that the ISOM in sub-stages 3.1.1.2, 3.1.2.2, and 3.1.3.2 depends on the iteration index of the external computational stage (index n) and improvisation/iteration index of the internal computational stage (index m), the aforementioned pseudocode must be redefined by adding indices n and m to all mathematical equations used in these pseudocodes.

Table 4.62 The pseudocode related to the update of the feasible ranges of continuous decision-making variables for each homogeneous musical group in the symphony orchestra in the proposed multi-objective SOSA

Algorithm 53: Pseudocode for the update of the feasible ranges of continuous decision-making variables for each homogeneous musical group in the symphony orchestra in the proposed multi-objective SOSA

Input: $\mathrm{x}^{1}_{n,m,g,p}$
Output: $x^{\min}_{n,m,g,v}$, $x^{\max}_{n,m,g,v}$

start main body

begin
for *homogeneous musical group g* $[g \in \Psi^{\mathrm{NHMG}}]$ **do**
  for *music player p of the musical homogeneous group g* $\left[p \in \Psi^{PN_g}\right]$ **do**
    set $\mathrm{x}^{\mathrm{best}}_{g,p} = \mathrm{x}^{1}_{n,m,g,p}$
  end for
end for
for *homogeneous musical group g* $[g \in \Psi^{\mathrm{NHMG}}]$ **do**
  for *continuous decision-making variable v* $[v \in \Psi^{\mathrm{NCDV}}]$ **do**
    $x^{\min}_{n,m,g,v} = \min\left(x^{\mathrm{best}}_{g,p,v}\right)\ ; \forall p \in \Psi^{PN_g}$
    $x^{\max}_{n,m,g,v} = \max\left(x^{\mathrm{best}}_{g,p,v}\right)\ ; \forall p \in \Psi^{PN_g}$
  end for
end for
terminate

end main body

Sub-stage 3.1.2.3 is related to the update of the feasible ranges of the pitches, continuous decision-making variables, for each homogeneous musical group in the next improvisation (only for random selection). This sub-stage is equivalent to sub-stage 4.3 of the single-objective SOSA, previously presented in Sect. 4.4.4.3 of this chapter.

In sub-stage 4.3 of the single-objective SOSA, the update of the feasible ranges of the continuous decision-making variables for each homogeneous musical group in the symphony orchestra is performed in accordance with the pseudocode displayed in Table 4.23.

Sub-stage 3.1.2.3 of the multi-objective SOSA is conceptually similar to sub-stage 4.3 of the single-objective SOSA. In the multi-objective SOSA, however, due to the existence of the external computational stage, this pseudocode must be restructured. As a result, Table 4.62 gives the pseudocode related to the update of the feasible ranges of continuous decision-making variables for each homogeneous musical group in the symphony orchestra in the multi-objective SOSA.

Sub-stage 3.1.3.3 is concerned with the process of updating the feasible ranges of pitches, continuous decision-making variables, for the inhomogeneous musical ensemble in the next improvisation (only for random selection). This sub-stage is equivalent to sub-stage 5.3 of the single-objective SOSA, previously presented in

Table 4.63 The pseudocode pertaining to the update process of the continuous decision-making variables for the inhomogeneous musical ensemble in the proposed multi-objective SOSA

Algorithm 54: Pseudocode for the update process of the continuous decision-making variables for the inhomogeneous musical ensemble in the proposed multi-objective SOSA

Input: $\mathrm{x}^{1}_{n,m,g,p}$

Output: $x^{\min}_{n,m,g,v}$, $x^{\max}_{n,m,g,v}$

start main body

begin
for *homogeneous musical group g* $[g \in \Psi^{\mathrm{NHMG}}]$ **do**
  for *music player p of the homogeneous musical group g* $\left[p \in \Psi^{PN_g}\right]$ **do**
    set $\mathrm{x}^{\mathrm{best}}_{g,p} = \mathrm{x}^{1}_{n,m,g,p}$
  end for
end for
for *homogeneous musical group g* $[g \in \Psi^{\mathrm{NHMG}}]$ **do**
  for *continuous decision-making variable v* $[v \in \Psi^{\mathrm{NCDV}}]$ **do**
    $x^{\min}_{n,m,g,v} = \min\left(x^{\mathrm{best}}_{g,p,v}\right) \; ; \forall p \in \Psi^{PN_g}$
    $x^{\max}_{n,m,g,v} = \max\left(x^{\mathrm{best}}_{g,p,v}\right) \; ; \forall p \in \Psi^{PN_g}$
  end for
end for
for *continuous decision-making variable v* $[v \in \Psi^{\mathrm{NCDV}}]$ **do**
  $x^{\min}_{n,m,v} = \min\left(x^{\min}_{g,v}\right); \forall g \in \Psi^{\mathrm{NHMG}}$
  $x^{\max}_{n,m,v} = \max\left(x^{\max}_{g,v}\right); \forall g \in \Psi^{\mathrm{NHMG}}$
end for
terminate

end main body

Sect. 4.4.5.3 of this chapter. In sub-stage 5.3 of the single-objective SOSA, the update of the feasible ranges of the continuous decision-making variables for the inhomogeneous musical ensemble is carried out according to the pseudocode presented in Table 4.24. Sub-stage 3.1.3.3 of the multi-objective SOSA is conceptually similar to sub-stage 5.3 of the single-objective SOSA.

In the multi-objective SOSA, however, due to the existence of the external computational stage, this pseudocode must be restructured. As a result, Table 4.63 presents the pseudocode pertaining to the process of updating the continuous decision-making variables for the inhomogeneous musical ensemble in the multi-objective SOSA. Sub-stages 3.1.1.3, 3.1.2.4, and 3.1.3.4 are relevant to the check of the stopping criterion of the SISS, GISSHMG, and GISSIME, respectively. These sub-stages of the multi-objective SOSA are similar to sub-stages 3.3, 4.4, and 5.4 of the single-objective SOSA, respectively, previously described in Sects. 4.4.3.3, 4.4.4.4, and 4.4.5.4 of this chapter, respectively. Only in sub-stages 3.1.1.3, 3.1.2.4, and 3.1.3.4 of the multi-objective SOSA the MNI-SIS, the

MNI-GISHMG, and the MNI-GISIME parameters must be replaced with the MNI-SISS, the MNI-GISSHMG, and the MNI-GISSIME parameters, respectively. As a result, in sub-stage 3.1.1.3 of the multi-objective SOSA, the computational efforts of the SISS are terminated when its stopping criterion—the MNI-SISS—is satisfied. Otherwise, sub-stages 3.1.1.1 and 3.1.1.2 are repeated.

In sub-stage 3.1.2.4 of the multi-objective SOSA, the computational efforts of the GISSHMG are also terminated when its stopping criterion—the MNI-GISSHMG—is satisfied. Otherwise, sub-stages 3.1.2.1, 3.1.2.2, and 3.1.2.3 are repeated.

Besides, in sub-stage 3.1.3.4 of the multi-objective SOSA, the computational efforts of the GISSIME are terminated when its stopping criterion—the MNI-GISSIME—is satisfied. Otherwise, sub-stages 3.1.3.1, 3.1.3.2, and 3.1.3.3 are repeated.

Sub-stage 3.2 of the multi-objective SOSA is concerned with the integration and separation procedures of melody vectors. The main goal of the implementation of this sub-stage is to construct the ISOM for the next iteration of the external computational stage or the new ISOM. The construction of the new ISOM requires the ISOM in the current iteration of the external computational stage and two other memories, referred to as the OSOM and HSOM. Implementation of this sub-stage in iteration n of the external computational stage begins by determining the $ISOM_n$ and $OSOM_n$. First, the ISOM for the internal computational sub-stage (sub-stage 3.1) in iteration n of the external computational stage is considered as the $ISOM_n$. The $ISOM_n$ obtained after the completion of the internal computational sub-stage (sub-stage 3.1) is also taken into account as the $OSOM_n$. Then, the $HSOM_n$ must be created by applying the integration procedure on the $ISOM_n$ and $OSOM_n$. All existing players in all homogeneous musical groups in the symphony orchestra are involved with this procedure. To clarify, consider player p in homogeneous musical group g in the symphony orchestra. This player integrates the $IPM_{n,g,p}$ and $OPM_{n,g,p}$ to form the $HPM_{n,g,p}$. Since all of the melody vectors available in the $IPM_{n,g,p}$ and $OPM_{n,g,p}$ are transferred to the $HPM_{n,g,p}$, elitism for player p in homogeneous musical group g in the symphony orchestra is ensured. This transfer also makes the size of the $HPM_{n,g,p}$ twice as large as the $IPM_{n,g,p}$ and $OPM_{n,g,p}$. That is say to that, unlike the $IPM_{n,g,p}$ and $OPM_{n,g,p}$ that have a size equal to $\{PMS_{g,p}\} \cdot \{\text{NDV} + \text{A}\}$, the $HPM_{n,g,p}$ has a size of $\{2 \cdot PMS_{g,p}\} \cdot \{\text{NDV} + \text{A}\}$. This procedure is also carried out by other existing players in homogeneous musical group g and by all of the remaining players in other available homogeneous musical groups in the symphony orchestra in the same way. After implementing the integration procedure by all players in homogeneous musical group g, all obtained $HPM_{n,g,p}$ are located next to each other and form the $HMM_{n,g}$. Consequently, after applying the integration procedure by all players in all homogeneous musical groups in the symphony orchestra, all obtained $HMM_{n,g}$ are located next to each other and form the $HSOM_n$. It is clear that the size of the $HMM_{n,g}$ will be twice the size of the $IMM_{n,g}$ and $OMM_{n,g}$. Put simply, unlike the $IMM_{n,g}$ and $OMM_{n,g}$ that have a size equal to $\left\{PMS_g^{\max}\right\} \cdot \{(\text{NDV+A}) \cdot PN_g\}$, the HMM_n has a size of $\left\{2 \cdot PMS_g^{\max}\right\} \cdot \{(\text{NDV+A}) \cdot PN_g\}$. Likewise, the size of the $HSOM_n$ will be twice the size of the $ISOM_n$ and $OSOM_n$. That is, unlike the $ISOM_n$ and $OSOM_n$ that have a size equal to $\left\{\sum_{g\in\Psi^{\text{NHMG}}} PMS_g^{\max}\right\} \cdot \{(\text{NDV}+\text{A}) \cdot \text{PN}^{\max}\}$, the $HSOM_n$ has a size of $\left\{2 \cdot \sum_{g\in\Psi^{\text{NHMG}}} PMS_g^{\max}\right\} \cdot \{(\text{NDV}+\text{A}) \cdot \text{PN}^{\max}\}$. Next, all melody vectors available in the $HSOM_n$ must be sorted according to their rank and crowding distance by employing the pseudocodes presented in Tables 4.57, 4.58, and 4.59. However, due to the fact that $HSOM_n$ in sub-stage 3.1.2 depends on the iteration index of the external computational stage (index n), the aforementioned pseudocode must be redefined by adding index n to all mathematical equations used in these pseudocodes. Afterwards, the new ISOM ($ISOM_{n+1}$) must be created by applying the separation procedure on the $HSOM_n$. Likewise, all players in all homogeneous musical groups in the symphony orchestra are involved with this procedure. Consider player p in homogeneous musical group g. This player employs the melody vectors stored in the $HPM_{n,g,p}$ in order to create the new $IPM_{g,p}$ ($IPM_{n+1,g,p}$). But considering that the size of the $HPM_{n,g,p}$ is twice the size of the $IPM_{n+1,g,p}$, only half of the melody vectors available in the $HPM_{n,g,p}$ can be transferred to the $IPM_{n+1,g,p}$. The melody vectors with lower—better—rank and higher—better—crowding distance available in the $HPM_{n,g,p}$ have a higher priority to transfer to the $IPM_{n+1,g,p}$. Thus, the first non-dominated front ($\mathcal{F}_{g,p,1}$) of the $HPM_{n,g,p}$ is selected. If the size of the $\mathcal{F}_{g,p,1}$ is smaller than the size of the $IPM_{n+1,g,p}$—$\{PMS_{g,p}\} \cdot \{\text{NDV} + \text{A}\}$—all melody vectors stored in the $\mathcal{F}_{g,p,1}$ are transferred to the $IPM_{n+1,g,p}$. Otherwise, the melody vectors with higher crowding distance of the $\mathcal{F}_{g,p,1}$ are chosen for transfer to the $IPM_{n+1,g,p}$. If the $\mathcal{F}_{g,p,1}$ fails to fill the $IPM_{n+1,g,p}$ completely, the subsequent non-dominated front ($\mathcal{F}_{g,p,2}$) of the $HPM_{n,g,p}$ will be selected to fill the remaining melody vectors of the $IPM_{n+1,g,p}$. In simple terms, the remaining melody vectors of the $IPM_{n+1,g,p}$ are selected from subsequent non-dominated front ($\mathcal{F}_{g,p,2}$) of the $HPM_{n,g,p}$. This process is repeated until the $IPM_{n+1,g,p}$ is completely filled. After implementing the separation procedure for all players in homogeneous musical group g, all obtained $IPM_{n+1,g,p}$ are located next to each other and form the $IMM_{n+1,g}$. Accordingly, after implementing the integration procedure for all players in all homogeneous musical groups in the symphony orchestra, all obtained $IMM_{n+1,g}$ are located next to each other and form the $ISOM_{n+1}$. The $ISOM_{n+1}$ is considered as the ISOM for the internal computational sub-stage in iteration $n + 1$ of the external computational stage. Table 4.64 gives the pseudocode related to the integration and separation procedures of melody vectors in the multi-objective SOSA.

Sub-stage 3.3 is related to the check of the stopping criterion of the external computational stage. In this sub-stage of the multi-objective SOSA, the computational efforts of the external computational stage are terminated when its stopping criterion (the MNI-E) is satisfied. Otherwise, sub-stages 3.1 and 3.2 are repeated. Stage 4 is related to selection of the final solution from the identified Pareto-optimal solution set.

This stage is equivalent to stage 4 of the single-objective SOSA. In the SOSA, the final optimal solution will be chosen using a simple and straightforward process. In this process, the best melody vector stored in the memory of each player in each homogeneous musical group is specified first. Then, the best melody vector is

Table 4.64 Pseudocode related to the integration and separation procedures of melody vectors in the proposed multi-objective SOSA

Algorithm 55: Pseudocode for the integration and separation procedures of melody vectors in the proposed multi-objective SOSA

Input: $ISOM_{n,1}$, $ISOM_{n,\text{MNI-I}}$
Output: $ISOM_{n+1}$

start main body

begin
set $ISOM_n = ISOM_{n,1}$
set $OSOM_n = ISOM_{n,\text{MNI-I}}$
set $PMS_0^{\max} = 0$
set $PMS_g^{\max} = \max(PMS_{g,p});\ \forall p \in \Psi^{PN_g}$
set $\text{PN}^{\max} = \max(PN_g);\ \forall g \in \Psi^{\text{NHMG}}$
construct the $HSOM_n$ with dimension $\left\{2 \cdot \sum_{g \in \Psi^{\text{NHMG}}} PMS_g^{\max}\right\} \cdot \{(\text{NDV} + \text{A}) \cdot \text{PN}^{\max}\}$ and zero initial value
for *musical homogeneous group g* $[g \in \Psi^{\text{NHMG}}]$ **do**
  $IMM_{n,g} = ISOM_n\left(1 + \sum_{u \in \Psi^{g-1}} PMS_u^{\max} : \sum_{z \in \Psi^g} PMS_z^{\max}, 1 : (\text{NDV} + \text{A}) \cdot PN_g\right)$
  $OMM_{n,g} = OSOM_n\left(1 + \sum_{u \in \Psi^{g-1}} PMS_u^{\max} : \sum_{z \in \Psi^g} PMS_z^{\max}, 1 : (\text{NDV} + \text{A}) \cdot PN_g\right)$
  for *music player p of the homogeneous musical group g* $\left[p \in \Psi^{PN_g}\right]$ **do**
    $IPM_{n,g,p} = IMM_{n,g}(1 : PMS_{g,p}, 1 + [(p-1) \cdot (\text{NDV} + \text{A})] : [p \cdot (\text{NDV} + \text{A})])$
    $OPM_{n,g,p} = OMM_{n,g}(1 : PMS_{g,p}, 1 + [(p-1) \cdot (\text{NDV} + \text{A})] : [p \cdot (\text{NDV} + \text{A})])$
    $HPM_{n,g,p} = IPM_{n,g,p} \cup OPM_{n,g,p}$
    $HMM_{n,g}\left(1 : 2 \cdot PMS_g^{\max}, 1 + [(p-1) \cdot (\text{NDV} + \text{A})] : [p \cdot (\text{NDV} + \text{A})]\right) = HPM_{n,g,p}$
  end for
  $HSOM_n\left(1 + 2 \cdot \sum_{u \in \Psi^{g-1}} PMS_u^{\max} : 2 \cdot \sum_{z \in \Psi^g} PMS_z^{\max}, 1 + [(p-1) \cdot (\text{NDV} + \text{A})] : [p \cdot (\text{NDV} + \text{A})]\right) = HMM_{n,g}$
end for
sort the $HSOM_n$
  Algorithm 48: Pseudocode for the MFNDSA used in order to determine the non-dominated front—rank—of each melody vector stored in the $HSOM_n$ under the proposed multi-objective SOSA, while the n index is added to all equations in this pseudocode
  Algorithm 49: Pseudocode for the MCCA used in order to determine the crowding distance of each melody vector stored in the $HSOM_n$ under the proposed multi-objective SOSA, while the n index is added to all equations in this pseudocode
  Algorithm 50: Pseudocode for sorting the melody vectors stored in the $HSOM_n$ under the proposed multi-objective SOSA, while the n index is added to all equations in this pseudocode
  construct the $ISOM_{n+1}$ with dimension $\left\{\sum_{g \in \Psi^{\text{NHMG}}} PMS_g^{\max}\right\} \cdot \{(\text{NDV} + \text{A}) \times \text{PN}^{\max}\}$ and zero initial value
for *musical homogeneous group g* $[g \in \Psi^{\text{NHMG}}]$ **do**
  $HMM_{n+1,g} = HSOM_n\left(1 + 2 \cdot \sum_{u \in \Psi^{g-1}} PMS_u^{\max} : 2 \cdot \sum_{z \in \Psi^g} PMS_z^{\max}, 1 : (\text{NDV} + \text{A}) \cdot PN_g\right)$
  for *music player p of the homogeneous musical group g* $\left[p \in \Psi^{PN_g}\right]$ **do**
    $IPM_{n+1,g,p} = HMM_{n,g}(1 : PMS_p, 1 + [(p-1) \cdot (\text{NDV} + \text{A})] : [p \cdot (\text{NDV} + \text{A})])$
    $IMM_{n+1,g}(1 : PMS_p, 1 + [(p-1) \cdot (\text{NDV} + \text{A})] : [p \cdot (\text{NDV} + \text{A})]) = IPM_{n+1,p}$
  end for
  $ISOM_{n+1}\left(1 + \sum_{u \in \Psi^{g-1}} PMS_u^{\max} : \sum_{z \in \Psi^g} PMS_z^{\max}, 1 : (\text{NDV} + \text{A}) \cdot PN_g\right) = IMM_{n+1,g}$
end for
terminate

end main body

selected from among the best melody vectors of the existing music players in each homogeneous musical group as the best melody vector of the corresponding homogeneous musical group. Eventually, the best melody vector is chosen from among the best available melody vectors in all homogeneous musical groups as the final optimal solution. In the multi-objective SOSA, however, the selection of the final optimal solution from the identified Pareto-optimal solution set is a complicated process. Here, the FSM is employed for choosing the most satisfactory solution from the identified Pareto-optimal solution set. The authors refer to Sect. 2.5 of Chap. 2 for a modern introduction to the FSM. The performance-driven architecture of the multi-objective SOSA is formed by placing the designed pseudocode in different stages and sub-stages of this algorithm in a regular sequence. Table 4.65 illustrates the pseudocode associated with the performance-driven architecture of the multi-objective SOSA. Sub-stages 3.3, 3.1.1.3, 3.1.2.4, and 3.1.3.4 (the check of the process of the stopping criterion of the external computational stage, the SISS, the GISSHMG, and the GISSIME) are defined by the first, second, third, and fourth WHILE loops in the pseudocode related to the performance-driven architecture of the multi-objective SOSA (see Table 4.65).

4.6 Conclusions

In the preceding chapter, a comprehensive review concerning the meta-heuristic music-inspired optimization algorithms was rigorously described, with a focus on the single-objective SS-HSA, single-objective SS-IHSA, and single-objective continuous TMS-MSA. According to what was represented, the enhancements accomplished on the single-objective SS-HSA, as an optimization algorithm with a single-stage computational structure and single-dimensional structure, could be broken down into three general categories, from the perspective of implementation: (1) enhancements applied on the single-objective SS-HSA from the standpoint of the parameter adjustments; (2) enhancements performed on the single-objective SS-HSA from the viewpoint of the combination of this algorithm with other meta-heuristic optimization algorithms; and (3) enhancements implemented on the single-objective SS-HSA from the perspective of architectural principles.

Also discussed was that the performance of most existing meta-heuristic optimization algorithms, even the single-objective SS-HSA and its enhanced versions, was highly affected by increasing an unbalanced number of dimensions of complicated, real-world, large-scale, non-convex, non-smooth optimization problems with big data. In this situation, most existing meta-heuristic optimization algorithms could not keep their desirable performance in the face of these optimization problems. Tenuous and vulnerable characteristics employed in the architecture of the existing meta-heuristic optimization algorithms—such as having only a single-stage computational structure; using single-dimensional structures; etc.—were the main reasons for this deficiency. It was then expressed that, in 2011, the single-objective continuous TMS-MSA, as a new version of architecture of the SS-HSA with a two-stage

Table 4.65 Pseudocode related to the performance-driven architecture of the proposed multi-objective SOSA

Algorithm 56: Pseudocode for the performance-driven architecture of the proposed multi-objective SOSA

Input: A, $BW_{g,p}^{\max}$, $BW_{g,p}^{\min}$, MNI-E, MNI-I, MNI-SISS, MNI-GISSHMG, MNI-GISSIME, NCDV, NDV, NDDV, $PAR_{g,p}^{\max}$, $PAR_{g,p}^{\min}$, $PMCR_{g,p}$, $PMS_{g,p}$, PN_g, $x_v^{\min}$, $x_v^{\max}$, $\{x_v(1), \ldots, x_v(w_v), \ldots, x_v(W_v)\}$

Output: Pareto optimal solution set and also x^{best}

start main body

begin

Stage 1—Definition stage: Definition of the MOOP and its parameters

Stage 2—Initialization stage

  Sub-stage 2.1: Initialization of the parameters of the multi-objective SOSA

  Sub-stage 2.2: Initialization of the of the ISOM

    Algorithm 47: Pseudocode for initialization of the entire set of PMs or entire set of MMs or SOM in the proposed multi-objective SOSA

    Algorithm 48: Pseudocode for the MFNDSA used in order to determine the non-dominated front—rank—of each melody vector stored in the IPMs or IMMs or ISOM under the proposed multi-objective SOSA

    Algorithm 49: Pseudocode for the MCCA used in order to determine the crowding distance of each melody vector stored in the IPMs or IMMs or ISOM under the proposed multi-objective SOSA

    Algorithm 50: Pseudocode for sorting the melody vectors stored in the IPMs or IMMs or ISOM under the proposed multi-objective SOSA

Stage 3—External computational stage

  set *iteration of the external computational stage* $n = 1$

  set $ISOM_n$ = ISOM

  while $n \leq$ MNI-E **do**

    Sub-stage 3.1—Internal computational sub-stage

      set *improvisation/iteration of the internal computational sub-stage* $m = 1$

      set $ISOM_{n,\,m} = ISOM_n$

      Sub-stage 3.1.1—single computational sub-stage or the SISS

      while $m \leq$ (MNI-SISS) **do**

        Sub-stage 3.1.1.1: Improvisation of a new melody vector by each existing player in the symphony orchestra

          Algorithm 51: Pseudocode for improvisation of a new melody vector by each existing player in the symphony orchestra in the proposed multi-objective SOSA

        Sub-stage 3.1.1.2: Update of each available IPM in the symphony orchestra

          Algorithm 52: Pseudocode for the update of the memory of all existing players in the symphony orchestra or the update of the $ISOM_{n,m}$ in the proposed multi-objective SOSA

          Algorithm 48: Pseudocode for the MFNDSA used in order to determine the non-dominated front—rank—of each melody vector stored in the $ISOM_{n,m}$ under the proposed multi-objective SOSA, while the n and m indices are added to all equations in this pseudocode

          Algorithm 49: Pseudocode for the MCCA used in order to determine the crowding distance of each melody vector stored in the $ISOM_{n,m}$ under the proposed multi-objective SOSA, while the n and m indices are added to all equations in this pseudocode

          Algorithm 50: Pseudocode for sorting the melody vectors stored in the $ISOM_{n,m}$ under the proposed multi-objective SOSA, while the n and m indices are added to all equations in this pseudocode

        set *improvisation/iteration of the internal computational sub-stage* $m = m + 1$

      end while

      Sub-stage 3.1.2—Group computational sub-stage for each homogeneous musical group or the GISSHMG

(continued)

Table 4.65 (continued)

while $m >$ (MNI-GISS) & $m \leq$ (MNI-GISS) + (MNI-GISSHMG) **do**
Sub-stage 3.1.2.1: Improvisation of a new melody vector by each existing player in the symphony orchestra taking into account the feasible ranges of the updated pitches for each homogeneous musical group
Algorithm 51: Pseudocode for improvisation of a new melody vector by each existing player in the symphony orchestra in the proposed multi-objective SOSA
Sub-stage 3.1.2.2: Update of each available IPM in the symphony orchestra
Algorithm 52: Pseudocode for the update of the memory of all existing players in the symphony orchestra or the update of the $ISOM_{n,m}$ in the proposed multi-objective SOSA
Algorithm 48: Pseudocode for the MFNDSA used in order to determine the non-dominated front—rank—of each melody vector stored in the $ISOM_{n,m}$ under the proposed multi-objective SOSA, while the n and m indices are added to all equations in this pseudocode
Algorithm 49: Pseudocode for the MCCA used in order to determine the crowding distance of each melody vector stored in the $ISOM_{n,m}$ under the proposed multi-objective SOSA, while the n and m indices are added to all equations in this pseudocode
Algorithm 50: Pseudocode for sorting the melody vectors stored in the $ISOM_{n,m}$ under the proposed multi-objective SOSA, while the n and m indices are added to all equations in this pseudocode
Sub-stage 3.1.2.3: Update of the feasible ranges of the pitches—continuous decision-making variables—for each homogeneous musical group in the next improvisation—only for random selection
Algorithm 53: Pseudocode for the update of the feasible ranges of continuous decision-making variables for each homogeneous musical group in the symphony orchestra in the proposed multi-objective SOSA
set *improvisation/iteration of the internal computational sub-stage* $m = m + 1$
end while
while $m >$ (MNI-GISS) + (MNI-GISSHMG) & $m \leq$ (MNI-GISS) + (MNI-GISSHMG) + (MNI-GISSIME) **do**
Sub-stage 3.1.3.1: Improvisation of a new melody vector by each existing player in the symphony orchestra taking into account the feasible ranges of the updated pitches for the inhomogeneous musical ensemble
Algorithm 51: Pseudocode for improvisation of a new melody vector by each existing player in the symphony orchestra in the proposed multi-objective SOSA
Sub-stage 3.1.3.2: Update of each available IPM in the symphony orchestra
Algorithm 52: Pseudocode for the update of the memory of all existing players in the symphony orchestra or the update of the $ISOM_{n,m}$ in the proposed multi-objective SOSA
Algorithm 48: Pseudocode for the MFNDSA used in order to determine the non-dominated front—rank—of each melody vector stored in the $ISOM_{n,m}$ under the proposed multi-objective SOSA, while the n and m indices are added to all equations in this pseudocode
Algorithm 49: Pseudocode for the MCCA used in order to determine the crowding distance of each melody vector stored in the $ISOM_{n,m}$ under the proposed multi-objective SOSA, while the n and m indices are added to all equations in this pseudocode
Algorithm 50: Pseudocode for sorting the melody vectors stored in the $ISOM_{n,m}$ under the proposed multi-objective SOSA, while the n and m indices are added to all equations in this pseudocode
Sub-stage 3.1.3.3: Update of the feasible ranges of the pitches—continuous decision-making variables—for the inhomogeneous musical ensemble in the next improvisation—only for random selection

(continued)

Table 4.65 (continued)

Algorithm 54: Pseudocode for the update process of the continuous decision-making variables for the inhomogeneous musical ensemble in the proposed multi-objective SOSA
set *improvisation/iteration of the internal computational sub-stage* $m = m + 1$
end while
Sub-stage 3.2—Integration and separation procedures of melody vectors
Algorithm 55: Pseudocode for the integration and separation procedures of melody vectors in the proposed multi-objective SOSA
set *iteration of the external computational stage* $n = n + 1$
end while
Stage 4—Selection stage: Selection of the final optimal solution—the best melody
Step-by-step process of the FSM presented in Sect. 2.5.3 of Chap. 2
terminate
end main body

computational multi-dimensional and single-homogenous structure, was developed in order to overcome the shortcomings in the architecture of the single-objective SS-HSA and its enhanced versions. Although organizing of the single-objective continuous TMS-MSA brought about an innovative direction in the architecture of the meta-heuristic algorithms in order to solve complicated, real-world, large-scale, non-convex, non-smooth optimization problems, this optimization algorithm was solely addressed for optimization problems with continuous decision-making variables. With that in mind, the single-objective continuous TMS-MSA could not be employed in dealing with optimization problems with a concurrent combination of the continuous and discrete decision-making variables.

In Sect. 4.2 of this chapter, therefore, the authors proposed a new single-objective continuous/discrete TMS-MSA in order to solve the complicated, real-world, large-scale, non-convex, non-smooth optimization problems with a simultaneous combination of the continuous and discrete decision-making variables. In order to enhance efficiency and efficacy of the performance of this optimization algorithm, the authors also developed a new and improved version of the single-objective continuous/discrete TMS-MSA, referred to as the single-objective TMS-EMSA, in Sect. 4.3 of this chapter. On the other hand, it was demonstrated that modern engineering challenges with multilevel dimensions in different branches of the engineering sciences, particularly electrical engineering, have been widely encountered in recent years. Consequently, these challenges could not be addressed in the form of single-level optimization problems, as traditional optimization problems, and had to be developed in the form of nontraditional multilevel optimization problems. In dealing with these multilevel optimization problems, many of the meta-heuristic optimization algorithms, even the single-objective continuous TMS-MSA, the proposed single-objective continuous/discrete TMS-MSA, and the single-objective TMS-EMSA, may not be able to show the most favorable performance. This is because of two reasons: (1) significant increase of data involved in the optimization process at different levels of the multilevel optimization problem and its interdependency and (2) having a single-homogeneous structure in the architecture of the aforementioned optimization algorithms.

In Sect. 4.4 of this chapter, then, the authors illustrated an innovative architectural version of the proposed single-objective TMS-EMSA, referred to as either the MMM-EMSA or the MMS-EMSA or the SOSA, to conquer the deficiencies in the architecture of the aforementioned optimization algorithms. The unique characteristics pondered in the architecture of the single-objective SOSA can result in a high degree of performance, flexibility, and robustness, not only in solving complicated, real-world, large-scale, non-convex, non-smooth optimization problems, but also in adjusting its computational burden. Furthermore, the well-designed architecture of the newly developed single-objective SOSA can provide parallel processing capability.

The single-objective SS-HSA, single-objective SS-IHSA, single-objective continuous TMS-MSA, single-objective continuous/discrete TMS-MSA, proposed single-objective TMS-EMSA, and proposed single-objective SOSA can be used to deal with single-objective optimization problems. In other words, the architecture of these optimization algorithms has been developed to be suitable only for solving SOOPs and is practically incapable of solving MOOPs. Technically speaking, if an optimization problem consists of multiple conflicting, noncommensurable and correlated objective functions, the most reasonable strategy is to take advantage of the multi-objective optimization process in order to deal with such optimization problems. In Sect. 4.5 of this chapter, then, the authors addressed the well-suited multi-objective versions of the aforementioned single-objective optimization algorithms with the aim of overcoming its inability to deal with MOOPs. Thus, a new multi-objective strategy was described to modify the architecture of the meta-heuristic music-inspired optimization algorithms with a single-stage computational and single-dimensional structure (i.e., the single-objective SS-HSA and single-objective SS-IHSA). Then, a new multi-objective strategy was expressed to alter the architecture of the meta-heuristic music-inspired optimization algorithms with a two-stage computational multi-dimensional and single-homogeneous structure (i.e., the single-objective continuous TMS-MSA, single-objective continuous/discrete TMS-MSA, and proposed single-objective TMS-EMSA). Eventually, the authors developed an innovative multi-objective strategy to amend the architecture of the meta-heuristic music-inspired optimization algorithm with a multi-stage computational multi-dimensional multiple-homogeneous—or multi-stage computational multi-dimensional single-inhomogeneous—structure (i.e., the single-objective SOSA).

Strictly speaking, these proposed optimization algorithms can bring about an innovative direction in the design and architecture of powerful and flexible meta-heuristic optimization algorithms in order to tackle the complexities of complicated real-world optimization problems. These algorithms can also be widely employed by specialists and researchers in different sciences, especially engineering sciences, in order to deal with a very wide range of optimization problems with different structures and characteristics. As a result, in this chapter, the main intention of the authors was to provide a powerful range of meta-heuristic music-inspired optimization algorithms in such a way that they have reasonable and applicable performance, flexibility, and robustness in dealing with highly complicated optimization problems of power systems. These optimization problems are mostly represented as a multilevel, large-scale, non-convex, non-smooth optimization problem having a nonlinear, mixed-integer nature with big data, the most pivotal of which will be addressed in Chaps. 5–7.

Appendix 1: List of Abbreviations and Acronyms

AIP	Alternative improvisation procedure
BW	Bandwidth
CDVs	Continuous decision-making variables
DDVs	Discrete decision-making variables
EAIP	Enhanced alternative improvisation procedure
FSM	Fuzzy satisfying method
GIS	Group improvisation stage
GISS	Group improvisation sub-stage
GISHMG	Group improvisation stage for each homogeneous musical group
GISSHMG	Group improvisation sub-stage for each homogeneous musical group
GISIME	Group improvisation stage for inhomogeneous musical ensemble
GISSIME	Group improvisation sub-stage for inhomogeneous musical ensemble
HHM	Hybrid harmony memory
HM	Harmony memory
HMCR	Harmony memory considering rate
HMM	Hybrid melody memory
HMMs	Hybrid melody memories
HMS	Harmony memory size
HPMs	Hybrid player memories
HSOM	Hybrid symphony orchestra memory
IHM	Input harmony memory
IMM	Input melody memory
IMMs	Input melody memories
IPMs	Input player memories
ISOM	Input symphony orchestra memory
MCCA	Modified crowded-comparison approach
MFNDSA	Modified fast non-dominated sorting approach
MM	Melody memory
MNI	Maximum number of improvisations/iterations
MNI-E	Maximum number of improvisations/iterations of the external computational stage
MNI-I	Maximum number of improvisations/iterations of the internal computational sub-stage
MNI-PGIS	Maximum number of improvisations/iterations of the pseudo-group improvisation stage

(continued)

MNI-PGISS	Maximum number of improvisations/iterations of the pseudo-group improvisation sub-stage
MNI-GIS	Maximum number of improvisations/iterations of the group improvisation stage
MNI-GISS	Maximum number of improvisations/iterations of the group improvisation sub-stage
MNI-GISHMG	Maximum number of improvisations/iterations of the group improvisation stage for each homogeneous musical group
MNI-GISSHMG	Maximum number of improvisations/iterations of the group improvisation sub-stage for each homogeneous musical group
MNI-GISIME	Maximum number of improvisations/iterations of the group improvisation stage for the inhomogeneous musical ensemble
MNI-GISSIME	Maximum number of improvisations/iterations of the group improvisation sub-stage for the inhomogeneous musical ensemble
MNI-SIS	Maximum number of improvisations/iterations of the single improvisation stage
MNI-SISS	Maximum number of improvisations/iterations of the single improvisation sub-stage
MMM-EMSA	Multi-stage computational multi-dimensional multiple-homogeneous enhanced melody search algorithm
MMS-EMSA	Multi-stage computational multi-dimensional single-inhomogeneous enhanced melody search algorithm
MOOPs	Multi-objective optimization problems
MOOAs	Multi-objective optimization algorithms
NCDV	Number of continuous decision-making variables
NDDV	Number of discrete decision-making variables
NDV	Number of decision-making variables including continuous and discrete decision-making variable
NHMG	Number of homogeneous musical groups
NIP	Novel improvisation procedure
NSGA-II	Non-dominated sorting genetic algorithm II
OHM	Output harmony memory
OMM	Output melody memory
OMMs	Output melody memories
OPMs	Output player memories
OSOM	Output symphony orchestra memory
PAR	Pitch adjusting rate
PGIS	Pseudo-group improvisation stage
PGISS	Pseudo-group improvisation sub-stage
PMCR	Player memory considering rate
PMs	Player memories
PMS	Player memory size
PN	Player number
SIS	Single improvisation stage
SISS	Single improvisation sub-stage
SOSA	Symphony orchestra search algorithm
SOM	Symphony orchestra memory
SOOPs	Single-objective optimization problems

(continued)

SS-HSA	Single-stage computational, single-dimensional harmony search algorithm
SS-IHSA	Single-stage computational, single-dimensional improved harmony search algorithm
TMS-EMSA	Two-stage computational, multi-dimensional, single-homogeneous enhanced melody search algorithm
TMS-MSA	Two-stage computational, multi-dimensional, single-homogeneous melody search algorithm

Appendix 2: List of Mathematical Symbols

Index:

a	Index for objective functions running from 1 to A
b	Index for equality constraints running from 1 to B
e	Index for inequality constraints running from 1 to E
m	Index for improvisations/iterations running from 1 to MNI under the SS-HSA and under the SS-IHSA, an index for improvisations/iterations running from 1 to (MNI-SIS) + (MNI-PGIS) under the continuous/discrete TMS-MSA, an index for improvisations/iterations running from 1 to (MNI-SIS) + (MNI-GIS) in the TMS-EMSA, and also an index for improvisations/iterations running from 1 to (MNI-SIS) + (MNI-GISHMG) + (MNI-GISIME) in the SOSA
n	Index for iterations of the external computational stage running from 1 to MNI-E
p	Index for existing players in a musical group running from 1 to PN under the continuous/discrete TMS-MSA and under the TMS-EMSA and also an index for existing players in homogeneous musical group g in the symphony orchestra running from 1 to PN_g under the SOSA
s, s^*	Index for harmony vectors stored in HM running from 1 to HMS under the SS-HSA and under the SS-IHSA, an index for melody vectors stored in each PM running from 1 to PMS under the continuous/discrete TMS-MSA, an index for melody vectors stored in the memory of player p in the musical group running from 1 to PMS_p under the TMS-EMSA, and also an index for melody vectors stored in the memory of player p in homogeneous musical group g in the symphony orchestra running from 1 to $PMS_{g,\,p}$ under the SOSA
v	Index for decision-making variables, including the continuous and discrete decision-making variables, running from 1 to NDV
w_v	Index for candidate permissible values of discrete decision-making variable v running from 1 to W_v

Set:

Ψ^{A}	Set of indices of objective functions
Ψ^{B}	Set of indices of equality constraints
Ψ^{E}	Set of indices of inequality constraints
Ψ^{HMS}	Set of indices of harmony vectors stored in the HM under the SS-HSA and under the SS-IHSA

(continued)

Ψ^{MNI}	Set of indices of improvisations/iterations in the SS-HSA and in the SS-IHSA
$\Psi^{\text{(MNI-SIS)+(MNI-PGIS)}}$	Set of indices of improvisations/iterations in the continuous/discrete TMS-MSA
$\Psi^{\text{(MNI-SIS)+(MNI-GIS)}}$	Set of indices of improvisations/iterations in the TMS-EMSA
$\Psi^{\text{(MNI-SIS)+(MNI-GISHMG)+(MNI-GISIME)}}$	Set of indices of improvisations/iterations in the SOSA
$\Psi^{\text{MNI-E}}$	Set of indices of iterations of the external computational stage
Ψ^{NDV}	Set of indices of decision-making variables, including the continuous and discrete decision-making variables
Ψ^{NCDV}	Set of indices of continuous decision-making variables
Ψ^{NDDV}	Set of indices of discrete decision-making variables
Ψ^{PMS}	Set of indices of melody vectors stored in each PM
Ψ^{PMS_p}	Set of indices of melody vectors stored in memory of player p in the musical group
$\Psi^{PMS_{g,p}}$	Set of indices of melody vectors stored in memory of player p in homogeneous musical group g in the symphony orchestra
Ψ^{PN}	Set of indices of existing players in a musical group
Ψ^{PN_g}	Set of indices of existing players in homogeneous musical group g in the symphony orchestra
W_v	Set of indices of candidate permissible values of discrete decision-making variable v

Parameters:

BW	Bandwidth
$\text{BW}^{\max}$	Maximum bandwidth
$BW_p^{\max}$	Maximum bandwidth of player p in the musical group
$BW_{g,p}^{\max}$	Maximum bandwidth of player p in homogeneous musical group g in the symphony orchestra
$\text{BW}^{\min}$	Minimum bandwidth
$BW_p^{\min}$	Minimum bandwidth of player p in the musical group
$BW_{g,p}^{\min}$	Minimum bandwidth of player p in homogeneous musical group g in the symphony orchestra
HMCR	Harmony memory considering rate
HMS	Harmony memory size
MNI	Maximum number of improvisations/iterations in the SS-HAS and in the SS-IHSA
MNI-E	Maximum number of iterations of the external computational stage
MNI-GIS	Maximum number of improvisations/iterations of the GIS in the TMS-EMSA
MNI-GISS	Maximum number of improvisations/iterations of the GISS in the multi-objective TMS-EMSA
MNI-GISHMG	Maximum number of improvisations/iterations of the GISHMG in the SOSA
MNI-GISSHMG	Maximum number of improvisations/iterations of the GISSHMG in the multi-objective SOSA

(continued)

MNI-GISIME	Maximum number of improvisations/iterations of the GISIME in the SOSA
MNI-GISSIME	Maximum number of improvisations/iterations of the GISSIME in the multi-objective SOSA
MNI-I	Maximum number of improvisations/iterations of the internal computational sub-stage
MNI-PGIS	Maximum number of improvisations/iterations of the PGIS in the continuous/ discrete TMS-MSA
MNI-PGISS	Maximum number of improvisations/iterations of the PGISS in the multi-objective continuous/discrete TMS-MSA
MNI-SIS	Maximum number of improvisations/iterations of the SIS in the continuous/ discrete TMS-MSA and in the TMS-EMSA and in the SOSA
MNI-SISS	Maximum number of improvisations/iterations of the SISS in the multi-objective continuous/discrete TMS-MSA and in the multi-objective TMS-EMSA and in the multi-objective SOSA
PAR	Pitch adjusting rate
$\mathrm{PAR}^{\max}$	Maximum pitch adjusting rate
$PAR_p^{\max}$	Maximum pitch adjusting rate of player p in the musical group
$PAR_{g,p}^{\max}$	Maximum pitch adjusting rate of existing players in homogeneous musical group g in the symphony orchestra
$\mathrm{PAR}^{\min}$	Minimum pitch adjusting rate
$PAR_p^{\min}$	Minimum pitch adjusting rate of player p in the musical group
$PAR_{g,p}^{\min}$	Minimum pitch adjusting rate of existing players in homogeneous musical group g in the symphony orchestra
PMCR	Player memory considering rate
$PMCR_p$	Player memory considering rate of player p in the musical group
$PMCR_{g,p}$	Player memory considering rate of player p in homogeneous musical group g in the symphony orchestra
PMS	Player memory size
PMS_p	Player memory size of player p in the musical group
$PMS_{g,p}$	Player memory size of player p in homogeneous musical group g in the symphony orchestra
PN	Number of existing players in the musical group
PN_g	Number of existing players in homogeneous musical group g in the symphony orchestra
$x_v^{\max}$	Upper bound on the decision-making variable v
$x_v^{\min}$	Lower bound on the decision-making variable v
X	Nonempty feasible decision-making space
Z	Feasible objective space

Variables:

BW_m	Bandwidth in improvisation/iteration m of the SS-IHSA and in improvisation/ iteration m of the continuous/discrete TMS-MSA
$BW_{m,p}$	Bandwidth of player p in the musical group under improvisation/iteration m of the TMS-EMSA
$BW_{m,g,p}$	Bandwidth of player p in homogeneous musical group g in the symphony orchestra under improvisation/iteration m of the SOSA

(continued)

$d_{\mathbf{x}^s}$	Number of harmony vectors available in the IHM, which dominate the harmony vector s in the multi-objective SS-HSA and in the multi-objective SS-IHSA
$d_{\mathbf{x}_p^s}$	Number of melody vectors available in the memory of player p in the musical group, which dominates the melody vector s of this player in the multi-objective continuous/discrete TMS-MSA and in the multi-objective TMS-EMSA
$d_{\mathbf{x}_{g,p}^s}$	Number of melody vectors available in the memory of player p in homogeneous musical group g in the symphony orchestra, which dominate the melody vector s of this player in the multi-objective SOSA
$\text{distance}_{\mathrm{x}^s}$	Crowding distance of harmony vector s stored in the IHM under the proposed multi-objective SS-HSA and under the multi-objective SS-IHSA
$\text{distance}_{\mathrm{x}_p^s}$	Crowding distance of melody vector s stored in the memory of player p in the musical group under the proposed multi-objective continuous/discrete TMS-MSA and under the multi-objective TMS-EMSA
$\text{distance}_{\mathrm{x}_{g,p}^s}$	Crowding distance of melody vector s stored in the memory of player p in homogeneous musical group g in the symphony orchestra under the proposed multi-objective SOSA
$f(\mathrm{x})$	Objective function of the optimization problem
$f_a(\mathrm{x})$	Objective function a of the optimization problem or component a of the vector of objective functions
$f(\mathrm{x}^s)$	Value of the objective function—fitness function—derived from the harmony vector s stored in the HM matrix
$f(\mathrm{x}_a^s)$	Value of the objective function a—fitness function—derived from the harmony vector s stored in the IHM matrix
$f\left(\mathrm{x}_p^s\right)$	Value of the objective function—fitness function—derived from the melody vector s stored in memory submatrix relevant to player p in the musical group
$f\left(\mathrm{x}_{p,a}^s\right)$	Value of the objective function a—fitness function—derived from the melody vector s stored in input memory submatrix relevant to player p in the musical group
$f\left(\mathrm{x}_{g,p}^s\right)$	Value of the objective function—fitness function—derived from the melody vector s stored in memory submatrix relevant to player p in homogeneous musical group g in the symphony orchestra
$f\left(\mathrm{x}_{g,p,a}^s\right)$	Value of the objective function a—fitness function—derived from the melody vector s stored in input memory submatrix relevant to player p in homogeneous musical group g in the symphony orchestra
$f(\mathrm{x}_m^{\text{new}})$	Value of the objective function—fitness function—derived from the new harmony vector under improvisation/iteration m of the SS-HSA and under improvisation/iteration m of the SS-IHSA
$f\left(\mathrm{x}_{m,p}^{\text{new}}\right)$	Value of the objective function—fitness function—derived from the new melody vector played by player p in the musical group under improvisation/iteration m of the continuous/discrete TMS-MSA and under improvisation/iteration m of the TMS-EMSA
$f\left(\mathrm{x}_{m,g,p}^{\text{new}}\right)$	Value of the objective function—fitness function—derived from the new melody vector played by player p in homogenous musical group g in the symphony orchestra under improvisation/iteration m of the SOSA
$\mathrm{F}(\mathrm{x})$	Vector of objective functions of the optimization problem
$\mathcal{F}_r$	Non-dominated front r in the IHM of the multi-objective SS-HSA or in the IHM of the multi-objective SS-IHSA

(continued)

$\mathcal{F}_{p,r_p}$	Non-dominated front r in the input memory of player p in the musical group under the proposed multi-objective continuous/discrete TMS-MSA and under the multi-objective TMS-EMSA
$\mathcal{F}_{g,p,r_{g,p}}$	Non-dominated front r in the input memory of player p in homogeneous musical group g in the symphony orchestra under the proposed multi-objective SOSA
$g_b(\mathrm{x})$	Equality constraint b of the optimization problem or component b of the vector of equality constraints
$\mathrm{G(x)}$	Vector of equality constraints of the optimization problem
$h_e(\mathrm{x})$	Inequality constraint e of the optimization problem or component e of the vector of inequality constraints
$\mathrm{H(x)}$	Vector of inequality constraints of the optimization problem
HM	Harmony memory matrix
HM_m	Harmony memory matrix in improvisation/iteration m of the SS-HSA and in improvisation/iteration m of the SS-IHSA
HHM	Hybrid harmony memory matrix
HMM	Hybrid melody memory matrix
HMM_g	Hybrid melody memory matrix relevant to existing homogeneous musical group g in the symphony orchestra
HPM_p	Hybrid player memory matrix relevant to player p in the musical group
$HPM_{g,p}$	Hybrid player memory matrix relevant to player p in homogeneous musical group g in the symphony orchestra
HSOM	Hybrid symphony orchestra memory matrix
IHM	Input harmony memory matrix
IMM	Input melody memory matrix
IMM_g	Input melody memory matrix relevant to existing homogeneous musical group g in the symphony orchestra
IPM_p	Input player memory matrix relevant to player p in the musical group
$IPM_{g,p}$	Input player memory matrix relevant to player p in homogeneous musical group g in the symphony orchestra
ISOM	Input symphony orchestra memory matrix
k	Random integer with a uniform distribution through the set $\{1, 2, \ldots, \mathrm{NCDV}\}$
l	Random integer with a uniform distribution through the set $\{1, 2, \ldots, \mathrm{NDDV}\}$
MM	Melody memory matrix
MM_m	Melody memory matrix in improvisation/iteration m of the continuous/discrete TMS-MSA or in improvisation/iteration m of the TMS-EMSA
MM_g	Melody memory matrix of the existing homogeneous musical group g in the symphony orchestra
OHM	Output harmony memory matrix
OMM	Output melody memory matrix
OMM_g	Output melody memory matrix relevant to existing homogeneous musical group g in the symphony orchestra
OPM_p	Output player memory matrix relevant to player p in the musical group
$OPM_{g,p}$	Output player memory matrix relevant to player p in homogeneous musical group g in the symphony orchestra
OSOM	Output symphony orchestra memory matrix
PAR_m	Pitch adjusting rate in improvisation/iteration m of the SS-IHSA and in improvisation/iteration m of the continuous/discrete TMS-MSA

(continued)

$PAR_{m,p}$	Pitch adjusting rate of player p in the musical group under improvisation/iteration m of the TMS-EMSA
$PAR_{m,g,p}$	Pitch adjusting rate of player p in homogeneous musical group g in the symphony orchestra under improvisation/iteration m of the SOSA
PM_p	Memory submatrix relevant to player p in the musical group
$PM_{m,p}$	Memory submatrix relevant to player p in the musical group in improvisation/iteration m of the continuous/discrete TMS-MSA and in improvisation/iteration m of the TMS-EMSA
$PM_{g,p}$	Memory submatrix relevant to player p in homogeneous musical group g in the symphony orchestra
$PM_{m,g,p}$	Memory submatrix relevant to player p in homogeneous musical group g in the symphony orchestra in improvisation/iteration m of the SOSA
r	Random integer with a uniform distribution through the set $\{1, 2, \ldots, \text{HMS}\}$ in the SS-HSA and random integer with a uniform distribution through the set $\{1, 2, \ldots, \text{PMS}\}$ in the continuous/discrete TMS-MSA
r_p	Random integer with a uniform distribution through the set $\{1, 2, \ldots, PMS_p\}$ in the TMS-EMSA
$r_{g,p}$	Random integer with a uniform distribution through the set $\{1, 2, \ldots, PMS_{g,p}\}$ in the SOSA
$\text{rank}_{\mathbf{x}^s}$	Non-dominated front—rank—of harmony vector s stored in the IHM under the proposed multi-objective SS-HSA and under the multi-objective SS-IHSA
$\text{rank}_{\mathbf{x}_p^s}$	Non-dominated front—rank—of melody vector s stored in the memory of player p in the musical group under the proposed multi-objective continuous/discrete TMS-MSA and under the multi-objective TMS-EMSA
$\text{rank}_{\mathbf{x}_{g,p}^s}$	Non-dominated front—rank—of melody vector s stored in the memory of player p in homogeneous musical group g in the symphony orchestra under the proposed multi-objective SOSA
SOM	Symphony orchestra memory matrix
SOM_m	Symphony orchestra memory matrix in improvisation/iteration m of the SOSA
$S_{\mathbf{x}^s}$	Set of harmony vectors available in the IHM that the harmony vector s dominates under the multi-objective SS-HSA and under the multi-objective SS-IHSA
$S_{\mathbf{x}_p^s}$	Set of melody vectors available in the memory of player p in the musical group that the melody vector s of this player dominates under the multi-objective continuous/discrete TMS-MSA and under the multi-objective TMS-EMSA
$S_{\mathbf{x}_{g,p}^s}$	Set of melody vectors available in the memory of the of player p in homogeneous musical group g in the symphony orchestra that the melody vector s of this player dominates under the multi-objective SOSA
t	Random integer with a uniform distribution through the set $\{-1, +1\}$ in the SS-HSA and in the SS-IHSA
U(0, 1)	Random number with a uniform distribution between 0 and 1
x_v	Decision-making variable v or component v of the vector of decision-making variable
$x_{m,p,v}^{\text{best}}$	Element v of the best melody vector stored in the memory submatrix relevant to player p in the musical group under improvisation/iteration m of the continuous/discrete TMS-MSA and under improvisation/iteration m of the TMS-EMSA
$x_{m,g,p,v}^{\text{best}}$	Element v of the best melody vector stored in the memory submatrix relevant to player p in homogeneous musical group g in the symphony orchestra under improvisation/iteration m of SOSA

(continued)

$x^{\text{best}}_{m,\text{best},v}$	Element v of the best melody vector stored in the memory submatrix relevant to the best existing player in the musical group under improvisation/iteration m of the TMS-EMSA
$x^{\text{best}}_{m,g,\text{best},v}$	Element v of the best melody vector stored in the memory submatrix relevant to the best existing player in the homogeneous musical group g in the symphony orchestra under improvisation/iteration m of the SOSA
$x^{\text{best}}_{m,\text{best},\text{best},v}$	Element v of the best melody vector stored in the memory submatrix relevant to the best existing player in the inhomogeneous musical ensemble or in the symphony orchestra under improvisation/iteration m of the SOSA
$x^{\max}_{m,v}$	Variable upper bound on the decision-making variable v under improvisation/iteration m of the PGIS in the continuous/discrete TMS-MSA and under improvisation/iteration m of the GIS in the TMS-EMSA and under improvisation/iteration m of the GISIME in the SOSA
$x^{\min}_{m,v}$	Variable lower bound on the decision-making variable v under improvisation/iteration m of the PGIS in the continuous/discrete TMS-MSA and under improvisation/iteration m of the GIS in the TMS-EMSA and under improvisation/iteration m of the GISIME in the SOSA
$x^{\max}_{m,g,v}$	Variable upper bound on the decision-making variable v under improvisation/iteration m of the GISHMG in the SOSA
$x^{\min}_{m,g,v}$	Variable lower bound on the decision-making variable v under improvisation/iteration m of the GISHMG in the SOSA
$x^{\text{new}}_{m,v}$	Element v of the new harmony vector under improvisation/iteration m of the SS-HSA and under improvisation/iteration m of the SS-IHSA
$x^{\text{new}}_{m,p,v}$	Element v of the new melody vector played by player p in the musical group under improvisation/iteration m of the continuous/discrete TMS-MSA and under in improvisation/iteration m of the TMS-EMSA
$x^{\text{new}}_{m,g,p,v}$	Element v of the new melody vector played by player p in homogeneous musical group g in the symphony orchestra under improvisation/iteration m of the SOSA
x^{s}_{v}	Element v of the harmony vector s stored in the HM matrix
$x^{s}_{p,v}$	Element v of the melody vector s stored in the memory submatrix relevant to player p in the musical group
$x^{s}_{g,p,v}$	Element v of the melody vector s stored in the memory submatrix relevant to player p in homogeneous musical group g in the symphony orchestra
$x_v(w_v)$	Candidate permissible value w of discrete decision-making variable v
x	Vector of decision-making variables
$\mathrm{x}^{\text{new}}_{m}$	New harmony vector in improvisation/iteration m of the SS-HSA and in improvisation/iteration m of the SS-IHSA
$\mathrm{x}^{\text{new}}_{m,p}$	New melody vector played by player p in the musical group in improvisation/iteration m of the continuous/discrete TMS-MSA and in improvisation/iteration m of the TMS-EMSA
$\mathrm{x}^{\text{new}}_{m,g,p}$	New melody vector played by player p in homogeneous musical group g in the symphony orchestra under improvisation/iteration m of the SOSA
x^{best}	Best harmony vector stored in the HM matrix under the SS-HSA and under the SS-IHSA, best melody vector stored in the MM matrix under the continuous/discrete TMS-MSA and under TMS-EMSA, and also best melody vector stored in the SOM matrix under the SOSA
$\mathrm{x}^{\text{best}}_{p}$	Best melody vector stored in the memory submatrix relevant to player p in the musical group

(continued)

$x_{g,p}^{best}$	Best melody vector stored in the memory submatrix relevant to player p in homogeneous musical group g in the symphony orchestra
x^{s}	Harmony vector s stored in the HM matrix
x_{p}^{s}	Melody vector s stored in the memory submatrix relevant to player p in the musical group
$x_{g,p}^{s}$	Melody vector s stored in the memory submatrix relevant to player p in homogeneous musical group g in the symphony orchestra
x^{worst}	Worst harmony vector stored in the HM matrix
x_{p}^{worst}	Worst melody vector stored in the memory submatrix relevant to player p in the musical group
$x_{g,p}^{worst}$	Worst melody vector stored in the memory submatrix relevant to player p in homogeneous musical group g in the symphony orchestra
y	Random integer with a uniform distribution through the set $\{x_v(1), \ldots, x_v(w_v), \ldots, x_v(W_v)\}$
z	Vector of the objective function

Note:

• Corresponding indices, sets, parameters, and variables of each single-objective meta-heuristic music-inspired optimization algorithm are valid in its multi-objective version.

• Corresponding variables of each single-objective meta-heuristic music-inspired optimization algorithm that have the index m are considered for its multi-objective version by adding the index n. For example, element v of the new melody vector played by player p in homogeneous musical group g in the symphony orchestra under improvisation/iteration m of the single-objective SOSA ($x_{m,g,p,v}^{new}$) is converted to the element v of the new melody vector played by player p in homogeneous musical group g in the symphony orchestra under iteration n of the external computational stage and under improvisation/iteration m of the internal computational sub-stage of the multi-objective SOSA ($x_{n,m,g,p,v}^{new}$) by adding the index n.

• Corresponding input, output, and hybrid memory relevant to each multi-objective meta-heuristic music-inspired optimization algorithm are rewritten by adding the index n and by simultaneously adding the indices n and m. For example, the IHM in the multi-objective SS-HSA is rewritten as the input harmony memory in iteration n of the external computational stage (IHM_n) by adding the index n and is rewritten as the input harmony memory in iteration n of the external computational stage and in improvisation/iteration m of the internal computational sub-stage (IHM_n) by simultaneously adding the indices n and m.

References

1. Z.W. Geem, G.H. Kim, G.V. Loganathan, A new heuristic optimization algorithm: harmony search. Simulation **76**(2), 60–68 (2001)
2. S.M. Ashrafi, A.B. Dariane, A novel and effective algorithm for numerical optimization: melody search (MS), in *11th International Conference on Hybrid Intelligence Systems (HIS)*, 2011
3. N. Rimsky-Korsakov, *Principles of Orchestration: With Musical Examples Drawn from His Own Works* (Kalmus, New York, 1912)
4. K. Deb, A. Pratap, S. Agarwal, T. Meyarivan, A fast and elitist multiobjective genetic algorithm: NSGA-II. IEEE Trans. Evol. Comput. **6**(2), 182–197 (2002)
5. M. Mahdavi, M. Fesanghary, E. Damangir, An improved harmony search algorithm for solving optimization problems. Appl. Math. Comput. **188**(2), 1567–1579 (2007)

Part II
Power Systems Operation and Planning Problems

Chapter 5
Power Systems Operation

5.1 Introduction

In a traditionally regulated electrical power industry, operation, planning, management, and control—as major actions of power systems in generation, transmission, and distribution networks—are the responsibilities associated with a vertically integrated utility (VIU). In such a situation, customers only deal with this monopolist VIU. There are, therefore, only two active entities in traditionally regulated power systems: (1) the utility and (2) customer. The existence of a natural monopoly and lack of competition in the conventionally regulated environments can bring about inefficiency and low productivity in the actions of the power systems. The first attempt to end the monopoly and develop competition in power systems in the 1990s was conducted in the United Kingdom [1]. Since then, many developed and developing countries throughout the world have been urged to change the structure of their power systems from the conventionally regulated environments to the competitive deregulated environments such as Norway, Sweden, the United States, Spain, the Netherlands, Chile, Argentina, and Australia, among others [2]. The reasons and implementation processes of deregulation in power systems can be varied from one country to another. As a general result, though, the deregulation process throughout these countries gives rise to decomposition of the three components of power systems: (1) generation, (2) transmission, and, (3) distribution, each of which can act independently. In the broadest sense, this process can lead to unbundling of the transmission actions of the VIU from the generation process and/or the distribution actions from the transmission process. Accordingly, the role of traditional entities concerned with the VIU has been changed, such that new independent entities are created that are classified into generation companies (GENCOs), transmission companies (TRANSCO), distribution companies (DISCOs), retail energy service companies (RESCOs), independent system operator (ISO), and so on [1]. The focus of this chapter is on the GENCOs and DISCOs as main market participants, and also the

M. Kiani-Moghaddam et al., *Modern Music-Inspired Optimization Algorithms for Electric Power Systems*, Power Systems,
https://doi.org/10.1007/978-3-030-12044-3_5

ISO as market operator. The actions of power systems have undergone major changes in open competitive environments. To illustrate, the power generation units are owned by the VIU, and all customers are forced to buy their electrical energy from this monopolist VIU. In this circumstance, the VIU schedules its power generation units with the aim of minimizing total generation costs, subject to technical and operational constraints of the power system equipment. In open competitive environments, however, the market participants should compete with each other to sell and buy the energy within the framework of a competitive electricity market. In this regard, profit maximization is more important than the minimization of total generation costs; and, therefore, the conventional least-cost approaches alone cannot be used by the newly created entities in the operation actions of the deregulated power systems. In recent years, then, most technical investigations have been dedicated to dealing with the new challenges of the operation actions in open competitive environments [3–6]. In other words, modeling and implementing of operation actions under the framework of a competitive electricity market, especially bidding strategy studies, are absolutely necessary. In doing so, there are a lot of technical publications that only assess strategic behaviors of the GENCOs as a generation-side Scheme [7]. These studies, therefore, are not able to reflect either the real behavior of market participants or the actual condition of the deregulated power systems. In addition, the schemes reported in the literature are modeled as a bi-level optimization problem, of either a nonlinear or non-convex nature. There is a need, then, for powerful optimization algorithms to overcome these optimization problems. In this chapter, the authors focus on three targets in the field of operation actions of deregulated power systems, as follows:

- Target 1: Presenting a two-level computational-logical framework for a bilateral bidding mechanism (BBM) within a competitive electricity market.
- Target 2: Considering the practical features and limitations in the proposed two-level computational-logical framework.
- Target 3: Solving the proposed framework by using the modern single-objective music-inspired algorithms that were addressed in Chap. 4 and that compare the obtained results with powerful state-of-the-art optimization algorithms.

In this chapter, the authors do not address the elementary details of power system operation, especially in the electricity market, as it is assumed that the readers are already familiar with the fundamental concepts of power system operation. Where appropriate, though, the reader will be referred to related studies that cover the elementary details of power system operation.

The rest of this chapter is arranged as follows: First, the main concepts of game theory and its modeling in the electricity markets are reviewed briefly in Sect. 5.2. In Sect. 5.3, the proposed framework for the BBM in the competitive security-constrained (CSC) electricity market is discussed in detail, including mathematical model of the proposed bi-level computational-logical framework, solution method, simulation results, case studies, and discussion of the results. Finally, the chapter ends with a brief summary and some concluding remarks in Sect. 5.4.

5.2 A Brief Review of Game Theory

In this section, we are going to discuss briefly the main concepts of game theory that form the main structure of the competition concept in the electricity markets.

The concept of game theory was introduced in the late 1920s by John von Neumann, who developed it in the mid-1940s along with Oskar Morgenstern as a powerful decision-making tool for the economic sciences [8]. Game theory is a branch of applied mathematics that deals with decision-making in conflict, cooperation, and competitive situations. In this theory, the decision maker evaluates its performance with respect to the rivals' strategies. In such circumstances, the decision of each decision maker will be affected by the decision of the others. Each game consists of a number of elements, as follows:

- Participants/players: Participants are decision makers in each game. Each participant may be a person, group, company, etc., who competes with each of the other participants in an uncertain situation.
- Move/choice: The action taken by each participant during the game.
- Outcome: The result of the completion of one or more moves by all participants.
- Strategy: A plan or action designed to achieve a major or overall goal—actually, a pure strategy, in contrast to mixed and totally mixed strategies.
- Payoff: A value or expected reward from a given outcome. Depending on the nature of each game, the payoff can be categorized into two major types: (1) cardinal payoff—a continuous measurement of the profit, quantity, utility, etc., and (2) ordinal payoffs—desirability of the outcomes.
- Rules: The specified conditions for the participants, moves, strategies, outcomes, and payoffs.

In each game, the participants influence each other, from which some will be benefited and others will be harmed. Each participant within a game seeks to maximize his/her profit, regardless of the actions of the other participants. The adopted strategies by each participant in each game have, therefore, two major features: (1) the strategies are constrained by the rules and conventions of the game and (2) the strategies are compatible with the benefits and privileges that a participant expects to achieve during the game.

5.2.1 Classifications of the Game

In the game theory literature, there are different types of games that can help to analyze the diverse kinds of the problems [9]. It is really a challenging task to classify these games systematically. Obviously, the classifications can largely depend on criteria, and there is no easy guideline to set out the criteria in the literature. As criteria may vary, detailed classifications can be an impossible task for a research work. Here, a classification of the games from some important points of the view has been tabulated in Table 5.1.

Table 5.1 Classification of the games

No.	Points of the view	Categories	Descriptions
1	Game interactions	Cooperative Noncooperative (√)	In the cooperative game, the participants may cooperate with each other. By contrast, in the non-cooperative game, each participant ignores possible cooperation with rivals in order to make his/her decisions
2	Game identities	Symmetric (√) Asymmetric	If the change in the identity of a participant does not affect his/her adopted strategy, then the game is considered symmetric; otherwise, the game is asymmetric
3	Game descriptions	Normal form (√) Extensive form	If the game description—The payoff and strategies—can be represented in tabular form, then it is considered a normal game. Unlike a normal game, the extensive form of the game is presented using a decision tree
4	Game benefit or outcome	Constant sum Zero sum (√) Nonzero sum	If the total benefit of all game participants remains constant, or zero, for every strategy profile, then the game is considered a constant-sum or zero-sum game, respectively. A zero-sum game is a special kind of constant-sum game. In contrast to the constant- and zero-sum games, in the nonzero-sum game, the total benefit of all game participants for every strategy profile can be different for each strategy profile
5	Choose moves and strategies	Simultaneous (√) Sequential	If each game participant selects moves and strategies without knowledge of the strategies adopted by rival participants, then the game is simultaneous; otherwise, the game is sequential
6	Game information	Perfect information Imperfect information (√)	If each participant has full knowledge of the moves and strategies previously made by all other participants, then the game information is perfect; otherwise, the game information is imperfect
7	Game features	Discrete (√) Continuous	A discrete game has a finite number of participants, choices, outcomes, etc. A continuous game continuously updates its strategies for the participants
8	Game participants	Many player (√) Population	In game theory, if the number of participants is arbitrary but limited, the game is called a multiplayer game; otherwise, the game is called a population-based game
9	Game moves	Finitely long (√) Infinitely long	If the outcomes and payoffs of all the participants are determined after a finite number of moves, then the game is a finite game; otherwise, the game is infinite
10	Game dimension	Combinatorial (√) Non-combinatorial	If there are so many possible moves such that finding an optimal strategy is complex, then the game is combinatorial; otherwise, the game is non-combinatorial
11	Game variables	Differential (√) Non-differential	If the state variables of the participants of the game evolve over time according to a differential equation, then the game is differential; otherwise, the game is non-differential
12	Game uncertainty	Stochastic outcomes	In stochastic outcomes, the related game can add an arbitrary acting player, who constructs chance choices and moves

5.2.2 The Concept of Nash Equilibrium

Nash equilibrium is a term used in game theory to describe a situation in which each participant's strategy is optimal, given the strategies of all other participants. A Nash equilibrium exists when there is no unilateral profitable deviation from any of the participants involved. In other words, no participant in the game would take a different action as long as every other participant's decision remains the same.

In game theory, it is assumed that the participants in the game will behave rationally. In other words, the selected strategies by participants are collinear with their benefits and privileges. In this theory, the adopted strategies, in the interests of the participants, depend on the selected strategies by rival participants. Hence, each participant should take into account the potential strategies adopted by rival participants in order to achieve a major or overall profit. An important issue that must be highlighted is that each participant is unaware of the strategies adopted by the other participants. However, each participant should continuously assess his/her decisions with respect to the strategic actions of rival participants in order to choose the best strategy. The Nash equilibrium point is achieved when two major conditions are met. First, a participant embraces a strategic action that will achieve maximum payoff after considering what the rival participants' strategies are believed to be. Second, each participant's beliefs are rational, and rival participants' strategies are adopted according to these beliefs. In game theory, then, the strategies embraced by the participants under this procedure are called Nash equilibrium strategies [9].

Let us consider a noncooperative game (S,u) with participant i and perfect information. Let S_i be the strategy set of participant i, $S = \{S_1, \ldots, S_i, \ldots, S_I\}$ be the set of possible strategy profiles, and $u(x) = (u_1(s), \ldots, u_i(s), \ldots, u_I(s))$ be the set of payoff profiles calculated at strategy profile $s \in S$. Furthermore, let s_i be a strategy of participant i, and s_{-i} be a strategy profile for all participants except participant i. When participant i selects strategy s_i, resulting in strategy profile $s = (s_1, \ldots, s_i, \ldots, s_I)$, then participant i obtains payoff $u_i(s)$. The obtained payoff by participants i depends not only on the selected strategy itself, but also on the selected strategies by all of the other participants. A strategy profile $s^* \in S$ is a Nash equilibrium point, if each unilateral deviation and/or move in the embraced strategy of participant i results in a disadvantage for the intended participant. In other words, there is no participant at the Nash equilibrium point who receives an advantage by changing his/her strategy. Given this, the strategy profile s^* is a Nash equilibrium point if and only if it satisfies Eq. (5.1):

$$u_i\left(s_i^*, s_{-i}^*\right) \geq u_i\left(s_i, s_{-i}^*\right); \quad \forall\left\{s_i \in S_i, i, -i \in \Psi^{\mathrm{I}}, s_i \neq s_i^*\right\} \tag{5.1}$$

Equation (5.1) shows that, at the Nash equilibrium point, if participant i is the only participant to change a strategy, then participant i will most likely be damaged by the change. In the Nash equilibrium point, therefore, the adopted strategies by all participants will be the best strategies in the game.

5.2.3 *Modeling of Game Theory in the Electricity Markets with Imperfect Competition*

In an electricity market with perfect competition, the participants have no control on market price; each participant must then increase his/her own power generation as far as the marginal cost of its generation is equal to the electricity market price. In an electricity market with imperfect competition, however, participants must evaluate the effects of their own power electrical generation on market price. In other words, each market participant must consider how his/her offered price impacts electrical power sales. Therefore, some of the market participants, the price-maker participants, could affect the market price by their decisions. The main characteristics of the game in real-world electricity markets are specified with a "√" sign in Table 5.1. Imperfect competition and strategic interactions among market participants in the electricity markets could be modeled and scrutinized by the Cournot, Stackelberg, Bertrand, and supply function equilibrium (SFE) models [10].

5.2.3.1 Cournot-Based Model and/or Playing with Quantities

The Cournot-based model is an economic model that can be used as a streamlined tool to describe the competition among several market participants [9]. In this model, each participant selects his/her quantity as a reaction to the known demand and costs, and the unknown quantities chosen by rival participants. In the Cournot-based model, the market participants make decisions regarding their own electrical power generation and allow the price be determined by the electricity market. Some of the main characteristics of the Cournot-based model within an electricity market can be described as follows:

- There are different market participants in the electricity market so that all market participants generate a homogeneous quantity (i.e., electrical power).
- There is no cooperation among market participants; that is, there is no collusion in this model.
- The decisions embraced by each market participant can affect the electricity market price, which means that all market participants have market power.
- The number of market participants in the electricity market is fixed.
- The market participants act simultaneously and compete with their rivals for generated electrical power.

The market participants seek to maximize their expected payoff, which means that they have economical and rational behavior. As previously mentioned, each market participant decides on the quantity of his/her own generated electrical power in the Cournot-based model. In simple terms, in this model, the quantity of the generated electrical power is considered as a strategic option for each market participant, as given by Eq. (5.2):

$$x_i = \rho_i; \quad \forall\{i \in \Psi^{\mathrm{I}}\} \tag{5.2}$$

As a result, the market clearing price and the maximum accepted price will be determined by using the inverse function of the demand in the electricity market. This function states the electricity market price as a function of the total traded electrical power, as given by Eq. (5.3):

$$\lambda_e = \lambda_e\left(\rho_i + \sum_{\substack{-i\in\Psi^{\mathrm{I}}\\ -i\neq i}} \rho_{-i}\right) = \lambda_e(Q); \quad \forall\{i, -i \in \Psi^{\mathrm{I}}, -i \neq i\} \tag{5.3}$$

Thus, if participant i assumes that rival participants will not adjust the quantity of their electrical power generation under the electricity market conditions, then his/her revenue will be given by Eq. (5.4):

$$\vartheta_i = \lambda_e \cdot \rho_i = \lambda_e\left(\rho_i + \sum_{\substack{-i\in\Psi^{\mathrm{I}}\\ -i\neq i}} \rho_{-i}\right) \cdot \rho_i; \quad \forall\{i, -i \in \Psi^{\mathrm{I}}, -i \neq i\} \tag{5.4}$$

The derivative of Eq. (5.4) with respect to the generated electrical power yields the marginal revenue of participant i, as given by Eq. (5.5):

$$\tilde{\vartheta}_i = \frac{\partial\left(\lambda_e\left(\rho_i + \sum_{\substack{-i\in\Psi^{\mathrm{I}}\\ -i\neq i}} \rho_{-i}\right) \cdot \rho_i\right)}{\partial \rho_i} = \lambda_e\left(\rho_i + \sum_{\substack{-i\in\Psi^{\mathrm{I}}\\ -i\neq i}} \rho_{-i}\right) + \frac{\partial\left(\lambda_e\left(\rho_i + \sum_{\substack{-i\in\Psi^{\mathrm{I}}\\ -i\neq i}} \rho_{-i}\right)\right)}{\partial \rho_i} \cdot \rho_i; \{i, -i \in \Psi^{\mathrm{I}}, -i \neq i\} \tag{5.5}$$

The Cournot-based model states that the market participants should be able to control the price such that the price is higher than the marginal cost of the generation. The difference between the price and marginal cost depends on the price elasticity of demand. Numerical results obtained from the Cournot-based model are very sensitive to the price elasticity, especially for commodities like electrical energy that has very little price elasticity. In this regard, the achieved equilibrium price via the Cournot-based model is often more than the price obtained in an actual electricity market [10].

5.2.3.2 Stackelberg Leadership-Based Model

The Stackelberg leadership-based model is a strategic economic game that was introduced by Heinrich Freiherr von Stackelberg in 1934 to develop the Cournot-based model [9]. In this model, there are two different sets of market participants: (1) leaders and (2) followers. First, the leaders simultaneously decide on the quantity of their own generated electrical power. Then, the followers analyze the adopted leaders' decisions and simultaneously select their decisions on the quantity of their own generated electrical power. In the Stackelberg leadership-based model, the quantity of the generated electrical power is considered as a strategic option for each market participant, similar to the Cournot-based model. The Stackelberg leadership-based model, however, is a sequential game, while the Cournot-based model is a simultaneous game.

5.2.3.3 Bertrand-Based Model and Playing with Prices

The Bertrand-based model is an economic model that is used to analyze interaction relationships among market participants and customers [10]. In this model, the regulatory strategy of the market participants is the offered prices; the customers, then, embrace quantities corresponding to the offered prices. The Bertrand- and Cournot-based models of the electricity market are duals of one another. In simple terms, one can go from the Cournot-based model to the Bertrand-based model by exchanging the role of the quantity (i.e., generated electrical power) and price (i.e., market price). Other characteristics of the Bertrand-based model are similar to the Cournot-based model. The offered price of each market participant is considered as a decision-making variable in the Bertrand-based model. In simple terms, in this model, the price of the generated electrical power is taken into account as a strategic option for each participant, as given by Eq. (5.6):

$$x_i = \lambda_i; \quad \forall \{ i \in \Psi^{\mathrm{I}} \} \tag{5.6}$$

Therefore, the amount of the electrical power sold by participant i is a function of its offered price and the rival participants' offered prices, according to Eq. (5.7):

$$\rho_i = \rho_i(\lambda_i, \lambda_{-i}); \quad \forall \{ i, -i \in \Psi^{\mathrm{I}}, i \neq -i \} \tag{5.7}$$

Hence, the revenue of participant i is determined using Eq. (5.8):

$$\vartheta_i = \lambda_e \cdot \rho_i = \lambda_e \cdot \rho_i(\lambda_i, \lambda_{-i}); \quad \forall \{i, -i \in \Psi^{\mathrm{I}}, i \neq -i\} \tag{5.8}$$

Participant i behaves in such a way that the rivals do not change their offered price by altering its offered price. This market participant holds the status quo until the offered price is lower than the price presented by the rivals; thus, he/she can sell any amount of electrical power according to Eq. (5.9):

$$\rho_i(\lambda_i, \lambda_{-i}) = \begin{cases} \rho_i; \text{ if } \lambda_i \leq \lambda_{-i} \\ 0; \text{ otherwise} \end{cases}; \quad \forall \{i, -i \in \Psi^{\mathrm{I}}, i \neq -i\} \tag{5.9}$$

In the uniform clearing mechanism-based electricity markets, the market price is determined in accordance with the marginal cost of the generation of the most efficient participant (i.e., the market participant with the lowest marginal price). In this situation, each of the market participants, who offer a price lower than the marginal cost of the generation of the most efficient participant, falls into a loser group. In addition, a higher price will not be tolerated, because it is beaten by the most efficient participant [10].

5.2.3.4 The Supply Function Equilibrium-Based Model

Although the Cournot-based model provides a good view of the imperfect electricity market, its performance in the electricity market brings about predictably irrational fluctuation of the prices at a high level. In the Bertrand-based model, the electricity market price is also adjusted on the marginal cost of the most efficient participant. In this regard, each participant's offer/bid of a price lower than the electricity market price encounters losses and a price higher than the electricity market price, which, again, is beaten by the most efficient participant. As a result, the SFE based model is presented in order to achieve more realistic prices, when compared to other models. Unlike the Cournot- and Bertrand-based models, in the SFE-based model, it is assumed that the delivered electrical power of participant i is a function of the electricity market price. In simple terms, the market participants are able to link their offered price with generated electrical power; therefore, the SFE-based model is the nearest model to the real behavior of the participants in an electricity market. Thus, the quantity of the electrical power that participant i intends to deliver is a function of the electricity market price, and can be described by the supply function of Eq. (5.10):

$$\rho_i = \rho_i(\lambda_e); \quad \forall \{i \in \Psi^{\mathrm{I}}\} \tag{5.10}$$

In the Cournot- and Bertrand-based models, the strategies embraced by the participants are limited to the quantity and the price of the electrical power generation, respectively. In contrast, in the SFE-based model, participant i competes with rival participants through decisions about the quantity and price of the generated electrical power simultaneously [10]. In simple terms, in this model, the quantity and the price of the generated electrical power are considered as a strategic option for each market participant. The SFE model is, therefore, the most widely used model for analyzing electricity markets, due to the ability to make connections between quantity and price of the generated electrical power and also to create a larger space of strategies for all market participants. In the Nash equilibrium point under the SFE-based model, the total electrical power demand is equal to the sum of the quantity of the generated electrical power by all electricity market participants, as given by Eq. (5.11):

$$Q(\lambda_e) = \sum_{i \in \Psi^{\mathrm{I}}} \rho_i(\lambda_e); \quad \forall\{i \in \Psi^{\mathrm{I}}\} \tag{5.11}$$

The revenue and profit of participant i are determined using Eqs. (5.12) and (5.13), respectively:

$$\vartheta_i = \lambda_e \cdot \rho_i(\lambda_e) = \lambda_e \cdot \left(Q(\lambda_e) - \sum_{\substack{-i \in \Psi^{\mathrm{I}} \\ -i \neq i}} \rho_{-i}(\lambda_e) \right); \quad \forall\{i, -i \in \Psi^{\mathrm{I}}, i \neq -i\} \tag{5.12}$$

$$\begin{aligned} \upsilon_i = \vartheta_i - \varphi_i(\rho_i(\lambda_e)) = \lambda_e \cdot \left(Q(\lambda_e) - \sum_{\substack{-i \in \Psi^{\mathrm{I}} \\ -i \neq i}} \rho_{-i}(\lambda_e) \right) \\ -\varphi_i \left(Q(\lambda_e) - \sum_{\substack{-i \in \Psi^{\mathrm{I}} \\ -i \neq i}} \rho_{-i}(\lambda_e) \right); \quad \forall\{i, -i \in \Psi^{\mathrm{I}}, i \neq -i\} \end{aligned} \tag{5.13}$$

The derivative of Eq. (5.13), with respect to the electricity market price, yields the optimal conditions, as determined using Eq. (5.14):

$$\begin{aligned} \rho_i(\lambda_e) = \left(\lambda_e - \frac{\partial \varphi_i(\rho_i)}{\partial \rho_i} \right) \\ \cdot \left(-\frac{\partial Q(\lambda_e)}{\partial \lambda_e} + \sum_{\substack{-i \in \Psi^{\mathrm{I}} \\ -i \neq i}} \frac{\partial \rho_{-i}(\lambda_e)}{\partial \lambda_e} \right); \quad \forall\{i, -i \in \Psi^{\mathrm{I}}, i \neq -i\} \end{aligned} \tag{5.14}$$

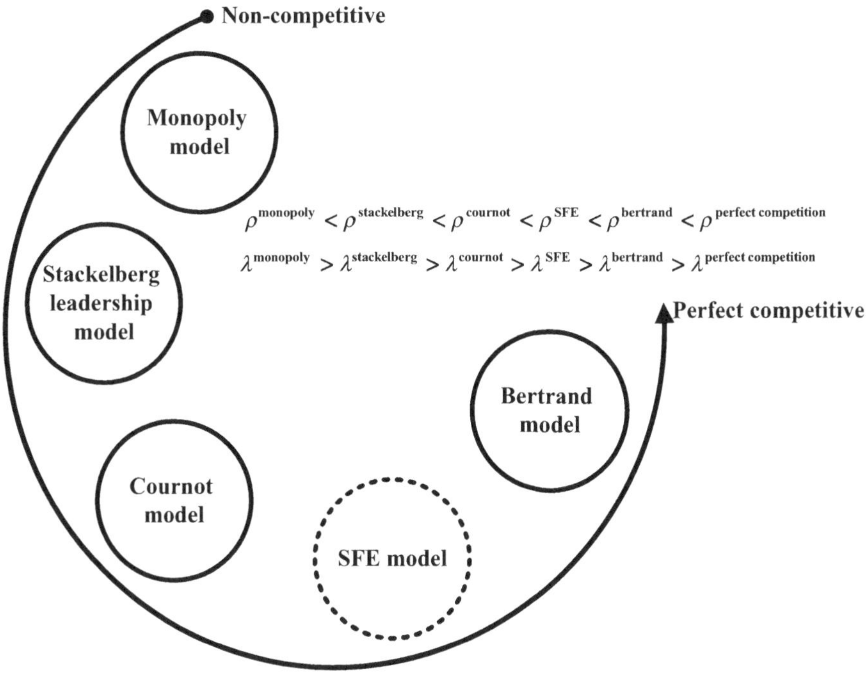

Fig. 5.1 Strategic interactions of the competition models

The answer to these equations is the equilibrium point at which all market participants will simultaneously maximize their profit.

Figure 5.1 shows a comparison of the strategic interactions of these competition models. As can be seen from this figure, the Bertrand-based model would be a more suitable model for the electricity markets, in that there is a high level of competition among market participants.

On the other hand, the Cournot-based model would be a more suitable model for the electricity markets, in that the number of the market participants is limited and there is no substantial competition among them. In the Stackelberg leadership-based model, compared with the Cournot-based model, there is also more competition among market participants. In addition, it can be seen that the SFE-based model has a mediocre level of competition among the market participants; therefore, this model, in terms of the amount of competition, would fall between the Cournot- and Bertrand-based models. This model would also be a more suitable model for the centralized electricity markets, where each market participant offers a supply/demand curve [10]. For a comprehensive overview of the application of game theory concepts, especially strategic interactions and competition models in power systems studies, please refer to the work by Kirschen and Strbac [10].

5.3 A Bilateral Bidding Mechanism in the Competitive Security-Constrained Electricity Market: A Bi-Level Computational-Logical Framework

In this section, a bi-level computational-logical framework is provided to analyze the BBM in the CSC electricity market.

The offered framework is considered as the central core for strategic tri-level generation expansion planning, strategic tri-level transmission expansion planning, and strategic quad-level coordination generation and transmission expansion planning, all of which are discussed in the next chapter. In the proposed bi-level computational-logical framework, the first level addresses the BBM problem, with the aim of maximizing the CSC market participants' profit; the second level solves the ISO's CSC electricity market clearing problem, with the aim of maximizing the community welfare function (CWF) along with minimizing the atmospheric emission cost function (AECF). The CSC electricity market serves as a platform for the participants to freely trade electrical power. The CSC market participants are comprised of two main groups: (1) the GENCOs that compete together to sell the electrical energy and (2) the DISCOs that compete with each other to buy electrical energy. Figure 5.2 illustrates a conceptual-view diagram of the proposed bi-level computational-logical framework. The proposed bi-level computational-logical framework is modeled by taking the following assumptions into account:

- Hypothesis 1: Electrical power is considered as a homogeneous commodity. This means that there is no difference in quality, variety, and so on.
- Hypothesis 2: Both GENCOs and DISCOs compete with their competitors to sell and buy the electrical energy, respectively.
- Hypothesis 3: Both GENCOs and DISCOs behave wisely. This means that they try to maximize their profits by implementation of optimal bidding strategies.
- Hypothesis 4: The GENCO's generation cost function and DISCO's benefit function represent private information; that is, the related functions are unknown by their competitors.
- Hypothesis 5: Both GENCOs and DISCOs provide their bidding strategies based on the SFE model.

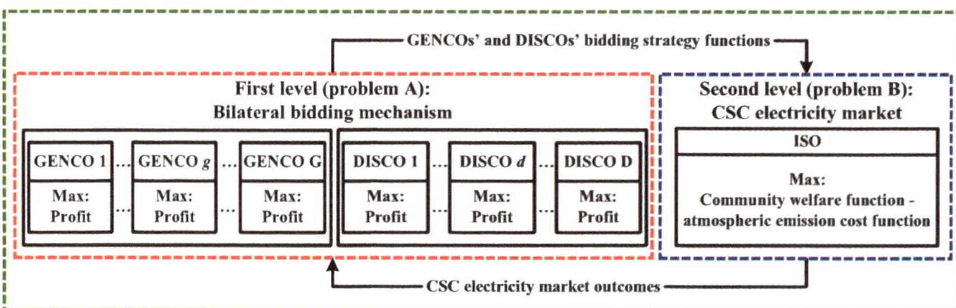

Fig. 5.2 Conceptual-view diagram of the proposed bi-level computational-logical framework

- Hypothesis 6: The operation of the CSC electricity market is based on the electrical power pool-based structure.
- Hypothesis 7: The adopted structure for the ISO is the maximal ISO (Max ISO).
- Hypothesis 8: The proposed CSC electricity market structure is cleared based on the locational marginal pricing mechanism.
- Hypothesis 9: A direct current (DC) optimal power flow method is widely employed for power flow analysis.
- Hypothesis 10: Transmission line loss is considered and applied in the power flow analysis by using the idea of an artificial load at each bus.
- Hypothesis 11: The proposed bi-level computational-logical framework considers a single hour of operation.

Table 5.2 was created as an attribute table in order to compare the most important features of the previous bidding strategy frameworks reported in the literature and the proposed bi-level computational-logical framework proposed in this chapter.

The authors have focused on most important and relevant articles in the literature. For further work on the application of bidding strategy concepts in power system studies, please refer to a review paper by Wang et al. [7].

5.3.1 Bilateral Bidding Strategy Model: First Level (Problem A)

In the BBM problem, both GENCOs and DISCOs generally submit their price bid to the ISO in terms of an hourly bid that is based on the marginal generation cost and the marginal benefit functions [$/MWh], respectively. In the CSC electricity market, the most competitive, the number of GENCOs and DISCOs is not important, as leaving a GENCO or DISCO has no effect on its performance or price. In such an environment, GENCOs and DISCOs should offer their marginal prices; otherwise, they will be removed from the CSC electricity market and/or they will obtain a smaller share of it. In real-world CSC electricity markets, however, due to the limited number of GENCOs and DISCOs and also the specific circumstances of power system operation, leaving of a GENCO or DISCO in the CSC electricity market will severely affect the CSC electricity market price and its performance. In such a structure, it is possible that neither GENCOs nor DISCOs intend to submit their real marginal generation cost and marginal benefit functions to the ISO, in order to obtain higher interest rates. In other words, GENCOs and DISCOs offer higher and lower prices than their marginal prices, respectively. In most related studies (see Table 5.2), the bidding mechanism in the CSC electricity market was designed and investigated only for GENCOs—named unilateral bidding mechanism (UBM). However, the BBM, under the operating conditions of the CSC electricity market, has not been so thoroughly investigated.

Table 5.2 Attributes of previous bidding strategy frameworks reported in the literature and the bi-level framework proposed in this study

Reference	Multilevel framework	Bidding strategy framework	Competition model	Number of strategic parameters for each participant	Modeling of real behavior of the participants	Modeling of market power as a restriction	Environmental considerations	Settlement mechanism of the CSC electricity market	Security constraint considerations	Loss considerations	Solution method
[11]	Bi-level	Unilateral: Only GENCOs	Cournot	One	No	No	No	Locational marginal pricing	Yes	No	General algebraic modeling system (GAMS)
[12]	Bi-level	Unilateral: Only GENCOs	Cournot	One	Yes	Yes	No	Locational marginal pricing	Yes	No	Particle swarm optimization (PSO)
[13]	Bi-level	Unilateral: Only GENCOs	SFE	One	Yes	No	No	Locational marginal pricing	Yes	No	GAMS
[14]	Bi-level	Unilateral: Only GENCOs	Semi-Bertrand Semi-Cournot	One	No	No	No	Uniform pricing	No	No	GAMS
[15]	Single-level	Unilateral: Only GENCOs	SFE	One	No	No	No	Uniform pricing	No	No	PSO
[16]	Bi-level	Unilateral: Only GENCOs	SFE	One	No	No	No	Locational marginal pricing	Yes	No	GAMS
[17]	Bi-level	Unilateral: Only GENCOs	Cournot	One	No	No	No	Locational marginal pricing	Yes	No	Genetic algorithm
[18]	Bi-level	Unilateral: Only GENCOs	SFE	One	No	No	No	Locational marginal pricing	Yes	No	Artificial bee colony algorithm
[19]	Bi-level	Unilateral: Only GENCOs	Cournot	Two	Yes	No	No	Locational marginal pricing	Yes	No	PSO

(continued)

Table 5.2 (continued)

Reference	Multilevel framework	Bidding strategy framework	Competition model	Number of strategic parameters for each participant	Modeling of real behavior of the participants	Modeling of market power as a restriction	Environmental considerations	Settlement mechanism of the CSC electricity market	Security constraint considerations	Loss considerations	Solution method
[20]	Single level	Unilateral: Only GENCOs	Cournot	One	No	No	No	No	No	No	Shuffled frog leaping algorithm
[21]	Bi-level	Unilateral: Only GENCOs	Cournot	One	No	No	No	Locational marginal pricing	Yes	No	GAMS
[22]	Bi-level	Unilateral: Only GENCOs	Cournot	One	No	No	No	Uniform pricing	No	No	GAMS
Proposed framework	Bi-level	Bilateral: GENCOs and DISCOs	SFE	Two	Yes	Yes	Yes: SO_2, CO_2, and NO_x atmospheric emissions	Locational marginal pricing	Yes	Yes	Three new single-objective music-inspired optimization algorithms (see Chap. 4): (a) continuous/discrete TMS-MSA, (b) TMS-EMSA, and (c) SOSA

To illustrate, most of the real-world electricity markets—such as California, New Zealand, and Spain—consider the BBM that enables DISCOs to actively participate in electricity trading by changing their normal pattern of consumption. In the CSC electricity markets with BBM, each DISCO tries to be active and compete with its competitors; it also tries to determine its competitors' bidding strategies by evaluating their strategic behaviors. As a result, it is necessary to develop a new, well-designed bilateral mechanism for bidding strategies of both GENCOs and DISCOs in the CSC electricity market.

5.3.1.1 Mathematical Model of Bidding Strategies for GENCOs

Here, it is assumed that all GENCOs are thermal electrical power generation units. In general, the fuel cost function of GENCO g is typically represented by a quadratic (i.e., second-order) polynomial function of its active power output, as shown in Eq. (5.15):

$$\hat{\varphi}_g(\rho_g) = \frac{1}{2} \cdot \hat{\alpha}_g \cdot \rho_g^2 + \hat{\beta}_g \cdot \rho_g + \hat{\gamma}_g; \quad \forall \{g \in \Psi^{\mathrm{G}}\} \tag{5.15}$$

Numerical results of Eq. (5.15) are given in terms of \$/h. In the economical operation of electrical power generation units, considering the cost of fuel as the cost of generation can lead to unrealistic profit for the GENCOs. This is due to the fact that the generation cost is composed of the other segments apart from the fuel cost, which is known as the additional costs, such as cost of labor, supplies, and maintenance [23]. These additional costs can be considered as a fixed percentage of the fuel cost. Hence, the quadratic polynomial generation cost function at operating point (ρ_g,ρ_d) can be considered for GENCO g using Eq. (5.16):

$$\varphi_g(\rho_g) = \frac{1}{2} \cdot \alpha_g \cdot \rho_g^2 + \beta_g \cdot \rho_g + \gamma_g; \quad \forall \{g \in \Psi^{\mathrm{G}}\} \tag{5.16}$$

It is worth noting that the generation cost function coefficients include the fuel cost function coefficients plus additional costs [17]. The generation cost function expresses the total payment, in terms of \$/h, for generating a certain amount of the electrical power, in terms of MW. The derivative of the quadratic generation cost function of GENCO g, with respect to ρ_g, yields the linear marginal generation cost function of this GENCO, as given by Eq. (5.17):

$$\lambda_g(\rho_g) = \alpha_g \cdot \rho_g + \beta_g; \quad \forall \{g \in \Psi^{\mathrm{G}}\} \tag{5.17}$$

This function describes the price, in terms of \$/MWh, of generating one more unit of electrical power in terms of MW. Figures 5.3a, b show graphical representations of the quadratic generation cost and linear marginal generation cost

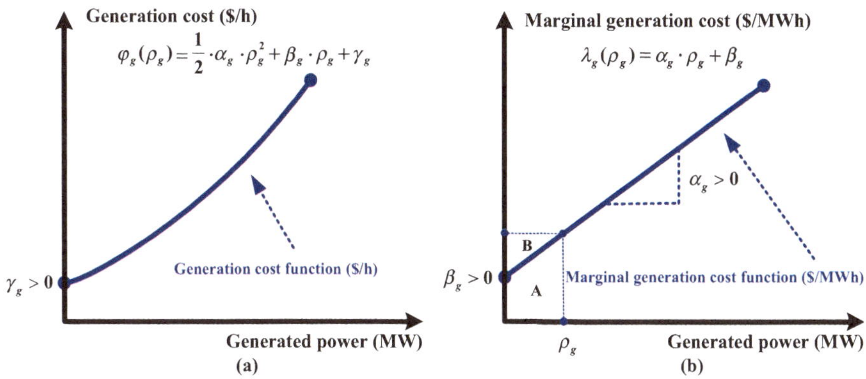

Fig. 5.3 (**a**) The quadratic generation cost function of GENCO *g*. (**b**) The linear marginal generation cost function of GENCO *g*

functions of GENCO *g*, respectively. If GENCO *g* sells a certain amount of the electrical power (i.e., ρ_g MW) at a specified price (i.e., marginal generation cost λ_g) at hour *t*, then the revenue of this GENCO at hour *t* can be calculated using Eq. (5.18):

$$\vartheta_g(\rho_g) = \rho_g \cdot [\lambda_g(\rho_g)] = \rho_g \cdot [\alpha_g \cdot \rho_g + \beta_g]; \quad \forall\{g \in \Psi^G\} \qquad (5.18)$$

This revenue is equal to the area labeled "A" plus the area labeled "B" in Fig. 5.3b. Thus, by subtracting the generation cost of GENCO *g* at hour *t* [the area labeled "A" in Fig. 5.3b, and Eq. (5.16)] from its revenue at this hour [the area labeled "A" plus area labeled "B" in Fig. 5.3b, and Eq. (5.18)], the profit of this GENCO at hour *t* can be calculated using Eq. (5.19):

$$\begin{aligned} \upsilon_g(\rho_g) &= \vartheta_g(\rho_g) - \varphi_g(\rho_g) \\ &= \rho_g \cdot [\alpha_g \cdot \rho_g + \beta_g] - \left[\frac{1}{2} \cdot \alpha_g \cdot \rho_g^2 + \beta_g \cdot \rho_g + \gamma_g\right]; \quad \forall\{g \in \Psi^G\} \end{aligned} \qquad (5.19)$$

After summarizing Eq. (5.19), the profit of GENCO *g* at hour *t* [the area labeled "B" in Fig. 5.3b] can be represented using Eq. (5.20):

$$\upsilon_g(\rho_g) = \frac{1}{2} \cdot \alpha_g \cdot \rho_g^2 - \gamma_g; \quad \forall\{g \in \Psi^G\} \qquad (5.20)$$

In this case, regardless of the bidding strategy, the profit obtained for GENCO *g* is pure profit. In simple terms, this profit is only a result of the participation in the CSC electricity market. Also, this GENCO can change its income and, consequently, its profit only by changing the amount of its generated electrical power. To determine the bidding strategy, each GENCO should take its competitors' behavior and power system operating conditions into account. As stated previously, in the CSC electricity markets, each GENCO has complete information about its bidding strategy, and has incomplete information about the strategic behaviors of its competitors that are formed based on their generation cost functions. Hence, each GENCO must estimate the generation cost function of its competitors in order to predict their strategic behaviors. In the process of the bidding strategy, GENCO g tries to hide data from its generation cost function from its competitors. Also, this GENCO tries to model the approximate generation cost function coefficients of its competitors by considering the type and capacity of the power plant, the type of fuel, statistical and empirical information from the history of the electricity market, etc. In simple terms, GENCO g simulates the strategic behavior of its competitors and then determines its own bidding strategy. In the proposed BBM, GENCO g modifies its marginal generation cost function with two bidding strategy parameters: 1) slope (i.e.,$\xi_{1,g} > 1$) and 2) intercept (i.e., $\xi_{2,g} > 1$). Therefore, the bidding strategy function of GENCO g is formed by multiplying its marginal generation cost function coefficients in its two strategic parameters using Eq. (5.21):

$$\tilde{\lambda}_g(\rho_g) = \xi_{1,g} \cdot \alpha_g \cdot \rho_g + \xi_{2,g} \cdot \beta_g; \quad \forall \{g \in \Psi^{G}\} \tag{5.21}$$

Figure 5.4 illustrates the actual marginal generation cost function—the blue solid line—and the bidding strategy function—the red dotted line—of GENCO g. If GENCO g sells a certain amount of electrical power (i.e., ρ_g MW) at a specified price (i.e., $\tilde{\lambda}_g$) at hour t, then the revenue of this GENCO at hour t can be calculated using Eq. (5.22):

$$\tilde{\vartheta}_g(\rho_g) = \rho_g \cdot \tilde{\lambda}_g(\rho_g) = \rho_g \cdot [\xi_{1,g} \cdot \alpha_g \cdot \rho_g + \xi_{2,g} \cdot \beta_g]; \quad \forall \{g \in \Psi^{G}\} \tag{5.22}$$

This revenue is equal to the sum or the areas labeled "A" through "D" in Fig. 5.4.

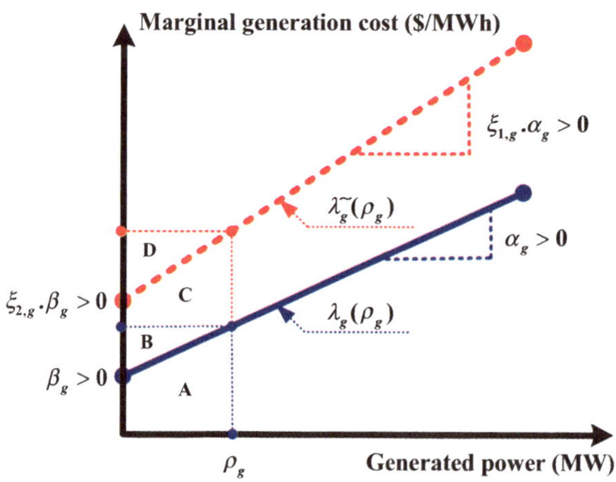

Fig. 5.4 Actual marginal generation cost function and bidding strategy function of GENCO g

Thus, by subtracting the generation cost of GENCO g at hour t [the area labeled "A" in Fig. 5.4, and Eq. (5.16)] from its revenue at this hour [the area labeled "A" plus area labeled "B" plus area labeled "C" plus area labeled "D" in Fig. 5.4, and Eq. (5.22)], the profit of this GENCO at hour t can be calculated using Eq. (5.23):

$$\begin{aligned}\tilde{\upsilon}_g(\rho_g) = \tilde{\vartheta}_g(\rho_g) - \varphi_g(\rho_g) = \rho_g \cdot \left[\xi_{1,g} \cdot \alpha_g \cdot \rho_g + \xi_{2,g} \cdot \beta_g\right] \\ - \left[\frac{1}{2} \cdot \alpha_g \cdot \rho_g^2 + \beta_g \cdot \rho_g + \gamma_g\right]; \quad \forall \{g \in \Psi^{\mathrm{G}}\}\end{aligned} \tag{5.23}$$

After summarizing Eq. (5.23), the profit of GENCO g at hour t—the sum of the areas labeled "B" through "D" in Fig. 5.4—can be expressed using Eq. (5.24):

$$\tilde{\upsilon}_g(\rho_g) = \frac{(2\xi_{1,g} - 1)}{2} \cdot \alpha_g \cdot \rho_g^2 + (\xi_{2,g} - 1) \cdot \beta_g \cdot \rho_g - \gamma_g; \quad \forall \{g \in \Psi^{\mathrm{G}}\} \tag{5.24}$$

In this case, considering the bidding strategy, the obtained profit for GENCO g is a mixed profit. In simple terms, a mixed profit arises from participation in the CSC electricity market and adoption of the optimal bidding strategies. Also, this GENCO can alter its income and, consequently, its profit by changing the amount of its generated electrical power and its two strategic bidding parameters.

5.3.1.2 Mathematical Model of a Bidding Strategy for DISCOs

Similarly, the quadratic (i.e., second-order) benefit function at operating point (ρ_g, ρ_d) can be considered for DISCO d, as given by Eq. (5.25):

$$\varphi_d(\rho_d) = \frac{1}{2} \cdot \alpha_d \cdot \rho_d^2 + \beta_d \cdot \rho_d + \gamma_d; \quad \forall \{d \in \Psi^{\mathrm{D}}\} \tag{5.25}$$

The derivative of the quadratic benefit function of DISCO d, with respect to ρ_d, yields the linear marginal benefit function of this DISCO, as given by Eq. (5.26):

$$\lambda_d(\rho_d) = \alpha_d \cdot \rho_d + \beta_d; \quad \forall \{d \in \Psi^{\mathrm{D}}\} \tag{5.26}$$

Figures 5.5a, b show graphical representations of the quadratic benefit and linear marginal benefit functions of DISCO d, respectively. For the same reason, if DISCO d purchases a certain amount of the electrical power (ρ_d MW) at a specified price (λ_d) at hour t, then the payment of this DISCO at hour t can be calculated using Eq. (5.27):

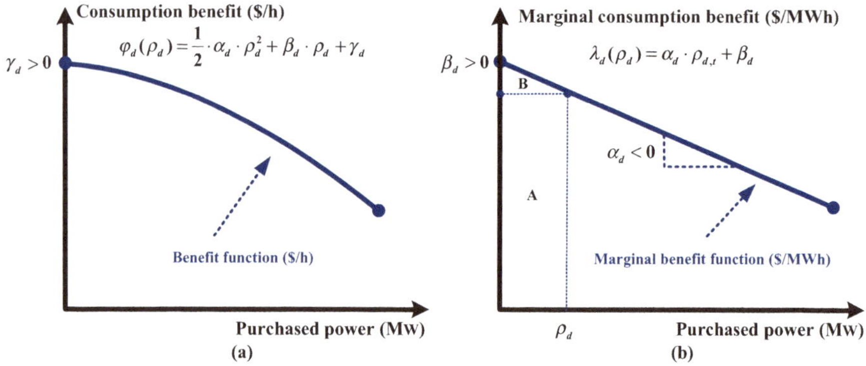

Fig. 5.5 (**a**) The quadratic benefit function of DISCO *d*. (**b**) The linear marginal benefit function of DISCO *d*

$$\varsigma_d(\rho_d) = \rho_d \cdot \lambda_d(\rho_d) = \rho_d \cdot [\alpha_d \cdot \rho_d + \beta_d]; \quad \forall\{d \in \Psi^{\mathrm{D}}\} \tag{5.27}$$

This payment is equal to the area labeled "A" in Fig. 5.5b. Thus, by subtracting the payment of DISCO *d* at hour *t* [the area labeled "A" in Fig. 5.5b, and Eq. (5.27)] from its benefit at this hour [the area labeled "A" plus the area labeled "B" in Fig. 5.5b, and Eq. (5.25)], the profit of this DISCO at hour *t* can be calculated using Eq. (5.28):

$$\begin{aligned} \upsilon_d(\rho_d) &= \varphi_d(\rho_d) - \varsigma_d(\rho_d) \\ &= \left[\frac{1}{2} \cdot \alpha_d \cdot \rho_d^2 + \beta_d \cdot \rho_d + \gamma_d\right] - \rho_d \cdot [\alpha_d \cdot \rho_d + \beta_d]; \quad \forall\{d \in \Psi^{\mathrm{D}}\} \end{aligned} \tag{5.28}$$

After summarizing Eq. (5.28), the profit of DISCO *d* at hour *t* can be expressed using Eq. (5.29):

$$\upsilon_d(\rho_d) = -\frac{1}{2} \cdot \alpha_d \cdot \rho_d^2 + \gamma_d; \quad \forall\{d \in \Psi^{\mathrm{D}}\} \tag{5.29}$$

In this case, and regardless of the bidding strategy, the profit obtained for DISCO *d* is pure profit. In simple terms, this profit is only a result of the participation in the CSC electricity market. Also, this DISCO can change its payment and, consequently, its profit simply by changing the amount of its purchased electrical power. Similar to the previous discussion for the GENCOs, in order to determine the bidding strategy, each DISCO also simulates the behavior of its competitors and then determines its own bidding strategy function. Thus, each DISCO tries to model the strategic behavior of its competitors in the CSC electricity market by considering their marginal benefit functions. In the proposed

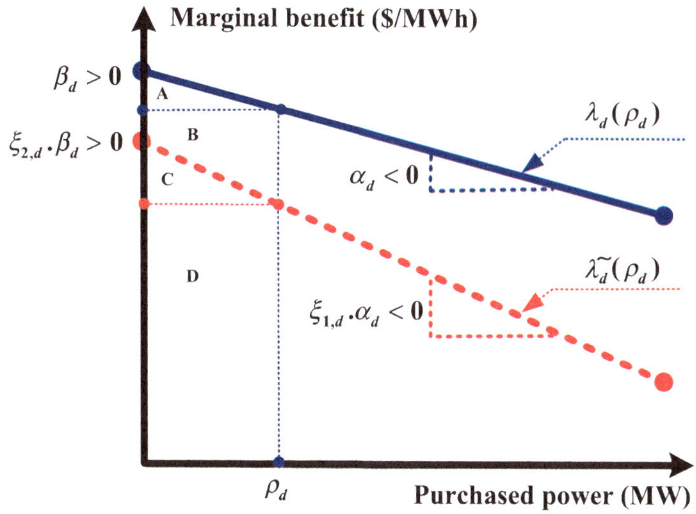

Fig. 5.6 Actual marginal benefit function and bidding strategy function of DISCO *d*

BBM, DISCO d modifies its marginal benefit function with two bidding strategy parameters: 1) slope ($\xi_{1,\ d} > 1$) and 2) intercept ($\xi_{2,\ d} < 1$).

Hence, the bidding strategy function of DISCO d is formed by multiplying its marginal benefit function with its strategic bidding parameters, as given by Eq. (5.30):

$$\tilde{\lambda}_d(\rho_d) = \xi_{1,d} \cdot \alpha_d \cdot \rho_d + \xi_{2,d} \cdot \beta_d; \quad \forall \{d \in \Psi^{\mathrm{D}}\} \tag{5.30}$$

Figure 5.6 illustrates the actual marginal benefit function—the blue solid line—and the bidding strategy function—the red dotted line—of DISCO d.

If DISCO d purchases a certain amount of electrical power (ρ_d MW) at a specific price ($\tilde{\lambda}_d$) at hour t, then the payment of this DISCO at hour t can be calculated using Eq. (5.31):

$$\tilde{\varsigma}_d(\rho_d) = \rho_d \cdot \tilde{\lambda}_d(\rho_d) = \rho_d \cdot \left[\xi_{1,d} \cdot \alpha_d \cdot \rho_d + \xi_{2,d} \cdot \beta_d\right]; \quad \forall \{d \in \Psi^{\mathrm{D}}\} \tag{5.31}$$

This payment is equal to the area labeled “D” in Fig. 5.6. Thus, by subtracting the payment of DISCO d at hour t [the area labeled “D” in Fig. 5.6, and Eq. (5.31)] from its benefit at this hour [the sum of the areas labeled “A” through “D” in Fig. 5.6, and Eq. (5.25)], the profit of this DISCO at hour t can be calculated using Eq. (5.32):

$$\tilde{\upsilon}_d(\rho_d) = \varphi_d(\rho_d) - \tilde{\varsigma}_d(\rho_d) = \left[\frac{1}{2}\cdot\alpha_d\cdot\rho_d^2 + \beta_d\cdot\rho_d + \gamma_d\right] \\ -\rho_d\cdot\left[\xi_{1,d}\cdot\alpha_d\cdot\rho_d + \xi_{2,d}\cdot\beta_d\right];\quad \forall\{d\in\Psi^{\mathrm{D}}\} \tag{5.32}$$

After summarizing Eq. (5.32), the profit of DISCO d at hour t can be described using Eq. (5.33):

$$\tilde{\upsilon}_d(\rho_d) = \frac{\left(1-2\xi_{1,d}\right)}{2}\cdot\alpha_d\cdot\rho_d^2 + \left(1-\xi_{2,d}\right)\cdot\beta_d\cdot\rho_d + \gamma_d;\quad \forall\{d\in\Psi^{\mathrm{D}}\} \tag{5.33}$$

In this case, considering the bidding strategy, the profit obtained for DISCO d is a mixed profit. In simple terms, a mixed profit results from participation in the CSC electricity market and adoption of optimal bidding strategies. Also, this DISCO can alter its payment and, consequently, its profit by changing the amount of its purchased electrical power and its two bidding strategy parameters.

5.3.2 Security-Constrained Electricity Market Model: Second Level (Problem B)

There is not a recognizable single pattern used by all countries for electricity market structure, due to the differences in regulatory policies and economic rules. However, all implemented electricity markets in the various countries can be classified into three main structures, as follows:

- Electrical power pool-based structure and/or centralized electricity market: In this structure, making connections between GENCOs and DISCOs in the electricity market is the responsibility of the ISO, and they cannot directly interact with each other.
- Bilateral contract structure and/or decentralized electricity market: In this structure, both GENCOs and DISCOs, freely and independently of the ISO, enter into bilateral contracts to sell and buy electrical energy. However, ISO confirmation is required in order to complete the bilateral contracts. This means that the ISO must confirm whether or not there is enough transmission capacity to fulfill the transactions and keep the security constraints of the power systems.
- Hybrid structure: In this structure, different characteristics of two previous structures can be employed simultaneously. In the hybrid structure, parallel with the electrical power pool-based structure, GENCOs and DISCOs can be entered into bilateral contracts. In simple terms, the operation of the electricity market, based on the electrical power pool-based structure, is not obligatory, and each GENCO or DISCO would be authorized to negotiate agreements directly with each other and/or to choose to accept the nodal price of the centralized electricity market in order to sell or buy electrical energy, respectively.

In the proposed bi-level computational-logical framework, the operation of the CSC electricity market is carried out based on the electrical power pool-based structure. A discussion of the general features of the different structures of the electricity market is outside the scope of this book. For a full overview of this field, please refer to the studies by Shahidehpour et al. [24] and Kirschen and Strbac [10]. Also, depending on the ISO's targets and authority, two possible structures can be considered for the ISO: (1) MaxISO and (2) minimal ISO (MinISO). As stated previously, in the proposed bi-level computational-logical framework, the adopted structure for the ISO is the MaxISO. For a thorough discussion on this topic, please refer the work by Shahidehpour et al. [24]. In the proposed bi-level computational-logical framework, the locational marginal pricing mechanism—also called nodal pricing or spot pricing—is adopted for CSC electricity market settlement. This mechanism is one of the most efficient mechanisms for determining the price of electrical power and scheduling the electrical power dispatch in the deregulated power system environment, which operates as a pool. This mechanism is complex but critical to ensuring a CSC electricity market's smooth and healthy operation. Most recently, many of the deregulated electricity markets—such as the Pennsylvania/New Jersey/Maryland (PJM) interconnection; the electric reliability council of Texas (ERCOT); the New York ISO (NYISO); ISO New England (ISO-NE); the California ISO (CAISO); the Midcontinent ISO (MISO) in the United States; the national electricity market management company limited (NEMMCO); New Zealand; and Singapore—have employed the locational marginal pricing mechanism in their power systems. Due to successful implementation of this mechanism, other electricity markets in different areas of the world are paving the way for the utilization of locational marginal pricing in their power systems. Nodal pricing, or locational marginal pricing (LMP), represents the lowest cost of the next—not the previous—electrical power supplied in terms of MW at a specified location—node—of the power system, taking into account both GENCOs' and DISCOs' bids and operational constraints. In general, this value consists of three major components: (1) the cost of the generated electrical power; (2) the cost of losses; and (3) the cost of congestion. For a comprehensive overview of the locational marginal pricing mechanism in the electricity market, please refer the studies by Shahidehpour et al. [24] and Litvinov [25]. In the power system literature, the locational marginal pricing mechanism can be broadly derived into two distinct models [24]: (1) the alternating current (AC) optimal power flow model and (2) the DC optimal power flow model. Many of the electricity markets in different countries have already implemented the DC optimal power flow model in order to handle the electricity market clearing process. Here, to limit computational burden to an acceptable level without losing the generality of the proposed bi-level computational-logical framework, the DC optimal power flow model is widely employed for power flow analysis. Please refer to the work by Christie et al. [26] for a comprehensive discussion of DC optimal power flow. In addition, transmission line loss is considered and applied in the power flow analysis by using the idea of an artificial load at each bus. Readers refer to [24, 27] for a thorough discussion on the artificial load concept. In the proposed bi-level computational-logical framework, both

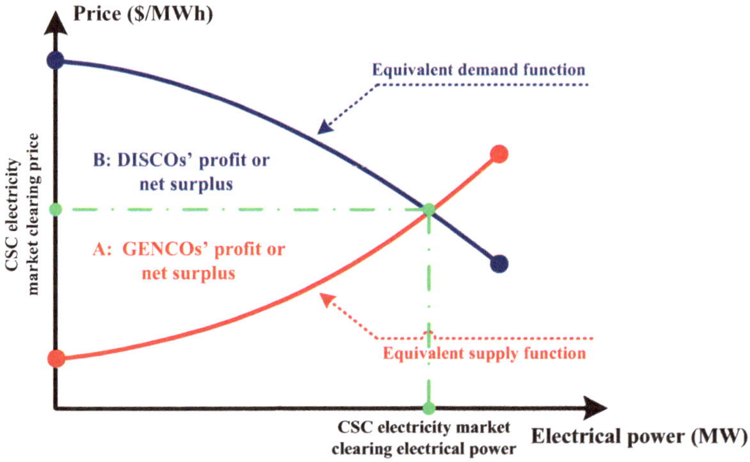

Fig. 5.7 Equivalent supply and demand functions

GENCOs and DISCOs submit their bidding strategy functions to the ISO. After the ISO receives these functions, it aggregates the bidding strategy functions to form an equivalent supply function and an equivalent demand function (see Fig. 5.7). Then, the ISO clears the CSC electricity market with the aim of maximizing CWF and minimizing AECF, subject to the technical constraints of the power system, which yields the LMPs. The CWF represents the profit of the CSC electricity market participants, which has been formed by two main parts: the GENCOs' profit or net surplus, and the DISCOs' profit or net surplus, as shown in Fig. 5.6. Hence, the mathematical expression of the CWF is illustrated by Eq. (5.34):

$$\begin{aligned}\mathrm{OF}_1^{\text{CSC-EM}} &= \mathrm{CWF} = \sum_{g\in\Psi^{\mathrm{G}}} \tilde{\upsilon}_g(\rho_g) + \sum_{d\in\Psi^{D}} \tilde{\upsilon}_d(\rho_d)\\ &= \sum_{g\in\Psi^{\mathrm{G}}} \frac{(2\xi_{1,g}-1)}{2}\cdot\alpha_g\cdot\rho_g^2 + (\xi_{2,g}-1)\cdot\beta_g\cdot\rho_g - \gamma_g\\ &+ \sum_{d\in\Psi^{\mathrm{D}}} \frac{(1-2\xi_{1,d})}{2}\cdot\alpha_d\cdot\rho_d^2 + (1-\xi_{2,d})\cdot\beta_d\cdot\rho_d + \gamma_d;\quad \forall\{g\in\Psi^{\mathrm{G}}, d\in\Psi^{\mathrm{D}}\}\end{aligned} \tag{5.34}$$

Electrical power generation by the thermal units can bring about significant destructive environmental impacts by entering the atmosphere through emissions such as sulfur dioxide (SO_2), carbon dioxide (CO_2), nitrogen oxide (NO_x), solid particles, and mercury. In recent years, power system planners and regulators take into account environmental issues during the planning process by estimating the costs associated with destructive environmental influences. In contrast, though, these issues are overlooked in the operational process. This negligence can lead to nonoptimal results that do not meet predetermined atmospheric emission levels, according to national and international environmental protocols. Hence, the authors intend to address environmental goals in the upper level of the proposed bi-level computational-logical framework. The SO_2 atmospheric emission function (AEF) of GENCO g represents the atmospheric emission of SO_2 in the electrical power generation process at hour t, in terms of ton/h. The SO_2 AEF is directly related to the amount of fuel consumed by GENCO g at this hour and, therefore, can be formulated as a quadratic polynomial function. In simple terms, this would be the same form as the fuel cost function [28–30]. Then, the SO_2 AECF of this GENCO is formed by multiplying the AEF by the price of SO_2 atmospheric emissions, in terms of \$/ton. Hence, the SO_2 AECF of GENCOs can be expressed using Eq. (5.35):

$$\mathrm{AECF}_{\mathrm{SO_2}} = \sum_{g \in \Psi^{\mathrm{G}}} \pi_g^{\mathrm{SO_2}} \cdot \left(\alpha_g^{\mathrm{SO_2}} \cdot \rho_g^2 + \beta_g^{\mathrm{SO_2}} \cdot \rho_g + \gamma_g^{\mathrm{SO_2}} \right); \quad \forall \{ g \in \Psi^{\mathrm{G}} \} \tag{5.35}$$

The CO_2 AEF of GENCO g describes the atmospheric emission of CO_2 in the electrical power generation process at hour t, in terms of ton/h. Similar to the SO_2, the CO_2 AEF is also directly dependent upon fuel consumption of GENCO g at this hour and, therefore, can be formulated as a quadratic polynomial function. Then, the CO_2 AECF of this GENCO is formed by multiplying the AEF with the price of CO_2 atmospheric emissions, in terms of \$/ton. Hence, the CO_2 AECF of GENCOs can be formulated using Eq. (5.36). For a comprehensive overview of the influence of CO_2 on generation scheduling, please refer to the work by Kockar et al. [31].

$$\mathrm{AECF}_{\mathrm{CO_2}} = \sum_{g \in \Psi^{\mathrm{G}}} \pi_g^{\mathrm{CO_2}} \cdot \left(\alpha_g^{\mathrm{CO_2}} \cdot \rho_g^2 + \beta_g^{\mathrm{CO_2}} \cdot \rho_g + \gamma_g^{\mathrm{CO_2}} \right); \quad \forall \{ g \in \Psi^{\mathrm{G}} \} \tag{5.36}$$

The NO_x AEF of GENCO g indicates the atmospheric emission of NO_x in the electrical power generation process at hour t, in terms of ton/h. Unlike the atmospheric emission of SO_2 and CO_2, the atmospheric emission of NO_x is primarily related to several factors involved in the combustion process; these include boiler temperature, air content, etc. Hence, the atmospheric emission of NO_x generally cannot be modeled by a quadratic polynomial function. Instead, an exponential function is adopted in order to model the atmospheric emission of NO_x [28–30]. In this regard, the NO_x AECF of GENCO g is formed by multiplying its AEF by the price of the atmospheric emission of NO_x, in terms of \$/ton. The NO_x AECF of GENCOs can be described using Eq. (5.37):

$$\mathrm{AECF}_{\mathrm{NO_x}} = \sum_{g \in \Psi^{\mathrm{G}}} \pi_g^{\mathrm{NO_x}} \cdot \left(\chi_g^{\mathrm{NO_x}} \cdot \exp\left(\delta_g^{\mathrm{NO_x}} \cdot \rho_g \right) \right); \quad \forall \{ g \in \Psi^{\mathrm{G}} \} \tag{5.37}$$

Finally, total AECF is determined using Eq. (5.38):

$$\mathrm{OF}_2^{\mathrm{CSC\text{-}EM}} = \mathrm{AECF} = \mathrm{W}_{\mathrm{SO}_2} \cdot \mathrm{AECF}_{\mathrm{SO}_2} + \mathrm{W}_{\mathrm{CO}_2} \cdot \mathrm{AECF}_{\mathrm{CO}_2} + \mathrm{W}_{\mathrm{NO}_x} \cdot \mathrm{AECF}_{\mathrm{NO}_x} \tag{5.38}$$

It is worth noting that it is possible that some of the atmospheric emissions have higher importance from a regulator's point of view. Hence, in the AECF, a weighting coefficient is considered for each atmospheric emission; that way, the determination of these coefficients is the responsibility of the regulator.

5.3.3 Overview of the Bi-Level Computational-Logical Framework

According to the equations identified in the previous sections, the proposed bi-level computational-logical framework can be summarized as follows:

$$\underset{x_{\mathrm{BBM}}}{\mathrm{Max}} : \tilde{\upsilon}_g(\rho_g) = \frac{(2\xi_{1,g} - 1)}{2} \cdot \alpha_g \cdot \rho_g^2 + (\xi_{2,g} - 1) \cdot \beta_g \cdot \rho_g - \gamma_g; \quad \forall\{g \in \Psi^{\mathrm{G}}\} \tag{5.39}$$

$$\underset{x_{\mathrm{BBM}}}{\mathrm{Max}} : \tilde{\upsilon}_d(\rho_d) = \frac{(1 - 2\xi_{1,d})}{2} \cdot \alpha_d \cdot \rho_d^2 + (1 - \xi_{2,d}) \cdot \beta_d \cdot \rho_d + \gamma_d; \quad \forall\{d \in \Psi^{\mathrm{D}}\} \tag{5.40}$$

Subject to:

$$\sum_{g \in \Psi^{\mathrm{G}}} \rho_g = \sum_{l \in \Psi^{\mathrm{L}}} \kappa_l + \sum_{d \in \Psi^{\mathrm{D}}} \rho_d; \quad \forall\{g \in \Psi^{\mathrm{G}}, l \in \Psi^{\mathrm{L}}, d \in \Psi^{\mathrm{D}}\} \tag{5.41}$$

$$\rho_g^{\min} \le \rho_g \le \rho_g^{\max}; \quad \forall\{g \in \Psi^{\mathrm{G}}\} \tag{5.42}$$

$$\rho_d^{\min} \le \rho_d \le \rho_d^{\max}; \quad \forall\{d \in \Psi^{\mathrm{D}}\} \tag{5.43}$$

$$\xi_{1,g}^{\min} \le \xi_{1,g} \le \xi_{1,g}^{\max}; \quad \forall\{g \in \Psi^{\mathrm{G}}\} \tag{5.44}$$

$$\xi_{2,g}^{\min} \le \xi_{2,g} \le \xi_{2,g}^{\max}; \quad \forall\{g \in \Psi^{\mathrm{G}}\} \tag{5.45}$$

$$\xi_{1,d}^{\min} \le \xi_{1,d} \le \xi_{1,d}^{\max}; \quad \forall\{d \in \Psi^{\mathrm{D}}\} \tag{5.46}$$

$$\xi_{2,d}^{\min} \le \xi_{2,d} \le \xi_{2,d}^{\max}; \quad \forall\{d \in \Psi^{\mathrm{D}}\} \tag{5.47}$$

$$\begin{aligned}
&\underset{x_{\text{CSC-EM}}}{\text{Max}}:\left\{\text{OF}^{\text{CSC-EM}}\right\}=\underset{x_{\text{CSC-EM}}}{\text{Max}}:\left\{\text{W}_{\text{OF}_1}^{\text{CSC-EM}}\cdot\text{OF}_1^{\text{CSC-EM}}-\text{W}_{\text{OF}_2}^{\text{CSC-EM}}\times\text{OF}_2^{\text{CSC-EM}}\right\}\\
&=\underset{x_{\text{CSC-EM}}}{\text{Max}}:\left\{\text{W}_{\text{OF}_1}^{\text{CSC-EM}}\cdot\left(\sum_{g\in\Psi^{\text{G}}}\tilde{\upsilon}_g(\rho_g)+\sum_{d\in\Psi^{\text{D}}}\tilde{\upsilon}_d(\rho_d)\right)\right.\\
&\left.-\text{W}_{\text{OF}_2}^{\text{CSC-EM}}\cdot\left(\text{W}_{\text{SO}_2}\cdot\text{AECF}_{\text{SO}_2}+\text{W}_{\text{CO}_2}\cdot\text{AECF}_{\text{CO}_2}+\text{W}_{\text{NO}_x}\cdot\text{AECF}_{\text{NO}_x}\right)\right\};\\
&\forall\left\{g\in\Psi^{\text{G}},d\in\Psi^{\text{D}}\right\}\\
&=\underset{x_{\text{CSC-EM}}}{\text{Max}}:\left\{\text{W}_{\text{OF}_1}^{\text{CSC-EM}}\cdot\left(\sum_{g\in\Psi^{\text{G}}}\frac{(2\xi_{1,g}-1)}{2}\cdot\alpha_g\cdot\rho_g^2+(\xi_{2,g}-1)\cdot\beta_g\cdot\rho_g-\gamma_g\right.\right.\\
&\left.+\sum_{d\in\Psi^{\text{D}}}\frac{(1-2\xi_{1,d})}{2}\cdot\alpha_d\cdot\rho_d^2+(1-\xi_{2,d})\cdot\beta_d\cdot\rho_d+\gamma_d\right)\\
&-\text{W}_{\text{OF}_2}^{\text{CSC-EM}}\cdot\left(\text{W}_{\text{SO}_2}\cdot\left(\sum_{g\in\Psi^{\text{G}}}\pi_g^{\text{SO}_2}\cdot\left(\alpha_g^{\text{SO}_2}\cdot\rho_g^2+\beta_g^{\text{SO}_2}\cdot\rho_g+\gamma_g^{\text{SO}_2}\right)\right)\right.\\
&-\text{W}_{\text{CO}_2}\cdot\left(\sum_{g\in\Psi^{\text{G}}}\pi_g^{\text{CO}_2}\cdot\left(\alpha_g^{\text{CO}_2}\cdot\rho_g^2+\beta_g^{\text{CO}_2}\cdot\rho_g+\gamma_g^{\text{CO}_2}\right)\right)\\
&\left.\left.-\text{W}_{\text{NO}_x}\cdot\left(\sum_{g\in\Psi^{\text{G}}}\pi_g^{\text{NO}_x}\cdot\left(\chi_g^{\text{NO}_x}\cdot\exp\left(\delta_g^{\text{NO}_x}\cdot\rho_g\right)\right)\right)\right)\right\};\\
&\forall\left\{g\in\Psi^{\text{G}},d\in\Psi^{\text{D}}\right\}
\end{aligned}\tag{5.48}$$

Subject to:

$$\sum_{g\in\Psi^{\text{G}}}\rho_g=\sum_{l\in\Psi^{\text{L}}}\kappa_l+\sum_{d\in\Psi^{\text{D}}}\rho_d;\quad\forall\left\{g\in\Psi^{\text{G}},l\in\Psi^{\text{L}},d\in\Psi^{\text{D}}\right\}\tag{5.49}$$

$$\rho_g^{\min}\leq\rho_g\leq\rho_g^{\max};\quad\forall\left\{g\in\Psi^{\text{G}}\right\}\tag{5.50}$$

$$\rho_d^{\min}\leq\rho_d\leq\rho_d^{\max};\quad\forall\left\{d\in\Psi^{\text{D}}\right\}\tag{5.51}$$

$$f_l^{\min}\leq f_l\leq f_l^{\max};\quad\forall\left\{l\in\Psi^{\text{L}}\right\}\tag{5.52}$$

Equations (5.39) and (5.40) illustrate the profit of GENCO g and the profit of DISCO d—objectives of the first level—respectively. The total generated electrical power by GENCOs must cover the total purchased electrical power by DISCOs plus network electrical power losses. Hence, Eq. (5.41) represents the network electrical power balance. The generated electrical power by each GENCO must satisfy the predetermined bounds. Similarly, the electrical power purchased by each DISCO must meet the predetermined bounds. Hence, Eqs. (5.42) and (5.43) show the lower and upper bounds related to the generated electrical power by each GENCO and the purchased electrical power by each DISCO, respectively. It is important to note that, in most real-world electricity markets, to prevent the application of market power in the excessive rise in prices, the price ceiling is considered for the bidding strategy parameters of both GENCOs and DISCOs. Hence, Eqs. (5.44) and (5.45) show the lower and upper bounds related to the slope and intercept parameters of the bidding strategy function of GENCO g, respectively. Similarly, Eqs. (5.46) and (5.47) illustrate the lower and upper bounds related to the slope and intercept parameters of the bidding strategy function of DISCO d, respectively. Equation (5.48) illustrates the objective of the second level, which has four parts: (1) the CWF; (2) the SO_2 AECF; (3) the CO_2 AECF; and, (4) the NO_x AECF. Equations (5.49) to (5.51) are similar to Eqs. (5.41) to (5.43) in the first level of the proposed bi-level computational-logical framework. Occurrence of an overload situation in the transmission lines can lead to system collapse in an extreme case. In this regards, Eq. (5.52) demonstrates the security restriction imposed on the transmission lines.

5.3.4 Solution Method and Implementation Considerations

In the proposed bi-level computational-logical framework, the decision-making variables of the solution vector of the single-objective optimization algorithm are (1) the generated electrical power by GENCOs; (2) the slope and intercept parameters of the bidding strategy function of GENCOs; (3) the purchased electrical power by DISCOs; and, (4) the slope and intercept parameters of the bidding strategy function of DISCOs. As previously discussed, in the first level of the bi-level computational-logical framework, both GENCOs and DISCOs try to choose the optimal bidding strategies in the CSC electricity market in order to maximize their profit. In the beginning, the marginal generation cost function of each GENCO and the marginal benefit function of each DISCO are set as their initial bidding strategy functions, respectively. This means that the bidding strategy parameters for all CSC electricity market participants are set to 1, though only for the first iteration. Then, using Eqs. (5.39) to (5.47), each CSC electricity market participant solves the bidding strategy problem to calculate and modify its bidding strategy parameters through the single-objective optimization algorithm while keeping the bidding strategy parameters of rivals. This process is repeated until the Nash equilibrium point is computed. In general, two stopping criteria can be considered for the first level of the bi-level computational-logical framework: (1) tolerance of profit changes and (2) calculation steps of the bidding strategy parameters. Here, to provide an appropriate convergence time, the stopping criterion is the tolerance of profit changes which is considered to be 0.1. A game theory problem may reach one Nash equilibrium point, multiple Nash equilibrium points, or no Nash equilibrium point. In view of real-world power system limitations, a Nash equilibrium may fail to exist. These limitations give rise to discontinuity in optimal behavior of a market participant relative to its rivals' bids. If the proposed bi-level computational-logical framework has no Nash equilibrium point, the adopted stopping criterion will not converge. On the other hand, binding mechanism restrictions may lead to multiple Nash equilibrium points. When a stopping criterion is fulfilled and the first Nash equilibrium point is obtained, the process will continue for a certain number of iterations. If the Nash equilibrium point does not change, the bi-level computational-logical framework will have only one Nash equilibrium point.

Otherwise, the framework will have multiple Nash equilibrium points, wherein all Nash equilibrium points are computed by repeating the process. If there are multiple Nash equilibrium points in the bi-level computational-logical framework, the profit of each CSC electricity market participant in each Nash equilibrium point will be different from the others. In this condition, among all obtained Nash equilibrium points, the Nash equilibrium point with the highest level of profit for the CSC electricity market participants is selected and the other Nash equilibrium points are discarded. After calculation of the Nash equilibrium point, the optimal bidding strategy functions of both GENCOs and DISCOs will be achieved. These optimal bidding strategy functions are then transferred to the second level. According to Fig. 5.8, the ISO aggregates these functions to form an equivalent supply function and an equivalent demand function. If the obtained optimal bidding strategies by both GENCOs and DISCOs generate a violation in the network constraints, the ISO solves the CSC electricity market clearing problem, with the aim of maximizing CWF and minimizing AECF to yield the LMPs. After determining the LMPs, the profit of the CSC electricity market participants, including GENCOs and DISCOs, is modified. Figure 5.8 shows the flowchart of the bi-level computational-logical framework. In the figure, the circles labeled "A" and "B" are the input and the output of the single-objective optimization algorithms for solving the optimization problem, respectively.

5.3.5 *Simulation Results and Case Studies*

Figure 5.9 shows the modified six-machine eight-bus test network that was chosen to test the proposed bi-level computational-logical framework.

This is a widely used test network in power system studies, especially for bidding strategy problems [11, 17, 24, 32]. The full data for this test network is given in Appendix 3 of this chapter. Table 5.14 shows a tabulation of the information for the transmission lines of the test network. Tables 5.15 and 5.16 show the parameters associated with GENCOs and DISCOs, respectively. Table 5.17 gives the coefficients of the SO_2, CO_2, and NO_x atmospheric emission functions, which were taken

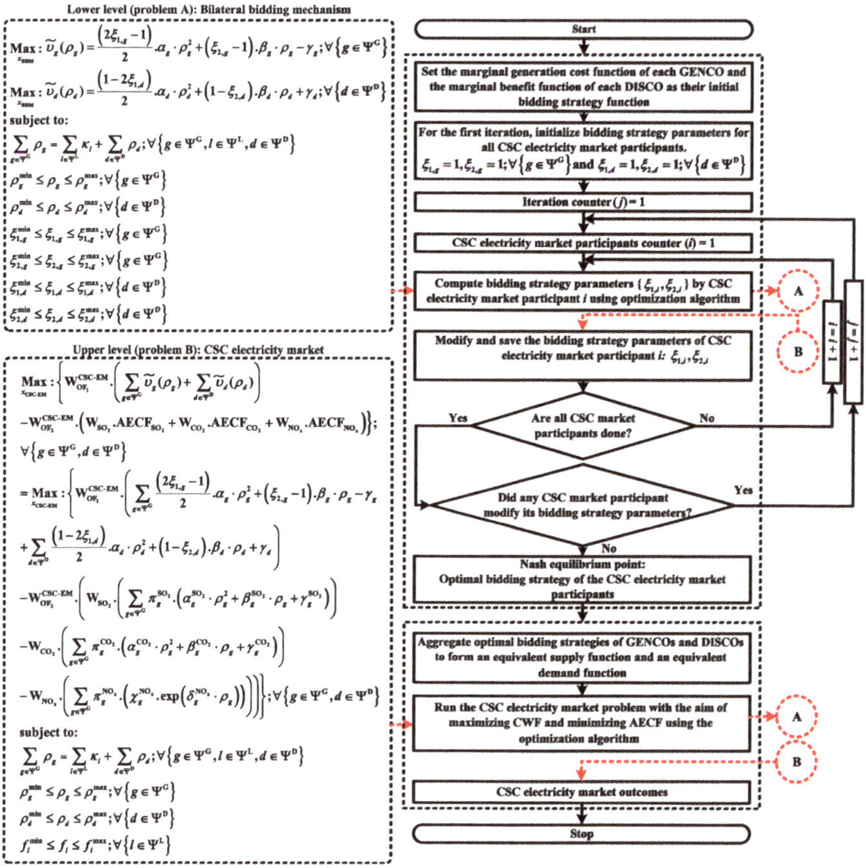

Fig. 5.8 Flowchart of the bi-level computational-logical framework

from the work by Abido [28, 29], Wang and Singh [30], and Kockar et al. [31] and modified for the aforementioned test network.

To prevent the application of market power, the price ceiling is considered for the CSC electricity market participant. Therefore, each CSC electricity market participant can change his/her bidding strategy parameters between predetermined minimum and maximum values of these parameters. Table 5.18 gives information related to the bidding strategy parameters of both GENCOs and DISCOs. Table 5.19 shows the parameters associated with weighting coefficients. To analyze the performance of the proposed bi-level computational-logical framework, two cases were defined and applied, as follows:

- First case: The bi-level computational-logical framework is run, while considering only a unilateral bidding mechanism, or generation-side bidding strategy, under two scenarios, as follows:

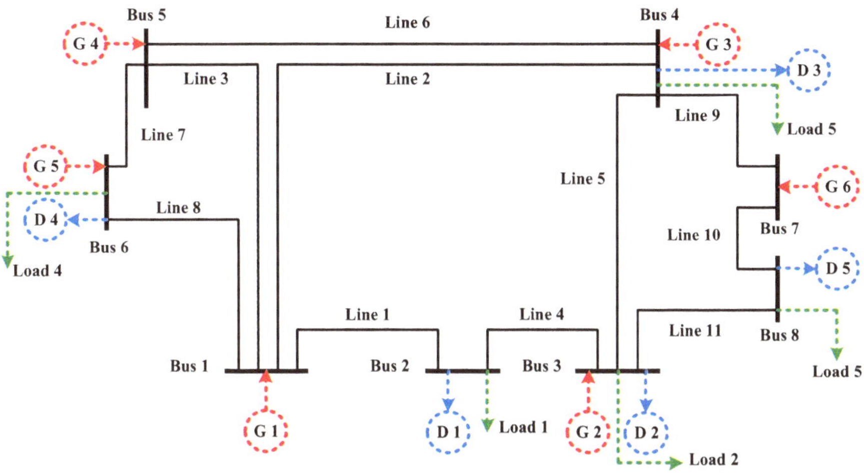

Fig. 5.9 The modified six-machine eight-bus test network

  - First scenario: The security constraint of the electricity market is disabled.
  - Second scenario: The security constraint of the electricity market is enabled.

- Second case: The bi-level computational-logical framework is run, while considering a bilateral bidding mechanism, or generation- and demand-side bidding strategies, under two scenarios, as follows:
  - First scenario: The security constraint of the electricity market is disabled.
  - Second scenario: The security constraint of the electricity market is enabled.

Two base scenarios were defined for the first and second cases for comparison targets. In these base scenarios, the strategic behavior of the CSC market participant in the lower level and the security constraint in the upper level of the proposed bi-level computational-logical framework were disabled. In simple terms, in the base scenario of the first case, each GENCO submits its true marginal generation cost function to the ISO. In the base scenario of the second case, each GENCO submits its true marginal generation cost function and each DISCO submits its true marginal benefit function to the ISO. In addition, the first and second cases, and the two base scenarios, were implemented by using only the proposed single-objective multi-stage computational multi-dimensional multi-homogeneous enhanced melody search algorithm (MMM-EMSA) or single-objective multi-stage computational multi-dimensional single-inhomogeneous enhanced melody search algorithm (MMS-EMSA) called a single-objective symphony orchestra search algorithm (SOSA), which was addressed in Chap. 4. Table 5.20 shows the parameter adjustments of the SOSA.

5.3.5.1 First Case: Simulation Results and Discussion

Tables 5.3 and 5.4 show the optimal results related to the bidding strategy parameters of GENCOs for the lower level of the bi-level computational-logical framework under the first and the second scenarios (i.e., unconstrained and constrained electricity markets) for the first case (i.e., UBM), respectively. During the optimization process of bidding strategy parameters associated with GENCO g, the rest of the GENCOs' parameters are demonstrated only for the first and second iterations. In subsequent iterations, the final results of the bidding strategy parameters of all GENCOs are illustrated after the optimization process. Tables 5.5 and 5.6 show the optimal results of the upper level and the CSC electricity market outcomes for the bi-level computational-logical framework under the first case. Tables 5.3 and 5.4 indicate that, for the Nash equilibrium point, the bidding strategy parameters of all GENCOs are larger than the constant value 1. This means that each GENCO has adopted a strategic behavior in order to compete with its rivals and earn more profit by selling the electrical power. By analyzing the results presented in Tables 5.3, 5.4, 5.5, and 5.6, it can be seen that the larger bidding strategy parameters (i.e., slope and intercept) are related to the GENCO with less generated electrical power and vice versa.

In simple terms, the smaller GENCO—in terms of generated electrical power—tends to adopt larger bidding strategy parameters to get more profit from the electricity market, thereby compensating for low profit due to the shortage of its electrical power generation capacity.

The LMPs in the different buses of the network were increased, clearly in the case that the GENCOs took strategic behavior into account, comparing the first and second scenarios to the base scenario. Also, the parameters in the second scenario, while applying the security constraint of the network, have more values when compared to the first scenario, ignoring the security constraint of the network. As can be seen from these tables, the GENCOs increased their profits by employing the optimal bidding strategies, when comparing the first and second scenarios to the base scenario. For the second scenario, when the network security constraint is activated, the GENCOs' profits will increase in comparison with the situation in which this constraint is ignored. One of the main reasons for increasing the GENCOs' profits in the security-constrained network is to focus on the congestion phenomenon or thermal capacity of the network lines by the GENCOs.

From Table 5.6, it can be seen that the environmental aspects in the upper level of the bi-level computational-logical framework, i.e., the CSC electricity market, led to a reduction in the amount of SO_2, CO_2, and NO_x emitted by GENCOs, and, subsequently, the cost of atmospheric emissions. In other words, when the ISO simultaneously considers the CWF and AECF as objective functions in the CSC electricity market clearing process, the number of different types of AEFs and the costs associated with these functions, namely AECFs, are lower, when compared with when the ISO only considers the CWF as an objective function.

Table 5.3 Bidding strategy parameters of GENCOs under the first scenario of the first case

Iteration counter (j)	GENCO counter (i)	Bidding strategy parameters of GENCOs											
		GENCO 1		GENCO 2		GENCO 3		GENCO 4		GENCO 5		GENCO 6	
		$\xi_{1,1}$	$\xi_{2,1}$	$\xi_{1,2}$	$\xi_{2,2}$	$\xi_{1,3}$	$\xi_{2,3}$	$\xi_{1,4}$	$\xi_{2,4}$	$\xi_{1,5}$	$\xi_{2,5}$	$\xi_{1,6}$	$\xi_{2,6}$
0	0	1.000000	1.000000	1.000000	1.000000	1.000000	1.000000	1.000000	1.000000	1.000000	1.000000	1.000000	1.000000
1	1	1.095773	1.0179	1.000000	1.000000	1.000000	1.000000	1.000000	1.000000	1.000000	1.000000	1.000000	1.000000
	2	1.095773	1.0179	1.084263	1.00639	1.000000	1.000000	1.000000	1.000000	1.000000	1.000000	1.000000	1.000000
	3	1.095773	1.0179	1.084263	1.00639	1.111763	1.03389	1.000000	1.000000	1.000000	1.000000	1.000000	1.000000
	4	1.095773	1.0179	1.084263	1.00639	1.111763	1.03389	1.089363	1.01149	1.000000	1.000000	1.000000	1.000000
	5	1.095773	1.0179	1.084263	1.00639	1.111763	1.03389	1.089363	1.01149	1.090563	1.01269	1.000000	1.000000
	6	1.095773	1.0179	1.084263	1.00639	1.111763	1.03389	1.089363	1.01149	1.090563	1.01269	1.113163	1.03529
2	1	1.109563	1.03169	1.084263	1.00639	1.111763	1.111763	1.089363	1.01149	1.090563	1.01269	1.113163	1.03529
	2	1.109563	1.03169	1.085863	1.00799	1.111763	1.111763	1.089363	1.01149	1.090563	1.01269	1.113163	1.03529
	3	1.109563	1.03169	1.085863	1.00799	1.112563	1.03469	1.089363	1.01149	1.090563	1.01269	1.113163	1.03529
	4	1.109563	1.03169	1.085863	1.00799	1.112563	1.03469	1.090963	1.01309	1.090563	1.01269	1.113163	1.03529
	5	1.109563	1.03169	1.085863	1.00799	1.112563	1.03469	1.090963	1.01309	1.091863	1.01399	1.113163	1.03529
	6	1.109563	1.03169	1.085863	1.00799	1.112563	1.03469	1.090963	1.01309	1.091863	1.01399	1.117663	1.03979
3	1–6	1.109563	1.03169	1.085863	1.00799	1.112563	1.03469	1.090963	1.01309	1.091863	1.01399	1.117663	1.03979
4	1–6	1.109563	1.03169	1.085863	1.00799	1.112563	1.03469	1.090963	1.01309	1.091863	1.01399	1.117663	1.03979
5	1–6	1.109563	1.03169	1.085863	1.00799	1.112563	1.03469	1.090963	1.01309	1.091863	1.01399	1.117663	1.03979

Table 5.4 Bidding strategy parameters of GENCOs under the second scenario of the first case

Iteration counter (j)	GENCO counter (i)	Bidding strategy parameters of GENCOs											
		GENCO 1		GENCO 2		GENCO 3		GENCO 4		GENCO 5		GENCO 6	
		$\xi_{1,1}$	$\xi_{2,1}$	$\xi_{1,2}$	$\xi_{2,2}$	$\xi_{1,3}$	$\xi_{2,3}$	$\xi_{1,4}$	$\xi_{2,4}$	$\xi_{1,5}$	$\xi_{2,5}$	$\xi_{1,6}$	$\xi_{2,6}$
0	0	1.000000	1.000000	1.000000	1.000000	1.000000	1.000000	1.000000	1.000000	1.000000	1.000000	1.000000	1.000000
1	1	1.128592	1.050719	1.000000	1.000000	1.000000	1.000000	1.000000	1.000000	1.000000	1.000000	1.000000	1.000000
	2	1.128592	1.050719	1.117082	1.039209	1.000000	1.000000	1.000000	1.000000	1.000000	1.000000	1.000000	1.000000
	3	1.128592	1.050719	1.117082	1.039209	1.123382	1.045509	1.000000	1.000000	1.000000	1.000000	1.000000	1.000000
	4	1.128592	1.050719	1.117082	1.039209	1.123382	1.045509	1.122182	1.044309	1.000000	1.000000	1.000000	1.000000
	5	1.128592	1.050719	1.117082	1.039209	1.123382	1.045509	1.122182	1.044309	1.145982	1.068109	1.000000	1.000000
	6	1.128592	1.050719	1.117082	1.039209	1.123382	1.045509	1.122182	1.044309	1.145982	1.068109	1.144582	1.066709
2	1	1.142382	1.064509	1.117082	1.039209	1.123382	1.045509	1.122182	1.044309	1.145982	1.068109	1.144582	1.066709
	2	1.142382	1.064509	1.118682	1.040809	1.123382	1.045509	1.122182	1.044309	1.145982	1.068109	1.144582	1.066709
	3	1.142382	1.064509	1.118682	1.040809	1.124682	1.046809	1.122182	1.044309	1.145982	1.068109	1.144582	1.066709
	4	1.142382	1.064509	1.118682	1.040809	1.124682	1.046809	1.123782	1.045909	1.145982	1.068109	1.144582	1.066709
	5	1.142382	1.064509	1.118682	1.040809	1.124682	1.046809	1.123782	1.045909	1.150482	1.072609	1.144582	1.066709
	6	1.142382	1.064509	1.118682	1.040809	1.124682	1.046809	1.123782	1.045909	1.150482	1.072609	1.145382	1.067509
3	1–6	1.142382	1.064509	1.118682	1.040809	1.124682	1.046809	1.123782	1.045909	1.150482	1.072609	1.145382	1.067509
4	1–6	1.142382	1.064509	1.118682	1.040809	1.124682	1.046809	1.123782	1.045909	1.150482	1.072609	1.145382	1.067509
5	1–6	1.142382	1.064509	1.118682	1.040809	1.124682	1.046809	1.123782	1.045909	1.150482	1.072609	1.145382	1.067509

Table 5.5 The CSC electricity market outcomes under the first case

Parameters	Symbols	Scenarios under first case: Base scenario	First scenario	Second scenario
Generation (MW)	ρ_1	158.2326273	178.9867588	187.0935563
	ρ_2	331.8822066	345.3022197	377.0875824
	ρ_3	160.3698883	169.9768812	186.9940877
	ρ_4	307.8330734	340.6697005	358.8354212
	ρ_5	286.7597059	208.8405421	119.8008206
	ρ_6	158.7649451	160.8279367	175.9086412
Total artificial demand or network losses (MW)	$\sum_{l \in \Psi^L} \kappa_l$	3.842446604	4.604039123	5.720109439
LMP ($/MWh)	LMP_1	14.19869159	18.88840636	21.28966218
	LMP_2	21.76526476	27.13634575	29.62582008
	LMP_3	29.50871611	31.10770095	39.63063621
	LMP_4	35.8634742	39.78331853	41.02491989
	LMP_5	15.5121182	16.32019658	18.03281369
	LMP_6	21.08354336	29.19117594	33.83014015
	LMP_7	22.48457606	25.57065989	28.194779
	LMP_8	24.38940412	28.58655675	31.11229217
GENCOs' profit ($\tilde{\upsilon}_g$; $/h)	$\tilde{\upsilon}_1$	486.2439798	888.0827229	1113.702238
	$\tilde{\upsilon}_2$	2350.49468	3055.080544	4149.935311
	$\tilde{\upsilon}_3$	367.0367668	693.9408779	908.1270185
	$\tilde{\upsilon}_4$	3426.384439	5040.082222	6028.634352
	$\tilde{\upsilon}_5$	2581.646126	1660.762501	721.7181367
	$\tilde{\upsilon}_6$	757.3455415	1094.463732	1451.635767

In the results, a positive improvement rate or reduction rate was observed in all defined scenarios of this case (i.e., the base, the first, and the second scenarios). As further explanation, Fig. 5.10 illustrates the changes of the AECF with and without consideration of the environmental aspects in the upper level of the bi-level computational-logical framework and improvement rate under the first case for the base, first, and second scenarios.

5.3.5.2 Second Case: Simulation Results and Discussion

Tables 5.7 and 5.8 show the optimal results associated with the bidding strategy parameters of both GENCOs and DISCOs for the lower level of the proposed bi-level computational-logical framework under the first and the second scenarios (i.e., unconstrained and constrained electricity markets) in the second case (i.e., BBM), respectively. For the same reason, during the optimization process of the bidding strategy parameter associated with GENCO g or DISCO d, the rest of the GENCOs'

Table 5.6 The CSC electricity market outcomes under the first case

Condition	Type of atmospheric emission	Scenario		
		Base scenario	First scenario	Second scenario
Ignoring environmental aspects in the CSC electricity market	AEF_{SO_2} (ton/h)	1.505934	1.542884	1.620131
	$AECF_{SO_2}$ ($/h)	198.783359	203.660815	215.264330
	AEF_{CO_2} (ton/h)	1.825581	1.848944	1.950279
	$AECF_{CO_2}$ ($/h)	401.628004	406.767696	429.061512
	AEF_{NO_x} (ton/h)	0.033307	0.033870	0.034111
	$AECF_{NO_x}$ ($/h)	2.464786	2.506443	2.524256
	AECF ($/h)	602.876148	612.934954	646.850098
Considering environmental aspects in the CSC electricity market	AEF_{SO_2} (ton/h)	1.433634	1.452218	1.522680
	$AECF_{SO_2}$ ($/h)	189.239735	191.692790	200.993772
	AEF_{CO_2} (ton/h)	1.719468	1.741641	1.826104
	$AECF_{CO_2}$ ($/h)	378.283040	383.161186	401.742989
	AEF_{NO_x} (ton/h)	0.031691	0.031744	0.031820
	$AECF_{NO_x}$ ($/h)	2.345182	2.349056	2.354716
	AECF ($/h)	569.867957	577.203032	605.09147
Improvement rate or reduction rate of the AECF (%)		5.475119	5.829643	6.455688

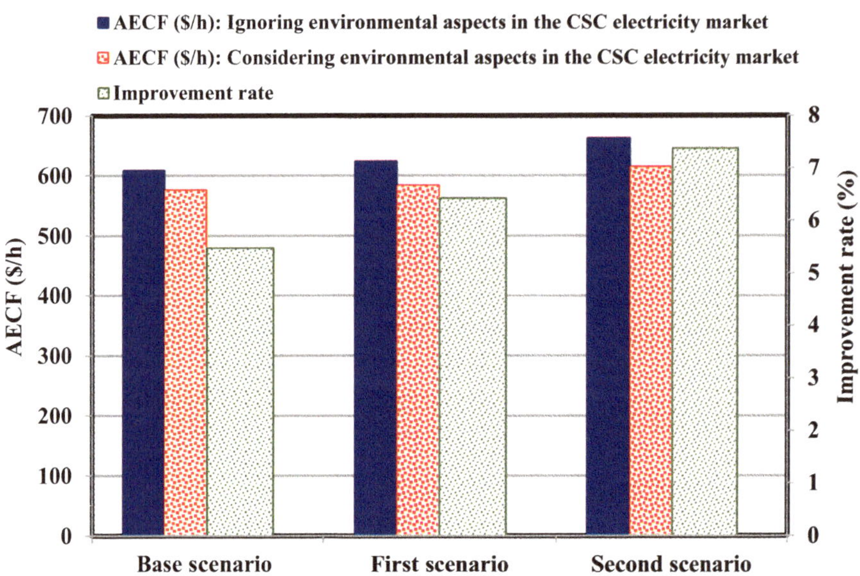

Fig. 5.10 Changes of the AECF with and without consideration of the environmental aspects and improvement rate under the first case for the base, first, and second scenarios

Table 5.7 Bidding strategy parameters of both GENCOs and DISCOs under the first scenario of the second case

Iteration counter (j)	GENCO + DISCO counter (i)	Bidding strategy parameters of GENCOs											
		GENCO 1		GENCO 2		GENCO 3		GENCO 4		GENCO 5		GENCO 6	
		$\xi_{1,1}$	$\xi_{2,1}$	$\xi_{1,2}$	$\xi_{2,2}$	$\xi_{1,3}$	$\xi_{2,3}$	$\xi_{1,4}$	$\xi_{2,4}$	$\xi_{1,5}$	$\xi_{2,5}$	$\xi_{1,6}$	$\xi_{2,6}$
0	0	1.000000	1.000000	1.000000	1.000000	1.000000	1.000000	1.000000	1.000000	1.000000	1.000000	1.000000	1.000000
1	1	1.614300	1.483237	1.000000	1.000000	1.000000	1.000000	1.000000	1.000000	1.000000	1.000000	1.000000	1.000000
	2	1.614300	1.483237	1.610200	1.479469	1.000000	1.000000	1.000000	1.000000	1.000000	1.000000	1.000000	1.000000
	3	1.614300	1.483237	1.610200	1.479469	1.635200	1.502440	1.000000	1.000000	1.000000	1.000000	1.000000	1.000000
	4	1.614300	1.483237	1.610200	1.479469	1.635200	1.502440	1.610800	1.480021	1.000000	1.000000	1.000000	1.000000
	5	1.614300	1.483237	1.610200	1.479469	1.635200	1.502440	1.610800	1.480021	1.612300	1.481399	1.000000	1.000000
	6	1.614300	1.483237	1.610200	1.479469	1.635200	1.502440	1.610800	1.480021	1.612300	1.481399	1.638300	1.505288
	7	1.614300	1.483237	1.610200	1.479469	1.635200	1.502440	1.610800	1.480021	1.612300	1.481399	1.638300	1.505288
	8	1.614300	1.483237	1.610200	1.479469	1.635200	1.502440	1.610800	1.480021	1.612300	1.481399	1.638300	1.505288
	9	1.614300	1.483237	1.610200	1.479469	1.635200	1.502440	1.610800	1.480021	1.612300	1.481399	1.638300	1.505288
	10	1.614300	1.483237	1.610200	1.479469	1.635200	1.502440	1.610800	1.480021	1.612300	1.481399	1.638300	1.505288
	11	1.614300	1.483237	1.610200	1.479469	1.635200	1.502440	1.610800	1.480021	1.612300	1.481399	1.638300	1.505288
2	1	1.618800	1.487371	1.610200	1.479469	1.635200	1.6352	1.610800	1.480021	1.612300	1.481399	1.638300	1.505288
	2	1.618800	1.487371	1.614000	1.482961	1.635200	1.6352	1.610800	1.480021	1.612300	1.481399	1.638300	1.505288
	3	1.618800	1.487371	1.614000	1.482961	1.640300	1.507126	1.610800	1.480021	1.612300	1.481399	1.638300	1.505288
	4	1.618800	1.487371	1.614000	1.482961	1.640300	1.507126	1.614200	1.483145	1.612300	1.481399	1.638300	1.505288
	5	1.618800	1.487371	1.614000	1.482961	1.640300	1.507126	1.614200	1.483145	1.615400	1.484247	1.638300	1.505288
	6	1.618800	1.487371	1.614000	1.482961	1.640300	1.507126	1.614200	1.483145	1.615400	1.484247	1.64400	1.510525
	7	1.618800	1.487371	1.614000	1.482961	1.640300	1.507126	1.614200	1.483145	1.615400	1.484247	1.64400	1.510525
	8	1.618800	1.487371	1.614000	1.482961	1.640300	1.507126	1.614200	1.483145	1.615400	1.484247	1.64400	1.510525
	9	1.618800	1.487371	1.614000	1.482961	1.640300	1.507126	1.614200	1.483145	1.615400	1.484247	1.64400	1.510525
	10	1.618800	1.487371	1.614000	1.482961	1.640300	1.507126	1.614200	1.483145	1.615400	1.484247	1.64400	1.510525
	11	1.618800	1.487371	1.614000	1.482961	1.640300	1.507126	1.614200	1.483145	1.615400	1.484247	1.64400	1.510525
3	1–11	1.618800	1.487371	1.614000	1.482961	1.640300	1.507126	1.614200	1.483145	1.615400	1.484247	1.644000	1.510525
4	1–11	1.618800	1.487371	1.614000	1.482961	1.640300	1.507126	1.614200	1.483145	1.615400	1.484247	1.644000	1.510525
5	1–11	1.618800	1.487371	1.614000	1.482961	1.640300	1.507126	1.614200	1.483145	1.615400	1.484247	1.644000	1.510525

(continued)

Table 5.7 (continued)

Iteration counter (j)	GENCO + DISCO counter (i)	Bidding strategy parameters of DISCOs									
		DISCO 1		DISCO 2		DISCO 3		DISCO 4		DISCO 5	
		$\xi_{1,1}$	$\xi_{2,1}$	$\xi_{1,2}$	$\xi_{2,2}$	$\xi_{1,3}$	$\xi_{2,3}$	$\xi_{1,4}$	$\xi_{2,4}$	$\xi_{1,5}$	$\xi_{2,5}$
0	0	1.000000	1.000000	1.000000	1.000000	1.000000	1.000000	1.000000	1.000000	1.000000	1.000000
1	1	1.000000	1.000000	1.000000	1.000000	1.000000	1.000000	1.000000	1.000000	1.000000	1.000000
	2	1.000000	1.000000	1.000000	1.000000	1.000000	1.000000	1.000000	1.000000	1.000000	1.000000
	3	1.000000	1.000000	1.000000	1.000000	1.000000	1.000000	1.000000	1.000000	1.000000	1.000000
	4	1.000000	1.000000	1.000000	1.000000	1.000000	1.000000	1.000000	1.000000	1.000000	1.000000
	5	1.000000	1.000000	1.000000	1.000000	1.000000	1.000000	1.000000	1.000000	1.000000	1.000000
	6	1.000000	1.000000	1.000000	1.000000	1.000000	1.000000	1.000000	1.000000	1.000000	1.000000
	7	1.542200	0.934817	1.000000	1.000000	1.000000	1.000000	1.000000	1.000000	1.000000	1.000000
	8	1.542200	0.934817	1.542400	0.934732	1.000000	1.000000	1.000000	1.000000	1.000000	1.000000
	9	1.542200	0.934817	1.542400	0.934732	1.542400	0.934117	1.000000	1.000000	1.000000	1.000000
	10	1.542200	0.934817	1.542400	0.934732	1.542400	0.934117	1.542800	0.934517	1.000000	1.000000
	11	1.542200	0.934817	1.542400	0.934732	1.542400	0.934117	1.542800	0.934517	1.542600	0.934717
2	1	1.542200	0.934817	1.542400	0.934732	1.542400	0.934117	1.542800	0.934517	1.542600	0.934717
	2	1.542200	0.934817	1.542400	0.934732	1.542400	0.934117	1.542800	0.934517	1.542600	0.934717
	3	1.542200	0.934817	1.542400	0.934732	1.542400	0.934117	1.542800	0.934517	1.542600	0.934717
	4	1.542200	0.934817	1.542400	0.934732	1.542400	0.934117	1.542800	0.934517	1.542600	0.934717
	5	1.542200	0.934817	1.542400	0.934732	1.542400	0.934117	1.542800	0.934517	1.542600	0.934717
	6	1.542200	0.934817	1.542400	0.934732	1.542400	0.934117	1.542800	0.934517	1.542600	0.934717
	7	1.542200	0.934512	1.542400	0.934732	1.542400	0.934117	1.542800	0.934517	1.542600	0.934717
	8	1.542200	0.934512	1.542400	0.934110	1.542400	0.934117	1.542800	0.934517	1.542600	0.934717
	9	1.542200	0.934512	1.542400	0.934110	1.542700	0.933715	1.542800	0.934517	1.542600	0.934717
	10	1.542200	0.934512	1.542400	0.934110	1.542700	0.933715	1.543700	0.932114	1.542600	0.934717
	11	1.542200	0.934512	1.542400	0.934110	1.542700	0.933715	1.543700	0.932114	1.542754	0.932870
3	1–11	1.542200	0.934512	1.542400	0.934110	1.542700	0.933715	1.543700	0.932114	1.542754	0.932870
4	1–11	1.542200	0.934512	1.542400	0.934110	1.542700	0.933715	1.543700	0.932114	1.542754	0.932870
5	1–11	1.542200	0.934512	1.542400	0.934110	1.542700	0.933715	1.543700	0.932114	1.542754	0.932870

Table 5.8 Bidding strategy parameters of both GENCOs and DISCOs under the second scenario of the second case

Iteration counter (j)	GENCO + DISCO counter (i)	Bidding strategy parameters of GENCOs											
		GENCO 1		GENCO 2		GENCO 3		GENCO 4		GENCO 5		GENCO 6	
		$\xi_{1,1}$	$\xi_{2,1}$	$\xi_{1,2}$	$\xi_{2,2}$	$\xi_{1,3}$	$\xi_{2,3}$	$\xi_{1,4}$	$\xi_{2,4}$	$\xi_{1,5}$	$\xi_{2,5}$	$\xi_{1,6}$	$\xi_{2,6}$
0	0	1.000000	1.000000	1.000000	1.000000	1.000000	1.000000	1.000000	1.000000	1.000000	1.000000	1.000000	1.000000
1	1	1.663598	1.509200	1.000000	1.000000	1.000000	1.000000	1.000000	1.000000	1.000000	1.000000	1.000000	1.000000
	2	1.663598	1.509200	1.653198	1.499765	1.000000	1.000000	1.000000	1.000000	1.000000	1.000000	1.000000	1.000000
	3	1.663598	1.509200	1.653198	1.499765	1.659698	1.505662	1.000000	1.000000	1.000000	1.000000	1.000000	1.000000
	4	1.663598	1.509200	1.653198	1.499765	1.659698	1.505662	1.655198	1.501580	1.000000	1.000000	1.000000	1.000000
	5	1.663598	1.509200	1.653198	1.499765	1.659698	1.505662	1.655198	1.501580	1.682098	1.525983	1.000000	1.000000
	6	1.663598	1.509200	1.653198	1.499765	1.659698	1.505662	1.655198	1.501580	1.682098	1.525983	1.684298	1.527979
	7	1.663598	1.509200	1.653198	1.499765	1.659698	1.505662	1.655198	1.501580	1.682098	1.525983	1.684298	1.527979
	8	1.663598	1.509200	1.653198	1.499765	1.659698	1.505662	1.655198	1.501580	1.682098	1.525983	1.684298	1.527979
	9	1.663598	1.509200	1.653198	1.499765	1.659698	1.505662	1.655198	1.501580	1.682098	1.525983	1.684298	1.527979
	10	1.663598	1.509200	1.653198	1.499765	1.659698	1.505662	1.655198	1.501580	1.682098	1.525983	1.684298	1.527979
	11	1.663598	1.509200	1.653198	1.499765	1.659698	1.505662	1.655198	1.501580	1.682098	1.525983	1.684298	1.527979
2	1	1.665598	1.511015	1.653198	1.499765	1.659698	1.505662	1.655198	1.50158	1.688498	1.531789	1.684298	1.527979
	2	1.665598	1.511015	1.656398	1.502668	1.659698	1.505662	1.655198	1.50158	1.688498	1.531789	1.684298	1.527979
	3	1.665598	1.511015	1.656398	1.502668	1.665198	1.509116	1.655198	1.50158	1.688498	1.531789	1.684298	1.527979
	4	1.665598	1.511015	1.656398	1.502668	1.665198	1.509116	1.660698	1.50657	1.688498	1.531789	1.684298	1.527979
	5	1.665598	1.511015	1.656398	1.502668	1.665198	1.509116	1.660698	1.50657	1.689624	1.532820	1.684298	1.527979
	6	1.665598	1.511015	1.656398	1.502668	1.665198	1.509116	1.660698	1.50657	1.689624	1.532820	1.687798	1.531969
	7	1.665598	1.511015	1.656398	1.502668	1.665198	1.509116	1.660698	1.50657	1.689624	1.532820	1.687798	1.531969
	8	1.665598	1.511015	1.656398	1.502668	1.665198	1.509116	1.660698	1.50657	1.689624	1.532820	1.687798	1.531969
	9	1.665598	1.511015	1.656398	1.502668	1.665198	1.509116	1.660698	1.50657	1.689624	1.532820	1.687798	1.531969
	10	1.665598	1.511015	1.656398	1.502668	1.665198	1.509116	1.660698	1.50657	1.689624	1.532820	1.687798	1.531969
	11	1.665598	1.511015	1.656398	1.502668	1.665198	1.509116	1.660698	1.50657	1.689624	1.532820	1.687798	1.531969
3	1–11	1.665598	1.511015	1.656398	1.502668	1.665198	1.509116	1.660698	1.50657	1.689624	1.532820	1.687798	1.531969
4	1–11	1.665598	1.511015	1.656398	1.502668	1.665198	1.509116	1.660698	1.50657	1.689624	1.532820	1.687798	1.531969
5	1–11	1.665598	1.511015	1.656398	1.502668	1.665198	1.509116	1.660698	1.50657	1.689624	1.532820	1.687798	1.531969

(continued)

Table 5.8 (continued)

		Bidding strategy parameters of DISCOs									
		DISCO 1		DISCO 2		DISCO 3		DISCO 4		DISCO 5	
Iteration counter (j)	GENCO + DISCO counter (i)	$\xi_{1,1}$	$\xi_{2,1}$	$\xi_{1,2}$	$\xi_{2,2}$	$\xi_{1,3}$	$\xi_{2,3}$	$\xi_{1,4}$	$\xi_{2,4}$	$\xi_{1,5}$	$\xi_{2,5}$
0	0	1.000000	1.000000	1.000000	1.000000	1.000000	1.000000	1.000000	1.000000	1.000000	1.000000
1	1	1.000000	1.000000	1.000000	1.000000	1.000000	1.000000	1.000000	1.000000	1.000000	1.000000
	2	1.000000	1.000000	1.000000	1.000000	1.000000	1.000000	1.000000	1.000000	1.000000	1.000000
	3	1.000000	1.000000	1.000000	1.000000	1.000000	1.000000	1.000000	1.000000	1.000000	1.000000
	4	1.000000	1.000000	1.000000	1.000000	1.000000	1.000000	1.000000	1.000000	1.000000	1.000000
	5	1.000000	1.000000	1.000000	1.000000	1.000000	1.000000	1.000000	1.000000	1.000000	1.000000
	6	1.000000	1.000000	1.000000	1.000000	1.000000	1.000000	1.000000	1.000000	1.000000	1.000000
	7	1.580998	0.955538	1.000000	1.000000	1.000000	1.000000	1.000000	1.000000	1.000000	1.000000
	8	1.580998	0.955538	1.581598	0.945338	1.000000	1.000000	1.000000	1.000000	1.000000	1.000000
	9	1.580998	0.955538	1.581598	0.945338	1.581898	0.945112	1.000000	1.000000	1.000000	1.000000
	10	1.580998	0.955538	1.581598	0.945338	1.581898	0.945112	1.584098	0.945022	1.000000	1.000000
	11	1.580998	0.955538	1.581598	0.945338	1.581898	0.945112	1.584098	0.945022	1.582998	0.945027
2	1	1.580998	0.955538	1.581598	0.945338	1.581898	0.945112	1.584098	0.945022	1.582998	0.945027
	2	1.580998	0.955538	1.581598	0.945338	1.581898	0.945112	1.584098	0.945022	1.582998	0.945027
	3	1.580998	0.955538	1.581598	0.945338	1.581898	0.945112	1.584098	0.945022	1.582998	0.945027
	4	1.580998	0.955538	1.581598	0.945338	1.581898	0.945112	1.584098	0.945022	1.582998	0.945027
	5	1.580998	0.955538	1.581598	0.945338	1.581898	0.945112	1.584098	0.945022	1.582998	0.945027
	6	1.580998	0.955538	1.581598	0.945338	1.581898	0.945112	1.584098	0.945022	1.582998	0.945027
	7	1.580368	0.951138	1.581598	0.945938	1.581898	0.945112	1.584098	0.945022	1.582998	0.945027
	8	1.580368	0.951138	1.581612	0.945144	1.581898	0.945112	1.584098	0.945022	1.582998	0.945027
	9	1.580368	0.951138	1.581612	0.945144	1.581924	0.944815	1.584098	0.945022	1.582998	0.945027
	10	1.580368	0.951138	1.581612	0.945144	1.581924	0.944815	1.584110	0.939110	1.582998	0.945027
	11	1.580368	0.951138	1.581612	0.945144	1.581924	0.944815	1.584110	0.939110	1.583016	0.942614
3	1–11	1.580368	0.951138	1.581612	0.945144	1.581924	0.944815	1.584110	0.939110	1.583016	0.942614
4	1–11	1.580368	0.951138	1.581612	0.945144	1.581924	0.944815	1.584110	0.939110	1.583016	0.942614
5	1–11	1.580368	0.951138	1.581612	0.945144	1.581924	0.944815	1.584110	0.939110	1.583016	0.942614

Table 5.9 The CSC electricity market outcomes under the second case

Parameters	Symbols	Scenarios under first case		
		Base scenario	First scenario	Second scenario
Generation (ρ_g;MW)	ρ_1	159.3456536	182.0401316	188.204892
	ρ_2	331.3288622	347.2390081	379.3430064
	ρ_3	159.4870209	170.9147665	190.1168826
	ρ_4	310.2335239	341.5523488	360.9722961
	ρ_5	288.1196686	210.2054134	121.1535897
	ρ_6	160.6876798	161.3936843	176.7897479
Consumption (ρ_d;MW)	ρ_1	468.3378947	471.456479	472.9619798
	ρ_2	254.5241593	261.4924088	268.779198
	ρ_3	231.5352562	255.8796893	262.176495
	ρ_4	192.0564724	199.8563836	187.0162575
	ρ_5	253.5462174	211.3150393	209.0660697
Total artificial demand or network losses (MW)	$\sum_{l \in \Psi^L} \kappa_l$	9.202409201	13.34535266	16.58041471
LMP ($/MWh)	LMP_1	14.22712934	22.67243543	23.24275854
	LMP_2	21.80283685	35.54553985	39.82725958
	LMP_3	29.55563608	36.80194143	47.20916641
	LMP_4	35.91806561	43.81002207	48.86606426
	LMP_5	15.54214152	21.07665092	29.80826203
	LMP_6	21.12029249	33.24608214	36.9448776
	LMP_7	22.52301651	29.27734773	34.6832035
	LMP_8	24.43014408	30.51852973	35.53537902
GENCOs' profit ($\tilde{\upsilon}_g$; $/h)	$\tilde{\upsilon}_1$	566.3538586	2857.947045	3245.647434
	$\tilde{\upsilon}_2$	2328.644521	9950.712702	11922.58918
	$\tilde{\upsilon}_3$	314.4797027	3601.278367	4209.324314
	$\tilde{\upsilon}_4$	3511.309057	11188.9591	13020.34538
	$\tilde{\upsilon}_5$	2632.364191	5140.343483	2408.599022
	$\tilde{\upsilon}_6$	757.3455415	3471.387178	4182.631796
DISCOs' profit ($\tilde{\upsilon}_d$; $/h)	$\tilde{\upsilon}_1$	5986.204657	12881.91878	13345.33597
	$\tilde{\upsilon}_2$	447.8165597	1160.094564	1219.455022
	$\tilde{\upsilon}_3$	874.5509105	2414.408139	2566.039375
	$\tilde{\upsilon}_4$	982.4479033	2530.385619	2279.780898
	$\tilde{\upsilon}_5$	1111.510634	1956.441751	1928.053275

or DISCOs' parameters are demonstrated only for first and second iterations. In subsequent iterations, only the final results of the bidding strategy parameters of all GENCOs and DISCOs are illustrated after the optimization process. Tables 5.9 and 5.10 show the optimal results of the upper level and the CSC electricity market outcomes for the bi-level computational-logical framework under the second case. Tables 5.7 and 5.8 present evidence that at the Nash equilibrium point, the bidding strategy parameters of GENCOs and DISCOs are not equal to the constant value of 1.

Table 5.10 The CSC electricity market outcomes under the second case

Condition	Type of atmospheric emission	Scenario		
		Base scenario	First scenario	Second scenario
Ignoring environmental aspects in the CSC electricity market	AEF_{SO_2} (ton/h)	1.534563	1.567605	1.666707
	$AECF_{SO_2}$ ($/h)	202.562392	206.923900	220.005403
	AEF_{CO_2} (ton/h)	1.835315	1.881794	2.000693
	$AECF_{CO_2}$ ($/h)	403.769477	413.994686	440.152618
	AEF_{NO_x} (ton/h)	0.033467	0.033877	0.034372
	$AECF_{NO_x}$ ($/h)	2.476568	2.506923	2.543590
	AECF ($/h)	608.808437	623.425509	662.701611
Considering environmental aspects in the CSC electricity market	AEF_{SO_2} (ton/h)	1.447701	1.467795	1.544678
	$AECF_{SO_2}$ ($/h)	191.096597	193.748970	203.897500
	AEF_{CO_2} (ton/h)	1.736344	1.760331	1.852494
	$AECF_{CO_2}$ ($/h)	381.9957212	387.272859	407.548721
	AEF_{NO_x} (ton/h)	0.031692	0.031750	0.031826
	$AECF_{NO_x}$ ($/h)	2.345235	2.349506	2.355176
	AECF ($/h)	575.437553	583.371335	613.801397
Improvement rate or reduction rate of the AECF (%)		5.481343	6.424853	7.378918

This means that both GENCOs and DISCOs have adopted a strategic behavior in order to compete with their competitors and earn more profit from selling and buying electrical power, respectively. By evaluating the results presented in Tables 5.7, 5.8, and 5.9, it can be seen that the larger bidding strategy parameters—slope and intercept—are related to the GENCO with less generated electrical power and vice versa. In addition, the larger bidding strategy parameter—slope—and smaller bidding strategy parameter—intercept—are related to the DISCO with less purchased electrical power and vice versa.

The fluctuations of the LMPs in different scenarios of this case are similar to the fluctuations in the first case. This means that the LMPs in the different buses of the network were clearly increased, as the strategic behavior was adopted by both GENCOs and DISCOs, when comparing the first and second scenarios to the base scenario. Also, the LMPs in the second scenario, taking into account the security constraint of the network, have more values in comparison with the first scenario, ignoring the security constraint of the network.

Due to the adoption of strategic behaviors by the GENCOs and DISCOs, their profits under the first and second scenarios are much higher than those under the base scenarios. Also, for the second scenario, when the network security constraint is activated, positive changes can be observed in the total profits of GENCOs and DISCOs.

Similar to the first case, integration of the environmental aspects in the upper level of the bi-level computational-logical framework results in atmospheric emissions (i.e., SO_2, CO_2, and NO_x) being produced by GENCOs in all scenarios of this

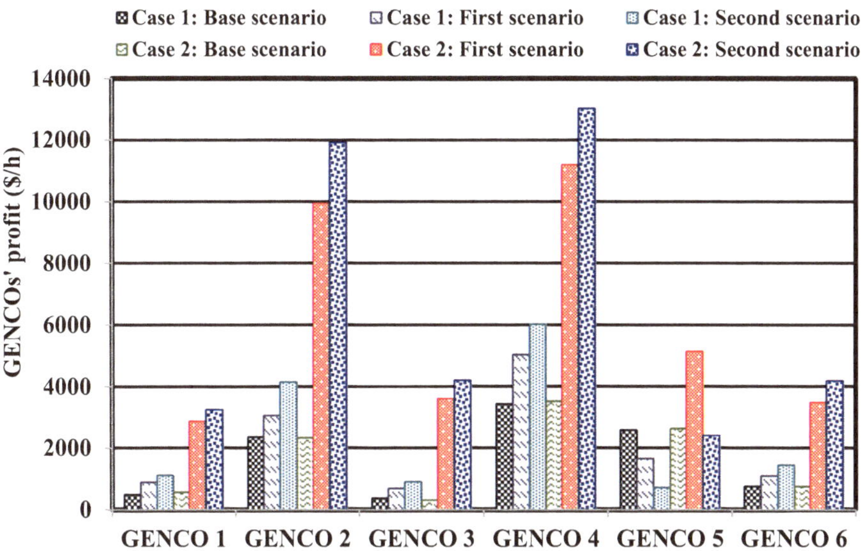

Fig. 5.11 Changes of the AECF with and without consideration of the environmental aspects and improvement rate under the second case for the base, first, and second scenarios

case (i.e., BBM), which are located in a more favorable situation, according to Table 5.10. Subsequently, the costs related to these atmospheric emissions are reduced to more favorable values. As further explanation, Fig. 5.11 shows the changes of the AECF with and without consideration of the environmental aspects in the upper level of the bi-level computational-logical framework, as well as improvement rates under the second case for the base, first, and second scenarios. Also, Fig. 5.12 shows the changes in GENCOs' profits under the first (i.e., UBM) and second (i.e., BBM) cases for the base, first, and second scenarios. From the optimal results illustrated in Fig. 5.12, it can be concluded that the GENCOs' profits in the second case increased in comparison with the first case. This increase in GENCOs' profits is due to the adoption of larger bidding strategy parameters (compare Table 5.3 with Table 5.7 for the first scenario of both cases, and Table 5.4 with Table 5.8 for the second scenario of both cases). In addition, Fig. 5.13 demonstrates the changes of the CSC electricity market prices, i.e., the LMPs under the first and second cases for the base, first, and second scenarios. As can be seen from Fig. 5.13, the LMPs in different buses of the network under the second case (i.e., BBM) are much higher than those under the first case (i.e., UBM), due to the adoption of strategic behaviors by GENCOs in generation side and by DISCOs in consumption side. As a result, the BBM can encourage and facilitate competition among all participants of the CSC electricity market in generation and consumption sides and, finally, improve efficiency of the CSC electricity market, which is the main goal of deregulating the power systems.

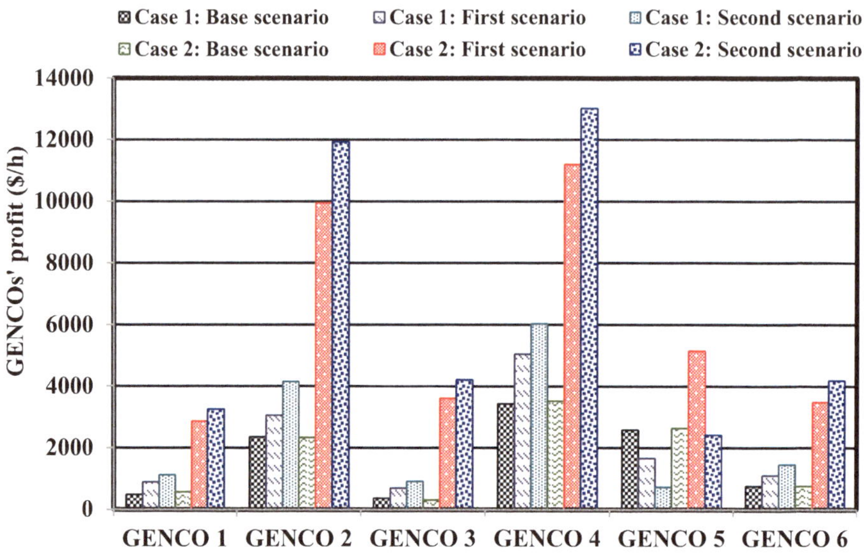

Fig. 5.12 Changes of GENCOs' profit under the first and second cases for the base, first, and second scenarios

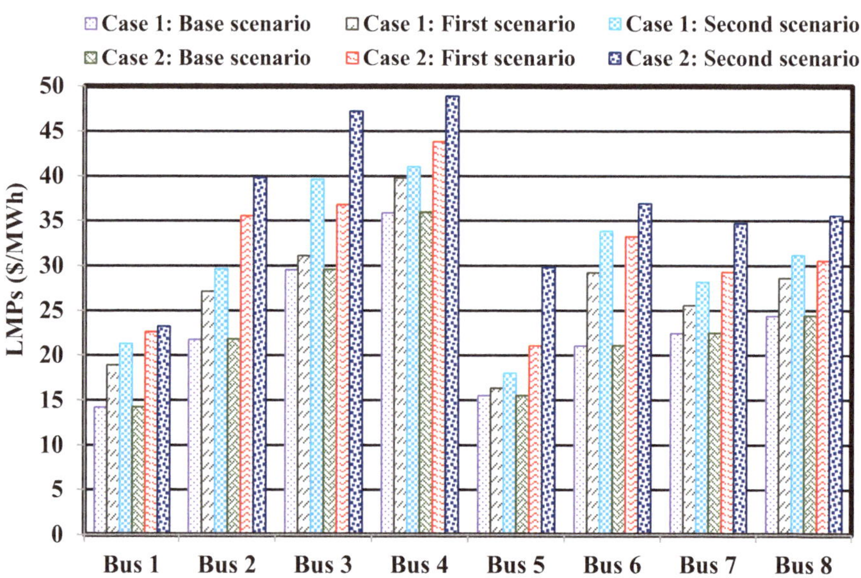

Fig. 5.13 Changes of the CSC electricity market prices under the first and second cases for the base, first, and second scenarios

5.3.5.3 Performance Evaluation of the Proposed Music-Inspired Optimization Algorithms

In this section, the performance of the proposed music-inspired optimization algorithms—which was addressed in Chap. 4 for the single-objective continuous/discrete two-stage computational multi-dimensional single-homogeneous melody search algorithm (TMS-MSA), single-objective two-stage computational multi-dimensional single-homogeneous enhanced melody search algorithm (TMS-EMSA), and single-objective SOSA—is compared with the single-objective single-stage computational single-dimensional harmony search algorithm (SS-HSA), single-objective single-stage computational single-dimensional improved harmony search algorithm (SS-IHSA), and non-dominated sorting genetic algorithm-II (NSGA-II) for the second case under the first and second scenarios.

It should be pointed out that the structure of the NSGI-II is modified to solve the proposed single-objective bi-level computational-logical framework. The detailed description of the NSGA-II can be found in [33]. Tables 5.21, 5.22, 5.23, 5.24, and 5.25 give the parameter adjustments of the single-objective TMS-EMSA, single-objective continuous/discrete TMS-MSA, single-objective SS-HSA, single-objective SS-IHSA, and NSGA-II, respectively. Table 5.11 shows the calculated values of both GENCOs' and DISCOs' profits by the proposed optimization algorithms. Bearing in mind the optimal results reported in Table 5.11, it clearly appears that the proposed bi-level computational-logical framework by the SOSA gives rise to more efficient results than other optimization algorithms. Also, an index of cost saving (ICS) is defined and employed to more clearly assess the performance of these optimization algorithms. The ICS is generally formulated using Eq. (5.53):

$$ICS_q^{n,\hat{n}} = \begin{cases} \left(1 - \left(\dfrac{I_q^n}{I_q^{\hat{n}}}\right)\right) \times 100; \text{ for minimization} \\ \left(\left(\dfrac{I_q^n}{I_q^{\hat{n}}}\right) - 1\right) \times 100; \text{ for maximization} \end{cases} ; \quad \forall \{n, \hat{n} \in \Psi^{\mathrm{N}}, q \subset \Psi^{\mathrm{Q}}\} \tag{5.53}$$

This index represents the amount of the superiority, or weakness, of optimization algorithm n compared with optimization algorithm $\hat{n}$, from the perspective of the obtained value of objective q by optimization algorithm $\hat{n}$. Tables 5.12 and 5.13 show the ICS for the proposed optimization algorithms under the first and second scenarios of the second case.

To clarify, the ICS shows 6.368757%, 9.675599%, 17.934617%, 13.815667%, and 22.282578% superiority—positive sign—of the SOSA performance compared with the performances of the TMS-EMSA, continuous/discrete TMS-MSA, SS-HSA, SS-IHSA, and NSGA-II, respectively, from the perspective of the obtained GENCOs' profits by corresponding optimization algorithms. Also, the ICS represents 8.190871%, 12.282241%, 19.116560%, 16.078196%, and 22.458994%

Table 5.11 Calculated values of both GENCOs' and DISCOs' profit by the proposed optimization algorithms for the second case under the first and second scenarios

Case No.	Scenario No.	Optimization algorithm	The calculated values of both GENCOs' and DISCOs' profits		
			GENCOs' profit ($/h)	DISCOs' profit ($/h)	GENCOs' + DISCOs' profit ($/h)
Second case	First scenario	SOSA	36210.627875	20943.248853	57153.876728
		TMS-EMSA	34042.541002	19357.685655	53400.226657
		Continuous/discrete TMS-MSA	33016.120235	18652.325219	51668.445454
		SS-HSA	30703.985474	17582.147115	48286.132589
		SS-IHSA	31815.152175	18042.362458	49857.514633
		NSGA-II	29612.254152	17102.254454	46714.508606
	Second scenario	SOSA	38989.137126	21338.664540	60327.801666
		TMS-EMSA	37011.212532	20018.597922	57029.810454
		Continuous/discrete TMS-MSA	34588.111855	18784.135565	53372.247420
		SS-HSA	32616.798558	17931.451253	50548.249811
		SS-IHSA	33437.024320	18314.929610	51751.953930
		NSGA-II	31812.769503	17245.687480	49058.456983

Table 5.12 The ICS for different objectives, based on various optimization algorithms for the second case under the first scenario

Scenario No.	Perspective	Optimization algorithm	Optimization algorithm					
			SOSA	TMS-EMSA	Continuous/discrete TMS-MSA	SS-HSA	SS-IHSA	NSGA-II
First scenario	GENCOs' profit	SOSA	0	6.368757	9.675599	17.934617	13.815667	22.282578
		TMS-EMSA	−5.987432	0	3.108847	10.873362	7.001031	14.960991
		Continuous/discrete TMS-MSA	−8.822016	−3.015112	0	7.530405	3.774830	11.494788
		SS-HSA	−15.207255	−9.807010	−7.003048	0	−3.492570	3.686755
		SS-IHSA	−12.138634	−6.542957	−3.637520	3.618965	0	7.439143
		NSGA-II	−18.222202	−13.013972	−10.309709	−3.555666	−6.924053	0
	DISCOs' profit	SOSA	0	8.190871	12.282241	19.116560	16.078196	22.458994
		TMS-EMSA	−7.570760	0	3.781622	10.098530	7.290193	13.187917
		Continuous/discrete TMS-MSA	−10.938721	−3.643826	0	6.086731	3.380725	9.063546
		SS-HSA	−16.048617	−9.172266	−5.737505	0	−2.550748	2.806019
		SS-IHSA	−13.851177	−6.794837	−3.270170	2.617515	0	5.496982
		NSGA-II	−18.340012	−11.651347	−8.310335	−2.729431	−5.210559	0
	GENCOs' + DISCOs' profit ($/h)	SOSA	0	7.029277	10.616598	18.364991	14.634428	22.347164
		TMS-EMSA	−6.567621	0	3.351719	10.591227	7.105673	14.311866
		Continuous/discrete TMS-MSA	−9.597653	−3.243022	0	7.004729	3.632212	10.604707
		SS-HSA	−15.515560	−9.576914	−6.546186	0	−3.151745	3.364316
		SS-IHSA	−12.766172	−6.634264	−3.504906	3.254313	0	6.728115
		NSGA-II	−18.265372	−12.520018	−9.587934	−3.254814	−6.303976	0

Table 5.13 The ICS for different objectives, based on various optimization algorithms for the second case under the second scenario

Scenario No.	Perspective	Optimization algorithm	Optimization algorithm					
			SOSA	TMS-EMSA	Continuous/discrete TMS-MSA	SS-HSA	SS-IHSA	NSGA-II
Second scenario	GENCOs' profit	SOSA	0	5.344122	12.724098	19.536983	16.604685	22.558135
		TMS-EMSA	−5.073014	0	7.005588	13.472855	10.689313	16.340743
		Continuous/discrete TMS-MSA	−11.287824	−6.546936	0	6.043858	3.442553	8.723988
		SS-HSA	−16.343882	−11.873196	−5.699395	0	−2.453046	2.527378
		SS-IHSA	−14.240153	−9.657041	−3.327986	2.514734	0	5.105669
		NSGA-II	−18.406069	−14.045589	−8.023977	−2.465076	−4.857653	0
	DISCOs' profit	SOSA	0	6.594201	13.599396	19.001324	16.509672	23.733336
		TMS-EMSA	−6.186266	0	6.571834	11.639585	9.302074	16.078862
		Continuous/discrete TMS-MSA	−11.971363	−6.166577	0	4.755244	2.561876	8.920769
		SS-HSA	−15.967322	−10.426038	−4.539385	0	−2.093802	3.976436
		SS-IHSA	−14.170216	−8.510427	−2.497884	2.138579	0	6.200055
		NSGA-II	−19.181037	−13.851671	−8.190145	−3.824362	−5.838090	0
	GENCOs' + DISCOs' profit ($/h)	SOSA	0	5.782925	13.032155	19.346964	16.571060	22.971257
		TMS-EMSA	−5.466785	0	6.852930	12.822522	10.198371	16.248683
		Continuous/discrete TMS-MSA	−11.529600	−6.413423	0	5.586736	3.130883	8.793163
		SS-HSA	−16.210688	−11.365215	−5.291134	0	−2.325910	3.036770
		SS-IHSA	−14.215416	−9.254557	−3.035835	2.381297	0	5.490382
		NSGA-II	−18.680185	−13.977520	−8.082459	−2.947268	−5.204628	0

superiority—positive sign—of the SOSA performance compared with the performances of the TMS-EMSA, continuous/discrete TMS-MSA, SS-HSA, SS-IHSA, and NSGA-II, respectively, from the standpoint of the obtained DISCOs' profits by corresponding optimization algorithms. In addition, the ICS displays 7.029277%, 10.616598%, 18.364991%, 14.634428%, and 22.347164% superiority—positive sign—of the SOSA performance compared with the performances of the TMS-EMSA, continuous/discrete TMS-MSA, SS-HSA, SS-IHSA, and NSGA-II, respectively, from the point of view of the obtained profits of all the CSC electricity markets including GENCOs and DISCOs by corresponding optimization algorithms. The ICS illustrates 5.987432% weakness—negative sign—of the TMS-EMSA performance compared with the performance of the SOSA, from the perspective of the obtained GENCOs' profits by the SOSA, and 3.108847%, 10.873362%, 7.001031%, and 14.960991% superiority—positive sign—of the TMS-EMSA performance compared with the performances of the continuous/discrete TMS-MSA, SS-HSA, SS-IHSA, and NSGA-II, respectively, from the perspective of the obtained GENCOs' profits by corresponding optimization algorithms.

Also, the ICS shows 7.570760% weakness—negative sign—of the TMS-EMSA performance compared with the performance of the SOSA, from the standpoint of the obtained DISCOs' profits by the SOSA, and 3.781622%, 10.098530%, 7.290193%, and 13.187917% superiority—positive sign—of the TMS-EMSA performance compared with the performances of the continuous/discrete TMS-MSA, SS-HSA, SS-IHSA, and NSGA-II, respectively, from the standpoint of the obtained DISCOs' profits by corresponding optimization algorithms. In addition, the ICS represents 6.567621% weakness—negative sign—of the TMS-EMSA performance compared with the performance of the SOSA, from the point of view of the obtained profits of all the CSC electricity markets including GENCOs and DISCOs by the SOSA, and 3.351719%, 10.591227%, 7.105673%, and 14.311866% superiority—positive sign—of the TMS-EMSA performance compared with the performances of the continuous/discrete TMS-MSA, SS-HSA, SS-IHSA, and NSGA-II, respectively, from the point of view of the obtained profits of all the CSC electricity markets including GENCOs and DISCOs by corresponding optimization algorithms. The results presented in Table 5.12 for the other optimization algorithms are analyzed in the same way. The results presented in Table 5.13 for all optimization algorithms are examined in the same way.

5.4 Conclusions

In this chapter, with a new point of view, the authors proposed a well-suited two-level computational-logical framework for a bilateral bidding mechanism within a competitive security-constrained electricity market. In the first level, the GENCOs and DISCOs maximized their profits. In the second level, the ISO cleared the competitive security-constrained electricity market by considering additional objectives containing the CO_2, SO_2, and NO_x emissions. The purpose of the proposed bi-level computational-logical framework is to provide a well-designed central core as a short-term operation slave problem for the strategic multilevel long-term expansion planning frameworks of the power system planning studies, which will be discussed in the next chapter. As a result, not only were the practical features and limitations of the proposed two-level computational-logical framework considered, but also a reduction in the production of atmospheric emissions was taken as a noteworthy short-term solution. In addition, to limit the computational burden to an acceptable level, without losing the generality of the bi-level computational-logical framework, the authors formulated a single hour of operation. The bi-level computational-logical framework can be, however, easily extended to the BBM in competitive day-ahead, security-constrained multi-period electricity markets.

The proposed two-level computational-logical framework was successfully implemented on the modified six-machine eight-bus test network. The results from the case studies demonstrate that the adoption of strategic behaviors by the GENCOs in the unilateral bidding mechanism of the CSC electricity market brought about a remarkable increase in their profits. In addition, increasing competition in the CSC electricity market, by importing DISCOs into the competition arena as a bilateral bidding mechanism, provided the expected positive results in CSC electricity market performance. As a consequence, the bilateral bidding mechanism encourages and facilitates competition among GENCOs and DISCOs and finally enhances the efficiency of the CSC electricity market, which is the most important target of the deregulation process of the power systems. It was also shown that the environmental aspects of both unilateral and bilateral bidding mechanisms gave rise to a significant reduction in the atmospheric emissions emitted by GENCOs, while at the same time the GENCOs' and DISCOs' profit increased.

The authors performed a thorough evaluation in which the SOSA, TMS-EMSA, and continuous/discrete TMS-MSA were compared with state-of-the-art optimization algorithms. Moreover, the performance of the optimization algorithms was thoroughly analyzed through a cost-saving index. The calculated results from the performance evaluation of the proposed optimization algorithms illustrate that the newly proposed SOSA is more economical and effective when compared to other optimization algorithms—especially with the state-of-the-art optimization algorithms. The newly proposed TMS-EMSA and continuous/discrete TMS-MSA, the SS-IHSA, the SS-HSA, and the NSGA-II are also ranked as the next most superior algorithms. As a result, the proposed modern music-inspired optimization algorithms addressed in Chap. 4, especially the SOSA, may be appropriate techniques for large-scale power system optimization problems with big data, in comparison to other optimization algorithms, when feasible regions of the solution space and/or dimensions of the problems with complexities—such as mixed-integer decision-making variables, multiple conflicting objective functions, nonlinearity, and discontinuity—increase.

Appendix 1: List of Abbreviations and Acronyms

AC	Alternative current
AEF	Atmospheric emission function
AECF	Atmospheric emission cost function
BBM	Bilateral bidding mechanism
CAISO	California independent system operator
CO_2	Carbon dioxide
CSC electricity market	Competitive security-constrained electricity market
CWF	Community welfare function
DC	Direct current
DISCOs	Distribution companies
ERCOT	Electric reliability Council of Texas
GAMS	General algebraic modeling system
GENCOs	Generation companies
ICS	Index of cost saving
ISO	Independent system operator
ISO-NE	Independent system operator-New England
LMP	Locational marginal price
MaxISO	Maximal independent system operator
MinISO	Minimal independent system operator
MISO	Midcontinent independent system operator
MMM-EMSA	Multi-stage computational multi-dimensional multiple-homogeneous enhanced melody search algorithm
MMS-EMSA	Multi-stage computational multi-dimensional single-inhomogeneous enhanced melody search algorithm
NEMMCO	National Electricity Market Management Company
NO_x	Nitrogen oxide
NSGA-II	Non-dominated sorting genetic algorithm-II
NYISO	New York independent system operator
PJM	Pennsylvania-New Jersey-Maryland
PSO	Particle swarm optimization
RESCOs	Retail energy service companies
SS-HSA	Single-stage computational single-dimensional harmony search algorithm
SS-IHSA	Single-stage computational single-dimensional improved harmony search algorithm
SO_2	Sulfur dioxide
SOSA	Symphony orchestra search algorithm
SFE	Supply function equilibrium
TMS-EMSA	Two-stage computational multi-dimensional single-homogeneous enhanced melody search algorithm
TMS-MSA	Two-stage computational multi-dimensional single-homogeneous melody search algorithm
TRANCOs	Transmission companies
UBM	Unilateral bidding mechanism
VIU	Vertically integrated utility

Appendix 2: List of Mathematical Symbols

Index:	
d	Index for DISCOs running from 1 to D
g	Index for GENCOs running from 1 to G
$i, -i$	Index for electricity market participants running from 1 to I
l	Index for transmission lines running from 1 to L
$n, \hat{n}$	Index for optimization algorithms running from 1 to N
q	Index for objective functions running from 1 to Q
Set:	
Ψ^{D}	Set of indices of DISCOs
Ψ^{G}	Set of indices of GENCOs
Ψ^{I}	Set of indices of electricity market participants
Ψ^{L}	Set of indices of transmission lines
Ψ^{N}	Set of indices of optimization algorithms
Ψ^{Q}	Set of indices of objective functions
Parameters:	
$f_l^{\max}$	Upper bound on the active electrical power flow in transmission line l [MW]
$f_l^{\min}$	Lower bound on the active electrical power flow in transmission line l [MW]
$\mathrm{W}_{\mathrm{CO_2}}$	Weighting coefficient of the CO_2 AECF in the second objective function of the CSC electricity market
$\mathrm{W}_{\mathrm{NO_x}}$	Weighting coefficient of the NO_x AECF in the second objective function of the CSC electricity market
$\mathrm{W}_{\mathrm{SO_2}}$	Weighting coefficient of the SO_2 AECF in the second objective function of the CSC electricity market
$\mathrm{W}_{\mathrm{OF_1}}^{\mathrm{CSC\text{-}EM}}$	Weighting coefficient of the first objective function of the CSC electricity market
$\mathrm{W}_{\mathrm{OF_2}}^{\mathrm{CSC\text{-}EM}}$	Weighting coefficient of the second objective function of the CSC electricity market
α_d	Benefit function coefficient of DISCO d [\$/MW^{2}h]
α_g	Generation cost function coefficient of GENCO g [\$/MW^{2}h]
$\hat{\alpha}_g$	Fuel cost function coefficient of GENCO g [\$/MW^{2}h]
$\alpha_g^{\mathrm{CO_2}}$	CO_2 atmospheric emission function coefficient of GENCO g [ton/MW^{2}h]
$\alpha_g^{\mathrm{SO_2}}$	SO_2 atmospheric emission function coefficient of GENCO g [ton/MW^{2}h]
β_d	Benefit function coefficient of DISCO d [\$/MWh]
β_g	Generation cost function coefficient of GENCO g [\$/MWh]
$\hat{\beta}_g$	Fuel cost function coefficient of GENCO g [\$/MWh]
$\beta_g^{\mathrm{CO_2}}$	CO_2 atmospheric emission function coefficient of GENCO g [ton/MWh]
$\beta_g^{\mathrm{SO_2}}$	SO_2 atmospheric emission function coefficient of GENCO g [ton/MWh]
γ_d	Benefit function coefficient of DISCO d [\$/h]
γ_g	Generation cost function coefficient of GENCO g [\$/h]
$\hat{\gamma}_g$	Fuel cost function coefficient of GENCO g [\$/h]
$\gamma_g^{\mathrm{CO_2}}$	CO_2 atmospheric emission function coefficient of GENCO g [ton/h]
$\gamma_g^{\mathrm{SO_2}}$	SO_2 atmospheric emission function coefficient of GENCO g [ton/h]

(continued)

$\delta_g^{NO_x}$	NO_x atmospheric emission coefficient of GENCO g [ton/MWh]
$\pi_g^{CO_2}$	Price of CO_2 atmospheric emission of GENCO g [$/ton]
$\pi_g^{NO_x}$	Price of NO_x atmospheric emission of GENCO g [$/ton]
$\pi_g^{SO_2}$	Price of SO_2 atmospheric emission of GENCO g [$/ton]
ρ_d^{max}	Upper bound on the purchased electrical power by DISCO d [MW]
ρ_d^{min}	Lower bound on the purchased electrical power by DISCO d [MW]
ρ_g^{max}	Upper bound on the generated electrical power by GENCO g [MW]
ρ_g^{min}	Lower bound on the generated electrical power by GENCO g [MW]
$\xi_{1,d}^{max}$	Upper bound of the bidding strategy parameter—slope—of DISCO d
$\xi_{2,d}^{max}$	Upper bound of the bidding strategy parameter—intercept—of DISCO d
$\xi_{1,d}^{min}$	Lower bound of the bidding strategy parameter—slope—of DISCO d
$\xi_{2,d}^{min}$	Lower bound of the bidding strategy parameter—intercept—of DISCO d
$\xi_{1,g}^{max}$	Upper bound of the bidding strategy parameter—slope—of GENCO g
$\xi_{2,g}^{max}$	Upper bound of the bidding strategy parameter—intercept—of GENCO g
$\xi_{1,g}^{min}$	Lower bound of the bidding strategy parameter—slope—of GENCO g
$\xi_{2,g}^{min}$	Lower bound of the bidding strategy parameter—intercept—of GENCO g
$\chi_g^{NO_x}$	NO_x atmospheric emission coefficient of GENCO g [ton/h]
Variables:	
$AECF_{CO_2}$	CO_2 atmospheric emission cost function [$/h]
$AECF_{NO_x}$	NO_x atmospheric emission cost function [$/h]
$AECF_{SO_2}$	SO_2 atmospheric emission cost function [$/h]
f_l	Active electrical power flow in transmission line l [MW]
OF^{CSC-EM}	The objective function of the CSC electricity market [$/h]
OF_1^{CSC-EM}	The first objective function of the CSC electricity market [$/h]
OF_2^{CSC-EM}	The second objective function of the CSC electricity market [$/h]
s_i	Strategy of electricity market participant i
s_i^*	Strategy of electricity market participant i at the Nash equilibrium point
s_{-i}^*	Strategy of the rivals of electricity market participant i at the Nash equilibrium point
u_i	Payoff profile of electricity market participant i
x_{BBM}	Decision-making variables of the first level, BBM, of the proposed bi-level computational-logical framework
x_{CSC-EM}	Decision-making variables of the second level, CSC electricity market, of the proposed bi-level computational-logical framework
x_i	Strategic option of electricity market participant i
κ_l	Active electrical power loss of transmission line l [MW]
$\lambda_d(\rho_d)$	Marginal benefit function of DISCO d [$/MWh]
λ_e	Electricity market price [$/MWh]
$\lambda_g(\rho_g)$	Marginal generation cost function of GENCO g [$/MWh]
λ_i	Offered price of the generated electrical power by electricity market participant i [$/MWh]
λ_{-i}	Offered price of the electrical power by the rivals of electricity market participant i [$/MWh]

(continued)

$\tilde{\lambda}_d(\rho_d)$	Bidding strategy function of DISCO d [$/MWh]
$\tilde{\lambda}_g(\rho_g)$	Bidding strategy function of GENCO g [$/MWh]
$\upsilon_d(\rho_d)$	Profit of DISCO d [$/h]
$\upsilon_g(\rho_g)$	Profit of GENCO g [$/h]
υ_i	Profit of electricity market participant i [$/h]
$\tilde{\upsilon}_d(\rho_d)$	Profit of DISCO d with consideration of the bidding strategy [$/h]
$\tilde{\upsilon}_g(\rho_g)$	Profit of GENCO g with consideration of the bidding strategy [$/h]
$\xi_{1,d}$	Bidding strategy parameter—slope—of DISCO d
$\xi_{2,d}$	Bidding strategy parameter—intercept—of DISCO d
$\xi_{1,g}$	Bidding strategy parameter—slope—of GENCO g
$\xi_{2,g}$	Bidding strategy parameter—intercept—of GENCO g
ρ_d	Purchased electrical power by DISCO d [MW]
ρ_g	Generated electrical power by GENCO g [MW]
ρ_i	Quantity of the generated electrical power by electricity market participant i [MW]
$\varsigma_d(\rho_d)$	Payment of DISCO d [$/h]
$\tilde{\varsigma}_d(\rho_d)$	Payment of DISCO d with consideration of the bidding strategy [$/h]
φ_i	Generation cost of electricity market participant i [$/h]
$\varphi_d(\rho_d)$	Benefit function of DISCO d [$/h]
$\varphi_g(\rho_g)$	Generation cost function of GENCO g [$/h]
$\hat{\varphi}_g(\rho_g)$	Fuel cost function of GENCO g [$/h]
ϑ_i	Revenue of electricity market participant i [$/h]
$\vartheta_g(\rho_g)$	Revenue of GENCO g [$/h]
$\tilde{\vartheta}_i$	Marginal revenue of electricity market participant i [$/MWh]
$\tilde{\vartheta}_g(\rho_g)$	Revenue of GENCO g with consideration of the bidding strategy [$/h]

Appendix 3: Input data

Table 5.14 Data of the transmission lines of the modified six-machine eight-bus test network

Line No.	From bus	To bus	R (p.u.)	X (p.u.)	Limit (MW)
1	1	2	0.0009	0.03	280
2	1	4	0.0009	0.03	140
3	1	5	0.000195	0.0065	380
4	2	3	0.000330	0.011	120
5	3	4	0.0009	0.03	230
6	4	5	0.0009	0.03	200
7	5	6	0.0006	0.02	300
8	6	1	0.00075	0.025	250
9	7	4	0.00045	0.015	250
10	7	8	0.00066	0.022	340
11	8	3	0.00054	0.018	240

Table 5.15 Data of the GENCOs of the modified six-machine eight-bus test network

		GENCOs' capacity		Generation cost function coefficients of GENCOs		
Bus No.	GENCO No.	ρ_g^{min} (MW)	ρ_g^{max} (MW)	α_g (\$/MW^{2}h)	β_g (\$/MWh)	γ_g(\$/h)
1	1	0	210	0.046	14	89.62
2	–	–	–	–	–	–
3	2	0	520	0.043	25	17.64
4	3	0	250	0.031	30	31.60
5	4	0	600	0.074	10	79.78
6	5	0	400	0.064	20	49.75
7	6	0	200	0.062	20	24.05
8	–	–	–	–	–	–

Table 5.16 Data associated with the demand and DISCOs of the modified six-machine eight-bus test network

		DISCOs' capacity		Benefit function coefficients of DISCOs			
Bus No.	DISCO No.	ρ_d^{min} (MW)	ρ_d^{max} (MW)	α_d (\$/MW^{2}h)	β_d (\$/MWh)	γ_d(\$/h)	Demand (MW)
1	–	–	–	–	–	–	0
2	1	0	600	−0.054	10	64.0143	300
3	2	0	450	−0.013	12	26.730	300
4	3	0	550	−0.031	15	43.618	300
5	–	–	–	–	–	–	0
6	4	0	400	−0.052	25	23.420	250
7	–	–	–	–	–	–	0
8	5	0	450	−0.034	25	18.654	250

Table 5.17 Data of the atmospheric emission functions related to the GENCOs of the modified six-machine eight-bus test network

GENCO No.	Bus No.	SO_2 atmospheric emission				CO_2 atmospheric emission				NO_x atmospheric emission		
		$\pi_g^{SO_2}$ ($/ton)	$\alpha_g^{SO_2}$ (ton/MW2 h)	$\beta_g^{SO_2}$ (ton/MWh)	$\gamma_g^{SO_2}$ (ton/h)	$\pi_g^{CO_2}$ ($/ton)	$\alpha_g^{CO_2}$ (ton/MW2 h)	$\beta_g^{CO_2}$ (ton/MWh)	$\gamma_g^{CO_2}$ (ton/h)	$\pi_g^{NO_x}$ ($/ton)	$\chi_g^{NO_x}$ (ton/h)	$\delta_g^{NO_x}$ (ton/MWh)
1	1	132	4.091×10^{-6}	-5.554×10^{-5}	0.6490×10^{-3}	220	4.91×10^{-6}	-6.66×10^{-5}	0.779×10^{-3}	74	2.857×10^{-3}	0.2×10^{-3}
2	3	132	2.543×10^{-6}	-6.047×10^{-5}	0.5638×10^{-3}	220	3.05×10^{-6}	-7.26×10^{-5}	0.677×10^{-3}	74	3.333×10^{-3}	$0. 5 \times 10^{-3}$
3	4	132	4.258×10^{-6}	-5.094×10^{-5}	0.4586×10^{-3}	220	5.11×10^{-6}	-6.11×10^{-5}	0.550×10^{-3}	74	8.000×10^{-3}	0. 001×10^{-3}
4	5	132	5.326×10^{-6}	-3.350×10^{-5}	0.3380×10^{-3}	220	6.39×10^{-6}	-4.26×10^{-5}	0.406×10^{-3}	74	2.000×10^{-3}	2×10^{-3}
5	6	132	4.258×10^{-6}	-5.094×10^{-5}	0.4586×10^{-3}	220	5.11×10^{-6}	-6.11×10^{-5}	0.550×10^{-3}	74	8.000×10^{-3}	0. 001×10^{-3}
6	7	132	6.131×10^{-6}	-5.555×10^{-5}	0.5151×10^{-3}	220	7.36×10^{-6}	-6.67×10^{-5}	0.618×10^{-3}	74	6.667×10^{-3}	0. 01×10^{-3}

Table 5.18 Data of the bidding strategy parameters associated with the CSC electricity market participants of the modified six-machine eight-bus test network

GENCO No.	GENCOs' bidding strategy parameters				DISCO No.	DISCOs' bidding strategy parameters			
	Slope		Intercept			Slope		Intercept	
	$\xi_{1,g}^{\min}$	$\xi_{1,g}^{\max}$	$\xi_{2,g}^{\min}$	$\xi_{2,g}^{\max}$		$\xi_{1,d}^{\min}$	$\xi_{1,d}^{\max}$	$\xi_{2,d}^{\min}$	$\xi_{2,d}^{\max}$
1	1	2.5	1	2.5	1	1	2.5	0	1
2	1	2.5	1	2.5	2	1	2.5	0	1
3	1	2.5	1	2.5	3	1	2.5	0	1
4	1	2.5	1	2.5	4	1	2.5	0	1
5	1	2.5	1	2.5	5	1	2.5	0	1
6	1	2.5	1	2.5	–	–	–	–	–

Table 5.19 Data of the weighting coefficients related to the proposed bi-level computational-logical framework

Weighting coefficients				
Weighting coefficients associated with the objective function		Weighting coefficients associated with the emission functions		
$W_{OF_1}^{CSC-EM}$	$W_{OF_2}^{CSC-EM}$	W_{SO_2}	W_{CO_2}	W_{NO_x}
1	1	1	1	1

Table 5.20 Parameter adjustments of the single-objective SOSA

No.	Single-objective SOSA parameters	Value
1	$BW_{g,p}^{\min}$	$0.4; \forall\{g \in [1],\ p \in [1]\}$
2	$BW_{g,p}^{\min}$	$0.35; \forall\{g \in [1,2],\ p \in [2]\}$
3	$BW_{g,p}^{\min}$	$0.30; \forall\{g \in [2],\ p \in [1]\}$
4	$BW_{g,p}^{\max}$	$0.95; \forall\{g \in [1],\ p \in [1]\}$
5	$BW_{g,p}^{\max}$	$0.9; \forall\{g \in [1,2],\ p \in [2]\}$
6	$BW_{g,p}^{\max}$	$0.85; \forall\{g \in [2],\ p \in [1]\}$
7	MNI-SIS	300
8	MNI-GISHMG	400
9	MNI-GISIME	500
10	NHMG	2
11	$PAR_{g,p}^{\min}$	$0.2; \forall\{g \in [1],\ p \in [1]\}$
12	$PAR_{g,p}^{\min}$	$0.3; \forall\{g \in [1,2],\ p \in [2]\}$
13	$PAR_{g,p}^{\min}$	$0.25; \forall\{g \in [2],\ p \in [1]\}$
14	$PAR_{g,p}^{\max}$	$2; \forall\{g \in [1],\ p \in [1]\}$

(continued)

Table 5.20 (continued)

No.	Single-objective SOSA parameters	Value
15	$PAR_{g,p}^{\max}$	$1.8; \forall\{g \in [1,2],\ p \in [2]\}$
16	$PAR_{g,p}^{\max}$	$1.6; \forall\{g \in [2],\ p \in [1]\}$
17	$PMCR_{g,\ p}$	$0.85; \forall\{g \in [1],\ p \in [1]\}$
18	$PMCR_{g,\ p}$	$0.8; \forall\{g \in [1,2],\ p \in [2]\}$
19	$PMCR_{g,\ p}$	$0.9; \forall\{g \in [2],\ p \in [1]\}$
20	$PMS_{g,\ p}$	$220; \forall\{g \in [1],\ p \in [1]\}$
21	$PMS_{g,\ p}$	$160; \forall\{g \in [1,2],\ p \in [2]\}$
22	$PMS_{g,\ p}$	$180; \forall\{g \in [2],\ p \in [1]\}$
23	PN_g	$2; \forall\{g \in [1,2]\}$
24	–	–

Table 5.21 Parameter adjustments of the single-objective TMS-EMSA

No.	Single-objective TMS-EMSA parameters	Value
1	$BW_p^{\min}$	$0.4; \forall\{\ p \in [1,2]\}$
2	$BW_p^{\min}$	$0.35; \forall\{\ p \in [3]\}$
3	$BW_p^{\min}$	$0.30; \forall\{\ p \in [4]\}$
4	$BW_p^{\max}$	$0.9; \forall\{\ p \in [1,2]\}$
5	$BW_p^{\max}$	$0.85; \forall\{\ p \in [3]\}$
6	$BW_p^{\max}$	$0.8; \forall\{\ p \in [4]\}$
7	MNI-SIS	500
8	MNI-GIS	700
9	$PAR_p^{\min}$	$0.2; \forall\{\ p \in [1,2]\}$
10	$PAR_p^{\min}$	$0.35; \forall\{\ p \in [3]\}$
11	$PAR_p^{\min}$	$0.30; \forall\{\ p \in [4]\}$
12	$PAR_p^{\max}$	$2; \forall\{\ p \in [1,2]\}$
13	$PAR_p^{\max}$	$1.75; \forall\{\ p \in [3]\}$
14	$PAR_p^{\max}$	$1.9; \forall\{\ p \in [4]\}$
15	$PMCR_p$	$0.85; \forall\{\ p \in [1,2]\}$
16	$PMCR_p$	$0.8; \forall\{\ p \in [3]\}$
17	$PMCR_p$	$075; \forall\{\ p \in [4]\}$
18	PMS_p	$220; \forall\{\ p \in [1,2]\}$
19	PMS_p	$180; \forall\{\ p \in [3]\}$
20	PMS_p	$160; \forall\{\ p \in [4]\}$
21	PN	4
22	–	–

Table 5.22 Parameter adjustments of the single-objective continuous/discrete TMS-MSA

No.	Single-objective continuous/discrete TMS-MSA parameters	Value
1	BW^{min}	0.4
2	BW^{max}	0.9
3	MNI-SIS	500
4	MNI-PGIS	700
5	PAR^{min}	0.2
6	PAR^{max}	2
7	PMCR	0.85
8	PMS	250
9	PN	4

Table 5.23 Parameter adjustments of the single-objective SS-HSA

No.	Single-objective SS-HSA parameters	Value
1	BW	0.75
2	HMCR	0.85
3	HMS	300
4	MNI	1200
5	PAR	0.9

Table 5.24 Parameter adjustments of the single-objective SS-IHSA

No.	Single-objective SS-IHSA parameters	Value
1	BW_{min}	0.4
2	BW_{max}	0.9
3	HMCR	0.85
4	HMS	300
5	MNI	1200
6	PAR_{min}	0.2
7	PAR_{max}	2

Table 5.25 Parameter adjustments of the NSGA-II

No.	NSGA-II parameters	Value
1	Crossover probability	0.95
2	Mutation probability	0.01
3	Population	400
4	MNI	1200

References

1. L.L. Lai, *Power System Restructuring and Deregulation, Trading Performance and Information Technology* (John Wiley and Sons, UK, 2001)
2. M. Shahidehpour, M. Alomoush, *Restructured Electrical Power Systems, Operation Trading and Volatility* (Dekker, New York, 2001)
3. F.L. Denny, D.E. Dismukes, *Power System Operation and Electricity Market* (CRC Press, Boca Raton, FL, 2002)
4. D. Gan, D. Feng, J. Xie, *Electricity Markets and Power System Economics*, 1st edn. (CRC Press, Boca Raton, FL, 2013)
5. E. Vaahedi, *Practical Power System Operation* (John Wiley and Sons, Hoboken, NJ, 2014)
6. M. Shivaie, M.T. Ameli, K. Jenab, Towards new competitive electricity markets based on information transparency: hypaethral market. Int. Trans. Electr. Energy Syst. **25**(8), 1621–1643 (2015)
7. J. Wang, Z. Zhou, A. Botterud, Modeling methods for GenCo bidding strategy optimization in the liberalized electricity spot market-A state-of-the-art review. Energy **36**, 4686–4700 (2011)
8. J. von Neumann, O. Morgenstern, *Theory of Games and Economic Behavior* (John Wiley and Sons, New York, 1944)
9. M.J. Osborne, A. Rubinstein, *A Course in Game Theory* (The MIT Press, Cambridge, MA, 1994)
10. D.S. Kirschen, G. Strbac, *Fundamentals of Power System Economics* (John Wiley and Sons, Chichester, 2004)
11. T. Li, M. Shahidehpour, Strategic bidding of transmission-constrained GENCOs with incomplete information. IEEE Trans. Power Syst. **20**(1), 437–447 (2005)
12. Y. Ma, C. Jiang, Z. Hou, C. Wang, The formulation of the optimal strategies for the electricity producers based on the particle swarm optimization algorithm. IEEE Trans. Power Syst. **21**(4), 1663–1671 (2006)
13. E. Bompard, Y. Ma, R. Napoli, G. Abrate, The demand elasticity impacts on the strategic bidding behavior of the electricity producers. IEEE Trans. Power Syst. **22**(1), 188–197 (2007)
14. M. Fampa, L.A. Barroso, D. Candal, L. Simonetti, Bilevel optimization applied to strategic pricing in competitive electricity markets. Comput. Optim. Appl. **39**(2), 121–142 (2008)
15. A.D. Yucekaya, J. Valenzuela, G. Dozier, Strategic bidding in electricity markets using particle swarm optimization. Electr. Pow. Syst. Res. **79**, 335–345 (2009)
16. E. Bompard, W. Lu, R. Napoli, X. Jiang, A supply function model for representing the strategic bidding of the producers in constrained electricity markets. Int. J. Electr. Power Energy Syst. **32**, 678–687 (2010)
17. J. Wang, Z. Zhou, A. Botterud, An evolutionary game approach to analyzing bidding strategies in electricity markets with elastic demand. Energy **36**, 3459–3467 (2011)
18. A.K. Jain, S.C. Srivastava, S.N. Singh, L. Srivastava, Strategic bidding in transmission constrained electricity markets using artificial bee colony algorithm. Electr. Power Compon. Syst. **40**, 1768–1788 (2012)
19. X. Luo, C. Chung, H. Yang, X. Tong, Optimal bidding strategy for generation companies under CVaR constraint. Int. Trans. Electr. Energy Syst. **24**, 1369–1384 (2014)
20. E.G. Kardakos, C.K. Simoglou, A.G. Bakirtzis, Optimal bidding strategy in transmission-constrained electricity markets. Electr. Pow. Syst. Res. **109**, 141–149 (2014)
21. F. Gao, G.B. Sheble, K.W. Hedman, C.N. Yu, Optimal bidding strategy for GENCOs based on parametric linear programming considering incomplete information. Int. J. Electr. Power Energy Syst. **66**, 272–279 (2015)
22. S. Koschker, D. Möst, Perfect competition vs. strategic behaviour models to derive electricity prices and the influence of renewables on market power. OR Spectr. **38**(3), 661–686 (2016)
23. A.J. Wood, B. Wollenberg, *Power Generation, Operation and Control*, 2nd edn. (John Wiley and Sons, New York, 1996)

24. M. Shahidehpour, H. Yamin, Z. Li, *Market Operations in Electric Power Systems* (John Wiley and Sons, New York, 2002)
25. E. Litvinov, Design and operation of the locational marginal prices-based electricity markets. IET Gener. Transm. Distrib. **4**(2), 315–323 (2010)
26. R.D. Christie, B.F. Wollenberg, I. Wangensteen, Transmission management in the deregulated environment. Proc. IEEE **88**, 170–195 (2000)
27. M. Shivaie, M.T. Ameli, A stochastic framework for multi-stage generation expansion planning under environmental and techno-economic constraints. Electr. Power Compon. Syst. **44**(17), 1917–1934 (2016)
28. M.A. Abido, Environmental/economic power dispatch using multi-objective evolutionary algorithms. IEEE Trans. Power Syst. **18**(4), 1529–1537 (2003)
29. M.A. Abido, Multi-objective evolutionary algorithms for power dispatch problem. IEEE Trans. Evol. Comput. **10**(3), 315–329 (2006)
30. L. Wang, C. Singh, Environmental/economic power dispatch using a fuzzified multi-objective particle swarm optimization algorithm. Electr. Pow. Syst. Res. **77**, 1654–1664 (2007)
31. I. Kockar, A.J. Conejo, J.R. McDonald, Influence of the emissions trading scheme on generation scheduling. Int. J. Electr. Power Energy Syst. **31**, 465–473 (2009)
32. H. Li, Y. Li, Z. Li, A multiperiod energy acquisition model for a distribution company with distributed generation and interruptible load. IEEE Trans. Power Syst. **22**(2), 588–596 (2007)
33. K. Deb, A. Pratap, S. Agarwal, T. Meyarivan, A fast and elitist multiobjective genetic algorithm: NSGA-II. IEEE Trans. Evol. Comput. **6**(2), 182–197 (2002)

Chapter 6
Power Systems Planning

6.1 Introduction

During the last decade, the remarkable growth in worldwide energy demand and emerging issues in power systems—including unbundling, deregulation, integration of renewable resources, and global environmental policies—have not only introduced new challenges and uncertainties, but also escalated existing ones. These issues, especially unbundling and deregulation, have given rise to the need for determining power system adequacy and receiving proper expansion planning signals, which, normally, are only possible after running operation actions under the framework of a competitive electricity market. In this way, well-adapted design and planning are increasingly needed to optimize expansion of the power systems, particularly the generation and transmission systems, as their pivotal components. In addition, it is widely known that optimization problems associated with power system planning studies are large-scale, complicated, and tough combinatorial problems having a non-convex, nonlinear, mixed-integer nature. This means that the number of local optimal points is greatly increased by enlarging the size of the power system under study. Traditional optimization algorithms encounter, therefore, various difficulties in finding the global optimal point.

For the reasons identified above, there are many studies being conducted that concentrate on the emerging challenges in power system planning studies [1–10]. The reported strategies in these studies can be classified as two different strategies. The first strategy considers only a simplified framework that corresponds to power system planning with features such as being a single-objective model, disregarding uncertainties, taking into account the traditional structure, lacking a consideration of practical aspects, and so forth [1–5]. As a result, the first strategy deals only with theoretical aspects and not practical applications. The second strategy, however, regards a more complex and realistic framework. To do so, the second strategy, with some simplifications (e.g., linearization of the nonlinear models), considers the

M. Kiani-Moghaddam et al., *Modern Music-Inspired Optimization Algorithms for Electric Power Systems*, Power Systems,
https://doi.org/10.1007/978-3-030-12044-3_6

solution process; therefore, its outputs are not based on actual conditions [6–10]. To illustrate, the electricity market participants have nonlinear behaviors in the market-clearing process. If these behaviors are linearized, it could bring about an unrealistic and unnecessary result. This destructive assumption is one of the most important reasons for the crisis in California's electricity market [11]. It is, consequently, essential to consider holistic strategies for power system planning studies without taking simplification of the vital aspects into account.

In this chapter, then, the authors focus on not only developing existing frameworks of power system planning studies by tackling the emerging issues, but also raising new optimization algorithms for solving combinatorial problems concerned with these studies. In this regard, the following five targets are contemplated and scrutinized in the field of power system expansion planning studies:

- Target 1: Incorporating a short-term, bi-level computational-logical operation framework, which was addressed in Chap. 5 in the power system planning studies, as a well-adapted long-term, multilevel computational-logical planning framework.
- Target 2: Providing a flexible strategy for handling severe uncertainties and figuring out the effects of risk-aversion and risk-taking decision-making policies in power system planning studies.
- Target 3: Developing a well-suited mechanism for value-based reliability assessment at both hierarchical level I (HL-I) and hierarchical level II (HL-II) in power system planning studies.
- Target 4: Considering the practical techno-economic and environmental sustainability characteristics in power system planning studies.
- Target 5: Studying the feasibility of implementing the proposed frameworks under real-world, large-scale test systems by using the offered modern meta-heuristic music-inspired optimization algorithms, which were addressed in Chap. 4, and comparing the results with powerful state-of-the-art multi-objective optimization algorithm.

In this chapter, the authors do not address the preliminary details of power system planning and design. It is assumed that readers are already familiar with these fundamental concepts. If required, readers should reference related studies that cover the preliminary details of power system planning and design.

The authors provide the following reading map of this chapter. In Sect. 6.2, an overview of power system planning studies is briefly described. In Sect. 6.3, the proposed framework for pseudo-dynamic generation expansion planning (PD-GEP) is discussed in detail, including mathematical modeling, solution method, numerical results, and discussion of the results. In Sect. 6.4, the offered framework for pseudo-dynamic transmission expansion planning (PD-TEP) is addressed in detail, including mathematical modeling, solution method, case studies, and analyses on the determined results. Section 6.5 represents the detail of the proposed framework for coordination of pseudo-dynamic generation and transmission expansion planning (PD-G&TEP), including mathematical modeling, solution method, numerical results, and discussion of the calculated results. In Sect. 6.6, the proposed framework for pseudo-dynamic distribution expansion planning (PD-DEP) is described in detail, including mathematical modeling, solution method, case studies, and investigation into the determined results. Finally, the chapter ends with a brief summary and some concluding remarks in Sect. 6.7.

6.2 A Brief Review of Power System Planning Studies

In this section, the authors discuss the different perspectives of power system planning studies that are involved in the proposed PD-GEP, PD-TEP, PD-G&TEP, and PD-DEP frameworks.

6.2.1 Why Do the Power Systems Need the Expansion Planning?

Nowadays, most facets of human life are dependent on electrical energy. Therefore, reliable, secure, and affordable energy provided for consumers at the consumption centers has been one of the key factors in achieving better standards of life. This is one of the most significant tasks of the relevant institutions in each country, in addition to its undeniable duty in economic development and social prosperity, and which could play an important role in national security. Electrical energy is generally generated at central power plants and then is transferred to the consumption centers through transmission and distribution systems. In other words, electrical energy is oftentimes generated long distances away from the consumption centers. On the other hand, population growth, industrial development, advancement of technology, climate changes, and so on have recently brought about considerable growth in worldwide energy consumption. Therefore, not only should power plants have the ability to generate more electrical energy, but also transmission and distribution systems should have sufficient capacity to transfer and distribute the generated electrical energy. In this way, if even one of these sections does not have the necessary ability, the power system will encounter a wide-scale blackout. As a consequence, given that sections associated with the power systems are generally operated within nominal capacity ranges, well-planned design and expansion are increasingly required to control and overcome the sizeable global energy consumption growth.

6.2.2 A Brief Review of Power System Planning Structure

In the broadest sense, power system planning studies can be divided into four components: (1) electrical power demand forecasting (EPDF), (2) generation expansion planning (GEP), (3) transmission expansion planning (TEP), and, (4) distribution expansion planning (DEP). Detailed descriptions of these components are provided in the following subsections. In the power system literature, the EPDF process is employed by the electrical power-providing companies to anticipate the future demand for a given horizon with regard to statistical data, behaviors of consumers, available system information, and so forth. This process helps these companies to make the right decisions when buying and selling electrical energy, developing infrastructure, etc. The EPDF process is also the first crucial step for power system planning. From the standpoint of the time period, the EPDF process can be classified into three different classes, as follows:

- Class I: Short-term EPDF that has a time period from 1 h up to 1 day.
- Class II: Mid-term EPDF that has a time period from 1 day to 7 years.
- Class III: Long-term EPDF that has a time period greater than 7 years.

In the power system literature, many techniques have been employed to puzzle out the EPDF process over the last few decades. These techniques can be generally categorized into three different types: (1) conventional techniques, (2) modified conventional techniques, and, (3) soft computing techniques. Readers interested in a comprehensive discussion on this topic are referred to the work by Alfares and Nazeeruddin [12] and Singh et al. [13]. After completing the EPDF process, power system planning will be performed at all levels—i.e., generation, transmission, and distribution levels. Generally speaking, the GEP studies are the first step in power system planning. The main purpose of these studies is to satisfy the adequacy of the generation systems for supplying the demand growth over the planning horizon. The GEP studies ensure that the generation system facilities have sufficient capacity to meet increasing demand, given that none of the techno-economic and environmental restrictions are violated in different operating modes of a power system. In a similar definition, the TEP studies ensure that the transmission system facilities have sufficient capacity to transfer the electrical energy from the central power plants to consumption centers over the planning horizon, while considering demand growth and provided that none of the techno-economic and environmental limitations are violated in different operating modes. In other words, in TEP studies, the planner seeks to maintain or restore the adequacy of the transmission facilities over the planning horizon. The DEP studies are also the final step in power system planning that provides the power delivery infrastructure to the consumption centers. These studies are performed in order to find an optimal strategy for constructing new distribution facilities and reinforcing the existing ones while simultaneously satisfying techno-economic constraints.

6.2.3 Power System Planning Issues

In general, power system planning studies can be classified from different point of views. Here, however, these classifications are addressed from the perspective of some important characteristics.

6.2.3.1 From the Standpoint of Power System Structure

Power system planning studies, from the standpoint of power system structure, have two main categories: (1) regulated and (2) deregulated/unbundled structures. In the regulated structure, a vertically integrated utility has the task of supplying the demand. This utility is responsible for planning and operation actions associated with the generation, transmission, distribution, and end-user levels in a specified geographic area. This structure can also be called a "monopoly at all levels." In this monopolistic environment, planning, operation, and coordination among these different levels are somewhat simple and ineffective. The utility predicts the demand growth and then expands the power system to supply electrical energy with the aim of minimizing investment costs subject to technical and operational constraints of the power system equipment. In simple terms, the utility concentrates on selecting the least-cost strategies to expand the power system. In the regulated structure, power system expansion planning at each level is generally composed of three main steps: (1) generate a candidate plan, (2) perform a financial analysis, and, (3) implement a technical analysis. These three steps are carried out separately and in order; this ensures the feasibility of the expansion plan. On the other hand, to end the monopoly in power systems, the power industries in many countries are forced to experience fundamental changes in their regulated structure. In this case, and in order to create a deregulated/unbundled structure, the vertically integrated utility is deconstructed and opened for competition with different dynamic participants. Some of the pioneering countries in the field of deregulation in power systems are the United States, the United Kingdom, Norway, Finland, Sweden, and some countries in South America. The aim of power system deregulation/unbundling is to facilitate competition and enhance efficiency and productivity at all levels of the power systems. Increasing the competition will, therefore, affect the quality and cost of different services in power systems. Broadly speaking, there are many other reasons that can be different for various regions and countries. The authors will not dwell on the reasons for power system deregulation or the changes resulting from this process and related topics, due to noncompliance with the targets of the chapter. For a detailed description of this topic, please refer to the work by Lai [14].

The deregulated structure has created fundamental changes in power system planning and, therefore, forced the adoption of strategies in regulated structures that are not practically applicable in this new circumstance. In the regulated structure, as stated earlier, the main concern of the utility is to find the expansion plans with the lowest investment and operation costs. In the deregulated structure, however, the planner should not only minimize the investment and operation costs, but also maximize the participants'/investors'/stakeholders' profits. In simple terms, in the deregulated structure, expansion planning decisions are made by taking into account the participants'/investors'/stakeholders' gain and interests, as a profit-driven framework. Please refer to the work by Mazer [15] for a comprehensive discussion of the differences of regulated and deregulated planning. As a result, it is necessary to develop new strategies in order to provide a well-designed model in deregulated power systems.

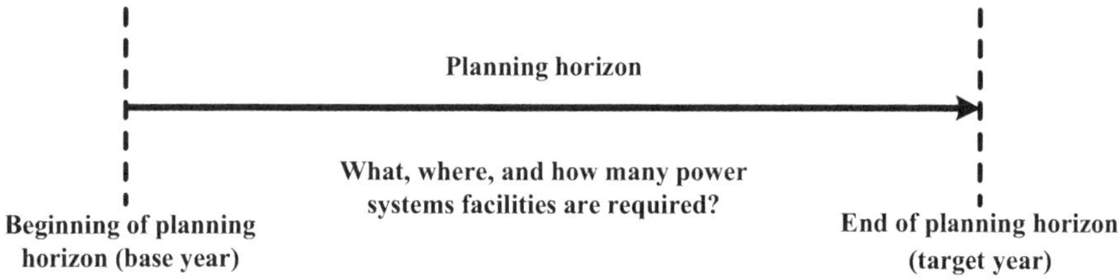

Fig. 6.1 Static model of power system planning

6.2.3.2 From the Standpoint of the Planning Horizon

Power system planning studies, from the standpoint of the planning horizon, have three main categories: (1) static or single-period model, (2) dynamic or multi-period model, and, (3) pseudo-dynamic or pseudo-multi-period model. In the static model, the entire planning horizon is represented by a single period. Figure 6.1 shows how the planning is done for this single-period planning horizon.

This single-period planning horizon is composed of a base year (i.e., the beginning of the planning horizon) and a target year (i.e., the end of the planning horizon). The power system requirements, or demand, are illustrated for the target year, and the optimal expansion plan is made for the base year. In this model, the planner does not consider the time aspect when simplifying the optimization problem associated with power system planning. In simple terms, the planner expands the power system based on techno-economic indices/constraints in order to satisfy three questions: (1) what, (2) where, and (3) how many new power system facilities or expansion options are required. One of the great advantages of using the static model—especially for the multilevel, large-scale, and real-world optimization problems—is to reduce computational burden and to decrease runtime of the program. This model, however, has two major disadvantages, which, given those reasons, is why it is less often used in power system planning. First, it is apparent that the demand in the target year has a much higher value compared to the base year, due to the fact that power system planning is generally a long-term optimization problem. It is possible, therefore, that a large fraction of the capacity corresponds to the optimal expansion plan not required and operated until the target year. Second, the forecasting of the power system conditions usually contains errors and is not accurate. Thus, if the actual conditions of the power system in the target year do not match the forecasted conditions (e.g., the actual demand is lower than the predicted one), then changes applied to the expansion plan will not be possible, since this plan has already been built. As a consequence, the static model leads to impractical results for real-world, large-scale optimization problems in power system planning. One way to cope with these disadvantages is to use the dynamic model. In the dynamic model, the entire planning horizon is split into several time periods, and power system planning is performed simultaneously for all time periods so that the expansion plans will be determined for all time periods simultaneously, as shown in Fig. 6.2. The power system requirements are illustrated for the target year of each time period, and the

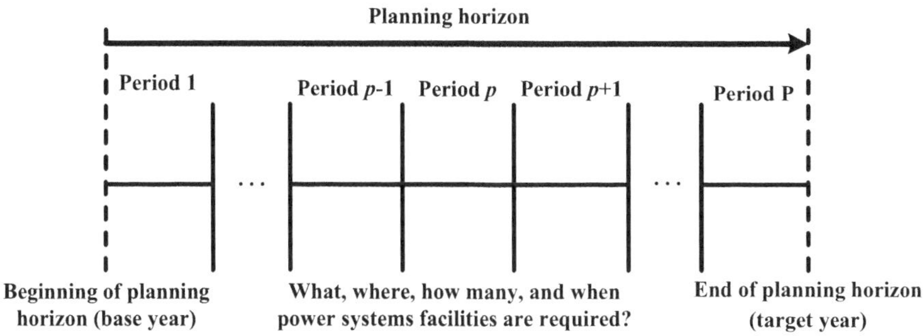

Fig. 6.2 Dynamic model of power system planning

expansion plans are made for the base year of each time period. The main goal of dynamic power system planning is to determine what, where, how many, and when new expansion options should be added to an existing power system in which different objectives subject to techno-economic limitations are satisfied. The utilization of the dynamic model in power system planning—especially for the multilevel, large-scale, and real-world optimization problems—leads to massive complexities.

As a result, finding the optimal expansion plan in a reasonable amount of time is extremely difficult and requires both enormous computational efforts and very powerful computing equipment. One of the simplest ways to restrict the computational burden to an acceptable level without losing the generality of the dynamic model is utilization of the pseudo-dynamic model. The only difference between the dynamic and pseudo-dynamic models is in the implementation process. As indicated earlier, power system planning is done simultaneously for all time periods of the planning horizon in the dynamic model and performed consecutively for all time periods of the planning horizon in the pseudo-dynamic model. In simple terms, in the pseudo-dynamic model, the synchronization of time periods is ignored and time periods are accomplished in a sequential order. In both dynamic and pseudo-dynamic models, the results from previous time periods also affect the calculated results of the subsequent time periods. For a comprehensive overview of power system planning from the standpoint of the planning horizon, please refer to the work by Seifi and Sepasian [16].

6.2.3.3 From the Standpoint of the Uncertainties

Power system planning studies from the standpoint of uncertainties have two main categories: (1) deterministic and (2) nondeterministic models. In the deterministic model, power system planning is modeled for the most critical condition, irrespective of its probability of occurrence. In this case, the power systems will be planned for the maximum value of each parameter, each of which is considered a deterministic parameter (i.e., demand, financial resources, fuel availability, and so forth). In the nondeterministic model, however, power system planning is modeled for all possible cases of the power systems that may occur in the future, taking into account the probability of their occurrence. It is a challenging task to classify the uncertainty parameters in power system planning in an orderly manner. Obviously, the classifications can largely depend on criteria, and there is no easy guideline for setting out the criteria in the literature. As the criteria may vary, detailed classifications can be an impossible task for a research study. Of studies found in the literature on power systems, the most technical investigations only attempt to focus on one dimension of the characteristics of the uncertainties—the nature of the uncertainty parameter. However, different dimensions associated with the characteristic of the uncertainty parameters can be useful to the planner and model developers for their classifications. Therefore, with inspiration from Walker et al. [17], the authors concentrated on three dimensions associated with the characteristics of the uncertainty parameters. According to Walker et al. [17], the uncertainty parameters can be classified based on three different dimensions, as follows:

- Location of the uncertainty parameter: This dimension refers to where the uncertainty parameter reveals itself within the model complex.
- Level of the uncertainty parameter: This dimension is related to the extent of the planner's ignorance of the uncertainty.
- Nature of the uncertainty parameter: This dimension depends on the answer to the question of whether the uncertainty is a result of insufficient knowledge of the planner or due to its inherent variability.

More detailed descriptions of these dimensions are beyond the scope of this chapter, but the interested reader may look to the work by Walker et al. [17] for a thorough discussion of these aspects. In order to create an analytical tool—i.e., conceptual framework—to categorize and report the different dimensions of the uncertainty parameters in power system planning, the authors have used the concept of an uncertainty matrix. The uncertainty matrix is an efficient tool that helps the model developers and users—and especially the planners—to identify, evaluate, and prioritize the basic characteristics of the uncertainty parameters in a systematic manner. In simple terms, this matrix provides a systematic and graphical overview of the basic characteristics of the uncertainty parameters. The location, level, and nature of the uncertainty parameters are employed in order to construct the uncertainty matrix. Table 6.1 shows a sample uncertainty matrix. The first column of the uncertainty matrix explains the source of the uncertainty parameters. The next five columns represent the location type of the uncertainty parameters. The next three columns describe the level of the uncertainty parameters, while the last two columns represent the nature of the uncertainty parameters for each location, respectively. According to the original uncertainty matrix [17], the first dimension, or location, of the uncertainty parameter is composed of five distinct forms: (1) context, (2) model uncertainty, (3) inputs, (4) parameter uncertainty, and, (5) model outcome uncertainty. The second dimension, or level, of the uncertainty parameter consists of three different forms: (1) statistical uncertainty, (2) scenario uncertainty, and, (3) recognized ignorance. The third dimension, or nature, of the uncertainty parameter is

Table 6.1 The constructed uncertainty sample matrix

Source of the uncertainty parameter	Location of the uncertainty parameter					Level of the uncertainty parameter			Nature of the uncertainty parameter	
	Context	Model uncertainty	Input	Parameter uncertainty	Model outcome uncertainty	Statistical uncertainty	Scenario uncertainty	Recognized ignorance	Epistemic uncertainty	Variability uncertainty
Inflation rate	–	–	√	–	–	√	–	–	–	√
Fuel availability	–	–	√	–	–	–	√	–	–	√
Solvers (local and global optimum)	–	√	–	–	–	–	–	√	√	–
Environmental policies	–	–	√	–	–	–	√	–	–	√
Strategic behavior of the market participants	–	–	–	–	√	–	–	√	–	√

composed of two distinct forms: (1) epistemic uncertainty and (2) variability uncertainty. To illustrate, the authors employ the characteristics of the uncertainty concerned with the inflation rate, the fuel availability, the solvers, the environmental policies, and the strategic behavior of the electricity market participants to construct a sample uncertainty matrix (see again Table 6.1). The uncertainty matrix can be used by the model developers and planners to classify the basic characteristics of the existing uncertainty parameters in their mathematical models in a systematic manner.

6.2.3.4 From the Standpoint of the Solving Algorithms

In this section, the authors will not investigate all of the characteristics of the solving algorithms again, as these features were already examined in the early chapters of the book. Here, the main goal is only to provide a classification for the solving algorithms employed in power system planning.

Power system planning studies from the standpoint of the solving algorithm can be divided into three main categories: (1) mathematical optimization algorithms, (2) heuristic optimization algorithms, and, (3) meta-heuristic optimization algorithms. Mathematical optimization algorithms employ a derivative-based calculation procedure to solve the planning problem. The exact mathematical formulation of this planning problem is, hence, required. In this case, power system planning is expressed as an optimization problem that contains one or more objective functions subject to a set of constraints. This problem is intrinsically a large-scale, non-convex combinatorial problem having a nonlinear, mixed-integer nature; as such, the mathematical optimization algorithms are faced with some great practical obstacles to finding the optimal expansion plan, especially when feasible regions of the solution space and/or dimensions of the planning problem increase. As a result, these algorithms may not be efficient in solving planning problems. The heuristic optimization algorithms are, then, considered as an alternative. Unlike the mathematical optimization algorithms, the heuristic algorithms employ a step-by-step process to generate, evaluate, and select the expansion plans in power system planning. This process can be accomplished by considering and ignoring developer's/user's/planner's help as interactive and noninteractive implementations, respectively. The heuristic optimization algorithms were inspired by heuristic principles such as logical rules, statistical rules, empirical rules, sensitivity analysis, and so on. In these optimization algorithms, the optimization process is continued until the given stopping index is satisfied. The stopping index is defined as a condition in which the algorithm fails to get a better solution for power system planning. Utilization of the heuristic optimization algorithms is very attractive for solving small-scale problems, because these algorithms have a poor computational structure. Moreover, these algorithms not only have deficient scalability and flexibility in real and binary solution space, but also are too weak to guarantee finding a near-optimal and/or optimal solution. In order to cope with drawbacks corresponding to existing traditional algorithms, the modern meta-heuristic optimization algorithms were developed. The modern meta-heuristic optimization algorithms are more applicable and have more effective techniques for solving a wide range of optimization problem types than do existing traditional algorithms. These algorithms are capable of finding the optimal solution in an appropriate computational time owing to an exploitation and exploration ability. In addition, they are flexible in solving large-scale problems, efficient in cases where existing traditional algorithms get stuck at local points, and are effective in parallel processing situations. For a full overview of power system planning from the standpoint of the solving algorithm, please refer to the work by Seifi and Sepasian [16].

6.3 Pseudo-Dynamic Generation Expansion Planning: A Strategic Tri-level Computational-Logical Framework

In this section, a well-designed strategic tri-level computational-logical framework is presented for the PD-GEP in deregulated environments. The proposed strategic tri-level computational-logical framework is divided into a long-term planning master problem and a short-term operational slave problem. The main task of the short-term operational slave problem is providing the competitive security-constrained (CSC) electricity market outcomes for the long-term planning master problem by considering both generation companies' (GENCOs) and distribution companies' (DISCOs) bids. The main task of the long-term planning master problem is also providing the optimal generation expansion plans compatible with the results of the short-term operational slave problem. In simple terms, the proposed strategic tri-level computational-logical framework evaluates the impacts of the generation expansion plans on strategic behaviors of both GENCOs and DISCOs and vice versa. In the first level, the bidding strategy is employed by both GENCOs and DISCOs to maximize their profit. In the second level, the CSC electricity market-clearing process is carried out from the perspective of the independent system operator (ISO) with the aim of maximizing the community welfare function (CWF). The third level of the proposed framework is also broken down into two different layers: (1) local and (2) global. In the local layer, each GENCO separately optimizes the expected profit, the capacity payment, and the investment cost, as a self-optimization approach. In the global layer, however, the obtained optimal generation expansion plans are aggregated and then evaluated by the ISO, with regard to the reserve margin of generation, the competitive conditions, and the reliability requirements.

Figure 6.3 illustrates a conceptual-view diagram of the proposed strategic tri-level computational-logical framework. The following assumptions are employed in the mathematical description process of the proposed strategic tri-level computational-logical framework:

- Hypotheses 1–10, discussed in Sect. 5.3 of Chap. 5, are considered here.
- Hypothesis 11: Severe twofold uncertainties associated with the demand and price parameters are considered and applied in the proposed strategic tri-level computational-logical framework.

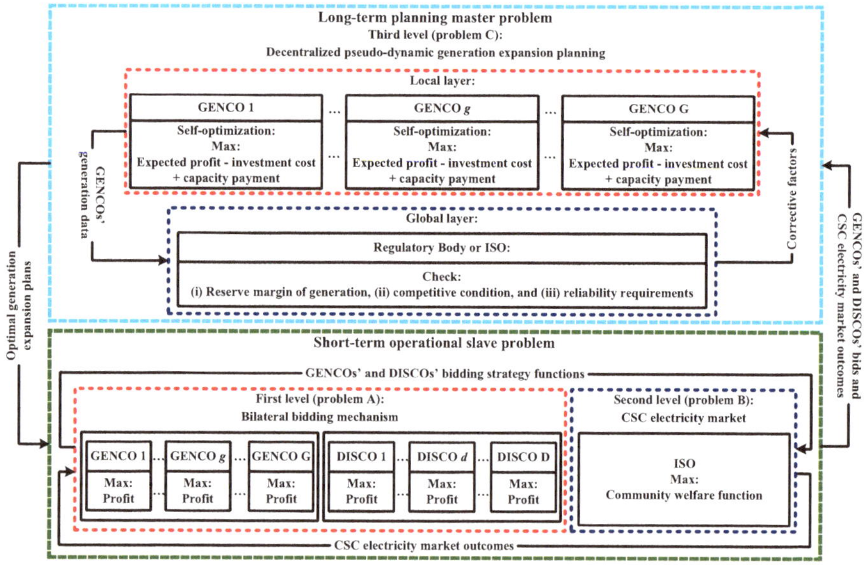

Fig. 6.3 Conceptual-view diagram of the proposed strategic tri-level computational-logical framework

- Hypothesis 12: A well-founded information-gap decision theory (IGDT) is widely employed to handle risks of the PD-GEP problem arising from severe twofold uncertainties.

The attribute table of Table 5.2 was created to compare the most important features of the previous GEP framework reported in the literature and the PD-GEP framework proposed in this chapter. Only the most important and relevant journal papers from the literature are reported here. Additional technical publications on generation expansion planning can be found in a review paper by Sadeghi et al. [29]. Additionally, Table 5.2 provides features related to the bi-level computational-logical framework (i.e., short-term operational slave problem). Finally, due to the fact that most technical studies in this area do not use a tri-level framework, the authors refrained from mentioning them in the attribute table of Table 6.2.

6.3.1 Mathematical Model of the Deterministic Strategic Tri-level Computational-Logical Framework

In this section, the mathematical model of the proposed strategic tri-level computational-logical framework is developed within a deterministic environment and discussed more in depth.

Table 6.2 Attributes of previous frameworks reported in the GEP literature and the strategic tri-level computational-logical framework proposed in this study

Reference	Type of expansion planning	Multilevel framework	Static or multi-period framework	Nondeterministic framework	Environmental constraints	System security constraint	Loss consideration	Uncertainty consideration	Strategic behaviors of CSC electricity market participants	Impacts of expansion plans on strategic behavior and vice versa	Solution method
[18]	Only capacity expansion	Bi-level	Static	No	No	Yes	No	Onefold: unit outage	No	No	Coevolutionary computation approach
[19]	Only capacity expansion	Single-level	Multi-period	No	Yes: only CO_2 atmospheric emission	No	No	No	No	No	Mathematical programming
[20]	Only capacity expansion	Single-level	Multi-period	No	Yes: CO_2 and NO_x atmospheric emissions	No	No	No	No	No	General Algebraic Modeling System (GAMS)
[21]	Only capacity expansion	Bi-level	Static	No	No	Yes	No	No	No	No	GAMS
[22]	Only capacity expansion	Bi-level	Multi-period	Yes:	No	No	No		No	No	GAMS
[23]	Only capacity expansion	Single-level	Static	No	Yes: only CO_2 emission	No	No	No	No	No	Mathematical programming
[24]	Only capacity expansion	Single-level	Multi-period	No	Yes: only CO_2 atmospheric emission	No	No	No	No	No	Mathematical Programming: CPLEX
[25]	Only capacity expansion	Bi-level	Static	Nondeterministic framework: Scenario generation	No	Yes	No	Threefold: rival producer investment, rival producer offering, and electrical power demand	No	No	GAMS

(continued)

Table 6.2 (continued)

Reference	Type of expansion planning	Multilevel framework	Static or multi-period framework	Nondeterministic framework	Environmental constraints	System security constraint	Loss consideration	Uncertainty consideration	Strategic behaviors of CSC electricity market participants	Impacts of expansion plans on strategic behavior and vice versa	Solution method
[26]	Only capacity expansion	Bi-level	Multi-period	No	Yes: only CO_2 atmospheric emission	No	No	No	No	No	Genetic algorithm (GA)
[27]	Only capacity expansion	Bi-level	Multi-period	No	No	No	No	No	No	No	GAMS
[28]	Only capacity expansion	Bi-level	Multi-period	Nondeterministic framework: cluster based	No	No	No	Twofold: electrical power demand growth and fuel prices	No	No	Heuristic algorithm
Proposed framework	Allocation and capacity expansion	Tri-level	Multi-period	Nondeterministic framework: stochastic modeling by the IGDT (risk-averse and risk-taker frameworks)	Yes: SO_2, CO_2, and NO_x atmospheric emissions	Yes	Yes	Severe twofold: price of the CSC electricity market and electrical power demand	Yes: bilateral bidding mechanism	Yes	Five new multi-objective music-inspired optimization algorithms (see Chap. 4): (1) SS-HSA; (2) SS-IHSA; (3) continuous/discrete TMS-MSA; (4) TMS-EMSA; and (5) SOSA

6.3.1.1 Bilateral Bidding Mechanism: First Level (Problem A)

In the first level of the proposed strategic tri-level computational-logical framework, both GENCOs and DISCOs try to maximize their profit by implementation of optimal bidding strategies. The implementation process of the bilateral bidding mechanism (BBM) is discussed in detail in Sect. 5.3.1 of Chap. 5. Here, the BBM modeling has been adapted by considering assumptions of the proposed strategic tri-level computational-logical framework. Therefore, the BBM (i.e., the first level) is modified to apply the index of period (p) and the index of the pattern (k), as given by Eqs. (6.1) and (6.2):

$$\begin{aligned}\underset{x_{\mathrm{BBM}}}{\mathrm{Max}}: \tilde{\upsilon}_{p,k,g}\left(\rho_{p,k,g}\right) &= \frac{\left(2\xi_{1,p,k,g}-1\right)}{2}\cdot\alpha_{p,k,g}\cdot\rho_{p,k,g}^{2}+\left(\xi_{2,p,k,g}-1\right)\\ &\cdot\beta_{p,k,g}\cdot\rho_{p,k,g}-\gamma_{p,k,g};\quad \forall\left\{p\in\Psi^{\mathrm{P}},k\in\Psi^{\mathrm{K}},g\in\Psi^{\mathrm{G}}\right\}\end{aligned} \tag{6.1}$$

$$\begin{aligned}\underset{x_{\mathrm{BBM}}}{\mathrm{Max}}: \tilde{\upsilon}_{p,k,d}\left(\rho_{p,k,d}\right) &= \frac{\left(1-2\xi_{1,p,k,d}\right)}{2}\cdot\alpha_{p,k,d}\cdot\rho_{p,k,d}^{2}+\left(1-\xi_{2,p,k,d}\right)\\ &\cdot\beta_{p,k,d}\cdot\rho_{p,k,d}+\gamma_{p,k,d};\quad \forall\left\{p\in\Psi^{\mathrm{P}},k\in\Psi^{\mathrm{K}},d\in\Psi^{\mathrm{D}}\right\}\end{aligned} \tag{6.2}$$

and subject to Eqs. (6.3) through (6.9):

$$\sum_{g\in\Psi^{\mathrm{G}}}\rho_{p,k,g}=\sum_{l\in\Psi^{\mathrm{L}}}\kappa_{p,k,l}+\sum_{d\in\Psi^{\mathrm{D}}}\rho_{p,k,d};\ \forall\left\{p\in\Psi^{\mathrm{P}},k\in\Psi^{\mathrm{K}},g\in\Psi^{\mathrm{G}},l\in\Psi^{\mathrm{L}},d\in\Psi^{\mathrm{D}}\right\} \tag{6.3}$$

$$\rho_{p,k,g}^{\min}\leq\rho_{p,k,g}\leq\rho_{p,k,g}^{\max};\quad \forall\left\{p\in\Psi^{\mathrm{P}},k\in\Psi^{\mathrm{K}},g\in\Psi^{\mathrm{G}}\right\} \tag{6.4}$$

$$\rho_{p,k,d}^{\min}\leq\rho_{p,k,d}\leq\rho_{p,k,d}^{\max};\quad \forall\left\{p\in\Psi^{\mathrm{P}},k\in\Psi^{\mathrm{K}},d\in\Psi^{\mathrm{D}}\right\} \tag{6.5}$$

$$\xi_{1,p,k,g}^{\min}\leq\xi_{1,p,k,g}\leq\xi_{1,p,k,g}^{\max};\quad \forall\left\{p\in\Psi^{\mathrm{P}},k\in\Psi^{\mathrm{K}},g\in\Psi^{\mathrm{G}}\right\} \tag{6.6}$$

$$\xi_{2,p,k,g}^{\min}\leq\xi_{2,p,k,g}\leq\xi_{2,p,k,g}^{\max};\quad \forall\left\{p\in\Psi^{\mathrm{P}},k\in\Psi^{\mathrm{K}},g\in\Psi^{\mathrm{G}}\right\} \tag{6.7}$$

$$\xi_{1,p,k,d}^{\min}\leq\xi_{1,p,k,d}\leq\xi_{1,p,k,d}^{\max};\quad \forall\left\{p\in\Psi^{\mathrm{P}},k\in\Psi^{\mathrm{K}},d\in\Psi^{\mathrm{D}}\right\} \tag{6.8}$$

$$\xi_{2,p,k,d}^{\min}\leq\xi_{2,p,k,d}\leq\xi_{2,p,k,d}^{\max};\quad \forall\left\{p\in\Psi^{\mathrm{P}},k\in\Psi^{\mathrm{K}},d\in\Psi^{\mathrm{D}}\right\} \tag{6.9}$$

6.3.1.2 Competitive Security-Constrained Electricity Market: Second Level (Problem B)

In the second level of the proposed strategic tri-level computational-logical framework, the ISO's CSC electricity market-clearing problem is performed with the aim of maximizing the CWF. The implementation process of the CSC electricity market settlement is discussed in detail in Sect. 5.3.2 of Chap. 5. In a similar manner, in the proposed strategic tri-level computational-logical framework, the CSC electricity market problem (i.e., the second level) is modified to apply the index of the period (p) and the index of the pattern (k), as given by Eq. (6.10):

$$\begin{aligned}
&\underset{x_{\text{CSC-EM}}}{\text{Max}}:CWF_{p,k}=\left(\sum_{g\in\Psi^{\text{G}}}\tilde{\upsilon}_{p,k,g}\left(\rho_{p,k,g}\right)+\sum_{d\in\Psi^{\text{D}}}\tilde{\upsilon}_{p,k,d}\left(\rho_{p,k,d}\right)\right);\\
&\forall\left\{p\in\Psi^{\text{P}},k\in\Psi^{\text{K}},g\in\Psi^{\text{G}},d\in\Psi^{\text{D}}\right\}\\
&\quad=\underset{x_{\text{CSC-EM}}}{\text{Max}}:\left(\sum_{g\in\Psi^{\text{G}}}\frac{\left(2\xi_{1,p,k,g}-1\right)}{2}\cdot\alpha_{p,k,g}\cdot\rho_{p,k,g}^{2}+\left(\xi_{2,p,k,g}-1\right)\cdot\beta_{p,k,g}\cdot\rho_{p,k,g}-\gamma_{p,k,g}\right.\\
&\quad\left.+\sum_{d\in\Psi^{\text{D}}}\frac{\left(1-2\xi_{1,p,k,d}\right)}{2}\cdot\alpha_{p,k,d}\cdot\rho_{p,k,d}^{2}+\left(1-\xi_{2,p,k,d}\right)\cdot\beta_{p,k,d}\cdot\rho_{p,k,d}+\gamma_{p,k,d}\right);\\
&\forall\left\{p\in\Psi^{\text{P}},k\in\Psi^{\text{K}},g\in\Psi^{\text{G}},d\in\Psi^{\text{D}}\right\}
\end{aligned}\tag{6.10}$$

and subject to Eqs. (6.11) through (6.14):

$$\sum_{g\in\Psi^{\text{G}}}\rho_{p,k,g}=\sum_{l\in\Psi^{\text{L}}}\kappa_{p,k,l}+\sum_{d\in\Psi^{\text{D}}}\rho_{p,k,d};\ \forall\left\{p\in\Psi^{\text{P}},k\in\Psi^{\text{K}},g\in\Psi^{\text{G}},l\in\Psi^{\text{L}},d\in\Psi^{\text{D}}\right\}\tag{6.11}$$

$$\rho_{p,k,g}^{\min}\le\rho_{p,k,g}\le\rho_{p,k,g}^{\max};\quad\forall\left\{p\in\Psi^{\text{P}},k\in\Psi^{\text{K}},g\in\Psi^{\text{G}}\right\}\tag{6.12}$$

$$\rho_{p,k,d}^{\min}\le\rho_{p,k,d}\le\rho_{p,k,d}^{\max};\quad\forall\left\{p\in\Psi^{\text{P}},k\in\Psi^{\text{K}},d\in\Psi^{\text{D}}\right\}\tag{6.13}$$

$$f_{p,k,l}^{\min}\le f_{p,k,l}\le f_{p,k,l}^{\max};\quad\forall\left\{p\in\Psi^{\text{P}},k\in\Psi^{\text{K}},l\in\Psi^{\text{L}}\right\}\tag{6.14}$$

6.3.1.3 Pseudo-Dynamic Generation Expansion Planning: Third Level (Problem C)

The third level of the strategic tri-level computational-logical framework consists of two layers: (1) the local layer and (2) the global layer. Presented here are the details of the mathematical model of these two layers. In the local layer, each GENCO separately makes its own optimal generation expansion plans over the planning horizon by solving the PD-GEP problem. Expected profit (EP), generation investment cost (GIC), and capacity payment (CP) are considered as objective functions in the offered PD-GEP problem. At the same time, the PD-GEP problem is subject to a set of techno-economic and environmental constraints.

In the regulated structure, the objective of the PD-GEP problem is to minimize the total GIC, while in the deregulated structure the most important objective of the PD-GEP problem is to maximize the EP of the GENCOs. In this structure, each GENCO tries to maximize its EP, while satisfying the ISO criteria such as reserve margin of generation, competitive condition, and reliability requirements. Therefore,

Eqs. (6.15) and (6.16) show that the first target of the proposed PD-GEP problem is formulated with the goal of maximizing the EP index:

$$\mathrm{OF}_{1,g}^{\text{PD-GEP}} = EP_{y,g} = \sum_{p \in \Psi^{\mathrm{P}}} \left(\frac{1}{(1+\mathrm{Ir})^{p-p_0}} \right) \cdot EP_{y,g,p}; \quad \forall \{p \in \Psi^{\mathrm{P}}\} \tag{6.15}$$

$$\begin{aligned} EP_{y,g,p} = & \sum_{k \in \Psi^{\mathrm{K}}} \sum_{eu \in \Psi_g^{\mathrm{EU}}} \Delta t_k \cdot \left[\left(p\lambda_{p,k,b_{eu}^g} \cdot \rho_{p,k,g_{eu}} \right) \right. \\ & \left. - \left(\frac{1}{2} \cdot \alpha_{p,k,g_{eu}} \cdot \rho_{p,k,g_{eu}}^2 + \beta_{p,k,g_{eu}} \cdot \rho_{p,k,g_{eu}} + \gamma_{p,k,g_{eu}} \right) \right] \\ & + \sum_{k \in \Psi^{\mathrm{K}}} \sum_{iu \in \Psi_g^{\mathrm{IU}}} \Delta t_k \cdot \left[\left(p\lambda_{p,k,b_{iu}^g} \cdot \rho_{p,k,g_{iu}} \right) \right. \\ & \left. - \left(\frac{1}{2} \cdot \alpha_{p,k,g_{iu}} \cdot \rho_{p,k,g_{iu}}^2 + \beta_{p,k,g_{iu}} \cdot \rho_{p,k,g_{iu}} + \gamma_{p,k,g_{iu}} \right) \right]; \\ & \forall \left\{ p \in \Psi^{\mathrm{P}}, k \in \Psi^{\mathrm{K}}, eu \in \Psi_g^{\mathrm{EU}}, iu \in \Psi_g^{\mathrm{IU}}, b_{eu}^g \in \Psi_{g_{eu}}^{\mathrm{B}}, b_{iu}^g \in \Psi_{g_{iu}}^{\mathrm{B}} \right\} \end{aligned} \tag{6.16}$$

Equation (6.15) explains the net present value of the EP index of GENCO g over the planning horizon. The EP of this GENCO at period p has two main parts: (1) the EP for the existing generation units owned by this GENCO and (2) the EP for the newly installed generation units owned by this GENCO [see Eq. (6.16)].

Due to limited financial resources, one of the main challenges in the generation expansion planning process is to find the optimum value of the GIC index. Hence, the second target of the PD-GEP problem is considered in order to minimize the GIC index, as determined by Eqs. (6.17) through (6.19):

$$\mathrm{OF}_{2,g}^{\text{PD-GEP}} = GIC_{y,g} = \sum_{p \in \Psi^{\mathrm{P}}} \left(\frac{1}{(1+\mathrm{Ir})^{p-p_0}} \right) \cdot GIC_{y,g,p}; \quad \forall \{p \in \Psi^{\mathrm{P}}\} \tag{6.17}$$

$$GIC_{y,g,p} = \sum_{iu \in \Psi_g^{\mathrm{IU}}} \left(GIC_{p,g_{iu}}^{\mathrm{f}} + GIC_{p,g_{iu}}^{\mathrm{v}} \right); \quad \forall \left\{ p \in \Psi^{\mathrm{P}}, iu \in \Psi_g^{\mathrm{IU}} \right\} \tag{6.18}$$

$$GIC_{p,g_{iu}}^{\mathrm{v}} = GIC_{p,g_{iu}} \cdot IC_{p,g_{iu}}; \quad \forall \left\{ p \in \Psi^{\mathrm{P}}, iu \in \Psi_g^{\mathrm{IU}} \right\} \tag{6.19}$$

Equation (6.17) shows the net present value of the GIC index of GENCO g over the planning horizon. The generation investment cost for installation of the new generation units that are owned by this GENCO at period p has two main parts: (1) the fixed GIC and (2) the variable GIC [see Eq. (6.18)]. The variable GIC of GENCO g generally depends on the installed generation capacity of this GENCO [see Eq. (6.19)].

On the other hand, in deregulated structures, each GENCO, through expansion of its own generation, can play an effective role in order to enhance power system reliability. In this regard, the ISO can employ an incentive mechanism, called a capacity payment, to encourage GENCOs to invest in expanding their generation capacity in places where the construction of new generation units increases power system reliability. In addition to increasing power system reliability, this mechanism can also increase the profit of the GENCOs. In the capacity payment mechanism, each GENCO receives incentive credits from the ISO, according to the degree of participation in improving power system reliability. Therefore, the third target of the PD-GEP problem is maximization of the capacity payment, namely the CP index, as given by Eqs. (6.20) and (6.21):

$$\mathrm{OF}_{3,g}^{\text{PD-GEP}} = CP_{y,g} = \sum_{p \in \Psi^{\mathrm{P}}} CP_{y,g,p}; \quad \forall \{p \in \Psi^{\mathrm{P}}\} \tag{6.20}$$

$$CP_{y,g,p} = \sum_{k \in \Psi^{\mathrm{K}}} \sum_{iu \in \Psi^{\mathrm{IU}}} \Delta t_k \cdot \left(ECO_{p,k,g_{iu}}^{\mathrm{b}} - ECO_{p,k,g_{iu}}^{\mathrm{a}}\right) \cdot \left(\mathrm{ECOC} - p\lambda_{p,k,b_{iu}^{g}}\right); \quad \forall \left\{p \in \Psi^{\mathrm{P}}, k \in \Psi^{\mathrm{K}}, iu \in \Psi_g^{\mathrm{IU}}\right\} \tag{6.21}$$

Equation (6.20) represents the CP index of GENCO g over the planning horizon. Equation (6.21) represents the CP of this GENCO at period p.

As a result, the overall objective function of the PD-GEP problem in the third level is formulated using Eq. (6.22):

$$\underset{x_{\text{PD-GEP}}}{\mathrm{Max}} : \left\{\mathrm{OF}_g^{\text{PD-GEP}}\right\} = \underset{x_{\text{PD-GEP}}}{\mathrm{Max}} : \left\{\mathrm{W}_{\mathrm{OF}_1,g}^{\text{PD-GEP}} \cdot \mathrm{OF}_{1,g}^{\text{PD-GEP}} - \mathrm{W}_{\mathrm{OF}_2,g}^{\text{PD-GEP}} \cdot \mathrm{OF}_{2,g}^{\text{PD-GEP}} + \mathrm{W}_{\mathrm{OF}_3,g}^{\text{PD-GEP}} \cdot \mathrm{OF}_{3,g}^{\text{PD-GEP}}\right\} \tag{6.22}$$

It is possible that the objective functions of the third level of the proposed tri-level computational-logical framework have different degrees of importance from the planner's (GENCO g) point of view. Hence, in the overall objective function, multiple weighting coefficients are assigned to each objective function by the planner.

The constraints of the DP-GEP problem are divided into three distinct classes: (1) the operational restrictions on each GENCO, technical constraints; (2) the availability of financial resources for each GENCO, economic constraints; and, (3) the environmental restrictions imposed on each GENCO, environmental constraints.

The generated power by the existing and installed generation units belonging to GENCO g must satisfy the predetermined limits. Eqs. (6.23) and (6.24) show the lower and upper limits related to the power generation of the existing and installed generation units owned by this GENCO, respectively:

$$\rho_{p,k,g_{eu}}^{\min} \le \rho_{p,k,g_{eu}} \le \rho_{p,k,g_{eu}}^{\max}; \quad \forall \left\{p \in \Psi^{\mathrm{P}}, k \in \Psi^{\mathrm{K}}, eu \in \Psi_g^{\mathrm{EU}}\right\} \tag{6.23}$$

$$\rho_{p,k,g_{iu}}^{\min} \leq \rho_{p,k,g_{iu}} \leq \rho_{p,k,g_{iu}}^{\max}; \quad \forall \left\{ p \in \Psi^{\mathrm{P}}, k \in \Psi^{\mathrm{K}}, iu \in \Psi_g^{\mathrm{IU}} \right\} \tag{6.24}$$

The capacity of each installed generation unit owned by GENCO g must be able to meet the maximum amount of generated power in all patterns. Therefore, the capacity of each installed generation unit owned by this GENCO can be determined using Eq. (6.25):

$$IC_{p,g_{iu}} = \max\left\{ \rho_{p,k,g_{iu}} \right\}; \quad \forall \left\{ p \in \Psi^{\mathrm{P}}, k \in \Psi^{\mathrm{K}}, iu \in \Psi_g^{\mathrm{IU}} \right\} \tag{6.25}$$

To prevent the exercise of the market power by GENCOs, each GENCO must limit its installed generation capacity. Therefore, the capacity of GENCO g at period p must satisfy Eq. (6.26):

$$\sum_{iu \in \Psi_g^{\mathrm{IU}}} IC_{p,g_{iu}} \leq IC_{p,g}^{\max}; \quad \forall \left\{ p \in \Psi^{\mathrm{P}}, k \in \Psi^{\mathrm{K}}, iu \in \Psi_g^{\mathrm{IU}} \right\} \tag{6.26}$$

The fuel consumption rate for GENCO g in each period of the planning horizon must meet the conditions specified by Eq. (6.27):

$$\sum_{k \in \Psi^{\mathrm{K}}} \sum_{eu \in \Psi_g^{\mathrm{EU}}} \Delta t_k \cdot F_{p,k,g_{eu}} \cdot \rho_{p,k,g_{eu}} + \sum_{k \in \Psi^{\mathrm{K}}} \sum_{iu \in \Psi_g^{\mathrm{IU}}} \Delta t_k \cdot F_{p,k,g_{iu}} \cdot \rho_{p,k,g_{iu}} \leq F_{p,g}^{\max};$$
$$\forall \left\{ p \in \Psi^{\mathrm{P}}, k \in \Psi^{\mathrm{K}}, eu \in \Psi_g^{\mathrm{EU}}, iu \in \Psi_g^{\mathrm{IU}} \right\} \tag{6.27}$$

In addition, generation expansion planning is a very costly process and each GENCO is faced with financial resource limitations to expand its generation capacity. Hence, the GIC of GENCO g in period p and in all periods must satisfy the financial constraints in accordance with Eqs. (6.28) and (6.29), respectively:

$$\sum_{iu \in \Psi_g^{\mathrm{IU}}} GIC_{p,g_{iu}} \leq GIC_{p,g}^{\max}; \quad \forall \left\{ p \in \Psi^{\mathrm{P}}, iu \in \Psi_g^{\mathrm{IU}} \right\} \tag{6.28}$$

$$\sum_{p \in \Psi^{\mathrm{P}}} \sum_{iu \in \Psi_g^{\mathrm{IU}}} GIC_{p,g_{iu}} \leq GIC_g^{\max}; \quad \forall \left\{ p \in \Psi^{\mathrm{P}}, iu \in \Psi_g^{\mathrm{IU}} \right\} \tag{6.29}$$

On the other hand, to comply with the regulations of the environmental protection agency (EPA), the emitted atmospheric emissions by each GENCO must be limited to acceptable levels. Hence, emitted sulfur dioxide (SO_2), carbon dioxide (CO_2), and nitrogen oxide (NO_x) atmospheric emissions by GENCO g are limited according to Eqs. (6.30) through (6.32), respectively:

$$\sum_{k\in\Psi^{\mathrm{K}}}\sum_{eu\in\Psi_g^{\mathrm{EU}}}\Delta t_k\cdot E_{p,k,g_{eu}}^{\mathrm{SO_2}}\cdot\rho_{p,k,g_{eu}}+\sum_{k\in\Psi^{\mathrm{K}}}\sum_{iu\in\Psi_g^{\mathrm{IU}}}\Delta t_k\cdot E_{p,k,g_{iu}}^{\mathrm{SO_2}}\cdot\rho_{p,k,g_{iu}}\leq E_{p,g}^{\mathrm{SO_2},\max};$$
$$\forall\left\{p\in\Psi^{\mathrm{P}},k\in\Psi^{\mathrm{K}},eu\in\Psi_g^{\mathrm{EU}},iu\in\Psi_g^{\mathrm{IU}}\right\} \tag{6.30}$$

$$\sum_{k\in\Psi^{\mathrm{K}}}\sum_{eu\in\Psi_g^{\mathrm{EU}}}\Delta t_k\cdot E_{p,k,g_{eu}}^{\mathrm{CO_2}}\cdot\rho_{p,k,g_{eu}}+\sum_{k\in\Psi^{\mathrm{K}}}\sum_{iu\in\Psi_g^{\mathrm{IU}}}\Delta t_k\cdot E_{p,k,g_{iu}}^{\mathrm{CO_2}}\cdot\rho_{p,k,g_{iu}}\leq E_{p,g}^{\mathrm{CO_2},\max};$$
$$\forall\left\{p\in\Psi^{\mathrm{P}},k\in\Psi^{\mathrm{K}},eu\in\Psi_g^{\mathrm{EU}},iu\in\Psi_g^{\mathrm{IU}}\right\} \tag{6.31}$$

$$\sum_{k\in\Psi^{\mathrm{K}}}\sum_{eu\in\Psi_g^{\mathrm{EU}}}\Delta t_k\cdot E_{p,k,g_{eu}}^{\mathrm{NO_x}}\cdot\rho_{p,k,g_{eu}}+\sum_{k\in\Psi^{\mathrm{K}}}\sum_{iu\in\Psi_g^{\mathrm{IU}}}\Delta t_k\cdot E_{p,k,g_{iu}}^{\mathrm{NO_x}}\cdot\rho_{p,k,g_{iu}}\leq E_{p,g}^{\mathrm{NO_x},\max};$$
$$\forall\left\{p\in\Psi^{\mathrm{P}},k\in\Psi^{\mathrm{K}},eu\in\Psi_g^{\mathrm{EU}},iu\in\Psi_g^{\mathrm{IU}}\right\} \tag{6.32}$$

Once GENCOs' optimal generation expansion plans are obtained in the local layer, the ISO aggregates these optimal plans in the global layer and then evaluates the established restrictions concerning the entire power system. If the obtained optimal plans violate one or more of these restrictions, the ISO sends corrective factors related to the violated restriction to the GENCOs in order to modify their expansion plans. In the global layer, three different global restrictions are investigated for the entire power system: (1) generation reserve margin, (2) competitive conditions, and, (3) reliability requirements. First, the generation reserve margin corresponds to the surplus of the existing and installed generation capacities, as a buffer for unforeseen demand fluctuations. Hence, the generation reserve margin in each period of the planning horizon must meet the conditions specified by Eq. (6.33):

$$\left(1+RM_p^{\min}\right)\cdot\left(\sum_{k\in\Psi^{\mathrm{K}}}\sum_{b\in\Psi^{\mathrm{B}}}p\rho_{p,k,b}^{\mathrm{p}}\right)\leq\left(1+\eta_p^{\mathrm{res}}\right)\cdot$$
$$\left(\sum_{g\in\Psi^{\mathrm{G}}}\sum_{eu\in\Psi_g^{\mathrm{EU}}}EC_{p,g_{eu}}+\sum_{g\in\Psi^{\mathrm{G}}}\sum_{iu\in\Psi_g^{\mathrm{IU}}}IC_{p,g_{iu}}\right)\leq\left(1+RM_p^{\max}\right)\cdot\left(\sum_{k\in\Psi^{\mathrm{K}}}\sum_{b\in\Psi^{\mathrm{B}}}p\rho_{p,k,b}^{\mathrm{p}}\right);$$
$$\forall\left\{g\in\Psi^{\mathrm{G}},p\in\Psi^{\mathrm{P}},k\in\Psi^{\mathrm{K}},b\in\Psi^{\mathrm{B}},eu\in\Psi_g^{\mathrm{EU}},iu\in\Psi_g^{\mathrm{IU}}\right\} \tag{6.33}$$

Second, to prevent the application of market power, the regulatory body—in this case, the ISO—should determine pre-established requirements to limit the share of the installed generation capacity owned by each GENCO regarding the total amount of the installed generation capacity by all GENCOs in the power system. The obtained generation capacity for GENCO g in the local layer, then, must satisfy pre-established requirements in the global layer using Eq. (6.34):

$$\left(1+\eta_{p,g}^{\mathrm{cap}}\right)\cdot\sum_{iu\in\Psi_g^{\mathrm{IU}}} IC_{p,g_{iu}} \le \frac{PERC_{p,g}^{\max}}{100}\cdot\sum_{g\in\Psi^{\mathrm{G}}}\sum_{iu\in\Psi_g^{\mathrm{IU}}} IC_{p,g_{iu}};$$
$$\forall\left\{g\in\Psi^{\mathrm{G}}, p\in\Psi^{\mathrm{P}}, g\in\Psi^{\mathrm{G}}, iu\in\Psi_g^{\mathrm{IU}}\right\} \tag{6.34}$$

Third, the obtained optimal expansion plans by GENCOs in the local layer must be able to satisfy the pre-established reliability requirements in the global layer. Therefore, to achieve a reliable power system, an adequacy index corresponding to the expected customer outage (ECO) is defined in this layer. The ECO at each period of the planning horizon should be smaller than a specified value by the regulatory body, in accordance with Eq. (6.35):

$$\left(1+\eta_p^{\mathrm{rel}}\right)\cdot ECO_p \le ECO^{\max};\quad \forall\left\{p\in\Psi^{\mathrm{P}}\right\} \tag{6.35}$$

The ECO index is determined at the HL-I by applying a well-founded linear optimization problem. The reliability evaluation at the HL-I is defined as an adequacy assessment process for the generation system. For a thorough discussion on the HL-I assessment studies at the planning and operational levels of power systems, please refer to the work by Billinton and Allan [30]. The calculation process of the ECO index at the HL-I studies is investigated in Sect. 6.4.1.3 of this chapter.

If the specified optimal expansion plans by the GENCOs violate the global layer restrictions, the ISO determines the amount of violation of each restriction and sends the corrective factors to the GENCOs. Using these received factors from the global layer, the PD-GEP problem—i.e., the local layer—should be solved again by each GENCO separately in order to find the optimal generation expansion plans. This iterative process is repeated until the determined optimal generation expansion plans by all GENCOs—i.e., the local layer outcomes—satisfy all restrictions in the global layer.

6.3.2 *Overview of the Deterministic Strategic Tri-level Computational-Logical Framework*

According to the equations identified in the previous sections, the proposed deterministic strategic tri-level computational-logical framework can be formulated as follows:

$$\underset{x_{\text{PD-GEP}}}{\text{Max}} : \left\{\text{OF}_{g}^{\text{PD-GEP}}\right\} = \underset{x_{\text{PD-GEP}}}{\text{Max}} : \left\{\text{W}_{\text{OF}_1,g}^{\text{PD-GEP}} \cdot \text{OF}_{1,g}^{\text{PD-GEP}} - \text{W}_{\text{OF}_2,g}^{\text{PD-GEP}} \cdot \text{OF}_{2,g}^{\text{PD-GEP}} + \text{W}_{\text{OF}_3,g}^{\text{PD-GEP}} \cdot \text{OF}_{3,g}^{\text{PD-GEP}}\right\}$$

(6.36) Third level

where,

$$\text{OF}_{1,g}^{\text{PD-GEP}} = EP_{y,g} = \sum_{p\in\Psi^{\text{P}}} \left(\frac{1}{(1+\text{Ir})^{p-p_0}}\right) \cdot EP_{y,g,p}; \quad \forall\{p\in\Psi^{\text{P}}\} \tag{6.37}$$

$$\begin{aligned} EP_{y,g,p} = & \sum_{k\in\Psi^{\text{K}}} \sum_{eu\in\Psi_g^{\text{EU}}} \Delta t_k \cdot \left[\left(p\lambda_{p,k,b_{eu}^g} \cdot \rho_{p,k,g_{eu}}\right)\right. \\ & \left.-\left(\frac{1}{2}\cdot \alpha_{p,k,g_{eu}} \cdot \rho_{p,k,g_{eu}}^2 + \beta_{p,k,g_{eu}} \cdot \rho_{p,k,g_{eu}} + \gamma_{p,k,g_{eu}}\right)\right] \\ & + \sum_{k\in\Psi^{\text{K}}} \sum_{iu\in\Psi_g^{IU}} \Delta t_k \cdot \\ & \left[\left(p\lambda_{p,k,b_{iu}^g} \cdot \rho_{p,k,g_{iu}}\right) - \left(\frac{1}{2}\cdot \alpha_{p,k,g_{iu}} \cdot \rho_{p,k,g_{iu}}^2 + \beta_{p,k,g_{iu}} \cdot \rho_{p,k,g_{iu}} + \gamma_{p,k,g_{iu}}\right)\right]; \\ & \forall\left\{p\in\Psi^{\text{P}}, k\in\Psi^{\text{K}}, eu\in\Psi_g^{\text{EU}}, iu\in\Psi_g^{\text{IU}}, b_{eu}^g\in\Psi_{g_{eu}}^{\text{B}}, b_{iu}^g\in\Psi_{g_{iu}}^{\text{B}}\right\} \end{aligned} \tag{6.38}$$

$$\text{OF}_{2,g}^{\text{PD-GEP}} = GIC_{y,g} = \sum_{p\in\Psi^{\text{P}}} \left(\frac{1}{(1+\text{Ir})^{p-p_0}}\right) \cdot GIC_{y,g,p}; \quad \forall\{p\in\Psi^{\text{P}}\} \tag{6.39}$$

$$GIC_{y,g,p} = \sum_{iu\in\Psi_g^{\text{IU}}} \left(GIC_{p,g_{iu}}^{\text{f}} + GIC_{p,g_{iu}}^{\text{v}}\right); \quad \forall\left\{p\in\Psi^{\text{P}}, iu\in\Psi_g^{\text{IU}}\right\} \tag{6.40}$$

$$GIC_{p,g_{iu}}^{\text{v}} = GIC_{p,g_{iu}} \cdot IC_{p,g_{iu}}; \quad \forall\left\{p\in\Psi^{\text{P}}, iu\in\Psi_g^{\text{IU}}\right\} \tag{6.41}$$

$$\text{OF}_{3,g}^{\text{PD-GEP}} = CP_{y,g} = \sum_{p\in\Psi^{\text{P}}} CP_{y,g,p}; \quad \forall\{p\in\Psi^{\text{P}}\} \tag{6.42}$$

$$\begin{aligned} CP_{y,g,p} = & \sum_{k\in\Psi^{\text{K}}} \sum_{iu\in\Psi^{\text{IU}}} \Delta t_k \cdot \left(ECO_{p,k,g_{iu}}^{\text{b}} - ECO_{p,k,g_{iu}}^{\text{a}}\right) \cdot \left(\text{ECOC} - p\lambda_{p,k,b_{iu}^g}\right); \\ & \forall\left\{p\in\Psi^{\text{P}}, k\in\Psi^{\text{K}}, iu\in\Psi_g^{\text{IU}}\right\} \end{aligned} \tag{6.43}$$

and subject to Eqs. (6.44) through (6.56):

$$\rho_{p,k,g_{eu}}^{\min} \leq \rho_{p,k,g_{eu}} \leq \rho_{p,k,g_{eu}}^{\max}; \quad \forall \left\{ p \in \Psi^{\mathrm{P}}, k \in \Psi^{\mathrm{K}}, eu \in \Psi_g^{\mathrm{EU}} \right\} \tag{6.44}$$

$$\rho_{p,k,g_{iu}}^{\min} \leq \rho_{p,k,g_{iu}} \leq \rho_{p,k,g_{iu}}^{\max}; \quad \forall \left\{ p \in \Psi^{\mathrm{P}}, k \in \Psi^{\mathrm{K}}, iu \in \Psi_g^{\mathrm{IU}} \right\} \tag{6.45}$$

$$IC_{p,g_{iu}} = \max\left\{ \rho_{p,k,g_{iu}} \right\}; \quad \forall \left\{ p \in \Psi^{\mathrm{P}}, k \in \Psi^{\mathrm{K}}, iu \in \Psi_g^{\mathrm{IU}} \right\} \tag{6.46}$$

$$\sum_{iu \in \Psi_g^{\mathrm{IU}}} IC_{p,g_{iu}} \leq IC_{p,g}^{\max}; \quad \forall \left\{ p \in \Psi^{\mathrm{P}}, k \in \Psi^{\mathrm{K}}, iu \in \Psi_g^{\mathrm{IU}} \right\} \tag{6.47}$$

$$\sum_{k \in \Psi^{\mathrm{K}}} \sum_{eu \in \Psi_g^{\mathrm{EU}}} \Delta t_k \cdot F_{p,k,g_{eu}} \cdot \rho_{p,k,g_{eu}} + \sum_{k \in \Psi^{\mathrm{K}}} \sum_{iu \in \Psi_g^{\mathrm{IU}}} \Delta t_k \cdot F_{p,k,g_{iu}} \cdot \rho_{p,k,g_{iu}} \leq F_{p,g}^{\max}; \\ \forall \left\{ p \in \Psi^{\mathrm{P}}, k \in \Psi^{\mathrm{K}}, eu \in \Psi_g^{\mathrm{EU}}, iu \in \Psi_g^{\mathrm{IU}} \right\} \tag{6.48}$$

$$\sum_{iu \in \Psi_g^{\mathrm{IU}}} GIC_{p,g_{iu}} \leq GIC_{p,g}^{\max}; \quad \forall \left\{ p \in \Psi^{\mathrm{P}}, iu \in \Psi_g^{\mathrm{IU}} \right\} \tag{6.49}$$

$$\sum_{p \in \Psi^{\mathrm{P}}} \sum_{iu \in \Psi_g^{\mathrm{IU}}} GIC_{p,g_{iu}} \leq GIC_g^{\max}; \quad \forall \left\{ p \in \Psi^{\mathrm{P}}, iu \in \Psi_g^{\mathrm{IU}} \right\} \tag{6.50}$$

$$\sum_{k \in \Psi^{\mathrm{K}}} \sum_{eu \in \Psi_g^{\mathrm{EU}}} \Delta t_k \cdot E_{p,k,g_{eu}}^{\mathrm{SO_2}} \cdot \rho_{p,k,g_{eu}} + \sum_{k \in \Psi^{\mathrm{K}}} \sum_{iu \in \Psi_g^{\mathrm{IU}}} \Delta t_k \cdot E_{p,k,g_{iu}}^{\mathrm{SO_2}} \cdot \rho_{p,k,g_{iu}} \leq E_{p,g}^{\mathrm{SO_2},\max}; \\ \forall \left\{ p \in \Psi^{\mathrm{P}}, k \in \Psi^{\mathrm{K}}, eu \in \Psi_g^{\mathrm{EU}}, iu \in \Psi_g^{\mathrm{IU}} \right\} \tag{6.51}$$

$$\sum_{k \in \Psi^{\mathrm{K}}} \sum_{eu \in \Psi_g^{\mathrm{EU}}} \Delta t_k \cdot E_{p,k,g_{eu}}^{\mathrm{CO_2}} \cdot \rho_{p,k,g_{eu}} + \sum_{k \in \Psi^{\mathrm{K}}} \sum_{iu \in \Psi_g^{\mathrm{IU}}} \Delta t_k \cdot E_{p,k,g_{iu}}^{\mathrm{CO_2}} \cdot \rho_{p,k,g_{iu}} \leq E_{p,g}^{\mathrm{CO_2},\max}; \\ \forall \left\{ p \in \Psi^{\mathrm{P}}, k \in \Psi^{\mathrm{K}}, eu \in \Psi_g^{\mathrm{EU}}, iu \in \Psi_g^{\mathrm{IU}} \right\} \tag{6.52}$$

$$\sum_{k \in \Psi^{\mathrm{K}}} \sum_{eu \in \Psi_g^{\mathrm{EU}}} \Delta t_k \cdot E_{p,k,g_{eu}}^{\mathrm{NO_x}} \cdot \rho_{p,k,g_{eu}} + \sum_{k \in \Psi^{\mathrm{K}}} \sum_{iu \in \Psi_g^{\mathrm{IU}}} \Delta t_k \cdot E_{p,k,g_{iu}}^{\mathrm{NO_x}} \cdot \rho_{p,k,g_{iu}} \leq E_{p,g}^{\mathrm{NO_x},\max}; \\ \forall \left\{ p \in \Psi^{\mathrm{P}}, k \in \Psi^{\mathrm{K}}, eu \in \Psi_g^{\mathrm{EU}}, iu \in \Psi_g^{\mathrm{IU}} \right\} \tag{6.53}$$

$$\left(1 + RM_p^{\min}\right) \cdot \left(\sum_{k \in \Psi^{\mathrm{K}}} \sum_{b \in \Psi^{\mathrm{B}}} p\rho_{p,k,b}^{\mathrm{p}} \right) \leq \left(1 + \eta_p^{\mathrm{res}}\right) \cdot \\ \left(\sum_{g \in \Psi^{\mathrm{G}}} \sum_{eu \in \Psi_g^{\mathrm{EU}}} EC_{p,g_{eu}} + \sum_{g \in \Psi^{\mathrm{G}}} \sum_{iu \in \Psi_g^{\mathrm{IU}}} IC_{p,g_{iu}} \right) \leq \left(1 + RM_p^{\max}\right) \cdot \left(\sum_{k \in \Psi^{\mathrm{K}}} \sum_{b \in \Psi^{\mathrm{B}}} p\rho_{p,k,b}^{\mathrm{p}} \right); \\ \forall \left\{ g \in \Psi^{\mathrm{G}}, p \in \Psi^{\mathrm{P}}, k \in \Psi^{\mathrm{K}}, b \in \Psi^{\mathrm{B}}, eu \in \Psi_g^{\mathrm{EU}}, iu \in \Psi_g^{\mathrm{IU}} \right\} \tag{6.54}$$

$$\left(1+\eta_{p,g}^{\text{cap}}\right)\cdot\sum_{iu\in\Psi_g^{\text{IU}}} IC_{p,g_{iu}}\leq\frac{PERC_{p,g}^{\max}}{100}\cdot\sum_{g\in\Psi^{\text{G}}}\sum_{iu\in\Psi_g^{\text{IU}}} IC_{p,g_{iu}};\quad \forall\left\{g\in\Psi^{\text{G}},p\in\Psi^{\text{P}},g\in\Psi^{\text{G}},iu\in\Psi_g^{\text{IU}}\right\} \tag{6.55}$$

$$\left(1+\eta_p^{\text{rel}}\right)\cdot ECO_p\leq ECO^{\max};\quad\forall\left\{p\in\Psi^{\text{P}}\right\} \tag{6.56}$$

$$\underset{x_{\text{BBM}}}{\text{Max}}:\tilde{\upsilon}_{p,k,g}\left(\rho_{p,k,g}\right)=\frac{\left(2\xi_{1,p,k,g}-1\right)}{2}\cdot\alpha_{p,k,g}\cdot\rho_{p,k,g}^2+\left(\xi_{2,p,k,g}-1\right)\cdot\beta_{p,k,g}\cdot\rho_{p,k,g}-\gamma_{p,k,g};\quad\forall\left\{p\in\Psi^{\text{P}},k\in\Psi^{\text{K}},g\in\Psi^{\text{G}}\right\} \tag{6.57}$$
First level

$$\underset{x_{\text{BBM}}}{\text{Max}}:\tilde{\upsilon}_{p,k,d}\left(\rho_{p,k,d}\right)=\frac{\left(1-2\xi_{1,p,k,d}\right)}{2}\cdot\alpha_{p,k,d}\cdot\rho_{p,k,d}^2+\left(1-\xi_{2,p,k,d}\right)\cdot\beta_{p,k,d}\cdot\rho_{p,k,d}+\gamma_{p,k,d};\quad\forall\left\{p\in\Psi^{\text{P}},k\in\Psi^{\text{K}},d\in\Psi^{\text{D}}\right\} \tag{6.58}$$
First level

and subject to Eqs. (6.59) through (6.65):

$$\sum_{g\in\Psi^{\text{G}}}\rho_{p,k,g}=\sum_{l\in\Psi^{\text{L}}}\kappa_{p,k,l}+\sum_{d\in\Psi^{\text{D}}}\rho_{p,k,d};\quad\forall\left\{p\in\Psi^{\text{P}},k\in\Psi^{\text{K}},g\in\Psi^{\text{G}},l\in\Psi^{\text{L}},d\in\Psi^{\text{D}}\right\} \tag{6.59}$$

$$\rho_{p,k,g}^{\min}\leq\rho_{p,k,g}\leq\rho_{p,k,g}^{\max};\quad\forall\left\{p\in\Psi^{\text{P}},k\in\Psi^{\text{K}},g\in\Psi^{\text{G}}\right\} \tag{6.60}$$

$$\rho_{p,k,d}^{\min}\leq\rho_{p,k,d}\leq\rho_{p,k,d}^{\max};\quad\forall\left\{p\in\Psi^{\text{P}},k\in\Psi^{\text{K}},d\in\Psi^{\text{D}}\right\} \tag{6.61}$$

$$\xi_{1,p,k,g}^{\min}\leq\xi_{1,p,k,g}\leq\xi_{1,p,k,g}^{\max};\quad\forall\left\{p\in\Psi^{\text{P}},k\in\Psi^{\text{K}},g\in\Psi^{\text{G}}\right\} \tag{6.62}$$

$$\xi_{2,p,k,g}^{\min}\leq\xi_{2,p,k,g}\leq\xi_{2,p,k,g}^{\max};\quad\forall\left\{p\in\Psi^{\text{P}},k\in\Psi^{\text{K}},g\in\Psi^{\text{G}}\right\} \tag{6.63}$$

$$\xi_{1,p,k,d}^{\min}\leq\xi_{1,p,k,d}\leq\xi_{1,p,k,d}^{\max};\quad\forall\left\{p\in\Psi^{\text{P}},k\in\Psi^{\text{K}},d\in\Psi^{\text{D}}\right\} \tag{6.64}$$

$$\xi_{2,p,k,d}^{\min}\leq\xi_{2,p,k,d}\leq\xi_{2,p,k,d}^{\max};\quad\forall\left\{p\in\Psi^{\text{P}},k\in\Psi^{\text{K}},d\in\Psi^{\text{D}}\right\} \tag{6.65}$$

$$\underset{x_{\text{CSC-EM}}}{\text{Max}} : CWF_{p,k} = \left(\sum_{g \in \Psi^{\text{G}}} \tilde{\upsilon}_{p,k,g}\left(\rho_{p,k,g}\right) + \sum_{d \in \Psi^{\text{D}}} \tilde{\upsilon}_{p,k,d}\left(\rho_{p,k,d}\right) \right);$$

$$\forall \{ p \in \Psi^{\text{P}}, k \in \Psi^{\text{K}}, g \in \Psi^{\text{G}}, d \in \Psi^{\text{D}} \}$$

$$= \underset{x_{\text{CSC-EM}}}{\text{Max}} : \left(\sum_{g \in \Psi^{\text{G}}} \frac{\left(2\xi_{1,p,k,g} - 1\right)}{2} \cdot \alpha_{p,k,g} \cdot \rho_{p,k,g}^{2} + \left(\xi_{2,p,k,g} - 1\right) \cdot \beta_{p,k,g} \cdot \rho_{p,k,g} - \gamma_{p,k,g} \right.$$

$$\left. + \sum_{d \in \Psi^{\text{D}}} \frac{\left(1 - 2\xi_{1,p,k,d}\right)}{2} \cdot \alpha_{p,k,d} \cdot \rho_{p,k,d}^{2} + \left(1 - \xi_{2,p,k,d}\right) \cdot \beta_{p,k,d} \cdot \rho_{p,k,d} + \gamma_{p,k,d} \right);$$

$$\forall \{ p \in \Psi^{\text{P}}, k \in \Psi^{\text{K}}, g \in \Psi^{\text{G}}, d \in \Psi^{\text{D}} \}$$

(6.66) Second level

and subject to Eqs. (6.67) through (6.70):

$$\sum_{g \in \Psi^{\text{G}}} \rho_{p,k,g} = \sum_{l \in \Psi^{\text{L}}} \kappa_{p,k,l} + \sum_{d \in \Psi^{\text{D}}} \rho_{p,k,d}; \quad \forall \{ p \in \Psi^{\text{P}}, k \in \Psi^{\text{K}}, g \in \Psi^{\text{G}}, l \in \Psi^{\text{L}}, d \in \Psi^{\text{D}} \} \tag{6.67}$$

$$\rho_{p,k,g}^{\min} \le \rho_{p,k,g} \le \rho_{p,k,g}^{\max}; \quad \forall \{ p \in \Psi^{\text{P}}, k \in \Psi^{\text{K}}, g \in \Psi^{\text{G}} \} \tag{6.68}$$

$$\rho_{p,k,d}^{\min} \le \rho_{p,k,d} \le \rho_{p,k,d}^{\max}; \quad \forall \{ p \in \Psi^{\text{P}}, k \in \Psi^{\text{K}}, d \in \Psi^{\text{D}} \} \tag{6.69}$$

$$f_{p,k,l}^{\min} \le f_{p,k,l} \le f_{p,k,l}^{\max}; \quad \forall \{ p \in \Psi^{\text{P}}, k \in \Psi^{\text{K}}, l \in \Psi^{\text{L}} \} \tag{6.70}$$

6.3.3 Mathematical Model of the Risk-Driven Strategic Tri-level Computational-Logical Framework

In the literature related to power systems, different techniques have been employed to model the uncertainty parameters. In the broadest sense, these techniques can be classified into six main classes of techniques: (1) probabilistic, (2) possibilistic, (3) hybrid possibilistic-probabilistic, (4) interval analysis, (5) robust optimization, and, (6) IGDT. Each of these techniques has advantages and disadvantages. Choosing a suitable technique for modeling and handling the uncertainty parameters in the optimization problem depends on many factors, such as location, level, nature of the uncertainty parameters, and planner preferences. Addressing these perspectives is out of the scope of this chapter, but the interested reader may look to the work by Aien et al. [31] for a discussion of these aspects.

Here, though, the authors employ the IGDT technique for handling risks of the PD-GEP problem arising from severe twofold uncertainty parameters for market price and demand. The IGDT is a well-established powerful technique for making decisions under severe uncertainties, as proposed by Ben-Haim [32]. This technique is radically different from others. The difference arises from the fact that uncertainty is modeled in this technique instead of the probability density functions as an information gap—i.e., distance between what is known or predicted and what may happen in reality. Therefore, when using the IGDT, there is no need for the primary data and/or recognizing structures associated with uncertainty parameters, including the probability density functions (PDFs) and/or fuzzy membership functions. More details about this technique can be found in a study by Ben-Haim [32]. To classify the basic characteristics of these uncertainty parameters in the proposed strategic tri-level computational-logical framework in a systematic manner, an uncertainty matrix is constructed for them, according to Table 6.3. These uncertainty parameters have a direct impact on the indices and constraints across all levels of the strategic tri-level computational-logical framework. Here, the IGDT risk-averse decision-making and the IGDT risk-taker decision-making policies of the strategic tri-level computational-logical framework are going to be placed under extensive scrutiny.

6.3.3.1 The IGDT Severe Twofold Uncertainty Model

There are many different IGDT models. Some of the most important are (1) the energy-bound model, (2) the envelope-bound model, (3) the Minkowski-norm model, (4) the slope-bound model, (5) the Fourier-bound model, (6) the discrete info-gap model, and so forth. Interested readers are directed to the work by Ben-Haim [32] for a comprehensive discussion of the main concepts of these information gap models. In view of the available structural features of these models, the envelope-bound model is more consistent with power system problems and its conditions. Therefore, the envelope-bound IGDT model will be widely employed here to handle uncertainty parameters. In the proposed strategic tri-level computational-logical framework, the envelope-bound IGDT model of the uncertain market price is given by Eqs. (6.71) through (6.73):

$$p\lambda = [p\lambda_1 \ \ldots \ p\lambda_{b-1} \ p\lambda_b \ p\lambda_{b+1} \ \ldots \ p\lambda_{\mathrm{B}}]; \quad \forall\{b \in \Psi^{\mathrm{B}}\} \tag{6.71}$$

$$p\lambda_b = \begin{bmatrix} p\lambda_{1,1,b} & \cdots & p\lambda_{p-1,1,b} & p\lambda_{p,1,b} & p\lambda_{p+1,1,b} & \cdots & p\lambda_{\mathrm{P},1,b} \\ \vdots & & \vdots & \vdots & \vdots & & \vdots \\ p\lambda_{1,k-1,b} & \cdots & p\lambda_{p-1,k-1,b} & p\lambda_{p,k-1,b} & p\lambda_{p+1,k-1,b} & \cdots & p\lambda_{\mathrm{P},k-1,b} \\ p\lambda_{1,k,b} & \cdots & p\lambda_{p-1,k,b} & p\lambda_{p,k,b} & p\lambda_{p+1,k,b} & \cdots & p\lambda_{\mathrm{P},k,b} \\ p\lambda_{1,k+1,b} & \cdots & p\lambda_{p-1,k+1,b} & p\lambda_{p,k+1,b} & p\lambda_{p+1,k+1,b} & \cdots & p\lambda_{\mathrm{P},k+1,b} \\ \vdots & & \vdots & \vdots & \vdots & & \vdots \\ p\lambda_{1,\mathrm{K},b} & \cdots & p\lambda_{p-1,\mathrm{K},b} & p\lambda_{p,\mathrm{K},b} & p\lambda_{p+1,\mathrm{K},b} & \cdots & p\lambda_{\mathrm{P},\mathrm{K},b} \end{bmatrix}; \quad \forall\{p \in \Psi^{\mathrm{P}}, k \in \Psi^{\mathrm{K}}, b \in \Psi^{\mathrm{B}}\} \tag{6.72}$$

Table 6.3 The constructed uncertainty matrix for severe twofold uncertainty parameters for market price and demand

	Location of the uncertainty parameter					Level of the uncertainty parameter			Nature of the uncertainty parameter	
Source of the uncertainty parameter	Context	Model uncertainty	Input	Parameter uncertainty	Model outcome uncertainty	Statistical uncertainty	Scenario uncertainty	Recognized ignorance	Epistemic uncertainty	Variability uncertainty
Market price	–	–	–	–	√	√	–	–	–	√
Demand	–	–	√	–	–	√	–	–	–	√

$$\Pi^{\text{price}}_{p,k,b}\left(\Delta^{\text{price}}, p\lambda_{p,k,b}\right) = \left\{ a\lambda_{p,k,b} : \left| \frac{a\lambda_{p,k,b} - p\lambda_{p,k,b}}{p\lambda_{p,k,b}} \right| \leq \Delta^{\text{price}} \right\};$$
$$\forall\left\{\Delta^{\text{price}} > 0, a\lambda_{p,k,b} \in \Pi^{\text{price}}_{p,k,b}\left(\Delta^{\text{price}}, p\lambda_{p,k,b}\right), p \in \Psi^{\text{P}}, k \in \Psi^{\text{K}}, b \in \Psi^{\text{B}}\right\} \tag{6.73}$$

In Eq. (6.73), if the actual market price, $a\lambda_{p,\ k,\ b}$, is equal to the predicted market price, $p\lambda_{p,\ k,\ b}$, the interval of the market price uncertainty, Δ^{price}, will be equal to zero; otherwise, this parameter will be equal to a positive value. Figure 6.4 shows how the market price robustness region is defined using Eq. (6.74):

$$RR^{\text{price}}_{p,k,b} = \left[\left(1 - \Delta^{\text{price}}\right) \cdot p\lambda_{p,k,b} \quad \left(1 + \Delta^{\text{price}}\right) \cdot p\lambda_{p,k,b}\right];$$
$$\forall\left\{\Delta^{\text{price}} > 0, p \in \Psi^{\text{P}}, k \in \Psi^{\text{K}}, b \in \Psi^{\text{B}}\right\} \tag{6.74}$$

Similarly, the envelope-bound IGDT model of the uncertain demand is given by Eqs. (6.75) through (6.77):

$$p\rho = \left[p\rho_1 \quad \cdots \quad p\rho_{b-1} \quad p\rho_b \quad p\rho_{b+1} \quad \cdots \quad p\rho_{\text{B}}\right]; \quad \forall\left\{b \in \Psi^{\text{B}}\right\} \tag{6.75}$$

$$p\rho_b = \begin{bmatrix} p\rho_{1,1,b} & \cdots & p\rho_{p-1,1,b} & p\rho_{p,1,b} & p\rho_{p+1,1,b} & \cdots & p\rho_{\text{P},1,b} \\ \vdots & & \vdots & \vdots & \vdots & & \vdots \\ p\rho_{1,k-1,b} & \cdots & p\rho_{p-1,k-1,b} & p\rho_{p,k-1,b} & p\rho_{p+1,k-1,b} & \cdots & p\rho_{\text{P},k-1,b} \\ p\rho_{1,k,b} & \cdots & p\rho_{p-1,k,b} & p\rho_{p,k,b} & p\rho_{p+1,k,b} & \cdots & p\rho_{\text{P},k,b} \\ p\rho_{1,k+1,b} & \cdots & p\rho_{p-1,k+1,b} & p\rho_{p,k+1,b} & p\rho_{p+1,k+1,b} & \cdots & p\rho_{\text{P},k+1,b} \\ \vdots & & \vdots & \vdots & \vdots & & \vdots \\ p\rho_{1,\text{K},b} & \cdots & p\rho_{p-1,\text{K},b} & p\rho_{p,\text{K},b} & p\rho_{p+1,\text{K},b} & \cdots & p\rho_{\text{P},\text{K},b} \end{bmatrix};$$
$$\forall\left\{p \in \Psi^{\text{P}}, k \in \Psi^{\text{K}}, b \in \Psi^{\text{B}}\right\} \tag{6.76}$$

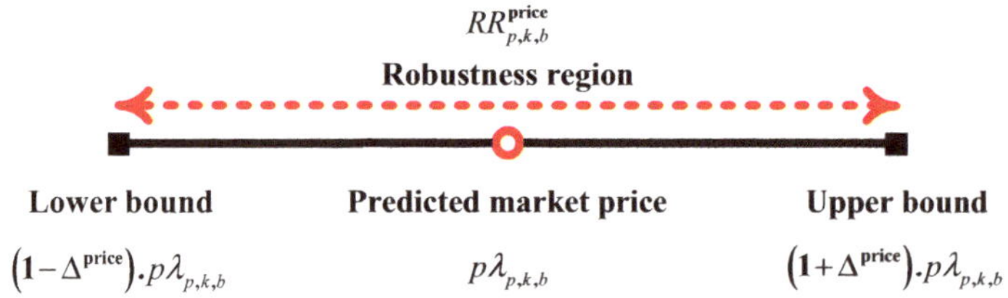

Fig. 6.4 The market price robustness region

$$\Pi_{p,k,b}^{\text{demand}}\left(\Delta^{\text{demand}}, p\rho_{p,k,b}\right) = \left\{ a\rho_{p,k,b} : \left| \frac{a\rho_{p,k,b} - p\rho_{p,k,b}}{p\rho_{p,k,b}} \right| \leq \Delta^{\text{demand}} \right\};$$
$$\forall \left\{ \Delta^{\text{demand}} > 0, a\rho_{p,k,b} \in \Pi_{p,k,b}^{\text{demand}}\left(\Delta^{\text{demand}}, p\rho_{p,k,b}\right), p \in \Psi^{\text{P}}, k \in \Psi^{\text{K}}, b \in \Psi^{\text{B}} \right\} \tag{6.77}$$

In Eq. (6.77), if the actual demand, $a\rho_{p,\ k,\ b}$, is equal to the predicted demand, $p\rho_{p,\ k,\ b}$, the interval of the demand uncertainty, Δ^{demand}, will be equal to zero; otherwise, this parameter will be equal to a positive value. Figure 6.5 shows how the demand robustness region is defined using Eq. (6.88):

$$RR_{p,k,b}^{\text{demand}} = \left[\left(1 - \Delta^{\text{demand}}\right) \cdot p\rho_{p,k,b} \quad \left(1 + \Delta^{\text{demand}}\right) \cdot p\rho_{p,k,b} \right];$$
$$\forall \left\{ \Delta^{\text{demand}} > 0, p \in \Psi^{\text{P}}, k \in \Psi^{\text{K}}, b \in \Psi^{\text{B}} \right\} \tag{6.78}$$

6.3.3.2 The IGDT Risk-Averse Decision-Making Policy: Robustness Function

In the IGDT-based strategic tri-level computational-logical framework, the robustness function addresses the destructive effects of the severe twofold uncertainties in the PD-GEP problem. In other words, this function describes the greatest level of uncertainty parameters, such that the minimum value of PD-GEP objectives associated with GENCO g ($\text{PD} - \text{GEPO}_g$) cannot be less than a predetermined critical profit. The robustness function for each GENCO is, therefore, the degree resistance against uncertainty parameters and immunity against smaller values of the $\text{PD} - \text{GEPO}_g$ at which defeat cannot arise.

This means that in the PD-GEP process by each GENCO, a large value of the robustness function is desirable. In the IGDT risk-averse decision-making policy, by using the predetermined critical profit, robust optimal generation expansion plans are determined by each GENCO. The robustness function for the proposed

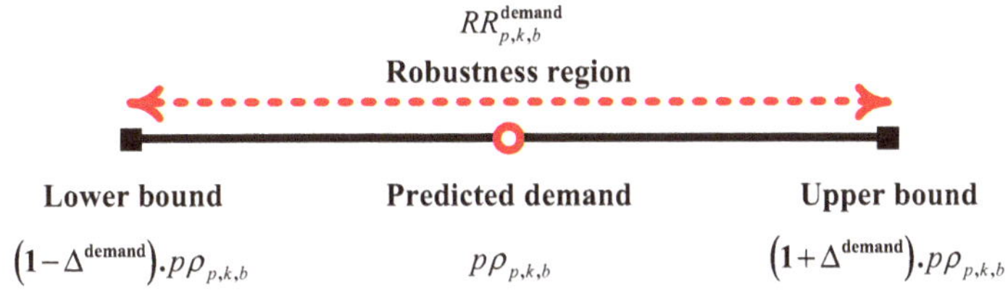

Fig. 6.5 The demand robustness region

strategic tri-level computational-logical framework can be expressed using Eq. (6.79):

$$\Upsilon_g^{\text{PD-GEP}}\left(x_{\text{PD-GEP}}, \varpi_{c,g}^{\text{PD-GEP}}\right) = \underset{\substack{\Delta_g^{\text{price}} \\ \Delta_g^{\text{demand}}}}{\text{Max}} \left\{ \left(\Delta_g^{\text{price}}, \Delta_g^{\text{demand}}\right) : \underset{\substack{x_{\text{PD-GEP}} \\ a\lambda_{p,k,b} \in \Pi_{p,k,g,b}^{\text{price}}\left(\Delta_g^{\text{price}}, p\lambda_{p,k,b}\right) \\ a\rho_{p,k,b} \in \Pi_{p,k,g,b}^{\text{demand}}\left(\Delta_g^{\text{demand}}, p\rho_{p,k,b}\right)}}{\text{Min}} \text{PD-GEPO}_g\left(x_{\text{PD-GEP}}, a\lambda_{p,k,g,b}, a\rho_{p,k,g,b}\right) \geq \varpi_{c,g}^{\text{PD-GEP}} \right\};$$
$$\forall\left\{\Delta_g^{\text{demand}} > 0, \Delta_g^{\text{price}} > 0, p \in \Psi^{\text{P}}, k \in \Psi^{\text{K}}, g \in \Psi^{\text{G}}, b \in \Psi^{\text{B}}\right\} \tag{6.79}$$

Therefore, the IGDT risk-averse decision-making policy for the proposed strategic tri-level computational-logical framework can be formulated as follows:

$$\Upsilon_g^{\text{PD-GEP}}\left(x_{\text{PD-GEP}}, \varpi_{c,g}^{\text{PD-GEP}}\right) = \underset{\substack{\Delta_g^{\text{price}} \\ \Delta_g^{\text{demand}}}}{\text{Max}} \left\{ \left(\Delta_g^{\text{price}}, \Delta_g^{\text{demand}}\right) : \underset{\substack{x_{\text{PD-GEP}} \\ a\lambda_{p,k,b} \in \Pi_{p,k,g,b}^{\text{price}}\left(\Delta_g^{\text{price}}, p\lambda_{p,k,b}\right) \\ a\rho_{p,k,b} \in \Pi_{p,k,g,b}^{\text{demand}}\left(\Delta_g^{\text{demand}}, p\rho_{p,k,b}\right)}}{\text{Min}} \text{PD-GEPO}_g\left(x_{\text{PD-GEP}}, a\lambda_{p,k,g,b}, a\rho_{p,k,g,b}\right) \geq \varpi_{c,g}^{\text{PD-GEP}} \right\};$$
$$\forall\left\{\Delta_g^{\text{demand}} > 0, \Delta_g^{\text{price}} > 0, p \in \Psi^{\text{P}}, k \in \Psi^{\text{K}}, g \in \Psi^{\text{G}}, b \in \Psi^{\text{B}}\right\} \tag{6.80) IGDT level}$$

and subject to Eqs. (6.81) through (6.86):

$$\underset{x_{\text{PD-GEP}}}{\text{Min}} \{\text{Eq.}(6.36)\} \geq \varpi_{c,g}^{\text{PD-GEP}};$$
$$\forall\left\{a\lambda_{p,k,g,b} = \left(1 - \Delta_g^{\text{price}}\right) \cdot p\lambda_{p,k,b}, a\rho_{p,k,g,b} = \left(1 - \Delta_g^{\text{demand}}\right) \cdot p\rho_{p,k,b}, p \in \Psi^{\text{P}}, k \in \Psi^{\text{K}}, g \in \Psi^{\text{G}}, b \in \Psi^{\text{B}}\right\} \tag{6.81) Third level}$$

$$a\lambda_{p,k,g,b} \leq \left(1 + \Delta_g^{\text{price}}\right) \cdot p\lambda_{p,k,b}; \quad \forall\left\{p \in \Psi^{\text{P}}, k \in \Psi^{\text{K}}, g \in \Psi^{\text{G}}, b \in \Psi^{\text{B}}\right\} \tag{6.82}$$

$$a\lambda_{p,k,g,b} \geq \left(1 - \Delta_g^{\text{price}}\right) \cdot p\lambda_{p,k,b}; \quad \forall\left\{p \in \Psi^{\text{P}}, k \in \Psi^{\text{K}}, g \in \Psi^{\text{G}}, b \in \Psi^{\text{B}}\right\} \tag{6.83}$$

$$a\rho_{p,k,g,b} \leq \left(1 + \Delta_g^{\text{demand}}\right) \cdot p\rho_{p,k,b}; \quad \forall\left\{p \in \Psi^{\text{P}}, k \in \Psi^{\text{K}}, g \in \Psi^{\text{G}}, b \in \Psi^{\text{B}}\right\} \tag{6.84}$$

$$a\rho_{p,k,g,b} \geq \left(1 - \Delta_g^{\text{demand}}\right) \cdot p\rho_{p,k,b};\quad \forall\left\{p \in \Psi^{\text{P}}, k \in \Psi^{\text{K}}, g \in \Psi^{\text{G}}, b \in \Psi^{\text{B}}\right\} \quad (6.85)$$

$$\{\text{Eqs.}(6.44) \text{ through } (6.56)\}|_{a\lambda_{p,k,g,b}, a\rho_{p,k,g,b}};$$
$$\forall\left\{a\lambda_{p,k,g,b} = \left(1 - \Delta_g^{\text{price}}\right) \cdot p\lambda_{p,k,b}, a\rho_{p,k,g,b} = \left(1 - \Delta_g^{\text{demand}}\right)\right.$$
$$\left.\cdot p\rho_{p,k,b}, p \in \Psi^{\text{P}}, g \in \Psi^{\text{G}}, k \in \Psi^{\text{K}}, b \in \Psi^{\text{B}}\right\} \quad (6.86)$$

$$\underset{x_{\text{BBM}}}{\text{Max}}\{\text{Eq.}(6.57)\}\Big|_{a\lambda_{p,k,b}, a\rho_{p,k,b}};\quad \forall\left\{a\lambda_{p,k,b} = \left(1 - \Delta^{\text{price}}\right) \cdot p\lambda_{p,k,b},\right.$$
$$\left.\text{and, } a\rho_{p,k,b} = \left(1 - \Delta^{\text{demand}}\right) \cdot p\rho_{p,k,b}, g \in \Psi^{\text{G}}, p \in \Psi^{\text{P}}, k \in \Psi^{\text{K}}, b \in \Psi^{\text{B}}\right\}$$
$$\underset{x_{\text{BBM}}}{\text{Max}}\{\text{Eq.}(6.58)\}\Big|_{a\lambda_{p,k,b}, a\rho_{p,k,b}};\quad \forall\left\{a\lambda_{p,k,b} = \left(1 - \Delta^{\text{price}}\right) \cdot p\lambda_{p,k,b},\right.$$
$$\left.\text{and, } a\rho_{p,k,b} = \left(1 - \Delta^{\text{demand}}\right) \cdot p\rho_{p,k,b}, d \in \Psi^{\text{D}}, p \in \Psi^{\text{P}}, k \in \Psi^{\text{K}}, b \in \Psi^{\text{B}}\right\}$$

(6.87) First level

and subject to Eq. (6.88):

$$\{\text{Eqs.}(6.59) \text{ through } (6.65)\}|_{a\lambda_{p,k,b}, a\rho_{p,k,b}};$$
$$\forall\left\{a\lambda_{p,k,b} = \left(1 - \Delta^{\text{price}}\right) \cdot p\lambda_{p,k,b}, a\rho_{p,k,b} = \left(1 - \Delta^{\text{demand}}\right) \cdot p\rho_{p,k,b}, p \in \Psi^{\text{P}}, k \in \Psi^{\text{K}}, b \in \Psi^{\text{B}}\right\} \quad (6.88)$$

$$\underset{x_{\text{CSCEM}}}{\text{Max}}\{\text{Eq.}(6.66)\}\Big|_{a\lambda_{p,k,b}, a\rho_{p,k,b}};\quad \forall\left\{a\lambda_{p,k,b} = \left(1 - \Delta^{\text{price}}\right) \cdot p\lambda_{p,k,b},\right.$$
$$\left.\text{and, } a\rho_{p,k,b} = \left(1 - \Delta^{\text{demand}}\right) \cdot p\rho_{p,k,b}, g \in \Psi^{\text{G}}, d \in \Psi^{\text{D}}, p \in \Psi^{\text{P}}, k \in \Psi^{\text{K}}, b \in \Psi^{\text{B}}\right\}$$

(6.89) Second level

and subject to Eq. (6.90):

$$\{\text{Eqs.}(6.67) \text{ through } (6.70)\}|_{a\lambda_{p,k,b}, a\rho_{p,k,b}};$$
$$\forall\left\{a\lambda_{p,k,b} = \left(1 - \Delta^{\text{price}}\right) \cdot p\lambda_{p,k,b}, a\rho_{p,k,b} = \left(1 - \Delta^{\text{demand}}\right) \cdot p\rho_{p,k,b}, p \in \Psi^{\text{P}}, k \in \Psi^{\text{K}}, b \in \Psi^{\text{B}}\right\} \quad (6.90)$$

In the proposed framework under the IGDT risk-averse decision-making policy, the minimum values of the PD - GEPO$_g$ are achieved for the highest level of the uncertainties in the robust region (see Figs. 6.4 and 6.5). The solution of the IGDT risk-averse decision-making policy will give optimal robust expansion plans of the

GENCOs, based on the defined values of their critical profits. The predetermined critical profit of GENCO g is determined using Eq. (6.91):

$$\varpi_{c,g}^{\text{PD-GEP}} = \left(1 - \sigma_{c,g}^{\text{PD-GEP}}\right) \cdot \varpi_{b,g}^{\text{PD-GEP}}; \quad \forall\{g \in \Psi^{\text{G}}\} \tag{6.91}$$

In Eq. (6.91), the IGDT base profit for each GENCO, $\varpi_{\text{b},g}^{\text{PD-GEP}}$, is calculated by solving the deterministic strategic tri-level computational-logical framework presented in Eqs. (6.36)–(6.70). The deterministic framework of the proposed strategic tri-level computational-logical framework is also named as a risk-neutral decision-making policy. In general, the critical profit of GENCO g is smaller than its base profit.

6.3.3.3 The IGDT Risk-Taker Decision-Making Strategy: Opportunity Function

In the IGDT-based strategic tri-level computational-logical framework, the opportunity function addresses the propitious effects of the severe twofold uncertainties in the PD-GEP problem. In simple terms, this function represents the smallest level of uncertainty parameters, such that the maximum value of PD-GEPO_g associated with GENCO g can possibly be as high as the predetermined target profit. The opportunity function for each GENCO is, therefore, the immunity against windfall rewards at which sweeping success can occur. This means that in the PD-GEP process for each GENCO, a small value of the opportunity function is desirable.

In the IGDT risk-taker decision-making policy, by using the predetermined target profit, opportunistic optimal generation expansion plans are determined by each GENCO. The opportunity function for the proposed strategic tri-level computational-logical framework can be defined using Eq. (6.92):

$$\Gamma_g^{\text{PD-GEP}}\left(x_{\text{PD-GEP}}, \varpi_{t,g}^{\text{PD-GEP}}\right) = \underset{\substack{\Delta_g^{\text{price}} \\ \Delta_g^{\text{demand}}}}{\text{Min}} \left\{ \left(\Delta_g^{\text{price}}, \Delta_g^{\text{demand}}\right) : \underset{\substack{x_{\text{PD-GEP}} \\ a\lambda_{p,k,g,b} \in \Pi_{p,k,g,b}^{\text{price}}\left(\Delta_g^{\text{price}}, p\lambda_{p,k,b}\right) \\ a\rho_{p,k,g,b} \in \Pi_{p,k,g,b}^{\text{demand}}\left(\Delta_g^{\text{demand}}, p\rho_{p,k,b}\right)}}{\text{Max}} \text{PD-GEPO}_g\left(x_{\text{PD-GEP}}, a\lambda_{p,k,g,b}, a\rho_{p,k,g,b}\right) \geq \varpi_{t,g}^{\text{PD-GEP}} \right\};$$

$$\forall\left\{\Delta_g^{\text{demand}} > 0, \Delta_g^{\text{price}} > 0, p \in \Psi^{\text{P}}, k \in \Psi^{\text{K}}, g \in \Psi^{\text{G}}, b \in \Psi^{\text{B}}\right\} \tag{6.92}$$

Therefore, the IGDT risk-taker decision-making policy for the proposed strategic tri-level computational-logical framework can be formulated as follows:

$$\Gamma_g^{\text{PD-GEP}}\left(x_{\text{PD-GEP}}, \varpi_{\text{t},g}^{\text{PD-GEP}}\right) = \underset{\substack{\Delta_g^{\text{price}} \\ \Delta_g^{\text{demand}}}}{\text{Min}} \left\{ \left(\Delta_g^{\text{price}}, \Delta_g^{\text{demand}}\right) : \underset{\substack{x_{\text{PD-GEP}} \\ a\lambda_{p,k,g,b} \in \Pi_{p,k,g,b}^{\text{price}}\left(\Delta_g^{\text{price}}, p\lambda_{p,k,b}\right) \\ a\rho_{p,k,g,b} \in \Pi_{p,k,g,b}^{\text{demand}}\left(\Delta_g^{\text{demand}}, p\rho_{p,k,b}\right)}}{\text{Max}} \text{PD-GEPO}_g\left(x_{\text{PD-GEP}}, a\lambda_{p,k,g,b}, a\rho_{p,k,g,b}\right) \geq \varpi_{\text{t},g}^{\text{PD-GEP}} \right\};$$

$$\forall\left\{\Delta_g^{\text{demand}} > 0, \Delta_g^{\text{price}} > 0, p \in \Psi^{\text{P}}, k \in \Psi^{\text{K}}, g \in \Psi^{\text{G}}, b \in \Psi^{\text{B}}\right\} \quad (6.93)\ \text{IGDT level}$$

and subject to Eqs. (6.94) through (6.99):

$$\underset{x_{\text{PD-GEP}}}{\text{Max}} \{\text{Eq.}(6.36)\} \geq \varpi_{\text{t},g}^{\text{PD-GEP}};$$
$$\forall\left\{a\lambda_{p,k,g,b} = \left(1 + \Delta_g^{\text{price}}\right) \cdot p\lambda_{p,k,b}, a\rho_{p,k,g,b} = \left(1 + \Delta_g^{\text{demand}}\right) \cdot p\rho_{p,k,b}, p \in \Psi^{\text{P}}, k \in \Psi^{\text{K}}, g \in \Psi^{\text{G}}, b \in \Psi^{\text{B}}\right\} \quad (6.94)\ \text{Third level}$$

$$a\lambda_{p,k,g,b} \leq \left(1 + \Delta_g^{\text{price}}\right) \cdot p\lambda_{p,k,b}; \quad \forall\left\{p \in \Psi^{\text{P}}, k \in \Psi^{\text{K}}, g \in \Psi^{\text{G}}, b \in \Psi^{\text{B}}\right\} \quad (6.95)$$

$$a\lambda_{p,k,g,b} \geq \left(1 - \Delta_g^{\text{price}}\right) \cdot p\lambda_{p,k,b}; \quad \forall\left\{p \in \Psi^{\text{P}}, k \in \Psi^{\text{K}}, g \in \Psi^{\text{G}}, b \in \Psi^{\text{B}}\right\} \quad (6.96)$$

$$a\rho_{p,k,g,b} \leq \left(1 + \Delta_g^{\text{demand}}\right) \cdot p\rho_{p,k,b}; \quad \forall\left\{p \in \Psi^{\text{P}}, k \in \Psi^{\text{K}}, g \in \Psi^{\text{G}}, b \in \Psi^{\text{B}}\right\} \quad (6.97)$$

$$a\rho_{p,k,g,b} \geq \left(1 - \Delta_g^{\text{demand}}\right) \cdot p\rho_{p,k,b}; \quad \forall\left\{p \in \Psi^{\text{P}}, k \in \Psi^{\text{K}}, g \in \Psi^{\text{G}}, b \in \Psi^{\text{B}}\right\} \quad (6.98)$$

$$\{\text{Eq.}(6.44) \text{ through } (6.56)\}|_{a\lambda_{p,k,g,b}, a\rho_{p,k,g,b}};$$
$$\forall\left\{a\lambda_{p,k,g,b} = \left(1 + \Delta_g^{\text{price}}\right) \cdot p\lambda_{p,k,b}, a\rho_{p,k,g,b} = \left(1 + \Delta^{\text{demand}}\right) \cdot p\rho_{p,k,b}, p \in \Psi^{\text{P}}, k \in \Psi^{\text{K}}, g \in \Psi^{\text{G}}, b \in \Psi^{\text{B}}\right\} \quad (6.99)$$

$$\underset{x_{\text{BBM}}}{\text{Max}}\,\{\text{Eq.}(6.57)\}\Big|_{a\lambda_{p,k,b},a\rho_{p,k,b}};\quad \forall\big\{a\lambda_{p,k,b}=\big(1+\Delta^{\text{price}}\big)\cdot p\lambda_{p,k,b},$$
$$\text{and}, a\rho_{p,k,b}=\big(1+\Delta^{\text{demand}}\big)\cdot p\rho_{p,k,b}, g\in\Psi^{\text{G}}, p\in\Psi^{\text{P}}, k\in\Psi^{\text{K}}, b\in\Psi^{\text{B}}\big\}$$
$$\underset{x_{\text{BBM}}}{\text{Max}}\,\{\text{Eq.}(6.58)\}\Big|_{a\lambda_{p,k,b},a\rho_{p,k,b}};\quad \forall\big\{a\lambda_{p,k,b}=\big(1+\Delta^{\text{price}}\big)\cdot p\lambda_{p,k,b},$$
$$\text{and}, a\rho_{p,k,b}=\big(1+\Delta^{\text{demand}}\big)\cdot p\rho_{p,k,b}, d\in\Psi^{\text{D}}, p\in\Psi^{\text{P}}, k\in\Psi^{\text{K}}, b\in\Psi^{\text{B}}\big\}$$
(6.100) First level

and subject to Eq. (6.101):

$$\{\text{Eqs.}(6.59)\text{ through }(6.65)\}\big|_{a\lambda_{p,k,b},a\rho_{p,k,b}};$$
$$\forall\big\{a\lambda_{p,k,b}=\big(1+\Delta^{\text{price}}\big)\cdot p\lambda_{p,k,b}, a\rho_{p,k,b}=\big(1+\Delta^{\text{demand}}\big)\cdot p\rho_{p,k,b}, p\in\Psi^{\text{P}}, k\in\Psi^{\text{K}}, b\in\Psi^{\text{B}}\big\} \tag{6.101}$$

$$\underset{x_{\text{CSCEM}}}{\text{Max}}\,\{\text{Eq.}(6.66)\}\Big|_{a\lambda_{p,k,b},a\rho_{p,k,b}};\quad \forall\big\{a\lambda_{p,k,b}=\big(1+\Delta^{\text{price}}\big)\cdot p\lambda_{p,k,b},$$
$$\text{and}, a\rho_{p,k,b}=\big(1+\Delta^{\text{demand}}\big)\cdot p\rho_{p,k,b}, g\in\Psi^{\text{G}}, d\in\Psi^{\text{D}}, p\in\Psi^{\text{P}}, k\in\Psi^{\text{K}}, b\in\Psi^{\text{B}}\big\}$$

(6.102) Second level

and subject to Eq. (6.103):

$$\{\text{Eqs.}(6.67)\text{ through }(6.70)\}\big|_{a\lambda_{p,k,b},a\rho_{p,k,b}};$$
$$\forall\big\{a\lambda_{p,k,b}=\big(1+\Delta^{\text{price}}\big)\cdot p\lambda_{p,k,b}, a\rho_{p,k,b}=\big(1+\Delta^{\text{demand}}\big)\cdot p\rho_{p,k,b}, p\in\Psi^{\text{P}}, k\in\Psi^{\text{K}}, b\in\Psi^{\text{B}}\big\} \tag{6.103}$$

In the proposed framework under the IGDT risk-taker decision-making policy, maximum values of the PD-GEPO$_g$ are achieved for the lowest level of uncertainties in the robust region (see Figs. 6.4 and 6.5). The solution of the IGDT risk-taker decision-making policy will give the optimal opportunistic expansion plans of the GENCOs, based on the defined values of their target profits. The predetermined target profit is also determined by GENCO g using Eq. (6.104):

$$\varpi_{\text{t},g}^{\text{PD-GEP}}=\left(1+\sigma_{\text{t},g}^{\text{PD-GEP}}\right)\cdot\varpi_{\text{b},g}^{\text{PD-GEP}};\quad\forall\big\{g\in\Psi^{\text{G}}\big\} \tag{6.104}$$

Generally, the target profit of GENCO g is greater than its base profit. Similarly, from Eq. (6.104), the IGDT base profit for each GENCO, $\varpi_{\text{b},g}^{\text{PD-GEP}}$, is calculated by solving the deterministic strategic tri-level computational-logical framework presented in Eqs. (6.36)–(6.70):

6.3.4 Solution Method and Implementation Considerations

In the proposed strategic tri-level computational-logical framework, the decision-making variables in the short-term operation slave problem are (1) the generated electrical power by the GENCOs; (2) the slope and intercept parameters of the bidding strategy function associated with the GENCOs; (3) the purchased electrical power by the DISCOs; and, (4) the slope and intercept parameters of the bidding strategy function associated with the DISCOs. At the same time, the decision-making variables in the long-term planning master problem are (1) the time of the augmented and newly installed generation units by the GENCOs over the planning horizon; (2) the location of the newly installed generation units by the GENCOs; and, (3) the capacity of the augmented and newly installed generation units by the GENCOs. The proposed strategic tri-level computational-logical framework starts from the first level, or the BBM, where market participants, including the GENCOs and DISCOs, compete with their competitors to buy and sell energy, based on the supply function equilibrium (SFE) model. In this level, each market participant solves the bidding strategy problem to calculate and modify its bidding strategy parameters [see Eqs. (6.57)–(6.65)] using an optimization algorithm, while the bidding strategy parameters of the rivals are kept constant. This process is repeated until the Nash equilibrium point is reached. After the Nash equilibrium point has been determined, the optimal bidding strategy functions of both GENCOs and DISCOs will have been achieved. Then, these optimal bidding strategy functions are transferred to the second level—the CSC electricity market problem. In this level, the ISO aggregates these functions to form equivalent supply and demand functions. If the obtained optimal bidding strategies by both the GENCOs and DISCOs generate a violation in the network constraints, the ISO solves the CSC electricity market-clearing problem [see Eqs. (6.66)–(6.70)] using an optimization algorithm, with the aim of maximizing CWF. The first and second levels are repeated for all patterns considered in period p of the planning horizon. After performing the short-term operational slave problem in all of the patterns of period p, the information related to the GENCOs, DISCOs, CSC electricity market outcomes, etc. is transferred to the third level—the long-term planning master problem. As noted previously, the third level is composed of both local and global layers. In the local layer, each GENCO solves the PD-GEP problem in order to determine the augmented and newly installed generation units [see Eqs. (6.36)–(6.53)] using an optimization algorithm. In other words, each GENCO competes with its rivals to determine the augmented and newly installed generation units under a Cournot-based model by an iterative process until the Nash equilibrium point is reached. After completing the local layer, the GENCOs' generation data, or optimal generation expansion plans, are transferred to the global layer. In this layer, the ISO receives these data and checks the established global constraints [see Eqs. (6.54)–(6.56)] for the entire power system. If the obtained optimal expansion plans by the GENCOs in the local layer violate any of these constraints, the ISO sends the corrective factors related to the violated constraints to the GENCOs. Then, the GENCOs update their expansion plans according to the received corrective factors from the global layer. After this modification and update, the GENCOs resubmit their optimal expansion plans to the global layer. This iterative process between the two layers is repeated until the obtained optimal expansion plans by the GENCOs in the local layer satisfy all established constraints concerned with the global layer. Finally, the obtained optimal expansion plans are approved for period p and considered as operational equipment in the next period of planning.

Figure 6.6 shows the flowchart of the proposed strategic tri-level computational-logical framework under the IGDT risk-neutral, the IGDT risk-averse, and the IGDT risk-taker decision-making policies. To implement the proposed framework under each of these decision-making policies, the block associated with this decision-making policy is active, while all other blocks are inactive in the flowchart.

6.3.5 Simulation Results and Case Studies

Figure 6.7 shows that the proposed strategic tri-level computational-logical framework was implemented and tested on a modified real-world, large-scale 46-bus power grid in southern Brazil [33–36]. Network data for this modified test system are tabulated in Appendix 3. Table 6.95 presents the data of the transmission lines associated with the modified test system.

Table 6.96 gives the parameters of the GENCOs' and candidate generation units. Table 6.97 gives the demand and parameters of the DISCOs. In addition, parameters associated with the SO_2, the CO_2, and the NO_x atmospheric emissions and fuel consumption for all GENCOs were taken from the studies by Tekiner et al. [20], Careri et al. [24], Chen et al. [37], Zhou et al. [38], and Kaplan [39] and modified for this current test system, according to Table 6.98. Since it is assumed that all units use the same technology (i.e., oil/steam, the value of emitted CO_2, SO_2, and NO_x atmospheric emissions), the fuel consumption rate for power generation of 1 MWh for all GENCOs is considered equal. However, the maximum permissible value of the emitted atmospheric emissions and the maximum value of fuel consumption are considered different for each GENCO, due to the fact that the GENCOs have different capacities. Table 6.99 gives the parameters of the available financial resources and the construction costs of the generation units for all GENCOs. Practically speaking, the available financial resources are not equal for the GENCOs. In this study, then, these values are taken into account differently for each GENCO. Moreover, the cost of building a generation unit

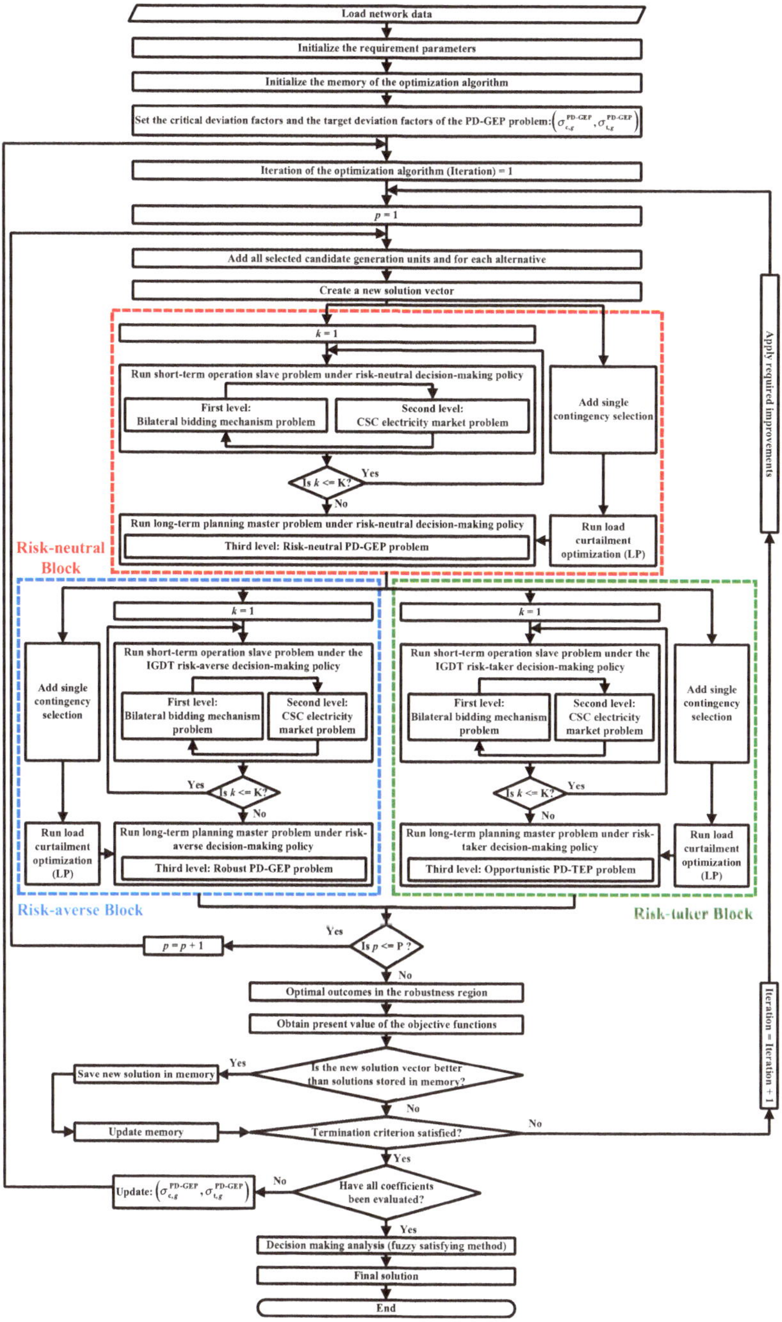

Fig. 6.6 Flowchart of the strategic tri-level computational-logical framework under the IGDT risk-neutral, the IGDT risk-averse, and the IGDT risk-taker decision-making policies

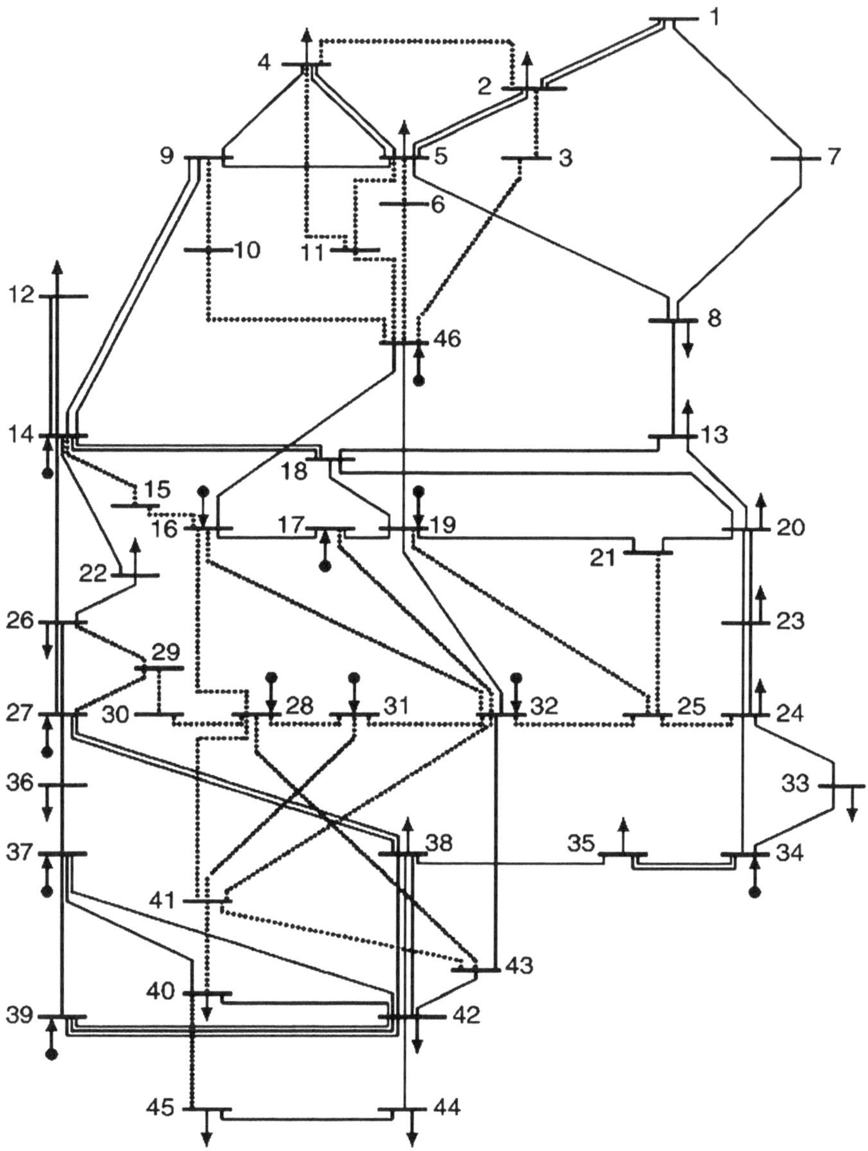

Fig. 6.7 The modified 46-bus network under study in southern Brazil

varies in different places of the power network. These values, therefore, are regarded differently for GENCO.

In order to prevent the application of market power, a price ceiling is defined and used for the market participants, including GENCOs and DISCOs. Consequently, each market participant can change its bidding strategy parameters between predetermined minimum and maximum values of these parameters. In this regard, data of the bidding strategy parameters of both GENCOs and DISCOs are given in Table 6.100. Table 6.101 shows the parameters of the weighting coefficients for the objective functions associated with the PD-GEP problem. The planning horizon is set at 10 years, divided into 1-year periods. Solving the proposed strategic tri-level computational-logical framework for all hours of a day and for all days of a year would be excessively difficult, complicated, and time consuming. This high computational burden can be reduced by using different time patterns, such as daily and seasonal patterns. In here, the time span of the network load duration curve (LDC) can be split into a number of time steps, using Eq. (6.105) [40]:

$$\Delta t_k = round\left(2^{0.5.(1+k)}\right); \quad \forall\left\{k \in \Psi^{\mathrm{K}}\right\} \tag{6.105}$$

In this exponential scheme, the number of patterns is equal to 22 for the LDC with a time interval of 8760 h. Time intervals for the last two patterns are regarded as equal. The load step in pattern k is also considered as the average load value during this pattern. Therefore, instead of solving the short-term operation slave problem in 8760 operational points, this slave problem is solved at 22 operational points. The expected costumer outage cost (ECOC) is equal to 250 \$/MWh. Here, it is assumed that the installed capacity by each GENCO in period p should not exceed 50% of its existing capacity in period $p - 1$. The annual interest rate is set at 5%. The annual demand growth is also considered to be 4%. In addition, maximum and minimum reserve margins are considered to be 40% and 10% of the average of the peak demand, respectively. The maximum value of the ECO is set at 3% of the average of the peak demand. Furthermore, the maximum percentage of installed generation capacity that can be installed by GENCO g at each period of the planning horizon is 70%. Table 6.102 provides a summary of these assumptions.

To investigate and analyze the performance of the proposed strategic tri-level computational-logical framework, two different cases were defined and applied, as follows:

- First case: The proposed strategic tri-level computational-logical framework is run under the IGDT risk-averse decision-making policy while considering the following two scenarios:
  - The strategic behavior of the market participants (i.e., first level) is ignored.
  - The strategic behavior of the market participants (i.e., first level) is considered.
- Second case: The proposed tri-level computational-logical framework is run under the IGDT risk-taker decision-making policy while considering the following two scenarios:
  - The strategic behavior of the market participants (i.e., first level) is disregarded.
  - The strategic behavior of the market participants (i.e., first level) is regarded.

The proposed strategic tri-level computational-logical framework under the IGDT risk-averse and the IGDT risk-taker decision-making policies is implemented and solved by using the proposed multi-objective multi-stage computational, multi-dimensional, multiple-homogeneous enhanced melody search algorithm (MMM-EMSA), or multi-objective multi-stage computational, multi-dimensional, single-inhomogeneous enhanced melody search algorithm (MMS-EMSA), or multi-objective symphony orchestra search algorithm (SOSA), which was addressed in Chap. 4. Table 6.103 gives the parameter adjustments of the newly developed multi-objective SOSA.

In the proposed strategic tri-level computational-logical framework under the IGDT risk-averse and the IGDT risk-taker decision-making policies, the GENCOs need to have access to the obtained results from the proposed framework under a risk-neutral/deterministic decision-making policy in order to calculate their critical and target profits: $\varpi_{t,g}^{PD-GEP}$ and $\varpi_{c,g}^{PD-GEP}$. At first, the proposed deterministic strategic tri-level computational-logical framework [Eqs. (6.36)–(6.70)] is solved based on the predicted demand and market price by the offered single-objective SOSA. Then, the obtained results are placed at the disposal of the GENCOs to determine their optimal expansion plans under the IGDT risk-averse and the IGDT risk-taker decision-making policies. Since the strategic tri-level computational-logical framework under the IGDT risk-averse and the IGDT risk-taker decision-making policies is examined for first and second scenarios, the proposed deterministic strategic tri-level computational-logical framework should also be solved based on these two scenarios.

After solving the proposed framework under the risk-neutral/deterministic decision-making policy, the optimal base profits of the GENCOs over the planning horizon under the first scenario (ignoring the BBM) and the second scenario (considering the BBM) are obtained according to Table 6.4. It is necessary to mention that the profit presented in Table 6.4 is the net profit that consists of the EP minus the GIC plus the CP [see Eq. (6.22)]. From the optimal results presented in Table 6.4, it can be concluded that the profit of the GENCOs over the planning horizon under the second scenario is much higher than that under the first scenario, due to the adoption of strategic behavior by the market participants.

6.3.5.1 First Case: Simulation Results and Discussion

The strategic tri-level computational-logical framework under the IGDT risk-averse decision-making policy is solved by optimization of Eqs. (6.80) through (6.91) under different values of the critical profit deviation factor, $\sigma_{c,g}^{PD-GEP} \in (0, 0.8)$, which leads to the different critical profits, $\varpi_{c,g}^{PD-GEP}$. It should be noted that all GENCOs are involved in the process of solving of the framework under the IGDT risk-averse decision-making policy. In other words, for each value of the critical profit deviation factor, $\sigma_{c,g}^{PD-GEP} \in (0, 0.8)$, GENCO g tries to find the maximum

Table 6.4 Optimal base profits of the GENCOs over the planning horizon under the first and second scenarios

GENCO No.	Base profit ($\varpi_{b,g}^{PD-GEP}$; M$) First scenario: Ignoring the BBM	Second scenario: Considering the BBM
GENCO 1	282.786547	379.429696
GENCO 2	323.304161	432.011602
GENCO 3	287.964564	372.153910
GENCO 4	306.263997	403.399222
GENCO 5	102.405250	137.569162
GENCO 6	178.743738	211.271669
GENCO 7	169.835064	200.286671
GENCO 8	138.424512	160.864376
GENCO 9	192.145473	236.769605
GENCO 10	129.647592	168.557316
GENCO 11	187.134549	217.478392
GENCO 12	206.884127	257.618738
Total	2505.539573	3177.410359

values of Δ_g^{price} and Δ_g^{demand} by solving Eqs. (6.80) through (6.86) in a multi-objective manner using a Pareto optimality concept.

This process brings about the organization of an optimal solution set, refereed to as the Pareto-optimal solution set instead of a single optimal solution. After determination of the Pareto-optimal solution set, only one solution for a specific value of the critical cost deviation factor, $\sigma_{c,g}^{PD-GEP}$, should be selected from the Pareto-optimal solution set by this GENCO. This means that a well-suited compromise among different objectives, namely Δ_g^{price} and Δ_g^{demand}, is established in this process. In this way, two main questions are raised: (1) which solution must be chosen by this GENCO and (2) how to select it. GENCO g selects the best solution based on its requirements and preferences. In the relevant literature, many approaches have been reported on choosing a trade-off solution among the Pareto-optimal solution set (see Chap. 2). Here, though, the authors employ fuzzy satisfying method (FSM) based on the conservative methodology—the min-max formulation—to select the best solution. The reason for choosing this method is its simplicity and resemblance to human ratiocination. This approach is explicitly addressed in Chap. 2. Consider the first GENCO's performance (i.e., GENCO 1). If this GENCO chooses a value of zero for its critical profit deviation factor, $\sigma_{c,g}^{PD-GEP} = 0$, the critical profit of this GENCO is identical to its base profit, which is obtained from the deterministic framework as $\varpi_{c,g}^{PD-GEP} = \varpi_{b,g}^{PD-GEP} = \$282.786547$M and $\varpi_{c,g}^{PD-GEP} = \varpi_{b,g}^{PD-GEP} = \$379.429696$ M for the first and second scenarios, respectively. It should be noted that adopting the risk-neutral decision-making policy by this GENCO results in a value of zero for its critical profit deviation factor. Under these conditions, the robustness and risk level of the obtained optimal generation expansion plans by this GENCO will be zero. In other words, this GENCO is not ready to deal with demand and market price uncertainties; therefore, its profit [see Eq. (6.22)] is extremely vulnerable to these uncertainties. Suppose that GENCO 1 chooses a critical profit deviation factor of 0.4 (i.e., $\sigma_{c,g}^{PD-GEP} = 0.4$). In this situation, Eqs. (6.80) through (6.86) are solved by this GENCO in a multi-objective manner for this value of the critical profit deviation factor with the aim of maximizing Δ_g^{price} and Δ_g^{demand} according to the first and second scenarios. Figure 6.8 shows the Pareto-optimal solution set for this value of the critical profit deviation factor related to GENCO 1 under the first and second scenarios. For illustration, consider the performance of GENCO 1 under the first scenario (ignoring the BBM).

For example, in the solution marked with a square in Fig. 6.8, a critical profit of $\varpi_{c,g}^{PD-GEP} = (1 - 0.4) \times \varpi_{b,g}^{PD-GEP} = \$169.671928\text{M}$ is guaranteed for this GENCO, provided that Δ_g^{price} and Δ_g^{demand} do not exceed 36.396412% and 3.999482%, respectively. Put simply, if there is a difference of less than 36.396412% between the actual market price, $a\lambda_{p,\ k,\ b}$, and the predicted market price, $p\lambda_{p,\ k,\ b}$, and also a difference of less than 3.999482% between actual demand, $a\rho_{p,\ k,\ b}$, and predicted demand, $p\rho_{p,\ k,\ b}$, the obtained profit by this GENCO will be at least $\varpi_{c,g}^{PD-GEP} = \$169.671928\text{M}$. To clarify, if the actual market price, $a\lambda_{p,\ k,\ b}$, falls within the market price robustness region, $RR_{p,k,b}^{price}$, and actual demand, $a\rho_{p,\ k,\ b}$, falls within the demand robustness region, $RR_{p,k,b}^{demand}$, then the profit obtained by this GENCO will be at least $\varpi_{c,g}^{PD-GEP} = \$169.671928\text{M}$. Note that if the actual market price and/or demand are not at least 36.396412% and 3.999482% lower than their corresponding predicted values, respectively, it will not be possible to achieve the critical profit by this GENCO. Put another way, if the actual market price and/or actual demand are not within the market price and demand robustness regions, respectively, it is not possible to achieve critical profit by this GENCO. By decreasing the critical profit by GENCO 1, $\varpi_{c,g}^{PD-GEP}$, due to enlarging its critical profit deviation factor, $\sigma_{c,g}^{PD-GEP}$, the robustness parameters of this GENCO become larger. This means that lower critical profits of this GENCO in the larger market price robustness region, $RR_{p,k,b}^{price}$, and larger demand robustness region, $RR_{p,k,b}^{demand}$, can be guaranteed and vice versa. In other words, better robustness of the calculated generation expansion plans by this GENCO results in a lower critical profit value of this GENCO, $\varpi_{c,g}^{PD-GEP}$, and/or in a larger value of the critical profit deviation factor of this GENCO, $\sigma_{c,g}^{PD-GEP}$. The rest of the solutions provided in Fig. 6.8 have the same interpretations. For the same reason, the Pareto-optimal solution set related to the other critical profit deviation factors of GENCO 1 has the same interpretations. It is relevant to note that these solutions are produced because of the adoption of different critical profit deviation factors by this GENCO. This analysis is also true for other GENCOs under the first scenario. It is also true for all GENCOs under the second scenario. It should be noted that there are a number of the GENCOs that may choose different amounts for their critical profit deviation factors—$\sigma_{c,g}^{PD-GEP}$. This condition leads to the formation of different critical profits and, consequently, different robustness values for the GENCOs. However, as previously mentioned,

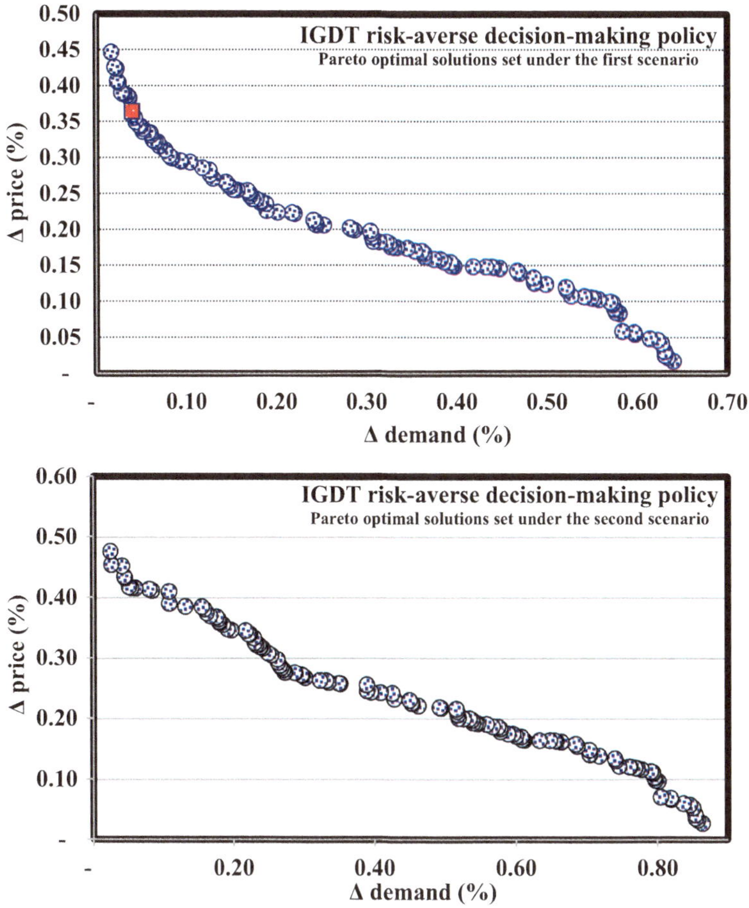

Fig. 6.8 The Pareto-optimal solution of robustness related to GENCO 1 against market price and demand uncertainties under the first and second scenarios

the third level of the strategic tri-level computational-logical framework is composed of two different layers: (1) local and (2) global. The existence of the iterative process between the local and global layers leads to the convergence of the interval of the market price uncertainty, Δ^{price}, and the interval of the demand uncertainty, Δ^{demand}, at a point of equilibrium. After solving the proposed framework under the first and the second scenarios of the first case—the IGDT risk-averse decision-making policy—by the newly proposed multi-objective SOSA, the optimal values for Δ^{price}, Δ^{demand}, and relevant critical profit deviation factors are calculated by the FSM in accordance with Table 6.5.

Tables 6.6 shows the optimal results related to the bidding strategy parameters of the market participants for the first level (i.e., the BBM) of the proposed strategic tri-level computational-logical framework under the second scenario in the first case (i.e., the IGDT risk-averse decision-making policy). Only the results for the first, fifth, and tenth periods of the planning horizon are presented here. These results are also related to the peak load. Since in the first scenario of the first case the strategic behaviors of the market participants are ignored, these participants lack the bidding strategy parameters. In simple terms, in the first scenario of the first case, the market participants, including GENCOs and DISCOs, submit their real marginal generation cost and marginal benefit functions to the ISO, respectively. As shown in Table 6.6, in the Nash equilibrium point obtained in the first, fifth, and tenth periods, the bidding strategy parameters (i.e., slope and intercept) of the market participants are not equal to the constant value 1. This means that all GENCOs and DISCOs, as market participants, have adopted a strategic behavior in order to compete with their rivals and earn more profit from selling and buying the electrical energy.

The optimal results of the second level (i.e., the CSC electricity market) for the proposed strategic tri-level computational-logical framework under the first and second scenarios of the first case (i.e., the IGDT risk-averse decision-making policy) are also presented in Tables 6.7 and 6.8. Again, only the results for the first, fifth, and tenth periods of the planning horizon are presented here. These results are also related to the peak load. By analyzing the results presented in Table 6.7, it can be seen that there is a slight difference between the power generated by the GENCOs and the power purchased by the DISCOs during the periods of the planning horizon. This is due to the fact that transmission line loss is considered and applied in the power flow analysis using the idea of an artificial load at each bus. For more information on the idea of artificial load, please refer to Sect. 5.3.2 of Chap. 5. From the results presented in Table 6.7, it can also be seen that the amount of power generated by the GENCOs, the amount of the power purchased by the DISCOs, and the amount of transmission line loss in the network have increased due to growth in demand during the planning horizon.

By assessing the results presented in Tables 6.6 and 6.7, it is clear that the larger bidding strategy parameters (i.e., slope and intercept) are related to the GENCO with less generated power and vice versa. In other words, the smaller GENCO—in terms of generated power—tends to adopt larger bidding strategy parameters in order to get more profit from the electricity market, and compensate for low profit due to a shortage of its power generation capacity. It is clear that the larger slope and smaller intercept of the bidding strategy parameters are related to the DISCO with less

Table 6.5 Optimal values for Δ^{price}, Δ^{demand}, and relevant critical cost deviation factors under the first and second scenarios of the first case

Parameter	Optimal values	
	First scenario: Ignoring the BBM	Second scenario: Considering the BBM
Δ^{price} (%)	16.562321	29.202145
Δ^{demand} (%)	34.265895	27.322562
Critical cost deviation factor	0.30	0.30

Table 6.6 Bidding strategy parameters of the market participants under the second scenario of the first case

Period No.	The market participants	Bidding strategy parameters of the market participants		Period No.	The market participants	Bidding strategy parameters of the market participants		Period No.	The market participants	Bidding strategy parameters of the market participants	
		$\xi_{1,p,k,g}$, $\xi_{1,p,k,d}$	$\xi_{2,p,k,g}$, $\xi_{2,p,k,d}$			$\xi_{1,p,k,g}$, $\xi_{1,p,k,d}$	$\xi_{2,p,k,g}$, $\xi_{2,p,k,d}$			$\xi_{1,p,k,g}$, $\xi_{1,p,k,d}$	$\xi_{2,p,k,g}$, $\xi_{2,p,k,d}$
1	GENCO 1	1.307711	1.108158	5	GENCO 1	1.323562	1.121591	10	GENCO 1	1.328792	1.126023
	GENCO 2	1.202073	1.018641		GENCO 2	1.216644	1.030988		GENCO 2	1.210800	1.026036
	GENCO 3	1.360179	1.176712		GENCO 3	1.400613	1.223293		GENCO 3	1.330148	1.129127
	GENCO 4	1.191590	1.009758		GENCO 4	1.206034	1.021997		GENCO 4	1.221452	1.035063
	GENCO 5	1.517822	1.311460		GENCO 5	1.536220	1.327356		GENCO 5	1.542291	1.332602
	GENCO 6	1.320531	1.119022		GENCO 6	1.348279	1.142536		GENCO 6	1.333607	1.130051
	GENCO 7	1.383839	1.208643		GENCO 7	1.376666	1.190975		GENCO 7	1.382107	1.195682
	GENCO 8	1.429856	1.285825		GENCO 8	1.447188	1.301411		GENCO 8	1.452907	1.306554
	GENCO 9	1.332132	1.128853		GENCO 9	1.336537	1.132586		GENCO 9	1.341819	1.137062
	GENCO 10	1.467890	1.300191		GENCO 10	1.485683	1.315951		GENCO 10	1.491554	1.321151
	GENCO 11	1.408873	1.239008		GENCO 11	1.425950	1.254026		GENCO 11	1.431586	1.258982
	GENCO 12	1.357630	1.150461		GENCO 12	1.374086	1.164406		GENCO 12	1.379517	1.169007
	DISCO 1	1.215050	0.922176		DISCO 1	1.338156	0.923120		DISCO 1	1.224906	0.929656
	DISCO 2	1.214077	0.922939		DISCO 2	1.215426	0.923965		DISCO 2	1.375011	0.928915
	DISCO 3	1.336671	0.922096		DISCO 3	1.365463	0.922465		DISCO 3	1.401766	0.928907
	DISCO 4	1.405823	0.919759		DISCO 4	1.422761	0.920038		DISCO 4	1.434269	0.925376
	DISCO 5	1.428157	0.897129		DISCO 5	1.216400	0.923200		DISCO 5	1.222700	0.937406
	DISCO 6	1.425562	0.907023		DISCO 6	1.424310	0.918951		DISCO 6	1.437125	0.914380
	DISCO 7	1.210438	0.936015		DISCO 7	1.214210	0.930896		DISCO 7	1.223924	0.930425
	DISCO 8	1.424411	0.916933		DISCO 8	1.443371	0.781831		DISCO 8	1.453463	0.787298
	DISCO 9	1.363947	0.921441		DISCO 9	1.392033	0.922457		DISCO 9	1.417226	0.927219
	DISCO 10	1.422729	0.917931		DISCO 10	1.425994	0.917952		DISCO 10	1.347513	0.929575

(continued)

Table 6.6 (continued)

Period No.	The market participants	Bidding strategy parameters of the market participants		Period No.	The market participants	Bidding strategy parameters of the market participants		Period No.	The market participants	Bidding strategy parameters of the market participants	
		$\xi_{1,p,k,g}$, $\xi_{1,p,k,d}$	$\xi_{2,p,k,g}$, $\xi_{2,p,k,d}$			$\xi_{1,p,k,g}$, $\xi_{1,p,k,d}$	$\xi_{2,p,k,g}$, $\xi_{2,p,k,d}$			$\xi_{1,p,k,g}$, $\xi_{1,p,k,d}$	$\xi_{2,p,k,g}$, $\xi_{2,p,k,d}$
	DISCO 11	1.428169	0.880006		DISCO 11	1.429744	0.898126		DISCO 11	1.439741	0.904406
	DISCO 12	1.429308	0.783939		DISCO 12	1.430772	0.863366		DISCO 12	1.442094	0.789664
	DISCO 13	1.390488	0.921434		DISCO 13	1.407385	0.920781		DISCO 13	1.432709	0.926471
	DISCO 14	1.421182	0.919016		DISCO 14	1.427146	0.908031		DISCO 14	1.435965	0.924370
	DISCO 15	1.429184	0.862407		DISCO 15	1.429756	0.880984		DISCO 15	1.440901	0.790297
	DISCO 16	1.441769	0.780963		DISCO 16	1.440812	0.783397		DISCO 16	1.450887	0.788874
	DISCO 17	1.212863	0.929863		DISCO 17	1.211783	0.937055		DISCO 17	1.220256	0.943607
	DISCO 18	1.430491	0.783311		DISCO 18	1.430896	0.784810		DISCO 18	1.440776	0.869402
	DISCO 19	1.439213	0.782527		DISCO 19	1.432080	0.784181		DISCO 19	1.439753	0.887144

Table 6.7 The market outcomes under the first and second scenarios of the first case

Period No.	The market participants	Generation and consumption ($\rho_{p,k,g}, \rho_{p,k,d}$; MW) First scenario	Second scenario	Period No.	The market participants	Generation and consumption ($\rho_{p,k,g}, \rho_{p,k,d}$; MW) First scenario	Second scenario	Period No.	The market participants	Generation and consumption ($\rho_{p,k,g}, \rho_{p,k,d}$; MW) First scenario	Second scenario
1	GENCO 1	952.665681	962.999597	5	GENCO 1	1201.696115	1213.622306	10	GENCO 1	1229.623704	1249.115863
	GENCO 2	1115.149508	1066.800271		GENCO 2	1320.755500	1329.63052		GENCO 2	2491.775099	2418.726689
	GENCO 3	443.774677	448.192167		GENCO 3	553.935215	558.503201		GENCO 3	1042.909341	1046.775019
	GENCO 4	1232.456625	1225.051608		GENCO 4	1499.487167	1501.770113		GENCO 4	1654.175296	1668.079125
	GENCO 5	110.888787	112.379384		GENCO 5	139.376825	140.526186		GENCO 5	189.860642	143.362312
	GENCO 6	636.235394	644.787841		GENCO 6	700.506231	706.282904		GENCO 6	774.885914	790.537240
	GENCO 7	396.296695	399.539297		GENCO 7	579.300217	584.077374		GENCO 7	654.998446	695.865335
	GENCO 8	371.147002	386.136060		GENCO 8	466.497039	470.343972		GENCO 8	492.444344	499.836544
	GENCO 9	573.862025	580.407690		GENCO 9	748.440349	724.364914		GENCO 9	732.298890	738.984185
	GENCO 10	196.100705	198.736744		GENCO 10	246.480230	248.512811		GENCO 10	295.198944	293.528344
	GENCO 11	382.134449	388.999137		GENCO 11	492.458863	480.257556		GENCO 11	569.155910	589.950206
	GENCO 12	485.299865	491.302265		GENCO 12	608.680949	613.700391		GENCO 12	684.059535	696.086210
	DISCO 1	519.177989	510.740742		DISCO 1	603.794215	610.435682		DISCO 1	758.729495	759.550403
	DISCO 2	602.246467	590.859260		DISCO 2	700.401290	708.105392		DISCO 2	664.359663	665.078468
	DISCO 3	456.876630	450.651853		DISCO 3	531.338910	537.183401		DISCO 3	579.777064	580.404355
	DISCO 4	290.739673	290.414815		DISCO 4	338.124760	341.843982		DISCO 4	368.949041	369.348225
	DISCO 5	249.205434	252.355556		DISCO 5	609.401841	593.009127		DISCO 5	1419.186244	1420.721734
	DISCO 6	269.972554	269.385185		DISCO 6	313.972992	317.426555		DISCO 6	342.595538	342.966209
	DISCO 7	904.774533	983.915034		DISCO 7	1159.284895	1172.036511		DISCO 7	1264.968141	1266.336775
	DISCO 8	269.972554	269.385185		DISCO 8	116.148968	117.426555		DISCO 8	126.737392	126.874516
	DISCO 9	373.808152	370.533334		DISCO 9	434.731835	439.513691		DISCO 9	574.254974	574.876290
	DISCO 10	269.972554	271.385185		DISCO 10	313.972992	317.426555		DISCO 10	742.163225	742.966209
	DISCO 11	249.205434	248.355556		DISCO 11	289.821223	293.009127		DISCO 11	316.242035	316.584193

(continued)

Table 6.7 (continued)

Period No.	The market participants	Generation and consumption ($\rho_{p,\,k,\,g}$, $\rho_{p,\,k,\,d}$; MW)		Period No.	The market participants	Generation and consumption ($\rho_{p,\,k,\,g}$, $\rho_{p,\,k,\,d}$; MW)		Period No.	The market participants	Generation and consumption ($\rho_{p,\,k,\,g}$, $\rho_{p,\,k,\,d}$; MW)	
		First scenario	Second scenario			First scenario	Second scenario			First scenario	Second scenario
	DISCO 12	166.136956	172.237037		DISCO 12	193.214149	195.339418		DISCO 12	210.828023	211.056129
	DISCO 13	309.060422	308.084950		DISCO 13	359.431446	363.385031		DISCO 13	392.198095	392.622434
	DISCO 14	249.972554	271.385185		DISCO 14	313.972992	317.426555		DISCO 14	342.595538	342.966209
	DISCO 15	207.671195	210.296296		DISCO 15	241.517686	244.174273		DISCO 15	263.535029	263.820161
	DISCO 16	189.369837	150.207418		DISCO 16	169.062380	170.921991		DISCO 16	184.474520	184.674113
	DISCO 17	957.287500	936.362965		DISCO 17	1477.717393	1423.201657		DISCO 17	1647.588893	1637.710283
	DISCO 18	161.262951	168.237037		DISCO 18	193.214149	195.339418		DISCO 18	288.494551	288.806688
	DISCO 19	183.286603	155.207407		DISCO 19	174.479335	176.398530		DISCO 19	290.277220	290.591286
	$\sum_{l \in \Psi^{L}} \kappa_{p,k,l}$	16.011417	25.332065		$\sum_{l \in \Psi^{L}} \kappa_{p,k,l}$	24.011245	37.988795		$\sum_{l \in \Psi^{L}} \kappa_{p,k,l}$	33.431230	52.892389

Table 6.8 The market outcomes under the first and second scenarios of the first case

Period No.	Parameter	LMP ($/MWh)		Period No.	Parameter	LMP ($/MWh)		Period No.	Parameter	LMP ($/MWh)	
		First scenario: ignoring the BBM	Second scenario: considering the BBM			First scenario: ignoring the BBM	Second scenario: considering the BBM			First scenario: ignoring the BBM	Second scenario: considering the BBM
1	LMP_1	24.162519	27.843799	5	LMP_1	24.650894	28.814437	10	LMP_1	25.617395	29.996485
	LMP_2	19.543256	22.520768		LMP_2	19.938266	23.305844		LMP_2	20.719996	24.261915
	LMP_3	29.548379	34.050222		LMP_3	30.145614	35.237216		LMP_3	31.327549	36.682745
	LMP_4	21.284034	24.526763		LMP_4	21.714229	25.381768		LMP_4	22.565591	26.422999
	LMP_5	19.470461	22.436882		LMP_5	19.863999	23.219034		LMP_5	20.642818	24.171544
	LMP_6	26.393898	30.415140		LMP_6	26.927374	31.475415		LMP_6	27.983130	32.766624
	LMP_7	18.369644	21.168350		LMP_7	18.740933	21.906281		LMP_7	19.475718	22.804938
	LMP_8	21.797483	25.118438		LMP_8	22.238056	25.994070		LMP_8	23.109956	27.060419
	LMP_9	16.124213	18.580817		LMP_9	16.450117	19.228546		LMP_9	17.095086	20.017355
	LMP_{10}	26.855449	30.947010		LMP_{10}	27.398254	32.025826		LMP_{10}	28.472472	33.339615
	LMP_{11}	15.372034	17.714040		LMP_{11}	15.682735	18.331553		LMP_{11}	16.297617	19.083565
	LMP_{12}	12.705170	14.640866		LMP_{12}	12.961969	15.151248		LMP_{12}	13.470175	15.772795
	LMP_{13}	22.630815	26.078732		LMP_{13}	23.088231	26.987839		LMP_{13}	23.993464	28.094956
	LMP_{14}	37.644830	43.380207		LMP_{14}	38.405711	44.892446		LMP_{14}	39.911505	46.734059
	LMP_{15}	26.508432	30.547124		LMP_{15}	27.044223	31.612000		LMP_{15}	28.104561	32.908812
	LMP_{16}	13.528302	15.589406		LMP_{16}	13.801738	16.132855		LMP_{16}	14.342870	16.794669
	LMP_{17}	16.994602	19.583815		LMP_{17}	17.338099	20.266508		LMP_{17}	18.017883	21.097897
	LMP_{18}	35.909247	41.380199		LMP_{18}	36.635048	42.822717		LMP_{18}	38.071418	44.579424
	LMP_{19}	17.406168	20.058085		LMP_{19}	17.757983	20.757312		LMP_{19}	18.454230	21.608834
	LMP_{20}	19.368382	22.319252		LMP_{20}	19.759858	23.097303		LMP_{20}	20.534593	24.044819
	LMP_{21}	24.383309	28.098228		LMP_{21}	24.876147	29.077735		LMP_{21}	25.851479	30.270585
	LMP_{22}	16.705329	19.250469		LMP_{22}	17.042979	19.921542		LMP_{22}	17.711192	20.738780

(continued)

Table 6.8 (continued)

Period No.	Parameter	LMP ($/MWh) First scenario: ignoring the BBM	LMP ($/MWh) Second scenario: considering the BBM	Period No.	Parameter	LMP ($/MWh) First scenario: ignoring the BBM	LMP ($/MWh) Second scenario: considering the BBM	Period No.	Parameter	LMP ($/MWh) First scenario: ignoring the BBM	LMP ($/MWh) Second scenario: considering the BBM
	LMP_{23}	14.659425	16.892861		LMP_{23}	14.955723	17.481748		LMP_{23}	15.542100	18.198898
	LMP_{24}	21.925026	25.265412		LMP_{24}	22.368177	26.146167		LMP_{24}	23.245178	27.218756
	LMP_{25}	32.075702	36.962594		LMP_{25}	32.724019	38.251114		LMP_{25}	34.007047	39.820282
	LMP_{26}	21.708229	25.015585		LMP_{26}	22.146998	25.887631		LMP_{26}	23.015327	26.949614
	LMP_{27}	16.703345	19.248182		LMP_{27}	17.040954	19.919176		LMP_{27}	17.709088	20.736316
	LMP_{28}	35.978480	41.459980		LMP_{28}	36.705681	42.905280		LMP_{28}	38.144820	44.665374
	LMP_{29}	19.523100	22.497542		LMP_{29}	19.917703	23.281808		LMP_{29}	20.698627	24.236893
	LMP_{30}	22.296713	25.693728		LMP_{30}	22.747376	26.589414		LMP_{30}	23.639245	27.680186
	LMP_{31}	15.538325	17.905666		LMP_{31}	15.852387	18.529860		LMP_{31}	16.473920	19.290006
	LMP_{32}	21.368375	24.623954		LMP_{32}	21.800275	25.482347		LMP_{32}	22.655010	26.527704
	LMP_{33}	16.753013	19.305418		LMP_{33}	17.091627	19.978407		LMP_{33}	17.761747	20.797977
	LMP_{34}	13.836750	15.944847		LMP_{34}	14.116419	16.500686		LMP_{34}	14.669889	17.177591
	LMP_{35}	22.483430	25.908892		LMP_{35}	22.937867	26.812079		LMP_{35}	23.837205	27.911986
	LMP_{36}	15.686307	18.076193		LMP_{36}	16.003360	18.706331		LMP_{36}	16.630812	19.473717
	LMP_{37}	19.256288	22.190079		LMP_{37}	19.645498	22.963628		LMP_{37}	20.415749	23.905660
	LMP_{38}	27.616340	31.823827		LMP_{38}	28.174525	32.933209		LMP_{38}	29.279178	34.284221
	LMP_{39}	19.439973	22.401750		LMP_{39}	19.832896	23.182677		LMP_{39}	20.610495	24.133695
	LMP_{40}	21.419924	24.683356		LMP_{40}	21.852865	25.543820		LMP_{40}	22.709662	26.591699
	LMP_{41}	16.517489	19.034011		LMP_{41}	16.851343	19.697539		LMP_{41}	17.512042	20.505587
	LMP_{42}	20.219855	23.300450		LMP_{42}	20.628540	24.112706		LMP_{42}	21.437334	25.101877
	LMP_{43}	24.088975	27.759050		LMP_{43}	24.575864	28.726734		LMP_{43}	25.539423	29.905185
	LMP_{44}	21.894066	25.229735		LMP_{44}	22.336591	26.109247		LMP_{44}	23.212353	27.180321
	LMP_{45}	19.323539	22.267577		LMP_{45}	19.714109	23.043827		LMP_{45}	20.487050	23.989149
	LMP_{46}	13.521436	15.581494		LMP_{46}	13.794733	16.124667		LMP_{46}	14.335590	16.786146

purchased power and vice versa. In simple terms, the smaller DISCO—in terms of purchased electrical power—tends to adopt a larger slope and smaller intercept associated with bidding strategy parameters in order to get more profit from the electricity market, and compensate for low profit due to a shortage of its power purchasing capacity.

As set out in Table 6.8, the LMPs' fluctuations in the different buses of the network over the planning horizon clearly increased for the case where strategic behaviors were taken into account by market participants—i.e., the second scenario compared to the first ones. Table 6.9 shows the optimal robust expansion plans obtained by the GENGOs over the planning horizon under the first and second scenarios of the first case (i.e., the IGDT risk-averse decision-making policy). To illustrate the results presented in Table 6.9, consider, for example, period 6 of the planning horizon. In the first scenario of this period, GENCO 5 expands the capacity of its existing generation units located on bus 27 of the network. GENCO 3 installs a new generation unit on bus 32 of the network, while the existing generation units of this GENCO are located on bus 17 of the network. This is due to the fact that the type of expansion planning in the proposed strategic tri-level computational logical framework is allocation and capacity expansion.

In simple terms, in the proposed framework, each GENCO not only increases the capacity of its existing generation units, but also can decide to install new generation units in the candidate locations via optimal allocation. In the second scenario of this period, however, GENCO 3 and GENCO 12 expand the capacity of their existing generation units located on buses 17 and 46 of the network, respectively. For the remaining periods of the planning horizon, the same analysis can be done. In addition, the number of augmented generation units—plus the number of newly installed generation units over the planning horizon in the second scenario, where the strategic behaviors of the market participants are taken into account—is less than the number of them in the first scenario, where the strategic behaviors of the market participants are ignored. Table 6.10 presents the values of the EP, the GIC, and the CP of the GENCOs under the first and second scenarios of the first case (i.e., the IGDT risk-averse decision-making policy). Looking at Table 6.10, it can be seen that the EP of all GENCOs increased by employing optimal bidding strategies—i.e., the second scenario compared to the first. In other words, when the BBM is activated in the proposed framework, the GENCOs' profits will increase in comparison with the situation where the BBM is ignored. Investments by all GENCOs to expand generation capacity in the second scenario, taking into account the BBM, have less value compared to the first scenario without taking into account the BBM. Therefore, by modeling and applying strategic interactions among the market participants in the proposed framework, the generation units with the lowest capacity need to be installed in the network; lower costs, then, for the installation of these generation units are needed. In addition, the capacity payment made by the ISO to the GENCOs is decreased by considering the BBM in the second scenario versus ignoring the BBM, as in the first scenario.

Hence, utilizing the first level in the proposed framework can significantly reduce unreliability costs, owing to the adoption of strategic behavior by the market

Table 6.9 Robust generation expansion plans under the proposed framework under the first and second scenarios of the first case

Period No.	Optimal robust generation expansion plans	
	First scenario: Ignoring the BBM	Second scenario: Considering the BBM
1	–	–
2	GENCO 8 (32: I), GENCO 4 (19: I)	GENCO 6 (28: I), GENCO 9 (34: I)
3	–	GENCO 5 (27: I), GENCO 2 (16: I)
4	GENCO 6 (28: I), GENCO 1 (14: I)	GENCO 4 (19: I), GENCO 8 (32: I), GENCO 7 (16: C)
5	GENCO 9 (34: I), GENCO 3 (17: I)	–
6	GENCO 5 (27: I), GENCO 3 (32: C)	GENCO 3 (17: I), GENCO 12 (46: I)
7	GENCO 7 (31: I), GENCO 2 (16: C)	–
8	–	GENCO 1 (14: I), GENCO 11 (39: I)
9	GENCO 10 (37: I), GENCO 4 (46: C)	–
10	GENCO 2 (16: I), GENCO 12 (31: C)	GENCO 10 (37: I), GENCO 2 (16: I)
Number of augmented generation units over the planning horizon	10	12
Number of newly installed generation units over the planning horizon	4	1

participants. As a result, adoption of strategic behaviors by both GENCOs and DISCOs can effectively lead to (1) improved competition and increased efficiency of the CSC electricity market, (2) a reduction of the ECO, (3) a decrease in the required investment cost for generation expansion, (4) provision of more flexible generation expansion plans against severe uncertainties, and so on.

6.3.5.2 Second Case: Simulation Results and Discussion

The strategic tri-level computational-logical framework under the IGDT risk-taker decision-making policy is solved optimization of Eqs. (6.93) through (6.104) under different values of the target profit deviation factor, $\sigma_{t,g}^{PD-GEP} \in (0, 0.8)$, which leads to the different target profits, $\varpi_{t,g}^{PD-GEP}$. It should be highlighted that all GENCOs are involved in the process of solving of the framework under the IGDT risk-taker decision-making policy. In other words, for each value of the target profit deviation factor, $\sigma_{c,g}^{PD-GEP} \in (0, 0.8)$, GENCO g tries to find the minimum values of Δ_g^{price} and

Table 6.10 Changes of the objective functions of the third level of the proposed framework under the first and second scenarios of the first case

	The EP of all GENCOs		The GIC of all GENCOs		The CP of all GENCOs	
Period No.	First scenario: Ignoring the BBM	Second scenario: Considering the BBM	First scenario: Ignoring the BBM	Second scenario: Considering the BBM	First scenario: Ignoring the BBM	Second scenario: Considering the BBM
1	147.034746	204.553580	–	–	–	–
2	286.740614	309.491962	242.290586	214.221033	12.9177547	7.488880
3	301.139275	379.986034	–	237.804245	–	6.035881
4	386.955771	428.307065	218.258331	425.351187	11.2154538	14.883169
5	406.897299	421.140516	264.258870	–	10.2588789	–
6	479.402850	492.576620	237.472006	212.550077	15.7693428	7.760390
7	489.121270	480.012098	238.572833	–	14.0801494	–
8	468.139220	502.206477	–	187.939613	–	6.094699
9	495.005932	497.707536	229.071587	–	14.8793842	–
10	499.907061	561.340776	212.354982	203.186365	12.5335175	6.399873
Total	3960.344038	4277.322664	1642.279195	1481.052520	91.654481	48.662891

Δ_g^{demand} by solving Eqs. (6.93) through (6.99) in a multi-objective manner using a Pareto optimality concept. In a manner similar to the IGDT risk-averse decision-making policy, GENCO g uses the FSM, based on the conservative methodology—the min-max formulation—to select the best solution from among the Pareto-optimal solution set. This process addresses a well-suited compromise among different objectives, namely Δ_g^{price} and Δ_g^{demand}. In order to achieve a profit equal to the obtained profit from the deterministic framework by GENCO 1, $\varpi_{\text{t},g}^{\text{PD-GEP}} = \varpi_{\text{b},g}^{\text{PD-GEP}} = \$282.786547\text{M}$ and $\varpi_{\text{t},g}^{\text{PD-GEP}} = \varpi_{\text{b},g}^{\text{PD-GEP}} = \$379.429696$ M for the first and second scenarios, respectively, this GENCO must set its target profit deviation factor to zero, $\sigma_{\text{t},g}^{\text{PD-GEP}} = 0$. In this condition, the opportunity and risk level of the determined generation expansion plans by this GENCO will be zero. In other words, this GENCO is not ready to deal with demand and market price uncertainties; therefore, its profit [see Eq. (6.22)] is extremely vulnerable to these uncertainties. Suppose that GENCO 1 sets its target profit deviation factor to 0.4 (i.e., $\sigma_{\text{t},g}^{\text{PD-GEP}} = 0.4$).

In this situation, Eqs. (6.93) through (6.99) are solved by this GENCO in a multi-objective manner for this value of the target cost deviation factor with the aim of minimizing Δ_g^{price} and Δ_g^{demand} based on the first and second scenarios. Figure 6.9 illustrates the Pareto-optimal solution set for this value of target profit deviation factor related to GENCO 1 under the first and second scenarios. As further elucidation, consider the performance of GENCO 1 under the first scenario (ignoring the BBM). For example, in the solution marked with a square in Fig. 6.9, a target profit of $\varpi_{\text{t},g}^{\text{PD-GEP}} = (1 + 0.4) \times \varpi_{\text{b},g}^{\text{PD-GEP}} = \$395.901165\text{M}$ is guaranteed for this GENCO, provided that Δ_g^{price} and Δ_g^{demand} are not less than 28.553757% and 6.195080%, respectively.

That is, to reach a profit 40% higher than the base profit by GENCO 1, actual market price and demand must be at least 28.553757% and 6.195080% higher than their corresponding predicted values, respectively. Put another way, if there is a difference of more than 28.553757% between actual market price, $a\lambda_{p,\ k,\ b}$, and predicted market price, $p\lambda_{p,\ k,\ b}$, and if there is a difference of more than 6.195080% between actual demand, $a\rho_{p,\ k,\ b}$, and predicted demand, $p\rho_{p,\ k,\ b}$, the obtained profit by this GENCO will be at least $\varpi_{\text{t},g}^{\text{PD-GEP}} = \$395.901165\text{M}$. And, if the actual market price, $a\lambda_{p,\ k,\ b}$, falls within the market price robustness region, $RR_{p,k,b}^{\text{price}}$, and the actual demand, $a\rho_{p,\ k,\ b}$, falls within the demand robustness region, $RR_{p,k,b}^{\text{demand}}$, the obtained profit by this GENCO will be at least $\varpi_{\text{t},g}^{\text{PD-GEP}} = \$395.901165\text{M}$. Note that if the actual market price and/or demand are not at least 28.553757% and 6.195080% higher than their corresponding predicted values, respectively, it will not be possible to achieve the target profit by this GENCO. Put another way, if the actual market price and/or actual demand are not within the market price and demand robustness regions, respectively, it is not possible to achieve target profit by this GENCO. If this GENCO wants to have a larger target profit, $\varpi_{\text{t},g}^{\text{PD-GEP}}$, it must choose a larger target profit deviation factor, $\sigma_{\text{t},g}^{\text{PD-GEP}}$, which leads to a larger

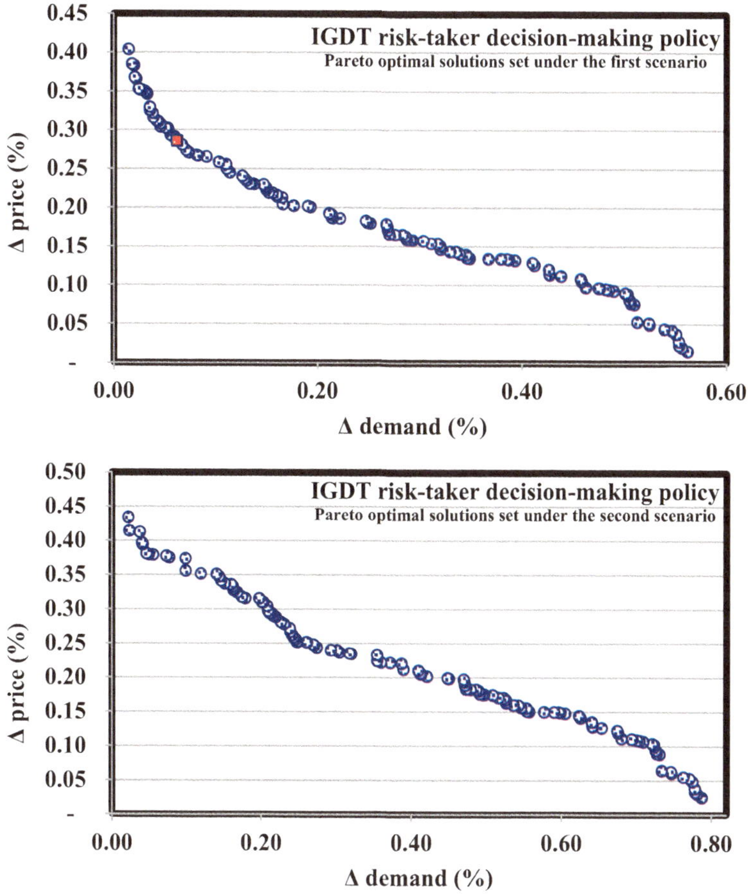

Fig. 6.9 The Pareto-optimal solutions of opportunity related to GENCO 1 against market price and demand uncertainties under the first and second scenarios

opportunity for this GENCO. This means that a larger target profit obtained by this GENCO in the larger market price robustness region, $RR_{p,k,b}^{\text{price}}$, and in the larger demand robustness region, $RR_{p,k,b}^{\text{demand}}$, can be guaranteed and vice versa. That is to say that a better opportunity of the obtained generation expansion plans by this GENCO is determined by a higher value of the target profit specified by this GENCO, $\varpi_{t,g}^{\text{PD-GEP}}$, and/or by a larger value of the target profit deviation factor of this GENCO, $\sigma_{t,g}^{\text{PD-GEP}}$. The rest of the solutions provided in Fig. 6.9 have the same interpretations. For the same reason, the Pareto-optimal solution set related to the other target profit deviation factors of GENCO 1 has the same interpretations. It should be noted that these solutions are produced because of the adoption of different target profit deviation factors by this GENCO. This analysis is also true for other GENCOs under the first scenario. It is also true for all GENCOs under the second scenario.

It should be highlighted that there are a number of GENCOs that may choose different amounts for their target profit deviation factors—$\sigma_{c,g}^{PD-GEP}$. This condition leads to the formation of different opportunities for the GENCOs. In a manner similar to the first case, the existence of the iterative process between the local and global layers leads to the convergence of the interval of the market price uncertainty, Δ^{price}, and the interval of the demand uncertainty, Δ^{demand}, at a point of equilibrium. After solving the proposed framework under the first and the second scenarios of the second case—the IGDT risk-taker decision-making policy—by the newly proposed multi-objective SOSA, the optimal values for Δ^{price}, Δ^{demand}, and relevant target profit deviation factors are calculated by the FSM in accordance with Table 6.11. Table 6.12 shows the optimal results associated with the bidding strategy parameters of both GENCOs and DISCOs for the first level of the proposed strategic tri-level computational-logical framework under the second scenario in the second case (i.e., the IGDT risk-taker decision-making policy). For the same reason, these results are only presented for the first, fifth, and tenth periods of the planning horizon. These results are also associated with the peak load. From Table 6.12, it is evident that in the Nash equilibrium point determined in the first, fifth, and tenth periods the bidding strategy parameters (i.e., slope and intercept) of the GENCOs and DISCOs are not equal to the constant value 1. This means that both GENCOs and DISCOs, as market participants, have adopted a strategic behavior in order to compete with their competitors and earn more profit by selling and buying electrical energy, respectively.

Tables 6.13 and 6.14 present the optimal results of the second level (i.e., the CSC electricity market) for the proposed strategic tri-level computational-logical framework under the first and second scenarios of the second case (i.e., the IGDT risk-taker decision-making policy). As before, these results are only presented for the first, fifth, and tenth periods of the planning horizon. These results are also related to the peak load. Because of transmission line loss considerations in the power flow analysis, using the idea of an artificial load at each bus, there is a very small difference between the power generated by the GENCOs and the power purchased by the DISCOs in the periods of the planning horizon (see Table 6.13). Furthermore,

Table 6.11 Optimal values for Δ^{price}, Δ^{demand}, and relevant critical cost deviation factors under the first and second scenarios of the second case

Parameter	Optimal values	
	First scenario: Ignoring the BBM	Second scenario: Considering the BBM
Δ^{price} (%)	21.032623	27.658547
Δ^{demand} (%)	16.154226	20.265233
Target cost deviation factor	0.25	0.25

Table 6.12 Bidding strategy parameters of the market participants under the second scenario of the second case

Period No.	The market participants	Bidding strategy parameters of the market participants		Period No.	The market participants	Bidding strategy parameters of the market participants		Period No.	The market participants	Bidding strategy parameters of the market participants	
		$\xi_{1,p,k,g}$, $\xi_{1,p,k,d}$	$\xi_{2,p,k,g}$, $\xi_{2,p,k,d}$			$\xi_{1,p,k,g}$, $\xi_{1,p,k,d}$	$\xi_{2,p,k,g}$, $\xi_{2,p,k,d}$			$\xi_{1,p,k,g}$, $\xi_{1,p,k,d}$	$\xi_{2,p,k,g}$, $\xi_{2,p,k,d}$
1	GENCO 1	1.312689	1.112377	5	GENCO 1	1.328735	1.125974	10	GENCO 1	1.335317	1.131552
	GENCO 2	1.206650	1.022519		GENCO 2	1.221399	1.035018		GENCO 2	1.216745	1.031074
	GENCO 3	1.365358	1.181192		GENCO 3	1.406087	1.228074		GENCO 3	1.413052	1.234157
	GENCO 4	1.196127	1.013602		GENCO 4	1.210748	1.025992		GENCO 4	1.227450	1.040145
	GENCO 5	1.523601	1.316453		GENCO 5	1.542224	1.332544		GENCO 5	1.549864	1.339145
	GENCO 6	1.325558	1.123283		GENCO 6	1.353549	1.147002		GENCO 6	1.360254	1.152684
	GENCO 7	1.389108	1.213244		GENCO 7	1.382047	1.195630		GENCO 7	1.388893	1.201553
	GENCO 8	1.435300	1.290721		GENCO 8	1.452844	1.306498		GENCO 8	1.460041	1.312970
	GENCO 9	1.337204	1.133151		GENCO 9	1.341761	1.137013		GENCO 9	1.348408	1.142645
	GENCO 10	1.473479	1.305141		GENCO 10	1.491490	1.321094		GENCO 10	1.498878	1.327638
	GENCO 11	1.414237	1.243725		GENCO 11	1.431524	1.258928		GENCO 11	1.438615	1.265164
	GENCO 12	1.362799	1.154841		GENCO 12	1.379457	1.168957		GENCO 12	1.386291	1.174747
	DISCO 1	1.222484	0.927817		DISCO 1	1.350468	0.931613		DISCO 1	1.248563	0.947611
	DISCO 2	1.221504	0.928585		DISCO 2	1.226608	0.932465		DISCO 2	1.401567	0.946856
	DISCO 3	1.344849	0.927737		DISCO 3	1.378026	0.930952		DISCO 3	1.428839	0.946848
	DISCO 4	1.414423	0.925386		DISCO 4	1.435851	0.928502		DISCO 4	1.461970	0.943249
	DISCO 5	1.436894	0.902618		DISCO 5	1.227591	0.931694		DISCO 5	1.246315	0.955510
	DISCO 6	1.434283	0.912572		DISCO 6	1.437414	0.927405		DISCO 6	1.464881	0.932040
	DISCO 7	1.217843	0.941741		DISCO 7	1.225381	0.939461		DISCO 7	1.247563	0.948395
	DISCO 8	1.433125	0.922543		DISCO 8	1.456650	0.789024		DISCO 8	1.481535	0.802503
	DISCO 9	1.372292	0.927078		DISCO 9	1.404840	0.930944		DISCO 9	1.444597	0.945127
	DISCO 10	1.431433	0.923547		DISCO 10	1.439113	0.926397		DISCO 10	1.373538	0.947528

(continued)

Table 6.12 (continued)

Period No.	The market participants	Bidding strategy parameters of the market participants		Period No.	The market participants	Bidding strategy parameters of the market participants		Period No.	The market participants	Bidding strategy parameters of the market participants	
		$\xi_{1,p,k,g}$, $\xi_{1,p,k,d}$	$\xi_{2,p,k,g}$, $\xi_{2,p,k,d}$			$\xi_{1,p,k,g}$, $\xi_{1,p,k,d}$	$\xi_{2,p,k,g}$, $\xi_{2,p,k,d}$			$\xi_{1,p,k,g}$, $\xi_{1,p,k,d}$	$\xi_{2,p,k,g}$, $\xi_{2,p,k,d}$
	DISCO 11	1.436906	0.885389		DISCO 11	1.442898	0.906389		DISCO 11	1.467548	0.921873
	DISCO 12	1.438052	0.788735		DISCO 12	1.443935	0.871309		DISCO 12	1.469946	0.804915
	DISCO 13	1.398994	0.927071		DISCO 13	1.420333	0.929252		DISCO 13	1.460380	0.944364
	DISCO 14	1.429876	0.924639		DISCO 14	1.440276	0.916385		DISCO 14	1.463698	0.942223
	DISCO 15	1.437927	0.867683		DISCO 15	1.442910	0.889089		DISCO 15	1.468730	0.805561
	DISCO 16	1.450589	0.785741		DISCO 16	1.454068	0.790604		DISCO 16	1.478908	0.804110
	DISCO 17	1.220282	0.935552		DISCO 17	1.222932	0.945676		DISCO 17	1.243824	0.961831
	DISCO 18	1.439242	0.788103		DISCO 18	1.444061	0.792030		DISCO 18	1.468602	0.886194
	DISCO 19	1.448018	0.787315		DISCO 19	1.445256	0.791396		DISCO 19	1.467560	0.904278

Table 6.13 The CSC electricity market outcomes under the first and second scenarios of the second case

Period No.	The market participants	Generation and consumption ($\rho_{p,k,g}, \rho_{p,k,d}$; MW) First scenario: Ignoring the BBM	Generation and consumption ($\rho_{p,k,g}, \rho_{p,k,d}$; MW) Second scenario: Considering the BBM	Period No.	The market participants	Generation and consumption ($\rho_{p,k,g}, \rho_{p,k,d}$; MW) First scenario: Ignoring the BBM	Generation and consumption ($\rho_{p,k,g}, \rho_{p,k,d}$; MW) Second scenario: Considering the BBM	Period No.	The market participants	Generation and consumption ($\rho_{p,k,g}, \rho_{p,k,d}$; MW) First scenario: Ignoring the BBM	Generation and consumption ($\rho_{p,k,g}, \rho_{p,k,d}$; MW) Second scenario: Considering the BBM
1	GENCO 1	952.665681	962.999597	5	GENCO 1	1218.139387	1219.691641	10	GENCO 1	1361.246923	1281.625364
	GENCO 2	1115.149508	1066.800271		GENCO 2	1337.341765	1339.047811		GENCO 2	2789.643749	2498.585964
	GENCO 3	443.774677	448.192167		GENCO 3	569.600507	570.316036		GENCO 3	633.653103	1078.878013
	GENCO 4	1232.456625	1225.051608		GENCO 4	1516.288095	1518.225012		GENCO 4	1691.141825	1701.430332
	GENCO 5	110.888787	112.379384		GENCO 5	154.544221	154.724257		GENCO 5	184.402571	173.650331
	GENCO 6	636.235394	644.787841		GENCO 6	716.347559	717.252417		GENCO 6	806.025446	722.125447
	GENCO 7	396.296695	399.539297		GENCO 7	594.995973	600.447462		GENCO 7	671.752621	725.263343
	GENCO 8	371.147002	386.136060		GENCO 8	482.057315	482.659899		GENCO 8	546.788512	530.840727
	GENCO 9	573.862025	580.407690		GENCO 9	764.339247	765.306022		GENCO 9	874.455271	770.468821
	GENCO 10	196.100705	198.736744		GENCO 10	266.109933	266.433907		GENCO 10	303.052433	324.118049
	GENCO 11	382.134449	388.999137		GENCO 11	508.050321	508.686440		GENCO 11	575.549193	620.321673
	GENCO 12	485.299865	491.302265		GENCO 12	624.411992	625.198238		GENCO 12	704.300790	729.484662
	DISCO 1	519.177989	511.743697		DISCO 1	613.294768	619.941741		DISCO 1	771.475251	776.2969665
	DISCO 2	602.246467	592.022689		DISCO 2	709.981931	717.692420		DISCO 2	685.012663	681.732174
	DISCO 3	456.876630	451.534454		DISCO 3	539.779396	546.628732		DISCO 3	598.346928	596.974835
	DISCO 4	290.739673	290.976470		DISCO 4	347.405070	351.127375		DISCO 4	383.311681	385.711258
	DISCO 5	249.205434	250.836975		DISCO 5	618.907042	602.500739		DISCO 5	1436.581164	1438.118164
	DISCO 6	269.972554	270.906722		DISCO 6	328.233279	335.411259		DISCO 6	358.932275	359.303311
	DISCO 7	904.774533	973.347899		DISCO 7	1169.245956	1182.008144		DISCO 7	1282.211480	1283.581459
	DISCO 8	269.972554	270.906722		DISCO 8	125.245257	126.523903		DISCO 8	139.861962	142.999221
	DISCO 9	373.808152	371.255462		DISCO 9	444.092233	448.878053		DISCO 9	590.819410	591.441337

(continued)

Table 6.13 (continued)

Period No.	The market participants	Generation and consumption ($\rho_{p,k,g}$, $\rho_{p,k,d}$; MW)		Period No.	The market participants	Generation and consumption ($\rho_{p,k,g}$, $\rho_{p,k,d}$; MW)		Period No.	The market participants	Generation and consumption ($\rho_{p,k,g}$, $\rho_{p,k,d}$; MW)	
		First scenario: Ignoring the BBM	Second scenario: Considering the BBM			First scenario: Ignoring the BBM	Second scenario: Considering the BBM			First scenario: Ignoring the BBM	Second scenario: Considering the BBM
	DISCO 10	269.972554	270.906722		DISCO 10	325.233279	329.689705		DISCO 10	758.892698	759.696472
	DISCO 11	249.205434	250.836975		DISCO 11	299.061488	302.252035		DISCO 11	335.552869	332.895364
	DISCO 12	166.136956	178.557983		DISCO 12	204.374326	206.501357		DISCO 12	223.035246	227.263576
	DISCO 13	309.060422	308.681998		DISCO 13	368.729418	372.686282		DISCO 13	408.583587	409.008343
	DISCO 14	249.972554	270.906722		DISCO 14	323.233279	326.689705		DISCO 14	361.932275	359.303311
	DISCO 15	207.671195	210.697479		DISCO 15	249.717907	253.376696		DISCO 15	279.794058	280.079472
	DISCO 16	189.369837	149.626400		DISCO 16	178.202535	180.063687		DISCO 16	204.655840	200.855629
	DISCO 17	957.287500	933.208404		DISCO 17	1487.942439	1433.381508		DISCO 17	1665.208310	1655.319991
	DISCO 18	161.262951	172.557988		DISCO 18	204.374326	204.501357		DISCO 18	304.778113	305.090557
	DISCO 19	183.286603	150.488235		DISCO 19	183.623981	185.544766		DISCO 19	309.562538	307.876908
	$\sum_{l \in \Psi^L} \kappa_{p,k,l}$	16.011417	25.332065		$\sum_{l \in \Psi^L} \kappa_{p,k,l}$	31.548403	42.589677		$\sum_{l \in \Psi^L} \kappa_{p,k,l}$	43.464089	63.244377

Table 6.14 The CSC electricity market outcomes under the first and second scenarios of the second case

Period No.	Parameter	LMP ($/MWh) First scenario: Ignoring the BBM	LMP ($/MWh) Second scenario: Considering the BBM	Period No.	Parameter	LMP ($/MWh) First scenario: Ignoring the BBM	LMP ($/MWh) Second scenario: Considering the BBM	Period No.	Parameter	LMP ($/MWh) First scenario: Ignoring the BBM	LMP ($/MWh) Second scenario: Considering the BBM
1	LMP_1	25.022536	28.834843	5	LMP_1	26.753925	31.103770	10	LMP_1	28.089942	32.720753
	LMP_2	20.238859	23.322350		LMP_2	21.830425	25.348690		LMP_2	22.911029	26.656546
	LMP_3	30.600095	35.262171		LMP_3	32.494512	37.813944		LMP_3	34.128329	39.791354
	LMP_4	22.041597	25.399744		LMP_4	23.685856	27.517503		LMP_4	24.862713	28.941855
	LMP_5	20.163473	23.235478		LMP_5	21.752836	25.257995		LMP_5	22.829415	26.560980
	LMP_6	27.333336	31.497706		LMP_6	29.132269	33.883816		LMP_6	30.591665	35.650125
	LMP_7	19.023474	21.921795		LMP_7	20.579516	23.886502		LMP_7	21.595228	25.115818
	LMP_8	22.573321	26.012479		LMP_8	24.233122	28.157203		LMP_8	25.438369	29.615915
	LMP_9	16.698122	19.242164		LMP_9	18.186196	21.088949		LMP_9	19.077751	22.167999
	LMP_{10}	27.811315	32.048507		LMP_{10}	29.624219	34.458857		LMP_{10}	31.109136	36.256053
	LMP_{11}	15.919171	18.344536		LMP_{11}	17.384476	20.151819		LMP_{11}	18.234441	21.180532
	LMP_{12}	13.157386	15.161979		LMP_{12}	14.541966	16.829208		LMP_{12}	15.244472	17.679452
	LMP_{13}	23.436314	27.006953		LMP_{13}	25.121339	29.195440		LMP_{13}	26.372664	30.709910
	LMP_{14}	38.984723	44.924239		LMP_{14}	41.124217	47.901208		LMP_{14}	43.205710	49.420441
	LMP_{15}	27.451947	31.634388		LMP_{15}	29.254347	34.026513		LMP_{15}	30.720076	35.800487
	LMP_{16}	14.009815	16.144280		LMP_{16}	15.419311	17.854737		LMP_{16}	16.167331	18.760066
	LMP_{17}	17.599491	20.280861		LMP_{17}	19.113911	22.173356		LMP_{17}	20.053593	23.310653
	LMP_{18}	37.187365	42.853045		LMP_{18}	39.274323	45.738866		LMP_{18}	41.259851	48.141953
	LMP_{19}	18.025706	20.772012		LMP_{19}	19.552584	22.686120		LMP_{19}	20.515022	23.850960
	LMP_{20}	20.057762	23.113661		LMP_{20}	21.644035	25.130817		LMP_{20}	22.714969	26.426971
	LMP_{21}	25.251185	29.098328		LMP_{21}	26.989258	31.378850		LMP_{21}	28.337482	33.010608
	LMP_{22}	17.299922	19.935651		LMP_{22}	18.805586	21.812954		LMP_{22}	19.729272	22.930893
	LMP_{23}	15.181198	17.494129		LMP_{23}	16.624933	19.263988		LMP_{23}	17.435496	20.245013
	LMP_{24}	22.705403	26.164684		LMP_{24}	24.369065	28.316106		LMP_{24}	25.581364	29.783353

(continued)

Table 6.14 (continued)

Period No.	Parameter	LMP ($/MWh) First scenario: Ignoring the BBM	LMP ($/MWh) Second scenario: Considering the BBM	Period No.	Parameter	LMP ($/MWh) First scenario: Ignoring the BBM	LMP ($/MWh) Second scenario: Considering the BBM	Period No.	Parameter	LMP ($/MWh) First scenario: Ignoring the BBM	LMP ($/MWh) Second scenario: Considering the BBM
	LMP_{25}	33.217372	38.278204		LMP_{25}	35.188291	40.962702		LMP_{25}	36.961851	43.109244
	LMP_{26}	22.480890	25.905965		LMP_{26}	24.137989	28.046002		LMP_{26}	25.338301	29.498740
	LMP_{27}	17.297867	19.933283		LMP_{27}	18.803470	21.810481		LMP_{27}	19.727047	22.928288
	LMP_{28}	37.259062	42.935666		LMP_{28}	39.348117	45.825123		LMP_{28}	41.337472	48.232843
	LMP_{29}	20.217986	23.298297		LMP_{29}	21.808942	25.323578		LMP_{29}	22.888432	26.630086
	LMP_{30}	23.090320	26.608245		LMP_{30}	24.765232	28.779186		LMP_{30}	25.998083	30.271307
	LMP_{31}	16.091381	18.542983		LMP_{31}	17.561720	20.358999		LMP_{31}	18.420879	21.398841
	LMP_{32}	22.128940	25.500394		LMP_{32}	23.775752	27.622583		LMP_{32}	24.957272	29.052578
	LMP_{33}	17.349303	19.992556		LMP_{33}	18.856410	21.872363		LMP_{33}	19.782734	22.993493
	LMP_{34}	14.329241	16.512372		LMP_{34}	15.748074	18.239028		LMP_{34}	16.513149	19.164998
	LMP_{35}	23.283683	26.831068		LMP_{35}	24.964247	29.011815		LMP_{35}	26.207422	30.516431
	LMP_{36}	16.244629	18.719580		LMP_{36}	17.719448	20.543367		LMP_{36}	18.586790	21.593112
	LMP_{37}	19.941677	22.979891		LMP_{37}	21.524557	24.991160		LMP_{37}	22.589294	26.279812
	LMP_{38}	28.599289	32.956533		LMP_{38}	30.435225	35.406842		LMP_{38}	31.962213	37.254958
	LMP_{39}	20.131900	23.199095		LMP_{39}	21.720340	25.220011		LMP_{39}	22.795234	26.520956
	LMP_{40}	22.182323	25.561911		LMP_{40}	23.830696	27.686806		LMP_{40}	25.015066	29.120251
	LMP_{41}	17.105397	19.711489		LMP_{41}	18.605374	21.578927		LMP_{41}	19.518675	22.684295
	LMP_{42}	20.939540	24.129783		LMP_{42}	22.551587	26.191656		LMP_{42}	23.669602	27.544791
	LMP_{43}	24.946374	28.747078		LMP_{43}	26.675538	31.012143		LMP_{43}	28.007488	32.624204
	LMP_{44}	22.673341	26.127738		LMP_{44}	24.336066	28.277534		LMP_{44}	25.546653	29.742709
	LMP_{45}	20.011322	23.060147		LMP_{45}	21.596238	25.074948		LMP_{45}	22.664693	26.368101
	LMP_{46}	14.002705	16.136086		LMP_{46}	15.411993	17.846183		LMP_{46}	16.159633	18.751052

owing to the growth of demand during the planning horizon, there is positive growth in the power generated by the GENCOs, the power purchased by the DISCOs, and the transmission line loss in the network (see Table 6.13).

By evaluating the results shown in Tables 6.12 and 6.13, it can be observed that the larger bidding strategy parameters (i.e., slope and intercept) are related to the GENCOs with less generated power and vice versa. In addition, the larger slope and smaller intercept of the bidding strategy parameters are related to the DISCOs with the lowest amount of purchased power and vice versa.

Looking at Table 6.14, it can be seen that the LMPs' fluctuations in the different buses of the network over the planning horizon clearly increased for the case where strategic behaviors were taken into account by market participants—i.e., the second scenario compared to the first. These LMP fluctuations in the first and second scenarios of the second case (i.e., the IGDT risk-taker decision-making policy) are similar to LMP fluctuations in the corresponding scenarios of the first case (i.e., the IGDT risk-averse decision-making policy).

Table 6.15 gives the optimal opportunistic expansion plans obtained by the GENGOs over the planning horizon under the first and second scenarios of the second case (i.e., the IGDT risk-taker decision-making policy). To describe the results presented in Table 6.15, consider for instance the eighth period of the planning horizon.

In the first scenario of this period, GENCO 5 and GENCO 12 expand the capacity of their existing generation units that are located on buses 27 and 46 of the network, respectively. In the second scenario of this period, however, GENCO 11 expands the capacity of its existing generation units located on bus 39 of the network. At the same time, GENCO 5 installs a new generation unit on bus 32 of the network, while the existing generation units of this GENCO are located on bus 27 of the network. This is because of the fact that, in the third level of the proposed strategic tri-level computational-logical framework, each GENCO decides to determine the location of the new installed generation units as well as their capacities. The same analysis can be done for the remaining periods of the planning horizon.

In addition, the number of augmented generation units plus the number of newly installed generation units over the planning horizon in the second scenario was decreased in comparison with the first scenario. This reduction in the number of augmented and installed generation units is due to the adoption of strategic behaviors by the market participants (see Table 6.15). Table 6.16 gives the values of the EP, GIC, and CP of all the GENCOs under the first and second scenarios of the second case (i.e., the IGDT risk-taker decision-making policy).

By evaluating the results presented in Table 6.16, it can be concluded that when the BBM is activated in the proposed strategic tri-level computational-logical framework, the EP of all GENCOs will increase, the GIC completed by all of the GENCOs will decrease, and the CP paid by the ISO to all GENCOs will also decrease, in comparison with the situation in which the BBM is ignored—i.e., the second scenario compared to the first. As a result, considering the BBM in the proposed framework can encourage and facilitate competition among all market participants,

Table 6.15 Opportunity generation expansion plans of the proposed framework under the first and second scenarios of the second case

Period No.	Optimal opportunity generation expansion plans	
	First scenario: Ignoring the BBM	Second scenario: Considering the BBM
1	–	–
2	GENCO 4 (19: I), GENCO 3 (17: I)	GENCO 1 (14: I), GENCO 12 (46: I)
3	GENCO 2 (16: I), GENCO 8 (32: I)	GENCO 6 (28: I), GENCO 8 (32: I)
4	GENCO 1 (14: I), GENCO 9 (34: I)	GENCO 4 (19: I), GENCO 9 (34: I)
5	GENCO 8 (32: I), GENCO 2 (16: I)	–
6	–	GENCO 2 (16: I), GENCO 7 (31: I), GENCO 6 (28: C)
7	GENCO 7 (31: I), GENCO 6 (28: C)	GENCO 3 (17: I), GENCO 10 (37: I)
8	GENCO 12 (46: I), GENCO 5 (27: I)	GENCO 11 (39: I), GENCO 5 (32: C)
9	GENCO 11 (39: I), GENCO 3 (31: C)	–
10	GENCO 10 (37: I), GENCO 12 (46: C)	GENCO 5 (27: I), GENCO 2 (16: C)
Number of augmented generation units associated with GENCOs	13	12
Number of newly installed generation units associated with GENCOs	3	3

improve efficiency of the CSC electricity market, decrease the required GIC for generation expansion, and provide more flexible generation expansion plans against severe uncertainties. As set out in the results, changes in the objective functions under the first scenario (i.e., considering the BBM) compared to the second scenario (i.e., ignoring the BBM) had a similar trend in both cases. In simple terms, in the second scenario of both cases in which the strategic behaviors of the market participants were adopted, it can be observed that not only did the profit increase, but investment and unreliability costs decreased, in comparison to the first scenario. As a consequence, the integration of the strategic behaviors of the market participants in the PD-GEP problem brings about more productivity and flexibility in the proposed framework.

Table 6.16 Changes in the objective functions of the third level of the proposed framework under the first and second scenarios of the second case

	Expected profit of all GENCOs		Investment costs of all GENCOs		Capacity payments of all GENCOs	
Period	First scenario: Ignoring the BBM	Second scenario: Considering the BBM	First scenario: Ignoring the BBM	Second scenario: Considering the BBM	First scenario: Ignoring the BBM	Second scenario: Considering the BBM
1	249.978461	273.331214	–	–	–	–
2	361.296444	399.440349	314.310111	325.239999	12.949833	7.507477
3	432.460372	497.183646	297.132921	288.225332	12.096569	6.050869
4	496.110447	521.932147	297.132921	307.344055	11.243304	7.779661
5	516.632479	507.503531	359.757218	–	10.284354	–
6	517.849213	654.878607	–	515.051099	–	14.920129
7	552.834388	683.824595	324.788715	331.055923	14.115114	8.472572
8	599.703267	703.338684	323.290069	271.757712	15.808502	6.109833
9	638.749114	718.044624	311.853891	–	14.916333	–
10	692.485563	899.805759	289.096209	293.804275	12.564641	6.415766
Total	5058.099751	5859.283160	2517.362059	2332.478397	103.978654	57.256310

6.3.5.3 Quantitative Verification of the Proposed IGDT Risk-Taker Decision-Making Policy in Comparison to a Robust Optimization Technique

In general, the robust optimization (RO) technique is employed to solve optimization problems under worst-case uncertainty situations. In this technique, the uncertainty data are modeled via a set of deterministic and bounded intervals. Then, a deterministic version of the optimization problem is solved for the worst condition of the uncertainty parameters in order to obtain an optimal solution that is immunized against data uncertainties. Please refer to the work by Ben-Tal et al. [41] for a comprehensive discussion of this robust optimization technique.

The worst-case uncertainty situation in the proposed strategic tri-level computational-logical framework is observed in the second case (i.e., the IGDT risk-taker decision-making policy). Hence, in this section, the performance of the proposed strategic tri-level computational-logical framework under the IGDT risk-taker decision-making policy is compared with the performance of this framework under the RO technique. Table 6.17 gives the results obtained under the IGDT risk-taker decision-making policy and RO technique. Table 6.18 summarizes the calculated optimal values of the objective functions of the third level related to the strategic tri-level computational-logical framework under the IGDT risk-taker decision-making policy and RO technique. These results are related to the second scenario (i.e., considering the BBM).

Table 6.18 shows that the number of augmented generation units plus the number of newly installed generation units under the IGDT risk-taker decision-making policy is also less than the number of those in the RO technique. As set out in Table 6.18, it can be seen that the EP of the GENCOs clearly increased in the IGDT risk-taker decision-making policy compared to the RO technique. Also, the proposed framework under the IGDT risk-taker decision-making policy leads to lower GIC for construction of new generation units and expansion of existing units. In addition, the CP paid by the ISO to the GENCOs in the IGDT risk-taker decision-making policy has a smaller value compared to that for the RO technique. Therefore, the unreliability costs decreased significantly in the proposed framework under the IGDT risk-taker decision-making policy compared with the proposed framework under the RO technique. As a result, the IGDT risk-taker decision-making policy provides more appropriate allocation and capacity expansion for the PD-GEP problem, thereby preventing unnecessary investments. For this reason, the IGDT risk-taker decision-making policy can be a suitable policy for handling severe uncertainties in large-scale optimization problems.

6.3.5.4 Performance Evaluation of the Proposed Optimization Algorithms: Simulation Results and Discussion

In this section, the performance of the proposed modern meta-heuristic music-inspired optimization algorithms addressed in Chap. 4, namely the multi-objective

Table 6.17 Optimal expansion plans of the proposed framework under the IGDT risk-taker decision-making policy and the RO technique

Period	Optimal robust generation expansion plans	
	The IGDT risk-averse decision-making policy	The RO technique
1	–	–
2	GENCO 1 (14: I), GENCO 12 (46: I)	GENCO 2 (16: I), GENCO 8 (32: I)
3	GENCO 6 (28: I), GENCO 8 (32: I)	GENCO 4 (19: I), GENCO 10 (37: I)
4	GENCO 4 (19: I), GENCO 9 (34: I)	GENCO 6 (28: I), GENCO 4 (46: C)
5	–	GENCO 1 (14: I), GENCO 8 (32: C)
6	GENCO 2 (16: I), GENCO 7 (31: I), GENCO 6 (28: C)	GENCO 7 (31: I), GENCO 9 (34: I),
7	GENCO 3 (17: I), GENCO 10 (37: I)	GENCO 3 (17: I), GENCO 2 (16: C)
8	GENCO 11 (39: I), GENCO 5 (32: C)	GENCO 11 (39: I), GENCO 5 (27: I)
9	–	–
10	GENCO 5 (27: I), GENCO 2 (16: C)	GENCO 12 (46: I), GENCO 10 (28: C)
Number of augmented generation units associated with GENCOs	12	12
Number of newly installed generation units associated with GENCOs	3	4

Table 6.18 Changes in the objective functions of the third level related to the strategic tri-level computational-logical framework under the IGDT risk-taker decision-making policy and RO technique

Objective	Uncertainty modeling technique	
	The IGDT risk-averse decision-making policy	The RO technique
EP of all GENCOs (M$)	5859.283160	5635.528226
GIC of all GENCOs (M$)	2332.478397	2509.532485
CP of all GENCOs (M$)	57.256310	61.984392

single-stage computational single-dimensional harmony search algorithm (SS-HSA), multi-objective single-stage computational single-dimensional improved harmony search algorithm (SS-IHSA), multi-objective continuous/discrete two-stage computational multi-dimensional single-homogeneou melody search algorithm

(TMS-MSA), multi-objective two-stage computational multi-dimensional single-homogeneou enhanced melody search algorithm (TMS-EMSA), and multi-objective SOSA, is compared with non-dominated sorting genetic algorithm-II (NSGA-II) for the first and second scenarios of the first and second cases. Tables 6.104, 6.105, 6.106, 6.107, and 6.108 illustrate the parameter adjustments of the multi-objective TMS-EMSA, multi-objective continuous/discrete TMS-MSA, multi-objective SS-HSA, multi-objective SS-IHSA, and NSGA-II, respectively. Please refer to the work by Deb et al. [42] for a detailed description of the NSGA-II.

Tables 6.19 and 6.20 give the calculated optimal results associated with the EP, GIC, and CP of all GENCOs over the planning horizon by the proposed multi-objective optimization algorithms under the first case (i.e., the IGDT risk-averse decision-making policy) and the second case (i.e., the IGDT risk-taker decision-making policy), respectively. Looking at these tables, it can be seen that the results of the proposed strategic tri-level computational-logical framework under the first and second scenarios of the first and second cases by the proposed multi-objective SOSA lead to more efficient results than other proposed meta-heuristic music-inspired optimization algorithms and NSGA-II. The accurate evaluation of the performance of these multi-objective optimization algorithms relative to each other is performed using an index of cost saving (ICS). A detailed description of the ICS is addressed in Sect. 5.3.5.3 of Chap. 5. Tables 6.21 and 6.22 present the ICS for the proposed multi-objective optimization algorithms under the first and second cases, respectively. To illustrate, consider the results of the proposed framework under the first scenario of the first case (the IGDT risk-averse decision-making policy) according to Table 6.21.

The ICS shows 7.273100%, 12.143418%, 20.198803%, 15.921764%, and 21.915521% superiority—positive sign—of the multi-objective SOSA performance compared with the performances of the multi-objective TMS-EMSA, multi-objective continuous/discrete TMS-MSA, multi-objective SS-HSA, multi-objective SS-IHSA, and NSGA-II, respectively, from the perspective of the obtained EP index by corresponding optimization algorithms. Also, the ICS represents 7.2638390%, 11.636653%, 19.616536%, 16.351328%, and 21.322670% superiority—positive sign—of the multi-objective SOSA performance compared with the performances of the multi-objective TMS-EMSA, multi-objective continuous/discrete TMS-MSA, multi-objective SS-HSA, multi-objective SS-IHSA, and NSGA-II, respectively, from the standpoint of the obtained GIC index by corresponding optimization algorithms. In addition, the ICS illustrates 5.419441%, 12.588433%, 20.460035%, 16.753005%, and 21.958834% superiority—positive sign—of the multi-objective SOSA performance compared with the performances of the multi-objective TMS-EMSA, multi-objective continuous/discrete TMS-MSA, multi-objective SS-HSA, multi-objective SS-IHSA, and NSGA-II, respectively, from the point of view of the obtained CP index by corresponding optimization algorithms. The ICS illustrates 6.779985% weakness—negative sign—of the multi-objective TMS-EMSA performance compared with the performance of the multi-objective SOSA, from the perspective of the obtained EP index by the multi-objective SOSA, and 4.540111%, 12.049342%, 8.062286%, and 13.649667% superiority—positive sign—of the multi-objective TMS-EMSA performance compared with the performances of the multi-objective

Table 6.19 The calculated values of the objective functions of the third level of the proposed framework by the proposed multi-objective optimization algorithms under the first case

	The IGDT risk-averse decision-making policy					
	First scenario: Ignoring the BBM			Second scenario: Considering the BBM		
Multi-objective optimization algorithm	EP of all GENCOs over the planning horizon (M$)	GIC of all GENCOs over the planning horizon (M$)	CP of all GENCOs over the planning horizon (M$)	EP of all GENCOs over the planning horizon (M$)	GIC of all GENCOs over the planning horizon (M$)	CP of all GENCOs over the planning horizon (M$)
SOSA	3960.344038	1642.279195	91.654481	4277.322668	1481.052524	48.662894
TM-EMSA	3691.833301	1770.915658	86.942674	4029.245588	1563.195467	46.229509
Continuous/discrete TMS-MSA	3531.499298	1858.552503	81.406658	3846.081609	1679.295016	43.938903
SS-HSA	3294.828178	2043.056023	76.087045	3575.246767	1833.081294	40.962958
SS-IHSA	3416.393842	1963.305773	78.502888	3669.719508	1762.476334	41.797525
NSGA-II	3248.433003	2087.360108	75.151982	3507.246010	1876.530009	40.199141

Table 6.20 The calculated values of the objective functions of the third level of the proposed framework by the proposed multi-objective optimization algorithms under the second case

	The IGDT risk-taker decision-making policy					
	First scenario: Ignoring the BBM			Second scenario: Considering the BBM		
Multi-objective optimization algorithm	EP of all GENCOs over the planning horizon (M$)	GIC of all GENCOs over the planning horizon (M$)	CP of all GENCOs over the planning horizon (M$)	EP of all GENCOs over the planning horizon (M$)	GIC of all GENCOs over the planning horizon (M$)	CP of all GENCOs over the planning horizon (M$)
SOSA	5058.099751	2517.478397	103.978654	5859.283160	2332.478397	57.256310
TM-EMSA	4769.723205	2647.209537	99.409889	5507.023288	2462.117222	54.524649
Continuous/discrete TMS-MSA	4462.988006	2771.619401	94.977939	5256.950538	2603.444989	52.028601
SS-HSA	4172.004574	3027.971019	86.842213	4897.323066	2865.500765	47.816126
SS-IHSA	4234.588088	2943.168077	89.823035	5034.997014	2796.637729	49.157529
NSGA-II	4115.949852	3076.156182	85.073146	4803.562960	2897.642599	46.961759

Table 6.21 The ICS for different objectives of the third level of the proposed framework, based on various multi-objective optimization algorithms under the first and second scenarios of the first case

Case No.	Scenario No.	Perspective	Multi-objective optimization algorithm	Multi-objective optimization algorithm: SOSA	TMS-EMSA	Continuous/ discrete TMS-MSA	SS-HSA	SS-IHSA	NSGA-II
First case	First scenario	EP of all GENCOs over the planning horizon	SOSA	0	7.273100	12.143418	20.198803	15.921764	21.915521
			TM-EMSA	−6.779985	0	4.540111	12.049342	8.062286	13.649667
			Continuous/discrete TMS-MSA	−10.828472	−4.342937	0	7.183109	3.369209	8.713933
			SS-HSA	−16.804496	−10.753603	−6.701717	0	−3.558303	1.428232
			SS-IHSA	−13.734923	−7.460777	−3.259393	3.689590	0	5.170518
			NSGA-II	−17.975989	−12.010301	−8.015470	−1.408121	−4.916319	0
		GIC of all GENCOs over the planning horizon	SOSA	0	7.2638390	11.636653	19.616536	16.351328	21.322670
			TM-EMSA	−7.832801	0	4.715327	13.320259	9.799294	15.160031
			Continuous/discrete TMS-MSA	−13.169095	−4.948674	0	9.030761	5.335555	10.961577
			SS-HSA	−24.403696	−15.367212	−9.927269	0	−4.062039	2.122493
			SS-IHSA	−19.547625	−10.863877	−5.636282	3.903478	0	5.943120
			NSGA-II	−27.101416	−17.868973	−12.311064	−2.168520	−6.318645	0
		CP of all GENCOs over the planning horizon	SOSA	0	5.419441	12.588433	20.460035	16.753005	21.958834
			TM-EMSA	−5.140836	0	6.800446	14.267381	10.750924	15.689129
			Continuous/discrete TMS-MSA	−11.180929	−6.367432	0	6.991483	3.698933	8.322702
			SS-HSA	−16.984915	−12.485961	−6.534616	0	−3.077393	1.244229
			SS-IHSA	−14.349099	−9.707299	−3.566993	3.175104	0	4.458839
			NSGA-II	−18.005119	−13.561455	−7.683248	−1.228938	−4.268513	0

(continued)

Table 6.21 (continued)

Case No.	Scenario No.	Perspective	Multi-objective optimization algorithm	Multi-objective optimization algorithm					
				SOSA	TMS-EMSA	Continuous/discrete TMS-MSA	SS-HSA	SS-IHSA	NSGA-II
	Second scenario	EP of all GENCOs over the planning horizon	SOSA	0	6.156911	11.212478	19.637131	16.557209	21.956733
			TM-EMSA	−5.799821	0	4.762352	12.698391	9.797099	14.883460
			Continuous/dis-crete TMS-MSA	−10.082032	−4.545862	0	7.575276	4.805874	9.661016
			SS-HSA	−16.413910	−11.267588	−7.041838	0	−2.574385	1.938864
			SS-IHSA	−14.205222	−8.922913	−4.585500	2.642411	0	4.632509
			NSGA-II	−18.003707	−12.955268	−8.809891	−1.901987	−4.427409	0
		GIC of all GENCOs over the planning horizon	SOSA	0	5.254809	11.805102	19.204209	15.967522	21.074935
			TM-EMSA	−5.546254	0	6.913588	14.723069	11.306867	16.697550
			Continuous/dis-crete TMS-MSA	−13.385244	−7.427065	0	8.389495	4.719570	10.510622
			SS-HSA	−23.768824	−17.265008	−9.157788	0	−4.006009	2.315375
			SS-IHSA	−19.001609	−12.748301	−4.953347	3.851709	0	6.077903
			NSGA-II	−26.702462	−20.044488	−11.745106	−2.370255	−6.471217	0
		CP of all GENCOs over the planning horizon	SOSA	0	5.263705	10.751272	18.797314	16.425300	21.054561
			TM-EMSA	−5.000493	0	5.213161	12.856862	10.603460	15.001235
			Continuous/dis-crete TMS-MSA	−9.707583	−4.954856	0	7.264966	5.123217	9.303089
			SS-HSA	−15.823013	−11.392184	-6.772916	0	−1.996689	1.900082
			SS-IHSA	−14.108016	−9.586915	−4.873535	2.037369	0	3.976164
			NSGA-II	−17.392622	−13.044412	−8.511277	−1.864652	−3.824111	0

Table 6.22 The ICS for different objectives of the third level of the proposed framework, based on various multi-objective optimization algorithms under the first and second scenarios of the second case

Case No.	Scenario No.	Perspective	Multi-objective optimization algorithm	Multi-objective optimization algorithm: SOSA	TMS-EMSA	Continuous/discrete TMS-MSA	SS-HSA	SS-IHSA	NSGA-II
Second case	First scenario	EP of all GENCOs over the planning horizon	SOSA	0	6.045980	13.334379	21.239074	19.447267	22.890218
			TM-EMSA	−5.701282	0	6.872866	14.326893	12.637241	15.883899
			Continuous/discrete TMS-MSA	−11.765520	−6.430880	0	6.974667	5.393674	8.431544
			SS-HSA	−17.518341	−12.531516	−6.519924	0	−1.477912	1.361890
			SS-IHSA	−16.281048	−11.219416	−5.117645	1.500082	0	2.882402
			NSGA-II	−18.626558	−13.706735	−7.775915	−1.343592	−2.801647	0
		GIC of all GENCOs over the planning horizon	SOSA	0	4.900675	9.169404	16.859230	14.463655	18.161554
			TM-EMSA	−5.153217	0	4.488706	12.574806	10.055781	13.944241
			Continuous/discrete TMS-MSA	−10.095062	−4.699660	0	8.466118	5.828708	9.899912
			SS-HSA	−20.277935	−14.383503	−9.249163	0	−2.881348	1.566408
			SS-IHSA	−16.909368	−11.180019	−6.189474	2.800652	0	4.323190
			NSGA-II	−22.191959	−16.203728	−10.987684	−1.591335	−4.518535	0
		CP of all GENCOs over the planning horizon	SOSA	0	4.595885	9.476637	19.732847	15.759453	22.222650
			TM-EMSA	−4.393945	0	4.666294	14.471851	10.673046	16.852254
			Continuous/discrete TMS-MSA	−8.656310	−4.458258	0	9.368400	5.738955	11.642678
			SS-HSA	−16.480729	−12.642279	−8.565911	0	−3.318549	2.079465
			SS-IHSA	−13.613966	−9.643762	−5.427475	3.432457	0	5.583300
			NSGA-II	−18.182105	−14.421847	−10.428519	−2.037104	−5.288052	0
	Second scenario		SOSA	0	6.396556	11.457833	19.642569	16.371134	21.977857

(continued)

Table 6.22 (continued)

Case No.	Scenario No.	Perspective	Multi-objective optimization algorithm	Multi-objective optimization algorithm					
				SOSA	TMS-EMSA	Continuous/discrete TMS-MSA	SS-HSA	SS-IHSA	NSGA-II
		EP of all GENCOs over the planning horizon	TM-EMSA	−6.011996	0	4.756992	12.449663	9.374906	14.644553
			Continuous/dis-crete TMS-MSA	−10.279971	−4.540978	0	7.343347	4.408215	9.438568
			SS-HSA	−16.417709	−11.071320	−6.840990	0	−2.734340	1.951886
			SS-IHSA	−14.068037	−8.571350	−4.222096	2.811208	0	4.817966
			NSGA-II	−18.017907	−12.773876	−8.624535	−1.914517	−4.596508	0
		GIC of all GENCOs over the planning horizon	SOSA	0	5.265339	10.408001	18.601368	16.597048	19.504275
			TM-EMSA	−5.557986	0	5.428490	14.077244	11.961524	15.030334
			Continuous/dis-crete TMS-MSA	−11.617110	−5.740090	0	9.145200	6.908035	10.152998
			SS-HSA	−22.852189	−16.383604	−10.065731	0	−2.462350	1.109240
			SS-IHSA	−19.899834	−13.586701	−7.420657	2.403176	0	3.485760
			NSGA-II	−24.230201	−17.689059	−11.300319	−1.121682	−3.611653	0
		CP of all GENCOs over the planning horizon	SOSA	0	5.009956	10.047760	19.742678	16.475158	21.921135
			TM-EMSA	−4.770934	0	4.797453	14.029833	10.918205	16.104358
			Continuous/dis-crete TMS-MSA	−9.130363	−4.577834	0	8.809737	5.840553	10.789293
			SS-HSA	−16.487587	−12.303651	−8.096460	0	−2.728784	1.819282
			SS-IHSA	−14.144783	−9.843474	−5.518257	2.805336	0	4.675655
			NSGA-II	−17.979767	−13.870589	−9.738570	−1.786775	−4.466803	0

continuous/discrete TMS-MSA, multi-objective SS-HSA, multi-objective SS-IHSA, and NSGA-II, respectively, from the perspective of the obtained EP index by corresponding optimization algorithms. Also, the ICS shows 7.832801% weakness—negative sign—of the multi-objective TMS-EMSA performance compared with the performance of the multi-objective SOSA, from the standpoint of the obtained GIC index by the multi-objective SOSA, and 4.715327%, 13.320259%, 9.799294%, and 15.160031% superiority—positive sign—of the multi-objective TMS-EMSA performance compared with the performances of the multi-objective continuous/discrete TMS-MSA, multi-objective SS-HSA, multi-objective SS-IHSA, and NSGA-II, respectively, from the standpoint of the obtained GIC index by corresponding optimization algorithms.

In addition, the ICS represents 5.140836% weakness—negative sign—of the multi-objective TMS-EMSA performance compared with the performance of the multi-objective SOSA, from the point of view of the obtained CP index by the multi-objective SOSA, and 6.800446%, 14.267381%, 10.750924%, and 15.689129% superiority—positive sign—of the multi-objective TMS-EMSA performance compared with the performances of the multi-objective continuous/discrete TMS-MSA, multi-objective SS-HSA, multi-objective SS-IHSA, and NSGA-II, respectively, from the point of view of the obtained CP index by corresponding optimization algorithms. By the same token, the optimal results presented in Table 6.18 for the other multi-objective optimization algorithms and in Table 6.22 for all multi-objective optimization algorithms are analyzed in the same way.

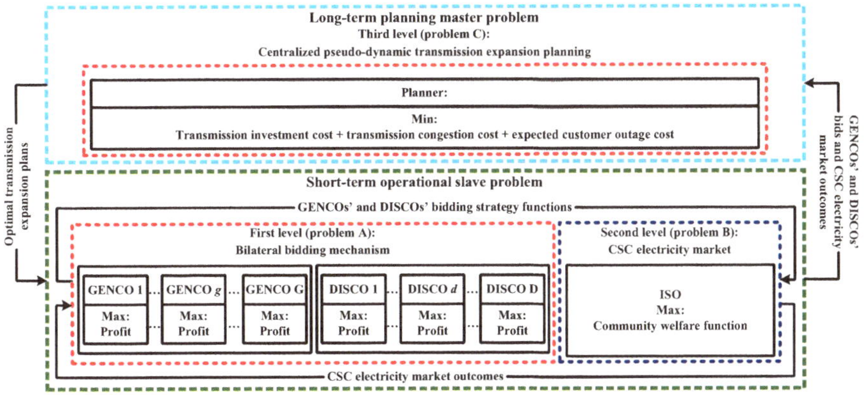

Fig. 6.10 Conceptual-view diagram of the proposed strategic tri-level computational-logical framework

6.4 Pseudo-Dynamic Transmission Expansion Planning: A Strategic Tri-level Computational-Logical Framework

In this section, a strategic tri-level computational-logical framework is addressed to the PD-TEP under deregulated environments. The proposed strategic tri-level computational-logical framework is broken down into a long-term planning master problem and a short-term operational slave problem. In the proposed framework, the outcomes of the short-term operational slave problem—including the GENCOs' and DISCOs' bids and the CSC electricity market outcomes—supply the long-term planning master problem. Mutually, the outcomes of the long-term planning master problem—including optimal transmission expansion plans—supply the short-term operational slave problem. In other words, the proposed framework evaluates the impact of the transmission expansion plans on strategic behaviors of the market participants—including the GENCOs and DISCOs—and vice versa. In the first level, each GENCO/DISCO tries to maximize its profit by employing a bidding mechanism. In the second level, the ISO performs the CSC electricity market-clearing process with the aim of maximizing the CWF. In the third level, however, the planner expands the existing transmission network using expansion options so that the energy supply of current and future customers is provided not only with minimum investment and congestion costs, but also with maximum service continuity (and/or minimum expected customer outage costs). Figure 6.10 shows a conceptual-view diagram of the strategic tri-level computational-logical framework. The assumptions used in the mathematical description process of the strategic tri-level computational-logical framework are as follows:

- Hypotheses 1–10, discussed in Sect. 5.3 of the Chap. 5, are considered in here.
- Hypothesis 11: Severe twofold uncertainties related to market price and demand parameters are considered and applied in the strategic tri-level computational-logical framework.
- Hypothesis 12: A well-founded IGDT is widely employed to handle risks of the PD-TEP problem arising from severe twofold uncertainties.
- Each electrical load is divided into S different sectors by existing customer classes.

Table 6.23 provides a summary of attributes of previous TEP frameworks reported in the literature, and the PD-TEP framework proposed in this chapter. The authors, though, will only focus on the most important and relevant journal papers found in the literature. Additional work on recent practices and new challenges on the topic of TEP can be found in the study by Lumbreras and Ramos [53]. Also, Table 5.2 (see Chap. 5) presents features related to the bi-level computational-logical framework (i.e., the short-term operational slave problem). However, due to the fact that most technical studies in this area do not use a tri-level framework, none are included in Table 6.23.

Table 6.23 Attributes of previous frameworks reported in the TEP literature and the strategic tri-level computational-logical framework proposed in this study

Reference	Multilevel framework	Static or multi-period framework	Nondeterministic framework	Reliability consideration	Type of modeling reliability criterion	System security constraint	Loss consideration	Uncertainty consideration	Strategic behaviors of CSC electricity market participants	Impact of expansion plans on strategic behavior and vice versa	Solution method
[43]	Single-level	Static	No	No	–	Yes	No	No	No	No	Particle swarm optimization (PSO)
[44]	Single-level	Static	Nondeterministic framework: scenario generation	No	–	Yes	No	Twofold: generation cost and electrical power demand	No	No	Stochastic optimization: recourse programming
[45]	Two-level	Static	Nondeterministic framework: scenario generation	No	–	Yes	No	Twofold: transmission line availability and electrical power demand	No	No	GAMS
[46]	Single-level	Multi-period	No	Yes: reliability assessment incompatible with the value-based approach	Deterministic model	Yes	No	No	No	No	Ant colony optimization
[47]	Two-level	Multi-period	No	No	–	Yes	No	No	No	No	Niche GA
[48]	Single-level	Multi-period	No	Yes: reliability constraint	–	Yes	Yes	No	No	No	GAMS
[49]	Single-level	Multi-period	Nondeterministic framework: fuzzy set concept	Yes: reliability constraint	–	Yes	No	Onefold: electrical power demand	No	No	PSO
[50]	Single-level	Static	Nondeterministic framework: scenario generation	No	–	Yes	No	Onefold: electrical power demand	No	No	Hybrid modified NSGA-II and Chu–Beasley algorithm
[51]	Single-level	Static	Nondeterministic framework: scenario generation	No	–	Yes	No	Onefold: electrical power generation	No	No	Strength Pareto evolutionary algorithm 2

(continued)

Table 6.23 (continued)

Reference	Multilevel framework	Static or multi-period framework	Nondeterministic framework	Reliability consideration	Type of modeling reliability criterion	System security constraint	Loss consideration	Uncertainty consideration	Strategic behaviors of CSC electricity market participants	Impact of expansion plans on strategic behavior and vice versa	Solution method
[52]	Single-level	Static	No	Yes: reliability constraint	–	Yes	No	No	No	No	Benders decomposition approach
Proposed framework	Tri-level	Multi-period	Nondeterministic framework: stochastic modeling by the IGDT (risk-averse and risk-taker frameworks)	Yes: value-based reliability approach compatible with the restructured environment	Probabilistic model based on sector customer damage function	Yes	Yes	Severe twofold: price of the CSC electricity market and electrical power demand	Yes: bilateral bidding mechanism	Yes	Five new multi-objective music-inspired optimization algorithms (see Chap. 4): (1) SS-HSA; (2) SS-IHSA; (3) continuous/discrete TMS-MSA; (4) TMS-EMSA; and (5) SOSA

6.4.1 Mathematical Model of the Deterministic Strategic Tri-level Computational-Logical Framework

In this section, the mathematical model of the proposed strategic tri-level computational-logical framework is developed within a deterministic environment and discussed more in depth.

6.4.1.1 Bilateral Bidding Mechanism: First Level (Problem A)

In the first level of the strategic tri-level computational-logical framework, the market participants, including GENCOs and DISCOs, try to choose optimal bidding strategies to maximize their profits. The first level of the strategic tri-level computational-logical framework related to the PD-TEP problem is virtually the same as the first level of the proposed PD-GEP problem in Sect. 6.3 of this chapter. Hence, the mathematical model of the BBM (i.e., the first level) is considered in accordance with Eqs. (6.1) through (6.9), presented in Sect. 6.3.1.1.

6.4.1.2 Competitive Security-Constrained Electricity Market: Second Level (Problem B)

In the second level of the strategic tri-level computational-logical framework, the ISO clears the CSC electricity market with the aim of maximizing the CWF. The second level of the strategic tri-level computational-logical framework related to the PD-TEP problem is virtually the same as the second level of the PD-GEP problem in Sect. 6.3. Hence, the mathematical model of the CSC electricity market-clearing problem (i.e., the second level) is taken into account based on Eqs. (6.10) through (6.14) presented in Sect. 6.3.1.2.

6.4.1.3 Pseudo-Dynamic Transmission Expansion Planning: Third Level (Problem C)

In the third level of the strategic tri-level computational-logical framework, a centralized PD-TEP problem is considered. In this problem, an entity is responsible for determining the optimal transmission expansion plans, or planner. The proposed centralized PD-TEP problem takes into account not only transmission investment costs (TIC) and transmission congestion costs (TCC), but also the ECOC in the optimization problem as three different objectives. Concurrently, the short-term and long-term constraints are considered in the optimization problem as two different classes of the constraints. In general, the TEP process required the huge financial resources, while the planner's budget for purchasing and installing the transmission network equipment is limited. Hence, any effort to reduce the TIC affords the saving of a significant amount of capital. The first target of the PD-TEP problem is, therefore, considered in order to minimize the TIC index, as determined by Eqs. (6.106) through (6.108):

$$\mathrm{OF}_1^{\text{PD-TEP}} = TIC_y = \sum_{p \in \Psi^{\mathrm{P}}} \left(\frac{1}{(1 + \mathrm{Ir})^{p - p_0}} \right) \cdot TIC_{y,p}; \quad \forall \{ p \in \Psi^{\mathrm{P}} \} \tag{6.106}$$

$$TIC_{y,p} = \sum_{(s,r) \in \Psi^{\mathrm{L}}} TIC_{p,(s,r)} \cdot \iota_{p,(s,r)}; \quad \forall \{ p \in \Psi^{\mathrm{P}}, (s,r) \in \Psi^{\mathrm{L}} \} \tag{6.107}$$

$$TIC_{p,(s,r)} = CCL^{\mathrm{b}} \cdot LE_{(s,r)} \cdot f_{p,(s,r)}; \quad \forall \{ p \in \Psi^{\mathrm{P}}, (s,r) \in \Psi^{\mathrm{L}} \} \tag{6.108}$$

Equation (6.106) shows the net present value of the TIC index over the planning horizon. The transmission investment costs for the installation of new transmission lines at period p and at corridor (s,r) depend on the number of transmission lines installed in this corridor [see Eq. (6.107)]. The transmission investment costs for installation of a new transmission line at corridor (s,r) also depend on the length of this corridor as well as the base construction costs and the thermal capacity of this corridor [see Eq. (6.108)].

The phenomenon of transmission congestion occurs when multiple market participants simultaneously intend to use an open-access transmission network. Under these circumstances, the electrical energy demand by consumers may not be provided by the low-priced suppliers. In this way, consumers inevitably purchase a portion of their demand from alternative sources at higher prices. Therefore, the possibility of using the lowest-priced resources to meet the demand is decreased when the transmission network is congested; and, consequently, the possibility of applying market power by the market participants is also increased. In the deregulated structure, hence, transmission congestion cost is one of the main indices that must be considered in the process of the PD-TEP problem. The second target of the PD-TEP problem is minimization of the TCC index—as a factor for encouraging competition in the CSC electricity market—as given by Eqs. (6.109) through (6.111):

$$\mathrm{OF}_2^{\text{PD-TEP}} = TCC_y = \sum_{p \in \Psi^{\mathrm{P}}} \left(\frac{1}{(1 + \mathrm{Ir})^{p - p_0}} \right) \cdot TCC_{y,p}; \quad \forall \{ p \in \Psi^{\mathrm{P}} \} \tag{6.109}$$

$$TCC_{y,p} = \sum_{k \in \Psi^{\mathrm{K}}} \Delta t_k \cdot TCC_{p,k}; \quad \forall \{ p \in \Psi^{\mathrm{P}}, k \in \Psi^{\mathrm{K}} \} \tag{6.110}$$

$$TCC_{p,k} = \sum_{l \in \Psi^{\mathrm{L}}} \left(p\lambda_{p,k,l_2} - p\lambda_{p,k,l_1} \right) \cdot f_{p,k,l_{1-2}}; \quad \forall \{ p \in \Psi^{\mathrm{P}}, k \in \Psi^{\mathrm{K}}, l \in \Psi^{\mathrm{L}} \} \tag{6.111}$$

Equation (6.109) shows the net present value of the TCC index over the planning horizon. The TCC index at period p depends on the congestion costs in all patterns [see Eq. (6.110)]. The value of the TCC index in each pattern also depends on the difference between the predicted prices at the sending and receiving buses and the active power flow in each transmission line [see Eq. (6.111)]. There is another approach to calculate the TCC. In this approach, the TCC index is equal to the difference between the values of the CWF for two distinct strategies of the CSC electricity market operation (i.e., with and without considering the security restriction imposed on the transmission lines). Therefore, the TCC index in each period of the planning horizon can be calculated using Eq. (6.112):

$$\begin{aligned} & TCC_{y,p} \\ & = \sum_{k \in \Psi^{\mathrm{K}}} \Delta t_k \cdot \left(\left[\underset{\rho_{p,k,g}, \rho_{p,k,d}}{\mathrm{Max}:} \quad \text{Eq.}(6.10); \text{subject to Eqs.}(6.11) \text{ through } (6.13) \right] \right. \\ & \qquad \left. - \left[\underset{\rho_{p,k,g}, \rho_{p,k,d}}{\mathrm{Max}:} \quad \text{Eq.}(6.10); \text{subject to Eqs.}(6.11) \text{ through } (6.14) \right] \right); \\ & \qquad \forall \{ p \in \Psi^{\mathrm{P}}, k \in \Psi^{\mathrm{K}}, g \in \Psi^{\mathrm{G}}, d \in \Psi^{\mathrm{D}} \} \end{aligned} \tag{6.112}$$

In order to evaluate reliability worth in the expansion plans for the transmission network, a well-defined value-based index must be developed for the PD-TEP problem. The third target of the PD-TEP problem is, thus, formulated with the aim of minimizing the expected costumer outage costs, namely the ECOC index, as shown in Eqs. (6.113) and (6.114):

$$\mathrm{OF}_3^{\text{PD-TEP}} = ECOC_y^{\text{tran}} = \sum_{p \in \Psi^{\mathrm{P}}} \left(\frac{1}{(1 + \mathrm{Ir})^{p - p_0}} \right) \cdot ECOC_{y,p}^{\text{tran}}; \quad \forall \{ p \in \Psi^{\mathrm{P}} \} \tag{6.113}$$

$$ECOC_{y,p}^{\text{tran}} = ECOC_{y,p}^{\text{HL-II}} - ECOC_{y,p}^{\text{HL-I}}; \quad \forall \{ p \in \Psi^{\mathrm{P}} \} \tag{6.114}$$

In the power system literature, a wide range of the reliability criteria for achieving predetermined reliability levels can be found. Focus on recognizing the various features of these reliability criteria is necessary to choose the appropriate reliability index. However, addressing these details is beyond the scope of this book. For a detailed description of the different types of reliability indices and their features, please refer to the work by Billinton and Allan [54]. Equation (6.113) represents the net present value of the ECOC index over the planning horizon. The ECOC index of the transmission network is equal to the calculated cost/worth of this index at HL-II minus the determined cost/worth at HL-I [see Eq. (6.114)]. A value-based reliability assessment at HL-I is defined as the adequacy assessment process of power generation capacity. In simple terms, the HL-I analysis is employed to assess the ability of the power generation network to meet the electrical load requirements. For the HL-I analysis, only generation network facilities are considered. The value-based reliability assessment at HL-II is also defined as the adequacy assessment process of power generation and transmission capacities. In other words, the HL-II analysis is employed to assess the integrated ability of the composite system (i.e., generation and transmission networks) to deliver energy to the customers. For the HL-II analysis, the generation and transmission network facilities are considered simultaneously. The ECOC at HL-II can be determined by the power load curtailment cost (LCC) index and state probability of all considered outage events, as given by Eq. (6.115):

$$ECOC_{y,p}^{\text{HL-II}} = \sum_{e \in \Psi^{\text{E}}} LCC_{p,e}^{\text{HL-II}} \cdot \Pr(e); \quad \forall \{p \in \Psi^{\text{P}}, e \in \Psi^{\text{E}}\} \tag{6.115}$$

In the calculation process of the ECOC at HL-II, a set of events, E, that include the outage of the generation and transmission network facilities and a combination of the simultaneous outage of generation and transmission network facilities are taken into account. In addition, a recently developed linear optimization approach is employed to calculate the LCC at HL-II, as given by Eqs. (6.116) through (6.122):

$$\begin{aligned} LCC_{p,e}^{\text{HL-II}} = \underset{x_{\text{ECOC}}}{\text{Min}} : & \sum_{b \in \Psi^{\text{B}}} \sum_{s \in \Psi^{\text{S}}} c\rho_{p,b,s,e} \cdot C_{p,b,s,e}(\Delta t_e) \\ & \cdot w_{p,b,s}; \quad \forall \{p \in \Psi^{\text{P}}, b \in \Psi^{\text{B}}, s \in \Psi^{\text{S}}, e \in \Psi^{\text{E}}\} \end{aligned} \tag{6.116}$$

subject to Eqs. (6.117) through (6.122):

$$\begin{aligned} & \sum_{g \in \Psi^{\text{G}}} \rho_{p,g,b} - \sum_{s \in \Psi^{\text{S}}} \left(\rho_{p,b,s}^{\max} - c\rho_{p,b,s,e}\right) = \sum_{b' \in \Psi^{\text{B}}} Y_{p,bb',e} \cdot \theta_{p,b',e}; \\ & \forall \{p \in \Psi^{\text{P}}, g \in \Psi^{\text{G}}, b, b' \in \Psi^{\text{B}}, s \in \Psi^{\text{S}}, e \in \Psi^{\text{E}}\} \end{aligned} \tag{6.117}$$

$$0 \leq \sum_{b \in \Psi^{\text{B}}} \sum_{s \in \Psi^{\text{S}}} c\rho_{p,b,s,e} \leq c\rho_p^{\max}; \quad \forall \{p \in \Psi^{\text{P}}, b \in \Psi^{\text{B}}, s \in \Psi^{\text{S}}, e \in \Psi^{\text{E}}\} \tag{6.118}$$

$$0 \leq \sum_{s \in \Psi^{\text{S}}} c\rho_{p,b,s,e} \leq c\rho_{p,b}^{\max}; \quad \forall \{p \in \Psi^{\text{P}}, b \in \Psi^{\text{B}}, s \in \Psi^{\text{S}}, e \in \Psi^{\text{E}}\} \tag{6.119}$$

$$0 \leq c\rho_{p,b,s,e} \leq c\rho_{p,b,s}^{\max}; \quad \forall \{p \in \Psi^{\text{P}}, b \in \Psi^{\text{B}}, s \in \Psi^{\text{S}}, e \in \Psi^{\text{E}}\} \tag{6.120}$$

$$\begin{aligned} & -Y_{p,bb',e} \cdot \left(\theta_{p,b,e} - \theta_{p,b',e}\right) \leq f_{p,l}^{\max}; \\ & \forall \{p \in \Psi^{\text{P}}, b, b' \in \Psi^{\text{B}}, e \in \Psi^{\text{E}}, l \in \Psi^{\text{L}}\} \end{aligned} \tag{6.121}$$

$$\begin{aligned} & -Y_{p,bb',e} \cdot \left(\theta_{p,b',e} - \theta_{p,b,e}\right) \leq f_{p,l}^{\max}; \\ & \forall \{p \in \Psi^{\text{P}}, b, b' \in \Psi^{\text{B}}, e \in \Psi^{\text{E}}, l \in \Psi^{\text{L}}\} \end{aligned} \tag{6.122}$$

Equation (6.116) shows the LCC index for outage event e in period p. Equation (6.117) represents the power balance constraint, which was adapted to the linear optimization approach. Eqs. (6.118) through (6.120) illustrate the restrictions imposed on load curtailment in each outage event, each bus, and each load sector of period p, respectively. Equations (6.121) and (6.122) demonstrate the restrictions associated with power flow in each transmission line at period p, which was adapted to the linear optimization approach.

In the proposed value-based reliability assessment methodology, each electrical load is split into S different sectors by existing customer classes. Each sector of the load has a specific sector customer damage function (SCDF). The SCDF is applied in order to determine the load curtailment costs of the relevant sector. The calculation process of the ECOC at HL-I is similar to the process at HL-II [see Eqs. (6.115) through (6.122)]. In this calculation process, however, only the outage of the generation network facilities is considered, and the outage of the transmission network facilities and its constraints are ignored in the proposed linear optimization approach [see Eqs. (6.121) and (6.122)].

As a result, the overall objective function of the PD-TEP problem in the third level is formulated according to Eq. (6.123):

$$\underset{x_{\text{PD-TEP}}}{\text{Min}} : \left\{\text{OF}^{\text{PD-TEP}}\right\} = \underset{x_{\text{PD-TEP}}}{\text{Min}} : \left\{\text{W}_{\text{OF}_1}^{\text{PD-TEP}} \cdot \text{OF}_1^{\text{PD-TEP}} + \text{W}_{\text{OF}_2}^{\text{PD-TEP}} \cdot \text{OF}_2^{\text{PD-TEP}} + \text{W}_{\text{OF}_3}^{\text{PD-TEP}} \cdot \text{OF}_3^{\text{PD-TEP}}\right\} \tag{6.123}$$

In the third level of the proposed tri-level computational-logical framework, objective functions have different degrees of importance from the perspective of a planner. In the overall objective function, then, the planner applies a weighting coefficient for each objective function.

The constraints of the PD-TEP problem are divided into two different classes: (1) short-term and (2) long-term constraints. The Kirchhoff's point and loop laws in the transmission network must be met at each pattern of period p, as determined by Eqs. (6.124) and (6.125), respectively:

$$\text{NB}_p \cdot \text{PF}_{p,k} + \text{GE}_{p,k} = \text{DE}_{p,k} + \text{ADE}_{p,k}; \quad \forall\left\{p \in \Psi^{\text{P}}, k \in \Psi^{\text{K}}\right\} \tag{6.124}$$

$$f_{p,k,(s,r)} - \left(y_{p,(s,r)}^{0} + \tau_{p,(s,r)} \cdot l_{p,(s,r)}\right) \cdot \left(\theta_{p,k,s} - \theta_{p,k,r}\right) = 0; \quad \forall\left\{p \in \Psi^{\text{P}}, k \in \Psi^{\text{K}}, (s,r) \in \Psi^{\text{L}}\right\} \tag{6.125}$$

The DC power flow limitation must be satisfied at each pattern of period p, in accordance with Eq. (6.126):

$$\left|f_{p,k,(s,r)}\right| \leq \left(y_{p,(s,r)}^{0} + \tau_{p,(s,r)} \cdot l_{p,(s,r)}\right) \cdot f_{p,(s,r)}^{\max} = 0; \quad \forall\left\{p \in \Psi^{\text{P}}, k \in \Psi^{\text{K}}, (s,r) \in \Psi^{\text{L}}\right\} \tag{6.126}$$

In general, there are multiple limitations in the required budget for purchasing and installing the transmission network equipment. Hence, the restrictions related to the

TIC index in period p and in all periods of the planning horizon are depicted by Eqs. (6.127) and (6.128):

$$TIC_{y,p} \leq TIC_{p}^{\max}; \quad \forall\{p \in \Psi^{\mathrm{P}}\} \tag{6.127}$$

$$\sum_{p \in \Psi^{\mathrm{P}}} TIC_{y,p} \leq TIC^{\max}; \quad \forall\{p \in \Psi^{\mathrm{P}}\} \tag{6.128}$$

Moreover, there are several limitations in the installation of the transmission network equipment from a practical perspective. That is, the number of installed transmission lines in each corridor of the transmission network must satisfy the predetermined standard levels, according to Eq. (6.129):

$$\begin{aligned} 0 \leq & \sum_{p \in \Psi^{\mathrm{P}}} \iota_{p,(s,r)} \\ & \leq \iota_{(s,r)}^{\max}; \quad \forall\{p \in \Psi^{\mathrm{P}}, (s,r) \in \Psi^{\mathrm{L}}, \iota_{p,(s,r)} \text{ is an integer variable}\} \end{aligned} \tag{6.129}$$

With the transmission network expansion in each period, the admittance matrix of the transmission network must be updated according to Eqs. (6.130) and (6.131):

\openup 5pt

$$Y_{p,sr} = -\left(y_{(p-1),(s,r)}^{0} + \iota_{p,(s,r)} \cdot \tau_{p,(s,r)}\right); \quad \forall\{p \in \Psi^{\mathrm{P}}, (s,r) \in \Psi^{\mathrm{L}}, s \neq r\} \tag{6.130}$$

$$\begin{aligned} Y_{p,ss} = & \; y_{(p-1),ss}^{0} \\ & + \sum_{r \in \Psi_{s}^{\mathrm{L}}} \left(y_{(p-1),(s,r)}^{0} + \iota_{p,(s,r)} \cdot \tau_{p,(s,r)}\right); \quad \forall\{p \in \Psi^{\mathrm{P}}, s \in \Psi^{\mathrm{L}}, r \in \Psi_{s}^{\mathrm{L}}, s \neq r\} \end{aligned} \tag{6.131}$$

6.4.2 *Overview of the Deterministic Strategic Tri-level Computational-Logical Framework*

According to the equations identified in the previous sections, the proposed PD-TEP problem can be formulated as a deterministic strategic tri-level computational-logical framework, as follows:

$$\underset{x_{\text{PD-TEP}}}{\text{Min}} : \left\{ \text{OF}^{\text{PD-TEP}} \right\} = \underset{x_{\text{PD-TEP}}}{\text{Min}} : \left\{ \text{W}_{\text{OF}_1}^{\text{PD-TEP}} \cdot \text{OF}_1^{\text{PD-TEP}} + \text{W}_{\text{OF}_2}^{\text{PD-TEP}} \cdot \text{OF}_2^{\text{PD-TEP}} + \text{W}_{\text{OF}_3}^{\text{PD-TEP}} \cdot \text{OF}_3^{\text{PD-TEP}} \right\}$$

(6.132) Third level

where

$$\text{OF}_1^{\text{PD-TEP}} = TIC_y = \sum_{p \in \Psi^{\text{P}}} \left(\frac{1}{(1+\text{Ir})^{p-p_0}} \right) \cdot TIC_{y,p}; \quad \forall \{ p \in \Psi^{\text{P}} \} \tag{6.133}$$

$$TIC_{y,p} = \sum_{(s,r) \in \Psi^{\text{L}}} TIC_{p,(s,r)} \cdot \iota_{p,(s,r)}; \quad \forall \{ p \in \Psi^{\text{P}}, (s,r) \in \Psi^{\text{L}} \} \tag{6.134}$$

$$TIC_{p,(s,r)} = CCL^{\text{b}} \cdot LE_{(s,r)} \cdot f_{(s,r)}; \quad \forall \{ p \in \Psi^{\text{P}}, (s,r) \in \Psi^{\text{L}} \} \tag{6.135}$$

$$\text{OF}_2^{\text{PD-TEP}} = TCC_y = \sum_{p \in \Psi^{\text{P}}} \left(\frac{1}{(1+\text{Ir})^{p-p_0}} \right) \cdot TCC_{y,p}; \quad \forall \{ p \in \Psi^{\text{P}} \} \tag{6.136}$$

$$TCC_{y,p} = \sum_{k \in \Psi^{\text{K}}} \Delta t_k \cdot TCC_{p,k}; \quad \forall \{ p \in \Psi^{\text{P}}, k \in \Psi^{\text{K}} \} \tag{6.137}$$

$$TCC_{p,k} = \sum_{l \in \Psi^{\text{L}}} \left(p\lambda_{p,k,l_2} - p\lambda_{p,k,l_1} \right) \cdot f_{p,k,l_{1-2}}; \quad \forall \{ p \in \Psi^{\text{P}}, k \in \Psi^{\text{K}}, l \in \Psi^{\text{L}} \} \tag{6.138}$$

$$\text{OF}_3^{\text{PD-TEP}} = ECOC_y^{\text{tran}} = \sum_{p \in \Psi^{\text{P}}} \left(\frac{1}{(1+\text{Ir})^{p-p_0}} \right) \cdot ECOC_{y,p}^{\text{tran}}; \quad \forall \{ p \in \Psi^{\text{P}} \} \tag{6.139}$$

$$ECOC_{y,p}^{\text{tran}} = ECOC_{y,p}^{\text{HL-II}} - ECOC_{y,p}^{\text{HL-I}}; \quad \forall \{ p \in \Psi^{\text{P}} \} \tag{6.140}$$

and subject to Eqs. (6.141) through (6.148):

$$\text{NB}_p \cdot \text{PF}_{p,k} + \text{GE}_{p,k} = \text{DE}_{p,k} + \text{ADE}_{p,k}; \quad \forall \{ p \in \Psi^{\text{P}}, k \in \Psi^{\text{K}} \} \tag{6.141}$$

$$f_{p,k,(s,r)} - \left(y_{p,(s,r)}^0 + \tau_{p,(s,r)} \cdot \iota_{p,(s,r)} \right) \cdot \left(\theta_{p,k,s} - \theta_{p,k,r} \right) = 0; \quad \forall \{ p \in \Psi^{\text{P}}, k \in \Psi^{\text{K}}, (s,r) \in \Psi^{\text{L}} \} \tag{6.142}$$

$$\left| f_{p,k,(s,r)} \right| \leq \left(y_{p,(s,r)}^0 + \tau_{p,(s,r)} \cdot \iota_{p,(s,r)} \right) \cdot f_{p,(s,r)}^{\max} = 0; \quad \forall \{ p \in \Psi^{\text{P}}, k \in \Psi^{\text{K}}, (s,r) \in \Psi^{\text{L}} \} \tag{6.143}$$

$$TIC_{y,p} \leq TIC_p^{\max}; \quad \forall \{p \in \Psi^{\mathrm{P}}\} \tag{6.144}$$

$$\sum_{p \in \Psi^{\mathrm{P}}} TIC_{y,p} \leq TIC^{\max}; \quad \forall \{p \in \Psi^{\mathrm{P}}\} \tag{6.145}$$

$$0 \leq \sum_{p \in \Psi^{\mathrm{P}}} \iota_{p,(s,r)} \leq \iota_{(s,r)}^{\max}; \quad \forall \{p \in \Psi^{\mathrm{P}}, (s,r) \in \Psi^{\mathrm{L}}, \iota_{p,(s,r)} \text{ is an integer variable}\} \tag{6.146}$$

$$Y_{p,sr} = -\left(y^0_{(p-1),(s,r)} + \iota_{p,(s,r)} \cdot \tau_{p,(s,r)}\right); \quad \forall \{p \in \Psi^{\mathrm{P}}, (s,r) \in \Psi^{\mathrm{L}}, s \neq r\} \tag{6.147}$$

$$Y_{p,ss} = y^0_{(p-1),ss} + \sum_{r \in \Psi^{\mathrm{L}}_s} \left(y^0_{(p-1),(s,r)} + \iota_{p,(s,r)} \cdot \tau_{p,(s,r)}\right); \quad \forall \{p \in \Psi^{\mathrm{P}}, s \in \Psi^{\mathrm{L}}, r \in \Psi^{\mathrm{L}}_s, s \neq r\} \tag{6.148}$$

$$\underset{x_{\mathrm{BBM}}}{\mathrm{Max}} : \tilde{\upsilon}_{p,k,g}\left(\rho_{p,k,g}\right) = \frac{\left(2\xi_{1,p,k,g} - 1\right)}{2} \cdot \alpha_{p,k,g} \cdot \rho^2_{p,k,g} + \left(\xi_{2,p,k,g} - 1\right) \cdot \beta_{p,k,g} \cdot \rho_{p,k,g} - \gamma_{p,k,g}; \forall \{p \in \Psi^{\mathrm{P}}, k \in \Psi^{\mathrm{K}}, g \in \Psi^{\mathrm{G}}\} \tag{6.149}$$

First level

$$\underset{x_{\mathrm{BBM}}}{\mathrm{Max}} : \tilde{\upsilon}_{p,k,d}\left(\rho_{p,k,d}\right) = \frac{\left(1 - 2\xi_{1,p,k,d}\right)}{2} \cdot \alpha_{p,k,d} \cdot \rho^2_{p,k,d} + \left(1 - \xi_{2,p,k,d}\right) \cdot \beta_{p,k,d} \cdot \rho_{p,k,d} + \gamma_{p,k,d}; \forall \{p \in \Psi^{\mathrm{P}}, k \in \Psi^{\mathrm{K}}, d \in \Psi^{\mathrm{D}}\} \tag{6.150}$$

First level

and subject to Eqs. (6.151) through (6.157):

$$\sum_{g \in \Psi^{\mathrm{G}}} \rho_{p,k,g} = \sum_{l \in \Psi^{\mathrm{L}}} \kappa_{p,k,l} + \sum_{d \in \Psi^{\mathrm{D}}} \rho_{p,k,d}; \quad \forall \{p \in \Psi^{\mathrm{P}}, k \in \Psi^{\mathrm{K}}, g \in \Psi^{\mathrm{G}}, l \in \Psi^{L}, d \in \Psi^{\mathrm{D}}\} \tag{6.151}$$

$$\rho^{\min}_{p,k,g} \leq \rho_{p,k,g} \leq \rho^{\max}_{p,k,g}; \quad \forall \{p \in \Psi^{\mathrm{P}}, k \in \Psi^{\mathrm{K}}, g \in \Psi^{\mathrm{G}}\} \tag{6.152}$$

$$\rho^{\min}_{p,k,d} \leq \rho_{p,k,d} \leq \rho^{\max}_{p,k,d}; \quad \forall \{p \in \Psi^{\mathrm{P}}, k \in \Psi^{\mathrm{K}}, d \in \Psi^{\mathrm{D}}\} \tag{6.153}$$

$$\xi^{\min}_{1,p,k,g} \leq \xi_{1,p,k,g} \leq \xi^{\max}_{1,p,k,g}; \quad \forall \{p \in \Psi^{\mathrm{P}}, k \in \Psi^{\mathrm{K}}, g \in \Psi^{\mathrm{G}}\} \tag{6.154}$$

$$\xi^{\min}_{2,p,k,g} \leq \xi_{2,p,k,g} \leq \xi^{\max}_{2,p,k,g}; \quad \forall \{p \in \Psi^{\mathrm{P}}, k \in \Psi^{\mathrm{K}}, g \in \Psi^{\mathrm{G}}\} \tag{6.155}$$

$$\xi_{1,p,k,d}^{\min} \leq \xi_{1,p,k,d} \leq \xi_{1,p,k,d}^{\max}; \quad \forall\{p \in \Psi^{\mathrm{P}}, k \in \Psi^{\mathrm{K}}, d \in \Psi^{\mathrm{D}}\} \tag{6.156}$$

$$\xi_{2,p,k,d}^{\min} \leq \xi_{2,p,k,d} \leq \xi_{2,p,k,d}^{\max}; \quad \forall\{p \in \Psi^{\mathrm{P}}, k \in \Psi^{\mathrm{K}}, d \in \Psi^{\mathrm{D}}\} \tag{6.157}$$

$$\underset{x_{\text{CSC-EM}}}{\text{Max}} : CWF_{p,k} = \left(\sum_{g \in \Psi^{\mathrm{G}}} \tilde{\upsilon}_{p,k,g}\left(\rho_{p,k,g}\right) + \sum_{d \in \Psi^{\mathrm{D}}} \tilde{\upsilon}_{p,k,d}\left(\rho_{p,k,d}\right) \right); \quad \forall\{p \in \Psi^{\mathrm{P}}, k \in \Psi^{\mathrm{K}}, g \in \Psi^{\mathrm{G}}, d \in \Psi^{\mathrm{D}}\}$$

(6.158) Second level

$$= \underset{x_{\text{CSC-EM}}}{\text{Max}} : \left(\sum_{g \in \Psi^{\mathrm{G}}} \frac{\left(2\xi_{1,p,k,g} - 1\right)}{2} \cdot \alpha_{p,k,g} \cdot \rho_{p,k,g}^{2} + \left(\xi_{2,p,k,g} - 1\right) \cdot \beta_{p,k,g} \cdot \rho_{p,k,g} - \gamma_{p,k,g} + \sum_{d \in \Psi^{\mathrm{D}}} \frac{\left(1 - 2\xi_{1,p,k,d}\right)}{2} \cdot \alpha_{p,k,d} \cdot \rho_{p,k,d}^{2} + \left(1 - \xi_{2,p,k,d}\right) \cdot \beta_{p,k,d} \cdot \rho_{p,k,d} + \gamma_{p,k,d} \right); \quad \forall\{p \in \Psi^{\mathrm{P}}, k \in \Psi^{\mathrm{K}}, g \in \Psi^{\mathrm{G}}, d \in \Psi^{\mathrm{D}}\}$$

and subject to Eqs. (6.159) through (6.162):

$$\sum_{g \in \Psi^{\mathrm{G}}} \rho_{p,k,g} = \sum_{l \in \Psi^{\mathrm{L}}} \kappa_{p,k,l} + \sum_{d \in \Psi^{\mathrm{D}}} \rho_{p,k,d}; \quad \forall\{p \in \Psi^{\mathrm{P}}, k \in \Psi^{\mathrm{K}}, g \in \Psi^{\mathrm{G}}, l \in \Psi^{\mathrm{L}}, d \in \Psi^{\mathrm{D}}\} \tag{6.159}$$

$$\rho_{p,k,g}^{\min} \leq \rho_{p,k,g} \leq \rho_{p,k,g}^{\max}; \quad \forall\{p \in \Psi^{\mathrm{P}}, k \in \Psi^{\mathrm{K}}, g \in \Psi^{\mathrm{G}}\} \tag{6.160}$$

$$\rho_{p,k,d}^{\min} \leq \rho_{p,k,d} \leq \rho_{p,k,d}^{\max}; \quad \forall\{p \in \Psi^{\mathrm{P}}, k \in \Psi^{\mathrm{K}}, d \in \Psi^{\mathrm{D}}\} \tag{6.161}$$

$$f_{p,k,l}^{\min} \leq f_{p,k,l} \leq f_{p,k,l}^{\max}; \quad \forall\{p \in \Psi^{\mathrm{P}}, k \in \Psi^{\mathrm{K}}, l \in \Psi^{\mathrm{L}}\} \tag{6.162}$$

6.4.3 *Mathematical Model of the Risk-Driven Strategic Tri-level Computational-Logical Framework*

In order to deal with risks of PD-TEP problem arising from the severe twofold uncertainty parameters of market price and demand, IGDT is employed. To see the basic characteristics of these uncertainty parameters in the proposed strategic tri-level computational-logical framework in a systematic manner, please refer to the uncertainty matrix in Table 6.3. These uncertainties directly influence the objectives and the constraints of all levels of the strategic tri-level computational-logical framework. Here, the IGDT risk-averse decision-making and risk-taker decision-making policies of the strategic tri-level computational-logical framework are developed within a stochastic environment. In the following sections, the authors will present an in-depth discussion about these decision-making policies.

6.4.3.1 The IGDT Severe Twofold Uncertainty Model

In the proposed PD-TEP problem, the envelope-bound IGDT model is developed to handle the uncertainty parameters. The envelope-bound IGDT model of uncertain market price and demand is fully described in Sect. 6.3.3.1.

6.4.3.2 The IGDT Risk-Averse Decision-Making Policy: Robustness Function

In the IGDT-based strategic tri-level computational-logical framework, the robustness function addresses the pernicious face of the severe twofold uncertainties in the PD-TEP problem. In other words, the robustness function expresses the greatest level of uncertainty parameters, so that the minimum value of the PD-TEP objectives (PD-TEPO) cannot be less than a predetermined critical cost. The robustness function is, therefore, the degree of resistance against uncertainty parameters and immunity against smaller values of the PD-TEPO at which defeat cannot arise. This means that a large value of the robustness function is desirable in PD-TEP. In the IGDT risk-averse decision-making policy, by using the predetermined critical cost, robust optimal transmission expansion plans are determined by the planner. The robustness function for the strategic tri-level computational-logical framework can be expressed using Eq. (6.163):

$$\begin{aligned}
&\Upsilon^{\text{PD-TEP}}\left(x_{\text{PD-TEP}}, \varpi_c^{\text{PD-TEP}}\right)\\
&= \underset{\substack{\Delta^{\text{price}}\\ \Delta^{\text{demand}}}}{\text{Max}} \left\{ \left(\Delta^{\text{price}}, \Delta^{\text{demand}}\right) : \underset{\substack{x_{\text{PD-TEP}}\\ a\lambda_{p,k,b} \in \Pi_{p,k,b}^{\text{price}}\left(\Delta^{\text{price}}, p\lambda_{p,k,b}\right)\\ a\rho_{p,k,b} \in \Pi_{p,k,b}^{\text{demand}}\left(\Delta^{\text{demand}}, p\rho_{p,k,b}\right)}}{\text{Min}} \text{PD-TEPO}\left(x_{\text{PD-TEP}}, a\lambda_{p,k,b}, a\rho_{p,k,b}\right) \geq \varpi_{\text{c}}^{\text{PD-TEP}} \right\};\\
&\forall\left\{\Delta^{\text{demand}} > 0, \Delta^{\text{price}} > 0, p \in \Psi^{\text{P}}, k \in \Psi^{\text{K}}, b \in \Psi^{\text{B}}\right\}
\end{aligned} \tag{6.163}$$

In the proposed policy, PD-TEPO, as a performance function, cannot exceed a predetermined critical value. In this condition, the robustness function addresses the greatest level of the uncertainty parameters, such that the maximum value of PD-TEPO cannot be greater than a predetermined critical cost. In this case, the expressed robustness function of Eq. (6.164) is modified, as follows:

$$\Upsilon^{\text{PD-TEP}}\left(x_{\text{PD-TEP}}, \varpi_{\text{c}}^{\text{PD-TEP}}\right) = \underset{\substack{\Delta^{\text{price}} \\ \Delta^{\text{demand}}}}{\text{Max}} \left\{ \left(\Delta^{\text{price}}, \Delta^{\text{demand}}\right) : \underset{\substack{x_{\text{PD-TEP}} \\ a\lambda_{p,k,b} \in \Pi_{p,k,b}^{\text{price}}\left(\Delta^{\text{price}}, p\lambda_{p,k,b}\right) \\ a\rho_{p,k,b} \in \Pi_{p,k,b}^{\text{demand}}\left(\Delta^{\text{demand}}, p\rho_{p,k,b}\right)}}{\text{Max}} \text{PD-TEPO}\left(x_{\text{PD-TEP}}, a\lambda_{p,k,b}, a\rho_{p,k,b}\right) \le \varpi_{\text{c}}^{\text{PD-TEP}} \right\};$$
$$\forall\left\{\Delta^{\text{demand}} > 0, \Delta^{\text{price}} > 0, p \in \Psi^{\text{P}}, k \in \Psi^{\text{K}}, b \in \Psi^{\text{B}}\right\} \tag{6.164}$$

Therefore, the IGDT risk-averse decision-making policy for the strategic tri-level computational-logical framework can be formulated from Eq. (6.165):

$$\Upsilon^{\text{PD-TEP}}\left(x_{\text{PD-TEP}}, \varpi_{\text{c}}^{\text{PD-TEP}}\right) = \underset{\substack{\Delta^{\text{price}} \\ \Delta^{\text{demand}}}}{\text{Max}} \left\{ \left(\Delta^{\text{price}}, \Delta^{\text{demand}}\right) : \underset{\substack{x_{\text{PD-TEP}} \\ a\lambda_{p,k,b} \in \Pi_{p,k,b}^{\text{price}}\left(\Delta^{\text{price}}, p\lambda_{p,k,b}\right) \\ a\rho_{p,k,b} \in \Pi_{p,k,b}^{\text{demand}}\left(\Delta^{\text{demand}}, p\rho_{p,k,b}\right)}}{\text{Max}} \text{PD-TEPO}\left(x_{\text{PD-TEP}}, a\lambda_{p,k,b}, a\rho_{p,k,b}\right) \le \varpi_{c}^{\text{PD-TEP}} \right\};$$
$$\forall\left\{\Delta^{\text{demand}} > 0, \Delta^{\text{price}} > 0, p \in \Psi^{\text{P}}, k \in \Psi^{\text{K}}, b \in \Psi^{\text{B}}\right\} \tag{6.165) IGDT level (}$$

and subject to Eqs. (6.166) through (6.171):

$$\underset{x_{\text{PD-TEP}}}{\text{Max}} \left\{\text{Eq.}(6.132)\right\} \le \varpi_{\text{c}}^{\text{PD-TEP}};$$
$$\forall\left\{a\lambda_{p,k,b} = \left(1 + \Delta^{\text{price}}\right) \cdot p\lambda_{p,k,b}, a\rho_{p,k,b} = \left(1 + \Delta^{\text{demand}}\right)\right.$$
$$\left. \cdot p\rho_{p,k,b}, p \in \Psi^{\text{P}}, k \in \Psi^{\text{K}}, b \in \Psi^{\text{B}}\right\} \tag{6.166) Third level (}$$

$$a\lambda_{p,k,b} \le \left(1 + \Delta^{\text{price}}\right) \cdot p\lambda_{p,k,b}; \quad \forall\left\{p \in \Psi^{\text{P}}, k \in \Psi^{\text{K}}, b \in \Psi^{\text{B}}\right\} \tag{6.167}$$

$$a\lambda_{p,k,b} \ge \left(1 - \Delta^{\text{price}}\right) \cdot p\lambda_{p,k,b}; \quad \forall\left\{p \in \Psi^{\text{P}}, k \in \Psi^{\text{K}}, b \in \Psi^{\text{B}}\right\} \tag{6.168}$$

$$a\rho_{p,k,b} \le \left(1 + \Delta^{\text{demand}}\right) \cdot p\rho_{p,k,b}; \quad \forall\left\{p \in \Psi^{\text{P}}, k \in \Psi^{\text{K}}, b \in \Psi^{\text{B}}\right\} \tag{6.169}$$

$$a\rho_{p,k,b} \ge \left(1 - \Delta^{\text{demand}}\right) \cdot p\rho_{p,k,b}; \quad \forall\left\{p \in \Psi^{\text{P}}, k \in \Psi^{\text{K}}, b \in \Psi^{\text{B}}\right\} \tag{6.170}$$

$$\left.\left\{\begin{array}{l}\text{Eqs.}(6.141)\text{ through }(6.143)\text{ and} \\ \text{Eqs.}(6.146)\text{ through }(6.148)\end{array}\right\}\right|_{a\lambda_{p,k,b}, a\rho_{p,k,b}};$$
$$\forall\left\{a\lambda_{p,k,b} = \left(1+\Delta^{\text{price}}\right) \cdot p\lambda_{p,k,b}, a\rho_{p,k,b} = \left(1+\Delta^{\text{demand}}\right) \cdot p\rho_{p,k,b}, p \in \Psi^{\text{P}}, k \in \Psi^{\text{K}}, b \in \Psi^{\text{B}}\right\} \tag{6.171}$$

$$\underset{x_{\text{BBM}}}{\text{Max}}\,\{\text{Eq.}(6.149)\}\Big|_{a\lambda_{p,k,b},a\rho_{p,k,b}};\quad \forall\left\{a\lambda_{p,k,b}=\left(1+\Delta^{\text{price}}\right)\cdot p\lambda_{p,k,b},\right.$$
$$\left.\text{and,}\quad a\rho_{p,k,b}=\left(1+\Delta^{\text{demand}}\right)\cdot p\rho_{p,k,b},g\in\Psi^{\text{G}},p\in\Psi^{\text{P}},k\in\Psi^{\text{K}},b\in\Psi^{\text{B}}\right\}$$
$$\underset{x_{\text{BBM}}}{\text{Max}}\,\{\text{Eq.}(6.150)\}\Big|_{a\lambda_{p,k,b},a\rho_{p,k,b}};\quad \forall\left\{a\lambda_{p,k,b}=\left(1+\Delta^{\text{price}}\right)\cdot p\lambda_{p,k,b},\right.$$
$$\left.\text{and,}\quad a\rho_{p,k,b}=\left(1+\Delta^{\text{demand}}\right)\cdot p\rho_{p,k,b},d\in\Psi^{\text{D}},p\in\Psi^{\text{P}},k\in\Psi^{\text{K}},b\in\Psi^{\text{B}}\right\}$$

(6.172) First level

and subject to Eq. (6.173):

$$\{\text{Eqs.}(6.151)\text{ through }(6.157)\}|_{a\lambda_{p,k,b},a\rho_{p,k,b}};$$
$$\forall\left\{a\lambda_{p,k,b}=\left(1+\Delta^{\text{price}}\right)\cdot p\lambda_{p,k,b},a\rho_{p,k,b}=\left(1+\Delta^{\text{demand}}\right)\cdot p\rho_{p,k,b},p\in\Psi^{\text{P}},k\in\Psi^{\text{K}},b\in\Psi^{\text{B}}\right\} \tag{6.173}$$

$$\underset{x_{\text{CSCEM}}}{\text{Max}}\,\{\text{Eq.}(6.158)\}\Big|_{a\lambda_{p,k,b},a\rho_{p,k,b}};\quad \forall\left\{a\lambda_{p,k,b}=\left(1+\Delta^{\text{price}}\right)\cdot p\lambda_{p,k,b},\right.$$
$$\left.\text{and,}\quad a\rho_{p,k,b}=\left(1+\Delta^{\text{demand}}\right)\cdot p\rho_{p,k,b},g\in\Psi^{\text{G}},d\in\Psi^{\text{D}},p\in\Psi^{\text{P}},k\in\Psi^{\text{K}},b\in\Psi^{\text{B}}\right\}$$

(6.174) Second level

and subject to Eq. (6.175):

$$\{\text{Eqs.}(6.159)\text{ through }(6.162)\}|_{a\lambda_{p,k,b},a\rho_{p,k,b}};$$
$$\forall\left\{a\lambda_{p,k,b}=\left(1+\Delta^{\text{price}}\right)\cdot p\lambda_{p,k,b},a\rho_{p,k,b}=\left(1+\Delta^{\text{demand}}\right)\cdot p\rho_{p,k,b},p\in\Psi^{\text{P}},k\in\Psi^{\text{K}},b\in\Psi^{\text{B}}\right\} \tag{6.175}$$

In the proposed framework under the IGDT risk-averse decision-making policy, the maximum values of PD-TEPO are achieved for the highest level of the uncertainties in the robust region (see Figs. 6.4 and 6.5). The solution of the IGDT risk-averse decision-making policy gives the optimal robust transmission expansion plans of the planner, based on the defined value of its critical cost. The predetermined critical cost is also determined using Eq. (6.176):

$$\varpi_{\text{c}}^{\text{PD-TEP}}=\left(1+\sigma_{\text{c}}^{\text{PD-TEP}}\right)\cdot\varpi_{\text{b}}^{\text{PD-TEP}} \tag{6.176}$$

From Eq. (6.176), the IGDT base cost, $\varpi_{\text{b}}^{\text{PD-TEP}}$, can be calculated by solving the deterministic strategic tri-level computational-logical framework as presented in

Eqs. (6.132) through (6.162). The deterministic framework of the strategic tri-level computational-logical framework is called a risk-neutral decision-making policy. In the PD-TEP problem, the determined critical cost by the planner is also larger than its base cost.

6.4.3.3 The IGDT Risk-Taker Decision-Making Policy: Opportunity Function

In the IGDT-based strategic tri-level computational-logical framework, the opportunity function addresses the propitious face of the severe twofold uncertainties in the PD-TEP problem. In simple terms, this function represents the smallest level of uncertainty parameters so that the maximum value of PD-TEPO can be potentially as large as a predetermined target cost. The opportunity function is, therefore, the immunity against a windfall reward, where sweeping success can occur. This means that a small value of the opportunity function is desirable in PD-TEP. In the IGDT risk-taker decision-making policy, by using the predetermined target cost, opportunistic optimal transmission expansion plans are determined by the planner. The opportunity function for the strategic tri-level computational-logical framework can be defined using Eq. (6.177):

$$
\begin{aligned}
&\Gamma^{\text{PD-TEP}}\left(x_{\text{PD-TEP}}, \varpi_{\text{t}}^{\text{PD-TEP}}\right) \\
&= \underset{\substack{\Delta^{\text{price}} \\ \Delta^{\text{demand}}}}{\text{Min}} \left\{ \left(\Delta^{\text{price}}, \Delta^{\text{demand}}\right) : \underset{\substack{x_{\text{PD-TEP}} \\ a\lambda_{p,k,b} \in \Pi_{p,k,b}^{\text{price}}\left(\Delta^{\text{price}}, p\lambda_{p,k,b}\right) \\ a\rho_{p,k,b} \in \Pi_{p,k,b}^{\text{demand}}\left(\Delta^{\text{demand}}, p\rho_{p,k,b}\right)}}{\text{Max}} \text{PD-TEPO}\left(x_{\text{PD-TEP}}, a\lambda_{p,k,b}, a\rho_{p,k,b}\right) \geq \varpi_{\text{t}}^{\text{PD-TEP}} \right\}; \\
&\forall \left\{\Delta^{\text{demand}} > 0, \Delta^{\text{price}} > 0, p \in \Psi^{\text{P}}, k \in \Psi^{\text{K}}, b \in \Psi^{\text{B}}\right\}
\end{aligned} \tag{6.177}
$$

Likewise, in the proposed policy, PD-TEPO, as a performance function, must not exceed a predetermined target cost. In this case, the opportunity function illustrates the smallest level of the uncertainty parameters, such that the minimum value of PD-TEPO can possibly be as small as a predetermined target cost. In this case, the expressed opportunity function in Eq. (6.177) is modified, yielding Eq. (6.178):

$$
\begin{aligned}
&\Gamma^{\text{PD-TEP}}\left(x_{\text{PD-TEP}}, \varpi_{\text{t}}^{\text{PD-TEP}}\right) \\
&= \underset{\substack{\Delta^{\text{price}} \\ \Delta^{\text{demand}}}}{\text{Min}} \left\{ \left(\Delta^{\text{price}}, \Delta^{\text{demand}}\right) : \underset{\substack{x_{\text{PD-TEP}} \\ a\lambda_{p,k,b} \in \Pi_{p,k,b}^{\text{price}}\left(\Delta^{\text{price}}, p\lambda_{p,k,b}\right) \\ a\rho_{p,k,b} \in \Pi_{p,k,b}^{\text{demand}}\left(\Delta^{\text{demand}}, p\rho_{p,k,b}\right)}}{\text{Min}} \text{PD-TEPO}\left(x_{\text{PD-TEP}}, a\lambda_{p,k,b}, a\rho_{p,k,b}\right) \leq \varpi_{\text{t}}^{\text{PD-TEP}} \right\}; \\
&\forall \left\{\Delta^{\text{demand}} > 0, \Delta^{\text{price}} > 0, p \in \Psi^{\text{P}}, k \in \Psi^{\text{K}}, b \in \Psi^{\text{B}}\right\}
\end{aligned} \tag{6.178}
$$

Therefore, the IGDT risk-taker decision-making policy for the strategic tri-level computational-logical framework can be formulated as indicated by Eq. (6.179):

$$\Gamma^{\text{PD-TEP}}\left(x_{\text{PD-TEP}}, \varpi_{\text{t}}^{\text{PD-TEP}}\right) = \underset{\substack{\Delta^{\text{price}} \\ \Delta^{\text{demand}}}}{\text{Min}} \left\{ \left(\Delta^{\text{price}}, \Delta^{\text{demand}}\right) : \underset{\substack{x_{\text{PD-TEP}} \\ a\lambda_{p,k,b} \in \Pi_{p,k,b}^{\text{price}}\left(\Delta^{\text{price}}, p\lambda_{p,k,b}\right) \\ a\rho_{p,k,b} \in \Pi_{p,k,b}^{\text{demand}}\left(\Delta^{\text{demand}}, p\rho_{p,k,b}\right)}}{\text{Min}} \text{PD-TEPO}\left(x_{\text{PD-TEP}}, a\lambda_{p,k,b}, a\rho_{p,k,b}\right) \le \varpi_{\text{t}}^{\text{PD-TEP}} \right\};$$
$$\forall\left\{\Delta^{\text{demand}} > 0, \Delta^{\text{price}} > 0, p \in \Psi^{\text{P}}, k \in \Psi^{\text{K}}, b \in \Psi^{\text{B}}\right\} \quad (6.179)$$

and subject to Eqs. (6.180) through (6.185):

$$\underset{x_{\text{PD-TEP}}}{\text{Min}} \left\{\text{Eq.}(6.132)\right\} \le \varpi_{\text{t}}^{\text{PD-TEP}};$$
$$\forall\left\{a\lambda_{p,k,b} = \left(1 - \Delta^{\text{price}}\right) \cdot p\lambda_{p,k,b}, a\rho_{p,k,b} = \left(1 - \Delta^{\text{demand}}\right) \right.$$
$$\left. \cdot p\rho_{p,k,b}, p \in \Psi^{\text{P}}, k \in \Psi^{\text{K}}, b \in \Psi^{\text{B}}\right\} \quad (6.180) \text{ Third level}$$

$$a\lambda_{p,k,b} \le \left(1 + \Delta^{\text{price}}\right) \cdot p\lambda_{p,k,b}; \quad \forall\left\{p \in \Psi^{\text{P}}, k \in \Psi^{\text{K}}, b \in \Psi^{\text{B}}\right\} \quad (6.181)$$

$$a\lambda_{p,k,b} \ge \left(1 - \Delta^{\text{price}}\right) \cdot p\lambda_{p,k,b}; \quad \forall\left\{p \in \Psi^{\text{P}}, k \in \Psi^{\text{K}}, b \in \Psi^{\text{B}}\right\} \quad (6.182)$$

$$a\rho_{p,k,b} \le \left(1 + \Delta^{\text{demand}}\right) \cdot p\rho_{p,k,b}; \quad \forall\left\{p \in \Psi^{\text{P}}, k \in \Psi^{\text{K}}, b \in \Psi^{\text{B}}\right\} \quad (6.183)$$

$$a\rho_{p,k,b} \ge \left(1 - \Delta^{\text{demand}}\right) \cdot p\rho_{p,k,b}; \quad \forall\left\{p \in \Psi^{\text{P}}, k \in \Psi^{\text{K}}, b \in \Psi^{\text{B}}\right\} \quad (6.184)$$

$$\left.\left\{\begin{array}{l} \text{Eqs.}(6.141) \text{ through } (6.143) \text{ and} \\ \text{Eqs.}(6.146) \text{ through } (6.148) \end{array}\right\}\right|_{a\lambda_{p,k,b}, a\rho_{p,k,b}};$$
$$\forall\left\{a\lambda_{p,k,b} = \left(1 - \Delta^{\text{price}}\right) \cdot p\lambda_{p,k,b}, a\rho_{p,k,b} = \left(1 - \Delta^{\text{demand}}\right) \cdot p\rho_{p,k,b}, p \in \Psi^{\text{P}}, k \in \Psi^{\text{K}}, b \in \Psi^{\text{B}}\right\}$$
$$(6.185)$$

$$\underset{x_{\text{BBM}}}{\text{Max}} \left.\left\{\text{Eq.}(6.149)\right\}\right|_{a\lambda_{p,k,b}, a\rho_{p,k,b}}; \quad \forall\left\{a\lambda_{p,k,b} = \left(1 - \Delta^{\text{price}}\right) \cdot p\lambda_{p,k,b}, \right.$$
$$\left. \text{and, } a\rho_{p,k,b} = \left(1 - \Delta^{\text{demand}}\right) \cdot p\rho_{p,k,b}, g \in \Psi^{\text{G}}, p \in \Psi^{\text{P}}, k \in \Psi^{\text{K}}, b \in \Psi^{\text{B}}\right\}$$
$$\underset{x_{\text{BBM}}}{\text{Max}} \left.\left\{\text{Eq.}(6.150)\right\}\right|_{a\lambda_{p,k,b}, a\rho_{p,k,b}}; \quad \forall\left\{a\lambda_{p,k,b} = \left(1 - \Delta^{\text{price}}\right) \cdot p\lambda_{p,k,b}, \right.$$
$$\left. \text{and, } a\rho_{p,k,b} = \left(1 - \Delta^{\text{demand}}\right) \cdot p\rho_{p,k,b}, d \in \Psi^{\text{D}}, p \in \Psi^{\text{P}}, k \in \Psi^{\text{K}}, b \in \Psi^{\text{B}}\right\}$$
$$(6.186) \text{ First level}$$

and subject to Eq. (6.187):

$$\{\text{Eqs.}(6.151)\,\text{through}\,(6.157)\}|_{a\lambda_{p,k,b},a\rho_{p,k,b}};$$
$$\forall\{a\lambda_{p,k,b}=\left(1-\Delta^{\text{price}}\right)\cdot p\lambda_{p,k,b},a\rho_{p,k,b}=\left(1-\Delta^{\text{demand}}\right)\cdot p\rho_{p,k,b},p\in\Psi^{\text{P}},k\in\Psi^{\text{K}},b\in\Psi^{\text{B}}\}$$

(6.187) Second level

$$\underset{x_{\text{CSCEM}}}{\text{Max}}\ \{\text{Eq.}(\boldsymbol{6.158})\}\Big|_{a\lambda_{p,k,b},a\rho_{p,k,b}};\quad \forall\{a\lambda_{p,k,b}=\left(1-\Delta^{\text{price}}\right)\cdot p\lambda_{p,k,b},$$
$$\text{and,}\ a\rho_{p,k,b}=\left(1-\Delta^{\text{demand}}\right)\cdot p\rho_{p,k,b},g\in\Psi^{\text{G}},d\in\Psi^{\text{D}},p\in\Psi^{\text{P}},k\in\Psi^{\text{K}},b\in\Psi^{\text{B}}\}\tag{6.188}$$

and subject to Eq. (6.189):

$$\{\text{Eqs.}(6.159)\,\text{through}\,(6.162)\}|_{a\lambda_{p,k,b},a\rho_{p,k,b}};$$
$$\forall\{a\lambda_{p,k,b}=\left(1-\Delta^{\text{price}}\right)\cdot p\lambda_{p,k,b},a\rho_{p,k,b}=\left(1-\Delta^{\text{demand}}\right)\cdot p\rho_{p,k,b},p\in\Psi^{\text{P}},k\in\Psi^{\text{K}},b\in\Psi^{\text{B}}\}\tag{6.189}$$

In the proposed framework under the IGDT risk-taker decision-making policy, the minimum values of PD-TEPO are achieved for the lowest level of the uncertainties in the robust region (see Figs. 6.4 and 6.5). The solution of the IGDT risk-taker decision-making policy gives the optimal opportunistic transmission expansion plans of the planner, based on the defined value of its target cost. The predetermined target cost can also be determined using Eq. (6.190):

$$\varpi_{\text{t}}^{\text{PD-TEP}}=\left(1-\sigma_{\text{t}}^{\text{PD-TEP}}\right)\cdot\varpi_{\text{b}}^{\text{PD-TEP}}\tag{6.190}$$

Similarly, in Eq. (6.190), the IGDT base cost, $\varpi_{\text{b}}^{\text{PD-TEP}}$, can be calculated by solving the deterministic strategic tri-level computational-logical framework, as presented in Eqs. (6.132) through (6.162). In the PD-TEP problem, the determined target cost by the planner is also smaller than its base cost.

6.4.4 Solution Method and Implementation Considerations

In the proposed strategic tri-level computational-logical framework, the decision-making variables of the solution vector in the short-term operational slave problem are (1) the generated power by GENCOs; (2) the slope and intercept parameters of the bidding strategy function of the GENCOs; (3) the purchased power by DISCOs; and, (4) the slope and intercept parameters of the bidding strategy function of the DISCOs. At the same time, the decision-making variables of the solution vector in the long-term planning master problem are (1) the time of the newly installed transmission lines over the planning horizon; (2) the location of the newly installed transmission lines; and, (3) the capacity of the newly installed transmission lines. The strategic tri-level computational-logical framework starts from the first level. At this level (the BBM), the market participants—including GENCOs and DISCOs—prepare their bidding strategy functions according to the SFE model. At this level, each participant individually solves the bidding strategy problem to calculate and modify its bidding strategy parameters [see Eqs. (6.149) through (6.157)] using an optimization algorithm while keeping the bidding strategy parameters of its rivals. This process will be continued iteratively until no market participant changes its own bidding strategy parameters (i.e., the Nash equilibrium point is achieved). After obtaining the Nash equilibrium point in the first level, the optimal bidding strategy functions of both GENCOs and DISCOs are transferred to the second level (i.e., the CSC electricity market problem). At this level, by aggregating these optimal bidding strategy functions by the ISO, equivalent supply and equivalent demand functions are created. If the obtained optimal bidding strategies by both GENCOs and DISCOs cause a violation in the network constraints, the ISO solves the CSC electricity market-clearing problem [see Eqs. (6.158) through (6.162)] using an optimization algorithm with the aim of maximizing CWF. The first and second levels are repeated for all considered patterns in period p of the planning horizon. After completing the short-term operational slave problem for period p, the information related to the GENCOs, DISCOs, CSC electricity market outcomes, etc. is transferred to the third level (i.e., long-term planning master problem). At the this level, the planner solves PD-TEP with the aim of minimizing the TIC, TCC, and ECOC, subject to the short-term and the long-term constraints in order to determine the newly installed transmission lines [see Eqs. (6.132) through (6.148)] using an optimization algorithm. Finally, the optimal transmission expansion plans are determined for period p and are considered as future operational equipment in the next period of planning. The flowchart of the strategic tri-level computational-logical framework under the IGDT risk-neutral, IGDT risk-averse, and IGDT risk-taker decision-making policies is depicted in Fig. 6.11. To implement the proposed framework under one of these decision-making policies, a block relevant to this decision-making policy is active, while other blocks are inactive in the flowchart.

6.4.5 Simulation Results and Case Studies

In this section, two test systems, including a modified IEEE 30-bus test system and a modified large-scale Iranian 400 kV transmission network, are taken into consideration to evaluate the performance of the proposed strategic tri-level computational-logical framework.

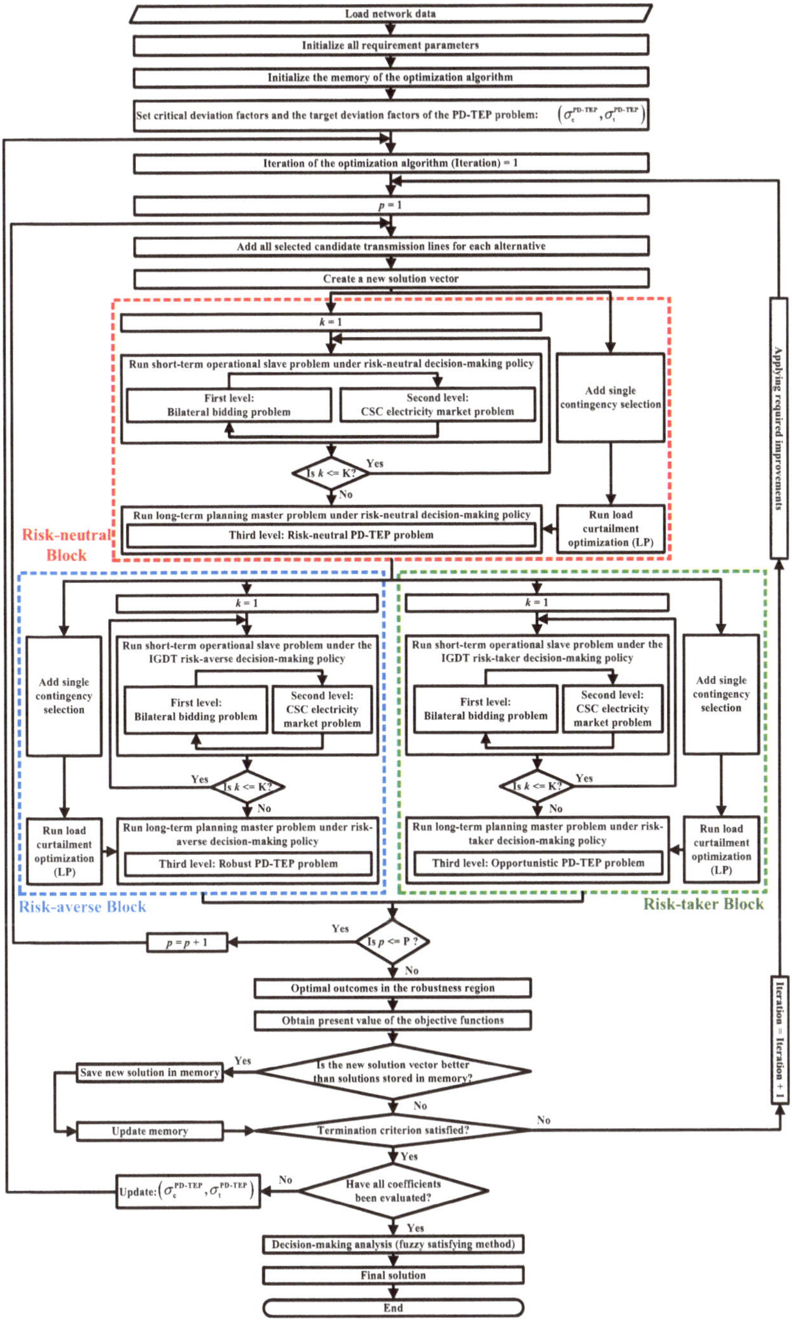

Fig. 6.11 Flowchart of the strategic tri-level computational-logical framework under the IGDT risk-neutral, IGDT risk-averse, and IGDT risk-taker decision-making policies

6.4.5.1 The Modified IEEE 30-Bus Test System

Figure 6.12 shows how the strategic tri-level computational-logical framework was implemented and tested on a modified IEEE 30-bus test system [55, 56]. This modified test system consists of 41 existing transmission lines, 30 buses, 12 GENCOs, and 19 DISCOs. The base apparent power and voltage for the entire modified IEEE 30-bus test system is equal to 100 MVA and 135 kV, respectively. Network data for this modified test network are tabulated in Appendix 3. Tables 6.109 and 6.110 show the data for the existing and candidate transmission lines associated with the modified IEEE 30-bus test system, respectively. Table 6.111 presents the parameters for the GENCOs. Table 6.112 presents the parameters for demand and DISCOs. In order to avoid market power, a price ceiling is specified and employed for the market participants, including GENCOs and DISCOs. In this condition, each

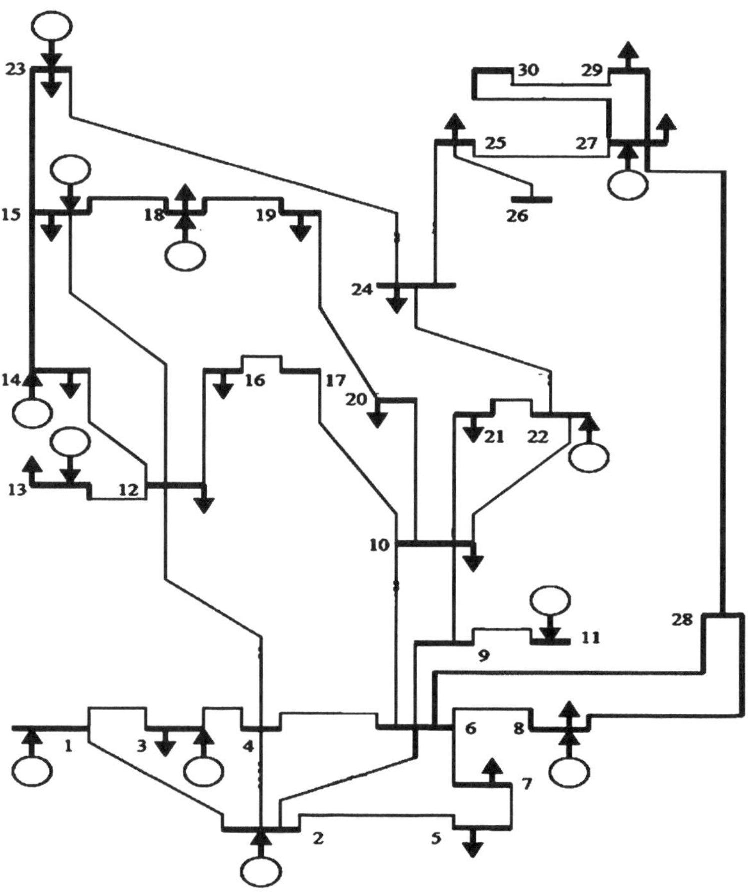

Fig. 6.12 The modified IEEE 30-bus test system

market participant can change its bidding strategy parameters between predetermined minimum and maximum values and, therefore, cannot grab exorbitant profit through high bidding strategy parameters.

In this regard, Table 6.113 presents the data of the bidding strategy parameters of both GENCOs and DISCOs. Parameters of weighting coefficients for the objective functions related to the PD-TEP problem are tabulated in Table 6.114. Each load in the modified IEEE 30-bus test system is split into three sectors, $S = 3$, by the existing customer's classes: (1) residential, (2) commercial, and, (3) industrial. The specified values for the residential, commercial, and industrial sectors of each load on each bus of the network are 60%, 20%, and 20% of the maximum load at the relevant bus, respectively. The SCDF data for different sectors of each load on each bus of the modified IEEE 30-bus test system are tabulated in Table 6.115. Tables 6.109 and 6.110 give the failure and repair rates for transmission lines. These values for GENCOs are also presented in Table 6.111. The maximum allowable amount of curtailed load at load sector s of bus b for period p is considered to be 30% of the maximum load at the relevant sector. The maximum allowable amount of curtailed load at bus b for period p is also considered to be 30% of the maximum load at the relevant bus. In addition, the maximum allowable amount of curtailed load at period p is assumed to be 30% of the maximum load for the relevant period. Outage cost-weight coefficients are considered similar to all sectors in each load of each bus.

The planner expands the transmission network for a 10-year planning horizon, $P = 10$, divided into 1-year periods. Solving the proposed strategic tri-level computational-logical framework for all hours of a day and for all days of a year is excessively difficult, complicated, and time consuming. Hence, similar to the PD-GEP problem presented in Sect. 6.3 of this chapter, by employing the daily and seasonal time patterns, the high computational burden of the PD-TEP problem is reduced to some extent. So, the 8760-h time span of the network LDC is modeled using 22 steps, $K = 22$, according to Eq. (6.105). The load step in pattern k is also regarded as the average load value during this pattern. The annual interest rate is set at 10%. The annual demand growth is considered to be 5%. The annual price growth is set to be 4%.

Candidate transmission lines can be installed in nine existing and nine new corridors of the network. However, a maximum number of installed transmission lines that can be installed in each corridor over the planning horizon is considered to be four transmission lines including existing ones. The basic construction cost for the transmission line is assumed to be 1200 \$/MW-km. The physical parameters of the candidate transmission lines are considered similar to existing ones. The maximum allowable value of the TIC index in period p as well as in all periods is set to \$90M and \$750M, respectively. Table 6.116 provides a summary of these assumptions.

Two different cases and two different scenarios are employed in order to investigate and analyze the performance of the strategic tri-level computational-logical framework. The distinction between the first and second cases is in the decision-making policy adopted by the planner to deal with the risks involved in the PD-TEP problem. Moreover, the difference between the first and second scenarios is in adopting or not adopting the strategic behavior by the market participants. The first and second cases and scenarios are as follows:

- First case: The strategic tri-level computational-logical framework is run under the IGDT risk-averse decision-making policy while considering the following two scenarios:
  - The strategic behavior of the market participants (i.e., first level) is ignored.
  - The strategic behavior of the market participants (i.e., first level) is considered.
- Second case: The tri-level computational-logical framework is run under the IGDT risk-taker decision-making policy while considering the following two scenarios:
  - The strategic behavior of the market participants (i.e., first level) is disregarded.
  - The strategic behavior of the market participants (i.e., first level) is regarded.

The strategic tri-level computational-logical framework under the IGDT risk-averse and IGDT risk-taker decision-making policies is implemented and solved by using the proposed multi-objective SOSA addressed in Chap. 4. Table 6.103 gives the parameter adjustments of the newly developed multi-objective SOSA.

In the strategic tri-level computational-logical framework under the IGDT risk-averse and IGDT risk-taker decision-making policies, the planner must have access to the results on the basis of the risk-neutral/deterministic decision-making policy in order to calculate its critical and target costs, $\varpi_{\mathrm{c}}^{\mathrm{PD-TEP}}$ and $\varpi_{\mathrm{t}}^{\mathrm{PD-TEP}}$. First, then, the proposed deterministic strategic tri-level computational-logical framework [see Eqs. (6.132) through (6.162)] is solved based on predicted market price and demand by the proposed single-objective SOSA. Then, the planner uses the optimal results from the deterministic strategic tri-level computational-logical framework to calculate the optimal transmission expansion plans under the IGDT risk-averse and IGDT risk-taker decision-making policies. Since the strategic tri-level computational-logical framework under the IGDT risk-averse and the IGDT risk-taker decision-making policies is examined for both the first and second scenarios, the deterministic strategic tri-level computational-logical framework should also be solved based on these two scenarios. After solving the proposed framework under the risk-neutral/deterministic decision-making policy, the optimal base costs of the planner over the planning horizon under the first scenario (ignoring the BBM) and the second scenario (considering the BBM) are calculated according to Table 6.24. The costs presented in Table 6.24 are mixed costs, consisting of the TIC, TCC, and ECOC [see Eq. (6.132)]. A striking point that must be highlighted in the results in Table 6.24 is the difference between the costs of the first scenario (ignoring the BBM) and second scenario (considering the BBM), which clearly illustrates the effect of adopting strategic behaviors by the market participants on cost reduction. In simple terms,

Table 6.24 Optimal base cost of the planner over the planning horizon under the first and second scenarios

Scenario no.	Base cost ($\varpi_{\mathrm{b}}^{\mathrm{PD-TEP}}$; M$)
First scenario	591.719878
Second scenario	573.968904

when the strategic behaviors are taken into account by the market participants, a lower value is determined for PD-TEPO, compared to situations in which strategic behaviors by the market participants are ignored.

6.4.5.1.1 First Case: Simulation Results and Discussion

The strategic tri-level computational-logical framework under the IGDT risk-averse decision-making policy is solved by optimization of Eqs. (6.165) through (6.176) under different values of the critical cost deviation factor, $\sigma_{\mathrm{c}}^{\mathrm{PD-TEP}} \in (0, 0.8)$, which leads to the different critical costs, $\varpi_{\mathrm{c}}^{\mathrm{PD-TEP}}$. In other words, for each value of the critical cost deviation factor, $\sigma_{\mathrm{c}}^{\mathrm{PD-TEP}} \in (0, 0.8)$, the planner tries to find the maximum values of Δ^{price} and Δ^{demand} by solving Eqs. (6.165) through (6.176) in a multi-objective manner using a Pareto optimality concept. This process gives rise to the formation of an optimal solutions set, called Pareto-optimal solution set instead of a single optimal solution. After calculation of the Pareto-optimal solution set, the planner is only allowed to choose one solution for a specific value of the critical cost deviation factor, $\sigma_{\mathrm{c}}^{\mathrm{PD-TEP}}$. This means that a well-suited compromise among different objectives, namely Δ^{price} and Δ^{demand}, is established in this process. In this way, two main questions are raised: (1) which solution must be chosen by the planner and (2) how to select it. The best solution is chosen based on the planner's requirements and preferences. In the relevant literature, many approaches have been reported on choosing a trade-off solution among the Pareto-optimal solution set (see Chap. 2). Here, though, the authors employ FSM based on the conservative methodology—the min-max formulation—to select the final solution.

The reason for choosing this method is its simplicity and resemblance to human ratiocination. This approach is explicitly addressed in Chap. 2. If the planner chooses a value of zero for its critical cost deviation factor, $\sigma_{\mathrm{c}}^{\mathrm{PD-TEP}} = 0$, the critical cost of the planner is identical to its base cost, which is obtained from the deterministic framework as $\varpi_{\mathrm{c}}^{\mathrm{PD-TEP}} = \varpi_{\mathrm{b}}^{\mathrm{PD-TEP}} = \$591.719878\mathrm{M}$ and $\varpi_{\mathrm{c}}^{\mathrm{PD-TEP}} = \varpi_{\mathrm{b}}^{\mathrm{PD-TEP}} = \$573.968904\mathrm{M}$ for the first and second scenarios, respectively. It should be noted that adopting the risk-neutral decision-making policy by the planner results in a value of zero for its critical cost deviation factor. In this condition, the robustness and risk level of the transmission expansion plans by the planner will be zero. In other words, the planner is not ready to deal with the severe uncertainties of market price and demand; therefore, the cost that the planner needs to invest in transmission

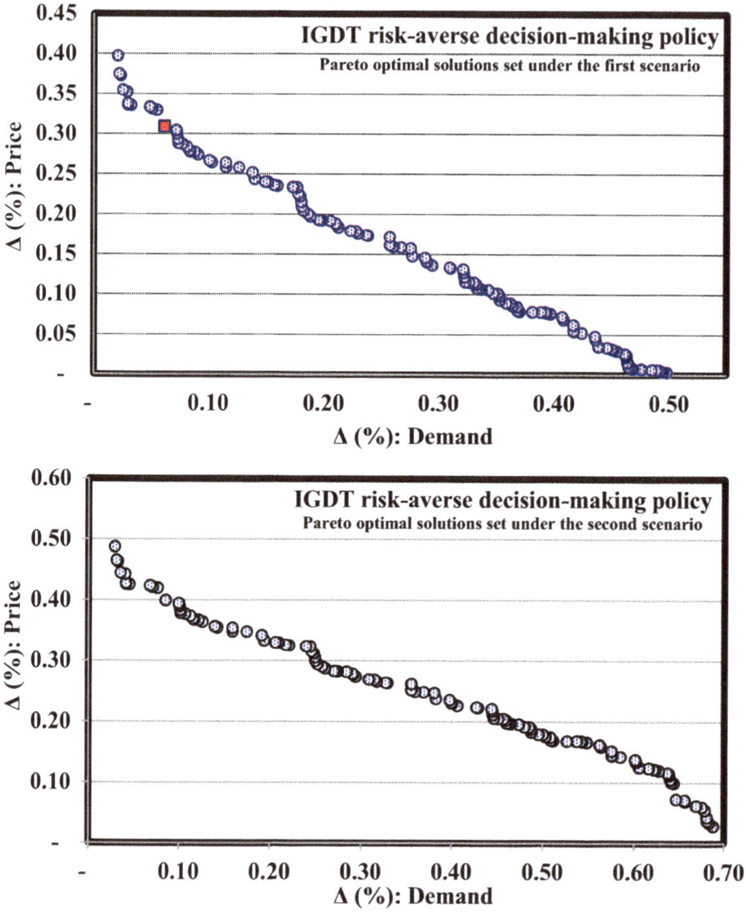

Fig. 6.13 The Pareto-optimal solutions of robustness against market price and demand uncertainties under the first and second scenarios

network expansion, namely PD-TEPO [see Eq. (6.132)], is severely affected by these uncertainties.

Suppose that the planner chooses a critical cost deviation factor of 0.4 (i.e., $\sigma_{c}^{PD-TEP} = 0.4$). In this situation, Eqs. (6.165) through (6.176) are solved in a multi-objective manner for this value of the critical cost deviation factor with the aim of maximizing Δ^{price} and Δ^{demand} according to the first and second scenarios. Figure 6.13 illustrates the Pareto-optimal solution set for this value of the critical cost deviation factor under the first and second scenarios. For illustration, consider the planner's performance under the first scenario (ignoring the BBM). For example, in the solution marked with a square in Fig. 6.13, a critical cost of $\varpi_{c}^{PD-TEP} = (1 + 0.4) \times \varpi_{b}^{PD-TEP} = \$828.407829\text{M}$ is guaranteed for the planner, provided that Δ^{price} and Δ^{demand} do not exceed 30.899748% and 6.180689%, respectively. Put simply, if there is a difference of less than 30.899748% between the actual market price, $a\lambda_{p,\ k,\ b}$, and the predicted market price, $p\lambda_{p,\ k,\ b}$, and also a difference of less than 6.180689% between the actual demand, $a\rho_{p,\ k,\ b}$, and the predicted demand, $p\rho_{p,\ k,\ b}$, the specified cost for transmission network expansion by the planner will not exceed $\varpi_{\mathrm{c}}^{\mathrm{PD-TEP}} = \$828.407829\mathrm{M}$. To clarify, if the actual market price, $a\lambda_{p,\ k,\ b}$, falls within the market price robustness region, $RR_{p,k,b}^{\mathrm{price}}$, and the actual demand, $a\rho_{p,\ k,\ b}$, falls within the demand robustness region, $RR_{p,k,b}^{\mathrm{demand}}$, then the cost specified for transmission network planning by the planner will be at most $\varpi_{\mathrm{c}}^{\mathrm{PD-TEP}} = \$828.407829\mathrm{M}$. Note that if the actual market price and/or actual demand are not within the market price and demand robustness regions, respectively, it is not possible for the planner to achieve a critical cost.

By increasing the critical cost by the planner, $\varpi_{\mathrm{c}}^{\mathrm{PD-TEP}}$, due to enlarging of its critical cost deviation factor, $\sigma_{\mathrm{c}}^{\mathrm{PD-TEP}}$, the robustness parameters of the planner become larger. This means that larger critical costs of the planner in the larger market price robustness region, $RR_{p,k,b}^{\mathrm{price}}$, and larger demand robustness region, $RR_{p,k,b}^{\mathrm{demand}}$, can be guaranteed and vice versa. In other words, better robustness of the calculated transmission expansion plans by the planner results in a higher critical cost value of the planner, $\varpi_{\mathrm{c}}^{\mathrm{PD-TEP}}$, and/or in a larger value of the critical cost deviation factor of the planner, $\sigma_{\mathrm{c}}^{\mathrm{PD-TEP}}$. The rest of the solutions provided in Fig. 6.13 have the same interpretations. For the same reason, the Pareto-optimal solution set related to the other critical cost deviation factors of the planner has the same interpretations. It is relevant to note that these solutions are produced because of the adoption of different critical cost deviation factors by the planner. In addition, this analysis is true for planner performance under the second scenario. After solving the proposed framework under the first and the second scenarios of the first case—the IGDT risk-averse decision-making policy—by the newly proposed multi-objective SOSA, the optimal values for Δ^{price}, Δ^{demand}, and relevant critical cost deviation factors are calculated by the FSM in accordance with Table 6.25. Table 6.26 gives the results associated with the bidding strategy parameters of the GENCOs and DISCOs, which were calculated in the first level (the BBM) of the strategic tri-level computational-logical framework, under the second scenario in the first case—the IGDT risk-averse decision-making policy. Only the results for the first, fifth, and tenth periods of the planning horizon are

Table 6.25 Optimal values for Δ^{price}, Δ^{demand}, and relevant critical cost deviation factors under the first and second scenarios of the first case

Parameter	Optimal values	
	First scenario: Ignoring the BBM	Second scenario: Considering the BBM
Δ^{price} (%)	10.273621	13.242882
Δ^{demand} (%)	33.871452	31.082576
Critical cost deviation factor	0.35	0.35

presented here. These results are also related to the peak load. Since in the first scenario of the first case the strategic behaviors of the market participants are ignored, the market participants—including GENCOs and DISCOs—lack the bidding strategy parameters. That is, in the first scenario of the first case, the GENCOs and DISCOs submit their real marginal generation cost and marginal benefit functions to the ISO, respectively.

For the Nash equilibrium point obtained in the first, fifth, and tenth periods, the bidding strategy parameters of slope and intercept of the market participants are not equal to a constant value of 1 (see Table 6.26). This means that all GENCOs and DISCOs, as market participants, have adopted a strategic behavior in order to compete with their rivals and earn more profit from selling and buying energy.

Tables 6.27 and 6.28 summarize the results of the CSC electricity market (i.e., the second level) for the strategic tri-level computational-logical framework under the first and second scenarios of the IGDT risk-averse decision-making policy (i.e., the first case). For the same reason, these results are only presented for the first, fifth, and tenth periods of the planning horizon. These results are also related to the peak load.

Due to the consideration of transmission line loss in the power flow analysis using the idea of an artificial load at each bus, a tiny difference between the power generated by the GENCOs and the power purchased by the DISCOs can be observed during the periods of the planning horizon (see Table 6.27). The main idea of the artificial load is presented in Sect. 5.3.2 of Chap. 5. Furthermore, demand growth over the planning horizon leads to an increase in the amount of power generated by the GENCOs, power purchased by the DISCOs, and transmission line loss in the network (see Table 6.27). The larger bidding strategy parameters of slope and intercept belong to the GENCO that has the lowest generated power and vice versa (see Tables 6.26 and 6.27). This expression means that the smaller GENCO—in terms of generated power—intends to adopt larger bidding strategy parameters in order to improve its profit in the electricity market and compensate for low profit owing to the shortage of its power generation capacity. It is evident that the larger slope and smaller intercept of the bidding strategy parameters are relevant to the DISCO with less purchased power and vice versa. In other words, the smaller DISCO—in terms of purchased electrical power—intends to adopt a larger slope and smaller intercept associated with bidding strategy parameters to enhance its profit in the electricity market and compensate for low profit, due to the shortage of its power purchasing capacity.

The LMP fluctuations on the different buses of the network over the planning horizon have clearly been increased in the case where the strategic behaviors are taken into account by the market participants—the second scenario compared to the first (see Table 6.28). Table 6.29 shows the robust transmission expansion plans by the planner over the planning horizon under the first and second scenarios of the first case—the IGDT risk-averse decision-making policy. In order to illustrate the results raised in Table 6.29, consider the first period of the planning horizon as an example. In the first scenario of this period, the planner installs two transmission lines in the

Table 6.26 Bidding strategy parameters of the GENCOs under the second scenario of the first case

Period No.	GENCO No.	Bidding strategy parameters of GENCOs		Period No.	GENCO No.	Bidding strategy parameters of GENCOs		Period No.	GENCO No.	Bidding strategy parameters of GENCOs	
		$\xi_{1,p,k,g}, \xi_{1,p,k,d}$	$\xi_{2,p,k,g}, \xi_{2,p,k,d}$			$\xi_{1,p,k,g}, \xi_{1,p,k,d}$	$\xi_{2,p,k,g}, \xi_{2,p,k,d}$			$\xi_{1,p,k,g}, \xi_{1,p,k,d}$	$\xi_{2,p,k,g}, \xi_{2,p,k,d}$
1	GENCO 1	1.321889	1.120173	5	GENCO 1	1.217896	1.122745	10	GENCO 1	1.355000	1.148232
	GENCO 2	1.333502	1.130014		GENCO 2	1.378083	1.252049		GENCO 2	1.212046	1.027092
	GENCO 3	1.519384	1.312809		GENCO 3	1.517800	1.304551		GENCO 3	1.433059	1.260278
	GENCO 4	1.203310	1.019689		GENCO 4	1.207275	1.023049		GENCO 4	1.222709	1.036128
	GENCO 5	1.359027	1.151644		GENCO 5	1.349666	1.228722		GENCO 5	1.383529	1.196912
	GENCO 6	1.309056	1.109299		GENCO 6	1.337912	1.203712		GENCO 6	1.380936	1.170210
	GENCO 7	1.192816	1.010797		GENCO 7	1.324923	1.192201		GENCO 7	1.330160	1.127182
	GENCO 8	1.469401	1.301529		GENCO 8	1.487211	1.302750		GENCO 8	1.343200	1.138232
	GENCO 9	1.410323	1.240283		GENCO 9	1.427418	1.273751		GENCO 9	1.493089	1.322511
	GENCO 10	1.385263	1.209886		GENCO 10	1.375500	1.247305		GENCO 10	1.407595	1.229391
	GENCO 11	1.361579	1.177923		GENCO 11	1.402054	1.255317		GENCO 11	1.454402	1.307899
	GENCO 12	1.431327	1.287148		GENCO 12	1.448677	1.295604		GENCO 12	1.543878	1.333973
	DISCO 1	1.214110	0.930820		DISCO 1	1.215459	0.931854		DISCO 1	1.223958	0.938370
	DISCO 2	1.365351	0.922389		DISCO 2	1.216676	0.924915		DISCO 2	1.225184	0.931382
	DISCO 3	1.422644	0.919962		DISCO 3	1.217652	0.924150		DISCO 3	1.376425	0.929871
	DISCO 4	1.429627	0.898052		DISCO 4	1.431227	0.881890		DISCO 4	1.441223	0.905337
	DISCO 5	1.211684	0.936978		DISCO 5	1.213030	0.938019		DISCO 5	1.221512	0.944578
	DISCO 6	1.338046	0.923044		DISCO 6	1.366868	0.923414		DISCO 6	1.403208	0.929863
	DISCO 7	1.427029	0.907957		DISCO 7	1.428614	0.908966		DISCO 7	1.438604	0.915321
	DISCO 8	1.429638	0.880911		DISCO 8	1.339533	0.924070		DISCO 8	1.226166	0.930612
	DISCO 9	1.425877	0.917876		DISCO 9	1.431215	0.899050		DISCO 9	1.418684	0.928173
	DISCO 10	1.440694	0.783332		DISCO 10	1.444856	0.782635		DISCO 10	1.442384	0.791111
	DISCO 11	1.430654	0.863295		DISCO 11	1.432244	0.864254		DISCO 11	1.442258	0.870297

(continued)

Table 6.26 (continued)

Period No.	GENCO No.	Bidding strategy parameters of GENCOs		Period No.	GENCO No.	Bidding strategy parameters of GENCOs		Period No.	GENCO No.	Bidding strategy parameters of GENCOs	
		$\xi_{1,p,k,g}, \xi_{1,p,k,d}$	$\xi_{2,p,k,g}, \xi_{2,p,k,d}$			$\xi_{1,p,k,g}, \xi_{1,p,k,d}$	$\xi_{2,p,k,g}, \xi_{2,p,k,d}$			$\xi_{1,p,k,g}, \xi_{1,p,k,d}$	$\xi_{2,p,k,g}, \xi_{2,p,k,d}$
	DISCO 12	1.430779	0.784745		DISCO 12	1.433554	0.784988		DISCO 12	1.452379	0.789686
	DISCO 13	1.216300	0.923125		DISCO 13	1.408833	0.921728		DISCO 13	1.437442	0.925321
	DISCO 14	1.424193	0.918875		DISCO 14	1.427461	0.918896		DISCO 14	1.441234	0.888056
	DISCO 15	1.215326	0.923889		DISCO 15	1.393465	0.923406		DISCO 15	1.435744	0.926329
	DISCO 16	1.431963	0.784117		DISCO 16	1.442295	0.784203		DISCO 16	1.454959	0.788108
	DISCO 17	1.443252	0.781767		DISCO 17	1.432368	0.785617		DISCO 17	1.443578	0.790477
	DISCO 18	1.407269	0.920705		DISCO 18	1.425775	0.919896		DISCO 18	1.434183	0.927424
	DISCO 19	1.391918	0.922382		DISCO 19	1.424225	0.920984		DISCO 19	1.348900	0.930531

Table 6.27 Market outcomes under the first and second scenarios of the first case

Period No.	The market participants	Generation and consumption ($\rho_{p,k,g}$, $\rho_{p,k,d}$; MW)		Period No.	The market participants	Generation and consumption ($\rho_{p,k,g}$, $\rho_{p,k,d}$; MW)		Period No.	The market participants	Generation and consumption ($\rho_{p,k,g}$, $\rho_{p,k,d}$; MW)	
		First scenario: Ignoring the BBM	Second scenario: Considering the BBM			First scenario: Ignoring the BBM	Second scenario: Considering the BBM			First scenario: Ignoring the BBM	Second scenario: Considering the BBM
1	GENCO 1	104.817340	106.942379	5	GENCO 1	185.302857	193.236002	10	GENCO 1	190.089650	195.056017
	GENCO 2	110.409626	105.986829		GENCO 2	124.231768	126.296612		GENCO 2	234.717571	247.487756
	GENCO 3	43.838594	44.519122		GENCO 3	77.268248	77.897793		GENCO 3	140.092318	146.001202
	GENCO 4	122.024064	119.708332		GENCO 4	209.162992	209.460890		GENCO 4	230.264467	236.658932
	GENCO 5	100.978967	103.026187		GENCO 5	148.814242	149.026710		GENCO 5	225.802556	153.055844
	GENCO 6	112.992909	115.283698		GENCO 6	176.272434	177.708618		GENCO 6	192.370621	194.677286
	GENCO 7	138.236864	139.572096		GENCO 7	184.692066	185.196849		GENCO 7	222.465740	231.835344
	GENCO 8	46.746823	47.694555		GENCO 8	82.779592	83.454041		GENCO 8	183.000044	205.540292
	GENCO 9	57.001069	57.740206		GENCO 9	114.165299	110.482051		GENCO 9	112.739926	112.712754
	GENCO 10	56.772214	57.923198		GENCO 10	125.532623	126.555403		GENCO 10	125.642216	149.480847
	GENCO 11	63.885651	64.652594		GENCO 11	115.637983	112.954388		GENCO 11	139.452999	138.456491
	GENCO 12	55.048962	56.115104		GENCO 12	97.274033	98.066576		GENCO 12	110.141052	111.232382
	DISCO 1	197.591022	197.961245		DISCO 1	283.234674	286.545505		DISCO 1	335.874755	335.938155
	DISCO 2	52.886745	52.985839		DISCO 2	163.765329	163.939005		DISCO 2	185.321303	185.401811
	DISCO 3	40.120979	40.196153		DISCO 3	108.373698	109.565325		DISCO 3	110.240442	110.319717
	DISCO 4	25.531532	25.579370		DISCO 4	30.783262	31.398790		DISCO 4	73.585885	73.595501
	DISCO 5	212.282770	212.680521		DISCO 5	338.178183	329.081299		DISCO 5	417.469895	417.521577
	DISCO 6	53.929851	54.030899		DISCO 6	95.022998	95.923034		DISCO 6	103.674245	103.786415
	DISCO 7	28.230142	27.920238		DISCO 7	43.600359	43.721722		DISCO 7	77.569898	77.613825
	DISCO 8	23.707851	23.752272		DISCO 8	99.574334	99.965376		DISCO 8	128.639951	128.709133
	DISCO 9	32.826256	32.887762		DISCO 9	39.578480	39.969873		DISCO 9	92.275137	92.304974
	DISCO 10	8.596851	8.612959		DISCO 10	10.365188	10.572445		DISCO 10	25.248904	25.262223
	DISCO 11	21.884170	21.925174		DISCO 11	26.385653	26.913249		DISCO 11	38.787901	38.829868
	DISCO 12	17.611647	17.644645		DISCO 12	21.234289	21.658880		DISCO 12	23.167538	23.182604
	DISCO 13	60.362383	60.475483		DISCO 13	72.778674	72.833923		DISCO 13	79.404716	79.460628

(continued)

Table 6.27 (continued)

Period No.	The market participants	Generation and consumption ($\rho_{p,k,g}$, $\rho_{p,k,d}$; MW)		Period No.	The market participants	Generation and consumption ($\rho_{p,k,g}$, $\rho_{p,k,d}$; MW)		Period No.	The market participants	Generation and consumption ($\rho_{p,k,g}$, $\rho_{p,k,d}$; MW)	
		First scenario: Ignoring the BBM	Second scenario: Considering the BBM			First scenario: Ignoring the BBM	Second scenario: Considering the BBM			First scenario: Ignoring the BBM	Second scenario: Considering the BBM
	DISCO 14	38.818851	38.891585		DISCO 14	47.803728	47.959589		DISCO 14	52.155957	52.212387
	DISCO 15	63.569808	63.688918		DISCO 15	76.645854	77.178429		DISCO 15	83.623978	83.714455
	DISCO 16	11.254666	11.275753		DISCO 16	13.569704	13.841037		DISCO 16	14.805140	14.821159
	DISCO 17	8.334321	8.349936		DISCO 17	23.365727	22.503722		DISCO 17	26.048908	25.121462
	DISCO 18	49.700447	49.793570		DISCO 18	59.923622	60.121827		DISCO 18	89.464243	89.491039
	DISCO 19	52.759698	51.347666		DISCO 19	67.707443	68.198169		DISCO 19	112.631045	112.702906
	$\sum_{l \in \Psi^L} \kappa_{p,k,l}$	12.753087	19.164305		$\sum_{l \in \Psi^L} \kappa_{p,k,l}$	19.242932	28.444727		$\sum_{l \in \Psi^L} \kappa_{p,k,l}$	36.789319	52.205306

Table 6.28 Market outcomes under the first and second scenarios of the first case

Period No.	Parameter	LMP ($/MWh)		Period No.	Parameter	LMP ($/MWh)		Period No.	Parameter	LMP ($/MWh)	
		First scenario: Ignoring the BBM	Second scenario: Considering the BBM			First scenario: Ignoring the BBM	Second scenario: Considering the BBM			First scenario: Ignoring the BBM	Second scenario: Considering the BBM
1	LMP_1	24.429321	28.151250	5	LMP_1	24.923089	29.132605	10	LMP_1	25.900262	30.327706
	LMP_2	19.759052	22.769442		LMP_2	20.158424	23.563187		LMP_2	20.948786	24.529815
	LMP_3	29.874652	34.426204		LMP_3	30.478482	35.626305		LMP_3	31.673468	37.087796
	LMP_4	21.519053	24.797587		LMP_4	21.953998	25.662034		LMP_4	22.814760	26.714762
	LMP_5	19.685453	22.684630		LMP_5	20.083338	23.475419		LMP_5	20.870756	24.438446
	LMP_6	26.685339	30.750983		LMP_6	27.224706	31.822966		LMP_6	28.292120	33.128433
	LMP_7	18.572481	21.402091		LMP_7	18.947870	22.148170		LMP_7	19.690769	23.056750
	LMP_8	22.038171	25.395796		LMP_8	22.483609	26.281096		LMP_8	23.365136	27.359220
	LMP_9	16.302257	18.785986		LMP_9	16.631759	19.440868		LMP_9	17.283850	20.238387
	LMP_{10}	27.151987	31.288727		LMP_{10}	27.700785	32.379455		LMP_{10}	28.786865	33.707751
	LMP_{11}	15.541772	17.909638		LMP_{11}	15.855904	18.533970		LMP_{11}	16.477575	19.294286
	LMP_{12}	12.845461	14.802531		LMP_{12}	13.105095	15.318549		LMP_{12}	13.618913	15.946958
	LMP_{13}	22.880704	26.366693		LMP_{13}	23.343171	27.285839		LMP_{13}	24.258400	28.405180
	LMP_{14}	38.060504	43.859211		LMP_{14}	38.829787	45.388148		LMP_{14}	40.352207	47.250097
	LMP_{15}	26.801138	30.884425		LMP_{15}	27.342845	31.961059		LMP_{15}	28.414891	33.272191
	LMP_{16}	13.677682	15.761544		LMP_{16}	13.954136	16.310994		LMP_{16}	14.501244	16.980116
	LMP_{17}	17.182257	19.800059		LMP_{17}	17.529546	20.490291		LMP_{17}	18.216837	21.330860
	LMP_{18}	36.305756	41.837119		LMP_{18}	37.039572	43.295565		LMP_{18}	38.491802	45.071670
	LMP_{19}	17.598367	20.279566		LMP_{19}	17.954067	20.986514		LMP_{19}	18.658002	21.847439
	LMP_{20}	19.582248	22.565701		LMP_{20}	19.978046	23.352344		LMP_{20}	20.761336	24.310322
	LMP_{21}	24.652550	28.408488		LMP_{21}	25.150829	29.398811		LMP_{21}	26.136931	30.604832
	LMP_{22}	16.889789	19.463033		LMP_{22}	17.231167	20.141516		LMP_{22}	17.906759	20.967777

(continued)

Table 6.28 (continued)

Period No.	Parameter	LMP ($/MWh)		Period No.	Parameter	LMP ($/MWh)		Period No.	Parameter	LMP ($/MWh)	
		First scenario: Ignoring the BBM	Second scenario: Considering the BBM			First scenario: Ignoring the BBM	Second scenario: Considering the BBM			First scenario: Ignoring the BBM	Second scenario: Considering the BBM
	LMP_{23}	14.821294	17.079392		LMP_{23}	15.120864	17.674782		LMP_{23}	15.713716	18.399851
	LMP_{24}	22.167122	25.544393		LMP_{24}	22.615166	26.434873		LMP_{24}	23.501851	27.519306
	LMP_{25}	32.429881	37.370735		LMP_{25}	33.085357	38.673483		LMP_{25}	34.382553	40.259977
	LMP_{26}	21.947931	25.291807		LMP_{26}	22.391545	26.173482		LMP_{26}	23.269462	27.247192
	LMP_{27}	16.887783	19.460720		LMP_{27}	17.229120	20.139123		LMP_{27}	17.904632	20.965286
	LMP_{28}	36.375754	41.917781		LMP_{28}	37.110985	43.379040		LMP_{28}	38.566015	45.158569
	LMP_{29}	19.738674	22.745959		LMP_{29}	20.137634	23.538886		LMP_{29}	20.927181	24.504517
	LMP_{30}	22.542913	25.977438		LMP_{30}	22.998553	26.883014		LMP_{30}	23.900269	27.985830

existing corridors, (2-6) and (12-16), of the network. The planner also establishes a new transmission line in the new corridor, (18-16), of the network. In the second scenario of this period, however, the planner installs two new transmission lines in the existing corridor, (10-22), and in the new corridor, (9-12), of the network. For the remaining periods of the planning horizon, the same analysis can be accomplished.

The number of newly installed transmission lines in the existing corridors of the network, plus the number of newly installed transmission lines in the new corridors of the network over the planning horizon in the second scenario where the strategic behaviors of the market participants are taken into account, is also less than that in the first scenario where the strategic behaviors of the market participants are ignored (see Table 6.29).

Table 6.30 summarizes the values for the TIC, TCC, and ECOC indices, PD-TEPO, under the first and second scenarios of the first case—the IGDT risk-averse decision-making policy. It is clear that considering the BBM (the first level) in the proposed framework can significantly reduce the TIC, TCC, and ECOC indices (see Table 6.30). In other words, the costs associated with investment, congestion, and unreliability in the second scenario, taking into account the BBM, have fewer values compared to the first scenario in which the BBM is not taken into account. As a result, the main benefits of adopting strategic behaviors by the market participants, including both GENCOs and DISCOs, are as follows: (1) it improves the competition and increases the efficiency of the CSC electricity market; (2) it reduces the required investment costs of transmission network expansion; (3) it decreases the costs caused by congestion phenomenon in the transmission network; (4) it reduces the unreliability costs imposed on the consumers; (5) it provides more flexible transmission expansion plans against severe uncertainties, and so on.

6.4.5.1.2 Second Case: Simulation Results and Discussion

The proposed strategic tri-level computational-logical framework under the IGDT risk-taker decision-making policy is solved by optimization of Eqs. (6.179) through (6.190) under different values of the target cost deviation factor, $\sigma_{\mathrm{t}}^{\mathrm{PD-TEP}} \in (0, 0.8)$, which leads to different target costs, $\varpi_{\mathrm{t}}^{\mathrm{PD-TEP}}$. In other words, for each value of the target cost deviation factor, $\sigma_{\mathrm{c}}^{\mathrm{PD-TEP}} \in (0, 0.8)$, the planner tries to find the minimum values of Δ^{price} and Δ^{demand} by solving Eqs. (6.165) through (6.176) in a multi-objective manner using a Pareto optimality concept. In a manner similar to the IGDT risk-averse decision-making policy, the planner uses the FSM, based on the conservative methodology—the min-max formulation—to select the optimal solution from among the Pareto-optimal solution set. This process addresses a well-suited compromise among different objectives, namely Δ^{price} and Δ^{demand}. In order to achieve a cost equal to the obtained cost from the deterministic framework, $\varpi_{\mathrm{c}}^{\mathrm{PD-TEP}} = \varpi_{\mathrm{b}}^{\mathrm{PD-TEP}} = \$591.719878\mathrm{M}$ and $\varpi_{\mathrm{c}}^{\mathrm{PD-TEP}} = \varpi_{\mathrm{b}}^{\mathrm{PD-TEP}} = \$573.968904$ M for the first and second scenarios, respectively, the planner must set its target cost deviation factor to zero, $\sigma_{\mathrm{t}}^{\mathrm{PD-TEP}} = 0$. In this condition, the opportunity and risk level of the transmission expansion plans by the planner will be zero. More precisely, the planner is not ready to deal with the severe uncertainties of market price and demand; therefore, the cost that the planner needs to invest in transmission network expansion—PD-TEPO [see Eq. (6.132)]—is severely affected by these uncertainties.

Suppose that the planner sets its target cost deviation factor to 0.4 (i.e., $\sigma_t^{PD-TEP} = 0.4$). In this situation, Eqs. (6.179) through (6.190) are solved in a multi-objective manner for this value of the target cost deviation factor with the aim of minimizing Δ^{price} and Δ^{demand} based on the first and second scenarios. Figure 6.14 illustrates the Pareto-optimal solution set for this value of target cost deviation factor under the first and second scenarios. As further elucidation, consider planner performance under the first scenario (ignoring the BBM). For example, in the solution marked with a square in Fig. 6.14, a target cost of $\varpi_t^{PD-TEP} = (1 - 0.4) \times \varpi_b^{PD-TEP} = \$355.031926M$ is guaranteed for the planner, provided that Δ^{price} and Δ^{demand} are not less than 21.014326% and 10.252934%, respectively. That is, to reach a cost 40% lower than the base cost by the planner, actual market price and demand must be at least

Table 6.29 Optimal robust transmission expansion plans under the first and second scenarios of the first case

Period No.	Optimal robust transmission expansion plans	
	First scenario: Ignoring the BBM	Second scenario: Considering the BBM
1	(2-6: I), (12-16: I), (18-16: C),	(9-12: C), (10-22: I)
2	(1-2: I), (2-8: C)	(2-6: I), (13-14: C), (23-25: C)
3	(6.8: I), (15-16: C)	(1-2: I), (27-30: I)
4	(1-3: I), (10-11: C)	(2-8: C), (6.8: I)
5	(27-30: I)	(19-24: C)
6	(4-12: I), (10-22: I)	(27-28: I), (4-12: I)
7	(27-28: I)	(12-16: I)
8	(13-14: C), (18-16: C)	(4-12: I), (15-16: C)
9	(9-12: C)	–
10	(10-12: C), (23-25: C)	(18-16: C)
Number of newly installed transmission lines in existing corridors over the planning horizon	9	9
Number of newly installed transmission lines in new corridors over the planning horizon	9	7

Table 6.30 Changes in the objective functions of the third level of the proposed framework under the first and second scenarios of the first case

Objectives	First scenario: Ignoring the BBM	Second scenario: Considering the BBM
TIC (M$)	608.323342	595.515927
TCC (M$)	27.133290	19.304774
ECOC (M$)	127.375258	103.262874
PD-TEPO (M$)	762.831890	718.083575

21.014326% and 10.252934% lower than their corresponding predicted values, respectively.

Put another way, if there is a difference of more than 21.014326% between actual market price, $a\lambda_{p,\ k,\ b}$, and predicted market price, $p\lambda_{p,\ k,\ b}$, and if there is a difference of more than 10.252934% between actual demand, $a\rho_{p,\ k,\ b}$, and predicted demand, $p\rho_{p,\ k,\ b}$, the specified cost by the planner will not be greater than $\varpi_{\mathrm{t}}^{\mathrm{PD-TEP}} = \$355.031926\mathrm{M}$. And, if the actual market price, $a\lambda_{p,\ k,\ b}$, falls within the market price robustness region, $RR_{p,k,b}^{\mathrm{price}}$, and actual demand, $a\rho_{p,\ k,\ b}$, falls within the demand robustness region, $RR_{p,k,b}^{\mathrm{demand}}$, the specified cost by the planner will not be greater than $\varpi_{\mathrm{t}}^{\mathrm{PD-TEP}} = \$355.031926\mathrm{M}$. Note that if actual market price and/or demand are not at least 21.014326% and 10.252934% lower than their corresponding predicted values, respectively, it will not be possible to achieve the target cost by the planner. If the planner wants to have a smaller target cost, $\varpi_{\mathrm{t}}^{\mathrm{PD-TEP}}$, it must choose a larger target cost deviation factor, $\sigma_{\mathrm{t}}^{\mathrm{PD-TEP}}$, which leads to a larger opportunity for the planner. This means that a lower target cost specified by the planner in the larger market price robustness region, $RR_{p,k,b}^{\mathrm{price}}$, and in the larger demand robustness region, $RR_{p,k,b}^{\mathrm{demand}}$, can be guaranteed and vice versa. That is to say that a better opportunity of the obtained transmission expansion plans by the planner is determined by a higher value of the target cost specified by the planner, $\varpi_{\mathrm{t}}^{\mathrm{PD-TEP}}$, and/or by a larger value of the target cost deviation factor of the planner, $\sigma_{\mathrm{t}}^{\mathrm{PD-TEP}}$. The rest of the solutions provided in Fig. 6.14 have the same interpretations. By the same token, solutions available in the Pareto-optimal solution set have the same interpretations. It should be noted that these solutions are generated owing to the adoption of different target cost deviation factors by the planner. Besides, this evaluation is true for planner performance under the second scenario. After solving the proposed framework under the first and the second scenarios of the second case—the IGDT risk-taker decision-making policy—by the newly proposed multi-objective SOSA, the optimal values for Δ^{price}, Δ^{demand}, and relevant target cost deviation factors are calculated by the FSM in accordance with Table 6.31. Table 6.32 represents the results related to the bidding strategy parameters of the GENCOs and DISCOs, which were calculated in the first level (the BBM) of the strategic tri-level computational-logical framework, under the second scenario in the first case—the IGDT risk-taker decision-making policy. For this reason, these results are only presented for the first, fifth, and tenth periods of the planning horizon. These results are also associated with the peak load. Due to the fact that in the first scenario of the second case the strategic behaviors of the market participants are ignored, the market participants—including GENCOs and DISCOs—lack the bidding strategy parameters. That is, in the first scenario of the second case, the GENCOs and DISCOs submit their real marginal generation costs and marginal benefit functions to the ISO, respectively. For the Nash equilibrium point obtained in the first, fifth, and tenth periods, the strategic parameters of slope and intercept of the market participants are not equal to a constant value of 1 (see Table 6.32). This reflects the fact that the market participants, including GENCOs and DISCOs, by adopting the strategic behaviors try to maximize their profit in selling and buying energy, respectively.

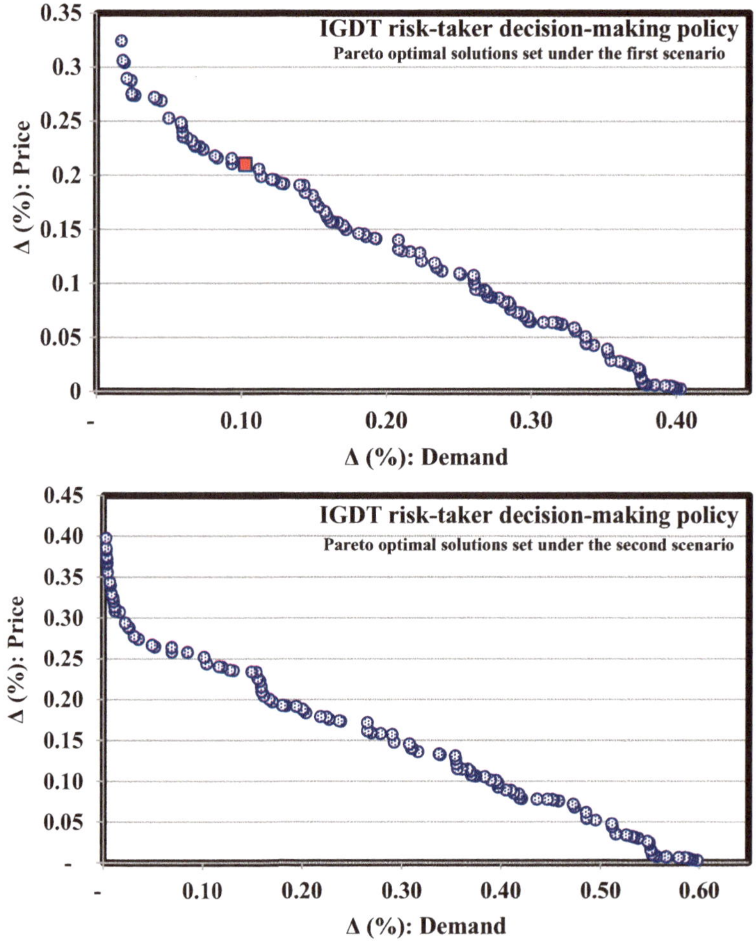

Fig. 6.14 Pareto-optimal solutions of opportunity against market price and demand uncertainties under the first and second scenarios

Tables 6.33 and 6.34 summarize the results of the CSC electricity market (the second level) for the strategic tri-level computational-logical framework under the first and second scenarios of the IGDT risk-taker decision-making policy (the second case). For this reason, these results are only presented for the first, fifth, and tenth periods of the planning horizon. These results are also related to the peak load. There is a difference between the power generated by the GENCOs and the power purchased by the DISCOs in the periods of the planning horizon, which reflects transmission network losses (see Table 6.33). A positive growth in the power generated by the GENCOs and the power purchased by the DISCOs, as well as transmission line losses in the network, can be observed as a result of demand growth (see Table 6.33). In addition, the larger bidding strategy parameters of slope and intercept belong to the GENCO that has the lowest generated power and vice versa (see Tables 6.32 and 6.33). This expression means that the smaller GENCO—in terms of generated power—intends to adopt larger bidding strategy parameters to enhance its profit in the electricity market and compensate for low profit, due to a shortage in its power generation capacity. Furthermore, it is evident that the larger slope and smaller intercept of the bidding strategy parameters are relevant to the DISCO with less purchased power and vice versa. In other words, the smaller DISCO—in terms of purchased power—intends to adopt a larger slope and smaller intercept associated with bidding strategy parameters in order to improve its profit in the electricity market and compensate for lower profit, owing to the shortage of its power purchasing capacity. The LMP's fluctuations on the different buses of the network over the planning horizon have clearly been increased in the case where the strategic behaviors are taken into account by the market participants—the second scenario compared to the first (see Table 6.34). Table 6.35 depicts the opportunistic transmission expansion plans by the planner over the planning horizon under the first and second scenarios of the second case—the IGDT risk-taker decision-making policy. To illustrate the results presented in Table 6.35, consider the first period of the planning horizon as an example. In the first scenario of this period, the planner installs only one transmission line in the existing corridor, (12-16), of the network.

In the second scenario of this period, however, the planner establishes two new transmission lines in the existing corridor, (10-22), and in the new corridor, (9-12), of the network. For the remaining periods of the planning horizon, the same analysis can be performed. The number of newly installed transmission lines in the existing

Table 6.31 Optimal values for Δ^{price}, Δ^{demand}, and relevant target cost deviation factors under the first and second scenarios of the second case

Parameter	Optimal values	
	First scenario: Ignoring the BBM	Second scenario: Considering the BBM
Δ^{price} (%)	9.247533	10.732155
Δ^{demand} (%)	27.365371	26.342581
Target cost deviation factor	0.3	0.3

Table 6.32 Bidding strategy parameters of the market participants under the second scenario of the second case

Period No.	The market participants	Bidding strategy parameters of the market participants		Period No.	The market participants	Bidding strategy parameters of the market participants		Period No.	The market participants	Bidding strategy parameters of the market participants	
		$\xi_{1,p,k,g}$, $\xi_{1,p,k,d}$	$\xi_{2,p,k,g}$, $\xi_{2,p,k,d}$			$\xi_{1,p,k,g}$, $\xi_{1,p,k,d}$	$\xi_{2,p,k,g}$, $\xi_{2,p,k,d}$			$\xi_{1,p,k,g}$, $\xi_{1,p,k,d}$	$\xi_{2,p,k,g}$, $\xi_{2,p,k,d}$
1	GENCO 1	1.319099	1.117809	5	GENCO 1	1.215325	1.120374	10	GENCO 1	1.342139	1.145808
	GENCO 2	1.330687	1.127629		GENCO 2	1.375173	1.299870		GENCO 2	1.209487	1.024924
	GENCO 3	1.516176	1.310038		GENCO 3	1.534554	1.341966		GENCO 3	1.380034	1.207617
	GENCO 4	1.200770	1.017537		GENCO 4	1.204726	1.020889		GENCO 4	1.220128	1.033940
	GENCO 5	1.356158	1.149213		GENCO 5	1.346817	1.225917		GENCO 5	1.280608	1.094386
	GENCO 6	1.306293	1.106957		GENCO 6	1.335088	1.191297		GENCO 6	1.338021	1.127740
	GENCO 7	1.190298	1.008663		GENCO 7	1.322126	1.189684		GENCO 7	1.327352	1.124802
	GENCO 8	1.466299	1.298781		GENCO 8	1.484072	1.338958		GENCO 8	1.348364	1.155829
	GENCO 9	1.407345	1.237665		GENCO 9	1.424404	1.331358		GENCO 9	1.419937	1.279719
	GENCO 10	1.382338	1.207332		GENCO 10	1.372596	1.254524		GENCO 10	1.464623	1.346796
	GENCO 11	1.358704	1.175436		GENCO 11	1.399094	1.302666		GENCO 11	1.391332	1.225138
	GENCO 12	1.428306	1.284431		GENCO 12	1.445618	1.333143		GENCO 12	1.460619	1.331157
	DISCO 1	1.211547	0.928855		DISCO 1	1.212894	0.929887		DISCO 1	1.221374	0.936389
	DISCO 2	1.362468	0.920442		DISCO 2	1.214108	0.922963		DISCO 2	1.222597	0.929416
	DISCO 3	1.419641	0.918020		DISCO 3	1.215081	0.922199		DISCO 3	1.373520	0.927908
	DISCO 4	1.426609	0.896157		DISCO 4	1.428205	0.880028		DISCO 4	1.438180	0.903425
	DISCO 5	1.209126	0.935000		DISCO 5	1.210469	0.936039		DISCO 5	1.218933	0.942584
	DISCO 6	1.335222	0.921096		DISCO 6	1.363982	0.921464		DISCO 6	1.400246	0.927900
	DISCO 7	1.424016	0.906040		DISCO 7	1.425598	0.907047		DISCO 7	1.435566	0.913389
	DISCO 8	1.426620	0.879052		DISCO 8	1.336705	0.922119		DISCO 8	1.223577	0.928648
	DISCO 9	1.422867	0.915939		DISCO 9	1.428194	0.897152		DISCO 9	1.415689	0.926214
	DISCO 10	1.437652	0.781679		DISCO 10	1.441806	0.780983		DISCO 10	1.441339	0.787440

	DISCO 11	1.427634	0.861472		DISCO 11	1.429220	0.862429		DISCO 11	1.439213	0.868460
	DISCO 12	1.427758	0.783089		DISCO 12	1.430527	0.783331		DISCO 12	1.449313	0.786019
	DISCO 13	1.213733	0.921176		DISCO 13	1.405859	0.919782		DISCO 13	1.436408	0.913368
	DISCO 14	1.421186	0.916936		DISCO 14	1.424447	0.916956		DISCO 14	1.438192	0.886182
	DISCO 15	1.212760	0.921938		DISCO 15	1.390523	0.921457		DISCO 15	1.432713	0.924373
	DISCO 16	1.428940	0.782461		DISCO 16	1.439250	0.782547		DISCO 16	1.451887	0.786444
	DISCO 17	1.440206	0.780116		DISCO 17	1.429344	0.783959		DISCO 17	1.440530	0.788808
	DISCO 18	1.404298	0.918762		DISCO 18	1.422765	0.917954		DISCO 18	1.431156	0.925466
	DISCO 19	1.388980	0.920434		DISCO 19	1.401218	0.919940		DISCO 19	1.346052	0.928567

corridors of the network, plus the number of newly installed transmission lines in the new corridors of the network over the planning horizon in the second scenario, where the strategic behaviors of the market participants are taken into account is less than that in the first scenario in which the strategic behaviors of the market participants are ignored (see Table 6.35).

Table 6.36 summarizes the values for the TIC, TCC, and ECOC indices, PD-TEPO, under the first and second scenarios of the second case—the IGDT risk-taker decision-making policy. It is clear that considering the BBM (the first level) in the proposed framework can significantly reduce the TIC, TCC, and ECOC indices (see Table 6.36). In other words, the costs associated with investment, congestion, and unreliability in the second scenario, taking into account the BBM, have fewer values compared to the first scenario, where the BBM is not considered. As a consequence, improving the competition and increasing the efficiency of the CSC electricity market; reducing costs related to investment, congestion phenomenon, and unreliability; and, finally, providing more flexible transmission expansion plans against severe uncertainties are the principal advantages of adopting strategic behaviors by the GENCOs and DISCOs as market participants. By comparing the results from the first case (the IGDT risk-averse decision-making policy) and second case (the IGDT risk-taker decision-making policy), it can be seen that when the strategic behaviors of the market participants (the BBM) are activated in the strategic tri-level computational-logical framework, a higher reduction in PD-TEPO is observed.

As a result, the integration of the strategic behaviors of the market participants in the PD-TEP problem can clearly enhance productivity and flexibility of the proposed framework.

6.4.5.2 Large-Scale Iranian 400 kV Transmission Network

In this section, the proposed strategic tri-level computational-logical framework has been implemented in a modified, large-scale Iranian 400 kV transmission network [40, 57], as shown in Fig. 6.15. In this figure, the existing and candidate 400 kV transmission lines/substations are illustrated with solid and dotted lines/circles, respectively. The Iranian 400 kV transmission network configuration demonstrated in Fig. 6.15 is in accordance with the network configuration from 2007. The owner of this network is the state-owned TAVANIR Company that is also responsible for network planning. In this real-world network, the Iran Grid Management Company (IGMC), which was established and became active in the second half of 2005, plays the role of ISO. The IGMC provides conditions that all market participants can freely participate without discrimination in electrical energy exchanges. According to the network configuration in 2007, the IGMC manages the power generation section with an installed capacity of about 35 GW, which is connected by an interconnected transmission network. The Iranian 400 kV transmission network consists of two voltage levels of 400 and 230 kV, 52 buses, and 102 transmission lines. In addition, in the power generation section of this real-world network, thermal, combined cycle,

Table 6.33 Market outcomes under the first and second scenarios of the second case

Period No.	The market participants	Generation and consumption ($\rho_{p,k,g}$, $\rho_{p,k,d}$; MW) First scenario: Ignoring the BBM	Second scenario: Considering the BBM	Period No.	The market participants	Generation and consumption ($\rho_{p,k,g}$, $\rho_{p,k,d}$; MW) First scenario: Ignoring the BBM	Second scenario: Considering the BBM	Period No.	The market participants	Generation and consumption ($\rho_{p,k,g}$, $\rho_{p,k,d}$; MW) First scenario: Ignoring the BBM	Second scenario: Considering the BBM
1	GENCO 1	111.328122	107.588137	5	GENCO 1	150.040260	154.437743	10	GENCO 1	163.546318	164.0487087
	GENCO 2	104.864073	106.788548		GENCO 2	103.272524	104.938560		GENCO 2	200.351219	205.6831808
	GENCO 3	45.211019	44.435835		GENCO 3	64.232258	65.257339		GENCO 3	104.952288	108.2889681
	GENCO 4	114.740322	119.484381		GENCO 4	173.874931	174.404971		GENCO 4	200.379369	201.4647411
	GENCO 5	103.816920	102.833445		GENCO 5	123.707669	123.904869		GENCO 5	198.399665	198.9449512
	GENCO 6	116.923612	115.068024		GENCO 6	146.533367	147.027974		GENCO 6	171.580780	170.7435643
	GENCO 7	147.907644	139.310984		GENCO 7	152.532516	148.012705		GENCO 7	178.423466	179.3584818
	GENCO 8	39.648522	47.605328		GENCO 8	68.813779	69.697994		GENCO 8	143.222899	144.9738385
	GENCO 9	55.231630	57.632185		GENCO 9	94.904378	95.299274		GENCO 9	98.555918	98.9509355
	GENCO 10	55.104159	57.814835		GENCO 10	104.353911	105.145390		GENCO 10	83.713153	83.851145
	GENCO 11	64.546403	64.531642		GENCO 11	96.128604	96.275211		GENCO 11	104.382061	104.8662516
	GENCO 12	53.417887	56.010123		GENCO 12	80.645125	83.477140		GENCO 12	98.237909	98.69071301
	DISCO 1	197.393254	197.746045		DISCO 1	231.266065	231.850354		DISCO 1	258.932871	260.3954829
	DISCO 2	52.833811	52.928239		DISCO 2	135.616985	135.805511		DISCO 2	124.938353	125.4698264
	DISCO 3	40.080822	40.152457		DISCO 3	89.746189	89.946170		DISCO 3	98.115418	98.18597392
	DISCO 4	25.505978	25.551563		DISCO 4	25.492167	25.025731		DISCO 4	65.492388	65.50094729
	DISCO 5	212.070297	212.449320		DISCO 5	280.051374	280.681212		DISCO 5	330.237976	331.5995981
	DISCO 6	53.875873	53.972163		DISCO 6	78.690236	78.812869		DISCO 6	92.271417	92.37125062
	DISCO 7	28.201887	27.889887		DISCO 7	36.106234	36.885608		DISCO 7	69.038211	69.07730713
	DISCO 8	23.684122	23.726452		DISCO 8	82.459279	82.908083		DISCO 8	104.491218	105.5527919
	DISCO 9	32.793400	32.852010		DISCO 9	32.775644	32.948756		DISCO 9	82.126064	82.15261971
	DISCO 10	9.589144	9.690677		DISCO 10	8.583596	8.603014		DISCO 10	22.471851	22.4837055
	DISCO 11	21.862266	21.901340		DISCO 11	21.850429	21.936341		DISCO 11	34.521733	34.55908448
	DISCO 12	17.594019	17.625464		DISCO 12	17.584493	17.963165		DISCO 12	20.619408	20.63281762
	DISCO 13	60.301966	60.409741		DISCO 13	60.269316	54.417219		DISCO 13	70.671223	65.53935833

(continued)

Table 6.33 (continued)

Period No.	The market participants	Generation and consumption ($\rho_{p,k,g}$, $\rho_{p,k,d}$; MW)		Period No.	The market participants	Generation and consumption ($\rho_{p,k,g}$, $\rho_{p,k,d}$; MW)		Period No.	The market participants	Generation and consumption ($\rho_{p,k,g}$, $\rho_{p,k,d}$; MW)	
		First scenario: Ignoring the BBM	Second scenario: Considering the BBM			First scenario: Ignoring the BBM	Second scenario: Considering the BBM			First scenario: Ignoring the BBM	Second scenario: Considering the BBM
	DISCO 14	38.779997	38.849307		DISCO 14	39.587118	39.654720		DISCO 14	46.419475	46.46969936
	DISCO 15	63.506182	63.619683		DISCO 15	63.471796	63.641628		DISCO 15	74.426420	74.50694658
	DISCO 16	11.243401	11.263496		DISCO 16	11.237313	11.910063		DISCO 16	12.176766	12.19102309
	DISCO 17	8.325979	8.340859		DISCO 17	19.349575	19.714558		DISCO 17	23.183864	23.3584261
	DISCO 18	49.650702	49.739440		DISCO 18	49.623818	50.471678		DISCO 18	79.624332	79.64818089
	DISCO 19	52.706891	51.291847		DISCO 19	56.069739	56.654686		DISCO 19	100.243085	100.3070427
	$\sum_{l \in \Psi^{L}} \kappa_{p,k,l}$	12.740322	19.103472		$\sum_{l \in \Psi^{L}} \kappa_{p,k,l}$	19.207951	28.04780025		$\sum_{l \in \Psi^{L}} \kappa_{p,k,l}$	35.742969	49.86339713

Table 6.34 Market outcomes under the first and second scenarios of the second case

Period No.	Parameter	LMP ($/MWh)		Period No.	Parameter	LMP ($/MWh)		Period No.	Parameter	LMP ($/MWh)	
		First scenario	Second scenario			First scenario	Second scenario			First scenario	Second scenario
1	LMP_1	24.377748	28.091819	5	LMP_1	24.870474	29.071103	10	LMP_1	25.845584	30.263681
	LMP_2	19.717339	22.721373		LMP_2	20.115867	23.513443		LMP_2	20.904561	24.478030
	LMP_3	29.811583	34.353526		LMP_3	30.414138	35.551094		LMP_3	31.606602	37.009499
	LMP_4	21.473623	24.745237		LMP_4	21.907650	25.607858		LMP_4	22.766595	26.658364
	LMP_5	19.643895	22.636740		LMP_5	20.040940	23.425859		LMP_5	20.826695	24.386854
	LMP_6	26.629004	30.686065		LMP_6	27.167232	31.755784		LMP_6	28.232392	33.058495
	LMP_7	18.533272	21.356909		LMP_7	18.907869	22.101413		LMP_7	19.649200	23.008075
	LMP_8	21.991646	25.342183		LMP_8	22.436143	26.225614		LMP_8	23.315809	27.301462
	LMP_9	16.267841	18.746327		LMP_9	16.596648	19.399826		LMP_9	17.247362	20.195661
	LMP_{10}	27.094666	31.222673		LMP_{10}	27.642306	32.311099		LMP_{10}	28.726093	33.636590
	LMP_{11}	15.508962	17.871829		LMP_{11}	15.822430	18.494843		LMP_{11}	16.442789	19.253553
	LMP_{12}	12.818343	14.771281		LMP_{12}	13.077428	15.286209		LMP_{12}	13.590162	15.913292
	LMP_{13}	22.832400	26.311030		LMP_{13}	23.293891	27.228236		LMP_{13}	24.207187	28.345214
	LMP_{14}	37.980154	43.766620		LMP_{14}	38.747813	45.292329		LMP_{14}	40.267019	47.150346
	LMP_{15}	26.744558	30.819224		LMP_{15}	27.285122	31.893586		LMP_{15}	28.354904	33.201950
	LMP_{16}	13.648807	15.728270		LMP_{16}	13.924678	16.276559		LMP_{16}	14.470630	16.944269
	LMP_{17}	17.145983	19.758259		LMP_{17}	17.492539	20.447034		LMP_{17}	18.178379	21.285828
	LMP_{18}	36.229111	41.748796		LMP_{18}	36.961377	43.204163		LMP_{18}	38.410542	44.976519
	LMP_{19}	17.561215	20.236753		LMP_{19}	17.916164	20.942209		LMP_{19}	18.618613	21.801317
	LMP_{20}	19.540908	22.518062		LMP_{20}	19.935871	23.303044		LMP_{20}	20.717507	24.259000
	LMP_{21}	24.600505	28.348515		LMP_{21}	25.097733	29.336747		LMP_{21}	26.081753	30.540222
	LMP_{22}	16.854133	19.421944		LMP_{22}	17.194790	20.098995		LMP_{22}	17.868956	20.923512
	LMP_{23}	14.790005	17.043336		LMP_{23}	15.088942	17.637468		LMP_{23}	15.680542	18.361006

(continued)

Table 6.34 (continued)

Period No.	Parameter	LMP ($/MWh)		Period No.	Parameter	LMP ($/MWh)		Period No.	Parameter	LMP ($/MWh)	
		First scenario	Second scenario			First scenario	Second scenario			First scenario	Second scenario
	LMP_{24}	22.120324	25.490466		LMP_{24}	22.567423	26.379066		LMP_{24}	23.452236	27.461209
	LMP_{25}	32.361418	37.291841		LMP_{25}	33.015510	38.591839		LMP_{25}	34.309967	40.174984
	LMP_{26}	21.901596	25.238413		LMP_{26}	22.344274	26.118227		LMP_{26}	23.220337	27.189670
	LMP_{27}	16.852131	19.419637		LMP_{27}	17.192748	20.096607		LMP_{27}	17.866833	20.921026
	LMP_{28}	36.298961	41.829288		LMP_{28}	37.032639	43.287462		LMP_{28}	38.484598	45.063234
	LMP_{29}	19.697004	22.697940		LMP_{29}	20.095121	23.4891931		LMP_{29}	20.883002	24.452785
	LMP_{30}	22.495322	25.922597		LMP_{30}	22.950000	26.8262615		LMP_{30}	23.849813	27.926749

gas turbines, and hydroelectric power plants have been used as the main sources for supplying demand. Due to the abundance of natural gas resources in Iran, natural gas is supplied at a subsidized price to most GENCOs as the main fuel. In general, most of the natural gas resources are located and explored in the southern part of Iran, especially along the shores of the Persian Gulf. Hence, with the construction of new power generation plants in this part of the country, the cost of gas transfer is significantly reduced. In this study, most new capacities in the power generation section will be, therefore, created by GENCOs that are located in the southern part of the country.

For more information related to this network, interested readers should look at the annual generation and transmission reports by Tavanir [58]. Network data for this modified test network are tabulated in Appendix 3. Table 6.117 presents the data for existing and candidate transmission lines associated with this network. Table 6.118 provides the parameters for the GENCOs. Table 6.119 provides the demand and the parameters for the DISCOs. Table 6.120 provides the data for the bidding strategy parameters of both the GENCOs and the DISCOs. Table 6.121 provides the parameters of the weighting coefficients for the objective functions concerned with the PD-TEP problem. Since the required information related to the value-based reliability assessment methodology is unavailable for the Iranian 400 kV transmission network, this information has been taken from the IEEE reliability test system (IEEE-RTS) [59] and modified for this network. In this real-world network, each load on each bus is categorized into three sectors: (1) residential, (2) commercial,

Table 6.35 Optimal opportunity transmission expansion plans under the first and second scenarios of the second case

	Optimal opportunity transmission expansion plans	
Period No.	First scenario: Ignoring the BBM	Second scenario: Considering the BBM
1	(12-16: I)	(9-12: C), (10-22: I)
2	(18-16: C)	(18-16: C)
3	–	(13-14: C)
4	(1-3: I)	(2-8: C)
5	(10-22: I)	–
6	(4-12: I)	(27-28: I)
7	(27-28: I), (10-11: C)	(12-16: I)
8	(13-14: C)	(2-6: I)
9	(10-12: C)	–
10	(23-25: C)	(15-16: C)
Number of newly installed transmission lines in existing corridors over the planning horizon	5	4
Number of newly installed transmission lines in new corridors over the planning horizon	5	5

Table 6.36 Changes in the objective functions of the third level of the proposed framework under the first and second scenarios of the second case

Objectives	First scenario: Ignoring the BBM	Second scenario: Considering the BBM
TIC (M$)	298.547285	273.5713297
TCC (M$)	19.219646	18.0777638
ECOC (M$)	94.052889	86.68458189
PD-TEPO (M$)	411.819820	378.333675

and, (3) industrial. In addition, 60%, 20%, and 20% of the maximum load at each bus is allocated to the residential, commercial, and industrial sectors, respectively. The SCDF and its duration at each load sector of each load are given according to Table 6.122. Tables 6.117 and 6.118 present the failure and repair rates for the transmission lines and GENCOs, respectively. The maximum allowable amount of curtailed load at load sector s of bus b for period p is considered to be 30% of the maximum load at the relevant sector. The maximum allowable amount of curtailed load at bus b for period p is considered to be 30% of the maximum load on the relevant bus. The maximum allowable amount of curtailed load at period p is also taken to be 30% of the maximum load at the relevant period. All sectors of each load on each bus have the same outage cost weight coefficient. The planning horizon is considered to be 10-year periods, where the 8760-h time span of the network LDC is modeled using 22 steps (K = 22), according to Eq. (6.105). The load step in pattern k is also considered to be an average load value during this period. The annual interest rate and annual demand growth are considered to be 10% and 5%, respectively. The price growth is also set at 4%. Candidate transmission lines can be installed in all of the existing and 27 new corridors of the network. However, the maximum number of transmission lines that can be installed in each corridor over the planning horizon is considered to be three, including existing ones. The basic construction costs for three bundles and two bundles of 400 kV transmission lines are assumed to be 350 and 200 $/MW-km, respectively. Physical parameters of the candidate transmission lines are considered similar to existing ones. The maximum allowable value of the TIC index in period p and in all remaining periods is set at $60M and $500M, respectively. Table 6.123 shows a summary of these assumptions.

In order to analyze the sufficiency of the tri-level computational-logical framework on the real-world, large-scale Iranian 400 kV transmission network, two different cases and two distinct scenarios are defined and applied. These cases and scenarios are quite similar to the cases and scenarios defined in the IEEE 30-bus test system. The strategic tri-level computational-logical framework under two different scenarios of the first case (the IGDT risk-averse decision-making policy) and the second case (the IGDT risk-taker decision-making policy) is implemented and solved by using the proposed multi-objective SOSA. The parameter adjustments of the multi-objective SOSA are considered in accordance with Table 6.103.

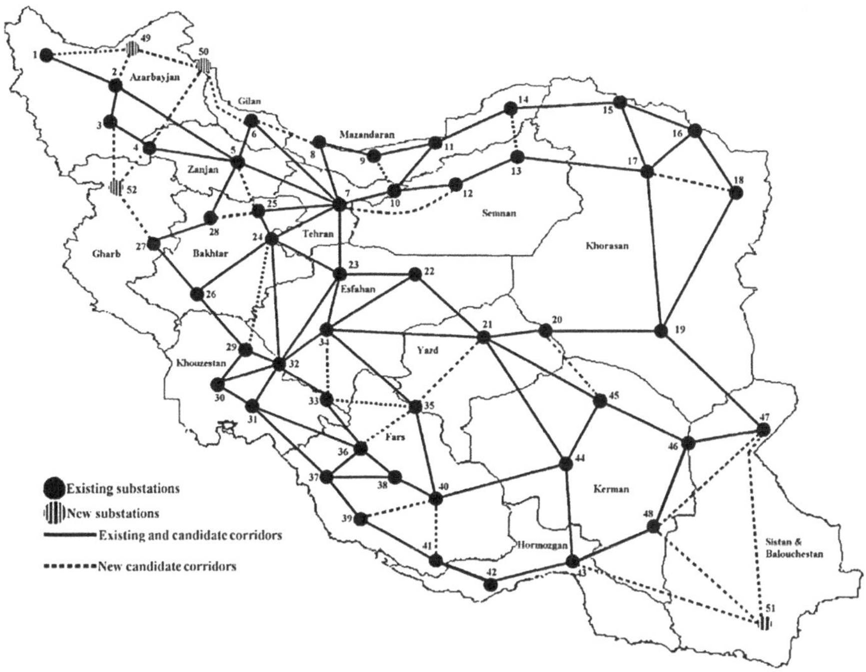

Fig. 6.15 Modified, real-world, large-scale Iranian 400 kV transmission grid

Initially, the planner solves the strategic tri-level computational-logical framework under the risk-neutral/deterministic decision-making policy [Eqs. (6.132) though (6.162)] by the SOSA, based on the first and second scenarios. Then, the optimal results of the proposed framework, according to the first and second scenarios, are employed by the planner so as to calculate the optimal transmission expansion plans under the IGDT risk-averse and IGDT risk-taker decision-making policies. The results presented in Table 6.37 illustrate the transmission expansion plans of the proposed framework under the risk-neutral/deterministic decision-making policy on the basis of the first and second scenarios. The values for the different objectives of the PD-TEP problem [see Eq. (6.132)]—the third level of the proposed strategic tri-level computational-logical framework—that collectively constitute the base costs are also itemized in Table 6.38. As can be seen from Table 6.37, when the proposed framework is solved using the strategic behaviors of the market participants (the second scenario), the number of newly installed transmission lines in the existing corridors of the network plus the number of the newly installed transmission lines in the new corridors of the network over the planning horizon is less than the condition in which the strategic behaviors of the market participants are ignored (the first scenario). Consequently, the planner invests less in the installation of new transmission lines (see Table 6.38). This trend is also seen in the results presented in Table 6.38. Simply put, by activating the strategic behaviors of market participants in the proposed framework (the second scenario) a tangible decrease in the values obtained for the TIC, TCC, and ECOC indices is observed, in comparison with the situation where the strategic behaviors of the market participants in the proposed framework are ignored (the first scenario).

6.4.5.2.1 First Case: Simulation Results and Discussion

The implementation process of the strategic tri-level computational-logical framework under the IGDT risk-averse decision-making policy is described in Sect. 6.4.5.1.1. After solving the proposed framework under the first and the second scenarios of the first case (the IGDT risk-averse decision-making policy) by the multi-objective SOSA, the optimal values for Δ^{price}, Δ^{demand}, and relevant critical cost deviation factors are determined by the FSM in accordance with Table 6.39. Table 6.40 gives the robust transmission expansion plans by the planner over the planning horizon under the first and second scenarios corresponding to the first case—the IGDT risk-averse decision-making policy. Looking at the results presented in Table 6.40, it is evident that the number of newly installed transmission lines in the existing corridors of the network plus the number of the newly installed transmission lines in the new corridors of the network over the planning horizon in the second scenario, where the strategic behaviors of the market participants are taken into account, is less than that for the first scenario, where the strategic behaviors of the market participants are ignored. Table 6.41 summarizes the values for the TIC, TCC, ECOC, and PD-TEPO indices under the first and second scenarios of the first case—the IGDT risk-averse decision-making policy. It is obvious that, considering the BBM (the first level) in the proposed framework can significantly reduce the TIC, TCC, and ECOC indices. In other words, the cost corresponding to investment in the transmission network, congestion, and unreliability in the second scenario (taking into account the BBM) is lower, when compared to that of the first scenario (without accounting for the BBM.

6.4.5.2.2 Second Case: Simulation Results and Discussion

The implementation process of the strategic tri-level computational-logical framework under the IGDT risk-taker decision-making policy is described in Sect. 6.4.5.1.2. After solving the framework under the first and the second scenarios of the second case (the IGDT risk-taker decision-making policy) by the multi-objective SOSA, the optimal values for Δ^{price}, Δ^{demand}, and corresponding target cost deviation factors are calculated by the FSM according to Table 6.42. Table 6.43 illustrates the opportunity transmission expansion plans by the planner over the planning horizon under the first and second scenarios corresponding to the second case (the IGDT risk-taker decision-making policy).

Table 6.43 shows that when the market participants adopt the strategic behaviors, the number of newly installed transmission lines in the existing and the new

Table 6.37 Optimal transmission expansion plans of the proposed framework under the risk-neutral/deterministic decision-making policy, based on the first and second scenarios

Period No.	Optimal transmission expansion plans	
	First scenario: Ignoring the BBM	Second scenario: Considering the BBM
1	(10-12: I), (2-49: C)	(27-52: C), (7-23: I)
2	(24-29: C), (7-23: I)	(33-34: C)
3	(17-19: I)	(7-12: C)
4	(7-12: C), (33-34: C)	(33-35: C), (49-50: C)
5	(42-43: I)	(2-49: C), (4-52: C)
6	(4-52: C), (40-41: C)	(7-24: I), (6.50: C)
7	(49-50: C)	(40-41: C), (43-51: C)
8	(13-14: C), (43-51: C)	(13-14: C), (36.37: I)
9	(4-50: C), (36.37: I)	(48-51: C)
10	(27-52: C), (47-51: C)	(17-19: I)
Number of newly installed transmission lines in existing corridors over the planning horizon	5	4
Number of newly installed transmission lines in new corridors over the planning horizon	12	12

Table 6.38 Changes in the objective functions of the third level along with the base cost under the first and second scenarios

Objectives	First scenario: Ignoring the BBM	Second scenario: Considering the BBM
TIC (M$)	297.592846	285.799452
TCC (M$)	15.791226	13.361885
ECOC (M$)	81.475942	76.186319
Base cost ($\varpi_b^{\text{PD-TEP}}$; M$)	394.860014	375.347655

corridors of the network is reduced, when compared to the condition where the strategic behaviors are ignored by the market participants.

Table 6.44 illustrates the values for the TIC, TCC, ECOC, and PD-TEPO indices under the first and second scenarios of the second case (the IGDT risk-taker decision-making policy). It can be seen from Table 6.44 that there is a significant difference in the results obtained under the first scenario (ignoring the BBM) and second scenario (considering the BBM). Simply put, when the market participants adopt the strategic behaviors, the required investment to expand the transmission network, the congestion cost created in the transmission network, and the unreliability cost imposed on the consumers decrease in comparison with the situation where the market participants lack the strategic behaviors (the second scenario compared to the first).

Table 6.39 Optimal values for Δ^{price}, Δ^{demand}, and relevant critical cost deviation factors under the first and second scenarios of the first case

Parameter	Optimal values	
	First scenario: Ignoring the BBM	Second scenario: Considering the BBM
Δ^{price} (%)	21.736723	16.084485
Δ^{demand} (%)	44.382903	47.677700
Critical cost deviation factor	0.55	0.55

6.4.5.2.3 Investigation of the Effects of Volatility in Market Price and Demand Uncertainties

In this section, the performance of the proposed strategic tri-level computational-logical framework is investigated by considering the volatilities in market price and demand uncertainties. Two different trials are defined and applied to the first and second scenarios of the first case (the IGDT risk-averse decision-making policy) as follows:

- Trial 1: The uncertainty values of market price and demand are assumed to be higher than the predicted values and within the permissible range of $RR_{p,k,b}^{\text{price}}$ and $RR_{p,k,b}^{\text{demand}}$, which are controllable volatilities.
- Trial 2: The uncertainty values of market price and demand are assumed to be higher than the predicted values and outside the permissible range of $RR_{p,k,b}^{\text{price}}$ and $RR_{p,k,b}^{\text{demand}}$, which are uncontrollable volatilities.

Table 6.39 provides the upper limits of the permissible range of $RR_{p,k,b}^{\text{price}}$ and $RR_{p,k,b}^{\text{demand}}$. The volatilities in market price and demand uncertainties in each trial are also set at 50% of the upper limit.

Moreover, two different trials are described and implemented for the first and second scenarios of the second case (the IGDT risk-taker decision-making policy) as follows:

- Trial 3: The uncertainty values of market price and demand are assumed to be lower than the predicted values and within the permissible range of $RR_{p,k,b}^{\text{price}}$ and $RR_{p,k,b}^{\text{demand}}$, which are controllable volatilities.
- Trial 4: The uncertainty values of market price and demand are assumed to be lower than the predicted values and outside the permissible range of $RR_{p,k,b}^{\text{price}}$ and $RR_{p,k,b}^{\text{demand}}$, which are uncontrollable volatilities.

The lower limits of the permissible range of $RR_{p,k,b}^{\text{price}}$ and $RR_{p,k,b}^{\text{demand}}$ are considered on the basis of the optimal values in Table 6.42. The volatilities in market price and demand uncertainties in each trial are also set at 50% of the lower limit. Table 6.45

Table 6.40 Optimal robust transmission expansion plans of the proposed framework under the first and second scenarios of the first case

Period No.	Optimal robust transmission expansion plans	
	First scenario: Ignoring the BBM	Second scenario: Considering the BBM
1	(15-17: I), (30-32: I)	(1-49: C), (7-10: I), (15-17: I)
2	(6.50: C), (7-23: I), (2-49: C)	(13-14: C), (30-32: I)
3	(36.37: I), (10-11: I), (29-32: I)	(24-29: C), (27-52: C), (7-23: I)
4	(35-36: C), (33-34: C), (43-51: C)	(36.37: I), (24-32: I)
5	(47-51: C), (7-10: I), (33-36: I)	(4-50: C), (43-51: C) , (21-35: C)
6	(4-50: C), (33-35: C), (7-24: I)	(17-19: I), (49-50: C)
7	(4-52: C), (1-49: C)	(7-24: I), (48-51: C) , (37-39: I)
8	(39-40: C), (39-41: I), (24-29: C)	(4-52: C), (39-40: C)
9	(13-14: C), (17-19: I), (27-52: C)	(36.37: I), (33-34: C)
10	(7-24: I), (31-36: I)	(7-24: I), (10-12: I)
Number of newly installed transmission lines in existing corridors over the planning horizon	13	12
Number of newly installed transmission lines in new corridors over the planning horizon	14	12

Table 6.41 Changes in the objective functions of the third level of the proposed framework under the first and second scenarios of the first case

Objectives	First scenario: Ignoring the BBM	Second scenario: Considering the BBM
TIC (M$)	531.267686	479.297893
TCC (M$)	12.273694	10.005164
ECOC (M$)	64.385489	60.986187
PD-TEPO (M$)	607.926869	550.289243

provides detailed descriptions for the values of the uncertainty parameters under the first and second scenarios of the first case (the IGDT risk-averse decision-making policy) and the second case (the IGDT risk-taker decision-making policy). Tables 6.46 and 6.47 show the transmission expansion plans and calculated values of the objective

Table 6.42 Optimal values for Δ^{price}, Δ^{demand}, and corresponding target cost deviation factors under the first and second scenarios of the second case

Parameter	Optimal values	
	First scenario: Ignoring the BBM	Second scenario: Considering the BBM
Δ^{price} (%)	14.245120	12.362311
Δ^{demand} (%)	23.164476	27.214550
Target cost deviation factor	0.35	0.35

Table 6.43 Optimal opportunity transmission expansion plans of the proposed framework under the first and second scenarios of the second case

Period No.	Optimal opportunity transmission expansion plans	
	First scenario: Ignoring the BBM	Second scenario: Considering the BBM
1	(33-34: C)	(4-50: C)
2	(1-49: C)	(13-14: C)
3	(4-50: C)	(7-23: I)
4	(7-24: I)	(31-32: I)
5	(30-32: I)	(36.37: I)
6	(13-14: C)	(48-51: C)
7	(37-36: I)	(1-49: C)
8	(7-12: C), (13-17: I)	(27-52: C)
9	(43-51: C)	(7-10: I)
10	(27-52: C)	(39-40: C)
Number of newly installed transmission lines in existing corridors over the planning horizon	4	4
Number of newly installed transmission lines in new corridors over the planning horizon	7	6

Table 6.44 Changes in the objective functions of the third level of the proposed framework under the first and second scenarios of the second case

Objectives	First scenario: Ignoring the BBM	Second scenario: Considering the BBM
TIC (M$)	194.645129	181.693654
TCC (M$)	11.032460	10.853425
ECOC (M$)	47.930716	42.986253
PD-TEPO (M$)	253.608305	235.533332

functions of the third level of the strategic tri-level computational-logical framework for trials 1 and 2 under the first and second scenarios corresponding to the first case (the IGDT risk-averse decision-making policy), respectively. Consider trial 1 in which the values of uncertain market price and demand are higher than the predicted values and lower than the upper limit of the permissible range. In this trial, the number of newly required transmission lines to be installed on the network in order to answer the demand for growth and the various desires of the stakeholders is more than the newly required transmission lines in the risk-neutral/deterministic decision-making policy and lower than the newly required transmission lines in the risk-averse decision-making policy (see Tables 6.37, 6.40, and 6.46). This trend is observed in trial 1 for both scenarios of the first case—the IGDT risk-averse decision-making policy.

In trial 1, the value for the PD-TEPO is also more than the value for it in the risk-neutral/deterministic decision-making policy (the base cost) and lower than the obtained optimal value for it in the risk-averse decision-making policy (the critical cost). This is due to the fact that the changes in market price and demand uncertainties are within the relevant permissible range for $RR_{p,k,b}^{\text{price}}$ and $RR_{p,k,b}^{\text{demand}}$ (see Tables 6.38, 6.41, and 6.47). That is, when the volatilities in market price and demand uncertainties are controllable, the value for the PD-TEPO will be higher than the base cost, and, consequently, lower than the critical cost. This trend is observed for trial 1 of both scenarios of the first case—the IGDT risk-averse decision-making policy. Now consider trial 2 in which the values of uncertain market price and demand are higher than both the predicted values and the upper limit of the permissible range. In this trial, more new transmission lines must be added to the network to answer the demand growth and various desires of stakeholders in comparison with the risk-neutral/deterministic and risk-averse decision-making policies (see Tables 6.37, 6.40, and 6.46). This trend can be seen in trial 2 for both scenarios of the first case—the IGDT risk-averse decision-making policy. In trial 2, the value for the PD-TEPO is also higher than the value for it in the risk-neutral/deterministic decision-making policy (the base cost) and the value for it in the risk-averse decision-making policy (the critical cost). This is because of the fact that changes in market price and demand uncertainties are outside the relevant permissible range of $RR_{p,k,b}^{\text{price}}$ and $RR_{p,k,b}^{\text{demand}}$ (see Tables 6.38, 6.41, and 6.47). In other words, when the volatilities in market price and demand uncertainties are uncontrollable, the value for the PD-TEPO will be higher than the base cost and, consequently, higher than the critical cost. This trend is true in trial 2 for both scenarios of the first case—the IGDT risk-averse decision-making policy. Tables 6.48 and 6.49 provide the transmission expansion plans and values of the objective functions for the third level of the strategic tri-level computational-logical framework for trials 3 and 4 under the first and second scenarios related to the second case (the IGDT risk-taker decision-making policy), respectively.

Consider trial 3 in which the values of uncertain market price and demand are lower than the predicted values and within the limits of the permissible range. In this trial, the number of newly required transmission lines to be installed on the network is lower than the newly required transmission lines in the risk-neutral/deterministic decision-making policy and more than the newly required transmission lines in the risk-taker decision-making policy (see Tables 6.37, 6.43, and 6.48).

This trend can be seen in trial 3 of both scenarios of the second case—the IGDT risk-taker decision-making policy. In trial 3, the value for the PD-TEPO is lower

Table 6.45 Detailed descriptions for the values of the uncertainty parameters under the first and second scenarios of the first and second cases

	The IGDT risk-averse decision-making policy				The IGDT risk-taker decision-making policy			
	First scenario: Ignoring the BBM		Second scenario: Considering the BBM		First scenario: Ignoring the BBM		Second scenario: Considering the BBM	
Parameter	Trial 1	Trial 2	Trial 1	Trial 2	Trial 3	Trial 4	Trial 3	Trial 4
Δ^{price} (%)	10.868381	32.605084	8.042242	24.126727	7.122560	21.367680	6.181155	18.543466
Δ^{demand} (%)	22.191451	66.574354	23.838850	71.516549	11.582238	34.746713	13.607275	40.821824

than the value for it in the risk-neutral/deterministic decision-making policy (the base cost) and more than the value for it in risk-averse decision-making policy (the target cost). This is due to the fact that the changes in market price and demand uncertainties are within the relevant permissible range of $RR_{p,k,b}^{\text{price}}$ and $RR_{p,k,b}^{\text{demand}}$ (see Tables 6.38, 6.44, and 6.49). That is to say that when the volatilities in market price and demand uncertainties are controllable, the value for the PD-TEPO will be lower than the base cost and, therefore, higher than the target cost. This trend is observed in trial 3 of both scenarios of the second case—the IGDT risk-taker decision-making policy.

Consider trial 4 in which the values of uncertain market price and demand are lower than the predicted values and below the lower limit of the permissible range. In this trial, fewer new transmission lines need to be established in the network, in comparison with the risk-neutral/deterministic and risk-taker decision-making policies (see Tables 6.37, 6.43, and 6.48). This trend can be seen in trial 4 of both scenarios of the second case—the IGDT risk-taker decision-making policy. In trial 4, the value for the PD-TEPO is also lower than the value for it in the risk-neutral/deterministic decision-making policy (the base cost) and the value for it in the risk-taker decision-making policy (the target cost). This is due to the fact that the changes in market price and demand uncertainties are outside the permissible range of $RR_{p,k,b}^{\text{price}}$ and $RR_{p,k,b}^{\text{demand}}$ (see Tables 6.38, 6.44, and 6.49). In other words, when the volatilities in market price and demand uncertainties are uncontrollable, the value for the PD-TEPO will be lower than the base cost and, therefore, lower than the target cost. This trend is true in trial 4 of both scenarios of the second case—the IGDT risk-taker decision-making policy.

6.4.5.2.4 Quantitative Verification of the Proposed IGDT Risk-Averse Decision-Making Policy in Comparison to the Robust Optimization Technique

In this section, the performance of the proposed strategic tri-level computational-logical framework under the IGDT risk-averse decision-making policy is compared with the performance of this framework under the RO technique; due to this fact the worst-case uncertainty situation in the proposed strategic tri-level computational-logical framework is observed in the first case (i.e., the IGDT risk-averse decision-making policy). For a comprehensive discussion of the RO technique, please refer to the work by Ben-Tal et al. [41].

Table 6.50 provides the optimal robust transmission expansion plans obtained under the IGDT risk-averse decision-making policy and RO technique. Table 6.51 summarizes the calculated optimal values of the objective functions of the third level associated with the strategic tri-level computational-logical framework under the IGDT risk-averse decision-making policy and RO technique. These optimal results are related to the second scenario (considering the BBM). Table 6.50 indicates that the number of newly installed transmission lines in both the existing

Table 6.46 Optimal robust transmission expansion plans of the proposed framework under trials 1 and 2 of the first and second scenarios of the first case

	Optimal robust transmission expansion plans			
	First scenario: Ignoring the BBM		Second scenario: Considering the BBM	
Period No.	Trial 1	Trial 2	Trial 1	Trial 2
1	(4-50: C), (13-14: C), (37-39: I)	(5-7: I), (13-14: C), (30-32: I), (12-13: I)	(13-14: C), (29-32: I)	(2-49: C), (31-32: I), (36.37: I)
2	(30-32: I), (7-12: C), (24-32: I)	(7-24: I), (1-49: C), (37-39: I)	(1-49: C), (7-12: C)	(6.50: C), (13-14: C), (37-39: I)
3	(43-51: C), (6.50: C)	(13-17: I), (39-40: C), (37-37: I)	(37-39: I), (39-40: C)	(13-17: I), (11-14: I), (7-12: C)
4	(27-52: C), (13-17: I), (7-23: I)	(27-52: C), (40-44: I), (43-51: C)	(4-50: C), (7-24: I)	(39-40: C), (42-43: I), (9-10: C)
5	(33-34: C), (1-49: C)	(4-50: C), (7-12: C), (29-30: I)	(7-23: I), (36.37: I)	(29-30: I), (4-50: C), (1-49: C)
6	(18-19: I), (47-51: C)	(42-43: I), (35-36: C), (4-50: C) , (26.29: I)	(42-43: I), (43-51: C)	(43-51: C), (47-51: C), (7-24: I)
7	(24-29: C), (42-43: I)	(17-19: I), (47-51: C), (7-23: I)	(4-52: C), (49-50: C)	(35-36: C), (3-52: C), (27-52: C)
8	(11-14: I), (3-52: C)	(33-34: C), (35-21: C), (36.37: I)	(17-19: I), (47-51: C)	(14-15: I), (18-19: I)
9	(29-30: I), (36.37: I)	(14-15: I), (49-50: C), (24-29: C)	(29-30: I), (27-52: C)	(21-35: C), (29-24: C), (7-23: I)
10	(9-10: C), (2-49: C)	(21-34: I), (9-10: C), (7-24: I),	(24-29: C)	(33-34: C), (20-21: I)
Number of newly installed transmission lines in existing corridors over the planning horizon	10	17	8	12
Number of newly installed transmission lines in new corridors over the planning horizon	13	15	11	16

and new corridors of the network under the IGDT risk-averse decision-making policy is less than that under the RO technique. According to Table 6.51, there are fewer optimal values obtained for the objectives of the third level of the tri-level computational-logical framework than for the IGDT risk-averse decision-making

Table 6.47 Changes in the objective functions of the third level of the proposed framwork under trials 1 and 2 of the first and second scenarios of the first case

	First scenario: Ignoring the BBM		Second scenario: Considering the BBM	
Objective	Trial 1	Trial 2	Trial 1	Trial 2
TIC (M$)	439.564238	617.337896	362.984555	536.974819
TCC (M$)	13.730237	12.897757	11.920536	11.262744
ECOC (M$)	76.320025	78.348025	71.383510	70.532041
PD-TEPO (M$)	529.614500	708.085836	446.288601	618.769604

Table 6.48 Optimal opportunity transmission expansion plans of the proposed framework under trials 3 and 4 of the first and second scenarios of the second case

	Optimal opportunity transmission expansion plans			
	First scenario: Ignoring the BBM		Second scenario: Considering the BBM	
Period No.	Trial 3	Trial 4	Trial 3	Trial 4
1	(13-14: C)	(36.37: I)	(29-30: I)	–
2	(36.37: I), (7-12: C)	(29-30: I)	(13-14: C), (7-12: C)	(13-14: C)
3	(33-34: C)	–	(36.37: I), (1-49: C)	(36.37: I)
4	(29-30: I), (27-52: C)	(13-14: C)	(24-29: C)	(30-29: I)
5	(13-17: I)	(1-49: C)	(7-24: I)	–
6	(1-49: C)	(7-12: C)	(49-50: C)	(7-12: C)
7	(37-39: I), (39-40: C)	–	(4-50: C)	(13-17: I)
8	(7-23: I), (49-50: C)	(29-24: C)	(3-52: C)	(24-29: C)
9	(4-50: C)	(7-24: I)	(27-52: C)	–
10	(4-52: C)	(2-49: C)	(13-17: I)	(7-24: I)
Number of newly installed transmission lines in existing corridors over the planning horizon	5	3	4	4
Number of newly installed transmission lines in new corridors over the planning horizon	9	5	8	3

policy compared to the RO technique. In other words, the framework under the IGDT risk-averse decision-making policy affords a lower TIC for construction of new transmission lines. The TCC in the IGDT risk-averse decision-making policy also has a smaller value than the RO technique. In addition, the ECOC has been

Table 6.49 Changes in the objective functions of the third level of the proposed framework under trials 3 and 4 of the first and second scenarios of the second case

	First scenario: Ignoring the BBM		Second scenario: Considering the BBM	
Objectives	Trial 3	Trial 4	Trial 3	Trial 4
TIC (M$)	239.118988	134.743024	216.746553	121.729559
TCC (M$)	12.411843	11.784685	10.957655	8.938515
ECOC (M$)	51.703329	47.449089	43.086286	38.580649
PD-TEPO (M$)	303.234160	193.976798	270.790494	169.248723

significantly decreased in the proposed framework under the IGDT risk-averse decision-making policy compared with the RO technique. As a result, the IGDT risk-averse decision-making policy provides a more well-planned expansion plan for transmission network and, thus, prevents unnecessary investments. For this reason, the IGDT risk-averse decision-making policy can be a suitable policy for handling severe uncertainties in large-scale optimization problems.

6.4.5.2.5 Performance Evaluation of the Proposed Optimization Algorithms: Simulation Results and Discussion

In this section, the performance of the proposed modern meta-heuristic music-inspired optimization algorithms addressed in Chap. 4, namely the multi-objective SS-HSA, multi-objective SS-IHSA, multi-objective continuous/discrete TMS-MSA, multi-objective TMS-EMSA, and multi-objective SOSA, is compared with that of the NSGA-II for the first and second scenarios of the first case (the IGDT risk-averse decision-making policy) and second case (the IGDT risk-taker decision-making policy). Tables 6.104, 6.105, 6.106, 6.107, and 6.108 illustrate the parameter adjustments of the multi-objective TMS-EMSA, multi-objective continuous/discrete TMS-MSA, multi-objective SS-HSA, multi-objective SS-IHSA, and NSGA-II, respectively. The calculated optimal results related to the TIC, TCC, and ECOC over the planning horizon by the multi-objective optimization algorithms under the first case (the IGDT risk-averse decision-making policy) and the second case (the IGDT risk-taker decision-making policy) are given in Tables 6.52 and 6.53, respectively. These tables clearly show that the proposed strategic tri-level computational-logical framework under the first and second scenarios of the first and second cases by the proposed multi-objective SOSA leads to more efficient results than other proposed meta-heuristic music-inspired optimization algorithms and NSGA-II. The accurate evaluation of the performance of these multi-objective optimization algorithms relative to each other is performed using the ICS. Please refer to Sect. 5.3.5.3 of Chap. 5 for more information on the ICS. Tables 6.54 and 6.55 present the ICS for the multi-objective optimization algorithms under the first and second cases, respectively.

To clarify, consider the optimal results of the proposed framework under the first scenario of the first case (the IGDT risk-averse decision-making policy) according to Table 6.54.

The ICS shows 8.384936%, 11.726171%, 18.828557%, 16.758274%, and 20.631733% superiority—positive sign—of the multi-objective SOSA performance compared with the performances of the multi-objective TMS-EMSA, multi-objective continuous/discrete TMS-MSA, multi-objective SS-HSA, multi-objective SS-IHSA, and NSGA-II, respectively, from the perspective of the obtained TIC index by corresponding optimization algorithms. Also, the ICS represents 9.177692%, 12.132572%, 19.778299%, 17.226608%, and 20.677813% superiority—positive sign—of the multi-objective SOSA performance compared with the performances of the multi-objective TMS-EMSA, multi-objective continuous/discrete TMS-MSA, multi-objective SS-HSA, multi-objective SS-IHSA, and NSGA-II, respectively, from the standpoint of the obtained TCC index by corresponding optimization algorithms.

Table 6.50 Optimal robust transmission expansion plans of the proposed framework under the IGDT risk-averse decision-making policy and RO technique

Period	Uncertainty modeling technique	
	The IGDT risk-averse decision-making policy	The RO technique
1	(1-49: C), (7-10: I), (15-17: I)	(4-52: C), (7-12: C), (36.37: I)
2	(13-14: C), (30-32: I)	(21-35: C), (13-14: C)
3	(24-29: C), (27-52: C), (7-23: I)	(31-32: I), (7-24: I), (1-49: C)
4	(36.37: I), (24-32: I)	(33-34: C), (37-39: I), (9-10: C)
5	(4-50: C), (43-51: C) , (21-35: C)	(27-52: C), (42-43: I)
6	(17-19: I), (49-50: C)	(2-49: C), (7-23: C), (7-10: I)
7	(7-24: I), (48-51: C) , (37-39: I)	(15-14: I), (43-51: C)
8	(4-52: C), (39-40: C)	(17-19: I), (47-51: C) , (39-40: C)
9	(36.37: I), (33-34: C)	(23-34: I), (4-50: C), (4-5: I)
10	(7-24: I), (10-12: I)	(29-30: I), (6.50: C) , (26.29: I)
Number of newly installed transmission lines in existing corridors over the planning horizon	12	12
Number of newly installed transmission lines in new corridors over the planning horizon	12	15

Table 6.51 Changes in the objective functions of the third level of the proposed framework under the IGDT risk-averse decision-making policy and RO technique

Objectives	Uncertainty modeling technique	
	The IGDT risk-averse decision-making policy	The RO technique
TIC (M$)	479.297893	507.373359
TCC (M$)	10.005164	12.209900
ECOC (M$)	60.986187	67.239945
PD-TEPO (M$)	550.289243	586.823204

In addition, the ICS illustrates 5.059116%, 12.784405%, 18.648607%, 16.368394%, and 21.076515% superiority—positive sign—of the multi-objective SOSA performance compared with the performances of the multi-objective TMS-EMSA, multi-objective continuous/discrete TMS-MSA, multi-objective SS-HSA, multi-objective SS-IHSA, and NSGA-II, respectively, from the point of view of the obtained ECOC index by corresponding optimization algorithms. The ICS illustrates 9.152355% weakness—negative sign—of the multi-objective TMS-EMSA performance compared with the performance of the multi-objective SOSA, from the perspective of the obtained TIC index by the multi-objective SOSA, and 3.647036%, 11.399457%, 9.139695%, and 13.367666% superiority—positive sign—of the multi-objective TMS-EMSA performance compared with the performances of the multi-objective continuous/discrete TMS-MSA, multi-objective SS-HSA, multi-objective SS-IHSA, and NSGA-II, respectively, from the perspective of the obtained TIC index by corresponding optimization algorithms. Also, the ICS shows 10.105108% weakness—negative sign—of the multi-objective TMS-EMSA performance compared with the performance of the multi-objective SOSA, from the standpoint of the obtained TCC index by the multi-objective SOSA, and 3.253473%, 11.671809%, 8.862267%, and 12.662221% superiority—positive sign—of the multi-objective TMS-EMSA performance compared with the performances of the multi-objective continuous/discrete TMS-MSA, multi-objective SS-HSA, multi-objective SS-IHSA, and NSGA-II, respectively, from the standpoint of the obtained TCC index by corresponding optimization algorithms. In addition, the ICS represents 5.328702% weakness—negative sign—of the multi-objective TMS-EMSA performance compared with the performance of the multi-objective SOSA, from the point of view of the obtained ECOC index by the multi-objective SOSA, and 8.136946%, 14.313633%, 11.911915%, and 16.870918% superiority—positive sign—of the multi-objective TMS-EMSA performance compared with the performances of the multi-objective continuous/discrete TMS-MSA, multi-objective SS-HSA, multi-objective SS-IHSA, and NSGA-II, respectively, from the point of view of the obtained ECOC index by corresponding optimization algorithms. By the same token, the optimal results presented in Table 6.54 for the other multi-objective optimization algorithms, and in Table 6.55 for all multi-objective optimization algorithms, are analyzed in the same way.

Table 6.52 Calculated values for the objective functions of the third level of the proposed framework by the proposed multi-objective optimization algorithms under the first case

Multi-objective optimization algorithm	The IGDT risk-averse decision-making policy					
	First scenario: Ignoring the BBM			Second scenario: Considering the BBM		
	TIC over the planning horizon (M$)	TCC over the planning horizon (M$)	ECOC over the planning horizon (M$)	TIC over the planning horizon (M$)	TCC over the planning horizon (M$)	ECOC over the planning horizon (M$)
SOSA	531.267686	12.273694	64.385489	479.297893	10.005164	60.986187
TMS-EMSA	579.891195	13.513964	67.816400	518.308583	10.722802	65.183788
Continuous/ disrete TMS-MSA	601.840540	13.968423	73.823368	549.795465	11.409725	68.795356
SS-HSA	654.500729	15.299718	79.144913	587.839519	12.569679	75.799924
SS-IHSA	638.222815	14.828067	76.987030	572.456939	12.189094	72.828946
NSGA-II	669.370400	15.473217	81.579633	607.960019	12.907194	77.958896

Table 6.53 Calculated values for the objective functions of the third level of the proposed framework by the proposed multi-objective optimization algorithms under the second case

Multi-objective optimization algorithm	The IGDT risk-taker decision-making policy					
	First scenario: Ignoring the BBM			Second scenario: Considering the BBM		
	TIC over the planning horizon (M$)	TCC over the planning horizon (M$)	ECOC over the planning horizon (M$)	TIC over the planning horizon (M$)	TCC over the planning horizon (M$)	ECOC over the planning horizon (M$)
SOSA	194.645129	11.032460	47.930716	181.693654	10.853425	42.986253
TMS-EMSA	210.160495	11.659777	50.681844	194.160495	11.759777	45.681844
Continuous/ disrete TMS-MSA	221.361624	12.364377	53.674062	203.361624	12.394377	48.974062
SS-HSA	237.109202	13.618579	58.028571	223.109202	13.338579	52.028571
SS-IHSA	229.089539	13.279675	56.390297	218.089539	12.869675	50.390297
NSGA-II	241.444868	13.890429	59.610378	227.444868	13.960429	54.610378

Table 6.54 The ICS for different objectives of the third level of the proposed framework, based on various multi-objective optimization algorithms under the first and second scenarios of the first case

Case No.	Scenario No.	Perspective	Multi-objective optimization algorithms	Multi-objective optimization algorithms: SOSA	TMS-EMSA	Continuous/discrete TMS-MSA	SS-HSA	SS-IHSA	NSGA-II
First case	First scenario	TIC over the planning horizon	SOSA	0	8.384936	11.726171	18.828557	16.758274	20.631733
			TM-EMSA	−9.152355	0	3.647036	11.399457	9.139695	13.367666
			Continuous/discrete TMS-MSA	−13.283859	−3.785079	0	8.045856	5.700560	10.088563
			SS-HSA	−23.196036	−12.866126	−8.749857	0	−2.550506	2.221441
			SS-IHSA	−20.132059	−10.059062	−6.045168	2.487073	0	4.653265
			NSGA-II	−25.994939	−15.430343	−11.220556	−2.271910	−4.880362	0
		TCC over the planning horizon	SOSA	0	9.177692	12.132572	19.778299	17.226608	20.677813
			TM-EMSA	−10.105108	0	3.253473	11.671809	8.862267	12.662221
			Continuous/discrete TMS-MSA	−13.807815	−3.362884	0	8.701434	5.797411	9.725152
			SS-HSA	−24.654549	−13.214139	−9.530746	0	−3.180799	1.121285
			SS-IHSA	−20.811770	−9.724038	−6.154195	3.082743	0	4.169462
			NSGA-II	−26.068134	−14.497988	−10.772826	−1.134001	−4.350870	0
		ECOC over the planning horizon	SOSA	0	5.059116	12.784405	18.648607	16.368394	21.076515
			TM-EMSA	−5.328702	0	8.136946	14.313633	11.911915	16.870918
			Continuous/discrete TMS-MSA	−14.658395	−8.857692	0	6.723799	4.109344	9.507599
			SS-HSA	−22.923526	−16.704680	−7.208483	0	−2.802917	2.984470
			SS-IHSA	−19.572020	−13.522732	−4.285447	2.726496	0	5.629595
			NSGA-II	−26.704999	−20.294844	−10.506517	−3.076281	−5.965424	0

	Second scenario	TIC over the planning horizon	SOSA	0	7.526537	12.822508	18.464499	16.273546	21.162925
			TM-EMSA	−8.139132	0	5.727017	11.828217	9.458939	14.746271
			Continuous/discrete TMS-MSA	−14.708509	−6.074929	0	6.471843	3.958633	9.567167
			SS-HSA	−22.645964	−13.414969	−6.919674	0	−2.687115	3.309510
			SS-IHSA	−19.436565	−10.447127	−4.121800	2.616799	0	5.839706
			NSGA-II	−26.843875	−17.296922	−10.579307	−3.422787	−6.201877	0
		TCC over the planning horizon	SOSA	0	6.692635	12.310209	20.402390	17.917082	22.483817
			TM-EMSA	−7.172676	0	6.020504	14.693111	12.029540	16.923833
			Continuous/discrete TMS-MSA	−14.038361	−6.406189	0	9.228191	6.393986	11.601816
			SS-HSA	−25.631914	−17.223828	−10.166362	0	−3.122340	2.614937
			SS-IHSA	−21.828028	−13.674522	−6.830743	3.027802	0	5.563564
			NSGA-II	−29.005322	−20.371466	−13.124496	−2.685152	−5.891332	0
		ECOC over the planning horizon	SOSA	0	6.439639	11.351302	19.543208	16.261060	21.771356
			TM-EMSA	−6.882871	0	5.249726	14.005470	10.497416	16.386979
			Continuous/discrete TMS-MSA	−12.804816	−5.540592	0	9.240864	5.538443	11.754322
			SS-HSA	−24.290315	−16.286466	−10.181745	0	−4.079391	2.769372
			SS-IHSA	−19.418756	−11.728618	−5.863171	3.919499	0	6.580326
			NSGA-II	−27.830415	−19.598597	−13.319997	−2.848250	−7.043833	0

6.5 Coordination of Pseudo-Dynamic Generation and Transmission Expansion Planning: A Strategic Quad-Level Computational-Logical Framework

In previous sections, a strategic tri-level computational-logical framework was presented for each of the PD-GEP and the PD-TEP problems. However, if there is no coordination between these two problems, it is possible that the obtained optimal expansion plans for the generation and transmission networks may be economically undesirable. In other words, ignoring coordination of these problems could lead to a reduction in the profit of all stakeholders and unnecessary investments in generation and transmission expansion planning. By applying the coordination between the PD-GEP and PD-TEP problems, it is possible to avoid overinvestment, to earn more profit by all stakeholders, to achieve more flexible generation and transmission expansion plans, and so on. In this section, the authors present a strategic quad-level computational-logical framework for the coordination of pseudo-dynamic generation and transmission expansion-planning (PD-G&TEP) problems within deregulated environments. The proposed strategic quad-level computational-logical framework is split into a long-term planning master problem and a short-term operation slave problem. Providing CSC electricity market outcomes for the long-term planning master problem by taking into consideration both GENCOs' and DISCOs' bids is the main task of the short-term operation slave problem. In addition, providing coordination of optimal generation and transmission expansion plans compatible with the results of the short-term operation slave problem is the main task of the long-term planning master problem. In simple terms, the proposed strategic quad-level computational-logical framework scrutinizes the impact of the coordination of optimal generation and transmission expansion plans on strategic behaviors of both GENCOs and DISCOs and vice versa.

In order to model the strategic behaviors of the CSC electricity market participants, including GENCOs and DISCOs, the BBM problem is considered in the proposed framework as the lowest level (i.e., the first level). At this level, each GENCO/DISCO seeks to maximize its profit by modeling its strategic behavior in the form of a bidding mechanism. The CSC electricity market-clearing problem is also included in the proposed framework as the intermediate-one level (i.e., the second level). At this level, the CSC electricity market-clearing problem is done by the ISO with the aim of the CWF maximization. The first and second levels of the proposed framework give rise to the formation of the short-term operation slave problem. A decentralized PD-GEP problem is considered in the proposed framework as the intermediate-two level (i.e., the third level). The decentralized PD-GEP problem considered for this level is divided into two layers: (1) local and (2) global. In the local layer, each GENCO separately optimizes the EP, GIC, and CP, as a self-optimization problem that is limited to a set of environmental/techno-economic constraints. In the global layer, the regulatory body, or ISO, checks whether or not the optimal results obtained at the local layer by the GENCOs violate the predetermined global restrictions for the entire power system. Due to common concerns about network reliability and the time-consuming process of transmission network expansion, a centralized PD-TEP problem is included into the proposed framework as the uppermost level (i.e., the fourth level). In the proposed centralized PD-TEP problem, the TIC, the TCC, and the ECOC are the three objectives considered in the optimization problem. At the same time, both short-term and the long-term restrictions are modeled as constraints in the proposed PD-TEP problem. The third and fourth levels of the proposed framework bring about the formation of the long-term planning master problem. Figure 6.16 depicts a conceptual view of the proposed strategic quad-level computational-logical framework. The following assumptions are employed in the description of the mathematical process of the proposed strategic quad-level computational-logical framework:

- Hypotheses 1–10, discussed in Sect. 5.3 of Chap. 5, are evaluated.
- Hypothesis 11: Severe twofold uncertainties associated with the market price and demand parameters are taken into account and applied in the proposed strategic quad-level computational-logical framework.
- Hypothesis 12: A well-founded IGDT is widely employed to handle risks of the PD-G&TEP problem arising from severe twofold uncertainties.
- Each electrical load is split into S different sectors by the existing customer classes.

Table 6.56 presents a summary of the attributes of previous coordination of generation and transmission expansion planning (G&TEP) frameworks reported in the literature, and the PD-G&TEP framework proposed in this chapter. The authors, though, will only focus on the most important and relevant journal papers found in the literature. Also, Table 5.2 (see Chap. 5) illustrates features related to the bi-level computational-logical framework (i.e., the short-term operational slave problem). However, due to the fact that most technical studies in this area do not deal with a quad-level framework, the authors did not include them in Table 6.56.

6.5.1 Mathematical Model of the Deterministic Strategic Quad-Level Computational-Logical Framework

In this section, the mathematical model of the proposed deterministic strategic quad-level computational-logical framework is presented.

6.5.1.1 Bilateral Bidding Mechanism: First Level (Problem A)

In the first level of the proposed strategic quad-level computational-logical framework, each GENCO/DISCO manipulates its bidding strategy parameters (i.e., slope and intercept) in order to earn more profit. The first level of the proposed strategic quad-level computational-logical framework related to the PD-G&TEP problem is

Table 6.55 The ICS for different objectives of the third level of the proposed framework, based on various multi-objective optimization algorithms under the first and second scenarios of the second case

Case No.	Scenario No.	Perspective	Multi-objective optimization algorithms	Multi-objective optimization algorithms					
				SOSA	TMS-EMSA	Continuous/discrete TMS-MSA	SS-HSA	SS-IHSA	NSGA-II
Second case	First scenario	TIC over the planning horizon	SOSA	0	7.382627	12.069162	17.909078	15.035348	19.383198
			TM-EMSA	−7.971104	0	5.060104	11.365525	8.262727	12.957149
			Continuous/discrete TMS-MSA	−13.725745	−5.329797	0	6.641487	3.373316	8.317941
			SS-HSA	−21.816149	−12.822917	−7.113960	0	−3.500667	1.795716
			SS-IHSA	−17.696004	−9.006946	−3.491081	3.382265	0	5.117246
			NSGA-II	−24.043622	−14.885943	−9.072595	−1.828552	−5.393231	0
		TCC over the planning horizon	SOSA	0	5.3801800	10.772212	18.989639	16.922213	20.575095
			TM-EMSA	−5.686102	0	5.698629	14.383306	12.198325	16.058913
			Continuous/discrete TMS-MSA	−12.072711	−6.042997	0	9.209492	6.892472	10.986356
			SS-HSA	−23.441000	−16.799652	−10.143673	0	−2.552050	1.957102
			SS-IHSA	−20.369119	−13.893044	−7.402702	2.488541	0	4.396941
			NSGA-II	−25.905093	−19.131172	−12.342328	−1.996170	−4.599163	0
		ECOC over the planning horizon	SOSA	0	5.428231	10.700412	17.401522	15.001838	19.593336
			TM-EMSA	−5.739801	0	5.574793	12.660534	10.123112	14.978153
			Continuous/discrete TMS-MSA	−11.982600	−5.903924	0	7.504077	4.816848	9.958527
			SS-HSA	−21.067607	−14.495776	−8.112873	0	−2.905240	2.653576
			SS-IHSA	−17.649603	−11.263309	−5.060610	2.823219	0	5.401879
			NSGA-II	−24.367802	−17.616829	−11.059934	−2.725910	−5.710345	0

	Second scenario	TIC over the planning horizon	SOSA	0	6.420894	10.654896	18.562904	16.688505	20.115298
			TM-EMSA	−6.861462	0	4.524515	12.975129	10.972119	14.634040
			Continuous/discrete TMS-MSA	−11.925551	−4.738929	0	8.851081	6.753150	10.588607
			SS-HSA	−22.794163	−14.909679	−9.710572	0	−2.301652	1.906249
			SS-IHSA	−20.031456	−12.324362	−7.242229	2.249868	0	4.113229
			NSGA-II	−25.180414	−17.142711	−11.842570	-1.943293	−4.289673	0
		TCC over the planning horizon	SOSA	0	7.707220	12.432670	18.631325	15.666673	22.255791
			TM-EMSA	−8.350838	0	5.120063	11.836358	8.624133	15.763498
			Continuous/discrete TMS-MSA	−14.197841	−5.396360	0	7.078730	3.693162	11.217792
			SS-HSA	−22.897417	−13.425441	−7.617986	0	−3.643479	4.454376
			SS-IHSA	−18.577085	−9.438087	−3.834787	3.515396	0	7.813183
			NSGA-II	−28.626945	−18.713382	−12.635181	−4.662040	−8.475381	0
		ECOC over the planning horizon	SOSA	0	5.900792	12.226490	17.379524	14.693392	21.285560
			TM-EMSA	−6.270821	0	6.722370	12.198541	9.343967	16.349518
			Continuous/discrete TMS-MSA	−13.929590	−7.206841	0	5.870830	2.810531	10.320961
			SS-HSA	−21.035371	−13.893324	−6.236993	0	−3.251169	4.727685
			SS-IHSA	−17.224214	−10.307055	−2.891806	3.148796	0	7.727617
			NSGA-II	−27.041494	−19.545038	−11.508777	−4.962286	-8.374788	0

virtually the same as the first level of the proposed PD-GEP and PD-TEP problems presented in Sects. 6.3 and 6.4 of this chapter, respectively.

Therefore, the mathematical model of the BBM (i.e., the first level) is considered in accordance with Eqs. (6.1) through (6.9), which were explained in Sect. 6.3.1.1 of this chapter.

6.5.1.2 Competitive Security-Constrained Electricity Market: Second Level (Problem B)

In the second level of the proposed strategic quad-level computational-logical framework, it is assumed that the CSC electricity market-clearing problem is solved by the ISO with the aim of maximizing the CWF. The second level of the proposed strategic quad-level computational-logical framework related to the PD-G&TEP problem is virtually the same as the second level of the proposed PD-GEP and PD-TEP problems presented in Sects. 6.3 and 6.4 of this chapter. Therefore, the mathematical model of the CSC electricity market-clearing problem (i.e., the second level) is taken into account using Eqs. (6.10) through (6.14), which were described in Sect. 6.3.1.2 of this chapter.

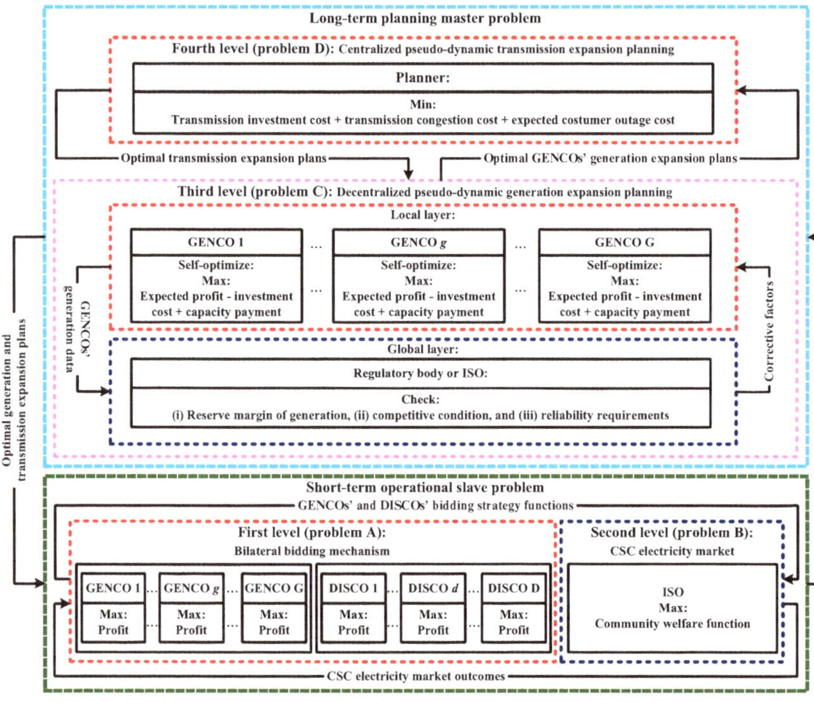

Fig. 6.16 Conceptual view of the proposed strategic quad-level computational-logical framework

6.5.1.3 Pseudo-Dynamic Generation Expansion Planning: Third Level (Problem C)

In the third level of the proposed strategic quad-level computational-logical framework, a decentralized PD-GEP problem is considered. This level of the proposed framework is divided into two layers: (1) local and (2) global. In the local layer, each GENCO separately solves the PD-GEP problem in order to determine its own optimal generation expansion plans over the planning horizon. The proposed PD-GEP problem takes into account the EP, the GIC, and the CP in the optimization as three objective functions. Concurrently, a set of environmental/techno-economic restrictions are employed as PD-GEP problem constraints. In the global layer, however, the specified optimal expansion plans by GENCOs are evaluated by the regulatory body, or ISO, to ensure that these optimal expansion plans do not violate any established global restrictions concerned with the entire power system, which include (1) generation reserve margin, (2) competitive condition, and (3) reliability requirements. The third level of the proposed strategic quad-level computational-logical framework related to the PD-G&TEP problem is virtually the same as the third level of the proposed PD-GEP problem in Sect. 6.3 of this chapter. Therefore, the mathematical model of the PD-GEP problem (i.e., the third level) is taken into account based on Eqs. (6.15) through (6.35), which were expressed in Sect. 6.3.1.3 of this chapter.

6.5.1.4 Pseudo-Dynamic Transmission Expansion Planning: Fourth Level (Problem D)

In the fourth level of the proposed strategic quad-level computational-logical framework, a centralized PD-TEP problem is used. The TIC, the TCC, and the ECOC are taken into account as three objective functions in the proposed PD-TEP problem. At the same time, the proposed PD-TEP problem is subjected to two different classes of constraints: (1) short term and (2) long term. The fourth level of the proposed strategic quad-level computational-logical framework associated with the PD-G&TEP problem is virtually the same as the third level of the proposed PD-TEP problem in Sect. 6.4 of this chapter. Therefore, the mathematical model of the PD-GEP problem (i.e., the fourth level) is considered on the basis of Eqs. (6.106) through (6.131), which were presented in Sect. 6.4.1.3 of this chapter.

6.5.2 Overview of the Deterministic Strategic Quad-Level Computational-Logical Framework

According to the equations identified in the previous sections, the proposed PD-G&TEP problem can be entirely formulated as a deterministic strategic quad-level computational-logical framework, as follows:

Table 6.56 Attributes of previous frameworks reported in the G&TEP literature and the strategic quad-level computational-logical framework proposed in this study

Reference	Multi-level framework	Type of generation expansion	Static or multi-period framework	Nondeterministic framework	Reliability consideration	Type of modeling reliability criterion	Environmental constraints	System security constraint	Loss consideration	Uncertainty consideration	Coordinated decision	Strategic behaviors of CSC electricity market participants	Impacts of expansion plans on strategic behavior and vice versa	Solution method
[60]	Tri-level	Only capacity expansion	Multi-period	No	No	–	No	Yes	No	No	Yes	No	No	Benders decomposition
[61]	Tri-level	Only capacity expansion	Multi-period	Nondeterministic framework: scenario generation	Yes: reliability constraint	–	No	Yes	No	Threefold: demand, outages of generation units, and outage of transmission lines	Yes	No	No	GAMS
[62]	Quad-level	Only capacity expansion	Multi-period	No	Yes: reliability assessment incompatible with the value-based approach	Deterministic model	No	Yes	No	No	Yes	Yes: one-sided	No	Q-learning and search-based technique
[63]	Bi-level	Only capacity expansion	Static	No	No	–	No	Yes	No	No	No	No	No	GAMS
[64, 65]	Tri-level	Only capacity expansion	Static	No	No	–	No	Yes	No	No	Yes	Yes: one-sided	No	GA
[66]	Bi-level	Only capacity expansion	Static	No	Yes: reliability assessment incompatible with the value-based approach	Deterministic model	Yes: SO_2, and NO_x atmospheric emissions	Yes	No	No	No	No	No	GAMS
[67]	Tri-level	Only capacity expansion	Static	Nondeterministic framework: scenario generation	No	–	No	Yes	No	Onefold: demand	Yes	No	No	GAMS

[68]	Tri-level	Only capacity expansion	Multi-period	Nondeterministic framework: Monte Carlo simulation	Yes: reliability constraint	–	No	Yes	No	Fourfold: demand, bid prices, outages of generation units, and outage of transmission lines	Yes	No	No	GAMS and iterative approach
[69]	Bi-level	Only capacity expansion	Static	No	No	–	No	Yes	No	No	Yes	No	No	GAMS
[70, 71]	Tri-level	Only capacity expansion	Static	No	No	–	No	Yes	No	No	Yes	Yes: one-sided	No	Iterative approach
[72]	Bi-level	Only capacity expansion	Static	No	Yes: reliability assessment incompatible with the value-based approach	Deterministic model	No	Yes	No	No	No	No	No	GAMS
[73]	Bi-level	Only capacity expansion	Multi-period	No	Yes: reliability assessment incompatible with the value-based approach	Deterministic model	Yes: SO_2, and NO_x atmospheric emissions	Yes	No	No	No	No	No	Benders decomposition
Proposed framework	Quad-level	Allocation and capacity expansion	Multi-period	Nondeterministic framework: stochastic modeling by the IGDT (risk-averse and risk-taker frameworks)	Yes: value-based reliability approach compatible with the restructured environment	Probabilistic model based on sector customer damage function	Yes: SO_2, CO_2, and NO_x atmospheric emissions	Yes	Yes	Severe twofold: price of the CSC electricity market and electrical power demand	Yes	Yes: bilateral bidding mechanism	Yes	Five new multi-objective music-inspired optimization algorithms (see Chap. 4): (1) SS-HSA; (2) SS-IHSA; (3) continuous/discrete TMS-MSA; (4) TMS-EMSA; and (5) SOSA

$$\underset{x_{\text{PD-TEP}}}{\text{Min}} : \left\{\text{OF}^{\text{PD-TEP}}\right\}$$
$$= \underset{x_{\text{PD-TEP}}}{\text{Min}} : \left\{\text{W}_{\text{OF}_1}^{\text{PD-TEP}} \cdot \text{OF}_1^{\text{PD-TEP}} + \text{W}_{\text{OF}_2}^{\text{PD-TEP}} \cdot \text{OF}_2^{\text{PD-TEP}} + \text{W}_{\text{OF}_3}^{\text{PD-TEP}} \cdot \text{OF}_3^{\text{PD-TEP}}\right\}$$

(6.191) Fourth level

where

$$\text{OF}_1^{\text{PD-TEP}} = TIC_y = \sum_{p \in \Psi^{\text{P}}} \left(\frac{1}{(1+\text{Ir})^{p-p_0}}\right) \cdot TIC_{y,p}; \quad \forall \left\{p \in \Psi^{\text{P}}\right\} \tag{6.192}$$

$$TIC_{y,p} = \sum_{(s,r) \in \Psi^{\text{L}}} TIC_{p,(s,r)} \cdot \iota_{p,(s,r)}; \quad \forall \left\{p \in \Psi^{\text{P}}, (s,r) \in \Psi^{\text{L}}\right\} \tag{6.193}$$

$$TIC_{p,(s,r)} = CCL^{\text{b}} \cdot LE_{(s,r)} \cdot f_{(s,r)}; \quad \forall \left\{p \in \Psi^{\text{P}}, (s,r) \in \Psi^{\text{L}}\right\} \tag{6.194}$$

$$\text{OF}_2^{\text{PD-TEP}} = TCC_y = \sum_{p \in \Psi^{\text{P}}} \left(\frac{1}{(1+\text{Ir})^{p-p_0}}\right) \cdot TCC_{y,p}; \quad \forall \left\{p \in \Psi^{\text{P}}\right\} \tag{6.195}$$

$$TCC_{y,p} = \sum_{k \in \Psi^{\text{K}}} \Delta t_k \cdot TCC_{p,k}; \quad \forall \left\{p \in \Psi^{\text{P}}, k \in \Psi^{\text{K}}\right\} \tag{6.196}$$

$$TCC_{p,k} = \sum_{l \in \Psi^{\text{L}}} \left(p\lambda_{p,k,l_2} - p\lambda_{p,k,l_1}\right) \cdot f_{p,k,l_{1-2}}; \quad \forall \left\{p \in \Psi^{\text{P}}, k \in \Psi^{\text{K}}, l \in \Psi^{\text{L}}\right\} \tag{6.197}$$

$$\text{OF}_3^{\text{PD-TEP}} = ECOC_y^{\text{tran}} = \sum_{p \in \Psi^{\text{P}}} \left(\frac{1}{(1+\text{Ir})^{p-p_0}}\right) \cdot ECOC_{y,p}^{\text{tran}}; \quad \forall \left\{p \in \Psi^{\text{P}}\right\} \tag{6.198}$$

$$ECOC_{y,p}^{\text{tran}} = ECOC_{y,p}^{\text{HL-II}} - ECOC_{y,p}^{\text{HL-I}}; \quad \forall \left\{p \in \Psi^{\text{P}}\right\} \tag{6.199}$$

and subject to Eqs. (6.200) through (6.207):

$$\text{NB}_p \cdot \text{PF}_{p,k} + \text{GE}_{p,k} = \text{DE}_{p,k} + \text{ADE}_{p,k}; \quad \forall \left\{p \in \Psi^{\text{P}}, k \in \Psi^{\text{K}}\right\} \tag{6.200}$$

$$f_{p,k,(s,r)} - \left(y_{p,(s,r)}^0 + \tau_{p,(s,r)} \cdot \iota_{p,(s,r)}\right) \cdot \left(\theta_{p,k,s} - \theta_{p,k,r}\right) = 0; \quad \forall \left\{p \in \Psi^{\text{P}}, k \in \Psi^{\text{K}}, (s,r) \in \Psi^{\text{L}}\right\} \tag{6.201}$$

$$\left|f_{p,k,(s,r)}\right| \leq \left(y_{p,(s,r)}^0 + \tau_{p,(s,r)} \cdot \iota_{p,(s,r)}\right) \cdot f_{p,(s,r)}^{\max} = 0; \quad \forall \left\{p \in \Psi^{\text{P}}, k \in \Psi^{\text{K}}, (s,r) \in \Psi^{\text{L}}\right\} \tag{6.202}$$

$$TIC_{y,p} \le TIC_{p}^{\max}; \quad \forall \{p \in \Psi^{\mathrm{P}}\} \tag{6.203}$$

$$\sum_{p \in \Psi^{\mathrm{P}}} TIC_{y,p} \le TIC^{\max}; \quad \forall \{p \in \Psi^{\mathrm{P}}\} \tag{6.204}$$

$$0 \le \sum_{p \in \Psi^{\mathrm{P}}} \iota_{p,(s,r)} \le \iota_{(s,r)}^{\max}; \quad \forall \{p \in \Psi^{\mathrm{P}}, (s,r) \in \Psi^{\mathrm{L}}, \iota_{p,(s,r)} \text{ is an integer variable}\} \tag{6.205}$$

$$Y_{p,sr} = -\left(y_{(p-1),(s,r)}^{0} + \iota_{p,(s,r)} \cdot \tau_{p,(s,r)}\right); \quad \forall \{p \in \Psi^{\mathrm{P}}, (s,r) \in \Psi^{\mathrm{L}}, s \ne r\} \tag{6.206}$$

$$Y_{p,ss} = y_{(p-1),ss}^{0} + \sum_{r \in \Psi_{s}^{\mathrm{L}}} \left(y_{(p-1),(s,r)}^{0} + \iota_{p,(s,r)} \cdot \tau_{p,(s,r)}\right); \quad \forall \{p \in \Psi^{\mathrm{P}}, s \in \Psi^{\mathrm{L}}, r \in \Psi_{s}^{\mathrm{L}}, s \ne r\} \tag{6.207}$$

$$\underset{x_{\text{PD-GEP}}}{\text{Max}} : \left\{\text{OF}_{g}^{\text{PD-GEP}}\right\} = \underset{x_{\text{PD-GEP}}}{\text{Max}} : \left\{\text{W}_{OF_{1},g}^{\text{PD-GEP}} \cdot \text{OF}_{1,g}^{\text{PD-GEP}} - \text{W}_{\text{OF}_{2},g}^{\text{PD-GEP}} \cdot \text{OF}_{2,g}^{\text{PD-GEP}} + \text{W}_{\text{OF}_{3},g}^{\text{PD-GEP}} \cdot \text{OF}_{3,g}^{\text{PD-GEP}}\right\}$$

(6.208) Third level

where

$$\text{OF}_{1,g}^{\text{PD-GEP}} = EP_{y,g} = \sum_{p \in \Psi^{\mathrm{P}}} \left(\frac{1}{(1+\text{Ir})^{p-p_0}}\right) \cdot EP_{y,g,p}; \quad \forall \{p \in \Psi^{\mathrm{P}}\} \tag{6.209}$$

$$\begin{aligned} EP_{y,g,p} = & \sum_{k \in \Psi^{\mathrm{K}}} \sum_{eu \in \Psi_{g}^{\mathrm{EU}}} \Delta t_{k} \cdot \\ & \left[\left(p\lambda_{p,k,b_{eu}^{g}} \cdot \rho_{p,k,g_{eu}}\right) - \left(\frac{1}{2} \cdot \alpha_{p,k,g_{eu}} \cdot \rho_{p,k,g_{eu}}^{2} + \beta_{p,k,g_{eu}} \cdot \rho_{p,k,g_{eu}} + \gamma_{p,k,g_{eu}}\right)\right] \\ & + \sum_{k \in \Psi^{\mathrm{K}}} \sum_{iu \in \Psi_{g}^{\mathrm{IU}}} \Delta t_{k} \cdot \\ & \left[\left(p\lambda_{p,k,b_{iu}^{g}} \cdot \rho_{p,k,g_{iu}}\right) - \left(\frac{1}{2} \cdot \alpha_{p,k,g_{iu}} \cdot \rho_{p,k,g_{iu}}^{2} + \beta_{p,k,g_{iu}} \cdot \rho_{p,k,g_{iu}} + \gamma_{p,k,g_{iu}}\right)\right]; \\ & \forall \left\{p \in \Psi^{\mathrm{P}}, k \in \Psi^{\mathrm{K}}, eu \in \Psi_{g}^{\mathrm{EU}}, iu \in \Psi_{g}^{\mathrm{IU}}, b_{eu}^{g} \in \Psi_{g_{eu}}^{\mathrm{B}}, b_{iu}^{g} \in \Psi_{g_{iu}}^{\mathrm{B}}\right\} \end{aligned} \tag{6.210}$$

$$\mathrm{OF}_{2,g}^{\text{PD-GEP}} = GIC_{y,g} = \sum_{p\in\Psi^{\mathrm{P}}} \left(\frac{1}{(1+\mathrm{Ir})^{p-p_0}}\right) \cdot GIC_{y,g,p}; \quad \forall\{p \in \Psi^{\mathrm{P}}\} \tag{6.211}$$

$$GIC_{y,g,p} = \sum_{iu\in\Psi_g^{\mathrm{IU}}} \left(GIC_{p,g_{iu}}^{\mathrm{f}} + GIC_{p,g_{iu}}^{\mathrm{v}}\right); \quad \forall\left\{p \in \Psi^{\mathrm{P}}, iu \in \Psi_g^{\mathrm{IU}}\right\} \tag{6.212}$$

$$GIC_{p,g_{iu}}^{\mathrm{v}} = GIC_{p,g_{iu}} \cdot IC_{p,g_{iu}}; \quad \forall\left\{p \in \Psi^{\mathrm{P}}, iu \in \Psi_g^{\mathrm{IU}}\right\} \tag{6.213}$$

$$\mathrm{OF}_{3,g}^{\text{PD-GEP}} = CP_{y,g} = \sum_{p\in\Psi^{\mathrm{P}}} CP_{y,g,p}; \quad \forall\{p \in \Psi^{\mathrm{P}}\} \tag{6.214}$$

$$CP_{y,g,p} = \sum_{k\in\Psi^{\mathrm{K}}} \sum_{iu\in\Psi^{\mathrm{IU}}} \Delta t_k \cdot \left(ECO_{p,k,g_{iu}}^{\mathrm{b}} - ECO_{p,k,g_{iu}}^{\mathrm{a}}\right) \cdot \left(\mathrm{ECOC} - p\lambda_{p,k,b_{iu}^g}\right); \quad \forall\left\{p \in \Psi^{\mathrm{P}}, k \in \Psi^{\mathrm{K}}, iu \in \Psi_g^{\mathrm{IU}}\right\} \tag{6.215}$$

and subject to Eqs. (6.216) through (6.228):

$$\rho_{p,k,g_{eu}}^{\min} \le \rho_{p,k,g_{eu}} \le \rho_{p,k,g_{eu}}^{\max}; \quad \forall\left\{p \in \Psi^{\mathrm{P}}, k \in \Psi^{\mathrm{K}}, eu \in \Psi_g^{\mathrm{EU}}\right\} \tag{6.216}$$

$$\rho_{p,k,g_{iu}}^{\min} \le \rho_{p,k,g_{iu}} \le \rho_{p,k,g_{iu}}^{\max}; \quad \forall\left\{p \in \Psi^{\mathrm{P}}, k \in \Psi^{\mathrm{K}}, iu \in \Psi_g^{\mathrm{IU}}\right\} \tag{6.217}$$

$$IC_{p,g_{iu}} = \max\left\{\rho_{p,k,g_{iu}}\right\}; \quad \forall\left\{p \in \Psi^{\mathrm{P}}, k \in \Psi^{\mathrm{K}}, iu \in \Psi_g^{\mathrm{IU}}\right\} \tag{6.218}$$

$$\sum_{iu\in\Psi_g^{\mathrm{IU}}} IC_{p,g_{iu}} \le IC_{p,g}^{\max}; \quad \forall\left\{p \in \Psi^{\mathrm{P}}, k \in \Psi^{\mathrm{K}}, iu \in \Psi_g^{\mathrm{IU}}\right\} \tag{6.219}$$

$$\sum_{k\in\Psi^{\mathrm{K}}} \sum_{eu\in\Psi_g^{\mathrm{EU}}} \Delta t_k \cdot F_{p,k,g_{eu}} \cdot \rho_{p,k,g_{eu}} + \sum_{k\in\Psi^{\mathrm{K}}} \sum_{iu\in\Psi_g^{\mathrm{IU}}} \Delta t_k \cdot F_{p,k,g_{iu}} \cdot \rho_{p,k,g_{iu}} \le F_{p,g}^{\max}; \quad \forall\left\{p \in \Psi^{\mathrm{P}}, k \in \Psi^{\mathrm{K}}, eu \in \Psi_g^{\mathrm{EU}}, iu \in \Psi_g^{\mathrm{IU}}\right\} \tag{6.220}$$

$$\sum_{iu\in\Psi_g^{\mathrm{IU}}} GIC_{p,g_{iu}} \le GIC_{p,g}^{\max}; \quad \forall\left\{p \in \Psi^{\mathrm{P}}, iu \in \Psi_g^{\mathrm{IU}}\right\} \tag{6.221}$$

$$\sum_{p\in\Psi^{\mathrm{P}}} \sum_{iu\in\Psi_g^{\mathrm{IU}}} GIC_{p,g_{iu}} \le GIC_g^{\max}; \quad \forall\left\{p \in \Psi^{\mathrm{P}}, iu \in \Psi_g^{\mathrm{IU}}\right\} \tag{6.222}$$

$$\sum_{k\in\Psi^{\mathrm{K}}} \sum_{eu\in\Psi_g^{\mathrm{EU}}} \Delta t_k \cdot E_{p,k,g_{eu}}^{\mathrm{SO_2}} \cdot \rho_{p,k,g_{eu}} + \sum_{k\in\Psi^{\mathrm{K}}} \sum_{iu\in\Psi_g^{\mathrm{IU}}} \Delta t_k \cdot E_{p,k,g_{iu}}^{\mathrm{SO_2}} \cdot \rho_{p,k,g_{iu}} \le E_{p,g}^{\mathrm{SO_2},\max}; \quad \forall\left\{p \in \Psi^{\mathrm{P}}, k \in \Psi^{\mathrm{K}}, eu \in \Psi_g^{\mathrm{EU}}, iu \in \Psi_g^{\mathrm{IU}}\right\} \tag{6.223}$$

$$\sum_{k\in\Psi^{\mathrm{K}}}\sum_{eu\in\Psi_g^{\mathrm{EU}}}\Delta t_k\cdot E_{p,k,g_{eu}}^{\mathrm{CO_2}}\cdot\rho_{p,k,g_{eu}}+\sum_{k\in\Psi^{\mathrm{K}}}\sum_{iu\in\Psi_g^{\mathrm{IU}}}\Delta t_k\cdot E_{p,k,g_{iu}}^{\mathrm{CO_2}}\cdot\rho_{p,k,g_{iu}}\le E_{p,g}^{\mathrm{CO_2},\max};$$
$$\forall\left\{p\in\Psi^{\mathrm{P}},k\in\Psi^{\mathrm{K}},eu\in\Psi_g^{\mathrm{EU}},iu\in\Psi_g^{\mathrm{IU}}\right\}\tag{6.224}$$

$$\sum_{k\in\Psi^{\mathrm{K}}}\sum_{eu\in\Psi_g^{\mathrm{EU}}}\Delta t_k\cdot E_{p,k,g_{eu}}^{\mathrm{NO_x}}\cdot\rho_{p,k,g_{eu}}+\sum_{k\in\Psi^{\mathrm{K}}}\sum_{iu\in\Psi_g^{\mathrm{IU}}}\Delta t_k\cdot E_{p,k,g_{iu}}^{\mathrm{NO_x}}\cdot\rho_{p,k,g_{iu}}\le E_{p,g}^{\mathrm{NO_x},\max};$$
$$\forall\left\{p\in\Psi^{\mathrm{P}},k\in\Psi^{\mathrm{K}},eu\in\Psi_g^{\mathrm{EU}},iu\in\Psi_g^{\mathrm{IU}}\right\}\tag{6.225}$$

$$\left(1+RM_p^{\min}\right)\cdot\left(\sum_{k\in\Psi^{\mathrm{K}}}\sum_{b\in\Psi^{\mathrm{B}}}p\rho_{p,k,b}^{\mathrm{p}}\right)\le\left(1+\eta_p^{\mathrm{res}}\right)\cdot$$
$$\left(\sum_{g\in\Psi^{\mathrm{G}}}\sum_{eu\in\Psi_g^{\mathrm{EU}}}EC_{p,g_{eu}}+\sum_{g\in\Psi^{\mathrm{G}}}\sum_{iu\in\Psi_g^{\mathrm{IU}}}IC_{p,g_{iu}}\right)\le\left(1+RM_p^{\max}\right)\cdot\left(\sum_{k\in\Psi^{\mathrm{K}}}\sum_{b\in\Psi^{\mathrm{B}}}p\rho_{p,k,b}^{\mathrm{p}}\right);$$
$$\forall\left\{g\in\Psi^{\mathrm{G}},p\in\Psi^{\mathrm{P}},k\in\Psi^{\mathrm{K}},b\in\Psi^{\mathrm{B}},eu\in\Psi_g^{\mathrm{EU}},iu\in\Psi_g^{\mathrm{IU}}\right\}\tag{6.226}$$

$$\left(1+\eta_{p,g}^{\mathrm{cap}}\right)\cdot\sum_{iu\in\Psi_g^{\mathrm{IU}}}IC_{p,g_{iu}}\le\frac{PERC_{p,g}^{\max}}{100}\cdot\sum_{g\in\Psi^{\mathrm{G}}}\sum_{iu\in\Psi_g^{\mathrm{IU}}}IC_{p,g_{iu}};\tag{6.227}$$
$$\forall\left\{g\in\Psi^{\mathrm{G}},p\in\Psi^{\mathrm{P}},g\in\Psi^{\mathrm{G}},iu\in\Psi_g^{\mathrm{IU}}\right\}$$

$$\left(1+\eta_p^{\mathrm{rel}}\right)\cdot ECO_p\le ECO^{\max};\quad\forall\left\{p\in\Psi^{\mathrm{P}}\right\}\tag{6.228}$$

$$\underset{x_{\mathrm{BBM}}}{\mathrm{Max}}:\tilde{\upsilon}_{p,k,g}\left(\rho_{p,k,g}\right)=\frac{\left(2\xi_{1,p,k,g}-1\right)}{2}\cdot\alpha_{p,k,g}\cdot\rho_{p,k,g}^2$$
$$+\left(\xi_{2,p,k,g}-1\right)\cdot\beta_{p,k,g}\cdot\rho_{p,k,g}$$
$$-\gamma_{p,k,g};\forall\left\{p\in\Psi^{\mathrm{P}},k\in\Psi^{\mathrm{K}},g\in\Psi^{\mathrm{G}}\right\}\quad(6.229)\ \text{First level}$$

$$\underset{x_{\mathrm{BBM}}}{\mathrm{Max}}:\tilde{\upsilon}_{p,k,d}\left(\rho_{p,k,d}\right)=\frac{\left(1-2\xi_{1,p,k,d}\right)}{2}\cdot\alpha_{p,k,d}\cdot\rho_{p,k,d}^2$$
$$+\left(1-\xi_{2,p,k,d}\right)\cdot\beta_{p,k,d}\cdot\rho_{p,k,d}$$
$$+\gamma_{p,k,d};\quad\forall\left\{p\in\Psi^{\mathrm{P}},k\in\Psi^{\mathrm{K}},d\in\Psi^{\mathrm{D}}\right\}\quad(6.230)\ \text{First level}$$

and subject to Eqs. (6.231) through (6.237):

$$\sum_{g \in \Psi^{G}} \rho_{p,k,g} = \sum_{l \in \Psi^{L}} \kappa_{p,k,l} + \sum_{d \in \Psi^{D}} \rho_{p,k,d}; \quad \forall \{p \in \Psi^{P}, k \in \Psi^{K}, g \in \Psi^{G}, l \in \Psi^{L}, d \in \Psi^{D}\} \tag{6.231}$$

$$\rho_{p,k,g}^{\min} \leq \rho_{p,k,g} \leq \rho_{p,k,g}^{\max}; \quad \forall \{p \in \Psi^{P}, k \in \Psi^{K}, g \in \Psi^{G}\} \tag{6.232}$$

$$\rho_{p,k,d}^{\min} \leq \rho_{p,k,d} \leq \rho_{p,k,d}^{\max}; \quad \forall \{p \in \Psi^{P}, k \in \Psi^{K}, d \in \Psi^{D}\} \tag{6.233}$$

$$\xi_{1,p,k,g}^{\min} \leq \xi_{1,p,k,g} \leq \xi_{1,p,k,g}^{\max}; \quad \forall \{p \in \Psi^{P}, k \in \Psi^{K}, g \in \Psi^{G}\} \tag{6.234}$$

$$\xi_{2,p,k,g}^{\min} \leq \xi_{2,p,k,g} \leq \xi_{2,p,k,g}^{\max}; \quad \forall \{p \in \Psi^{P}, k \in \Psi^{K}, g \in \Psi^{G}\} \tag{6.235}$$

$$\xi_{1,p,k,d}^{\min} \leq \xi_{1,p,k,d} \leq \xi_{1,p,k,d}^{\max}; \quad \forall \{p \in \Psi^{P}, k \in \Psi^{K}, d \in \Psi^{D}\} \tag{6.236}$$

$$\xi_{2,p,k,d}^{\min} \leq \xi_{2,p,k,d} \leq \xi_{2,p,k,d}^{\max}; \quad \forall \{p \in \Psi^{P}, k \in \Psi^{K}, d \in \Psi^{D}\} \tag{6.237}$$

$$\underset{x_{\text{CSC-EM}}}{\text{Max}} : CWF_{p,k} = \left(\sum_{g \in \Psi^{G}} \tilde{\upsilon}_{p,k,g}\left(\rho_{p,k,g}\right) + \sum_{d \in \Psi^{D}} \tilde{\upsilon}_{p,k,d}\left(\rho_{p,k,d}\right) \right); \quad \forall \{p \in \Psi^{P}, k \in \Psi^{K}, g \in \Psi^{G}, d \in \Psi^{D}\}$$

(6.238) Second level

$$= \underset{x_{\text{CSC-EM}}}{\text{Max}} : \left(\sum_{g \in \Psi^{G}} \frac{\left(2\xi_{1,p,k,g} - 1\right)}{2} \cdot \alpha_{p,k,g} \cdot \rho_{p,k,g}^{2} + \left(\xi_{2,p,k,g} - 1\right) \cdot \beta_{p,k,g} \cdot \rho_{p,k,g} - \gamma_{p,k,g} + \sum_{d \in \Psi^{D}} \frac{\left(1 - 2\xi_{1,p,k,d}\right)}{2} \cdot \alpha_{p,k,d} \cdot \rho_{p,k,d}^{2} + \left(1 - \xi_{2,p,k,d}\right) \cdot \beta_{p,k,d} \cdot \rho_{p,k,d} + \gamma_{p,k,d} \right); \quad \forall \{p \in \Psi^{P}, k \in \Psi^{K}, g \in \Psi^{G}, d \in \Psi^{D}\}$$

and subject to Eqs. (6.239) through (6.242):

$$\sum_{g \in \Psi^{G}} \rho_{p,k,g} = \sum_{l \in \Psi^{L}} \kappa_{p,k,l} + \sum_{d \in \Psi^{D}} \rho_{p,k,d}; \quad \forall \{p \in \Psi^{P}, k \in \Psi^{K}, g \in \Psi^{G}, l \in \Psi^{L}, d \in \Psi^{D}\} \tag{6.239}$$

$$\rho_{p,k,g}^{\min} \leq \rho_{p,k,g} \leq \rho_{p,k,g}^{\max}; \quad \forall \{p \in \Psi^{P}, k \in \Psi^{K}, g \in \Psi^{G}\} \tag{6.240}$$

$$\rho_{p,k,d}^{\min} \le \rho_{p,k,d} \le \rho_{p,k,d}^{\max}; \quad \forall \{p \in \Psi^{\mathrm{P}}, k \in \Psi^{\mathrm{K}}, d \in \Psi^{\mathrm{D}}\} \tag{6.241}$$

$$f_{p,k,l}^{\min} \le f_{p,k,l} \le f_{p,k,l}^{\max}; \quad \forall \{p \in \Psi^{\mathrm{P}}, k \in \Psi^{\mathrm{K}}, l \in \Psi^{\mathrm{L}}\} \tag{6.242}$$

6.5.3 *Mathematical Model of the Risk-Driven Strategic Quad-Level Computational-Logical Framework*

The IGDT is widely used to cope with the risks of the PD-G&TEP problem stemming from severe twofold uncertainty parameters of market price and demand. The basic characteristics of these uncertainty parameters, in the form of an uncertainty matrix, are presented in Table 6.3. The objective functions and constraints existing at all levels of the proposed strategic quad-level computational-logical framework are severely affected in the face of these uncertain parameters. To investigate these effects, the two decision-making policies of the proposed framework—the IGDT risk-averse and the risk-taker decision-making policies—are developed within a stochastic environment. Detailed descriptions of these decision-making policies are also provided in the following sections.

6.5.3.1 The IGDT Severe Twofold Uncertainty Model

In order to handle the uncertainty parameters involved in the proposed PD-G&TEP problem, the envelope-bound IGDT model is employed. Full details associated with the envelope-bound IGDT model of the uncertain market price and demand are indicated in Sect. 6.3.3.1.

6.5.3.2 The IGDT Risk-Averse Decision-Making Policy: Robustness Function

As the transmission expansion planning problem is located in the uppermost level (i.e., the fourth level) of the proposed strategic quad-level computational-logical framework and given its interdependencies with other levels, risk handling of the PD-G&TEP problem arising from the market price and demand uncertainties is applied from the perspective of this level. The robustness and opportunity functions for the proposed strategic quad-level computational-logical framework are, hence, modeled at this level (i.e., the PD-TEP problem). In general, the robustness function represents the destructive face of the severe twofold uncertainties in the IGDT-based strategic quad-level computational-logical framework. In other words, the robustness function addresses the greatest level of uncertainty parameters so that the maximum value of the PD-TEPO cannot be greater than a predetermined critical cost. Therefore, the robustness function for the proposed strategic quad-level computational-logical framework can be defined using Eq. (6.243):

$$\Upsilon^{\text{PD-TEP}}\left(x_{\text{PD-TEP}}, \varpi_{c}^{\text{PD-TEP}}\right) = \underset{\substack{\Delta^{\text{price}}\\ \Delta^{\text{demand}}}}{\text{Max}} \left\{ \left(\Delta^{\text{price}}, \Delta^{\text{demand}}\right) : \underset{\substack{x_{\text{PD-TEP}} \\ a\lambda_{p,k,b} \in \Pi_{p,k,b}^{\text{price}}\left(\Delta^{\text{price}}, p\lambda_{p,k,b}\right) \\ a\rho_{p,k,b} \in \Pi_{p,k,b}^{\text{demand}}\left(\Delta^{\text{demand}}, p\rho_{p,k,b}\right)}}{\text{Max}} \text{PD-TEPO}\left(x_{\text{PD-TEP}}, a\lambda_{p,k,b}, a\rho_{p,k,b}\right) \leq \varpi_{c}^{\text{PD-TEP}} \right\};$$

$$\forall\left\{\Delta^{\text{demand}} > 0, \Delta^{\text{price}} > 0, p \in \Psi^{\text{P}}, k \in \Psi^{\text{K}}, b \in \Psi^{\text{B}}\right\} \tag{6.243}$$

Hence, the IGDT risk-averse decision-making policy for the proposed strategic quad-level computational-logical framework can be formulated as follows:

$$\Upsilon^{\text{PD-TEP}}\left(x_{\text{PD-TEP}}, \varpi_{c}^{\text{PD-TEP}}\right) = \underset{\substack{\Delta^{\text{price}}\\ \Delta^{\text{demand}}}}{\text{Max}} \left\{ \left(\Delta^{\text{price}}, \Delta^{\text{demand}}\right) : \underset{\substack{x_{\text{PD-TEP}} \\ a\lambda_{p,k,b} \in \Pi_{p,k,b}^{\text{price}}\left(\Delta^{\text{price}}, p\lambda_{p,k,b}\right) \\ a\rho_{p,k,b} \in \Pi_{p,k,b}^{\text{demand}}\left(\Delta^{\text{demand}}, p\rho_{p,k,b}\right)}}{\text{Max}} \text{PD-TEPO}\left(x_{\text{PD-TEP}}, a\lambda_{p,k,b}, a\rho_{p,k,b}\right) \leq \varpi_{c}^{\text{PD-TEP}} \right\};$$

$$\forall\left\{\Delta^{\text{demand}} > 0, \Delta^{\text{price}} > 0, p \in \Psi^{\text{P}}, k \in \Psi^{\text{K}}, b \in \Psi^{\text{B}}\right\} \tag{6.244) IGDT level (}$$

and subject to Eqs. (6.245) through (6.250):

$$\underset{x_{\text{PD-TEP}}}{\text{Max}} \left\{\text{Eq.}(6.191)\right\} \leq \varpi_{c}^{\text{PD-TEP}};$$

$$\forall\left\{a\lambda_{p,k,b} = \left(1+\Delta^{\text{price}}\right)\cdot p\lambda_{p,k,b}, a\rho_{p,k,b} = \left(1+\Delta^{\text{demand}}\right) \cdot p\rho_{p,k,b}, p \in \Psi^{\text{P}}, k \in \Psi^{\text{K}}, b \in \Psi^{\text{B}}\right\} \tag{6.245) Fourth level (}$$

$$a\lambda_{p,k,b} \leq \left(1+\Delta^{\text{price}}\right)\cdot p\lambda_{p,k,b}; \quad \forall\left\{p \in \Psi^{\text{P}}, k \in \Psi^{\text{K}}, b \in \Psi^{\text{B}}\right\} \tag{6.246}$$

$$a\lambda_{p,k,b} \geq \left(1-\Delta^{\text{price}}\right)\cdot p\lambda_{p,k,b}; \quad \forall\left\{p \in \Psi^{\text{P}}, k \in \Psi^{\text{K}}, b \in \Psi^{\text{B}}\right\} \tag{6.247}$$

$$a\rho_{p,k,b} \leq \left(1+\Delta^{\text{demand}}\right)\cdot p\rho_{p,k,b}; \quad \forall\left\{p \in \Psi^{\text{P}}, k \in \Psi^{\text{K}}, b \in \Psi^{\text{B}}\right\} \tag{6.248}$$

$$a\rho_{p,k,b} \geq \left(1-\Delta^{\text{demand}}\right)\cdot p\rho_{p,k,b}; \quad \forall\left\{p \in \Psi^{\text{P}}, k \in \Psi^{\text{K}}, b \in \Psi^{\text{B}}\right\} \tag{6.249}$$

$$\left.\begin{Bmatrix} \text{Eqs.}(6.200)\text{ through }(6.202)\text{ and} \\ \text{Eqs.}(6.205)\text{ through }(6.207) \end{Bmatrix}\right|_{a\lambda_{p,k,b}, a\rho_{p,k,b}};$$

$$\forall\left\{a\lambda_{p,k,b} = \left(1+\Delta^{\text{price}}\right)\cdot p\lambda_{p,k,b}, a\rho_{p,k,b} = \left(1+\Delta^{\text{demand}}\right)\cdot p\rho_{p,k,b}, p \in \Psi^{\text{P}}, k \in \Psi^{\text{K}}, b \in \Psi^{\text{B}}\right\} \tag{6.250}$$

$$\underset{x_{\text{PD-GEP}}}{\text{Max}} \ \{\text{Eq.}(6.208)\}\Big|_{a\lambda_{p,k,b},a\rho_{p,k,b}}; \ \forall\{a\lambda_{p,k,b} = \left(1+\Delta^{\text{price}}\right)\cdot p\lambda_{p,k,b}, a\rho_{p,k,b} = \left(1+\Delta^{\text{demand}}\right) \cdot p\rho_{p,k,b}, p \in \Psi^{\text{P}}, k \in \Psi^{\text{K}}, b \in \Psi^{\text{B}}\} \quad (6.251) \text{ Third level}$$

and subject to Eq. (6.252):

$$\{\text{Eqs.}(6.216)\,\text{through}\,(6.228)\}\Big|_{a\lambda_{p,k,b},a\rho_{p,k,b}}; \ \forall\{a\lambda_{p,k,b} = \left(1+\Delta^{\text{price}}\right)\cdot p\lambda_{p,k,b}, a\rho_{p,k,b} = \left(1+\Delta^{\text{demand}}\right)\cdot p\rho_{p,k,b}, p\in\Psi^{\text{P}}, k\in\Psi^{\text{K}}, b\in\Psi^{\text{B}}\} \quad (6.252)$$

$$\underset{x_{\text{BBM}}}{\text{Max}}\,\{\text{Eq.}(6.229)\}\Big|_{a\lambda_{p,k,b},a\rho_{p,k,b}}; \ \forall\{a\lambda_{p,k,b} = \left(1+\Delta^{\text{price}}\right)\cdot p\lambda_{p,k,b}, \text{ and, } a\rho_{p,k,b} = \left(1+\Delta^{\text{demand}}\right)\cdot p\rho_{p,k,b}, g\in\Psi^{\text{G}}, p\in\Psi^{\text{P}}, k\in\Psi^{\text{K}}, b\in\Psi^{\text{B}}\}$$
$$\underset{x_{\text{BBM}}}{\text{Max}}\,\{\text{Eq.}(6.230)\}\Big|_{a\lambda_{p,k,b},a\rho_{p,k,b}}; \forall\{a\lambda_{p,k,b} = \left(1+\Delta^{\text{price}}\right)\cdot p\lambda_{p,k,b}, \text{ and, } a\rho_{p,k,b} = \left(1+\Delta^{\text{demand}}\right)\cdot \leq p\rho_{p,k,b}, d\in\Psi^{\text{D}}, p\in\Psi^{\text{P}}, k\in\Psi^{\text{K}}, b\in\Psi^{\text{B}}\} \quad (6.253) \text{ First level}$$

and subject to Eq. (6.254):

$$\{\text{Eqs.}(6.231)\,\text{through}\,(6.237)\}|_{a\lambda_{p,k,b},a\rho_{p,k,b}}; \ \forall\{a\lambda_{p,k,b} = \left(1+\Delta^{\text{price}}\right)\cdot p\lambda_{p,k,b}, a\rho_{p,k,b} = \left(1+\Delta^{\text{demand}}\right)\cdot p\rho_{p,k,b}, p\in\Psi^{\text{P}}, k\in\Psi^{\text{K}}, b\in\Psi^{\text{B}}\} \quad (6.254)$$

$$\underset{x_{\text{CSCEM}}}{\text{Max}}\,\{\text{Eq.}(6.238)\}\Big|_{a\lambda_{p,k,b},a\rho_{p,k,b}}; \ \forall\{a\lambda_{p,k,b} = \left(1+\Delta^{\text{price}}\right)\cdot p\lambda_{p,k,b}, \text{ and, } a\rho_{p,k,b} = \left(1+\Delta^{\text{demand}}\right)\cdot p\rho_{p,k,b}, g\in\Psi^{\text{G}}, d\in\Psi^{\text{D}}, p\in\Psi^{\text{P}}, k\in\Psi^{\text{K}}, b\in\Psi^{\text{B}}\} \quad (6.255) \text{ Second level}$$

and subject to Eq. (6.256):

$$\{\text{Eqs.}(6.239)\,\text{through}\,(6.242)\}|_{a\lambda_{p,k,b},a\rho_{p,k,b}}; \ \forall\{a\lambda_{p,k,b} = \left(1+\Delta^{\text{price}}\right)\cdot p\lambda_{p,k,b}, a\rho_{p,k,b} = \left(1+\Delta^{\text{demand}}\right)\cdot p\rho_{p,k,b}, p\in\Psi^{\text{P}}, k\in\Psi^{\text{K}}, b\in\Psi^{\text{B}}\} \quad (6.256)$$

In the proposed framework under the IGDT risk-averse decision-making policy, the maximum values of the PD-TEPO are achieved for the highest level of the uncertain market price and demand so that these uncertain parameters are allowed by the IGDT risk-averse decision-making policy at the relevant robust region (see Figs. 6.4 and 6.5). The solution of the IGDT risk-averse decision-making policy provides not only the optimal robust transmission expansion plans of the planner based on the defined value of its critical cost, but also the optimal generation expansion plans adapted to the optimal robust transmission expansion plans. The predetermined critical cost is also depicted by Eq. (6.257):

$$\varpi_{\mathrm{c}}^{\text{PD-TEP}} = \left(1 + \sigma_{\mathrm{c}}^{\text{PD-TEP}}\right) \cdot \varpi_{\mathrm{b}}^{\text{PD-TEP}} \tag{6.257}$$

In Eq. (6.257), the IGDT base cost, $\varpi_{\mathrm{b}}^{\text{PD-TEP}}$, is calculated by solving the deterministic strategic quad-level computational-logical framework using Eqs. (6.191) through (6.242). The deterministic framework of the proposed strategic quad-level computational-logical framework is regarded as a risk-neutral decision-making policy. Generally, in the fourth level (i.e., the proposed PD-TEP problem) of the proposed strategic quad-level computational-logical framework, the determined critical cost by the planner is larger than its base cost.

6.5.3.3 The IGDT Risk-Taker Decision-Making Policy: Opportunity Function

In general, the opportunity function illustrates the propitious face of the severe twofold uncertainties in the IGDT-based strategic quad-level computational-logical framework. In other words, the opportunity function expresses the smallest level of uncertainty parameters so that the minimum value of the PD-TEPO cannot be greater than a predetermined target cost. Therefore, the opportunity function for the proposed strategic quad-level computational-logical framework can be described by Eq. (6.258):

$$\begin{aligned}
&\Gamma^{\text{PD-TEP}}\left(x_{\text{PD-TEP}}, \varpi_{\mathrm{t}}^{\text{PD-TEP}}\right) \\
&= \underset{\substack{\Delta^{\text{price}} \\ \Delta^{\text{demand}}}}{\text{Min}} \left\{ \left(\Delta^{\text{price}}, \Delta^{\text{demand}}\right) : \underset{\substack{x_{\text{PD-TEP}} \\ a\lambda_{p,k,b} \in \Pi_{p,k,b}^{\text{price}}\left(\Delta^{\text{price}}, p\lambda_{p,k,b}\right) \\ a\rho_{p,k,b} \in \Pi_{p,k,b}^{\text{demand}}\left(\Delta^{\text{demand}}, p\rho_{p,k,b}\right)}}{\text{Min}} \text{PD-TEPO}\left(x_{\text{PD-TEP}}, a\lambda_{p,k,b}, a\rho_{p,k,b}\right) \leq \varpi_{\mathrm{t}}^{\text{PD-TEP}} \right\}; \\
&\forall \left\{\Delta^{\text{demand}} > 0, \Delta^{\text{price}} > 0, p \in \Psi^{\mathrm{P}}, k \in \Psi^{\mathrm{K}}, b \in \Psi^{\mathrm{B}}\right\}
\end{aligned} \tag{6.258}$$

Hence, the IGDT risk-taker decision-making policy for the proposed strategic quad-level computational-logical framework can be formulated as follows:

$$\Gamma^{\text{PD-TEP}}\left(x_{\text{PD-TEP}}, \varpi_{\text{t}}^{\text{PD-TEP}}\right) = \underset{\substack{\Delta^{\text{price}} \\ \Delta^{\text{demand}}}}{\text{Min}} \left\{ \left(\Delta^{\text{price}}, \Delta^{\text{demand}}\right) : \underset{\substack{x_{\text{PD-TEP}} \\ a\lambda_{p,k,b} \in \Pi_{p,k,b}^{\text{price}}\left(\Delta^{\text{price}}, p\lambda_{p,k,b}\right) \\ a\rho_{p,k,b} \in \Pi_{p,k,b}^{\text{demand}}\left(\Delta^{\text{demand}}, p\rho_{p,k,b}\right)}}{\text{Min}} \text{PD-TEPO}\left(x_{\text{PD-TEP}}, a\lambda_{p,k,b}, a\rho_{p,k,b}\right) \leq \varpi_{\text{t}}^{\text{PD-TEP}} \right\};$$
$$\forall\left\{\Delta^{\text{demand}} > 0, \Delta^{\text{price}} > 0, p \in \Psi^{\text{P}}, k \in \Psi^{\text{K}}, b \in \Psi^{\text{B}}\right\} \quad (6.259) \text{ IGDT level}$$

and subject to Eqs. (6.260) through (6.265):

$$\underset{x_{\text{PD-TEP}}}{\text{Min}} \{\text{Eq.}(6.191)\} \leq \varpi_{\text{t}}^{\text{PD-TEP}};$$
$$\forall\left\{a\lambda_{p,k,b} = \left(1 - \Delta^{\text{price}}\right) \cdot p\lambda_{p,k,b}, a\rho_{p,k,b} = \left(1 - \Delta^{\text{demand}}\right) \cdot p\rho_{p,k,b}, p \in \Psi^{\text{P}}, k \in \Psi^{\text{K}}, b \in \Psi^{\text{B}}\right\} \quad (6.260) \text{ Fourth level}$$

$$a\lambda_{p,k,b} \leq \left(1 + \Delta^{\text{price}}\right) \cdot p\lambda_{p,k,b}; \quad \forall\left\{p \in \Psi^{\text{P}}, k \in \Psi^{\text{K}}, b \in \Psi^{\text{B}}\right\} \tag{6.261}$$

$$a\lambda_{p,k,b} \geq \left(1 - \Delta^{\text{price}}\right) \cdot p\lambda_{p,k,b}; \quad \forall\left\{p \in \Psi^{\text{P}}, k \in \Psi^{\text{K}}, b \in \Psi^{\text{B}}\right\} \tag{6.262}$$

$$a\rho_{p,k,b} \leq \left(1 + \Delta^{\text{demand}}\right) \cdot p\rho_{p,k,b}; \quad \forall\left\{p \in \Psi^{\text{P}}, k \in \Psi^{\text{K}}, b \in \Psi^{\text{B}}\right\} \tag{6.263}$$

$$a\rho_{p,k,b} \geq \left(1 - \Delta^{\text{demand}}\right) \cdot p\rho_{p,k,b}; \quad \forall\left\{p \in \Psi^{\text{P}}, k \in \Psi^{\text{K}}, b \in \Psi^{\text{B}}\right\} \tag{6.264}$$

$$\left.\left\{\begin{matrix}\text{Eqs.}(6.200) \text{ through } (6.202) \text{ and} \\ \text{Eqs.}(6.205) \text{ through } (6.207)\end{matrix}\right\}\right|_{a\lambda_{p,k,b}, a\rho_{p,k,b}};$$
$$\forall\left\{a\lambda_{p,k,b} = \left(1 - \Delta^{\text{price}}\right) \cdot p\lambda_{p,k,b}, a\rho_{p,k,b} = \left(1 - \Delta^{\text{demand}}\right) \cdot p\rho_{p,k,b}, p \in \Psi^{\text{P}}, k \in \Psi^{\text{K}}, b \in \Psi^{\text{B}}\right\} \tag{6.265}$$

$$\underset{x_{\text{PD-GEP}}}{\text{Max}} \left.\{\text{Eq.}(6.208)\}\right|_{a\lambda_{p,k,b}, a\rho_{p,k,b}};$$
$$\forall\left\{a\lambda_{p,k,b} = \left(1 - \Delta^{\text{price}}\right) \cdot p\lambda_{p,k,b}, a\rho_{p,k,b} = \left(1 - \Delta^{\text{demand}}\right) \cdot p\rho_{p,k,b}, p \in \Psi^{\text{P}}, k \in \Psi^{\text{K}}, b \in \Psi^{\text{B}}\right\}$$
$$(6.266) \text{ Third level}$$

and subject to Eq. (6.267):

$$\left.\{\text{Eqs.}(6.216) \text{ through } (6.228)\}\right|_{a\lambda_{p,k,b}, a\rho_{p,k,b}};$$
$$\forall\left\{a\lambda_{p,k,b} = \left(1 - \Delta^{\text{price}}\right) \cdot p\lambda_{p,k,b}, a\rho_{p,k,b} = \left(1 - \Delta^{\text{demand}}\right) \cdot p\rho_{p,k,b}, p \in \Psi^{\text{P}}, k \in \Psi^{\text{K}}, b \in \Psi^{\text{B}}\right\} \tag{6.267}$$

$$\underset{x_{\mathrm{BBM}}}{\mathrm{Max}}\ \{\mathrm{Eq.}(6.229)\}\Big|_{a\lambda_{p,k,b},a\rho_{p,k,b}};\forall\left\{a\lambda_{p,k,b}=\left(1-\Delta^{\mathrm{price}}\right)\cdot p\lambda_{p,k,b},\right.$$
$$\left.\text{and,}\quad a\rho_{p,k,b}=\left(1-\Delta^{\mathrm{demand}}\right)\cdot p\rho_{p,k,b},g\in\Psi^{\mathrm{G}},p\in\Psi^{\mathrm{P}},k\in\Psi^{\mathrm{K}},b\in\Psi^{\mathrm{B}}\right\}$$
$$\underset{x_{\mathrm{BBM}}}{\mathrm{Max}}\ \{\mathrm{Equation}\ (6.230)\}\Big|_{a\lambda_{p,k,b},a\rho_{p,k,b}};\quad\forall\left\{a\lambda_{p,k,b}=\left(1-\Delta^{\mathrm{price}}\right)\cdot p\lambda_{p,k,b},\right.$$
$$\left.\text{and,}\ a\rho_{p,k,b}=\left(1-\Delta^{\mathrm{demand}}\right)\cdot p\rho_{p,k,b},d\in\Psi^{\mathrm{D}},p\in\Psi^{\mathrm{P}},k\in\Psi^{\mathrm{K}},b\in\Psi^{\mathrm{B}}\right\}$$

(6.268) First level

and subject to Eq. (6.269):

$$\{\mathrm{Eqs.}(6.231)\ \text{through}\ (6.237)\}|_{a\lambda_{p,k,b},a\rho_{p,k,b}};$$
$$\forall\left\{a\lambda_{p,k,b}=\left(1-\Delta^{\mathrm{price}}\right)\cdot p\lambda_{p,k,b},a\rho_{p,k,b}=\left(1-\Delta^{\mathrm{demand}}\right)\right.$$
$$\left.\cdot p\rho_{p,k,b},p\in\Psi^{\mathrm{P}},k\in\Psi^{\mathrm{K}},b\in\Psi^{\mathrm{B}}\right\} \tag{6.269}$$

$$\underset{x_{\mathrm{CSCEM}}}{\mathrm{Max}}\ \{\mathrm{Eq.}(6.238)\}\Big|_{a\lambda_{p,k,b},a\rho_{p,k,b}};\quad\forall\left\{a\lambda_{p,k,b}=\left(1-\Delta^{\mathrm{price}}\right)\cdot p\lambda_{p,k,b},\right.$$
$$\left.\text{and,}\ a\rho_{p,k,b}=\left(1-\Delta^{\mathrm{demand}}\right)\cdot p\rho_{p,k,b},g\in\Psi^{\mathrm{G}},d\in\Psi^{\mathrm{D}},p\in\Psi^{\mathrm{P}},k\in\Psi^{\mathrm{K}},b\in\Psi^{\mathrm{B}}\right\}$$

(6.270) Second level

and subject to Eq. (6.271):

$$\{\mathrm{Eqs.}(6.239)\ \text{through}\ (6.242)\}|_{a\lambda_{p,k,b},a\rho_{p,k,b}};$$
$$\forall\left\{a\lambda_{p,k,b}=\left(1-\Delta^{\mathrm{price}}\right)\cdot p\lambda_{p,k,b},a\rho_{p,k,b}=\left(1-\Delta^{\mathrm{demand}}\right)\cdot p\rho_{p,k,b},p\in\Psi^{\mathrm{P}},k\in\Psi^{\mathrm{K}},b\in\Psi^{\mathrm{B}}\right\} \tag{6.271}$$

In the proposed framework under the IGDT risk-taker decision-making policy, the minimum values of the PD-TEPO are achieved for the lowest level of the uncertain market price and demand so that these uncertainty parameters are allowed by the IGDT risk-averse decision-making policy at the relevant robust region (see Figs. 6.4 and 6.5). The solution of the IGDT risk-averse decision-making policy gives not only the optimal opportunistic transmission expansion plans of the planner according to the defined value of its target cost, but also the optimal generation expansion plans adapted to the optimal robust transmission expansion plans. The predetermined target cost is demonstrated by Eq. (6.272):

$$\varpi_{\mathrm{t}}^{\text{PD-TEP}}=\left(1-\sigma_{\mathrm{t}}^{\text{PD-TEP}}\right)\cdot\varpi_{\mathrm{b}}^{\text{PD-TEP}} \tag{6.272}$$

In a similar manner, in Eq. (6.272), the IGDT base cost, $\varpi_{\mathrm{b}}^{\mathrm{PD\text{-}TEP}}$, is calculated by solving the deterministic strategic quad-level computational-logical framework using Eqs. (6.191) through (6.242). Generally, in the fourth level (i.e., the proposed PD-TEP problem) of the proposed strategic quad-level computational-logical framework, the determined target cost by the planner is lower than its base cost.

6.5.4 Solution Method and Implementation Considerations

In the proposed strategic quad-level computational-logical framework, the following decision-making variables of the solution vector are taken into account: (1) the generated power by the GENCOs; (2) the slope and intercept parameters of the bidding strategy function of the GE.NCOs; (3) the purchased power by the DISCOs; and, (4) the slope and intercept parameters of the bidding strategy function of the DISCOs in the short-term operational slave problem. Concurrently considered are the following decision-making variables from the long-term planning master problem: (1) the time of the newly installed transmission lines over the planning horizon; (2) the time of the augmented and newly installed generation units by the GENCOs over the planning horizon; (3) the location of the newly installed transmission lines; (4) the location of the newly installed generation units by the GENCOs; (5) the capacity of the newly installed transmission lines; and, (6) the capacity of the augmented and newly installed generation units by the GENCOs. The solution process of the strategic quad-level computational-logical framework (i.e., the PD-G&TEP problem) begins at the first level. In this level, the BBM, the SFE model is employed by the market participants, including GENCOs and DISCOs preparing their bidding strategy functions. In this level, each market participant performs a self-optimization of the bidding strategy problem [see Eqs. (6.229) through (6.237)] to calculate and modify its bidding strategy parameters using an optimization algorithm, assuming that the bidding strategy parameters of its rivals are fixed. This process continues iteratively until the market participants are reluctant to change their own bidding strategy parameters. This means that the Nash equilibrium point of the first level is obtained. After completing the first level, the obtained optimal bidding strategy functions of both GENCOs and DISCOs at the Nash equilibrium point are transferred from the first to the second level (i.e., the CSC electricity market problem) of the proposed framework. In this level, the aggregation of these optimal functions is done by the ISO with the aim of creating an equivalent supply function and an equivalent demand function. If these optimal functions fail to satisfy the network constraints, the CSC electricity market-clearing problem [see Eqs. (6.238) through (6.242)] is carried out using an optimization algorithm by the ISO with the aim of maximizing the CWF. In the strategic quad-level computational-logical framework, the short-term operational slave problem is formed and organized by a transverse relationship between the first and second levels (see Fig. 6.16). The first and second levels are repeated for all considered patterns in period p of the planning horizon. After completing the short-term operational slave problem for period p, the information associated with the GENCOs, the DISCOs, the CSC electricity market outcomes, etc. is transferred to the third level (i.e., the PD-GEP problem). The third level is broken into two layers: (1) local and (2) global. In the local layer, each GENCO first computes its augmented and newly installed generation units by solving the PD-GEP problem [see Eqs. (6.208) through (6.225)] with the aim of maximizing the EP, GIC, and CP, subject to the environmental/techno-economic constraints using an optimization algorithm. To further clarify, in this layer, each GENCO competes with its rivals to determine the augmented and newly installed generation units under a Cournot-based model using an iterative process. This iterative process of solving the PD-GEP problem is repeated until the Nash equilibrium point is achieved. Then, the GENCO' generation data, or optimal generation expansion plans, are transferred from the local to the global layer for the regulatory body's or ISO's admission. After receiving these data from the regulatory body or the ISO in the global layer, the established global constraints [see Eqs. (6.226) through (6.228)] for the entire power system are investigated by this entity. If the optimal expansion plans by the GENCOs in the local layer cannot satisfy these constraints, the corrective factors associated with the violated constraints are sent by the regulatory body or the ISO to the GENCOs. Then, the GENCOs will modify their expansion plans by taking the received corrective factors from the global layer into account. After this modification and update, the GENCOs resubmit their optimal expansion plans to the global layer to obtain the regulatory body's or ISO's admission. This iterative process between the two layers is repeated until the expansion plans by the GENCOs in the local layer do not lead to any violations of the established constraints related to the global layer. After completion of the third level, the expansion plans by the GENCOs are transferred to the fourth level (i.e., PD-TEP problem). In this level, the ISO, as transmission network planner, solves the PD-TEP problem with the aim of minimizing the TIC, TCC, and ECOC, subject to the short- and long-term constraints to determine the newly installed transmission lines [see Eqs. (6.191) through (6.207)] by considering the expansion plans of the GENCOs in the third level using an optimization algorithm. The transmission expansion plans by the ISO are, then, transferred from the fourth level to the third level. In the third level, each GENCO updates its expansion plans, considering the transmission expansion plans obtained by the ISO in the fourth level. This iterative process between the third and fourth levels is repeated until an equilibrium point in coordination with generation and transmission expansion planning is achieved. In the strategic quad-level computational-logical framework, the long-term planning master problem is formed and organized by a longitudinal relationship between the third and fourth levels (see Fig. 6.16). Finally, the generation and transmission expansion plans are determined for period p and are applied as operational equipment in the next period of planning. Figure 6.17 shows the flowchart of the strategic quad-level computational-logical framework under the IGDT risk-neutral, the IGDT risk-averse, and the IGDT risk-taker decision-making policies.

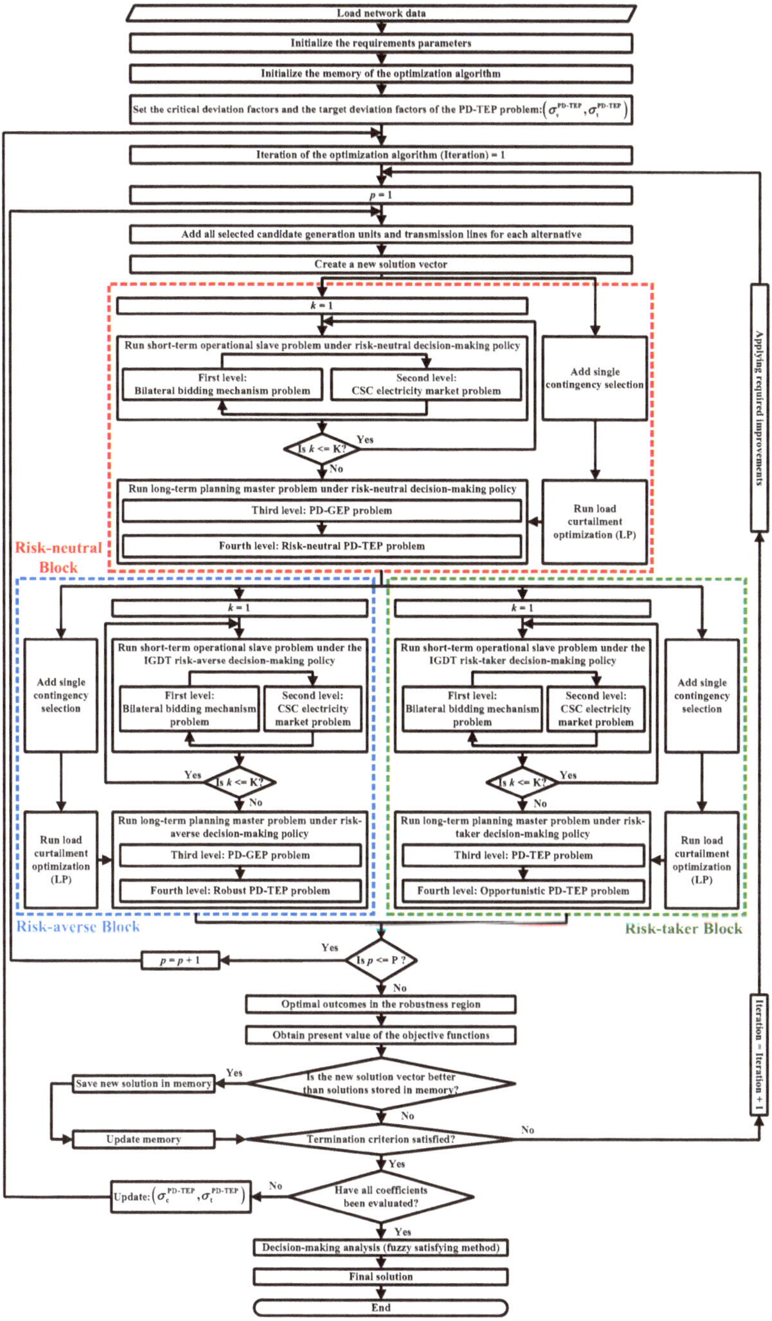

Fig. 6.17 Flowchart of the strategic quad-level computational-logical framework under the IGDT risk-neutral, the IGDT risk-averse, and the risk-taker decision-making policies

To implement the proposed framework under each of these decision-making policies, the block relevant to this decision-making policy is active, while other blocks in the flowchart are inactive.

6.5.5 Simulation Results and Case Studies

The strategic quad-level computational-logical framework is implemented in a modified, large-scale, 46-bus south Brazilian system [33–36]. Figure 6.7 shows the single-line diagram of this test system. In this figure, the existing and candidate transmission lines1 are shown as solid and dotted lines, respectively. This modified test system consists of 46 b11usses, 47 existing transmission lines, 32 candidate transmission lines, 12 GENCOs, and 19 DISCOs. The base apparent power and voltage for the entire modified large-scale 46-bus south 1Brazilian system are equal to 100 MVA and 135 kV,11 respectively. Full data of this test system are presented in Appendix 3.

Table 6.95 shows the existing and candidate transmission line data related to this test system. Table 6.96 presents the data related to the GENCOs and candidate generation units. Table 6.97 shows the data associated with the DISCOs and demand. The parameters of Table 6.98, corresponding to SO_2, CO_2, and NO_x atmospheric emissions as well as fuel consumption for all of the GENCOs, were taken from other studies [20, 24, 37–39] and modified for the aforementioned test system. Due to the fact that all generation units utilize the same technology (i.e., oil/steam), the value of emitted CO_2, SO_2, and NO_x atmospheric emissions as well as fuel consumption rate for power generation of 1 MWh for all GENCOs is considered equal. However, since GENCOs have different capacities, the maximum admissible value of the emitted atmospheric emissions and the maximum value of fuel consumption are taken into account differently for each GENCO. Table 6.99 presents the data related to the available financial resources and the construction costs of the generation units for all GENCOs. It should be noted that, in practical terms, the available financial resources are not equal for the GENCOs. In this study, then, these values are considered different for each GENCO. The cost of building a generation unit in various portions of the power network is also different. Therefore, these values are regarded differently for each GENCO. To deter the exercise of market power by the market participants, including GENCOs and DISCOs, a price ceiling is defined and employed for these participants, such that they can change their bidding strategy parameters within the predetermined admissible range. Table 6.100 shows the data for the bidding strategy parameters of both GENCOs and DISCOs. Tables 6.101 and 6.124 include the parameters of the weighting coefficients for the objective functions related to the PD-GEP and PD-TEP problems, respectively. In the modified, large-scale, 46-bus south Brazilian system, each load is divided into three sectors (i.e., $S = 3$) by the existing customer classes: (1) residential, (2) commercial, and, (3) industrial. The specified values for the residential, commercial, and industrial sectors of each load in each bus of the network are set at 60%, 20%, and 20% of the maximum load at the relevant bus, respectively. Table 6.125 presents the SCDF and its duration at each load sector of each load in each bus of the modified, large-scale, 46-bus south Brazilian system. Table 6.95 gives the failure and repair rates for the transmission lines. Table 6.96 shows the values for the GENCOs. The maximum admissible value of the curtailed load at load sector s of bus b in period p is considered to be 30% of the maximum load at the relevant sector. The maximum admissible value of the curtailed load at bus b of period p is taken to be 30% of the maximum load at the relevant bus. In addition, the maximum admissible value of the curtailed load at period p is considered to be 30% of the maximum load in the relevant period. All sectors of each load in each bus of the network have the same outage cost weighting coefficients. The proposed PD-G&TEP problem is solved for a 10-year planning horizon (i.e., $P = 10$) that is divided into 1-year periods.

Solving the proposed strategic quad-level computational-logical framework for all hours of a day and for all days of a year would be prohibitively difficult, complicated, and excessively time consuming. These complexities and the high computational burden of the proposed PD-G&TEP problem can be eliminated by the proper use of daily and seasonal time patterns. In this study, and without losing the generality of the proposed PD-G&TEP problem, the time span of the network LDC is broken down into a number of time steps, or patterns, using the proposed exponential scheme of Eq. (6.105). In the proposed exponential scheme, the number of patterns is set at 22 (i.e., $K = 22$) for the LDC with a time interval of 8760 h during which the time intervals for the last two patterns are considered to be equal. The load step for pattern k is also taken into account as the average load value during this pattern.

It is assumed that the installed capacity by each GENCO in period p should not exceed 50% of its existing capacity in period $p - 1$. Candidate transmission lines can also be installed in the 47 existing and 32 new corridors of the network. In addition, the maximum number of installed transmission lines that can be constructed in each corridor over the planning horizon is assumed to be four, including existing ones. The annual interest rate is considered to be 10%. The annual demand growth and the annual price growth are also considered to be 5% and 4%, respectively. The basic construction cost for the transmission line is set at 1200 \$/MW-km. It is important to point out that the physical parameters of the candidate transmission lines are considered similar to existing ones. In addition, the maximum and minimum reserve margins are set at 40% and 10% of the average of the peak demand, respectively. The maximum value of the ECO is regarded as 3% of the average of the peak demand. The ECOC is also equal to 250 \$/MWh. Furthermore, the maximum percentage of the installed generation capacity that can be installed by GENCO g at each period of the planning horizon is taken to be 70%. The maximum admissible values of the TIC index in period p and in all periods of the planning horizon are set at \$90M and \$750M, respectively. Table 6.126 summarizes these assumptions.

To investigate and analyze the performance of the proposed strategic quad-level computational-logical framework, two different cases and two different scenarios are employed, as follows:

- First case: The strategic quad-level computational-logical framework is run under the IGDT risk-averse decision-making policy while considering the following two scenarios:
  - The strategic behavior of the market participants (i.e., first level) is ignored.
  - The strategic behavior of the market participants (i.e., first level) is considered.
- Second case: The strategic quad-level computational-logical framework is run under the IGDT risk-taker decision-making policy while considering the following two scenarios:
  - The strategic behavior of the market participants (i.e., first level) is disregarded.
  - The strategic behavior of the market participants (i.e., first level) is regarded.

The strategic quad-level computational-logical framework under the IGDT risk-averse and the IGDT risk-taker decision-making policies is implemented and solved by using the proposed multi-objective SOSA that was addressed in Chap. 4. Table 6.103 gives the parameter adjustments of the newly developed multi-objective SOSA. To implement the strategic quad-level computational-logical framework under the IGDT risk-averse and the IGDT risk-taker decision-making policies, the planner must specify his/her critical and target costs (i.e., $\varpi_{\mathrm{c}}^{\text{PD-TEP}}$ and $\varpi_{\mathrm{t}}^{\text{PD-TEP}}$). Calculation of these costs by the planner requires that the results from the implementation of the proposed framework under the risk-neutral/deterministic decision-making policy be available. Hence, initially, the planner solves the deterministic strategic quad-level computational-logical framework [see Eqs. (6.191) through (6.242)] using the newly single-objective SOSA by considering the predicted market price and demand. Then, the results of the framework under the risk-neutral/deterministic decision-making policy are used to calculate the optimal generation and transmission expansion plans under the IGDT risk-averse and the IGDT risk-taker decision-making policies. It should be noted that the deterministic strategic quad-level computational-logical framework should be performed under the first and second scenarios, as the framework under the IGDT risk-averse decision-making and the IGDT risk-taker decision-making policies is implemented under these two scenarios. Table 6.57 shows the generation and transmission expansion plans of the framework under the risk-neutral/deterministic decision-making policy for the first and second scenarios. Table 6.58 provides the values for different objectives of the proposed PD-GEP problem [see Eq. (6.191)] (i.e., the third level of the strategic quad-level computational-logical framework) and the values for different objectives of the proposed PD-TEP problem [see Eq. (6.208)] (i.e., the fourth level of the proposed strategic quad-level computational-logical framework) for the first and second scenarios. To illustrate the results presented in Table 6.57, consider as an

Table 6.57 Optimal generation and transmission expansion plans of the proposed framework under the risk-neutral/deterministic decision-making policy, based on the first and second scenarios

	Optimal generation and transmission expansion plans			
	First scenario: Ignoring the BBM		Second scenario: Considering the BBM	
Period No.	Optimal generation expansion plans	Optimal transmission expansion plans	Optimal generation expansion plans	Optimal transmission expansion plans
1	GENCO 1 (14: I)	(9-14: I), (16.32: C)	GENCO 9 (34: I)	(24-34: I), (32-43: I)
2	GENCO 12 (32: C)	(37-40: I), (32-43: I), (14-18: I)	GENCO 3 (31: C)	(9-14: I), (31-41: C), (40-41: C)
3	GENCO 10 (37: I), GENCO 5 (27: I)	(26.27: I), (14-22: I), (13-18: I)	GENCO 6 (28: I)	(28-43: C), (42-43: I)
4	GENCO 7 (28: C)	(28-41: C), (40-41: C)	GENCO 1 (14: I)	(5-9: I), (13-18: I)
5	GENCO 6 (31: C)	(9-14: I), (31-32: C)	GENCO 4 (19: I)	(19-21: I)
6	GENCO 2 (16: I)	(5-9: I)	GENCO 5 (27: I)	(20-21: I), (26.27: I)
7	GENCO 4 (19: I), GENCO 9 (34: I)	(19-25: C), (24-25: C)	–	(37-42: I)
8	GENCO 11 (39: I)	(8-13: I), (19-21: I)	GENCO 10 (37: I)	(5-8: I)
9	GENCO 1 (14: I)	(24-34: I)	GENCO 12 (46: I)	(46.3: C), (2-3: C)
10	–	(40-42: I)	GENCO 8 (32: I)	(40-45: C)
Number of augmented generation units over the planning horizon	8	–	8	–
Number of newly installed generation units over the planning horizon	3	–	1	–
Number of newly installed transmission lines in existing corridors over the planning horizon	–	13	–	11
Number of newly installed transmission lines in new corridors over the planning horizon	–	6	–	6

example the second period of the planning horizon. In the first scenario of this period, GENCO 12 installs a new generation unit on bus 32 of the network, while the existing generation units of this GENCO are located on bus 46 of the network. In the first scenario of this period, the planner installs three new transmission lines in the existing corridors (37-40), (32-43), and (14-18) of the network.

In the second scenario of this period, however, GENCO 3 installs a new generation unit on bus 31 of the network, while the existing generation units of this GENCO are located on bus 17 of the network. In the second scenario of this period, the planner also installs one new transmission line in the existing corridor (9-14) and two new transmission lines in new corridors, (31-41) and (40-41), of the network.

An important point about the results presented in Table 6.57, which must be highlighted, is the difference in capacity expansion by the GENCOs. In the proposed strategic quad-level computational-logical framework, the GENCOs can install their new generation units in a location other than the location of their existing generation units. This is due to the fact that the type of generation expansion planning in the proposed framework is allocation and capacity expansion. More precisely, in the third level of the strategic quad-level computational-logical framework, each GENCO determines the location of the newly installed generation units as well as their capacities. For the remaining periods of the planning horizon, the same analysis can be done. The number of augmented generation units plus the number of newly installed generation units over the planning horizon in the second scenario—where the strategic behaviors of the market participants are taken into account—is less than the number of them in the first scenario, where the strategic behaviors of the market participants are ignored (see Table 6.57). In addition, the number of newly installed transmission lines in the existing corridors of the network plus the number of newly installed transmission lines in new corridors of the network over the planning horizon in the second scenario has been decreased in comparison with the first scenario. This reduction in the number of newly installed transmission lines in the existing and new corridors of the network is due to the adoption of strategic behaviors by the market participants (see Table 6.57). By evaluating the results presented in Table 6.58, it can be concluded that when the strategic behaviors of the market participants are activated in the strategic quad-level computational-logical framework, different objectives of the PD-GEP problem [see Eq. (6.191)] (i.e., the third level of the proposed framework) and different objectives of the PD-TEP problem [see Eq. (6.208)] (i.e., the fourth level of the proposed framework) have more favorable values in comparison to the situation where the strategic behaviors of the market participants are ignored (i.e., second scenario compared to the first). In other words, in the second scenario, where the strategic behaviors of the market participants are considered, an increase in the EP of all GENCOs, a decrease in the GIC and the CP of all GENCOs in the third level of the framework, and a decrease in the TIC, TCC, and ECOC in the fourth level of the framework are observed, in comparison with the first scenario in which the strategic behaviors of the market participants are ignored (see Table 6.58).

Table 6.58 Changes in the objective functions of the third and fourth levels along with the base costs of the proposed framework under the risk-neutral/deterministic decision-making policy, based on the first and second scenarios

Level	Objectives	First scenario: Ignoring the BBM	Second scenario: Considering the BBM
Third level	EP of all GENCOs (M$)	3726.770567	3902.332719
	GIC of all GENCOs (M$)	1230.686415	1106.982464
	CP of all GENCOs (M$)	84.482656	62.675434
Fourth level	TIC (M$)	788.010880	707.509198
	TCC (MS)	25.055357	21.486249
	ECOC (M$)	81.962370	74.551479
	Base cost ($\varpi_{b}^{PD\text{-}TEP}$; M$)	895.028607	803.546926

6.5.5.1 First Case: Simulation Results and Discussion

As noted earlier, due to the high importance of network reliability and the time-consuming process of transmission network expansion, the proposed PD-TEP problem is included in the highest level of the strategic quad-level computational-logical framework. Hence, treatment of the risks related to market price and demand that affect the performance of the strategic quad-level computational-logical framework must be handled at the topmost level of the framework (i.e., the PD-TEP problem). To implement the strategic quad-level computational-logical framework under the IGDT risk-averse decision-making policy, the planner must solve the optimization problem using Eqs. (6.244) through (6.257), using different values of the critical cost deviation factor, $\upsilon_{c}^{PD\text{-}TEP} \in (0, 0.8)$. The process of solving the strategic quad-level computational-logical framework under the IGDT risk-averse decision-making policy associated with the PD-G&TEP problem is quite similar to the process of solving the proposed strategic tri-level computational-logical framework related to the PD-TEP problem described in Sect. 6.4.5.1.1. After solving the proposed strategic quad-level computational-logical framework under the first and the second scenarios of the first case (i.e., the IGDT risk-averse decision-making policy) by the newly proposed multi-objective SOSA, the optimal values for Δ^{price} and Δ^{demand}, and the critical cost deviation factors related to them, are calculated using the FSM according to Table 6.59. Table 6.60 shows the optimal robust transmission expansion plans by the planner and the optimal expansion plans by the GENCOs over the planning horizon under the first and second scenarios related to the first case (i.e., the IGDT risk-averse decision-making policy). It is important to point out that the optimal expansion plans obtained by the GENCOs are coordinated and compatible with the optimal robust transmission expansion plans determined by the planner. To illustrate

Table 6.59 Optimal values for Δ^{price} and Δ^{demand} and the relevant critical cost deviation factors under the first and second scenarios of the first case

Parameter	Optimal value	
	First scenario: Ignoring the BBM	Second scenario: Considering the BBM
Δ^{price} (%)	0.147334	0.329587
Δ^{demand} (%)	0.479211	0.352871
Critical cost deviation factor	0.45	0.45

the results provided in Table 6.60, consider the second period of the planning horizon. In the first scenario of this period, GENCO 10 expands the capacity of its existing generation units located on bus 37 of the network. GENCO 11 also installs a new generation unit on bus 28 of the network, while the existing generation units of this GENCO are located on bus 39 of the network.

In the first scenario of this period, the planner installs one new transmission line in the existing corridor (37-42) and two new transmission lines in new corridors, (28-41) and (40-41), of the network. In the second scenario of this period, however, GENCO 7 expands the capacity of its existing generation units located on bus 31 of the network. In the second scenario of this period, the planner constructs one new transmission line in the existing corridor (13-18) and two new transmission lines in new corridors, (31-41) and (40-41), of the network. For the remaining periods of the planning horizon, the same analysis can be done. Looking at Table 6.60, it can be seen that the number of augmented generation units plus the number of newly installed generation units over the planning horizon has been decreased for the case where the strategic behaviors are taken into account by the market participants (i.e., the second scenario compared to the first). In addition, by adopting the strategic behaviors of the market participants, a significant reduction in the number of newly installed transmission lines in the existing corridors of the network plus the number of newly constructed transmission lines in new corridors of the network over the planning horizon is observed (i.e., the second scenario compared to the first).

Table 6.61 summarizes the optimal values for different objectives of the proposed PD-GEP problem (i.e., the third level of the strategic quad-level computational-logical framework) and the optimal values for different objectives of the proposed PD-TEP problem (i.e., the fourth level of the strategic quad-level computational-logical framework) for the first and second scenarios of the first case (i.e., the IGDT risk-averse decision-making policy). As set out in Table 6.61, by taking into account the strategic behaviors of the market participants, the strategic quad-level computational-logical framework leads to more favorable values for different objectives of the proposed PD-GEP problem [see Eq. (6.191)] (i.e., the third level of the proposed framework) and different objectives of the proposed PD-TEP problem [see Eq. (6.208)] (i.e., the fourth level of the proposed framework). More precisely, in the third level of the proposed framework, a significant increase in the EP of all

Table 6.60 Optimal generation expansion plans and optimal robust transmission expansion plans of the proposed framework under the first and second scenarios of the first case

Period No.	Optimal generation expansion plans and optimal robust transmission expansion plans			
	First scenario: Ignoring the BBM		Second scenario: Considering the BBM	
	Optimal generation expansion plans	Optimal robust transmission expansion plans	Optimal generation expansion plans	Optimal robust transmission expansion plans
1	GENCO 9 (34: I)	(4-9: I), (18-19: I), (33-34: I)	GENCO 5 (27: I)	(27-38: I), (9-14: I), (22-26: I)
2	GENCO 11 (28: C), GENCO 10 (37: I)	(37-42: I), (28-41: C), (40-41: C)	GENCO 7 (31: I)	(31-41: C), (40-41: C), (13-18: I)
3	GENCO 1 (14: I), GENCO 12 (46: I)	(9-14: I), (46.3: C), (2-3: C)	GENCO 9 (34: I), GENCO 12 (32: C)	(24-34: I), (32-43: I), (42-43: I)
4	GENCO 11 (39: I)	(24-34: I), (39-42: I)	GENCO 8 (46: C)	(46.6: C), (5-6: C), (40-45: C)
5	GENCO 4 (19: I)	(19-21: I), (20-21: I), (13-18: I)	GENCO 1 (14: I)	(12-14: I), (14-18: I), (19-21: I)
6	GENCO 6 (31: C), GENCO 9 (34: I)	(36.37: I), (40-41: C), (31-41: C), (40-45: C)	GENCO 2 (16: I)	(16.32: C), (46.16: I)
7	GENCO 8 (32: I)	(32-43: I), (42-43: I)	GENCO 10 (37: I)	(36.37: I), (37-42: I)
8	GENCO 2 (16: I)	(16.28: C), (16.32: C)	GENCO 6 (28: I)	(28-41: C), (20-21: I)
9	GENCO 3 (46: C)	(5-6: C), (46.6: C), (14-22: I)	GENCO 4 (19: I)	(18-19: I), (44-42: I)
10	GENCO 5 (27: I)	(27-38: I), (26.27: I)	GENCO 11 (39: I)	(39-42: I)
Number of augmented generation units over the planning horizon	10	–	9	–
Number of newly installed generation units over the planning horizon	3	–	2	–
Number of newly installed transmission lines in existing corridors over the planning horizon	–	16	–	17
Number of newly installed transmission lines in new corridors over the planning horizon	–	11	–	7

Table 6.61 Changes of the objective functions of the third and fourth levels of the proposed framework under the first and second scenarios of the first case

Level	Objectives	First scenario: Ignoring the BBM	Second scenario: Considering the BBM
Third level	EP of all GENCOs (M$)	4261.499839	4534.984813
	GIC of all GENCOs (M$)	1672.294725	1518.411324
	CP of all GENCOs (M$)	103.205914	87.675434
Fourth level	TIC (M$)	1147.379145	1038.327848
	TCC (MS)	23.453480	19.837990
	ECOC (M$)	76.108163	69.695342
	PD-TEPO (M$)	1246.940788	1097.861180

GENCOs, a significant reduction in the GIC of all GENCOs, and a significant reduction in the CP of all GENCOs are observed by activating the BBM in the proposed framework (i.e., the second scenario compared to the first). Integration of the BBM in the proposed framework also reduces the TIC, TCC and ECOC in the fourth level (i.e., the second scenario compared to the first).

6.5.5.2 Second Case: Simulation Results and Discussion

To implement the proposed strategic quad-level computational-logical framework under the IGDT risk-taker decision-making policy, the planner must solve the optimization problem in Eqs. (6.259) through (6.272) for different values of the target cost deviation factor, $\sigma_{\mathrm{t}}^{\mathrm{PD\text{-}TEP}} \in (0, 0.8)$.

Similarly, the process of solving the strategic quad-level computational-logical framework under the IGDT risk-taker decision-making policy associated with the PD-G&TEP problem is quite similar to the process of solving the proposed strategic tri-level computational-logical framework related to the PD-TEP problem that was addressed in Sect. 6.4.5.1.2.

After solving the strategic quad-level computational-logical framework under the first and second scenarios of the second case (i.e., the IGDT risk-taker decision-making policy) by the newly proposed multi-objective SOSA, the optimal values for Δ^{price} and Δ^{demand}, and the target cost deviation factors related to them, are calculated using the FSM according to Table 6.62. The optimal opportunistic transmission expansion plans by the planner and the optimal expansion plans by the GENCOs over the planning horizon under the first and second scenarios related to the second case (i.e., the IGDT risk-taker decision-making policy) are tabulated in Table 6.63.

The point to be made here is that the obtained optimal expansion plans by the GENCOs are coordinated and compatible with the determined optimal opportunistic transmission expansion plans. To describe the results presented in Table 6.63,

Table 6.62 Optimal values for Δ^{price} and Δ^{demand}, and the relevant target cost deviation factors, under the first and second scenarios of the second case

Parameter	Optimal value	
	First scenario: Ignoring the BBM	Second scenario: Considering the BBM
Δ^{price} (%)	0.140718	0.188745
Δ^{demand} (%)	0.318422	0.273125
Target cost deviation factor	0.40	0.40

consider this time the fourth period of the planning horizon. In the first scenario of this period, GENCO 5 expands the capacity of its existing generation units located on bus 27 of the network.

In the first scenario of this period, the planner installs one new transmission line in the existing corridor (22-26) of the network. In the second scenario of this period, however, GENCO 2 constructs a new generation unit on bus 28 of the network, while the existing generation units of this GENCO are located on bus 16 of the network. In the second scenario of this period, the planner installs two new transmission lines in new corridors, (40-41) and (28-41), of the network. For the remaining periods of the planning horizon, the same analysis can be done. From Table 6.63, it can be seen that adopting the strategic behaviors from the market participants gives rise to a reduction in the number of augmented generation units plus the number of newly installed generation units over the planning horizon (i.e., the second scenario compared to the first). The number of newly installed transmission lines in the existing corridors of the network plus the number of newly installed transmission lines in new corridors of the network over the planning horizon under the second scenario is fewer than that under the first scenario, due to the adoption of strategic behaviors by the market participants. Table 6.64 demonstrates the optimal values obtained for different objectives of the proposed PD-GEP problem (i.e., the third level of the strategic quad-level computational-logical framework) and the optimal values obtained for different objectives of the proposed PD-TEP problem (i.e., the fourth level of the strategic quad-level computational-logical framework) for the first and second scenarios of the second case (i.e., the IGDT risk-taker decision-making policy).

Looking at Table 6.64, one can see that different objectives of the proposed PD-GEP problem [see Eq. (6.191)] (i.e., the third level of the strategic quad-level computational-logical framework) and different objectives of the proposed PD-TEP problem [see Eq. (6.208)] (i.e., the fourth level of the strategic quad-level computational-logical framework) have been clearly improved in the case where the strategic behaviors are taken into account by the market participants (i.e., the second scenario compared to the first).

In other words, when the market participants adopt the strategic behaviors, the EP, GIC, and CP in the third level of the proposed framework and the TIC, TCC, and ECOC in the fourth level have more efficient values, in comparison with when the market participants lack these strategic behaviors. From the results of the

Table 6.63 Optimal generation expansion plans and optimal opportunistic transmission expansion plans of the proposed framework under the first and second scenarios of the second case

	Optimal generation expansion plans and optimal opportunistic transmission expansion plans			
	First scenario: Ignoring the BBM		Second scenario: Considering the BBM	
Period No.	Optimal generation expansion plans	Optimal opportunistic transmission expansion plans	Optimal generation expansion plans	Optimal opportunistic transmission expansion plans
1	–	(34-24, I)	GENCO 1 (14: I)	(9-14: I)
2	GENCO 1 (14: I)	(9-14: I), (13-18: I)	GENCO 10 (37: I)	(37-42: I)
3	GENCO 12 (32: C)	(32-43: I)	–	(18-19: I)
4	GENCO 5 (27: I)	(22-26: I)	GENCO 2 (28: C)	(40-41: C), (28-41: C)
5	GENCO 6 (28: I)	(40-41: C), (28-41: C)	GENCO 9 (34: I)	(24-34: I)
6	–	(5-9: I)	–	(13-18: I)
7	GENCO 10 (37: I)	(37-42: I), (42-44: I)	GENCO 8 (32: I)	(32-43: I)
8	GENCO 3 (17: I)	(17-32: C)	GENCO 7 (31: I)	(31-32: C)
9	GENCO 2 (31: I)	(31-41: C)	GENCO 5 (27: I)	(27-38: I)
10	GENCO 4 (19: I)	(18-19: I)	–	(5-9: I)
Number of augmented generation units over the planning horizon	7	–	6	–
Number of newly installed generation units over the planning horizon	1	–	1	–
Number of newly installed transmission lines in existing corridors over the planning horizon	–	9	–	8
Number of newly installed transmission lines in new corridors over the planning horizon	–	4	–	3

risk-neutral/deterministic decision-making policy, the IGDT risk-averse decision-making policy (i.e., first case), and the IGDT risk-taker decision-making policy (i.e., second case), it can be concluded that in all decision-making policies the adoption of strategic behaviors by the market participants results in a similar trend in the

Table 6.64 Changes of the objective functions of the third and fourth levels of the proposed framework under the first and second scenarios of the second case

Level	Objectives	First scenario: Ignoring the BBM	Second scenario: Considering the BBM
Third level	EP of all GENCOs (M$)	2396.773928	2502.743862
	GIC of all GENCOs (M$)	847.732641	802.475939
	CP of all GENCOs (M$)	62.451016	53.417199
Fourth level	TIC (M$)	451.979402	407.267978
	TCC (MS)	17.270095	13.948587
	ECOC (M$)	61.206425	52.944139
	PD-TEPO (M$)	530.455922	474.160704

obtained optimal results (i.e., second scenario compared to the first). More precisely, in the second scenario of all decision-making policies for which the strategic behaviors of the market participants are considered, there is an increase in the EP and a decrease in GIC and CP in the third level of the strategic quad-level computational-logical framework (i.e., the PD-GEP problem) and a decrease in the TIC, TCC, and ECOC in the fourth level of the strategic quad-level computational-logical framework (i.e., the PD-TEP problem is observed compared to the first scenario). As a result, the integration of the strategic behaviors of the market participants in the PD-G&TEP problem leads to increased productivity and flexibility of the proposed framework.

6.5.5.3 Investigation into the Performance of the Proposed Framework Under the Coordinated and Uncoordinated Decisions for thc PD-G&TEP Problem

In this section, the performance of the proposed strategic quad-level computational-logical framework is investigated under the coordinated and uncoordinated decisions for the PD-G&TEP problem. Two different trials are defined and applied to the second scenario (i.e., considering the BBM) of the first case (i.e., the IGDT risk-averse decision-making policy) and second case (i.e., the IGDT risk-averse decision-making policy), as follows:

- First trial: Coordinated decisions are considered between the third level (i.e., the PD-GEP problem) and fourth level (i.e., the PD-TEP problem, of the strategic quad-level computational-logical framework) that are the coordinated PD-G&TEP problem.
- Second trial: Coordinated decisions are ignored between the third level (i.e., the PD-GEP problem) and fourth level (i.e., the PD-TEP problem, of the strategic quad-level computational-logical framework) that are the uncoordinated PD-G&TEP problem.

Table 6.65 shows the robust transmission expansion plans by the planner and the expansion plans by the GENCOs over the planning horizon under the first and second trials of the second scenario associated with the first case (i.e., the IGDT risk-averse decision-making policy). Table 6.66 gives the opportunistic transmission expansion plans by the planner and the expansion plans by the GENCOs over the planning horizon under the first and second trials of the second scenario related to the second case (i.e., the IGDT risk-taker decision-making policy). As can be seen from Tables 6.65 and 6.66, when the coordinated decisions between the PD-GEP problem (i.e., the third level of the strategic quad-level computational-logical framework) and the PD-TEP problem (i.e., the fourth level of the strategic quad-level computational-logical framework) are considered, a significant reduction in the number of augmented generation units plus the number of newly installed generation units over the planning horizon in the third level of the framework is observed (i.e., the second trial compared to the first). Likewise, a significant reduction in the number of newly installed transmission lines in the existing corridors of the network plus the number of newly installed transmission lines in new corridors of the network over the planning horizon in the fourth level of the framework is discovered (i.e., the second trial compared to the first). In simple terms, in the second trial, not only must fewer augmented generation units plus newly installed generation units in the network over the planning horizon be constructed, but also fewer new transmission lines must be added in the existing and new corridors of the network over the planning horizon to answer the demand growth and various desires of stakeholders, compared to the situation where the coordinated decisions between the PD-GEP problem and the PD-TEP problem in the proposed framework are ignored (see Tables 6.65 and 6.66). The values for different objectives of the PD-GEP problem (i.e., the third level of the strategic quad-level computational-logical framework) and the values for different objectives of the PD-TEP problem (i.e., the fourth level of the strategic quad-level computational-logical framework) under the first and second trials of the second scenario concerned with the first case (i.e., the IGDT risk-averse decision-making policy) and second case (i.e., the IGDT risk-taker decision-making policy) are itemized in Tables 6.67 and 6.68, respectively.

As can be seen from Tables 6.67 and 6.68, considering coordinated decisions between the PD-GEP problem (i.e., the third level of the strategic quad-level computational-logical framework) and the PD-TEP problem (i.e., the fourth level of the strategic quad-level computational-logical framework) leads to more favorable values for different objectives of the PD-GEP problem [see Eq. (6.191)] and different objectives of the PD-TEP problem [see Eq. (6.208)]. In simple terms, by activating the coordinated decisions between the PD-GEP and the PD-TEP problems in the proposed framework, a considerable increase in the EP and a remarkable decrease in GIC and CP of the third level of the proposed framework are observed, in comparison with the situation where the coordinated decisions between the PD-GEP problem and the PD-TEP problem in the proposed framework are ignored (i.e., the second trial compared to the first). Likewise, a significant reduction in the TIC, TCC, and ECOC in the fourth level of the proposed framework is discovered (i.e., the

Table 6.65 Optimal generation expansion plans and robust transmission expansion plans of the proposed framework under the first and second trials of the second scenario related to the first case

	Optimal generation expansion plans and optimal robust transmission expansion plans			
	First trial: Ignoring coordinated decisions		Second trial: Considering coordinated decisions	
Period No.	Optimal generation expansion plans	Optimal robust transmission expansion plans	Optimal generation expansion plans	Optimal robust transmission expansion plans
1	GENCO 12 (28: C)	(9-14: I), (24-34: I), (32-43: I)	GENCO 5 (27: I)	(27-38: I), (9-14: I), (22-26: I)
2	GENCO 1 (14: I), GENCO 10 (37: I)	(28-41: C), (40-41: C), (37-42: I), (14-18: I)	GENCO 7 (31: I)	(31-41: C), (40-41: C), (13-18: I)
3	GENCO 3 (17: I)	(18-20: I), (17-32: C), (16.28: C)	GENCO 9 (34: I), GENCO 12 (32: C)	(24-34: I), (32-43: I), (42-43: I)
4	GENCO 5 (46: C)	(19-21: I), (42-43: I), (44-42: I)	GENCO 8 (46: C)	(46.6: C), (5-6: C), (40-45: C)
5	GENCO 9 (34: I)), GENCO 8 (32: I)	(46.6: C), (5-6: C), (22-26: I)	GENCO 1 (14: I)	(12-14: I), (14-18: I), (19-21: I)
6	GENCO 7 (31: I)	(23-24: I), (40-45: C)	GENCO 2 (16: I)	(16.32: C), (46.16: I)
7	GENCO 2 (16: I)	(31-32: C), (31-41: C), (39-42: I)	GENCO 10 (37: I)	(36.37: I), (37-42: I)
8	GENCO 11 (39: I)	(16.32: C), (32-43: I)	GENCO 6 (28: I)	(28-41: C), (20-21: I)
9	GENCO 5 (27: I)	(4-9: I), (12-14: I)	GENCO 4 (19: I)	(18-19: I), (44-42: I)
10	GENCO 4 (19: I)	(27-38: I), (46.19: I)	GENCO 11 (39: I)	(39-42: I)
Number of augmented generation units over the planning horizon	10	–	9	–
Number of newly installed generation units over the planning horizon	2	–	2	–
Number of newly installed transmission lines in existing corridors over the planning horizon	–	17	–	17
Number of newly installed transmission lines in new corridors over the planning horizon	–	10	–	7

Table 6.66 Optimal generation expansion plans and opportunistic transmission expansion plans of the proposed framework under the first and second trials of the second scenario related to the second case

Period No.	Optimal generation expansion plans and optimal opportunistic transmission expansion plans			
	First trial: Ignoring coordinated decisions		Second trial: Considering coordinated decisions	
	Optimal generation expansion plans	Optimal opportunistic transmission expansion plans	Optimal generation expansion plans	Optimal opportunistic transmission expansion plans
1	GENCO 6 (28: I)	(13-18: I), (32-43: I)	GENCO 1 (14: I)	(9-14: I)
2	GENCO 1 (14: I)	(28-41: C), (40-41: C)	GENCO 10 (37: I)	(37-42: I)
3	GENCO 7 (31: I)	(9-14: I)	–	(18-19: I)
4	–	(31-32: C)	GENCO 6 (28: I)	(40-41: C), (28-41: C)
5	GENCO 4 (19: I)	(19-21: I)	GENCO 9 (34: I)	(24-34: I)
6	GENCO 10 (37: I)	(37-42: I)	–	(13-18: I)
7	GENCO 12 (16: C)	(14-18: I), (5-9: I)	GENCO 8 (32: I)	(32-43: I)
8	–	(16.32: C)	GENCO 7 (31: I)	(31-32: C)
9	GENCO 9 (34: I)	(24-34: I)	GENCO 5 (27: I)	(27-38: I)
10	GENCO 5 (27: I)	(26.27: I)	–	(5-9: I)
Number of augmented generation units over the planning horizon	7	–	7	–
Number of newly installed generation units over the planning horizon	1	–	–	–
Number of newly installed transmission lines in existing corridors over the planning horizon	–	9	–	8
Number of newly installed transmission lines in new corridors over the planning horizon	–	4	–	3

second trial compared to the first). This trend is observed in both cases (see Tables 6.67 and 6.68).

As a result, creating coordination between the PD-GEP and the PD-TEP problems in the proposed framework can remarkably encourage and facilitate

Table 6.67 Changes of the objective functions of the third and fourth levels of the proposed framework under the first and second trials of the second scenario related to the first case

Level	Objectives	First trial: Ignoring coordinated decisions	Second trial: Considering coordinated decisions
Third level	EP of all GENCOs (M$)	4314.936125	4534.984813
	GIC of all GENCOs (M$)	1637.214520	1518.411324
	CP of all GENCOs (M$)	82.719250	74.675434
Fourth level	TIC (M$)	1152.812832	1038.327848
	TCC (MS)	21.177267	19.837990
	ECOC (M$)	74.024536	69.695342
	PD-TEPO (M$)	1249.014635	1097.861180

Table 6.68 Changes of the objective functions of the third and fourth levels of the proposed framework under the first and second trials of the second scenario related to the second case

Level	Objectives	First trial: Ignoring coordinated decisions	Second trial: Considering coordinated decisions
Third level	EP of all GENCOs (M$)	2374.210342	2502.743862
	GIC of all GENCOs (M$)	871.132540	802.475939
	CP of all GENCOs (M$)	51.340456	46.417199
Fourth level	TIC (M$)	451.216727	407.267978
	TCC (MS)	15.193727	13.948587
	ECOC (M$)	56.234305	52.944139
	PD-TEPO (M$)	522.644759	474.160704

competition among market participants, improve efficiency of the electricity market, avoid unnecessary investments in generation and transmission levels, and, finally, reduce unreliability costs in generation and transmission levels, which is the main target of power system planning.

6.5.5.4 Quantitative Verification of the Proposed IGDT Risk-Averse Decision-Making Policy in Comparison to the Robust Optimization Technique

In this section, the performance of the proposed strategic quad-level computational-logical framework under the IGDT risk-averse decision-making policy is compared with the performance of this framework under the RO technique; due to this fact the worst-case uncertainty situation in the proposed strategic quad-level computational-logical framework is observed in the first case (i.e., the IGDT risk-averse decision-making policy). Please refer to the work by Ben-Tal et al. [41] for a thorough discussion on the RO technique. Table 6.69 gives the robust transmission expansion plans by the planner and the expansion plans by the GENCOs over the planning horizon under the IGDT risk-averse decision-making policy and RO technique.

The calculated optimal values for different objectives of the PD-GEP problem (i.e., the third level) and the obtained optimal values for different objectives of the PD-TEP problem (i.e., the fourth level of the strategic quad-level computational-logical framework) under the IGDT risk-averse decision-making policy and RO technique are also tabulated in Table 6.70. These results are associated with the second scenario (i.e., considering the BBM). From Table 6.69, it can be seen that the number of augmented generation units plus the number of newly installed generation units over the planning horizon in the third level of the strategic quad-level computational-logical framework under the IGDT risk-averse decision-making policy is less than the number of those in the robust optimization technique. The number of newly installed transmission lines in the existing corridors of the network plus the number of newly installed transmission lines in new corridors of the network over the planning horizon in the fourth level of the proposed framework under the IGDT risk-averse decision-making policy is also less than the number of those in the robust optimization technique. As can be determined from Table 6.70, employing the IGDT risk-averse decision-making policy to handle the uncertainty parameters in the proposed framework brings about more efficient results in different objectives of the third level and different objectives of the fourth level, compared with using the RO technique to deal with uncertainty parameters. More precisely, the EP of all GENCOs has been clearly increased in the IGDT risk-averse decision-making policy compared to the RO technique. The required GIC for construction of new generation units and expansion of existing ones in the IGDT risk-averse decision-making policy is also lower than that in the RO technique. In addition, the CP paid by the ISO to the GENCOs in the IGDT risk-averse decision-making policy is lower than for the RO technique. The required TIC for construction of new transmission lines in existing and new corridors of the network in the IGDT risk-averse decision-making policy is less than that for the RO technique.

The proposed framework under the IGDT risk-averse decision-making policy leads to more efficient results in the TCC criterion compared to that in the RO technique. Similarly, the ECOC significantly decreased in the proposed framework under the IGDT risk-averse decision-making policy compared with the proposed framework under the RO technique.

As a result, it can be concluded that the IGDT risk-averse decision-making policy, compared to other methods, especially the RO technique, provides more appropriate allocation and capacity expansion for the PD-GEP and PD-TEP problems and, thus, prevents unnecessary investment in the generation and transmission levels. For this reason, the IGDT risk-averse decision-making policy can be a reasonable and suitable policy for handling severe uncertainties in large-scale optimization problems.

Table 6.69 Optimal generation expansion plans and optimal robust transmission expansion plans of the proposed framework under the IGDT risk-averse decision-making policy and RO technique

Period No.	Optimal generation expansion plans and optimal robust transmission expansion plans			
	The IGDT risk-averse decision-making policy		The RO technique	
	Optimal generation expansion plans	Optimal robust transmission expansion plans	Optimal generation expansion plans	Optimal robust transmission expansion plans
1	GENCO 5 (27: I)	(27-38: I), (9-14: I), (22-26: I)	GENCO 1 (14: I)	(14-18: I), (14-22: I), (9-14: I)
2	GENCO 7 (31: I)	(31-41: C), (40-41: C), (13-18: I)	GENCO 10 (37: I), GENCO 2 (32: C)	(32-43: I), (37-42: I), (37-40: I)
3	GENCO 9 (34: I), GENCO 12 (32: C)	(24-34: I), (32-43: I), (42-43: I)	GENCO 12 (46: I)	(46.6: C), (5-6: C)
4	GENCO 8 (46: C)	(46.6: C), (5-6: C), (40-45: C)	GENCO 9 (34: I), GENCO 11 (28: C)	(13-18: I), (24-34: I), (28-41: C), (40-41: C)
5	GENCO 1 (14: I)	(12-14: I), (14-18: I), (19-21: I)	GENCO 5 (27: I)	(26.27: I), (27-38: I), (40-45: C)
6	GENCO 2 (16: I)	(16.32: C), (46.16: I)	GENCO 8 (16: C)	(46.16: I), (16.28: C), (19-21: I)
7	GENCO 10 (37: I)	(36.37: I), (37-42: I)	GENCO 7 (31: I)	(31-32: C), (32-43: I), (31-41: C)
8	GENCO 6 (28: I)	(28-41: C), (20-21: I)	GENCO 4 (19: I)	(18-19: I), (18-20: I)
9	GENCO 4 (19: I)	(18-19: I), (44-42: I)	GENCO 3 (17: I)	(17-32: C), (20-21: I)
10	GENCO 11 (39: I)	(39-42: I)	GENCO 11 (39: I)	(39-42: I)
Number of augmented generation units over the planning horizon	9	–	9	–
Number of newly installed generation units over the planning horizon	2	–	3	–
Number of newly installed transmission lines in existing corridors over the planning horizon	–	17	–	17
Number of newly installed transmission lines in new corridors over the planning horizon	–	7	–	9

Table 6.70 Changes of the objective functions of the third and fourth levels of the proposed framework under the IGDT risk-averse decision-making policy and RO technique

Level	Objectives	The IGDT risk-averse decision-making policy	The RO technique
Third level	EP of all GENCOs (M$)	4534.984813	4392.952861
	GIC of all GENCOs (M$)	1518.411324	1604.431308
	CP of all GENCOs (M$)	87.675434	91.213416
Fourth level	TIC (M$)	1038.327848	1104.768245
	TCC (MS)	19.837990	21.008911
	ECOC (M$)	69.695342	72.941415
	PD-TEPO (M$)	1097.861180	1198.718571

6.5.5.5 Performance Evaluation of the Proposed Optimization Algorithms: Simulation Results and Discussion

In this section, the performance of the proposed modern meta-heuristic music-inspired optimization algorithms addressed in Chap. 4, namely the multi-objective SS-HSA, multi-objective SS-IHSA, multi-objective continuous/discrete TMS-MSA, multi-objective TMS-EMSA, and multi-objective SOSA, is compared with that of the NSGA-II for the first and second scenarios of the first case (i.e., the IGDT risk-averse decision-making policy) and second case (i.e., the IGDT risk-taker decision-making policy). Tables 6.104, 6.105, 6.106, 6.107, and 6.108 give the parameter adjustments of the multi-objective TMS-EMSA, multi-objective continuous/discrete TMS-MSA, multi-objective SS-HSA, multi-objective SS-IHSA, and NSGA-II, respectively. The calculated optimal results associated with the EP, GIC, and CP of all GENCOs in the PD-GEP problem (i.e., the third level of the strategic quad-level computational-logical framework) and the calculated optimal results concerned with the TIC, TCC, and ECOC in the PD-TEP problem (i.e., the fourth level of the strategic quad-level computational-logical framework) over the planning horizon by the multi-objective optimization algorithms under the first and second scenarios of the first case (i.e., the IGDT risk-averse decision-making policy) and the first and second scenarios of the second case (i.e., the IGDT risk-taker decision-making policy) are presented in Tables 6.71 and 6.72, respectively. As seen in these tables, the strategic quad-level computational-logical framework using the proposed multi-objective SOSA gives rise to more economical and effective results than other proposed meta-heuristic music-inspired optimization algorithms and NSGA-II. The ICS is also employed for accurate evaluation of the performance of these multi-objective optimization algorithms relative to each other. Please refer to Sect. 5.3.5.3 of Chap. 5 for a more detailed description of the ICS. Tables 6.73 and 6.74 show the ICS for the proposed multi-objective optimization algorithms under the first and second scenarios of the first case (i.e., the IGDT risk-averse decision-making policy), respectively. Meanwhile, Tables 6.75 and 6.76 present the ICS for the proposed multi-objective optimization algorithms under the first and second

Table 6.71 The calculated values of the objective functions of the third and fourth levels of the proposed framework by the proposed multi-objective optimization algorithms under the first and second scenarios of the first case

Policy	Scenario	Multi-objective optimization algorithm	Objective					
			EP of all GENCOs (M $)	GIC of all GENCOs (M $)	CP of all GENCOs (M$)	TIC (M$)	TCC (MS)	ECOC (M $)
The IGDT risk-averse decision-making policy	First scenario: Ignoring the BBM	SOSA	4261.499839	1672.294725	103.205914	1147.379145	23.453480	76.108163
		TMS-EMSA	4006.068612	1789.105702	97.023001	1251.581032	25.351892	80.569520
		Continuous/dis-crete TMS-MSA	3807.675373	1867.310081	92.108821	1298.120248	26.806564	85.822911
		SS-HSA	3601.039548	2077.313229	86.661466	1419.482598	29.271844	93.876946
		SS-IHSA	3673.142689	1984.143939	88.296827	1376.437733	28.737254	89.499040
		NSGA-II	3570.901218	2141.659758	85.529112	1487.867914	29.758651	94.912868
	Second scenario: Considering the BBM	SOSA	4534.984813	1518.411324	87.675434	1038.327848	19.837990	69.695342
		TMS-EMSA	4309.219448	1641.527386	83.219448	1101.303857	21.191175	73.725267
		Continuous/dis-crete TMS-MSA	4129.860315	1702.132049	80.660315	1178.196142	21.916315	75.962229
		SS-HSA	3874.724681	1862.317532	75.724681	1296.676451	24.104986	82.443032
		SS-IHSA	3978.248538	1796.383020	77.348538	1243.884849	23.021106	80.512492
		NSGA-II	3792.804016	1917.375009	74.104016	1347.828688	24.745351	84.325201

Table 6.72 The calculated values of the objective functions of the third and fourth levels of the proposed framework by the proposed multi-objective optimization algorithms under the first and second scenarios of the second case

Policy	Scenario	Multi-objective optimization algorithm	Objective					
			EP of all GENCOs (M $)	GIC of all GENCOs (M $)	CP of all GENCOs (M$)	TIC (M$)	TCC (MS)	ECOC (M $)
The IGDT risk-taker decision-making policy	First scenario: Ignoring the BBM	SOSA	2396.773928	847.732641	62.451016	451.979402	17.270095	61.206425
		TMS-EMSA	2278.532749	921.385537	59.144876	483.545004	18.645102	65.180301
		Continuous/discrete TMS-MSA	2191.914691	967.983607	56.623391	514.485258	19.719146	69.915173
		SS-HSA	2042.433088	1044.342987	52.456813	562.218561	21.683522	74.930066
		SS-IHSA	2087.977250	1019.368471	54.381848	536.978221	20.963052	72.879630
		NSGA-II	2001.940072	1082.335659	51.584739	590.321776	22.050269	76.815371
	Second scenario: Considering the BBM	SOSA	2502.743862	802.475939	53.417199	407.267978	13.948587	52.944139
		TMS-EMSA	2352.384165	861.875654	51.103842	439.844403	15.132007	55.842091
		Continuous/discrete TMS-MSA	2241.402796	912.750204	48.695381	458.901497	16.058387	59.036461
		SS-HSA	2093.879537	1007.208686	45.288719	507.976988	17.621434	65.082710
		SS-IHSA	2157.991978	951.730342	46.531113	486.648323	16.978772	63.349614
		NSGA-II	2052.075314	1023.422187	44.276111	524.869072	18.023233	66.551505

Table 6.73 The ICS for different objectives of the third and fourth levels of the proposed framework, based on various multi-objective optimization algorithms under the first scenario of the first case

Case No.	Scenario No.	Perspective	Multi-objective optimization algorithms	Multi-objective optimization algorithms					
				SOSA	TMS-EMSA	Continuous/discrete TMS-MSA	SS-HSA	SS-IHSA	NSGA-II
First case	First scenario	EP of all GENCOs over the planning horizon	SOSA	0	6.376107	11.918675	18.340823	16.017813	19.339617
			TM-EMSA	−5.993927	0	5.210350	11.247559	9.063789	12.186486
			Continuous/discrete TMS-MSA	−10.649407	−4.952317	0	5.738227	3.662604	6.630655
			SS-HSA	−15.498306	−10.110387	−5.426823	0	−1.962982	0.843997
			SS-IHSA	−13.806339	−8.310539	−3.533197	2.002286	0	2.863183
			NSGA-II	−16.205529	−10.862704	−6.218338	−0.836934	−2.783487	0
		GIC of all GENCOs over the planning horizon	SOSA	0	6.529014	10.443651	19.497228	15.717066	21.915947
			TM-EMSA	−6.985071	0	4.188076	13.874052	9.829843	16.461721
			Continuous/discrete TMS-MSA	−11.661542	−4.371143	0	10.109363	5.888376	12.810142
			SS-HSA	−24.219326	−16.109027	−11.246292	0	−4.695692	3.004516
			SS-IHSA	−18.647982	−10.901437	−6.256800	4.485086	0	7.354847
			NSGA-II	−28.067124	−19.705602	−14.692239	−3.097584	−7.938729	0
		CP of all GENCOs over the planning horizon	SOSA	0	6.372626	12.047807	19.090893	16.885189	20.667585
			TM-EMSA	−5.990851	0	5.335189	11.956334	9.882771	13.438569
			Continuous/discrete TMS-MSA	−10.752381	−5.064963	0	6.285786	4.317249	7.692946
			SS-HSA	−16.030523	−10.679462	−5.914042	0	−1.852117	1.323939
			SS-IHSA	−14.445962	−8.993923	−4.138576	1.887068	0	3.235991
			NSGA-II	−17.127703	−11.846561	−7.143408	−1.306640	−3.134557	0

(continued)

Table 6.73 (continued)

Case No.	Scenario No.	Perspective	Multi-objective optimization algorithms	Multi-objective optimization algorithms: SOSA	TMS-EMSA	Continuous/discrete TMS-MSA	SS-HSA	SS-IHSA	NSGA-II
		TIC over the planning horizon	SOSA	0	8.325620	11.612260	19.169199	16.641405	22.884341
			TM-EMSA	−9.081731	0	3.585123	11.828363	9.071002	15.880904
			Continuous/discrete TMS-MSA	−13.137863	−3.718434	0	8.549759	5.689867	12.752991
			SS-HSA	−23.715217	−13.415157	−9.349083	0	−3.127265	4.596195
			SS-IHSA	−19.963635	−9.975918	−6.033145	3.032433	0	7.489252
			NSGA-II	−29.675349	−18.879071	−14.617110	−4.817622	−8.095548	0
		TCC over the planning horizon	SOSA	0	7.488245	12.508443	19.876998	18.386495	21.187690
			TM-EMSA	−8.094372	0	5.426551	13.391544	11.780394	14.808329
			Continuous/discrete TMS-MSA	−14.296744	−5.737922	0	8.422018	6.718422	9.920096
			SS-HSA	−24.808105	−15.462167	−9.196553	0	−1.860268	1.635850
			SS-IHSA	−22.528742	−13.353488	−7.202303	1.826294	0	3.432269
			NSGA-II	−26.883733	−17.382367	−11.012552	−1.663055	−3.554261	0
		ECOC over the planning horizon	SOSA	0	5.537276	11.319527	18.927738	14.962034	19.812598
			TM-EMSA	−5.861864	0	6.121198	14.175393	9.977224	15.112121
			Continuous/discrete TMS-MSA	−12.764397	−6.520320	0	8.579353	4.107450	9.577159
			SS-HSA	−23.346751	−16.516700	−9.384481	0	−4.891567	1.091445
			SS-IHSA	−17.594535	−11.083000	−4.283388	4.663451	0	5.703997
			NSGA-II	−24.707869	−17.802449	−10.591527	−1.103489	−6.049034	0

scenarios of the second case (i.e., the IGDT risk-taker decision-making policy), respectively.

To illustrate, consider the optimal results of the proposed framework under the first scenario of the first case, according to Table 6.73. The ICS shows 6.376107%, 11.918675%, 18.340823%, 16.017813%, and 19.339617% superiority—positive sign—of the multi-objective SOSA performance compared with the performances of the multi-objective TMS-EMSA, multi-objective continuous/discrete TMS-MSA, multi-objective SS-HSA, multi-objective SS-IHSA, and NSGA-II, respectively, from the perspective of the obtained EP index by corresponding optimization algorithms. Also, the ICS represents 6.529014%, 10.443651%, 19.497228%, 15.717066%, and 21.915947% superiority—positive sign—of the multi-objective SOSA performance compared with the performances of the multi-objective TMS-MSA, multi-objective continuous/discrete TMS-MSA, multi-objective SS-HSA, multi-objective SS-IHSA, and NSGA-II, respectively, from the standpoint of the obtained GIC index by corresponding optimization algorithms. In addition, the ICS illustrates 6.372626%, 12.047807%, 19.090893%, 16.885189%, and 20.667585% superiority—positive sign—of the multi-objective SOSA performance compared with the performances of the multi-objective TMS-EMSA, multi-objective continuous/discrete TMS-MSA, multi-objective SS-HSA, multi-objective SS-IHSA, and NSGA-II, respectively, from the point of view of the obtained CP index by corresponding optimization algorithms. The ICS shows 8.325620%, 11.612260%, 19.169199%, 16.641405%, and 22.884341% superiority—positive sign—of the multi-objective SOSA performance compared with the performances of the multi-objective TMS-EMSA, multi-objective continuous/discrete TMS-MSA, multi-objective SS-HSA, multi-objective SS-IHSA, and NSGA-II, respectively, from the perspective of the obtained TIC index by corresponding optimization algorithms.

Also, the ICS represents 7.488245%, 12.508443%, 19.876998%, 18.386495%, and 21.187690% superiority—positive sign—of the multi-objective SOSA performance compared with the performances of the multi-objective TMS-EMSA, multi-objective continuous/discrete TMS-MSA, multi-objective SS-HSA, multi-objective SS-IHSA, and NSGA-II, respectively, from the standpoint of the obtained TCC index by corresponding optimization algorithms. In addition, the ICS illustrates 5.537276%, 11.319527%, 18.927738%, 14.962034%, and 19.812598% superiority—positive sign—of the multi-objective SOSA performance compared with the performances of the multi-objective TMS-EMSA, multi-objective continuous/discrete TMS-MSA, multi-objective SS-HSA, multi-objective SS-IHSA, and NSGA-II, respectively, from the point of view of the obtained ECOC index by corresponding optimization algorithms. The ICS illustrates 5.993927% weakness—negative sign—of the multi-objective TMS-EMSA performance compared with the performance of the multi-objective SOSA, from the perspective of the obtained EP index by the multi-objective SOSA, and 5.210350%, 11.247559%, 9.063789%, and 12.186486% superiority—positive sign—of the multi-objective TMS-EMSA performance compared with the performances of the multi-objective continuous/discrete TMS-MSA, multi-objective SS-HSA, multi-objective SS-IHSA, and NSGA-II, respectively, from the perspective of the obtained EP index by corresponding optimization algorithms. Also, the ICS shows 6.985071% weakness—negative sign—of the multi-objective TMS-EMSA performance compared with the performance of the multi-objective SOSA, from the standpoint of the obtained GIC index by the multi-objective SOSA, and 4.188076%, 13.874052%, 9.829843%, and 16.461721% superiority—positive sign—of the multi-objective TMS-EMSA performance compared with the performances of the multi-objective continuous/discrete TMS-MSA, multi-objective SS-HSA, multi-objective SS-IHSA, and NSGA-II, respectively, from the standpoint of the obtained GIC index by corresponding optimization algorithms. In addition, the ICS represents 5.990851% weakness—negative sign—of the multi-objective TMS-EMSA performance compared with the performance of the multi-objective SOSA, from the point of view of the obtained CP index by the multi-objective SOSA, and 5.335189%, 11.956334%, 9.882771%, and 13.438569% superiority—positive sign—of the multi-objective TMS-EMSA performance compared with the performances of the multi-objective continuous/discrete TMS-MSA, multi-objective SS-HSA, multi-objective SS-IHSA, and NSGA-II, respectively, from the point of view of the obtained CP index by corresponding optimization algorithms. The ICS illustrates 9.081731% weakness—negative sign—of the multi-objective TMS-EMSA performance compared with the performance of the multi-objective SOSA, from the perspective of the obtained TIC index by the multi-objective SOSA, and 3.585123%, 11.828363%, 9.071002%, and 15.880904% superiority—positive sign—of the multi-objective TMS-EMSA performance compared with the performances of the multi-objective continuous/discrete TMS-MSA, multi-objective SS-HSA, multi-objective SS-IHSA, and NSGA-II, respectively, from the perspective of the obtained TIC index by corresponding optimization algorithms. Also, the ICS shows 8.094372% weakness—negative sign—of the multi-objective TMS-EMSA performance compared with the performance of the multi-objective SOSA, from the standpoint of the obtained TCC index by the multi-objective SOSA, and 5.426551%, 13.391544%, 11.780394%, and 14.808329% superiority—positive sign—of the multi-objective TMS-EMSA performance compared with the performances of the multi-objective continuous/discrete TMS-MSA, multi-objective SS-HSA, multi-objective SS-IHSA, and NSGA-II, respectively, from the standpoint of the obtained TCC index by corresponding optimization algorithms. In addition, the ICS represents 5.861864% weakness—negative sign—of the multi-objective TMS-EMSA performance compared with the performance of the multi-objective SOSA, from the point of view of the obtained ECOC index by the multi-objective SOSA, and 6.121198%, 14.175393%, 9.977224%, and 15.112121% superiority—positive sign—of the multi-objective TMS-EMSA performance compared with the performances of the multi-objective continuous/discrete TMS-MSA, multi-objective SS-HSA, multi-objective SS-IHSA, and NSGA-II, respectively, from the point of view of the obtained ECOC index by corresponding optimization algorithms. The optimal results presented in Table 6.73 for the other multi-objective optimization algorithms, and Tables 6.74, 6.75, and 6.76 for all multi-objective optimization algorithms, are analyzed in the same way.

Table 6.74 The ICS for different objectives of the third and fourth levels of the proposed framework, based on various multi-objective optimization algorithms under the second scenario of the first case

Case No.	Scenario No.	Perspective	Multi-objective optimization algorithms	Multi-objective optimization algorithms: SOSA	TMS-EMSA	Continuous/discrete TMS-MSA	SS-HSA	SS-IHSA	NSGA-II
First case	Second scenario	EP of all GENCOs over the planning horizon	SOSA	0	5.239124	9.809641	17.040181	13.994507	19.568129
			TM-EMSA	−4.978304	0	4.342983	11.213564	8.319513	13.615663
			Continuous/discrete TMS-MSA	−8.933315	−4.162218	0	6.584613	3.811018	8.886731
			SS-HSA	−14.559258	−10.082911	−6.177827	0	−2.602247	2.159897
			SS-IHSA	−12.276475	−7.680530	−3.671111	2.671773	0	4.889377
			NSGA-II	−16.365673	−11.983966	−8.161445	−2.114231	−4.661461	0
		GIC of all GENCOs over the planning horizon	SOSA	0	7.500091	10.793564	18.466572	15.473965	20.807806
			TM-EMSA	−8.108215	0	3.560514	11.855666	8.620412	14.386732
			Continuous/discrete TMS-MSA	−12.099536	−3.691967	0	8.601405	5.246707	11.225918
			SS-HSA	−22.649081	−13.450287	−9.410872	0	−3.670403	2.871502
			SS-IHSA	−18.306745	−9.433630	−5.537230	3.540454	0	6.310293
			NSGA-II	−26.275073	−16.804326	−12.645491	−2.956395	−6.735311	0
		CP of all GENCOs over the planning horizon	SOSA	0	5.354500	8.697113	15.781846	13.351119	18.314011
			TM-EMSA	−5.082365	0	3.172728	9.897389	7.590201	12.300861
			Continuous/discrete TMS-MSA	−8.001236	−3.075162	0	6.517867	4.281628	8.847427
			SS-HSA	−13.630674	−9.006028	−6.119036	0	−2.099402	2.187013
			SS-IHSA	−11.778551	−7.054733	−4.105831	2.144422	0	4.378334
			NSGA-II	−15.479157	−10.953487	−8.128283	−2.140207	−4.194677	0
		TIC over the planning horizon	SOSA	0	5.718313	11.871393	19.923906	16.525404	22.962921

(continued)

Table 6.74 (continued)

Case No.	Scenario No.	Perspective	Multi-objective optimization algorithms	Multi-objective optimization algorithms: SOSA	TMS-EMSA	Continuous/ discrete TMS-MSA	SS-HSA	SS-IHSA	NSGA-II
			TM-EMSA	−6.065137	0	6.526272	15.067181	11.462555	18.290516
			Continuous/discrete TMS-MSA	−13.470533	−6.981931	0	9.137229	5.280931	12.585616
			SS-HSA	−24.881217	−17.740117	−10.056076	0	−4.244090	3.795158
			SS-IHSA	−19.796927	−12.946562	−5.575362	4.071301	0	7.711947
			NSGA-II	−29.807622	−22.384815	−14.397649	−3.944872	-8.356387	0
		TCC over the planning horizon	SOSA	0	6.385606	9.483003	17.701715	13.826946	19.831446
			TM-EMSA	−6.821179	0	3.308676	12.088001	7.948927	14.363004
			Continuous/discrete TMS-MSA	−10.476489	−3.421896	0	9.079743	4.799035	11.432595
			SS-HSA	−21.509215	−13.750115	−9.986491	0	−4.708201	2.587819
			SS-IHSA	−16.045557	−8.635344	−5.040952	4.496497	0	6.967955
			NSGA-II	−24.737189	−16.771962	−12.908356	−2.656566	−7.489844	0
		ECOC over the planning horizon	SOSA	0	5.466138	8.250004	15.462422	13.435368	17.349331
			TM-EMSA	−5.782201	0	2.944834	10.574289	8.430027	12.570303
			Continuous/discrete TMS-MSA	−8.991830	−3.034186	0	7.860946	5.651623	9.917523
			SS-HSA	−18.290591	−11.824663	−8.531612	0	−2.397814	2.232036
			SS-IHSA	−15.520621	−9.206104	−5.990165	2.341665	0	4.521434
			NSGA-II	−20.991157	−14.377613	−11.009382	−2.282993	−4.735549	0

Table 6.75 The ICS for different objectives of the third and fourth levels of the proposed framework, based on various multi-objective optimization algorithms under the first scenario of the second case

Case No.	Scenario No.	Perspective	Multi-objective optimization algorithms	Multi-objective optimization algorithms					
				SOSA	TMS-EMSA	Continuous/discrete TMS-MSA	SS-HSA	SS-IHSA	NSGA-II
Second case	First scenario	EP of all GENCOs over the planning horizon	SOSA	0	5.189356	9.346131	17.348957	14.789274	19.722561
			TM-EMSA	−4.933347	0	3.951707	11.559725	9.126320	13.816231
			Continuous/discrete TMS-MSA	−8.547290	−3.801484	0	7.318800	4.977901	9.489525
			SS-HSA	−14.784074	−10.361916	−6.819681	0	−2.181257	2.022688
			SS-IHSA	−12.883847	−8.363079	−4.741856	2.229897	0	4.297689
			NSGA-II	−16.473554	−12.139069	−8.667062	−1.982587	−4.120599	0
		GIC of all GENCOs over the planning horizon	SOSA	0	7.993710	12.422830	18.826223	16.837467	21.675624
			TM-EMSA	−8.688222	0	4.813931	11.773665	9.612121	14.870629
			Continuous/discrete TMS-MSA	−14.185011	−5.057391	0	7.311714	5.040852	10.565303
			SS-HSA	−23.192494	−13.344842	−7.888499	0	−2.449998	3.510248
			SS-IHSA	−20.246458	−10.634303	−5.308443	2.391409	0	5.817713
			NSGA-II	−27.674175	−17.468270	−11.813428	−3.637949	−6.177078	0
		CP of all GENCOs over the planning horizon	SOSA	0	5.589900	10.291903	19.052249	14.837980	21.064906
			TM-EMSA	−5.293973	0	4.453080	12.749655	8.758488	14.655762
			Continuous/discrete TMS-MSA	−9.331513	−4.263234	0	7.942872	4.121858	9.767718
			SS-HSA	−16.003267	−11.307933	−7.358404	0	−3.539848	1.690565
			SS-IHSA	−12.920795	−8.053154	−3.958687	3.669752	0	5.422357
			NSGA-II	−17.399680	−12.782404	−8.898534	−1.662460	−5.143460	0

(continued)

Table 6.75 (continued)

Case No.	Scenario No.	Perspective	Multi-objective optimization algorithms	Multi-objective optimization algorithms					
				SOSA	TMS-EMSA	Continuous/discrete TMS-MSA	SS-HSA	SS-IHSA	NSGA-II
		TIC over the planning horizon	SOSA	0	−6.527955	−12.149202	−19.607883	−15.829099	−23.435078
			TM-EMSA	6.983858	0	−6.013827	−13.993411	−9.950723	−18.087893
			Continuous/discrete TMS-MSA	13.829359	6.398629	0	−8.490168	−4.188803	−12.846640
			SS-HSA	24.390306	16.270162	9.277875	0	4.700440	−4.760660
			SS-IHSA	18.805905	11.050308	4.371935	−4.489417	0	−9.036352
			NSGA-II	30.608114	22.082075	14.740270	4.998628	9.934025	0
		TCC over the planning horizon	SOSA	0	7.374628	12.419660	20.353829	17.616504	21.678529
			TM-EMSA	−7.961780	0	5.446706	14.012576	11.057311	15.442745
			Continuous/discrete TMS-MSA	−14.180877	−5.760461	0	9.059303	5.933802	10.571857
			SS-HSA	−25.555314	−16.296076	−9.961770	0	−3.436856	1.663231
			SS-IHSA	−21.383536	−12.431951	−6.308112	3.322661	0	4.930629
			NSGA-II	−27.678909	−18.263064	−11.821622	−1.691362	−5.186348	0
		ECOC over the planning horizon	SOSA	0	6.096743	12.456163	18.315266	16.017102	20.320081
			TM-EMSA	−6.492579	0	6.772309	13.011819	10.564445	15.146799
			Continuous/discrete TMS-MSA	−14.228487	−7.264268	0	6.692764	4.067607	8.982834
			SS-HSA	−22.421896	−14.958146	−7.172824	0	−2.813455	2.454333
			SS-IHSA	−19.071862	−11.812355	−4.240076	2.736466	0	5.123637
			NSGA-II	−25.502136	−17.850592	−9.869385	−2.516086	−5.400330	0

Table 6.76 The ICS for different objectives of the third and fourth levels of the proposed framework, based on various multi-objective optimization algorithms under the second scenario of the second case

Case No.	Scenario No.	Perspective	Multi-objective optimization algorithms	Multi-objective optimization algorithms					
				SOSA	TMS-EMSA	Continuous/discrete TMS-MSA	SS-HSA	SS-IHSA	NSGA-II
Second case	Second scenario	EP of all GENCOs over the planning horizon	SOSA	0	6.391800	11.659710	19.526640	15.975586	21.961598
			TM-EMSA	−6.007794	0	4.951424	12.345725	9.008012	14.634397
			Continuous/discrete TMS-MSA	−10.442182	−4.717825	0	7.045451	3.865205	9.226146
			SS-HSA	−16.336643	−10.989048	−6.581737	0	−2.970930	2.037168
			SS-IHSA	−13.774957	−8.263624	−3.721366	3.061897	0	5.161441
			NSGA-II	−18.006978	−12.766148	−8.446829	−1.996496	−4.908112	0
		GIC of all GENCOs over the planning horizon	SOSA	0	6.891912	12.081538	20.326745	15.682425	21.588964
			TM-EMSA	−7.402055	0	5.573764	14.429287	9.441191	15.784935
			Continuous/discrete TMS-MSA	−13.741753	−5.902771	0	9.378243	4.095712	10.813912
			SS-HSA	−25.512634	−16.862413	−10.348776	0	−5.829208	1.584243
			SS-IHSA	−18.599237	−10.425481	−4.270624	5.508128	0	7.005109
			NSGA-II	−27.533068	−18.743600	−12.125111	−1.609745	−7.532789	0
		CP of all GENCOs over the planning horizon	SOSA	0	4.526777	9.696644	17.948134	14.798885	20.645643
			TM-EMSA	−4.330734	0	4.945974	12.840113	9.827250	15.420801
			Continuous/discrete TMS-MSA	−8.839508	−4.712876	0	7.522098	4.651227	9.981161
			SS-HSA	−15.216972	−11.379032	−6.995862	0	−2.670028	2.287030
			SS-IHSA	−12.891140	−8.947916	−4.444503	2.743274	0	5.093044
			NSGA-II	−17.112631	−13.360504	−9.075337	−2.235894	−4.846224	0

(continued)

Table 6.76 (continued)

Case No.	Scenario No.	Perspective	Multi-objective optimization algorithms	Multi-objective optimization algorithms: SOSA	TMS-EMSA	Continuous/discrete TMS-MSA	SS-HSA	SS-IHSA	NSGA-II
		TIC over the planning horizon	SOSA	0	7.406352	11.251547	19.825506	16.311644	22.405796
			TM-EMSA	−7.998769	0	4.152763	13.412533	9.617606	16.199214
			Continuous/discrete TMS-MSA	−12.678021	−4.332689	0	9.660967	5.701617	12.568386
			SS-HSA	−24.727947	−15.490156	−10.694123	0	−4.382767	3.218342
			SS-IHSA	−19.490937	−10.641017	−6.046357	4.198746	0	7.281958
			NSGA-II	−28.875605	−19.330624	−14.375105	−3.325364	−7.853874	0
		TCC over the planning horizon	SOSA	0	7.820641	13.138305	20.843065	17.846903	22.607741
			TM-EMSA	−8.484156	0	5.768823	14.127266	10.876905	16.041661
			Continuous/discrete TMS-MSA	−15.125546	−6.121990	0	8.870146	5.420798	10.901739
			SS-HSA	−26.331319	−16.451400	−9.733524	0	−3.785091	2.229339
			SS-IHSA	−21.723957	−12.204362	−5.731490	3.647047	0	5.795081
			NSGA-II	−29.211891	−19.106692	−12.235637	−2.280171	−6.151569	0
		ECOC over the planning horizon	SOSA	0	5.189547	10.319592	18.650991	16.425475	20.446368
			TM-EMSA	−5.473603	0	5.410842	14.198270	11.850937	16.091918
			Continuous/discrete TMS-MSA	−11.507075	−5.720362	0	9.290100	6.808491	11.292072
			SS-HSA	−22.927129	−16.547766	−10.241550	0	−2.735764	2.207004
			SS-IHSA	−19.653686	−13.444201	−7.305913	2.662913	0	4.811147
			NSGA-II	−25.701364	−19.178031	−12.729496	−2.256812	−5.054318	0

6.6 Pseudo-Dynamic Open-Loop Distribution Expansion Planning: A Techno-Economic Framework

In this section, a techno-economic framework is presented for the pseudo-dynamic open-loop distribution expansion planning (PD-DEP) problem in the presence of distributed generation resources (DGRs). To optimally plan the distribution network, the techno-economic framework modifies the topology of the distribution network by using one of the following two possible expansion options: reinforcement and/or installation of substation and branches, and augmentation and/or installation of DGRs. In doing so, the techno-economic framework is broken down into a short-term loadability-based optimal power flow (LBOPF) problem and a long-term planning problem. In the short-term LBOPF problem, the objective function is to maximize the loadability of the open-loop distribution network. At the same time, the constraints of the short-term LBOPF problem are divided into two blocks: (1) Kirchhoff's point and loop laws and (2) the operational limits on the distribution network equipment. In the long-term planning problem, however, not only are the distribution investment cost (DIC), the distribution operation cost (DOC), and the distribution maintenance cost (DMC) considered as economic objectives, but also the expected customer outage cost (ECOC) is regarded as a technical objective to reliability worth assessment. Concurrently, the constraints of the long-term planning problem are excided into three blocks: (1) limitations associated with the short-term LBOPF problem, (2) the availability of financial resources, and, (3) the logical restrictions. The following assumptions are employed in the mathematical description process of the proposed techno-economic framework:

- The PD-DEP is implemented from the perspective of the distribution network operator (DNO).
- In solving the PD-DEP problem, the DNO is authorized to optimally choose among the expansion options (i.e., reinforcement and/or installation of substations, branches, and DGRs) under an integrated framework.
- The required energy for the distribution network and its customers is supplied by purchasing the energy from the pool-based wholesale electricity market of the upstream network and/or by producing the energy of the DGRs.
- The daily load variations are modeled using a LDC throughout the planning horizon.
- In view of the unavailability of a continuous LDC, this curve is approximated by using discrete LDCs.
- The discrete LDC is split into M different load levels.
- As the pool-based wholesale electricity market price is competitively determined, each load level of the LDC has a distinct market price.
- The connection of a DGR to a bus on the distribution network is modeled as a negative P load.
- Uncertainties associated with the energy prices of the pool-based wholesale electricity market and demand/load are modeled and applied as a stochastic approach in the proposed techno-economic framework.

- The IGDT is widely employed to minimize risks of the PD-DEP problem arising from severe twofold uncertainties.
- Each electrical load is divided into S different sectors by existing customer classes.

For comparative purposes, the most important features of the previous DEP framework reported in the literature and the PD-DEP framework proposed in this chapter are employed to create the attribute table, Table 6.77. The authors have focused on the most important and relevant articles in the literature. For further work on advances in the DEP problem, please refer to the work by Ganguly et al. [87]. For a comprehensive overview of the DEP problem, please also refer to the work by Khator and Leung [88], Willis [89], and Ault et al. [90].

6.6.1 Mathematical Model of the Deterministic Techno-Economic Framework

In this section, the mathematical model of the proposed deterministic techno-economic framework is addressed from the perspective of the DNO in order to achieve an economical, reliable, and flexible plan. A number of planning indices are designed in order to evaluate the effects of each proposed plan on network costs and reliability. To do this, the methodology optimizes four different objective functions—the DIC, DOC, DMC, and ECOC. In general, the financial resources at the disposal of the DNO to purchase and install the distribution network equipment are usually restricted. The DNO must, then, try to optimally utilize and manage existing financial resources to reduce investment costs.

As a result, the first goal of the PD-DEP problem is formulated with the aim of minimizing investment costs (i.e., the DIC index), as given by Eqs. (6.273) and (6.274):

$$\mathrm{OF}_1^{\text{PD-DEP}} = DIC_y = \sum_{p \in \Psi^{\mathrm{P}}} \left(\frac{1}{(1+\mathrm{Ir})^{p-p_0}} \right) \cdot DIC_{y,p}; \quad \forall \{ p \in \Psi^{\mathrm{P}} \} \tag{6.273}$$

$$\begin{aligned} DIC_{y,p} = {} & \left(\frac{(1+\mathrm{Ir})^{\varepsilon_s}}{(1+\mathrm{Ir})^{\varepsilon_s}-1} \right) \cdot \left(\sum_{rs \in \Psi^{\mathrm{RS}}} IC_{p,rs}^{\mathrm{v}}(a) + \sum_{is \in \Psi^{\mathrm{IS}}} IC_{p,is}^{\mathrm{f}} + IC_{p,is}^{\mathrm{v}}(a) \right) \\ & + \left(\frac{(1+\mathrm{Ir})^{\varepsilon_b}}{(1+\mathrm{Ir})^{\varepsilon_b}-1} \right) \cdot \left(\sum_{rb \in \Psi^{\mathrm{RB}}} L_{rb} \cdot IC_{p,rb}(b) + \sum_{ib \in \Psi^{\mathrm{IB}}} L_{ib} \cdot IC_{p,ib}(b) \right) \\ & + \left(\frac{(1+\mathrm{Ir})^{\varepsilon_{dg}}}{(1+\mathrm{Ir})^{\varepsilon_{dg}}-1} \right) \cdot \left(\sum_{id \in \Psi^{\mathrm{ID}}} IC_{p,id} \cdot \left(\rho_{p,id}^{\mathrm{cap}} + \rho_{p,id}^{\mathrm{res}} \right) \right); \\ & \forall \{ p \in \Psi^{\mathrm{P}}, rs \in \Psi^{\mathrm{RS}}, is \in \Psi^{\mathrm{IS}}, rb \in \Psi^{\mathrm{RB}}, ib \in \Psi^{\mathrm{IB}}, id \in \Psi^{\mathrm{ID}} \} \end{aligned} \tag{6.274}$$

Table 6.77 Attributes of previous frameworks reported in the DEP literature and the techno-economic framework proposed in this study

Reference	Static or multi-period framework	Objectives	Nondeterministic framework	Reliability consideration	Type of modeling reliability criterion	DGR consideration	Open-loop configuration consideration	Uncertainty consideration	Economic and technical aspects	Solution method
[74]	Multi-period	Single objective: (1) investment cost	No	No	No	No	Yes: classic tree approach	No	No	Branch and bound algorithms
[75]	Static	Single objective: (1) investment cost	No	No	No	No	Yes: classic tree approach	No	No	Iterative solution approach
[76, 77]	Multi-period	Double objective: (1) investment cost and (2) operation cost	No	No	No	Yes	Yes: classic tree approach	No	No	Branch and bound algorithms
[78]	Multi-period	Single objective: (1) investment cost	No	No	No	No	Yes: classic tree approach	No	No	Dynamic programming
[79]	Multi-period	Double objective: (1) total cost and (2) dissatisfaction of technical constraints	No	No	No	Yes	No	No	No	Hybrid immune genetic algorithm
[80]	Multi-period	Four objectives: (1) investment cost; (2) operation cost; (3) maintenance cost; and (4) load shedding cost	No	Yes: reliability assessment incompatible with the value-based approach	Deterministic model	No	Yes: classic tree approach	No	No	GAMS

(continued)

Table 6.77 (continued)

Reference	Static or multi-period framework	Objectives	Nondeterministic framework	Reliability consideration	Type of modeling reliability criterion	DGR consideration	Open-loop configuration consideration	Uncertainty consideration	Economic and technical aspects	Solution method
[81]	Multi-period	Three objectives: (1) investment cost; (2) operation cost; and (3) customer outage cost	No	Yes: value-based reliability approach compatible with the restructured environment	Probabilistic model based on sector customer damage function	Yes	Yes: classic tree approach	No	No	GA
[82]	Multi-period	Five objectives: (1) investment cost; (2) operation and maintenance costs; (3) loss cost; (4) reliability cost; and (5) energy saving	No	Yes: reliability assessment incompatible with the value-based approach	Deterministic model	Yes	No	No	No	Modified discrete PSO
[83]	Multi-period	Double objective: (1) investment cost and (2) operation cost	No	No	No	Yes	No	No	No	Back-propagation approach
[84]	Static	Double objective: (1) investment cost and (2) operation cost	No	No	No	No	Yes: classic tree approach	No	No	Mathematical programming
[85]	Multi-period	Five objectives: (1) investment cost; (2) production cost; (3) maintenance cost; (4) energy losses; and	No	Yes: reliability assessment incompatible with the value-based approach	Deterministic model	Yes	Yes: classic tree approach	No	No	GAMS

(continued)

Table 6.77 (continued)

Reference	Static or multi-period framework	Objectives	Nondeterministic framework	Reliability consideration	Type of modeling reliability criterion	DGR consideration	Open-loop configuration consideration	Uncertainty consideration	Economic and technical aspects	Solution method
		(5) unserved energy								
[86]	Multi-period	Four objectives: (1) investment cost; (2) operation cost; (3) maintenance cost; and (4) interruption cost	No	Yes: reliability assessment incompatible with the value-based approach	Deterministic model	No	Yes: classic tree approach	No	No	GA
Proposed framework	Multi-period	Four objectives: (1) DIC; (2) DOC; (3) DMC; and (4) ECOC	Nondeterministic framework: stochastic modeling by the IGDT (risk-averse and risk-taker frameworks)	Yes: value-based reliability approach compatible with the restructured environment	Probabilistic model based on sector customer damage function	Yes	Yes: a new graph-based approach	Severe two-fold: pool-based wholesale and electrical power demand	Yes	Five new multi-objective music-inspired optimization algorithms (see Chap. 4): (i) SS-HSA; (2) SS-IHSA; (3) continuous/discrete TMS-MSA; (iv) TMS-EMSA; and (5) SOSA

Equation (6.273) indicates the net present value of the DIC index over the planning horizon. The DIC index of period p of the planning horizon has three main parts: (1) the investment cost for the reinforced and installed substations [i.e., the first term of Eq. (6.274)]; (2) the investment cost for the reinforced and installed branches [i.e., the second term of Eq. (6.274)]; and, (3) the investment cost for the installed DGRs [i.e., the third term of Eq. (6.274)].

In addition, the electrical efficiency of the distribution network is generally affected by increasing the losses in transformers, branches, and other equipment. To enhance the electrical efficiency of the distribution network, the DNO is obliged to use an appropriate index that minimizes the losses in the distribution equipment. The second goal of the PD-DEP problem is, therefore, considered in order to minimize the DOC index, as given by Eqs. (6.275) and (6.276):

$$\begin{aligned} &\mathrm{OF}_2^{\text{PD-DEP}} = DOC_y \\ &= \sum_{p \in \Psi^{\mathrm{P}}} \left(\left(\frac{1}{(1+\mathrm{Ir})^{p-p_0}} + \frac{(1+\mathrm{Ir})^{-\mathrm{P}}}{\mathrm{Ir}} \right) \cdot \left(\sum_{m \in \Psi^{\mathrm{M}}} 8760 \cdot \frac{\Delta t_m}{24} \cdot DOC_{y,p,m} \right) \right); \\ &\forall \{p \in \Psi^{\mathrm{P}}, m \in \Psi^{\mathrm{M}}\} \end{aligned} \tag{6.275}$$

$$\begin{aligned} DOC_{y,p,m} = {} & p\lambda_{p,m} \cdot \\ & \left(\sum_{es \in \Psi^{\mathrm{ES}}} R_{es} \cdot \left(I_{p,m,es}\right)^2 + \sum_{rs \in \Psi^{\mathrm{RS}}} R_{rs} \cdot \left(I_{p,m,rs}\right)^2 + \sum_{is \in \Psi^{\mathrm{IS}}} R_{is} \cdot \left(I_{p,m,is}\right)^2 \right) \\ & + p\lambda_{p,m} \cdot \left(\sum_{eb \in \Psi^{\mathrm{EB}}} R_{eb} \cdot \left(I_{p,m,eb}\right)^2 + \sum_{rb \in \Psi^{\mathrm{RB}}} R_{rb} \cdot \left(I_{p,m,rb}\right)^2 + \sum_{ib \in \Psi^{\mathrm{IB}}} R_{ib} \cdot \left(I_{p,m,ib}\right)^2 \right) \\ & + \left(\sum_{id \in \Psi^{\mathrm{ID}}} OC_{p,m,id} \cdot \rho_{p,m,id} \right) \\ & + \left(\sum_{es \in \Psi^{\mathrm{ES}}} p\lambda_{p,m} \cdot \rho_{p,m,es} + \sum_{rs \in \Psi^{\mathrm{RS}}} p\lambda_{p,m} \cdot \rho_{p,m,rs} + \sum_{is \in \Psi^{\mathrm{IS}}} p\lambda_{p,m} \cdot \rho_{p,m,is} \right); \\ & \forall \{p \in \Psi^{\mathrm{P}}, m \in \Psi^{\mathrm{M}}, es \in \Psi^{\mathrm{ES}}, rs \in \Psi^{\mathrm{RS}}, is \in \Psi^{\mathrm{IS}}, eb \in \Psi^{\mathrm{EB}}, rb \in \Psi^{\mathrm{RB}}, \\ & ib \in \Psi^{\mathrm{IB}}, id \in \Psi^{\mathrm{ID}}\} \end{aligned} \tag{6.276}$$

Equation (6.275) demonstrates the net present value of the DOC index over the planning horizon. The DOC index of period p of the planning horizon has four main parts: (1) the loss cost of existing, reinforced, and installed substations [i.e., the first term of Eq. (6.276)]; (2) the loss cost of existing, reinforced, and installed branches [i.e., the second term of Eq. (6.276)]; (3) the operational costs of installed DGRs [i.e., the third term of Eq. (6.276)]; and, (4) the cost of power purchased from the pool-based wholesale electricity market through existing, reinforced, and installed substations [i.e., the fourth term of Eq. (6.276)].

The maintenance budget limitation of the DNO, on the one hand, and increasing the complexity of the distribution network in view of the load growth and the proliferation of the network equipment, on the other hand, have greatly increased the importance of dealing with the maintenance cost in the PD-DEP problem. The third goal of the PD-DEP problem is, therefore, the minimization of the distribution maintenance cost (namely the DMC index), as given by Eqs. (6.277) through (6.281):

$$\begin{aligned}\mathrm{OF}_3^{\text{PD-DEP}} &= DMC_y \\ &= \sum_{p\in\Psi^{\mathrm{P}}}\left(\left(\frac{1}{(1+\mathrm{Ir})^{p-p_0}}+\frac{(1+\mathrm{Ir})^{-\mathrm{P}}}{\mathrm{Ir}}\right)\cdot DMC_{y,p}\right);\quad \forall\{p\in\Psi^{\mathrm{P}}\}\end{aligned} \tag{6.277}$$

$$\begin{aligned}DMC_{y,p} = {}&\left(\sum_{es\in\Psi^{\mathrm{ES}}} MC_{p,es}\cdot\rho_{p,es}^{\max}+\sum_{rs\in\Psi^{\mathrm{RS}}} MC_{p,rs}\cdot\rho_{p,rs}^{\max}+\sum_{is\in\Psi^{IS}} MC_{p,is}\cdot\rho_{p,is}^{\max}\right) \\ &+\left(\sum_{eb\in\Psi^{\mathrm{EB}}} MC_{p,eb}\cdot L_{eb}+\sum_{rb\in\Psi^{\mathrm{RB}}} MC_{p,rb}\cdot L_{rb}+\sum_{ib\in\Psi^{\mathrm{IB}}} MC_{p,ib}\cdot L_{ib}\right); \\ &\forall\{p\in\Psi^{\mathrm{P}}, es\in\Psi^{\mathrm{ES}}, rs\in\Psi^{\mathrm{RS}}, is\in\Psi^{\mathrm{IS}}, eb\in\Psi^{\mathrm{EB}}, rb\in\Psi^{\mathrm{RB}}, ib\in\Psi^{\mathrm{IB}}\}\end{aligned} \tag{6.278}$$

$$\rho_{p,es}^{\max} = \max\{\rho_{p,m,es}\};\quad \forall\{p\in\Psi^{\mathrm{P}}, m\in\Psi^{\mathrm{M}}, es\in\Psi^{\mathrm{ES}}\} \tag{6.279}$$

$$\rho_{p,rs}^{\max} = \max\{\rho_{p,m,rs}\};\quad \forall\{p\in\Psi^{\mathrm{P}}, m\in\Psi^{\mathrm{M}}, rs\in\Psi^{\mathrm{RS}}\} \tag{6.280}$$

$$\rho_{p,is}^{\max} = \max\{\rho_{p,m,is}\};\quad \forall\{p\in\Psi^{\mathrm{P}}, m\in\Psi^{\mathrm{M}}, is\in\Psi^{\mathrm{IS}}\} \tag{6.281}$$

Equation (6.277) shows the net present value of the DMC index over the planning horizon. The maintenance cost of period p of the planning horizon is divided into the maintenance cost of the substations and the branches. The first term of Eq. (6.278) represents the maintenance cost of existing, reinforced, and installed substations in the distribution network. This cost for a substation depends on the maximum operational injected power of the corresponding substation at different load levels [see Eqs. (6.279) through (6.281)]. It should be pointed out that the maximum operational injected power of a substation is equal to the maximum power supplied from the upstream network. The second term of Eq. (6.278), however, demonstrates the maintenance cost of existing, reinforced, and installed branches in the distribution network.

In general, unreliability has drastic and profound implications on different aspects of the PD-DEP problem. A well-suited index must be defined and applied to the PD-DEP problem with the aim of assessing the effects of each proposed plan on network reliability. Ignoring this index in the PD-DEP problem may give rise to the possibility that the proposed plans cannot effectively satisfy the predetermined reliability levels. Moreover, in view of the unique structural characteristics of the distribution network, the process of the reliability cost/worth evaluation in the PD-DEP problem is completely different from the PD-GEP and PD-TEP problems. The DNO must, therefore, consider an appropriate index to increase the degree of service continuity and to decrease the ECOC. The fourth goal of the PD-DEP problem is, accordingly, the minimization of the expected customer outage costs (namely the ECOC index), as given by Eqs. (6.282) and (6.283):

$$\begin{aligned}\mathrm{OF}_4^{\text{PD-DEP}} &= ECOC_y\\ &= \sum_{p\in\Psi^{\mathrm{P}}}\left(\left(\frac{1}{(1+\mathrm{Ir})^{p-p_0}}+\frac{(1+\mathrm{Ir})^{-\mathrm{P}}}{\mathrm{Ir}}\right)\cdot ECOC_{y,p}\right);\quad \forall\{p\in\Psi^{\mathrm{P}}\}\end{aligned} \tag{6.282}$$

$$\begin{aligned}&ECOC_{y,p}\\ &=\left(\sum_{m\in\Psi^{\mathrm{M}}}8760\cdot\frac{\Delta t_m}{24}\cdot\sum_{e\in\Psi^{\mathrm{E}}}\sum_{b\in\Psi^{\mathrm{B}}}\sum_{s\in\Psi^{\mathrm{S}}}p\rho_{p,m,e,b,s}\cdot C_{p,m,e,b,s}\left(\Delta t_{p,m,e,b,s}\right)\cdot FR_{p,m,e,b,s}\right);\\ &\forall\{p\in\Psi^{\mathrm{P}},m\in\Psi^{\mathrm{M}},e\in\Psi^{\mathrm{E}},b\in\Psi^{\mathrm{B}},s\in\Psi^{\mathrm{S}}\}\end{aligned} \tag{6.283}$$

Equation (6.282) explains the net present value of the ECOC index over the planning horizon. Equation (6.283) represents the ECOC of period p of the planning horizon.

The step-by-step procedure of reliability cost/worth evaluation in the PD-DEP problem can be summarized as follows:

1. Set the number of periods equal to P.
2. Set the period counter ($p = 1$).
3. Set the number of load levels equal to M.
4. Set the load level counter ($m = 1$).
5. Set the number of failure events equal to E.
6. Set the failure event counter ($e = 1$).
7. Find the average failure rate, $FR_{p,\ m,\ e}$, and outage duration time, $\Delta t_{p,\ m,\ e}$, of the failure event, e, at load level m of period p.
8. Find the affected buses using a direct search technique, according to the distribution network configuration.
9. Set the number of affected buses equal to B.
10. Set the affected bus counter ($b = 1$).
11. Set the number of load sectors equal to S.
12. Set the load sector counter ($s = 1$).
13. Find the average failure rate, $FR_{p,\ m,\ e,\ b,\ s}$, and outage duration time, $\Delta t_{p,\ m,\ e,\ b,\ s}$, at load sector s of affected bus b in failure event e at load level m of period p,

according to the distribution network configuration and transfer capability of alternate supplies (reserve feeder and/or DGRs).

14. Determine the cost of the outage based on the calculated outage duration time in the previous step, $\Delta t_{p,\ m,\ e,\ b,\ s}$, using the SCDF at load sector s of affected bus b at load level m of period p [i.e., $C_{p,\ m,\ e,\ b,\ s}(\Delta t_{p,\ m,\ e,\ b,\ s})$].
15. Calculate the ECOC at load sector s of affected bus b in failure event e at load level m of period p, according to $p\rho_{p,\ m,\ e,\ b,\ s} \cdot C_{p,\ m,\ e,\ b,\ s}(\Delta t_{p,\ m,\ e,\ b,\ s}) \cdot FR_{p,\ m,\ e,\ b,\ s}$.
16. $s = s + 1$: If $s \leq S$, go to step (13); otherwise go to the next step.
17. $b = b + 1$: If $b \leq B$, go to step (11); otherwise go to the next step.
18. $e = e + 1$: If $e \leq E$, go to step (7); otherwise go to the next step.
19. $m = m + 1$: If $m \leq M$, go to step (5); otherwise go to the next step.
20. $p = p + 1$: If $p \leq P$, go to step (3); otherwise go to the next step.
21. Stop.

Ignoring the fuse and breaker, the average failure rate and outage duration time at load sector s of affected bus b in failure event e at load level m of period p [see step (13)] is considered equal to the average failure rate and outage duration time of the failure event e at load level m of period p [see step (7)]. It is obvious that by taking into account the fuse and breaker these variables will be different from one another. For more information on this topic, please refer to the work by Billinton and Wang [91].

As a result, the overall objective function of the PD-DEP problem is formulated using Eq. (6.284):

$$\begin{aligned} &\underset{x_{\text{PD-DEP}}}{\text{Min}} : \left\{ \text{OF}^{\text{PD-DEP}} \right\} = \\ &\underset{x_{\text{PD-DEP}}}{\text{Min}} : \left\{ \text{W}_{\text{OF}_1}^{\text{PD-DEP}} \cdot \text{OF}_1^{\text{PD-DEP}} + \text{W}_{\text{OF}_2}^{\text{PD-DEP}} \cdot \text{OF}_2^{\text{PD-DEP}} \right. \\ &\left. + \text{W}_{\text{OF}_3}^{\text{PD-DEP}} \cdot \text{OF}_3^{\text{PD-DEP}} + \text{W}_{\text{OF}_4}^{\text{PD-DEP}} \cdot \text{OF}_4^{\text{PD-DEP}} \right\} \end{aligned} \tag{6.284}$$

In the proposed PD-DEP problem, the objective functions have different degrees of importance from the planner's (here, the DNO) point of view. Hence, in the overall objective function, multiple weighting coefficients are assigned to each objective function by the DNO. As stated previously, the constraints of the PD-DEP problem are divided into four distinct blocks: (1) the Kirchhoff's point and loop laws; (2) the operational limits on the distribution network equipment; (3) the availability of financial resources; and, (4) the logical restrictions.

The Kirchhoff's point and loop laws represent one of the most important equality constraints that deal with the power flow and potential difference in the distribution network. The Kirchhoff's point law on each bus of the distribution network must be, then, met at each load level of period p, as given by Eq. (6.285):

$$\mathrm{BBI}^{\mathrm{ex}} \cdot \left[\mathrm{PF}_{p,m}^{\mathrm{ex}}\right]^{\mathrm{Tr}} + \mathrm{BBI}^{\mathrm{re}} \cdot \left[\mathrm{PF}_{p,m}^{\boldsymbol{re}}\right]^{\mathrm{Tr}} + \mathrm{BBI}^{\mathrm{in}} \cdot \left[\mathrm{PF}_{p,m}^{\mathrm{in}}\right]^{\mathrm{Tr}} + g\rho_{p,m} + c\rho_{p,m} = p\rho_{p,m}; \forall\left\{p \in \Psi^{\mathrm{P}}, m \in \Psi^{\mathrm{M}}\right\} \tag{6.285}$$

The Kirchhoff's loop law must also be satisfied at each load level of period p. For the branches of the distribution network, refer to Eqs. (6.286) through (6.288):

$$Z_{eb} \cdot I_{p,m,eb} + \mathrm{V}_{p,m} \cdot \left[\mathrm{BBI}_{\mathrm{column}:eb}^{\mathrm{ex}}\right] = 0; \quad \forall\left\{p \in \Psi^{\mathrm{P}}, m \in \Psi^{\mathrm{M}}, eb \in \Psi^{\mathrm{EB}}\right\} \tag{6.286}$$

$$Z_{rb} \cdot I_{p,m,rb} + \mathrm{V}_{p,m} \cdot \left[\mathrm{BBI}_{\mathrm{column}:rb}^{\mathrm{re}}\right] = 0; \quad \forall\left\{p \in \Psi^{\mathrm{P}}, m \in \Psi^{\mathrm{M}}, rb \in \Psi^{\mathrm{RB}}\right\} \tag{6.287}$$

$$Z_{ib} \cdot I_{p,m,ib} + \mathrm{V}_{p,m} \cdot \left[\mathrm{BBI}_{\mathrm{column}:ib}^{\mathrm{in}}\right] = 0; \quad \forall\left\{p \in \Psi^{\mathrm{P}}, m \in \Psi^{\mathrm{M}}, ib \in \Psi^{\mathrm{IB}}\right\} \tag{6.288}$$

From a practical perspective, the distribution network equipment—such as transformers and branches—is characterized by a maximum/minimum current-carrying capacity, also referred to as thermal rating. The value of the injection current by a substation and the current passing through a branch of the distribution network should, therefore, satisfy the predetermined standards related to its capacity. In this way, the limits related to the injection currents at load level m of period p for the existing, reinforced, and installed substations are given by Eqs. (6.289) through (6.291):

$$0 \leq I_{p,m,es} \leq I_{es}^{\max}; \quad \forall\left\{p \in \Psi^{\mathrm{P}}, m \in \Psi^{\mathrm{M}}, es \in \Psi^{\mathrm{ES}}\right\} \tag{6.289}$$

$$0 \leq I_{p,m,rs} \leq I_{rs}^{\max}; \quad \forall\left\{p \in \Psi^{\mathrm{P}}, m \in \Psi^{\mathrm{M}}, rs \in \Psi^{\mathrm{RS}}\right\} \tag{6.290}$$

$$0 \leq I_{p,m,is} \leq I_{is}^{\max}; \quad \forall\left\{p \in \Psi^{\mathrm{P}}, m \in \Psi^{\mathrm{M}}, is \in \Psi^{\mathrm{IS}}\right\} \tag{6.291}$$

The range of acceptable values for the current passing through existing, reinforced, and installed branches at load level m of period p is presented by Eqs. (6.292) through (6.294):

$$0 \leq I_{p,m,eb} \leq I_{eb}^{\max}; \quad \forall\left\{p \in \Psi^{\mathrm{P}}, m \in \Psi^{\mathrm{M}}, eb \in \Psi^{\mathrm{EB}}\right\} \tag{6.292}$$

$$0 \leq I_{p,m,rb} \leq I_{rb}^{\max}; \quad \forall\left\{p \in \Psi^{\mathrm{P}}, m \in \Psi^{\mathrm{M}}, rb \in \Psi^{\mathrm{RB}}\right\} \tag{6.293}$$

$$0 \leq I_{p,m,ib} \leq I_{ib}^{\max}; \quad \forall\left\{p \in \Psi^{\mathrm{P}}, m \in \Psi^{\mathrm{M}}, ib \in \Psi^{\mathrm{IB}}\right\} \tag{6.294}$$

The voltage on each bus of the distribution network at load level m of period p must be kept within operating parameters. And since there are usually different types of buses in the distribution network, it is essential that the DNO determine the special limits for voltage on each type of bus. Eqs. (6.295) through (6.297) show the range of permissible values for voltage in the voltage conversion, electrical load, and generation buses of the distribution network, respectively.

The transfer buses (i.e., buses with no demand or generation) are considered load buses with zero demand:

$$V_{b_{vc}}^{\min} \leq V_{p,m,b_{vc}} \leq V_{b_{vc}}^{\max}; \quad \forall \{p \in \Psi^{P}, m \in \Psi^{M}, b_{vc} \in \Psi^{B_{vc}}\} \tag{6.295}$$

$$V_{b_{el}}^{\min} \leq V_{p,m,b_{el}} \leq V_{b_{el}}^{\max}; \quad \forall \{p \in \Psi^{P}, m \in \Psi^{M}, b_{el} \in \Psi^{B_{el}}\} \tag{6.296}$$

$$V_{b_{ge}}^{\min} \leq V_{p,m,b_{ge}} \leq V_{b_{ge}}^{\max}; \quad \forall \{p \in \Psi^{P}, m \in \Psi^{M}, b_{ge} \in \Psi^{B_{ge}}\} \tag{6.297}$$

The power generation of DGRs must be bounded within their capacity parameters. The range of acceptable values for the installed DGRs' power generation at load level m of period p is, then, presented by Eq. (6.298):

$$0 \leq \rho_{p,m,id} \leq \rho_{id}^{\max}; \quad \forall \{p \in \Psi^{P}, m \in \Psi^{M}, id \in \Psi^{ID}\} \tag{6.298}$$

The penetration limits of DGRs are often employed by the DNO to avoid reverse power flow from the downstream distribution network to the upstream network. The restrictions imposed by the DNO on the total installed capacity of the DGRs in period p and all other periods are determined using Eqs. (6.299) and (6.300):

$$\sum_{id \in \Psi^{ID}} \rho_{p,id}^{\text{cap}} + \rho_{p,id}^{\text{res}} \leq \rho_{p}^{\text{max-dg}}; \quad \forall \{p \in \Psi^{P}, id \in \Psi^{ID}\} \tag{6.299}$$

$$\sum_{p \in \Psi^{P}} \sum_{id \in \Psi^{ID}} \rho_{p,id}^{\text{cap}} + \rho_{p,id}^{\text{res}} \leq \rho^{\text{max-dg}}; \quad \forall \{p \in \Psi^{P}, id \in \Psi^{ID}\} \tag{6.300}$$

The DNO usually performs load curtailment with the aim of satisfying the power balance constraint and removing potential overload of the distribution network equipment. However, the amount of load curtailment on each bus of the distribution network must be restricted within its predetermined limits. The permissible range of curtailment load in load node b_{el} at load level m of period p is given by Eq. (6.301):

$$0 \leq c\rho_{p,m,b_{el}} \leq p\rho_{p,m,b_{el}}; \quad \forall \{p \in \Psi^{P}, m \in \Psi^{M}, b_{el} \in \Psi_{el}^{B}\} \tag{6.301}$$

To achieve easier and more reliable protection, the distribution network should be operated in an open-loop configuration. The open-loop configuration of the distribution network is considered as a constraint in the PD-DEP problem. Ignoring this constraint in the PD-DEP problem may result in the obtained expansion plans not being appropriate for real-world distribution networks. The implementation of this constraint is but one of the most complex limitations in the process of solving the PD-DEP problem. In this case, in order to maintain the open-loop configuration of the distribution network [i.e., meet Eq. (6.302)], a graph-based approach is employed:

$$\textit{Open-loop configuration of the distribution network} \tag{6.302}$$

In general, the distribution network configuration can be considered as a simple linear graph. Let G = (V, E) be a graph in which V = $\{v_1, v_2, ...\}$ represents a set of objects called vertices—corresponding with the distribution network buses—and E = $\{e_1, e_2, ...\}$ represents a set of arcs called edges—corresponding with the distribution network branches. According to graph theory, it is obvious that a tree is a connected graph G without any loop; thus, it is possible to compare the open-loop configuration of the distribution network with a tree. In other words, graph G is a tree if and only if it has two conditions: connectivity and no-loop. In addition, a disjoint union of trees is called a forest. In the proposed techno-economic framework, the configuration of the distribution network is demonstrated by a forest. This forest can contain one or multiple trees such that each tree has one and only one substation bus. It can, therefore, be stated that the configuration of the distribution network is open loop if and only if each sub-graph in the forest is a tree (i.e., satisfies these two conditions). The first condition is necessary but not sufficient; thus, the first condition should be coupled with the second to guarantee the open-loop configuration of the distribution network.

The first condition is evaluated for each sub-graph in the forest, as given by Eq. (6.303):

$$BBI_{sg} \neq \begin{bmatrix} BBI_{sg}^{1} & \vdots & 0 \\ \cdots & \cdots & \cdots \\ 0 & \vdots & BBI_{sg}^{2} \end{bmatrix}; \quad \forall \{sg \in \Psi^{SG}\} \tag{6.303}$$

This equation tells us that the sub-graph sg in the forest is connected if the bus-branch incidence matrix of this sub-graph cannot be written in a block-diagonal form as Equation (6.303). If condition 1 is false, condition 2 will not be evaluated. In this circumstance, the sub-graph is not a tree and there is no need to evaluate condition 2. If condition 1 is true, there is no guarantee that this sub-graph is a tree, so condition 2 should be evaluated.

Condition 2 is assessed for each sub-graph in the forest, according to Eq. (6.304):

$$\text{rank}\left(BBI_{sg}\right) = N_{sg}^{\text{bu}} - 1; \quad \forall \{sg \in \Psi^{SG}\} \tag{6.304}$$

This equation tells us that the rank of bus-branch incidence matrix for each sub-graph in the forest must be equal to the number of distribution network buses at this sub-graph minus 1. If both conditions are satisfied for a sub-graph, the sub-graph is a tree; otherwise, the sub-graph is not a tree. For a thorough discussion on the process of proving these conditions, please see the work by Deo [92].

The DNO's overall budget for purchasing and installing distribution network equipment is usually limited. The distribution investment costs in period p and in all other periods must, therefore, meet the conditions specified by Eqs. (6.305) and (6.306):

$$DIC_{y,p} \leq DIC_{p}^{\max}; \quad \forall \{p \in \Psi^{\mathrm{P}}\} \tag{6.305}$$

$$\sum_{p \in \Psi^{\mathrm{P}}} DIC_{y,p} \leq DIC^{\max}; \quad \forall \{p \in \Psi^{\mathrm{P}}\} \tag{6.306}$$

In the proposed techno-economic framework, to satisfy the rational and practical assumptions, a set of logical discrete or numerical constraints is considered with the aim of assessing the obtained distribution expansion plans. Here, some of the most important logical constraints are expressed. The number of installed substations on each candidate bus of the distribution network of period p and at all other periods must meet the conditions specified by Eqs. (6.307) and (6.308):

$$N_{p,cb^{\mathrm{i-s}}} \leq 1; \quad \forall \{p \in \Psi^{\mathrm{P}}, cb^{\mathrm{i-s}} \in \Psi^{\mathrm{CB_{i-s}}}\} \tag{6.307}$$

$$\sum_{p \in \Psi^{\mathrm{P}}} N_{p,cb^{\mathrm{i-s}}} \leq 1; \quad \forall \{p \in \Psi^{\mathrm{P}}, cb^{\mathrm{i-s}} \in \Psi^{\mathrm{CB_{i-s}}}\} \tag{6.308}$$

The candidate substation for reinforcement can only be reinforced once in period p and in all other periods in accordance with Eqs. (6.309) and (6.310):

$$N_{p,cb^{\mathrm{r-s}}} \leq 1; \quad \forall \{p \in \Psi^{\mathrm{P}}, cb^{\mathrm{r-s}} \in \Psi^{\mathrm{CB_{r-s}}}\} \tag{6.309}$$

$$\sum_{p \in \Psi^{\mathrm{P}}} N_{p,cb^{\mathrm{r-s}}} \leq 1; \quad \forall \{p \in \Psi^{\mathrm{P}}, cb^{\mathrm{r-s}} \in \Psi^{\mathrm{CB_{r-s}}}\} \tag{6.310}$$

The number of installed branches in each candidate location of period p and at all other periods must satisfy Eqs. (6.311) and (6.312):

$$N_{p,cl^{\mathrm{i-b}}} \leq 1; \quad \forall \{p \in \Psi^{\mathrm{P}}, cl^{\mathrm{i-b}} \in \Psi^{\mathrm{CL_{i-b}}}\} \tag{6.311}$$

$$\sum_{p \in \Psi^{\mathrm{P}}} N_{p,cl^{\mathrm{i-b}}} \leq 1; \quad \forall \{p \in \Psi^{\mathrm{P}}, cl^{\mathrm{i-b}} \in \Psi^{\mathrm{CL_{i-b}}}\} \tag{6.312}$$

The candidate branch for reinforcement can only be reinforced once in period p and in all other periods, according to Eqs. (6.313) and (6.314):

$$N_{p,cl^{\mathrm{r-b}}} \leq 1; \quad \forall \{p \in \Psi^{\mathrm{P}}, cl^{\mathrm{r-b}} \in \Psi^{\mathrm{CL_{r-b}}}\} \tag{6.313}$$

$$\sum_{p \in \Psi^{\mathrm{P}}} N_{p,cl^{\mathrm{r-b}}} \leq 1; \quad \forall \{p \in \Psi^{\mathrm{P}}, cl^{\mathrm{r-b}} \in \Psi^{\mathrm{CL_{r-b}}}\} \tag{6.314}$$

6.6.2 Mathematical Model of the Risk-Driven Techno-Economic Framework

In this section, to deal with risks of the PD-DEP problem arising from severe uncertainties of market price and demand, the IGDT is widely used. These uncertainties have drastic impacts on the objective functions and the constraints of the techno-economic framework. Here, the IGDT risk-averse decision-making and the IGDT risk-taker decision-making policies of the proposed techno-economic framework are examined.

6.6.2.1 The IGDT Severe Twofold Uncertainty Model

Similar to the PD-GEP, the PD-TEP, and the PD-G&TEP problems presented in previous sections, the envelope-bound IGDT model is also employed to handle the uncertainty parameters in the proposed PD-DEP problem. In the techno-economic framework, the envelope-bound IGDT model of the uncertain market price is presented according to Eqs. (6.315) through (6.317):

$$p\lambda = \begin{bmatrix} p\lambda_1 \\ \vdots \\ p\lambda_{m-1} \\ p\lambda_m \\ p\lambda_{m+1} \\ \vdots \\ p\lambda_{\mathrm{M}} \end{bmatrix}; \quad \forall\{m \in \Psi^{\mathrm{M}}\} \tag{6.315}$$

$$p\lambda_m = [p\lambda_{m,1} \quad \cdots \quad p\lambda_{m,p-1} \quad p\lambda_{m,p} \quad p\lambda_{m,p+1} \quad \cdots \quad p\lambda_{m,\mathrm{P}}]; \quad \forall\{m \in \Psi^{\mathrm{M}}, p \in \Psi^{\mathrm{P}}\} \tag{6.316}$$

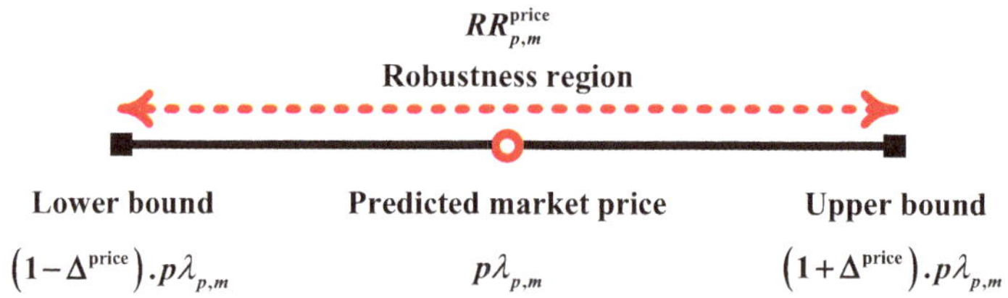

Fig. 6.18 The market price robustness region

$$\Pi_{p,m}^{\text{price}}\left(\Delta^{\text{price}}, p\lambda_{p,m}\right) = \left\{ a\lambda_{p,m} : \left| \frac{a\lambda_{p,m} - p\lambda_{p,m}}{p\lambda_{p,m}} \right| \leq \Delta^{\text{price}} \right\};$$
$$\forall \left\{ \Delta^{\text{price}} > 0, a\lambda_{p,m} \in \Pi_{p,m}^{\text{price}}\left(\Delta^{\text{price}}, p\lambda_{p,m}\right), p \in \Psi^{\text{P}}, m \in \Psi^{\text{M}} \right\} \tag{6.317}$$

In Eq. (6.317), if the actual market price, $a\lambda_{p,\ m}$, is equal to the predicted market price, $p\lambda_{p,\ m}$, the interval of the market price uncertainty, Δ^{price}, will be equal to zero. Otherwise, this parameter will be equal to a positive value. Hence, the market price robustness region is described according to Eqs. (6.318), as shown in Fig. 6.18:

$$RR_{p,m}^{\text{price}} = \left[\left(1 - \Delta^{\text{price}}\right) \cdot p\lambda_{p,m} \quad \left(1 + \Delta^{\text{price}}\right) \cdot p\lambda_{p,m} \right];$$
$$\forall \left\{ \Delta^{\text{price}} > 0, p \in \Psi^{\text{P}}, m \in \Psi^{\text{M}} \right\} \tag{6.318}$$

Similarly, the envelope-bound IGDT model of the uncertain demand is described by Eqs. (6.319) through (6.321):

$$p\rho = \left[p\rho_1 \quad \cdots \quad p\rho_{b-1} \quad p\rho_b \quad p\rho_{b+1} \quad \cdots \quad p\rho_{\text{B}} \right]; \quad \forall \left\{ b \in \Psi^{\text{B}} \right\} \tag{6.319}$$

$$p\rho_b = \begin{bmatrix} p\rho_{1,1,b} & \cdots & p\rho_{p-1,1,b} & p\rho_{p,1,b} & p\rho_{p+1,1,b} & \cdots & p\rho_{\mathbf{P},1,b} \\ \vdots & & \vdots & \vdots & \vdots & & \vdots \\ p\rho_{1,m-1,b} & \cdots & p\rho_{p-1,m-1,b} & p\rho_{p,m-1,b} & p\rho_{p+1,m-1,b} & \cdots & p\rho_{\mathbf{P},m-1,b} \\ p\rho_{1,m,b} & \cdots & p\rho_{p-1,m,b} & p\rho_{p,m,b} & p\rho_{p+1,m,b} & \cdots & p\rho_{\mathbf{P},m,b} \\ p\rho_{1,m+1,b} & \cdots & p\rho_{p-1,m+1,b} & p\rho_{p,m+1,b} & p\rho_{p+1,m+1,b} & \cdots & p\rho_{\mathbf{P},m+1,b} \\ \vdots & & \vdots & \vdots & \vdots & & \vdots \\ p\rho_{1,\text{M},b} & \cdots & p\rho_{p-1,\text{M},b} & p\rho_{p,\text{M},b} & p\rho_{p+1,\text{M},b} & \cdots & p\rho_{\text{P},\text{M},b} \end{bmatrix};$$
$$\forall \left\{ p \in \Psi^{\text{P}}, m \in \Psi^{\text{M}}, b \in \Psi^{\text{B}} \right\} \tag{6.320}$$

$$\Pi_{p,m,b}^{\text{demand}}\left(\Delta^{\text{demand}}, p\rho_{p,m,b}\right) = \left\{ a\rho_{p,m,b} : \left| \frac{a\rho_{p,m,b} - p\rho_{p,m,b}}{p\rho_{p,m,b}} \right| \leq \Delta^{\text{demand}} \right\};$$
$$\forall \left\{ \Delta^{\text{demand}} > 0, a\rho_{p,m,b} \in \Pi_{p,m,b}^{\text{demand}}\left(\Delta^{\text{demand}}, p\rho_{p,m,b}\right), p \in \Psi^{\text{P}}, m \in \Psi^{\text{M}}, b \in \Psi^{\text{B}} \right\} \tag{6.321}$$

In Eq. (6.321), if the actual power demand, $a\rho_{p,\ m,\ b}$, is equal to the predicted power demand, $p\rho_{p,\ m,\ b}$, the interval of the power demand uncertainty, Δ^{demand}, will be equal to zero. Otherwise, this parameter will be equal to a positive value. The power demand robustness region (see Fig. 6.19) can, then, be defined using Eq. (6.322):

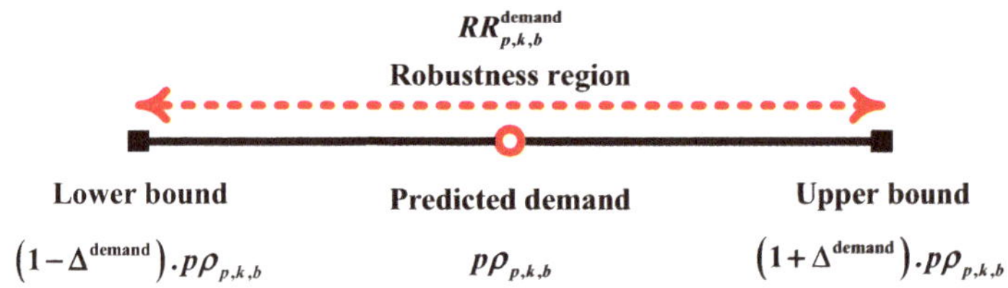

Fig. 6.19 The demand robustness region

$$RR^{\text{demand}}_{p,m,b} = \left[\left(1-\Delta^{\text{demand}}\right)\cdot p\rho_{p,m,b} \quad \left(1+\Delta^{\text{demand}}\right)\cdot p\rho_{p,m,b}\right]; \quad \forall\left\{\Delta^{\text{demand}} > 0, p\in\Psi^{\text{P}}, m\in\Psi^{\text{M}}, b\in\Psi^{\text{B}}\right\} \tag{6.322}$$

6.6.2.2 The IGDT Risk-Averse Decision-Making Model: Robustness Function

In general, the robustness function represents the destructive face of the severe twofold uncertainties in the IGDT-based techno-economic framework. In other words, the robustness function states that the greatest level of uncertainty parameters means that the maximum value of the PD-DEP objectives (PD-DEPO) cannot be greater than a predetermined critical cost. In the IGDT risk-averse decision-making policy, by using the predetermined IGDT critical cost, robust optimal distribution expansion plans are determined by the DNO. The robustness function for the techno-economic framework can be expressed using Eq. (6.323):

$$\Upsilon^{\text{PD-DEP}}\left(x_{\text{PD-DEP}}, \varpi_{\text{c}}^{\text{PD-DEP}}\right) = \underset{\substack{\Delta^{\text{price}}\\ \Delta^{\text{demand}}}}{\text{Max}}\left\{\left(\Delta^{\text{price}}, \Delta^{\text{demand}}\right) : \underset{\substack{x_{\text{PD-DEP}}\\ a\lambda_{p,m}\in\Pi^{\text{price}}_{p,m}\left(\Delta^{\text{price}}, p\lambda_{p,m}\right)\\ a\rho_{p,m,b}\in\Pi^{\text{demand}}_{p,m,b}\left(\Delta^{\text{demand}}, p\rho_{p,m,b}\right)}}{\text{Max}} \text{PD-DEPO}\left(x_{\text{PD-DEP}}, a\lambda_{p,m}, a\rho_{p,m,b}\right) \le \varpi_{\text{c}}^{\text{PD-DEP}}\right\};$$
$$\forall\left\{\Delta^{\text{price}} > 0, \Delta^{\text{demand}} > 0, p\in\Psi^{\text{P}}, m\in\Psi^{\text{M}}, b\in\Psi^{\text{B}}\right\} \tag{6.323}$$

The IGDT risk-averse decision-making policy of the techno-economic framework can be formulated as follows:

$$\Upsilon^{\text{PD-DEP}}\left(x_{\text{PD-DEP}}, \varpi_{\text{c}}^{\text{PD-DEP}}\right) = \underset{\substack{\Delta^{\text{price}}\\ \Delta^{\text{demand}}}}{\text{Max}}\left\{\left(\Delta^{\text{price}}, \Delta^{\text{demand}}\right) : \underset{\substack{x_{\text{PD-DEP}}\\ a\lambda_{p,m}\in\Pi^{\text{price}}_{p,m}\left(\Delta^{\text{price}}, p\lambda_{p,m}\right)\\ a\rho_{p,m,b}\in\Pi^{\text{demand}}_{p,m,b}\left(\Delta^{\text{demand}}, p\rho_{p,m,b}\right)}}{\text{Max}} \text{PD-DEPO}\left(x_{\text{PD-DEP}}, a\lambda_{p,m}, a\rho_{p,m,b}\right) \le \varpi_{\text{c}}^{\text{PD-DEP}}\right\};$$
$$\forall\left\{\Delta^{\text{price}} > 0, \Delta^{\text{demand}} > 0, p\in\Psi^{\text{P}}, m\in\Psi^{\text{M}}, b\in\Psi^{\text{B}}\right\} \tag{6.324}$$ IGDT level

and subject to Eqs. (6.325) through (6.330):

$$\underset{\mathrm{X_{PD\text{-}DEP}}}{\mathrm{Max}} \ \{\mathrm{Eq.}(6.284)\} \leq \varpi_{\mathrm{c}}^{\mathrm{PD\text{-}DEP}};$$
$$\forall\left\{a\lambda_{p,m} = \left(1+\Delta^{\mathrm{price}}\right)\cdot p\lambda_{p,m}, a\rho_{p,m,b} = \left(1+\Delta^{\mathrm{demand}}\right)\cdot p\rho_{p,m,b}, p\in\Psi^{\mathrm{P}}, m\in\Psi^{\mathrm{M}}, b\in\Psi^{\mathrm{B}}\right\}$$
$$(6.325) \ \text{PD-DEP level}$$

$$a\lambda_{p,m} \leq \left(1+\Delta^{\mathrm{price}}\right)\cdot p\lambda_{p,m}; \quad \forall\left\{p\in\Psi^{\mathrm{P}}, m\in\Psi^{\mathrm{M}}\right\} \tag{6.326}$$

$$a\lambda_{p,m} \geq \left(1-\Delta^{\mathrm{price}}\right)\cdot p\lambda_{p,m}; \quad \forall\left\{p\in\Psi^{\mathrm{P}}, m\in\Psi^{\mathrm{M}}\right\} \tag{6.327}$$

$$a\rho_{p,m,b} \leq \left(1+\Delta^{\mathrm{demand}}\right)\cdot p\rho_{p,m,b}; \quad \forall\left\{p\in\Psi^{\mathrm{P}}, m\in\Psi^{\mathrm{M}}, b\in\Psi^{\mathrm{B}}\right\} \tag{6.328}$$

$$a\rho_{p,m,b} \geq \left(1-\Delta^{\mathrm{demand}}\right)\cdot p\rho_{p,m,b}; \quad \forall\left\{p\in\Psi^{\mathrm{P}}, m\in\Psi^{\mathrm{M}}, b\in\Psi^{\mathrm{B}}\right\} \tag{6.329}$$

$$\left.\begin{Bmatrix}\text{Eqs.}(6.285)\ \text{through}\ (6.304)\ \text{and}\\ \text{Eqs.}(6.307)\ \text{through}\ (6.314)\end{Bmatrix}\right|_{a\lambda_{p,m}, a\rho_{p,m,b}};$$
$$\forall\left\{a\lambda_{p,m} = \left(1+\Delta^{\mathrm{price}}\right)\cdot p\lambda_{p,m}, a\rho_{p,m,b} = \left(1+\Delta^{\mathrm{demand}}\right)\cdot p\rho_{p,m,b}, p\in\Psi^{\mathrm{P}}, m\in\Psi^{\mathrm{M}}, b\in\Psi^{\mathrm{B}}\right\}$$
$$(6.330)$$

In the techno-economic framework under the IGDT risk-averse decision-making policy, the maximum values of the PD-DEPO are obtained for the highest level of the uncertain market price and demand so that these uncertain parameters are allowed by the IGDT risk-averse decision-making policy at the relevant robust region (see Figs. 6.18 and 6.19). The solution of the IGDT risk-averse decision-making policy yields the optimal robust distribution expansion plans of the DNO in accordance with the defined value of its critical cost. The predetermined critical cost can be calculated using Eq. (6.331):

$$\varpi_{\mathrm{c}}^{\mathrm{PD\text{-}DEP}} = \left(1+\sigma_{\mathrm{c}}^{\mathrm{PD\text{-}DEP}}\right)\cdot\varpi_{\mathrm{b}}^{\mathrm{PD\text{-}DEP}} \tag{6.331}$$

In Eq. (6.328), the IGDT base cost, $\varpi_{\mathrm{b}}^{\mathrm{PD-DEP}}$, is calculated by solving the deterministic techno-economic framework, as presented in Eqs. (6.273) through (6.314). The deterministic framework of the techno-economic framework is also called a risk-neutral decision-making policy. In general, in the techno-economic framework, the critical cost of the DNO is higher than its base cost.

6.6.2.3 The IGDT Risk-Taker Decision-Making Model: Opportunity Function

In general, the opportunity function illustrates the propitious face of the severe twofold uncertainties in the IGDT-based techno-economic framework. In other words, the opportunity function indicates that the lowest level of uncertainty parameters means that the minimum value of the PD-DEPO cannot be greater than a predetermined target cost. In the IGDT risk-taker decision-making policy, by using the predetermined IGDT target cost, opportunistic optimal distribution expansion plans are determined by the DNO. The opportunity function for the techno-economic framework can be defined using Eq. (6.332):

$$\Gamma^{\text{PD-DEP}}\left(x_{\text{PD-DEP}}, \varpi_{\text{t}}^{\text{PD-DEP}}\right) = \underset{\substack{\Delta^{\text{price}} \\ \Delta^{\text{demand}}}}{\text{Min}} \left\{ \left(\Delta^{\text{price}}, \Delta^{\text{demand}}\right) : \underset{\substack{x_{\text{PD-DEP}} \\ a\lambda_{p,m} \in \Pi_{p,m}^{\text{price}}\left(\Delta^{\text{price}}, p\lambda_{p,m}\right) \\ a\rho_{p,m,b} \in \Pi_{p,m,b}^{\text{demand}}\left(\Delta^{\text{demand}}, p\rho_{p,m,b}\right)}}{\text{Min}} \text{PD-DEPO}\left(x_{\text{PD-DEP}}, a\lambda_{p,m}, a\rho_{p,m,b}\right) \leq \varpi_{\text{t}}^{\text{PD-DEP}} \right\};$$

$$\forall\left\{\Delta^{\text{price}} > 0, \Delta^{\text{demand}} > 0, p \in \Psi^{\text{P}}, m \in \Psi^{\text{M}}, b \in \Psi^{\text{B}}\right\} \tag{6.332}$$

The IGDT risk-taker decision-making policy of the techno-economic framework can be formulated as follows:

$$\Gamma^{\text{PD-DEP}}\left(x_{\text{PD-DEP}}, \varpi_{\text{t}}^{\text{PD-DEP}}\right) = \underset{\substack{\Delta^{\text{price}} \\ \Delta^{\text{demand}}}}{\text{Min}} \left\{ \left(\Delta^{\text{price}}, \Delta^{\text{demand}}\right) : \underset{\substack{x_{\text{PD-DEP}} \\ a\lambda_{p,m} \in \Pi_{p,m}^{\text{price}}\left(\Delta^{\text{price}}, p\lambda_{p,m}\right) \\ a\rho_{p,m,b} \in \Pi_{p,m,b}^{\text{demand}}\left(\Delta^{\text{demand}}, p\rho_{p,m,b}\right)}}{\text{Min}} \text{PD-DEPO}\left(x_{\text{PD-DEP}}, a\lambda_{p,m}, a\rho_{p,m,b}\right) \leq \varpi_{\text{t}}^{\text{PD-DEP}} \right\};$$

$$\forall\left\{\Delta^{\text{price}} > 0, \Delta^{\text{demand}} > 0, p \in \Psi^{\text{P}}, m \in \Psi^{\text{M}}, b \in \Psi^{\text{B}}\right\} \tag{6.333 IGDT level}$$

and subject to Eqs. (6.334) through (6.339):

$$\underset{x_{\text{PD-DEP}}}{\text{Min}} \left\{\text{Eq.}(6.284)\right\} \leq \varpi_{\text{c}}^{\text{PD-DEP}};$$
$$\forall\left\{a\lambda_{p,m} = \left(1 - \Delta^{\text{price}}\right) \cdot p\lambda_{p,m}, a\rho_{p,m,b} = \left(1 - \Delta^{\text{demand}}\right) \cdot p\rho_{p,m,b}, p \in \Psi^{\text{P}}, m \in \Psi^{\text{M}}, b \in \Psi^{\text{B}}\right\}$$

$$\tag{6.334 PD-DEP level}$$

$$a\lambda_{p,m} \leq \left(1 + \Delta^{\text{price}}\right) \cdot p\lambda_{p,m}; \quad \forall\left\{p \in \Psi^{\text{P}}, m \in \Psi^{\text{M}}\right\} \tag{6.335}$$

$$a\lambda_{p,m} \geq \left(1 - \Delta^{\text{price}}\right) \cdot p\lambda_{p,m}; \quad \forall\left\{p \in \Psi^{\text{P}}, m \in \Psi^{\text{M}}\right\} \tag{6.336}$$

$$a\rho_{p,m,b} \leq \left(1 + \Delta^{\text{demand}}\right) \cdot p\rho_{p,m,b}; \quad \forall\left\{p \in \Psi^{\text{P}}, m \in \Psi^{\text{M}}, b \in \Psi^{\text{B}}\right\} \tag{6.337}$$

$$a\rho_{p,m,b} \geq \left(1 - \Delta^{\text{demand}}\right) \cdot p\rho_{p,m,b}; \quad \forall\left\{p \in \Psi^{\text{P}}, m \in \Psi^{\text{M}}, b \in \Psi^{\text{B}}\right\} \tag{6.338}$$

$$\left\{\begin{array}{l}\text{Eqs.}(6.285)\text{ through }(6.304)\text{ and}\\ \text{Eqs.}(6.307)\text{ through }(6.314)\end{array}\right\}\Bigg|_{a\lambda_{p,m},a\rho_{p,m,b}};$$
$$\forall\left\{a\lambda_{p,m}=\left(1-\Delta^{\text{price}}\right)\cdot p\lambda_{p,m},a\rho_{p,m,b}=\left(1-\Delta^{\text{demand}}\right)\cdot p\rho_{p,m,b},p\in\Psi^{\text{P}},m\in\Psi^{\text{M}},b\in\Psi^{\text{B}}\right\} \tag{6.339}$$

In the proposed techno-economic framework under the IGDT risk-taker decision-making policy, the minimum values of the PD-DEPO are obtained for the lowest level of the uncertain market price and demand so that these uncertain parameters are allowed by the IGDT risk-taker decision-making policy at a relevant robust region (see Figs. 6.18 and 6.19). The solution of the IGDT risk-taker decision-making policy yields the optimal opportunistic distribution expansion plans of the DNO, according to the defined value of its target cost. The predetermined target cost can be calculated using Eq. (6.340):

$$\varpi_{\text{t}}^{\text{PD-DEP}}=\left(1-\sigma_{\text{t}}^{\text{PD-DEP}}\right)\cdot\varpi_{\text{b}}^{\text{PD-DEP}} \tag{6.340}$$

Similarly, in Eq. (6.340), the IGDT base cost, $\varpi_{\text{b}}^{\text{PD-DEP}}$, is obtained by solving the deterministic techno-economic framework as described in Eqs. (6.273) through (6.314). In general, in the techno-economic framework, the determined target cost by the DNO is lower than its base cost.

6.6.3 Solution Method and Implementation Considerations

In the proposed techno-economic framework, the decision-making variables of the solution vector of the optimization algorithm are (1) the time of reinforced/installed equipment of the open-loop distribution network over the planning horizon; (2) the site, size, and power produced by the candidate installed DGRs; (3) the site and size of the candidate installed substations/branches; and, (4) the size of the candidate-reinforced substations/branches. The solution process of the techno-economic framework starts with the implementation of the short-term LBOPF problem. As previously mentioned, the proposed short-term LBOPF problem is solved with the aim of maximizing the loadability of the open-loop distribution network, subject to Kirchhoff's point and loop laws and the operational limits on the distribution network equipment. The short-term LBOPF problem is performed for all load levels in period p of the planning horizon. After completing the short-term LBOPF problem for period p, the LBOPF problem outcomes are transferred to the long-term planning problem (i.e., the PD-DEP problem). In the long-term planning problem, the DNO solves the PD-DEP problem with the aim of minimizing the DIC, DOC, DMC, and ECOC, subject to Kirchhoff's point and loop laws; the operational limits on the distribution network equipment; the availability of financial resources; and the logical limits in order to optimally determine the reinforced/

installed substations and branches as well as the newly installed DGRs [see Eqs. (6.273) through (6.214)]. Finally, the optimal distribution expansion plans are specified for period p and are considered as operational equipment in the next period of planning.

The flowchart of the techno-economic framework under the IGDT risk-neutral, the IGDT risk-averse, and the IGDT risk-taker decision-making policies is depicted in Fig. 6.20. In order to implement the proposed framework under each of these decision-making policies, the block relevant to this decision-making policy is active, while other blocks are inactive in the flowchart.

6.6.4 Simulation Results and Case Studies

The techno-economic framework is examined to solve the PD-DEP problem on a modified three-phase medium-voltage open-loop distribution test network [80]. The base apparent power and voltage for the entire test network are equal to 1 MVA and 13.8 kV, respectively. Figure 6.21 shows the primary single-line diagram of this modified distribution test network. According to this figure, this modified distribution test network has 24 load buses, 3 voltage conversions (i.e., substation, buses), and 39 branches. In Fig. 6.21, the circles indicate the load buses. The solid line squares denote the existing substations in the initial configuration of the open-loop distribution network that can be reinforced in period p, and dotted line squares represent candidate substations for installation in period p and reinforcement in period $p + 1$. In addition, the solid lines represent the existing branches in the initial configuration of the open-loop distribution network that cannot be reinforced, and double lines (i.e., one solid line plus one dotted line) depict the existing branches in the initial configuration of the open-loop distribution network that can be reinforced in period p. Candidate branches that can be installed in period p and candidate branches that can be installed in period p and reinforced in period $p + 1$ are depicted by dotted lines and double-dotted lines, respectively. Complete data for this modified open-loop distribution test network are given in Appendix 3. Data associated with the predicted power demand in each load level of period p are tabulated in Table 6.127. Data of branch lengths are provided in Table 6.127.

To describe the daily LDC in the proposed PD-DEP problem, period p of the planning horizon considers three different load levels (M = 3): (1) the first load level or the on-peak load level; (2) the second load level or mid-peak load level; and (3) the third load level or off-peak load level. Data for these load levels are given in Table 6.128. In this modified open-loop distribution test network, it is assumed that each electrical load on each bus is divided into three sectors: (1) residential, (2) commercial, and, (3) industrial. Table 6.129 gives the data for the customers and sector participation of each load. Table 6.130 gives the SCDF and its duration at each load sector on each load bus. For the sake of simplicity, the failure rate and repair times are determined in a similar manner for all branches. These values are considered to be 0.4 failures per year and 10 h, respectively. Economic lifetime data are provided

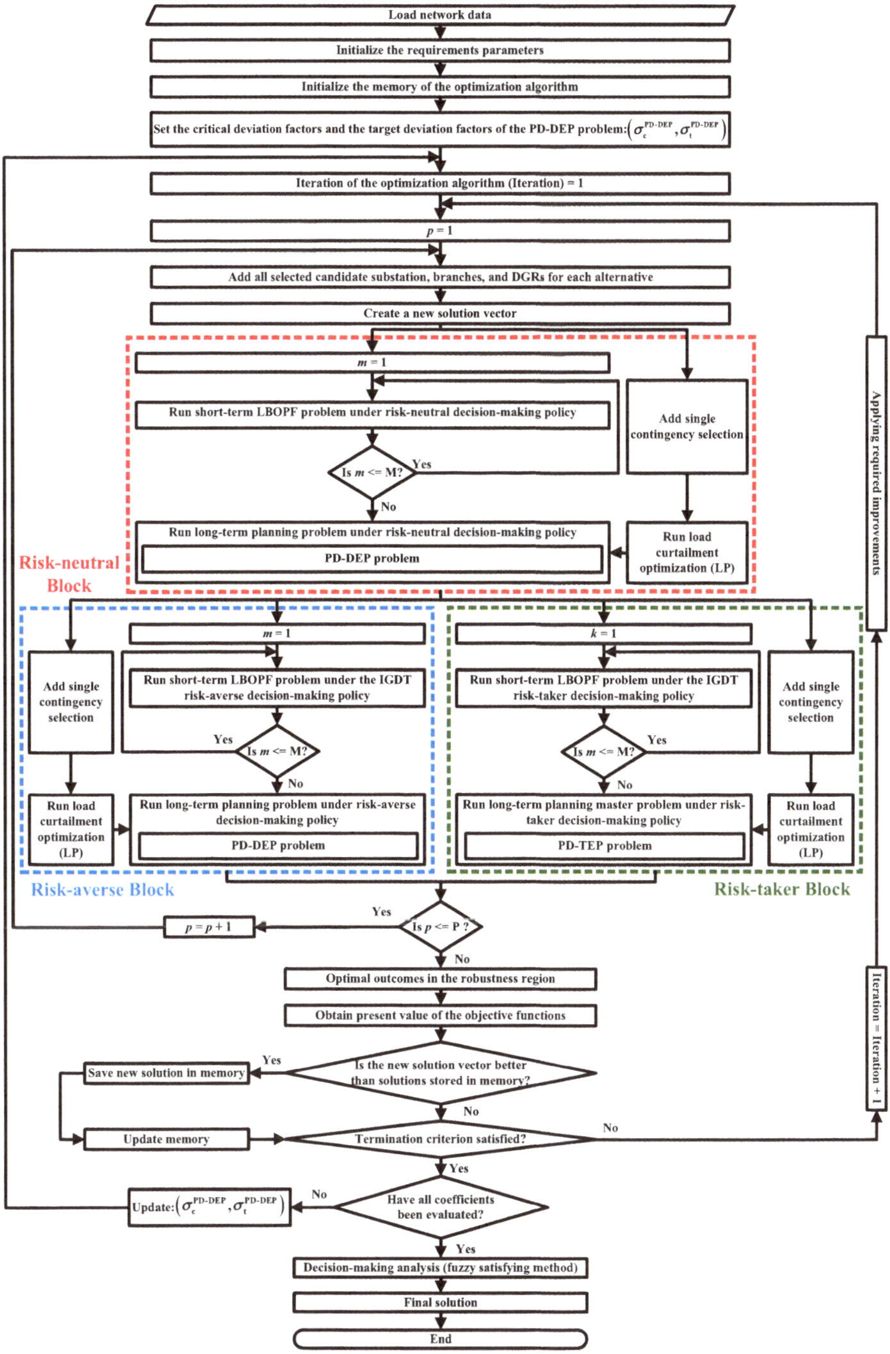

Fig. 6.20 Flowchart of the techno-economic framework under the IGDT risk-neutral, the IGDT risk-averse, and the IGDT risk-taker decision-making policies

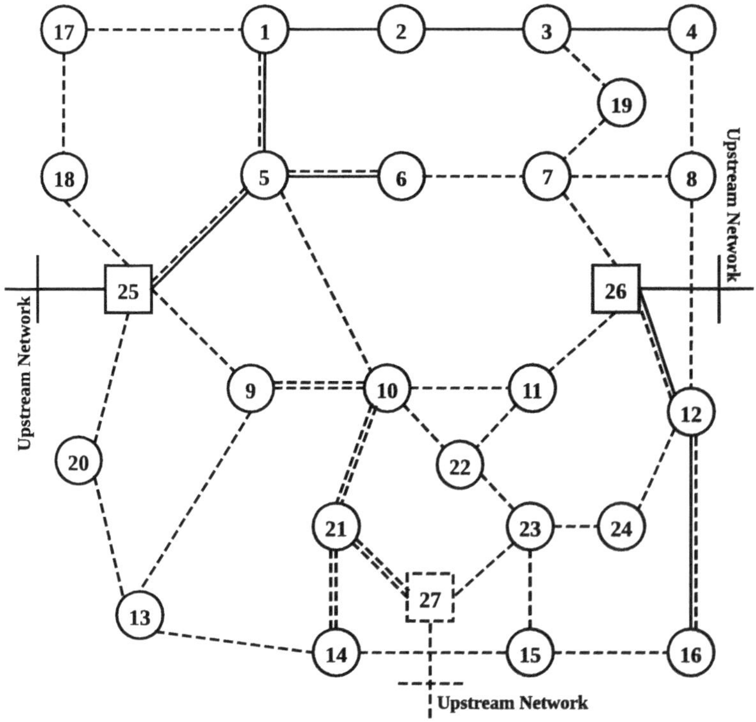

Fig. 6.21 Primary single-line diagram of the modified medium-voltage open-loop distribution test network

in Table 6.131. Table 6.132 presents the techno-economic features of the existing and candidate substations in the modified open-loop distribution test network. These characteristics for the existing and candidate branches in the modified open-loop distribution test network are given in Table 6.133. Table 6.134 gives the parameters of weighting coefficients for the objective functions related to the PD-DEP problem. The DNO expands the distribution network over a 3-year planning horizon divided into 1-year periods. The interest rate is considered to be 10%. The maximum admissible value of the DIC index in period p and in all other periods of the planning horizon is set to \$25M and \$70M, respectively. The voltage value on the voltage conversion buses of the modified open-loop distribution test network is fixed at 1.05 per unit. The lower and upper limits for voltage at electrical load and on generation buses of the modified open-loop distribution test network are considered to be 0.95 and 1.05 per unit, respectively. Each bus of the modified open-loop distribution network is regarded as a candidate bus for DGR installation. The sizes of the candidate DGRs are at multiples of 1 MW.

In the process of solving the PD-DEP problem, the DNO can install up to four DGRs along with one DGR in reserve on each candidate bus of the modified open-loop distribution network. The investment and operational costs of installed DGRs are set at \$0.89 M/MW and \$50 M/MW-h, respectively. The maximum admissible values for total installed capacity of the DGRs in period p and in all other periods of the planning horizon are considered to be 10 MW and 25 MW, respectively. The maximum admissible value of curtailment load on each electrical load bus of the modified open-loop distribution network is considered to be equal to the maximum predicted demand on the corresponding bus. Table 6.135 provides a summary of these assumptions.

To analyze the performance of the proposed strategic quad-level computational-logical framework, the following two cases and two scenarios are investigated:

- First case: The proposed techno-economic framework is run under the IGDT risk-averse decision-making policy while considering the following two scenarios:
  - DGRs are ignored.
  - DGRs are considered.
- Second case: The techno-economic framework is run under the IGDT risk-taker decision-making policy while considering the following two scenarios:
  - DGRs are ignored.
  - DGRs are considered.

The techno-economic framework under the IGDT risk-averse and the IGDT risk-taker decision-making policies is implemented and solved by using the proposed multi-objective SOSA, which was addressed in Chap. 4. Table 6.103 gives the parameter adjustments of the newly developed multi-objective SOSA. In the techno-economic framework under the IGDT risk-averse and the IGDT risk-taker decision-making policies, the DNO needs to have access to the obtained results from the framework under a risk-neutral/deterministic decision-making policy in order to calculate the critical, $\varpi_{\mathrm{c}}^{\mathrm{PD-DEP}}$, and target, $\varpi_{\mathrm{t}}^{\mathrm{PD-DEP}}$, costs. At first, the proposed risk-neutral/deterministic techno-economic framework [i.e., Eqs. (6.273) through (6.314)] is solved by the DNO, based on predicted demand and market price through the offered single-objective SOSA. Then, the calculated results under the risk-neutral/deterministic techno-economic framework are placed at the disposal of the DNO to specify the distribution expansion plans under the IGDT risk-averse and the IGDT risk-taker decision-making policies. Since the techno-economic framework under the IGDT risk-averse and the IGDT risk-taker decision-making policies is examined for first and second scenarios, the risk-neutral/deterministic techno-economic framework should also be solved based on these two scenarios. The distribution expansion plans of the proposed framework under the risk-neutral/deterministic decision-making policy for the first and second scenarios are tabulated in Table 6.78. Table 6.79 gives the values for different objectives of the proposed PD-DEP problem [see Eq. (6.191)] for the first and second scenarios. In the optimal results provided in Table 6.78, letters R and I are associated with the reinforced and installed distribution network equipment, respectively. The numbers following these letters illustrate the expansion options of the distribution network equipment, either 1 or 2. To clarify the results presented in Table 6.78, consider the second period of the planning horizon. In the first scenario of this period, the DNO installs a new substation of type 1 on bus 27 of the network. The DNO also installs a new branch of type 1 in corridors (12-24), (14-21), (15-23), and (13-14) of the distribution network, and a new branch of type 2 in corridors (21-27) and (23-27) of the distribution network. In the second scenario of this period, however, the DNO installs a new branch of type 1 in corridors (10-21), (13-20), (11-22), and (23-24) of the distribution network, and a new branch of type 2 in corridor (9-10) of the distribution network. The DNO reinforces the existing branch in corridor (12-26) of the distribution network with a new branch of type 2. In addition, the DNO installs two new DGRs on buses 19 and 21 of the distribution network. The optimal value for the size of these DGRs is 1 MW under operating conditions, plus 1 MW in reserve. For the remaining periods of the planning horizon, the same analysis can be performed.

From the results in Table 6.78, it is clear that the distribution network at the end of the planning horizon in both scenarios has an open-loop configuration. By analyzing the optimal results presented in Table 6.78, it can also be seen that the obtained configurations of the distribution network in each period of the planning horizon, for scenarios 1 and 2, are different. This is due to the fact that the different expansion options are available to the DNO in the first and second scenarios. When the DGRs are not used (i.e., scenario 1) in the techno-economic framework, candidate bus 27 for substation installation was expanded with a new substation of type 1. However, when DGRs are incorporated (i.e., scenario 2) in the techno-economic framework, there is no need to install and/or reinforce any substation. As a result, considering DGRs as one of the expansion options in the PD-DEP problem can provide a well-planned result to partially alleviate the budget restrictions related to the expensive equipment of the distribution networks, especially substations.

By evaluating the results presented in Table 6.79, it can be concluded that when DGRs are considered in the techno-economic framework, different objectives of the proposed PD-DEP problem [see Eq. (6.284)] have more favorable values in comparison with the situation in which DGRs are ignored (i.e., second scenario compared to the first). In simple terms, by considering DGRs in the techno-economic framework, the DIC, DOC, DMC, and ECOC have lower values than those where DGRs are not considered. As a consequence, keeping DGRs as an alternative expansion option in the PD-DEP problem can bring about optimal expansion plans for a distribution network with lower DIC, DOC, DMC, and ECOC.

6.6.4.1 First Case: Simulation Results and Discussion

In order to implement the proposed techno-economic framework under the IGDT risk-averse decision-making policy, the DNO must solve the optimization problem in Eqs. (6.324) through (6.331) for different values of the critical cost deviation factor [i.e., $\sigma_{c}^{PD-DEP} \in (0, 0.8)$]. These different values of the critical cost deviation

Table 6.78 Optimal distribution expansion plans of the proposed framework under the risk-neutral/deterministic decision-making policy, based on the first and second scenarios

	Optimal distribution expansion plans					
	First scenario: Ignoring DGRs			Second scenario: Considering DGRs		
Period No.	Substation	Branch	DGR	Substation	Branch	DGR
1	–	(5-25: R1), (11-26: I1), (12-26: R2), (8-12: I2), (7-8: I1), (7-19: I1), (9-25: I2), (20-25: I1)	–	–	(7-26: I2), (7-19: I1), (12:24: I1), (7:8: I1), (11-26: I1), (9-25: I2), (20-25: I1)	(11: 1+1 MW)
2	(27: I1)	(12-24: I1), (21-27: I2), (14-21: I1), (23-27: I2), (15-23: I1), (13-14: I1)	–	–	(12-26: R2), (9-10: I2), (10-21: I1), (13-20: I1), (11-22: I1), (23-24: I1)	(19: 1+1 MW), (21: 1 +1 MW)
3	–	(9-10: I2), (18-25: I2), (17-18: I1), (11-22: I1)	–	–	(18-25: I2), (17-18: I1), (14-21: I1), (15-16: I1)	(3: 1+1 MW)
Number of newly installed substations over the planning horizon	1	–	–	–	–	–
Number of reinforced substations over the planning horizon	–	–	–	–	–	–
Number of newly installed branches over the planning horizon	–	16	–	–	15	–
Number of reinforced branches over the planning horizon	–	2	–	–	1	–
Number of newly installed DGRs over the planning horizon	–	–	–	–	–	4

Table 6.79 Changes of the objective functions of the proposed framework under the risk-neutral/deterministic decision-making policy, based on the first and second scenarios

Objective	First scenario: Ignoring DGRs	Second scenario: Considering DGRs
DIC (M$)	38.562117	35.415087
DOC (M$)	8.332033	7.232227
DMC (M$)	1.504354	1.143652
ECOC (M$)	4.389782	3.727535
Base cost ($\varpi_{\mathrm{b}}^{\mathrm{PD-DEP}}$: M$)	52.788286	47.518501

factor lead to the formation of different values for critical cost, $\varpi_{\mathrm{c}}^{\mathrm{PD-DEP}}$. More precisely, from the DNO's perspective, Eqs. (6.324) through (6.331) were solved by using the Pareto optimality concept as a multi-objective optimization problem with the aim of maximizing Δ^{price} and Δ^{demand} in terms of each value of the critical cost deviation factor. In this case, the DNO is faced with a set of optimal solutions, called a Pareto-optimal solution set instead of a single optimal solution. After solving the optimization problem formed from Eqs. (6.324) through (6.331) by the DNO for different values of the critical cost deviation factor, $\sigma_{\mathrm{c}}^{\mathrm{PD-DEP}}$, multiple sets of the Pareto-optimal solutions are obtained. This way, each set of these Pareto-optimal solutions that corresponds to a value of the critical cost deviation factor indicates a compromise among different objectives of the optimization problem—Δ^{price} and Δ^{demand}. After obtaining the multiple sets of the Pareto-optimal solutions, the next step is to select the final optimal solution from the candidate solution sets available in the Pareto-optimal solutions. This means that the DNO is only allowed to choose one solution for a specific amount of the critical cost deviation factor from the Pareto-optimal solutions. The main question, then, is which solution must be chosen by the DNO and how to select it. The DNO chooses the best solution with respect to its requirements and preferences. There are many approaches for choosing a trade-off solution among the Pareto-optimal solution set that have been documented in the relevant literature. For more information on this topic, please refer to Chap. 2. The FSM, based on the conservative methodology—the min-max formulation—is employed by the authors to choose the final optimal solution. It should be noted that simplicity and proximity to human ratiocination are the main criteria in choosing this method. By selecting the zero value for the critical cost deviation factor (i.e., $\sigma_{\mathrm{c}}^{\mathrm{PD-DEP}} = 0$) by the DNO, the critical cost of the DNO is virtually the same as its base cost, which is obtained from the deterministic framework, $\varpi_{\mathrm{c}}^{\mathrm{PD-DEP}} = \varpi_{\mathrm{b}}^{\mathrm{PD-DEP}} = \$52.788286\mathrm{M}$ and $\varpi_{\mathrm{c}}^{\mathrm{PD-DEP}} = \varpi_{\mathrm{b}}^{\mathrm{PD-DEP}} = \$47.518501\mathrm{M}$ for the first and second scenarios, respectively. Because this condition is very similar to the risk-neutral decision-making policy, the robustness and risk level of the obtained optimal distribution expansion plans by the DNO will be zero. This means that the DNO lacks the readiness to deal with the severe uncertainties of market price and demand; therefore, the cost that the DNO needs to invest in distribution network expansion (i.e., namely the PD-DEPO) [see Eq. (6.284)] is significantly affected by these

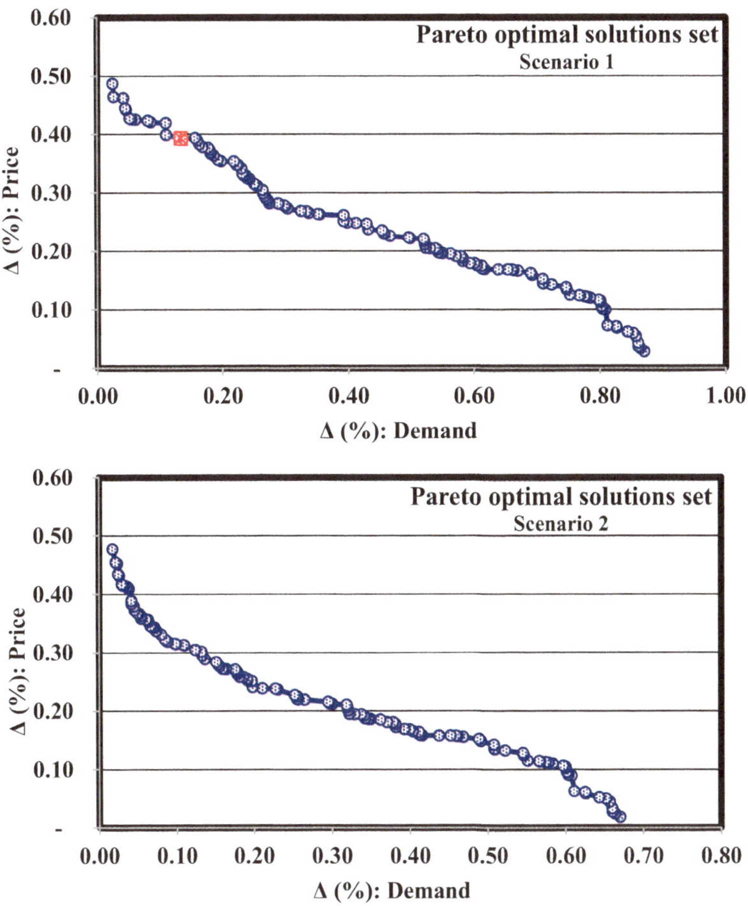

Fig. 6.22 The Pareto-optimal solution set of robustness against market price and demand uncertainties under the first and second scenarios

uncertainties. Suppose that the DNO sets its critical cost deviation factor value to 0.7 (i.e., $\sigma_{c}^{PD-DEP} = 0.7$). In this condition, the DNO solves the optimization problem using Eqs. (6.324) through (6.331) for this value of the critical cost deviation factor with the aim of maximizing Δ^{price} and Δ^{demand}, based on the first and second scenarios. The obtained Pareto-optimal solution sets for this value of the critical cost deviation factor under the first and second scenarios are depicted in Fig. 6.22. To illustrate, consider the DNO's performance under the first scenario (i.e., ignoring DGRs). For example, in the solution marked with a square in Fig. 6.22, a critical cost of $\varpi_{c}^{PD-DEP} = (1 + 0.7) \times \varpi_{b}^{PD-DEP} = \$89.740086M$ is guaranteed for the DNO, provided that $\Delta^{price}$and Δ^{demand} do not exceed 39.417668% and 13.182771%, respectively. In other words, if there is a difference of up to 39.417668% between actual market price, $a\lambda_{p,\ m,\ b}$, and predicted market price, $p\lambda_{p,\ m,\ b}$, and also a

Table 6.80 Optimal values for Δ^{price}, Δ^{demand}, and relevant critical cost deviation factors under the first and second scenarios of the first case

Parameter	Optimal value First scenario: Ignoring DGRs	 Second scenario: Considering DGRs
Δ^{price} (%)	11.614733	14.729587
Δ^{demand} (%)	21.449211	17.543287
Critical cost deviation factor	0.30	0.30

difference of up to 13.182771% between actual demand, $a\rho_{p,\ m,\ b}$, and predicted demand, $p\rho_{p,\ m,\ b}$, then the specified cost for distribution network expansion by the DNO will not exceed $\varpi_{\mathrm{c}}^{\mathrm{PD-DEP}} = \$89.740086\mathrm{M}$. So, if the actual market price falls within the market price robustness region, $RR_{p,m,b}^{\mathrm{price}}$, and the actual demand falls within the demand robustness region, $RR_{p,m,b}^{\mathrm{demand}}$, the specified cost for distribution network planning by the DNO will be at most $\varpi_{\mathrm{c}}^{\mathrm{PD-DEP}} = \$89.740086\mathrm{M}$. Note that if the actual market price is not within the market price robustness region and/or the actual demand is not within the demand robustness region, it is not possible to achieve critical cost by the DNO. By increasing the critical cost by the DNO (i.e., $\varpi_{\mathrm{c}}^{\mathrm{PD-DEP}}$), which is due to increasing its critical cost deviation factor (i.e., $\sigma_{\mathrm{c}}^{\mathrm{PD-DEP}}$), the robustness parameters of the DNO get larger. This means that the larger critical cost of the DNO in the larger market price robustness region and larger demand robustness region can be guaranteed and vice versa.

That is to say that better robustness of the distribution expansion plans obtained by the DNO is determined in the higher value of the critical cost of the DNO and/or in the larger value of the critical cost deviation factor of the DNO. The rest of the solutions provided in Fig. 6.22 have the same interpretations. All solutions available in all of the Pareto-optimal solution sets, which are due to the adoption of different critical cost deviation factors by the DNO, have the same interpretations. This analysis is also true for the DNO's performance under the second scenario.

After solving the techno-economic framework under the first and the second scenarios of the first case (i.e., the IGDT risk-averse decision-making policy) by the newly proposed multi-objective SOSA, the optimal values for Δ^{price} and Δ^{demand} and the relevant critical cost deviation factors are calculated by the FSM in accordance with Table 6.80. The optimal robust distribution expansion plans obtained by the DNO over the planning horizon under the first and second scenarios corresponding to the first case (i.e., the IGDT risk-averse decision-making policy) are tabulated in Table 6.81. To illustrate the optimal results presented in Table 6.78, consider the first period of the planning horizon. In the first scenario of this period, the DNO installs a new substation of type 2 on bus 27 of the distribution network. The DNO also installs a new branch of type 1 in corridors (11-26), (12-24), (7-19), and (8-12) of the distribution network and a new branch of type 2 in corridors (21-27), (7-26), (9-25), (20-25), and (18-25) of the distribution network. In addition, the DNO reinforces the

Table 6.81 Optimal robust distribution expansion plans of the proposed framework under the first and second scenarios of the first case

	Optimal robust distribution expansion plans					
	First scenario: Ignoring DGRs			Second scenario: Considering DGRs		
Period No.	Substation	Branch	DGR	Substation	Branch	DGR
1	(27: I2)	(5-25: R2), (21-27: I2), (11-26: I1), (7-26: I2), (12-24: I1), (7-19: I1), (8-12: I1), (9-25: I2), (20-25: I2), (18-25: I2)	–	–	(7-26: I2), (7-19: I1), (11-26: I1), (7-8: I1), (9-25: I2), (12-26: R1), (20-25: I1), (18-25: I2)	(7: 1+1 MW), (11: 2+1 MW)
2	–	(12-26: R2), (14-21: I2), (23-27: I2), (13-20: I1), (15-23: I1), (12-16: R1), (22-23: I1), (17-18: I1)	–	(25: R2)	(9-13: I2), (5-25: R2), (5-10: I1), (13-14: I1), (10-21: I1), (11-22: I1), (1-5: R1)	(13: 2+1 MW), (21: 1+1 MW)
3	(25: R1)	(9-10: I1), (21-27: R1), (1-5: R1), (5-6: R1)	–	–	(17-18: I1), (12-24: I2), (23-24: I1), (15-16: I2)	(23: 2+1 MW)
Number of newly installed substations over the planning horizon	1	–	–	–	–	–
Number of reinforced substations over the planning horizon	1	–	–	1	–	–
Number of newly installed branches over the planning horizon	–	16	–	–	16	–
Number of reinforced branches over the planning horizon	–	6	–	–	3	–
Number of newly installed DGRs over the planning horizon	–	–	–	–	–	5

existing branch in corridor (5-25) of the distribution network with a new branch of type 2. In the second scenario of this period, however, the DNO installs a new branch of type 1 in corridors (7-19), (11-26), (7-8), and (20-25) of the distribution network and a new branch of type 2 in corridors (7-26), (9-25), and (18-25) of the distribution network. The DNO also reinforces the existing branch in corridor (12-26) of the distribution network with a new branch of type 1. In addition, the DNO installs two new DGRs on buses 7 and 11 of the distribution network. The optimal value obtained for the size of the installed DGR on bus 7 of the distribution network is 1 MW under operating conditions plus 1 MW in reserve. This value for the installed DGR on bus 11 of the distribution network is 2 MW under operating conditions plus 1 MW in reserve. For the remaining periods of the planning horizon, the same analysis can be done. By analyzing the results in Table 6.81, it becomes clear that the configuration for the distribution network at the end of planning horizon in both scenarios is an open-loop configuration.

An important point outlined in optimal results presented in Table 6.81 is that the configurations of the distribution network in each period of the planning horizon for scenarios 1 and 2 are different. The existence of different expansion options available to the DNO in the first and second scenarios leads to this difference in the distribution network configuration under these scenarios. When DGRs are not used (i.e., scenario 1) in the techno-economic framework, candidate bus 27 for substation installation has been expanded with a new substation of type 2. The DNO also reinforces the existing substation on bus 25 of the distribution network with a new substation of type 1. To the contrary, when DGRs are incorporated (i.e., scenario 2) in the techno-economic framework, the DNO only reinforces the existing substation on bus 25 of the distribution network with a new substation of type 2. Here, a reduction in the number of installations and/or reinforcement of expensive equipment of the distribution network, which can lead to a well-planned result, is due to the use of DGRs as one of the expansion options in the PD-DEP problem. The optimal values obtained for the DIC, DOC, DMC, and ECOC indices (i.e., PD-DEPO) under the first and second scenarios of the first case (i.e., the IGDT risk-averse decision-making policy) are given in Table 6.82. As can be seen from Table 6.82, considering DGRs in the techno-economic framework gives rise to more favorable values for different objectives of the PD-DEP problem [see Eq. (6.284)]. Simply put, by incorporating DGRs in the techno-economic framework, a remarkable decrease in the DIC, DOC, DMC, and ECOC is observed in comparison with the situation in which DGRs are ignored (i.e., the second scenario compared to the first).

6.6.4.2 Second Case: Simulation Results and Discussion

In order to implement the proposed techno-economic framework under the IGDT risk-taker decision-making policy, the DNO must solve the formed optimization problem in Eqs. (6.333) through (6.340) under different values of the target cost deviation factor [i.e., $\sigma_t^{\mathrm{PD-DEP}} \in (0, 0.8)$]. Different values of the target cost

Table 6.82 Changes of the objective functions of the proposed framework under the first and second scenarios of the first case

Objective	First scenario: Ignoring DGRs	Second scenario: Considering DGRs
DIC (M$)	49.415087	43.898235
DOC (M$)	10.732227	8.951327
DMC (M$)	2.159612	1.970700
ECOC (M$)	3.755317	3.128792
PD-DEPO (M$)	66.062243	57.949054

deviation factor lead to the formation of different values for the target cost (i.e., ϖ_t^{PD-DEP}). From the DNO's perspective, Eqs. (6.333) through (6.340) are solved by using the Pareto optimality concept as a multi-objective optimization problem with the aim of minimizing Δ^{price} and Δ^{demand} in terms of each value of the target cost deviation factor. Similar to the IGDT risk-averse decision-making policy, in this case, the DNO employs the FSM based on the conservative methodology—the min-max formulation—to choose the optimal solution from among all of the Pareto-optimal solution sets. This optimal solution represents a compromise among different objectives (namely, Δ^{price} and Δ^{demand}).

To achieve a cost equal to the obtained cost from the deterministic framework (i.e., $\varpi_t^{PD-DEP} = \varpi_b^{PD-DEP} = \$52.788286M$ for the first scenario and $\varpi_t^{PD-DEP} = \varpi_b^{PD-DEP} = \$47.518501M$ for the second scenario), the DNO must set its target cost deviation factor to zero (i.e., $\sigma_t^{PD-DEP} = 0$). In this condition, the opportunity and risk level of the distribution expansion plans by the DNO will be zero. In other words, the DNO is not ready to deal with the severe uncertainties of market price and demand; therefore, the cost that the DNO needs to invest in distribution network expansion (i.e., the PD-DEPO) [see Eq. (6.284)] is seriously affected by these uncertainties. Suppose that the DNO sets its target cost deviation factor value to 0.7 (i.e., $\sigma_t^{PD-DEP} = 0.7$). In this condition, Eqs. (6.333) through (6.340) are solved in a multi-objective manner for this value of the target cost deviation factor with the aim of minimizing Δ^{price} and Δ^{demand} based on the first and second scenarios. Figure 6.23 illustrates the obtained Pareto-optimal solution set for this value of the target cost deviation factor under the first and second scenarios. As further elucidation, consider the DNO's performance under the first scenario (i.e., ignoring DGRs). For example, in the solution marked with a square in Fig. 6.23, a target cost of $\varpi_t^{PD-DEP} = (1 - 0.7) \times \varpi_b^{PD-DEP} = \$15.836485M$ is guaranteed for the DNO, provided that Δ^{price} and Δ^{demand} are not less than 38.196480% and 12.394319%, respectively. In order to reach a cost 50% lower than the base cost by the DNO, the actual market price and actual demand must be at least 38.196480% and 12.394319% lower than their corresponding predicted values, respectively. That is, if there is a difference of more than 38.196480% between actual market price and predicted market price, and also a difference of more than 12.394319% between actual demand and predicted demand, then the specified cost by the DNO will be at most $\varpi_t^{PD-DEP} = \$15.836485M$. In other words, if the actual market price falls within the market price robustness region, $RR_{p,m,b}^{price}$, and the actual demand falls within the

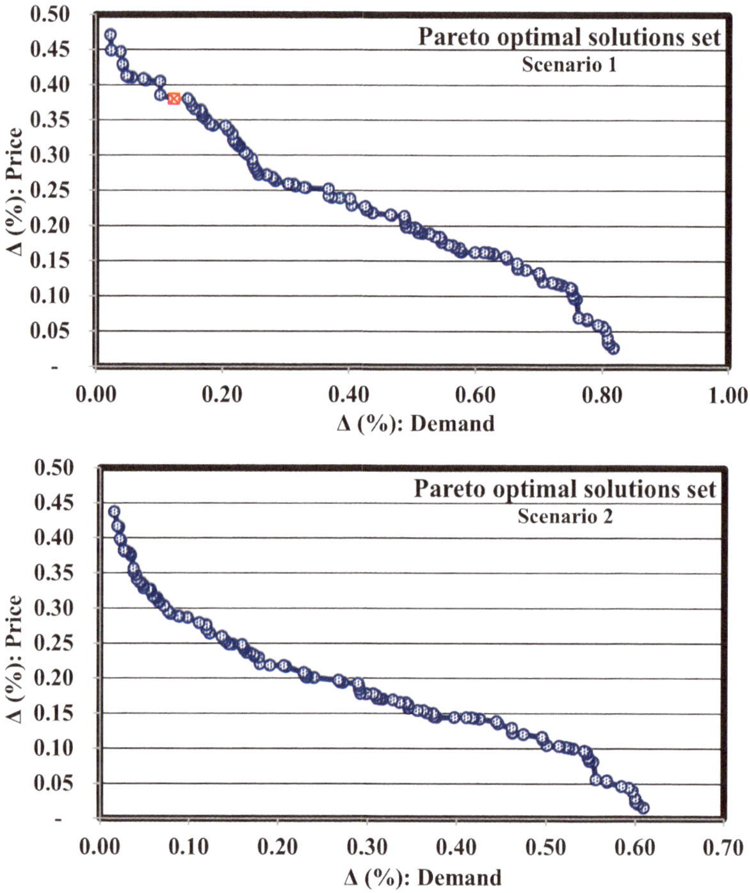

Fig. 6.23 The Pareto-optimal solution set of opportunity against market price and demand uncertainties under the first and second scenarios

demand robustness region, $RR_{p,m,b}^{\text{demand}}$, the specified cost by the DNO will be at most $\varpi_{\text{t}}^{\text{PD-DEP}} = \$15.836485\text{M}$. Note that if the actual market price and/or the actual demand are not at least 38.196480% and 12.394319% lower than their corresponding predicted values, respectively, it is not possible to achieve the target cost by the DNO. If the DNO wants to have a smaller target cost (i.e., $\varpi_{\text{t}}^{\text{PD-DEP}}$), it must choose a larger target cost deviation factor (i.e., $\sigma_{\text{t}}^{\text{PD-TEP}}$), which leads to a larger opportunity for the DNO. This means that a lower target cost of the DNO in the larger market price robustness region and in the larger demand robustness region can be guaranteed and vice versa. More precisely, a better opportunity of the distribution expansion plans obtained by the DNO is determined by a higher value of the target cost of the DNO (i.e., $\varpi_{\text{t}}^{\text{PD-DEP}}$) and/or a larger value of the target cost deviation factor of the DNO (i.e., $\sigma_{\text{t}}^{\text{PD-DEP}}$). The rest of the solutions provided in

Fig. 6.23 have the same interpretations. All solutions available in all of the Pareto-optimal solution sets, which are due to the adoption of different target cost deviation factors by the DNO, have the same interpretations. This analysis is also true for DNO performance under the second scenario.

After solving the proposed framework under the first and the second scenarios of the second case (i.e., the IGDT risk-taker decision-making policy) by the newly proposed multi-objective SOSA, the optimal values for Δ^{price} and Δ^{demand} and the relevant target cost deviation factors are calculated by the FSA according to Table 6.83. The opportunity distribution expansion plans obtained by the DNO over the planning horizon under the first and second scenarios corresponding to the second case (i.e., the IGDT risk-taker decision-making policy) are presented in Table 6.84.

To illustrate the optimal results presented in Table 6.84, consider the second period of the planning horizon. In the first scenario of this period, the DNO reinforces the existing substation on bus 25 of the distribution network with a new substation of type 1.

The DNO also installs a new branch of type 1 in corridors (18-25), (9-10), (11-22), (17-18), and (9-13) of the distribution network. In the second scenario of this period, however, the DNO installs a new branch of type 1 in corridors (13-20), (17-18), (5-10), (7-19), and (11-22) of the distribution network and a new branch of type 2 in corridor (18-25) of the distribution network. The DNO also installs a new DGR on bus 10 of the distribution network. The optimal value obtained for the size of the installed DGR is 1 MW under operating conditions plus 1 MW in reserve. For the remaining periods of the planning horizon, the same analysis can be done. As set out in Table 6.84, it can be seen that the obtained configuration for the distribution network at the end of planning horizon in both scenarios is an open-loop configuration. Another important point that can be seen from the results in Table 6.84 is that the configurations of the distribution network in each period of the planning horizon, for scenarios 1 and 2, are different. This is due to the fact that the different expansion options are available to the DNO in the first and second scenarios. When DGRs are not used (i.e., scenario 1) in the techno-economic framework, the DNO reinforces the existing substation on bus 25 of the distribution network with a new substation of type 1. When DGRs are incorporated (i.e., scenario 2) in the techno-economic framework, there is no need to install and/or reinforce any substation. In distribution network expansion planning studies, therefore, by integrating DGRs into the

Table 6.83 Optimal values for Δ^{price}, Δ^{demand}, and relevant critical cost deviation factors under the first and second scenarios of the second case

Parameter	Optimal value	
	First scenario: Ignoring DGRs	Second scenario: Considering DGRs
Δ^{price} (%)	4.590718	6.678745
Δ^{demand} (%)	11.408422	8.923125
Target cost deviation factor	0.15	0.15

Table 6.84 Optimal opportunity distribution expansion plans of the proposed framework under the first and second scenarios of the second case

	Optimal opportunity distribution expansion plans					
	First scenario: Ignoring DGRs			Second scenario: Considering DGRs		
Period No.	Substation	Branch	DGR	Substation	Branch	DGR
1	–	(6.7: I2), (7-19: I1), (8-12: I1), (20-25: I1), (9-25: I2), (11-26: I1), (12-24: I1)	–	–	(9-25: I2), (9-13: I1), (11-26: I2), (8-12: I2), (7-8: I2), (12-24: I1)	(24: 1+1 MW)
2	(25: R1)	(18-25: I1), (9-10: I1), (11-22: I1), (17-18: I1), (9-13: I1)	–	–	(13-20: I1), (18-25: I2), (17-18: I1), (5-10: I1), (7-19: I1), (11-22: I1)	(10: 1+1 MW)
3	–	(10-21: I1), (23-24: I1), (15-16: I2), (14-15: I1)	–	–	(13-14: I1), (15-16: I1), (10-21: I1), (22-23: I1)	
Number of newly installed substations over the planning horizon	–	–	–	–	–	–
Number of reinforced substations over the planning horizon	1	–	–	–	–	–
Number of newly installed branches over the planning horizon	–	16	–	–	16	–
Number of reinforced branches over the planning horizon	–	–	–	–	–	–
Number of newly installed DGRs over the planning horizon	–	–	–	–	–	2

Table 6.85 Changes in the objective functions of the proposed framework under the first and second scenarios of the second case

Objective	First scenario: Ignoring DGRs	Second scenario: Considering DGRs
DIC (M$)	32.269921	29.707316
DOC (M$)	7.160072	6.363647
DMC (M$)	1.239535	1.124115
ECOC (M$)	3.043652	2.730044
PD-DEPO (M$)	43.713180	39.925122

proposed techno-economic framework, a well-designed framework is created in which techno-economic restrictions correspond to the expense of equipment in the distribution networks, especially substations, being partially decreased. Table 6.85 provides the values for the DIC, DOC, DMC, and ECOC indices under the first and second scenarios of the second case (i.e., the IGDT risk-taker decision-making policy). From Table 6.85, it is obvious that considering DGRs in the techno-economic framework can significantly reduce the value of the different objectives of the PD-DEP problem [see Eq. (6.284)].

In other words, the investment performed by the DNO in order to expand the distribution network—the cost imposed on the DNO for the operation and maintenance of distribution network equipment, and also unreliability costs imposed on consumers in the second scenario taking into account DGRs—has less value compared to the first scenario without taking into account DGRs. With the overall evaluation of the results related to the proposed techno-economic framework under the risk-neutral/deterministic decision-making policy, the IGDT risk-averse decision-making policy (i.e., first case), and the IGDT risk-taker decision-making policy (i.e., second case), it can be concluded that, for all decision-making policies, the incorporation of DGRs in the techno-economic framework has given rise to a similar trend in the results (i.e., second scenario compared to the first). More precisely, in the second scenario of all decision-making policies in which DGRs are considered, a significant reduction in the DIC, DOC, DMC, and ECOC as the objective functions of the PD-DEP problem is observed compared to the first scenario. Therefore, considering DGRs in the techno-economic framework can avoid unnecessary investments in the distribution network by reducing the DNO's dependence on installing expensive equipment, especially substations, in the distribution network, decrease the operation and maintenance costs of the distribution network equipment, and alleviate unreliability costs imposed on the costumers. As a result, the integration of DGRs in the techno-economic framework can lead to more productivity and flexibility in the proposed techno-economic framework.

6.6.4.3 The Impact of the Presence of Distributed Generation Resources on the Voltage Profile

In this section, the impact of the presence of DGRs on the voltage profile of the distribution network is scrutinized. Table 6.86 shows the changes of the voltage profile on load buses of the distribution network in the final period of the planning horizon in the absence and presence of DGRs for the two scenarios of the first case (i.e., the IGDT risk-averse decision-making policy) and the second case (i.e., the IGDT risk-taker decision-making policy). The changes of the voltage profile of the distribution network are recorded at the end of the planning horizon. From the optimal results of both cases presented in Table 6.86, it can be seen that the presence of DGRs in the proposed techno-economic framework provides better voltage values on the majority of the load buses compared to conditions where DGRs are not considered.

Table 6.86 Changes of voltage profiles of the distribution network in the absence and presence of DGRs in the proposed framework under the first and second cases

First case: The IGDT risk-averse decision-making policy				First case: The IGDT risk-averse decision-making policy			
First scenario: Ignoring DGRs		Second scenario: Considering DGRs		First scenario: Ignoring DGRs		Second scenario: Considering DGRs	
Bus No.	Voltage (p.u.)	Bus No.	Voltage (p.u.)	Bus No.	Voltage (p.u.)	Bus No.	Voltage (p.u.)
1	0.993061	1	1.004152	1	1.020784	1	1.024045
2	0.973058	2	0.983420	2	0.998937	2	1.001039
3	0.964048	3	0.968720	3	0.981039	3	0.988047
4	0.951262	4	0.956200	4	0.967936	4	0.972037
5	1.025049	5	1.030360	5	1.035042	5	1.036061
6	1.013049	6	1.017512	6	1.011060	6	1.018038
7	1.019961	7	1.040214	7	0.992630	7	0.999031
8	1.012415	8	1.019854	8	1.008072	8	1.017039
9	1.024250	9	1.031028	9	1.032412	9	1.038958
10	0.999258	10	1.014471	10	1.006077	10	1.034066
11	1.022420	11	1.041553	11	1.027452	11	1.039048
12	1.026867	12	1.031661	12	1.026077	12	1.034028
13	0.994210	13	1.030263	13	1.011463	13	1.013828
14	0.993320	14	1.011845	14	0.974502	14	0.985065
15	0.982212	15	0.985414	15	0.989099	15	0.986219
16	1.006852	16	1.008432	16	1.004961	16	1.010029
17	0.996050	17	1.001758	17	1.010080	17	1.018252
18	1.023058	18	1.029885	18	1.033072	18	1.037141
19	1.004163	19	1.016325	19	0.978450	19	0.981039
20	1.016653	20	1.038029	20	1.034256	20	0.987561
21	1.020220	21	1.007321	21	0.989059	21	1.014064
22	0.976820	22	1.020325	22	1.012079	22	1.021033
23	0.998240	23	1.001263	23	0.991061	23	1.007032
24	1.015214	24	1.006874	24	1.009069	24	1.039362
25	1.050000	25	1.050000	25	1.050000	25	1.050000
26	1.050000	26	1.050000	26	1.050000	26	1.050000
27	1.050000	27	–	27	–	27	–

In other words, by incorporating DGRs into the techno-economic framework, the distribution network will have more favorable conditions in terms of voltage fluctuations.

6.6.4.4 Quantitative Verification of the Proposed IGDT Risk-Averse Decision-Making Policy in Comparison to the Robust Optimization Technique

In this section, the performance of the techno-economic framework under the IGDT risk-averse decision-making policy is compared with the performance of this framework under the RO technique; due to this fact the worst-case uncertainty situation in the proposed strategic tri-level computational-logical framework is observed in the first case (i.e., the IGDT risk-averse decision-making policy). For more information about the RO technique, please refer to the work by Ben-Tal et al. [41]. Table 6.87 shows the distribution expansion plans by the DNO across the planning horizon under the IGDT risk-averse decision-making policy and RO technique. These results are associated with the second scenario (i.e., considering DGRs). As set out in Table 6.87, it can be seen that there is less equipment installed on the distribution network over the planning horizon in the techno-economic framework under the IGDT risk-averse decision-making policy than for the RO technique. Table 6.88 gives the calculated optimal values for the DIC, DOC, DMC, and ECOC (i.e., the PD-DEPO) under the IGDT risk-averse decision-making policy and RO technique. These results are concerned with the second scenario (i.e., considering DGRs). By evaluating the results presented in Table 6.88, it can be concluded that employing the IGDT risk-averse decision-making policy to handle the uncertainty parameters leads to more efficient results for the different objectives of the PD-DEP problem compared with the RO technique. That is, the required DNO's budget (i.e., the DIC index) for purchasing and installing the new equipment in the distribution network under the IGDT risk-averse decision-making policy is lower than that under the RO technique.

The cost imposed on the DNO for the operation and maintenance of the distribution network (i.e., the DOC and DMC indices) has been clearly decreased under the IGDT risk-averse decision-making policy compared to the RO technique. Moreover, the unreliability cost imposed on the costumer (i.e., the ECOC index) under the IGDT risk-averse decision-making policy is less than that under the RO technique. As a result, it can be concluded that the IGDT risk-averse decision-making policy, compared to other methods, especially RO technique, provides a more suitable techno-economic expansion plan for the PD-DEP problem, thereby preventing unnecessary investment in the distribution network. For this reason, the IGDT risk-averse decision-making policy can be a reasonable and suitable policy for handling severe uncertainties in large-scale optimization problems.

Table 6.87 Optimal distribution expansion plans of the proposed framework under the IGDT risk-averse decision-making policy and RO technique

	Optimal distribution expansion plans					
	The IGDT risk-averse decision-making policy			The RO technique		
Period No.	Substation	Branch	DGR	Substation	Branch	DGR
1	–	(7-26: I2), (7-19: I1), (11-26: I1), (7-8: I1), (9-25: I2), (12-26: R1), (20-25: I1), (18-25: I2)	(7: 1+1 MW), (11: 2+1 MW)	(25: R2)	(7-26: I2), (7-19: I1), (11-26: I2), (8-12: I1), (9-25: I2), (18-25: I2), (5-25: R2), (20-25: I2), (13-20: I1)	(22: 1+1 MW), (13: 1+1 MW)
2	(25: R2)	(9-13: I2), (5-25: R2), (5-10: I1), (13-14: I1), (10-21: I1), (11-22: I1), (1-5: R1)	(13: 2+1 MW), (21: 1+1 MW)	–	(9-10: I2), (10-21: I1), (14-21: I1), (12-24: I1), (11-22: I1), (12-26: R2), (9-10: R1)	(21: 2+1 MW), (15: 1+1 MW)
3	–	(17-18: I1), (12-24: I2), (23-24: I1), (15-16: I2)	(23: 2+1 MW)	–	(17-18: I1), (23-24: I1), (15-16: I1), (1-5: R2)	(19: 1+1 MW), (4: 2+1 MW)
Number of newly installed substations over the planning horizon	–	–	–	–	–	–
Number of reinforced substations over the planning horizon	1	–	–	1	–	–
Number of newly installed branches over the planning horizon	–	16	–	–	16	–
Number of reinforced branches over the planning horizon	–	3	–	–	4	–
Number of newly installed DGRs over the planning horizon	–	–	5	–	–	5

Table 6.88 Changes of the objective functions of the proposed framework under the IGDT risk-averse decision-making policy and RO technique

Objective	The IGDT risk-averse decision-making policy	The RO technique
DIC (M$)	43.898235	45.728858
DOC (M$)	8.951327	9.313276
DMC (M$)	1.970700	2.087920
ECOC (M$)	3.128792	3.270700
PD-DEPO (M$)	57.949054	60.400754

6.6.4.5 Performance Evaluation of the Proposed Optimization Algorithms: Simulation Results and Discussion

In this section, the performance of the proposed modern meta-heuristic music-inspired optimization algorithms addressed in Chap. 4, namely the multi-objective SS-HSA, multi-objective SS-IHSA, multi-objective continuous/discrete TMS-MSA, multi-objective TMS-EMSA, and multi-objective SOSA, is compared with that of the NSGA-II for the first and second scenarios of the first case (i.e., the IGDT risk-averse decision-making policy) and second case (i.e., the IGDT risk-taker decision-making policy). Tables 6.104, 6.105, 6.106, 6.107, and 6.108 give the parameter adjustments of the multi-objective TMS-EMSA, multi-objective continuous/discrete TMS-MSA, multi-objective SS-HSA, multi-objective SS-IHSA, and NSGA-II, respectively. The calculated results associated with the DIC, DOC, DMC, and ECOC across the planning horizon by the proposed multi-objective optimization algorithms under the first and second scenarios of the first case (i.e., the IGDT risk-averse decision-making policy) and the second case (i.e., the IGDT risk-taker decision-making policy) are given in Tables 6.89 and 6.90, respectively. From these results, it can be seen that the techno-economic framework—in the absence of DGRs (i.e., scenario 1) and in the presence of DGRs (i.e., scenario 2) of both cases by the proposed multi-objective SOSA—leads to more efficient results than other proposed meta-heuristic music-inspired optimization algorithms and NSGA-II. The ICS is also used to evaluate the performance of these multi-objective optimization algorithms relative to each other. For more information about the ICS index, please refer to Sect. 5.3.5.3 of Chap. 5.

The ICS for the proposed optimization algorithms under the first and second scenarios of the first case is given in Tables 6.91 and 6.92, respectively. The ICS for the proposed optimization algorithms under the first and second scenarios of the second case is also presented in Tables 6.93 and 6.94, respectively. To illustrate, consider the results of the techno-economic framework under the first scenario of the first case according to Table 6.91. The ICS shows 7.025091%, 13.364386%, 19.094083%, 16.421875%, and 21.426192% superiority—positive sign—of the multi-objective SOSA performance compared with the performances of the multi-objective TMS-EMSA, multi-objective continuous/discrete TMS-MSA, multi-objective SS-HSA, multi-objective SS-IHSA, and NSGA-II, respectively, from the perspective of the obtained DIC index by corresponding optimization algorithms.

Table 6.89 The calculated values of the objective functions of the proposed framework by the multi-objective optimization algorithms under the first and second scenarios of the first case

Policy	Scenario	Multi-objective optimization algorithm	Objective			
			DIC over the planning horizon (M$)	DOC over the planning horizon (M$)	DMC over the planning horizon (M$)	ECOC over the planning horizon (M$)
The IGDT risk-averse decision-making policy	First scenario: Ignoring DGRs	SOSA	49.415087	10.732227	2.159612	3.755317
		TMS-EMSA	53.148842	11.334793	2.310936	3.989008
		Continuous/ discrete TMS-MSA	57.037845	12.140334	2.507013	4.249082
		SS-HSA	61.077223	13.245111	2.651192	4.618232
		SS-IHSA	59.124427	12.678985	2.581094	4.443821
		NSGA-II	62.890025	13.574028	2.715967	4.862905
	Second scenario: Considering DGRs	SOSA	43.898235	8.951327	1.970700	3.128792
		TMS-EMSA	46.245887	9.525016	2.117965	3.325903
		Continuous/ discrete TMS-MSA	48.878159	10.182520	2.235066	3.518653
		SS-HSA	53.982896	11.147240	2.429097	3.880551
		SS-IHSA	52.587076	10.652517	2.346153	3.716837
		NSGA-II	54.726259	11.440121	2.460041	3.957614

Table 6.90 The calculated values of the objective functions of the proposed framework by the multi-objective optimization algorithms under the first and second scenarios of the second case

Policy	Scenario	Multi-objective optimization algorithm	Objective			
			DIC over the planning horizon (M$)	DOC over the planning horizon (M$)	DMC over the planning horizon (M$)	ECOC over the planning horizon (M$)
The IGDT risk-taker decision-making policy	First scenario: Ignoring DGRs	SOSA	32.269921	7.160072	1.239535	3.043652
		TMS-EMSA	34.411216	7.653067	1.310445	3.292047
		Continuous/ discrete TMS-MSA	37.000705	8.164648	1.406411	3.486465
		SS-HSA	39.859079	8.789488	1.551522	3.814914
		SS-IHSA	38.919542	8.510572	1.482235	3.671193
		NSGA-II	40.682313	9.040519	1.597189	3.868156
	Second scenario: Considering DGRs	SOSA	29.707316	6.363647	1.124115	2.730044
		TMS-EMSA	31.513801	6.777079	1.200868	2.901242
		Continuous/ discrete TMS-MSA	33.173182	7.271283	1.291158	3.111450
		SS-HSA	36.767636	7.896969	1.400829	3.425166
		SS-IHSA	35.382096	7.539196	1.351879	3.340826
		NSGA-II	37.462973	8.091180	1.430062	3.491604

Table 6.91 The ICS for different objectives of the proposed framework, based on various multi-objective optimization algorithms under the first scenario of the first case

Case No.	Scenario No.	Perspective	Multi-objective optimization algorithm	Multi-objective optimization algorithm					
				SOSA	TMS-EMSA	Continuous/discrete TMS-MSA	SS-HSA	SS-IHSA	NSGA-II
First case	First scenario: Ignoring DGRs	DIC over the planning horizon	SOSA	0	7.025091	13.364386	19.094083	16.421875	21.426192
			TMS-EMSA	−7.555900	0	6.818285	12.980912	10.106795	15.489233
			Continuous/discrete TMS-MSA	−15.425973	−7.317192	0	6.613558	3.529136	9.305418
			SS-HSA	−23.600355	−14.917316	−7.081926	0	−3.302858	2.882495
			SS-IHSA	−19.648534	−11.243114	−3.658241	3.197257	0	5.987591
			NSGA-II	−27.268874	−18.328119	−10.260170	−2.968049	−6.368937	0
		DOC over the planning horizon	SOSA	0	5.316074	11.598585	18.972162	15.354210	20.935576
			TMS-EMSA	−5.614547	0	6.635245	14.422816	10.601731	16.496466
			Continuous/discrete TMS-MSA	−13.120362	−7.106799	0	8.341017	4.248376	10.562038
			SS-HSA	−23.414376	−16.853576	−9.100054	0	−4.465073	2.423134
			SS-IHSA	−18.139367	−11.858990	−4.436871	4.274226	0	6.593790
			NSGA-II	−26.479136	−19.755411	−11.809345	−2.483308	−7.059263	0
		DMC over the planning horizon	SOSA	0	6.548169	13.857167	18.541848	16.329587	20.484600
			TMS-EMSA	−7.006999	0	7.821140	12.834076	10.466802	14.912957
			Continuous/discrete TMS-MSA	−16.086269	−8.484743	0	5.438270	2.870139	7.693539
			SS-HSA	−22.762422	−14.723731	−5.751027	0	−2.715825	2.384970
			SS-IHSA	−19.516561	−11.690414	−2.954950	2.644018	0	4.965929
			NSGA-II	−25.761804	−17.526707	−8.334779	−2.443240	−5.225419	0

(continued)

Table 6.91 (continued)

Case No.	Scenario No.	Perspective	Multi-objective optimization algorithm	Multi-objective optimization algorithm					
				SOSA	TMS-EMSA	Continuous/discrete TMS-MSA	SS-HSA	SS-IHSA	NSGA-II
		ECOC over the planning horizon	SOSA	0	5.858373	11.620510	18.684964	15.493513	22.776262
			TMS-EMSA	−6.222936	0	6.120710	13.624781	10.234728	17.970678
			Continuous/discrete TMS-MSA	−13.148424	−6.519766	0	7.993318	4.382242	12.622557
			SS-HSA	−22.978486	−15.773946	−8.687758	0	−3.924798	5.031416
			SS-IHSA	−18.334111	−11.401656	−4.583084	3.776575	0	8.617976
			NSGA-II	−29.493862	−21.907627	−14.446014	−5.297979	−9.430712	0

Table 6.92 The ICS for different objectives of the proposed framework, based on various multi-objective optimization algorithms under the second scenario of the first case

Case No.	Scenario No.	Perspective	Multi-objective optimization algorithm	Multi-objective optimization algorithm					
				SOSA	TMS-EMSA	Continuous/discrete TMS-MSA	SS-HSA	SS-IHSA	NSGA-II
First case	Second scenario: Considering DGRs	DIC over the planning horizon	SOSA	0	5.076455	10.188444	18.681215	16.522768	19.785792
			TMS-EMSA	−5.347941	0	5.385374	14.332334	12.058455	15.495983
			Continuous/dis-crete TMS-MSA	−11.344247	−5.691905	0	9.456211	7.052905	10.686094
			SS-HSA	−22.972817	−16.730155	−10.443799	0	−2.654302	1.358329
			SS-IHSA	−19.793144	−13.711898	−7.588086	2.585670	0	3.908878
			NSGA-II	−24.666194	−18.337570	−11.964648	−1.377034	−4.067887	0
		DOC over the planning horizon	SOSA	0	6.022971	12.091240	19.699163	15.969840	21.754962
			TMS-EMSA	−6.408982	0	6.457183	14.552696	10.584362	16.740251
			Continuous/dis-crete TMS-MSA	−13.754307	−6.902917	0	8.654339	4.412074	10.992899
			SS-HSA	−24.531704	−17.031194	−9.474275	0	−4.644188	2.560121
			SS-IHSA	−19.004892	−11.837261	−4.615723	4.438076	0	6.884577
			NSGA-II	−27.803632	−20.106055	−12.350587	−2.627385	−7.393595	0
		DMC over the planning horizon	SOSA	0	6.9531366	11.828107	18.871086	16.002920	19.891579
			TMS-EMSA	−7.472725	0	5.239263	12.808545	9.726049	13.905296
			Continuous/dis-crete TMS-MSA	−13.414827	−5.528939	0	7.987783	4.734857	9.145172
			SS-HSA	−23.260618	−14.690138	−8.681220	0	−3.535319	1.257865
			SS-IHSA	−19.051758	−10.773926	−4.970188	3.414602	0	4.629516
			NSGA-II	−24.830822	−16.151164	−10.065698	−1.273889	−4.854244	0

(continued)

Table 6.92 (continued)

Case No.	Scenario No.	Perspective	Multi-objective optimization algorithm	Multi-objective optimization algorithm					
				SOSA	TMS-EMSA	Continuous/discrete TMS-MSA	SS-HSA	SS-IHSA	NSGA-II
		ECOC over the planning horizon	SOSA	0	5.926540	11.079836	19.372480	15.821113	20.942466
			TMS-EMSA	−6.299907	0	5.477948	14.293021	10.517921	15.961915
			Continuous/discrete TMS−MSA	−12.460432	−5.795418	0	9.325943	5.332060	11.091556
			SS-HSA	−24.027133	−16.676613	−10.285129	0	−4.404659	1.947208
			SS-IHSA	−18.794634	−11.754221	−5.632382	4.218833	0	6.083892
			NSGA-II	−26.490159	−18.993668	−12.475256	−1.985877	−6.478008	0

Table 6.93 The ICS for different objectives of the proposed framework, based on various multi-objective optimization algorithms under the first scenario of the second case

Case No.	Scenario No.	Perspective	Multi-objective optimization algorithm	Multi-objective optimization algorithm					
				SOSA	TMS-EMSA	Continuous/discrete TMS-MSA	SS-HSA	SS-IHSA	NSGA-II
Second case	First scenario: Ignoring DGRs	DIC over the planning horizon	SOSA	0	6.222665	12.785659	19.039973	17.085558	20.678253
			TMS-EMSA	−6.635575	0	6.998485	13.667809	11.583707	15.414799
			Continuous/dis-crete TMS-MSA	−14.660042	−7.525130	0	7.171199	4.930266	9.049652
			SS-HSA	−23.517746	−15.831649	−7.725187	0	−2.414049	2.023567
			SS-IHSA	−20.606251	−13.101327	−5.185947	2.357146	0	4.333015
			NSGA-II	−26.068834	−18.223991	−9.950102	−2.065361	−4.529269	0
		DOC over the planning horizon	SOSA	0	6.441796	12.303972	18.538235	15.868498	20.800210
			TMS-EMSA	−6.885335	0	6.265805	12.929319	10.075762	15.347039
			Continuous/dis-crete TMS-MSA	−14.030249	−6.684653	0	7.108946	4.064638	9.688282
			SS-HSA	−22.756978	−14.849223	−7.652993	0	−3.277288	2.776732
			SS-IHSA	−18.861542	−11.204723	−4.236851	3.173290	0	5.861909
			NSGA-II	−26.262962	−18.129359	−10.727602	−2.856036	−6.226925	0
		DMC over the planning horizon	SOSA	0	5.411138	11.865379	20.108448	16.373921	22.392716
			TMS-EMSA	−5.720693	0	6.823467	15.538097	11.589930	17.953041
			Continuous/dis-crete TMS-MSA	−13.462790	−7.323161	0	9.352816	5.115518	11.944610
			SS-HSA	−25.169681	−18.396575	−10.317823	0	−4.674494	2.859210
			SS-IHSA	−19.579923	−13.109287	−5.391311	4.465743	0	7.197269
			NSGA-II	−28.853885	−21.881422	−13.564882	−2.943367	−7.755450	0

(continued)

Table 6.93 (continued)

Case No.	Scenario No.	Perspective	Multi-objective optimization algorithm	Multi-objective optimization algorithm					
				SOSA	TMS-EMSA	Continuous/discrete TMS-MSA	SS-HSA	SS-IHSA	NSGA-II
		ECOC over the planning horizon	SOSA	0	7.545305	12.700916	20.217021	17.093653	21.315169
			TMS-EMSA	−8.161084	0	5.576364	13.705866	10.327596	14.893634
			Continuous/discrete TMS-MSA	−14.548739	−5.905687	0	8.609604	5.031824	9.867518
			SS-HSA	−25.340019	−15.882731	−9.420688	0	−3.914831	1.376418
			SS-IHSA	−20.618027	−11.517028	−5.298432	3.767345	0	5.091909
			NSGA-II	−27.089299	−17.500023	−10.947793	−1.395627	−5.365095	0

Table 6.94 The ICS for different objectives of the proposed framework, based on various multi-objective optimization algorithms under the second scenario of the second case

Case No.	Scenario No.	Perspective	Multi-objective optimization algorithm	Multi-objective optimization algorithm					
				SOSA	TMS-EMSA	Continuous/discrete TMS-MSA	SS-HSA	SS-IHSA	NSGA-II
Second case	Second scenario: Considering DGRs	DIC over the planning horizon	SOSA	0	5.732361	10.447794	19.202539	16.038563	20.702193
			TMS-EMSA	−6.080943	0	5.002176	14.289292	10.932916	15.880138
			Continuous/discrete TMS-MSA	−11.666709	−5.265569	0	9.776135	6.243027	11.450748
			SS-HSA	−23.766267	−16.671537	−10.835421	0	−3.915935	1.856064
			SS-IHSA	−19.102298	−12.274923	−6.658734	3.768368	0	5.554489
			NSGA-II	−26.106892	−18.877989	−12.931502	−1.891165	−5.881158	0
		DOC over the planning horizon	SOSA	0	6.100445	12.482473	19.416588	15.592498	21.350816
			TMS-EMSA	−6.496777	0	6.796654	14.181263	10.108730	16.241153
			Continuous/discrete TMS-MSA	−14.262828	−7.292286	0	7.923115	3.553601	10.133219
			SS-HSA	−24.095019	−16.524670	−8.604891	0	−4.745506	2.400280
			SS-IHSA	−18.472882	−11.245508	−3.684535	4.530510	0	6.822045
			NSGA-II	−27.146902	−19.390374	−11.275823	−2.459310	−7.321523	0
		DMC over the planning horizon	SOSA	0	6.391460	12.937456	19.753588	16.847957	21.393967
			TMS-EMSA	−6.827860	0	6.992947	14.274476	11.170452	16.026857
			Continuous/discrete TMS-MSA	−14.859956	−7.518728	0	7.829006	4.491600	9.713145
			SS-HSA	−24.616165	−16.651372	−8.494003	0	−3.620886	2.044177
			SS-IHSA	−20.261628	−12.575154	−4.702832	3.494359	0	5.467105
			NSGA-II	−27.216699	−19.085694	−10.758094	−2.086835	−5.783283	0

(continued)

Table 6.94 (continued)

Case No.	Scenario No.	Perspective	Multi-objective optimization algorithm	Multi-objective optimization algorithm					
				SOSA	TMS-EMSA	Continuous/discrete TMS-MSA	SS-HSA	SS-IHSA	NSGA-II
		ECOC over the planning horizon	SOSA	0	5.900852	12.258143	20.294549	18.282364	21.811179
			TMS-EMSA	−6.270887	0	6.755949	15.296309	13.157943	16.908045
			Continuous/discrete TMS-MSA	−13.970691	−7.245448	0	9.159147	6.865846	10.887660
			SS-HSA	−25.461934	−18.058610	−10.082630	0	−2.524525	1.902793
			SS-IHSA	−22.372606	−15.151579	−7.371996	2.462362	0	4.318301
			NSGA-II	−27.895521	−20.348595	−12.217904	−1.939701	−4.513195	0

Also, the ICS represents 5.316074%, 11.598585%, 18.972162%, 15.354210%, and 20.935576% superiority—positive sign—of the multi-objective SOSA performance compared with the performances of the multi-objective TMS-EMSA, multi-objective continuous/discrete TMS-MSA, multi-objective SS-HSA, multi-objective SS-IHSA, and NSGA-II, respectively, from the standpoint of the obtained DOC index by corresponding optimization algorithms. In addition, the ICS illustrates 6.548169%, 13.857167%, 18.541848%, 16.329587%, and 20.484600% superiority—positive sign—of the multi-objective SOSA performance compared with the performances of the multi-objective TMS-EMSA, multi-objective continuous/discrete TMS-MSA, multi-objective SS-HSA, multi-objective SS-IHSA, and NSGA-II, respectively, from the point of view of the obtained DMC index by corresponding optimization algorithms. Moreover, the ICS demonstrates 5.858373%, 11.620510%, 18.684964%, 15.493513%, and 22.776262% superiority—positive sign—of the multi-objective SOSA performance compared with the performances of the multi-objective TMS-EMSA, multi-objective continuous/discrete TMS-MSA, multi-objective SS-HSA, multi-objective SS-IHSA, and NSGA-II, respectively, from the perspective of the obtained ECOC index by corresponding optimization algorithms.

The ICS illustrates 7.555900% weakness—negative sign—of the multi-objective TMS-EMSA performance compared with the performance of the multi-objective SOSA, from the perspective of the obtained DIC index by the multi-objective SOSA, and 6.818285%, 12.980912%, 10.106795%, and 15.489233% superiority—positive sign—of the multi-objective TMS-EMSA performance compared with the performances of the multi-objective continuous/discrete TMS-MSA, multi-objective SS-HSA, multi-objective SS-IHSA, and NSGA-II, respectively, from the perspective of the obtained DIC index by corresponding optimization algorithms. Also, the ICS shows 5.614547% weakness—negative sign—of the multi-objective TMS-EMSA performance compared with the performance of the multi-objective SOSA, from the standpoint of the obtained DOC index by the multi-objective SOSA, and 6.635245%, 14.422816%, 10.601731%, and 16.496466% superiority—positive sign—of the multi-objective TMS-EMSA performance compared with the performances of the multi-objective continuous/discrete TMS-MSA, multi-objective SS-HSA, multi-objective SS-IHSA, and NSGA-II, respectively, from the standpoint of the obtained DOC index by corresponding optimization algorithms. In addition, the ICS demonstrates 7.006999% weakness—negative sign—of the multi-objective TMS-EMSA performance compared with the performance of the multi-objective SOSA, from the point of view of the obtained DMC index by the multi-objective SOSA, and 7.821140%, 12.834076%, 10.466802%, and 14.912957% superiority—positive sign—of the multi-objective TMS-EMSA performance compared with the performances of the multi-objective continuous/discrete TMS-MSA, multi-objective SS-HSA, multi-objective SS-IHSA, and NSGA-II, respectively, from the point of view of the obtained DMC index by corresponding optimization algorithms. Moreover, the ICS shows 6.222936% weakness—negative sign—of the multi-objective TMS-EMSA performance compared with the performance of the multi-objective SOSA, from the perspective of the obtained ECOC index by the multi-objective SOSA, and 6.120710%,

13.624781%, 10.234728%, and 17.970678% superiority—positive sign—of the multi-objective TMS-EMSA performance compared with the performances of the multi-objective continuous/discrete TMS-MSA, multi-objective SS-HSA, multi-objective SS-IHSA, and NSGA-II, respectively, from the perspective of the obtained ECOC index by corresponding optimization algorithms. The results presented in Table 6.91 for the other multi-objective optimization algorithms, and in Tables 6.92, 6.93, and 6.94 for all multi-objective optimization algorithms, are analyzed in the same way.

6.7 Conclusions

In this chapter, from a new perspective, the authors developed four innovative strategic multilevel computational-logical frameworks for the PD-GEP, PD-TEP, PD-G&TEP, and PD-DEP problems with the aim of optimally supplying, transferring, and distributing electric energy. In Sect. 6.2, different perspectives and aspects of power system planning were thoroughly investigated for utilization in the aforementioned frameworks. In Sect. 6.3, a strategic tri-level computational-logical framework was presented for the PD-GEP problem within the deregulated environments. The proposed framework related to the PD-GEP problem was also divided into a short-term operational slave problem and a long-term planning master problem. The short-term operational slave problem was organized by the BBM problem as the first level and the CSC electricity market problem as the second level. The short-term operational slave problem provided the CSC electricity market outcomes along with strategic GENCO and DISCO bids for the long-term planning master problem. The outputs of the long-term master planning problem serve as inputs to the short-term operational slave problem and vice versa.

More precisely, evaluating the impact of the generation expansion plans on strategic behaviors of both GENCOs and DISCOs and vice versa was one of the most important goals of the proposed framework associated with the PD-GEP problem. In the first level, by adopting strategic behaviors, the GENCOs and DISCOs tried to maximize their profits from participation in the CSC electricity market. In other words, each market participant sought to maximize its profit by solving the BBM problem. The BBM problem was bounded to four different types of constraints: (1) power balance restriction; (2) the GENCOs' generated power limitation; (3) the DISCOs' purchased power limitation; and, (4) the GENCOs' and the DISCOs' bidding parameter restrictions to avoid applying market power. In the second level, however, the CSC electricity market-clearing process was performed by the ISO with the aim of maximizing the CWF. The CSC electricity market problem was subject to four different types of constraints: (1) power balance restriction; (2) the GENCOs' generated power limitation; (3) the DISCOs' purchased power limitation; and, (4) a system security restriction imposed on the transmission lines. The third level of the proposed framework was composed of two different layers: local and global. In the local layer, each GENCO individually solved the PD-GEP problem to minimize the GIC and maximize the EP and CP while bounded to techno-economic and environmental constraints as well as limitations associated with financial resource availability. In the global layer, however, the ISO aggregated the GENCOs' optimal generation expansion plans obtained in the local layer, and then assessed these optimal plans to meet the restrictions concerned with the entire power system: (1) the generation reserve margin; (2) the competitive condition; and, (3) the reliability requirements.

In Sect. 6.4, a strategic tri-level computational-logical framework was addressed for the PD-TEP problem within the deregulated environments. The proposed framework corresponding to the PD-TEP problem was broken into two different problems: a short-term operational slave problem and a long-term planning master problem. These two problems serve each other. As previously mentioned, the first (i.e., the BBM problem) and second (i.e., CSC electricity market problem) levels formed the short-term operational slave problem. The outputs of the slave problem, which ultimately are inputs to the long-term master planning problem, are generated via an iterative feedback process between strategic bids of the GENCOs and DISCOs and outcomes of the CSC electricity market. Mutually, the outputs of the master problem are as inputs to the short-term operational slave problem.

Similar to the proposed framework related to the PD-GEP problem, assessing the effects of the transmission expansion plans on strategic behaviors of both GENCOs and DISCOs and vice versa was one of the most important targets of the proposed framework relevant to the PD-TEP problem. The first and second levels of the framework related to the PD-TEP problem were quite similar to the first and second levels of the framework concerned with the PD-GEP problem. In the third level of the framework pertaining to the PD-TEP problem, the TIC, TCC, and ECOC were minimized subject to two limitations: short- and long-term constraints.

The point to be made here is that ignoring coordination of the proposed frameworks concerned with the PD-GEP and PD-TEP problems could give rise to uneconomical expansion plans specified for the generation and transmission networks. In other words, lack of coordination of the proposed frameworks increased unnecessary investment costs in the generation and transmission networks and decreased the market participants' profit.

In Sect. 6.5, a strategic quad-level computational-logical framework was addressed for the PD-G&TEP problem within the deregulated environments. The proposed framework pertaining to the PD-G&TEP problem was divided into a short-term operational slave problem and a long-term planning master problem. The short-term operational slave problem was also organized by the BBM problem as the first level and the CSC electricity market problem as the second level. The long-term planning master problem was also organized by the PD-GEP problem as the third level and the PD-TEP problem as the fourth level. The outputs of the slave problem, which ultimately are inputs to the long-term master planning problem, are generated via an iterative feedback process between strategic bids of the GENCOs and DISCOs and outcomes of the CSC electricity market. Mutually, the outputs of the master problem, which ultimately are inputs to the short-term operational slave problem, are generated via an iterative feedback process between outcomes resulting from the generation expansion planning and outcomes arising from the transmission expansion planning. Simply put, evaluating the impact of the optimal generation and transmission expansion plans on the strategic behaviors of both GENCOs and DISCOs and vice versa was one of the most important goals of the proposed framework relevant to the PD-G&TEP problem. In the framework relevant to the PD-G&TEP problem, the first and second levels were borrowed from levels 1 and 2 of the frameworks concerned with the PD-GEP and/or the PD-TEP problems. In addition, the third and fourth levels of the framework related to the PD-G&TEP problem were also adapted from level 3 of the frameworks associated with the PD-GEP and PD-TEP problems, respectively.

In Sect. 6.6, a techno-economic framework was reported for the PD-DEP problem in the presence of DGRs. The proposed framework relevant to the PD-DEP problem was broken into an LBOPF and a long-term planning problem. In the LBOPF, the DNO sought to maximize the loadability of the open-loop distribution network subject to Kirchhoff's point and loop laws as well as the operational limits on the distribution network equipment. In the long-term planning problem, however, the PD-DEP was solved in order to minimize the DIC, DOC, DMC, and ECOC subject to limitations associated with the short-term LBOPF problem, the availability of financial resources, and the logical restrictions.

To minimize the risks of planning stemming from severe uncertainties of demand and market price, an IGDT under a twofold envelope-bound uncertainty model was widely employed in the frameworks associated with the PD-GEP, PD-TEP, PD-G&TEP, and PD-DEP problems. Hence, by applying the IGDT in the proposed frameworks related to the PD-GEP, PD-TEP, PD-G&TEP, and PD-DEP problems, they were rewritten in the form of two risk-averse and risk-taker decision-making policies.

A feasibility study of implementing the proposed frameworks relevant to the PD-GEP, PD-TEP, and PD-G&TEP problems under risk-neutral, risk-averse, and risk-taker policies by ignoring/considering the strategic behaviors of market participants (e.g., the first level of the frameworks) was thoroughly investigated. In this regard, the frameworks pertaining to the PD-GEP and PD-G&TEP problems were successfully implemented on a modified large-scale 46-bus south Brazilian system. The framework related to the PD-TEP problem was also successfully applied not only on a modified IEEE 30-bus test system but also on a modified large-scale Iranian 400 kV transmission network. Simulation results reflect the fact that the proposed frameworks pertaining to the PD-GEP, PD-TEP, and PD-G&TEP problems—by considering strategic behaviors of market participants (i.e., first level) under the IGDT risk-neutral, risk-averse, and risk-taker policies—give rise to achieving more appropriate techno-economic results compared to the proposed frameworks by ignoring strategic behaviors of market participants.

A feasibility study implementing the framework associated with the PD-DEP problem under the IGDT risk-neutral, risk-averse, and risk-taker policies by ignoring/considering DGRs was evaluated. The proposed framework related to the PD-DEP problem was also successfully implemented on a modified three-phase 13.8 kV 27-node open-loop distribution test network. The simulation results illustrated that the framework relevant to the PD-DEP problem in the presence of DGRs under the IGDT risk-neutral, risk-averse, and risk-taker policies yields better techno-economic outcomes compared to the PD-DEP framework in the absence of DGRs.

The performance of the proposed framework related to the PD-GEP problem under the IGDT risk-taker policy and the performance of the proposed frameworks associated with the PD-TEP, PD-G&TEP, and PD-DEP problems under the IGDT risk-averse policy were compared with the performance of those under the RO technique. The results demonstrated that the performance of the proposed framework relevant to the PD-GEP problem under the IGDT risk-taker policy and the performance of the proposed frameworks pertaining to the PD-TEP, PD-G&TEP, and PD-DEP problems under the IGDT risk-averse policy lead to more favorable techno-economic outputs when compared with the RO technique. As a result, decision-making policies according to the IGDT risk-averse and risk-taker policies were a more suitable approach for dealing with severe uncertainties in the proposed frameworks compared to decision-making policies based on the RO technique.

In addition, the efficiency of modern meta-heuristic music-inspired optimization algorithms including the multi-objective SOSA, multi-objective TMS-EMSA, multi-objective continuous/discrete TMS-MS, multi-objective SS-HSA, and multi-objective SS-IHSA, which were discussed in Chap. 4, was scrutinized on the proposed frameworks corresponding to the PD-GEP, PD-TEP, PD-G&TEP, and PD-DEP problems and compared to the current state-of-the-art multi-objective optimization algorithm making up the NSGA-II. To more clarify, the performance of these multi-objective optimization algorithms was exhaustively assessed by employing the cost-saving index. From the results, it can be demonstrated that the performance of the multi-objective optimization algorithms in the proposed frameworks concerned with the PD-GEP, PD-TEP, PD-G&TEP, and PD-DEP problems has a distinct trend in the ranking of these optimization algorithms. In other words, these results indicate that the newly developed multi-objective SOSA was more efficient and cost effective than other multi-objective optimization algorithms. As the results show, the multi-objective TMS-EMSA, multi-objective continuous/discrete TMS-MSA, multi-objective SS-IHSA, multi-objective SS-HSA, and NSGA-II were ranked as the next most efficient multi-objective optimization algorithms. Consequently, the modern meta-heuristic music-inspired optimization algorithms, which were discussed in Chap. 4, may be appropriate optimization algorithms for overcoming difficulties in solving the multi-objective, large-scale, non-convex, nonlinear, mixed-integer nature of real-world optimization problems with big data.

Appendix 1: List of Abbreviations and Acronyms

BBM	Bilateral bidding mechanism
CO_2	Carbon dioxide
CP	Capacity payment
CSC electricity market	Competitive security-constrained electricity market
CWF	Community welfare function

(continued)

DEP	Distribution expansion planning
DGRs	Distributed generation resources
DIC	Distribution investment cost
DISCO	Distribution companies
DMC	Distribution maintenance cost
DNO	Distribution network operator
DOC	Distribution operation cost
ECO	Expected customer outage
ECOC	Expected customer outage cost
EP	Expected profit
EPDF	Electrical power demand forecasting
FSM	Fuzzy satisfying method
GA	Genetic algorithm
GAMS	General algebraic modeling system
GENCOs	Generation companies
GEP	Generation expansion planning
GIC	Generation investment cost
G&TEP	Generation and transmission expansion planning
HL-I	Hierarchical level I
HL-II	Hierarchical level II
IGDT	Information gap decision theory
ISO	Independent system operator
LBOPF	Loadability-based optimal power flow
LCC	Load curtailment cost
LDC	Load duration curve
MMM-EMSA	Single-stage computational, multi-dimensional, multiple-homogeneous enhanced melody search algorithm
MMS-EMSA	Single-stage computational, multi-dimensional, single-inhomogeneous enhanced melody search algorithm
NO_x	Nitrogen oxide
NSGA-II	Non-dominated sorting genetic algorithm-II
PDF	Probability density function
PD-DEP	Pseudo-dynamic distribution expansion planning
PD-DEPO	Pseudo-dynamic distribution expansion planning objective function
PD-GEP	Pseudo-dynamic generation expansion planning
PD-GEPO	Pseudo-dynamic generation expansion planning objective function
PD-G&TEP	Pseudo-dynamic generation and transmission expansion planning
PD-TEP	Pseudo-dynamic transmission expansion planning
PD-TEPO	Pseudo-dynamic transmission expansion planning objective function
PSO	Particle swarm optimization
RO	Robust optimization
SCDF	Sector customer damage function
SFE	Supply function equilibrium
SO_2	Sulfur dioxide

(continued)

SOSA	Symphony orchestra search algorithm
SS-HSA	Single-stage computational, single-dimensional harmony search algorithm
SS-IHSA	Single-stage computational, single-dimensional improved harmony search algorithm
TCC	Transmission congestion cost
TEP	Transmission expansion planning
TIC	Transmission investment cost
TMS-EMSA	Two-stage computational, multi-dimensional single-homogeneous enhanced melody search algorithm
TMS-MSA	Two-stage computational, multi-dimensional single-homogeneous melody search algorithm
UK	United Kingdom
USA	United States of America

Appendix 2: List of Mathematical Symbols

Index	
b	Index for buses of the network running from 1 to B
b_{el}	Index for electrical load buses of the network running from 1 to B_{el}
b_{ge}	Index for generation buses of the network running from 1 to B_{ge}
b_{eu}^{g}	Index for the buses of the existing generation units owned by GENCO g running from 1 to B
b_{iu}^{g}	Index for the buses of the installed generation units owned by GENCO g running from 1 to B
b_{vc}	Index for voltage conversion buses of the network running from 1 to B_{vc}
$cb^{\mathrm{i\text{-}s}}$	Index for candidate buses of the network for installed substations running from 1 to $CB_{\mathrm{i\text{-}s}}$
$cb^{\mathrm{r\text{-}s}}$	Index for candidate buses of the network for reinforced substations running from 1 to $CB_{\mathrm{r\text{-}s}}$
$cl^{\mathrm{i\text{-}b}}$	Index for candidate locations in the network for installed branches running from 1 to $CL_{\mathrm{i\text{-}b}}$
$cl^{\mathrm{r\text{-}b}}$	Index for candidate locations in the network for reinforced branches running from 1 to $CL_{\mathrm{r\text{-}b}}$
d	Index for DISCOs running from 1 to D
e	Index for outage events running from 1 to E
eb	Index for existing branches in the network running from 1 to EB
es	Index for existing substations in the network running from 1 to ES
eu	Index for the existing generation units owned by GENCO g running from 1 to EU
g	Index for GENCOs running from 1 to G
ib	Index for installed branches in the network running from 1 to IB
id	Index for installed DGRs in the network running from 1 to ID

(continued)

is	Index for installed substations in the network running from 1 to IS
iu	Index for the installed generation units owned by GENCO *g* running from 1 to IU
k	Index for patterns running from 1 to K
l	Index for transmission lines running from 1 to L
p	Index for periods of planning horizon running from 1 to P
rb	Index for reinforced branches in the network running from 1 to RB
rs	Index for reinforced substations in the network running from 1 to RS
s	Index for load sectors in each load bus running from 1 to S
sg	Index for sub-graphs in the forest running from 1 to SG
(s, r)	Index for corridors of the network running from 1 to L
Set	
Ψ^{B}	Set of indices for the buses of the network
$\Psi^{\mathrm{B_{el}}}$	Set of indices of electrical load buses of the network
$\Psi^{\mathrm{B_{ge}}}$	Set of indices of generation buses of the network
$\Psi^{\mathrm{B_{vc}}}$	Set of indices of voltage conversion buses of the network
$\Psi^{\mathrm{B}}_{g_{eu}}$	Set of indices for the buses of the existing generation units owned by GENCO *g*
$\Psi^{\mathrm{B}}_{g_{iu}}$	Set of indices for the buses of the installed generation units owned by GENCO *g*
$\Psi^{\mathrm{CB_{i-s}}}$	Set of indices of candidate buses of the network for installed substations
$\Psi^{\mathrm{CB_{r-s}}}$	Set of indices of candidate buses of the network for reinforced substations
$\Psi^{\mathrm{CL_{i-b}}}$	Set of indices of candidate locations in the network for installed branches
$\Psi^{\mathrm{CL_{r-b}}}$	Set of indices of candidate locations in the network for reinforced branches
Ψ^{D}	Set of indices for the DISCOs
Ψ^{E}	Set of indices of outage events
Ψ^{EB}	Set of indices of existing branches in the network
Ψ^{ES}	Set of indices of existing substations in the network
Ψ^{EU}_{g}	Set of indices for the existing generation units owned by GENCO *g*
Ψ^{G}	Set of indices for the GENCOs
Ψ^{IB}	Set of indices of installed branches in the network
Ψ^{ID}	Set of indices of installed DGRs in the network
Ψ^{IS}	Set of indices of installed substations in the network
Ψ^{IU}_{g}	Set of indices for the installed generation units owned by GENCO *g*
Ψ^{K}	Set of indices for patterns
Ψ^{L}	Set of indices for transmission lines
Ψ^{L}_{s}	Set of transmission lines connected to the sending bus of corridor (s,r)
Ψ^{P}	Set of indices for periods

(continued)

Ψ^{RB}	Set of indices of reinforced branches in the network
Ψ^{RS}	Set of indices of reinforced substations in the network
Ψ^{S}	Set of indices of load sectors
Ψ^{SG}	Set of indices of sub-graphs in the forest
Parameters	
CCL^{b}	Basic construction cost for each transmission line [M$/MWkm]
$c\rho_{p}^{\max}$	Maximum allowed amount of the curtailed load in period p [MW]
$c\rho_{p,b}^{\max}$	Maximum allowed amount of the curtailed load on bus b in period p [MW]
$c\rho_{p,b,s}^{\max}$	Maximum allowed amount of the curtailed load at load sector s on bus b in period p [MW]
$DIC^{\max}$	Maximum budget available for distribution investment cost [M$]
$DIC_{p}^{\max}$	Maximum budget available for distribution investment cost of period p [M$]
ECOC	Expected customer outage cost [$/MWh]
$ECO^{\max}$	Maximum value of expected customer outage [MW]
$E_{p,k,g_{iu}}^{CO_2}$	Value of emitted CO_2 atmospheric emission of the installed generation unit iu owned by GENCO g at pattern k of period p [ton/MWh]
$E_{p,k,g_{eu}}^{CO_2}$	Value of emitted CO_2 atmospheric emission of the existing generation unit eu owned by GENCO g at pattern k of period p [ton/MWh]
$E_{p,g}^{CO_2,\max}$	Maximum permissible value of emitted CO_2 atmospheric emission in the atmosphere at period p [ton]
$E_{p,k,g_{iu}}^{NO_x}$	Value of emitted NO_x atmospheric emission of the installed generation unit iu owned by GENCO g at pattern k of period p [ton/MWh]
$E_{p,k,g_{eu}}^{NO_x}$	Value of emitted NO_x atmospheric emission of the existing generation unit eu owned by GENCO g at pattern k of period p [ton/MWh]
$E_{p,g}^{NO_x,\max}$	Maximum permissible value of emitted NO_x atmospheric emission in the atmosphere at period p [ton]
$E_{p,k,g_{iu}}^{SO_2}$	Value of emitted SO_2 atmospheric emission of the installed generation unit iu owned by GENCO g at pattern k of period p [ton/MWh]
$E_{p,k,g_{eu}}^{SO_2}$	Value of emitted SO_2 atmospheric emission of the existing generation unit eu owned by GENCO g at pattern k of period p [ton/MWh]
$E_{p,g}^{SO_2,\max}$	Maximum permissible value of emitted SO_2 atmospheric emission in the atmosphere at period p [ton]
$f_{p,k,l}^{\max}$	Upper limit on the active electrical power flow in transmission line m at pattern k of period p [MW]
$f_{p,k,l}^{\min}$	Lower limit on the active electrical power flow in transmission line l at pattern k of period p [MW]
$f_{p,(s,r)}^{\max}$	Maximum allowable active electrical power flow in corridor (s,r) in period p [MW]
$f_{p,l}^{\max}$	Maximum allowable active electrical power flow in transmission line l in period p [MW]

(continued)

$F_{p,k,g_{eu}}$	Fuel consumption rate of the existing generation unit *eu* owned by GENCO *g* at pattern *k* of period *p* [m3/MWh or L/MWh or kg/MWh]
$F_{p,k,g_{iu}}$	Fuel consumption rate of the installed generation unit *iu* owned by GENCO *g* at pattern *k* of period *p* [m3/MWh or L/MWh or kg/MWh]
$F_{p,g}^{\max}$	Maximum value of fuel consumption by GENCO *g* at period *p* [m3 or L or kg]
$FR_{p,\,m,\,e,\,b,\,s}$	Average failure rate in case of failure event *e*, at load sector *s* on affected load bus *b*, at load level *m* of period *p* [failure per year]
$g\rho_{p,\,m}$	Column vector of the bus injection at load level *m* of period *p*
$GIC_{p,\,g_{iu}}$	Generation investment cost for generating one unit of electrical power in MW [$/MW]
$GIC_{p,g_{iu}}^{\mathrm{f}}$	Fixed generation investment cost for the installed generation unit *iu* owned by GENCO *g* at period *p* [M$]
$GIC_{p,g}^{\max}$	Maximum value of the generation investment cost by GENCO *g* in period *p* [M$]
$GIC_{g}^{\max}$	Maximum value of the generation investment cost by GENCO *g* in all periods [M$]
$I_{eb}^{\max}$	Maximum current of existing branch *eb* [p.u.]
$I_{es}^{\max}$	Maximum current of existing substation *es* [p.u.]
$I_{ib}^{\max}$	Maximum current of installed branch *ib* [p.u.]
$I_{is}^{\max}$	Maximum current of installed substation *is* [p.u.]
$I_{rb}^{\max}$	Maximum current of reinforced branch *rb* [p.u.]
$I_{rs}^{\max}$	Maximum current of reinforced substation *rs* [p.u.]
Ir	Interest rate [%]
$IC_{p,g}^{\max}$	Maximum value of the installed generation capacity by GENCO *g* at period *p* [MW]
$IC_{p,is}^{\mathrm{f}}$	Fixed investment costs for the installed substation *is* of period *p* [M$]
$LE_{(s,\,r)}$	Length of corridor (*s*,*r*) [km]
L_{ib}	Length of installed branch *ib* [km]
L_{rb}	Length of reinforced branch *rb* [km]
p_0	Base year
$PERC_{p,g}^{\max}$	Maximum percentage of installed generation capacity that can be installed by GENCO *g* at period *p* [%]
R_{eb}	Resistance of existing branch *eb* [Ω]
R_{es}	Resistance of existing substation *es* [Ω]
R_{ib}	Resistance of installed branch *ib* [Ω]
R_{is}	Resistance of installed substation *is* [Ω]
R_{rb}	Resistance of reinforced branch *rb* [Ω]
R_{rs}	Resistance of reinforced substation *rs* [Ω]
$RM_{p}^{\max}$	Maximum reserve margin of generation in period *p* [%]
$RM_{p}^{\min}$	Minimum reserve margin of generation in period *p* [%]
$TIC_{p}^{\max}$	Maximum allowable value of the TIC index in period *p* [M$]
$TIC^{\max}$	Maximum allowable value of the TIC index over the planning horizon [M$]

(continued)

$V_{b_{\text{el}}}^{\max}$	Maximum voltage on electrical load bus b_{el} [p.u.]
$V_{b_{\text{el}}}^{\min}$	Minimum voltage on electrical load bus b_{el} [p.u.]
$V_{b_{\text{ge}}}^{\max}$	Maximum voltage on generation bus b_{ge} [p.u.]
$V_{b_{\text{ge}}}^{\min}$	Minimum voltage on generation bus b_{ge} [p.u.]
$V_{b_{\text{vc}}}^{\max}$	Maximum voltage on voltage conversion bus b_{vc} [p.u.]
$V_{b_{\text{vc}}}^{\min}$	Minimum voltage on voltage conversion bus b_{vc} [p.u.]
$w_{p,\, b,\, s}$	Outage cost weight coefficient for load sector s on bus b in period p
$\text{W}_{\text{OF}_1,g}^{\text{PD-GEP}}$	Weighting coefficient of the first objective function of the PD-GEP problem from the perspective of GENCO g
$\text{W}_{\text{OF}_2,g}^{\text{PD-GEP}}$	Weighting coefficient of the second objective function of the PD-GEP problem from the perspective of GENCO g
$\text{W}_{\text{OF}_3,g}^{\text{PD-GEP}}$	Weighting coefficient of the third objective function of the PD-GEP problem from the perspective of GENCO g
$\text{W}_{\text{OF}_1}^{\text{PD-TEP}}$	Weighting coefficient of the first objective function of the PD-TEP problem
$\text{W}_{\text{OF}_2}^{\text{PD-TEP}}$	Weighting coefficient of the second objective function of the PD-TEP problem
$\text{W}_{\text{OF}_3}^{\text{PD-TEP}}$	Weighting coefficient of the third objective function of the PD-TEP problem
$\text{W}_{\text{OF}_1}^{\text{PD-DEP}}$	Weighting coefficient of the first objective function of the PD-DEP problem
$\text{W}_{\text{OF}_2}^{\text{PD-DEP}}$	Weighting coefficient of the second objective function of the PD-DEP problem
$\text{W}_{\text{OF}_3}^{\text{PD-DEP}}$	Weighting coefficient of the third objective function of the PD-DEP problem
$\text{W}_{\text{OF}_4}^{\text{PD-DEP}}$	Weighting coefficient of the fourth objective function of the PD-DEP problem
$y_{p,(s,r)}^{0}$	Initial susceptance in corridor (s,r) in period p [mho]
$y_{(p-1),(s,r)}^{0}$	Initial susceptance in corridor (s,r) in period p-1 [mho]
$y_{(p-1),ss}^{0}$	Initial susceptance in the element at the sth row and sth column of the network admittance matrix in period p-1 [mho]
Z_{eb}	Impedance of existing branch eb [Ω]
Z_{ib}	Impedance of installed branch ib [Ω]
Z_{rb}	Impedance of reinforced branch rb [Ω]
$\alpha_{p,\, k,\, g}$	Generation cost function coefficient of GENCO g at pattern k of period p [\$/MW^{2}h]
$\alpha_{p,\, k,\, d}$	Benefit function coefficient of DISCO d at pattern k of period p [\$/MW^{2}h]
$\alpha_{p,k,g_{eu}}$	Generation cost function coefficient of the existing generation unit eu owned by GENCO g at pattern k of period p [\$/MW^{2}h]
$\alpha_{p,k,g_{iu}}$	Generation cost function coefficient of the installed generation unit iu owned by GENCO g at pattern k of period p [\$/MW^{2}h]
$\beta_{p,\, k,\, g}$	Generation cost function coefficient of GENCO g at pattern k of period p [\$/MWh]
$\beta_{p,\, k,\, d}$	Benefit function coefficient of DISCO d at pattern k of period p [\$/MWh]

(continued)

$\beta_{p,k,g_{eu}}$	Generation cost function coefficient of the existing generation unit *eu* owned by GENCO g at pattern k of period p [$/MWh]
$\beta_{p,k,g_{iu}}$	Generation cost function coefficient of the installed generation unit *iu* owned by GENCO g at pattern k of period p [$/MWh]
$\gamma_{p,\ k,\ g}$	Generation cost function coefficient of GENCO g at pattern k of period p [$/h]
$\gamma_{p,\ k,\ d}$	Benefit function coefficient of DISCO d at pattern k of period p [$/h]
$\gamma_{p,k,g_{eu}}$	Generation cost coefficient of the existing generation unit *eu* owned by GENCO g at pattern k of period p [$/h]
$\gamma_{p,k,g_{iu}}$	Generation cost function coefficient of the installed generation unit *iu* owned by GENCO g at pattern k of period p [$/h]
$\rho^{\text{max-dg}}$	Maximum capacity of DGRs that can be installed in the open-loop distribution network [MW]
$\rho_{id}^{\max}$	Maximum capacity of installed DGR *id* [MW]
$\rho_{p}^{\text{max-dg}}$	Maximum capacity of DGRs that can be installed in the open-loop distribution network of period p [MW]
$\rho_{p,k,g}^{\max}$	Upper limit on the generated electrical power by GENCO g at pattern k of period p [MW]
$\rho_{p,k,g_{eu}}^{\max}$	Upper limit on the generated electrical power by the existing generation unit *eu* owned by GENCO g at pattern k of period p [MW]
$\rho_{p,k,g_{iu}}^{\max}$	Upper limit on the generated electrical power by the installed generation unit *iu* owned by GENCO g at pattern k of period p [MW]
$\rho_{p,k,g}^{\min}$	Lower limit on the generated electrical power by GENCO g at pattern k of period p [MW]
$\rho_{p,k,g_{eu}}^{\min}$	Lower limit on the generated electrical power by the existing generation unit *eu* owned by GENCO g at pattern k of period p [MW]
$\rho_{p,k,g_{iu}}^{\min}$	Lower limit on the generated electrical power by the installed generation unit *iu* owned by GENCO g at pattern k of period p [MW]
$\rho_{p,k,d}^{\max}$	Upper limit on the purchased electrical power by DISCO d at pattern k of period p [MW]
$\rho_{p,k,d}^{\min}$	Lower limit on the purchased electrical power by DISCO d at pattern k of period p [MW]
$\xi_{1,p,k,g}^{\max}$	Upper level of the bidding strategy parameter (i.e., slope) of GENCO g at pattern k of period p
$\xi_{1,p,k,g}^{\min}$	Lower level of the bidding strategy parameter (i.e., slope) of GENCO g at pattern k of period p
$\xi_{2,p,k,g}^{\max}$	Upper level of the bidding strategy parameter (i.e., intercept) of GENCO g at pattern k of period p
$\xi_{2,p,k,g}^{\min}$	Lower level of the bidding strategy parameter (i.e., intercept) of GENCO g at pattern k of period p
$\xi_{1,p,k,d}^{\max}$	Upper level of the bidding strategy parameter (i.e., slope) of DISCO d at pattern k of period p
$\xi_{1,p,k,d}^{\min}$	Lower level of the bidding strategy parameter (i.e., slope) of DISCO d at pattern k of period p

(continued)

$\xi_{2,p,k,d}^{\max}$	Upper level of the bidding strategy parameter (i.e., intercept) of DISCO d at pattern k of period p
$\xi_{2,p,k,d}^{\min}$	Lower level of the bidding strategy parameter (i.e., intercept) of DISCO d at pattern k of period p
$\iota_{(s,r)}^{\max}$	Maximum allowable number of the transmission lines that can be added in corridor (s,r)
ε_b	Economic lifetime of the branches [year]
ε_{dg}	Economic lifetime of the DGRs [year]
ε_s	Economic lifetime of the substations [year]
Δt_k	Duration period of pattern k [h]
Δt_m	Durations of load level m [hour/day]
Variables	
$a\rho_{p,\ k,\ b}$	Actual demand on bus b of the network at pattern k of period p [MW]
$a\rho_{p,\ k,\ g,\ b}$	Actual demand on bus b of the network at pattern k of period p from the perspective of GENCO g [MW]
$a\rho_{p,\ m,\ b}$	Actual demand on bus b of the network at load level m of period p [MW]
$a\lambda_{p,\ k,\ b}$	Actual market price of bus b of the network at pattern k of period p [\$/MWh]
$a\lambda_{p,\ k,\ g,\ b}$	Actual market price of bus b of the network at pattern k of period p from the perspective of GENCO g [\$/MWh]
$a\lambda_{p,\ m}$	Actual market price at load level m of period p [\$/MWh]
$\mathrm{ADE}_{p,\ k}$	Artificial power demand matrix in pattern k in period p with dimension number of bus in 1
$\mathrm{BBI}^{\mathrm{ex}}$	Bus-branch incidence matrix for the existing open-loop distribution network
$\mathrm{BBI}^{\mathrm{ex}}_{\mathrm{column}:eb}$	Column eb of the bus-branch incidence matrix of the existing open-loop distribution network
$\mathrm{BBI}^{\mathrm{in}}$	Bus-branch incidence matrix for the installed open-loop distribution network
$\mathrm{BBI}^{\mathrm{in}}_{\mathrm{column}:ib}$	Column ib of the bus-branch incidence matrix of the existing open-loop distribution network
$\mathrm{BBI}^{\mathrm{re}}$	Bus-branch incidence matrix for the reinforced open-loop distribution network
$\mathrm{BBI}^{\mathrm{re}}_{\mathrm{column}:rb}$	Column rb of the bus-branch incidence matrix of the existing open-loop distribution network
BBI_{sg}	Bus-branch incidence matrix of sub-graph sg in the forest
$c\rho_{p,\ m}$	Column vector of bus-load shedding at load level m of period p
$c\rho_{p,\ b,\ s,\ e}$	Amount of power demand in case of outage event e, at load sector s on bus b in period p [MW]
$c\rho_{p,m,b_{el}}$	Bus-load shedding on electrical load bus b_{el} at load level m of period p [MW]
$C_{p,\ b,\ s,\ e}(\Delta t_e)$	Outage cost, based on the outage duration time, Δt_e, using the SCDF in case of outage event e, at load sector s on bus b in period p [\$/kW]
$C_{p,\ m,\ e,\ b,\ s}$	Outage costs, based on the outage duration time $\Delta t_{p,\ m,\ e,\ b,\ s}$, using the SCDF in case of failure event e, at load sector s on affected load bus b, at load level m of period p [\$/kW]

(continued)

$CP_{y,\,g}$	Capacity payment for the generation expansion plan y of GENCO g [M$]
$CP_{y,\,g,\,p}$	Capacity payment for the generation expansion plan y of GENCO g at period p [$]
$CWF_{p,\,k}$	Community welfare function at pattern k of period p [$/h]
$\mathrm{DE}_{p,\,k}$	Power demand matrix in pattern k in period p with dimension number of bus in 1
DIC_y	Distribution investment cost for distribution expansion plan y [M$]
$DIC_{y,\,p}$	Distribution investment cost for distribution expansion plan y of period p [M$]
DMC_y	Distribution maintenance cost for distribution expansion plan y [M$]
$DMC_{y,\,p}$	Distribution maintenance cost for distribution expansion plan y of period p [M$]
DOC_y	Distribution operation cost for distribution expansion plan y [M$]
$DOC_{y,\,p,\,m}$	Total operational costs for expansion plan y at load level m of period p [$/h]
ECO_p	Expected customer outage in period p [MW]
$ECO^{\mathrm{b}}_{p,k,g_{iu}}$	Expected customer outage before installation of new generation unit iu owned by GENCO g at pattern k of period p [MW]
$ECO^{\mathrm{a}}_{p,k,g_{iu}}$	Expected customer outage after installation of new generation unit iu owned by GENCO g at pattern k of period p [MW]
$ECOC_y$	Expected customer outage costs for distribution expansion plan y [M$]
$ECOC_{y,\,p}$	Expected customer outage costs for distribution expansion plan y of period p [M$]
$ECOC^{\mathrm{trans}}_{y}$	Expected costumer outage cost of a transmission network for transmission expansion plan y [M$]
$ECOC^{\mathrm{trans}}_{y,p}$	Expected costumer outage cost of a transmission network for transmission expansion plan y in period p [M$]
$ECOC^{\mathrm{HL\text{-}II}}_{y,p}$	Expected costumer outage cost at the HL-II for transmission expansion plan y in period p [M$]
$ECOC^{\mathrm{HL\text{-}I}}_{y,p}$	Expected costumer outage cost at the HL-I for transmission expansion plan y in period p [M$]
$EP_{y,\,g}$	Expected profit for the generation expansion plan y of GENCO g [M$]
$EP_{y,\,g,\,p}$	Expected profit for the generation expansion plan y of GENCO g at period p [M$]
$f_{p,\,k,\,l}$	Active electrical power flow in transmission line l at pattern k of period p [MW]
$f_{p,k,l_{1-2}}$	Active electrical power flow in transmission line l at pattern k in period p [MW]
$f_{(s,\,r)}$	Thermal capacity of corridor (s,r) [MW]
$f_{p,\,k,\,(s,\,r)}$	Active power in corridor (s,r) at pattern k in period p [MW]
$\mathrm{GE}_{p,\,k}$	Power generation matrix in pattern k in period p with dimension number of bus in 1
$GIC_{y,\,g}$	Generation investment cost for the generation expansion plan y of GENCO g [M$]
$GIC_{y,\,g,\,p}$	Generation investment cost for the generation expansion plan y of GENCO g at period p [M$]

(continued)

$GIC^{\mathrm{v}}_{p,g_{iu}}$	Variable generation investment cost for the installed generation unit *iu* owned by GENCO *g* at period *p* [M$]
$I_{p,\ m,\ eb}$	Current of existing branch *eb* at load level *m* of period *p* [p.u.]
$I_{p,\ m,\ es}$	Injected current of existing substation *es* at load level *m* of period *p* [p.u.]
$I_{p,\ m,\ ib}$	Current of installed branch *ib* at load level *m* of period *p* [p.u.]
$I_{p,\ m,\ is}$	Injected current of installed substation *is* at load level *m* of period *p* [p.u.]
$I_{p,\ m,\ rb}$	Current of reinforced branch *rb* at load level *m* of period *p* [p.u.]
$I_{p,\ m,\ rs}$	Injected current of reinforced substation *rs* at load level *m* of period *p* [p.u.]
$IC_{p,g_{iu}}$	Installed capacity of new generation unit *iu* owned by GENCO *g* at period *p* [MW]
$IC_{p,\ ib}(b)$	Investment costs for installed branch *ib* of period *p* associated with branch type *b* [M$/km]
$IC_{p,\ id}$	Investment costs for installed DGR *id* of period *p* [$/MW]
$IC^{v}_{p,is}(a)$	Variable investment costs for installed substation *is* of period *p* associated with substation type *a* [M$]
$IC_{p,\ rb}(b)$	Investment costs for reinforced branch *rb* of period *p* associated with branch type *b* [M$/km]
$IC^{\mathrm{v}}_{p,rs}(a)$	Investment costs for reinforced substation *rs* of period *p* associated with substation type *a* [M$]
$LCC^{\mathrm{HL\text{-}II}}_{p,e}$	Load curtailment cost at the HL-II for outage event *e* in period *p* [M$]
$MC_{p,\ eb}$	Maintenance costs of existing branch *eb* of period *p* [$/km]
$MC_{p,\ es}$	Maintenance costs of existing substation *es* of period *p* [$/MW]
$MC_{p,\ ib}$	Maintenance costs of installed branch *ib* of period *p* [$/km]
$MC_{p,\ is}$	Maintenance costs of installed substation *is* of period *p* [$/MW]
$MC_{p,\ rb}$	Maintenance costs of reinforced branch *rb* of period *p* [$/km]
$MC_{p,\ rs}$	Maintenance costs of reinforced substation *rs* of period *p* [$/MW]
N^{br}_{sg}	Number of branches in sub-graph *sg* in the forest
N^{bu}_{sg}	Number of buses in sub-graph *sg* in the forest
$N_{p,cl^{\mathrm{i-b}}}$	Number of installed branches in candidate location *cl* of period *p*
$N_{p,cl^{\mathrm{r-b}}}$	Number of reinforced branches in candidate location *cl* of period *p*
$N_{p,cb^{\mathrm{i-s}}}$	Number of installed substations on candidate bus *cb* of period *p*
$N_{p,cb^{\mathrm{r-s}}}$	Number of reinforced substations on candidate bus *cb* of period *p*
NB_p	Node-branch incidence matrix of the network in period *p* with dimensions of number of bus in number of line
$OC_{p,\ m,\ id}$	Operational costs of installed DGR *id* at load level *m* of period *p* [$/MWh]
$\mathrm{OF}^{\mathrm{PD\text{-}DEP}}$	Objective function of the PD-DEP problem [M$]
$\mathrm{OF}^{\mathrm{PD\text{-}GEP}}_{\mathrm{g}}$	Objective function of the PD-GEP problem from the perspective of GENCO *g* [M$]
$\mathrm{OF}^{\mathrm{PD\text{-}TEP}}$	Objective function of the PD-TEP problem [M$]
$\mathrm{OF}^{\mathrm{PD\text{-}GEP}}_{1,g}$	First objective function of the PD-GEP problem from the perspective of GENCO *g* [M$]

(continued)

$\mathrm{OF}_1^{\mathrm{PD\text{-}TEP}}$	First objective function of the PD-TEP problem [M$]
$\mathrm{OF}_1^{\mathrm{PD\text{-}DEP}}$	First objective function of the PD-DEP problem [M$]
$\mathrm{OF}_{2,g}^{\mathrm{PD\text{-}GEP}}$	Second objective function of the PD-GEP problem from the perspective of GENCO g [M$]
$\mathrm{OF}_2^{\mathrm{PD\text{-}TEP}}$	Second objective function of the PD-TEP problem [M$]
$\mathrm{OF}_2^{\mathrm{PD\text{-}DEP}}$	Second objective function of the PD-DEP problem [M$]
$\mathrm{OF}_{3,g}^{\mathrm{PD\text{-}GEP}}$	Third objective function of the PD-GEP problem from the perspective of GENCO g [M$]
$\mathrm{OF}_3^{\mathrm{PD\text{-}TEP}}$	Third objective function of the PD-TEP problem [M$]
$\mathrm{OF}_3^{\mathrm{PD\text{-}DEP}}$	Third objective function of the PD-DEP problem [M$]
$\mathrm{OF}_4^{\mathrm{PD\text{-}DEP}}$	Fourth objective function of the PD-DEP problem [M$]
PD - DEPO	Pseudo-dynamic distribution expansion planning objective function [M$]
PD - GEPO	Pseudo-dynamic generation expansion planning objective function [M$]
PD - TEPO	Pseudo-dynamic transmission expansion planning objective function [M$]
$\mathrm{PF}_{p,\,k}$	Active power flow matrix of the network in pattern k in period p with dimension number of line in 1
$\Pr(e)$	Probability of outage event e in period p [M$]
$p\rho$	Predicted demand on all buses of the network [MW]
$p\rho_b$	Predicted demand on bus b of the network [MW]
$p\rho_{p,\,k,\,b}$	Predicted demand on bus b of the network at pattern k of period p [MW]
$p\rho_{p,k,b}^{\mathrm{p}}$	Predicted average peak demand of bus b of the network at pattern k of period p [MW]
$p\rho_{p,\,m,\,e,\,b,\,s}$	Amount of predicted power demand in case of failure event e, at load sector s on affected load bus b, at load level m of period p [MW]
$p\rho_{p,m,b_{\mathrm{el}}}$	Predicted demand on electrical load bus b_{el} at load level m of period p [MW]
$p\rho_{p,\,m}$	Column vector of the bus load at load level m of period p
$p\lambda$	Predicted market price of all buses of the network [$/MWh]
$p\lambda_b$	Predicted market price of bus b of the network [$/MWh]
$p\lambda_{p,\,k,\,b}$	Predicted market price of bus b of the network at pattern k of period p [$/MWh]
$p\lambda_{p,k,b_{eu}^g}$	Predicted market price of the bus of the existing generation unit eu owned by GENCO g at pattern k of period p [$/MWh]
$p\lambda_{p,k,b_{iu}^g}$	Predicted market price of the bus of the installed generation unit iu owned by GENCO g at pattern k of period p [$/MWh]
$p\lambda_{p,k,l_2}$	Predicted market price on the receiving bus of transmission line l at pattern k in period p [$/MWh]
$p\lambda_{p,k,l_1}$	Predicted market price on the sending bus of transmission line l at pattern k in period p [$/MWh]
$p\lambda_{p,\,m}$	Predicted market price at load level m of period p [$/MWh]
$\mathrm{PF}_{p,m}^{\mathrm{ex}}$	Row vector of the current for existing branches at load level m of period p
$\mathrm{PF}_{p,m}^{\mathrm{in}}$	Row vector of the current for installed branches at load level m of period p

(continued)

$\mathrm{PF}^{\mathrm{re}}_{p,m}$	Row vector of the current for reinforced branches at load level m of period p
$RR^{\mathrm{demand}}_{p,k,b}$	Demand robustness region on bus b of the network at pattern k of period p
$RR^{\mathrm{price}}_{p,k,b}$	Market price robustness region on bus b of the network at pattern k of period p
TCC_y	Transmission congestion cost for the transmission expansion plan y [M$]
$TCC_{y,\,p}$	Transmission congestion cost for the transmission expansion plan y in period p [M$]
$TCC_{p,\,k}$	Transmission congestion cost at pattern k in period p [M$]
TIC_y	Transmission investment cost for the transmission expansion plan y [M$]
$TIC_{y,\,p}$	Transmission investment cost for the transmission expansion plan y in period p [M$]
$TIC_{p,\,(s,r)}$	Transmission investment cost for installation of a new transmission line in corridor (s,r) in period p [M$]
$\mathrm{V}_{p,\,m}$	Row vector of bus voltages at load level m of period p
$V_{p,m,b_{\mathrm{vc}}}$	Voltage on voltage conversion bus b_{vc} at load level m of period p [p.u.]
$V_{p,m,b_{\mathrm{el}}}$	Voltage on electrical load bus b_{el} at load level m of period p [p.u.]
$V_{p,m,b_{\mathrm{ge}}}$	Voltage on generation bus b_{ge} at load level m of period p [p.u.]
x_{BBM}	Decision-making variables of the first level (i.e., BBM) of the proposed strategic tri-level computational-logical framework
$x_{\mathrm{CSC\text{-}EM}}$	Decision-making variables of the second level (i.e., CSC electricity market) of the proposed strategic tri-level computational-logical framework
x_{ECOC}	Decision-making variables in the linear optimization approach of the value-based reliability assessment methodology
$x_{\mathrm{PD\text{-}DEP}}$	Decision-making variables of the pseudo-dynamic distribution expansion planning problem
$x_{\mathrm{PD\text{-}GEP}}$	Decision-making variables of the third level (i.e., pseudo-dynamic generation expansion planning problem) of the proposed strategic tri-level computational-logical framework
$x_{\mathrm{PD\text{-}TEP}}$	Decision-making variables of the third level (i.e., pseudo-dynamic transmission expansion planning problem) of the proposed strategic tri-level computational-logical framework
$Y_{p,\,bb',\,e}$	Element at the bth row and b'^{th} column in the network admittance matrix in case of outage event e in period p
$Y_{p,\,sr}$	Element at the sth row and r^{th} column in the network admittance matrix in period p
$Y_{p,\,ss}$	Element at the sth row and s^{th} column in the network admittance matrix in period p
$\rho_{p,\,k,\,g}$	Generated electrical power by GENCO g at pattern k of period p [MW]
$\rho_{p,k,g_{eu}}$	Generated electrical power by the existing generation unit eu owned by GENCO g at pattern k of period p [MW]
$\rho_{p,k,g_{iu}}$	Generated electrical power by the installed generation unit iu owned by GENCO g at pattern k of period p [MW]

(continued)

$\rho_{p,\ k,\ d}$	Purchased electrical power by DISCO d at pattern k of period p [MW]
$\rho_{p,\ g,\ b}$	Power generation by GENCO g located on bus b in period p [MW]
$\rho_{p,id}^{\text{cap}}$	Total operational capacity of installed DGR id of period p [MW]
$\rho_{p,id}^{\text{res}}$	Total reserve capacity of installed DGR id of period p [MW]
$\rho_{p,s,b}^{\max}$	Maximum power demand at load sector s on bus b in period p [MW]
$\rho_{p,\ m,\ es}$	Injected power of existing substation es at load level m of period p [MW]
$\rho_{p,\ m,\ is}$	Injected power of installed substation is at load level m of period p [MW]
$\rho_{p,\ m,\ id}$	Generated power of the installed DGR dg at load level m of period p [MW]
$\rho_{p,\ m,\ rs}$	Injected power of reinforced substation rs at load level m of period p [MW]
$\xi_{1,\ p,\ k,\ g}$	Bidding strategy parameter (i.e., slope) of GENCO g at pattern k of period p
$\xi_{2,\ p,\ k,\ g}$	Bidding strategy parameter (i.e., intercept) of GENCO g at pattern k of period p
$\xi_{1,\ p,\ k,\ d}$	Bidding strategy parameter (i.e., slope) of DISCO d at pattern k of period p
$\xi_{2,\ p,\ k,\ d}$	Bidding strategy parameter (i.e., intercept) of DISCO d at pattern k of period p
$\tilde{\upsilon}_{p,k,g}\left(\rho_{p,k,g}\right)$	Profit of GENCO g considering the bidding strategy at pattern k of period p [$/h]
$\tilde{\upsilon}_{p,k,d}\left(\rho_{p,k,d}\right)$	Profit of DISCO d considering the bidding strategy at pattern k of period p [$/h]
$\kappa_{p,\ k,\ l}$	Loss of transmission line l at pattern k of period p [MW]
η_{p}^{res}	Corrective factor related to the reserve margin of generation in period p [MW]
$\eta_{p,g}^{\text{cap}}$	Corrective factor related to the installed generation capacity by GENCO g in period p [MW]
η_{p}^{rel}	Corrective factor related to the reliability requirements in period p [MW]
Δ^{price}	Interval of market price uncertainty in bus b of the network at pattern k of period p [%]
$\Delta_{g}^{\text{price}}$	Interval of market price uncertainty in bus b of the network at pattern k of period p from the perspective of GENCO g [%]
Δ^{demand}	Interval of demand uncertainty in bus b of the network at pattern k of period p [%]
$\Delta_{g}^{\text{demand}}$	Interval of demand uncertainty in bus b of the network at pattern k of period p from the perspective of GENCO g [%]
Δt_{e}	Duration time of outage event e [hour]
$\Delta t_{p,\ m,\ e,\ b,\ s}$	Outage duration time in case of failure event e, at load sector s on affected load bus b, at load level m of period p [hour]
$\varpi_{\text{c}}^{\text{PD-DEP}}$	Predetermined IGDT critical cost of the PD-DEP problem
$\varpi_{\text{c},\ g}^{\text{PD-GEP}}$	Predetermined IGDT critical profit of the PD-GEP problem from the perspective of GENCO g
$\varpi_{\text{c}}^{\text{PD-TEP}}$	Predetermined IGDT critical cost of the PD-TEP problem

(continued)

$\varpi_{\mathrm{b}}^{\text{PD-DEP}}$	Predetermined IGDT base cost of the PD-DEP problem
$\varpi_{\mathrm{b},g}^{\text{PD-GEP}}$	Predetermined IGDT base profit of the PD-GEP problem from the perspective of GENCO g
$\varpi_{\mathrm{b}}^{\text{PD-TEP}}$	Predetermined IGDT base cost of the PD-TEP problem
$\varpi_{\mathrm{t}}^{\text{PD-DEP}}$	Predetermined IGDT target cost of the PD-DEP problem
$\varpi_{\mathrm{t},g}^{\text{PD-GEP}}$	Predetermined IGDT target profit of the PD-GEP problem from the perspective of GENCO g
$\varpi_{\mathrm{t}}^{\text{PD-TEP}}$	Predetermined IGDT target cost of the PD-TEP problem
$\sigma_{\mathrm{c}}^{\text{PD-DEP}}$	Critical cost deviation factor of the PD-DEP problem
$\sigma_{\mathrm{c},g}^{\text{PD-GEP}}$	Critical profit deviation factor of the PD-GEP problem from the perspective of GENCO g
$\sigma_{\mathrm{c}}^{\text{PD-TEP}}$	Critical cost deviation factor of the PD-TEP problem
$\sigma_{\mathrm{t}}^{\text{PD-DEP}}$	Target cost deviation factor of the PD-DEP problem
$\sigma_{\mathrm{t},g}^{\text{PD-GEP}}$	Target profit deviation factor of the PD-GEP problem from the perspective of GENCO g
$\sigma_{\mathrm{t}}^{\text{PD-TEP}}$	Target cost deviation factor of the PD-TEP problem
$\iota_{p,(s,r)}$	Number of new transmission lines added to corridor (s,r) in period p
$\tau_{p,(s,r)}$	Base susceptance of a new transmission line added to corridor (s,r) in period p [mho]
$\theta_{p,b',e}$	Phase angle of bus b' in case of outage event e in period p
$\theta_{p,b,e}$	Phase angle of bus b in case of outage event e in period p
$\theta_{p,k,s}$	Phase angle on the sending bus of corridor (s,r) at pattern k in period p
$\theta_{p,k,r}$	Phase angle on the receiving bus of corridor (s,r) at pattern k in period p
$\Pi_{p,k,b}^{\text{demand}}\left(\Delta^{\text{demand}}, p\rho_{p,k,b}\right)$	Envelope-bound IGDT model of uncertain demand on bus b of the network at pattern k of period p
$\Pi_{p,k,g,b}^{\text{demand}}\left(\Delta_g^{\text{demand}}, p\rho_{p,k,b}\right)$	Envelope-bound IGDT model of uncertain demand on bus b of the network at pattern k of period p from the perspective of GENCO g
$\Pi_{p,m,b}^{\text{demand}}\left(\Delta^{\text{demand}}, p\rho_{p,m,b}\right)$	Envelope-bound IGDT model of the uncertain demand on bus b of the network at load level m of period p
$\Pi_{p,k,b}^{\text{price}}\left(\Delta^{\text{price}}, p\lambda_{p,k,b}\right)$	Envelope-bound IGDT model of uncertain market price of bus b of the network at pattern k of period p
$\Pi_{p,m}^{\text{price}}\left(\Delta^{\text{price}}, p\lambda_{p,m}\right)$	Envelope-bound IGDT model of the uncertain market price at load level m of period p
$\Pi_{p,k,g,b}^{\text{price}}\left(\Delta_g^{\text{price}}, p\lambda_{p,k,b}\right)$	Envelope-bound IGDT model of uncertain market price of bus b of the network at pattern k of period p from the perspective of GENCO g
$\Upsilon^{\text{PD-DEP}}\left(x_{\text{PD-DEP}}, \varpi_{\mathrm{c}}^{\text{PD-DEP}}\right)$	IGDT robustness function for the PD-DEP problem
$\Upsilon_g^{\text{PD-GEP}}\left(x_{\text{PD-GEP}}, \varpi_{\mathrm{c},g}^{\text{PD-GEP}}\right)$	IGDT robustness function for the PD-GEP problem from the perspective of GENCO g
$\Upsilon^{\text{PD-TEP}}\left(x_{\text{PD-TEP}}, \varpi_{\mathrm{c}}^{\text{PD-TEP}}\right)$	IGDT robustness function for the PD-TEP problem
$\Gamma^{\text{PD-DEP}}\left(x_{\text{PD-DEP}}, \varpi_{\mathrm{t}}^{\text{PD-DEP}}\right)$	IGDT opportunity function for the PD-DEP problem
$\Gamma_g^{\text{PD-GEP}}\left(x_{\text{PD-GEP}}, \varpi_{\mathrm{t},g}^{\text{PD-GEP}}\right)$	IGDT opportunity function for the PD-GEP problem from the perspective of GENCO g
$\Gamma^{\text{PD-TEP}}\left(x_{\text{PD-TEP}}, \varpi_{\mathrm{t}}^{\text{PD-TEP}}\right)$	IGDT opportunity function for the PD-TEP problem

Appendix 3: Input Data

Table 6.95 Data of the transmission lines for the modified 46 buses in the south Brazilian network

Type	Corridor	Number of lines	R (p.u.)	X (p.u.)	Capacity (MW)	Type	Corridor	Number of lines	R (p.u.)	X (p.u.)	Capacity (MW)
I	01-07	1	0.001848	0.0616	270	I	44-45	1	0.005592	0.1864	200
I	01-02	2	0.003195	0.1065	270	I	19-32	1	0.000585	0.0195	1800
I	04-09	1	0.002772	0.0924	270	I	46-19	1	0.000666	0.0222	1800
I	05-09	1	0.003519	0.1173	270	I	46-16	1	0.000609	0.0203	1800
I	05-08	1	0.003396	0.1132	270	I	18-19	1	0.000375	0.0125	600
I	07-08	1	0.003069	0.1023	270	I	20-21	1	0.000375	0.0125	600
I	04-05	2	0.001698	0.0566	270	I	42-43	1	0.000375	0.0125	600
I	02-05	2	0.000972	0.0324	270	C	02-04	0	0.002646	0.0882	270
I	08-13	1	0.004044	0.1348	240	C	14-15	0	0.001122	0.0374	270
I	09-14	2	0.005268	0.1756	220	C	46-10	0	0.000243	0.0081	2000
I	12-14	2	0.002220	0.0740	270	C	04-11	0	0.006738	0.2246	240
I	14-18	2	0.004542	0.1514	240	C	05-11	0	0.002745	0.0915	270
I	13-18	1	0.005415	0.1805	220	C	46-06	0	0.000384	0.0128	2000
I	13-20	1	0.003219	0.1073	270	C	46-03	0	0.000609	0.0203	1800
I	18-20	1	0.005991	0.1997	200	C	16-28	0	0.000666	0.0222	1800
I	19-21	1	0.000834	0.0278	1500	C	16-32	0	0.000933	0.0311	1400
I	16-17	1	0.000234	0.0078	2000	C	17-32	0	0.000696	0.0232	1700
I	17-19	1	0.000183	0.0061	2000	C	19-25	0	0.000975	0.0325	1400
I	14-26	1	0.004842	0.1614	220	C	21-25	0	0.000522	0.0174	2000
I	14-22	1	0.002520	0.0840	270	C	25-32	0	0.000957	0.0319	1400
I	22-26	1	0.002370	0.0790	270	C	31-32	0	0.000138	0.0046	2000
I	20-23	2	0.002796	0.0932	270	C	28-31	0	0.000159	0.0053	2000

(continued)

Table 6.95 (continued)

Type	Corridor	Number of lines	R (p.u.)	X (p.u.)	Capacity (MW)	Type	Corridor	Number of lines	R (p.u.)	X (p.u.)	Capacity (MW)
I	23-24	2	0.002322	0.0774	270	C	28-30	0	0.000174	0.0058	2000
I	26-27	2	0.002496	0.0832	270	C	27-29	0	0.002994	0.0998	270
I	24-34	1	0.004941	0.1647	220	C	26-29	0	0.001623	0.0541	270
I	24-33	1	0.004344	0.1448	240	C	28-41	0	0.001017	0.0339	1300
I	33-34	1	0.003795	0.1265	270	C	28-43	0	0.001218	0.0406	1200
I	27-36	1	0.002745	0.0915	270	C	31-41	0	0.000834	0.0278	1500
I	27-38	2	0.00624	0.2080	200	C	32-41	0	0.000927	0.0309	1400
I	36-37	1	0.003171	0.1057	270	C	41-43	0	0.000417	0.0139	2000
I	34-35	2	0.001473	0.0491	270	C	40-45	0	0.006615	0.2205	180
I	35-38	1	0.005940	0.1980	200	C	15-16	0	0.000375	0.0125	600
I	37-39	1	0.000849	0.0283	270	C	46-11	0	0.000375	0.0125	600
I	37-40	1	0.003843	0.1281	270	C	24-25	0	0.000375	0.0125	600
I	37-42	1	0.006315	0.2105	200	C	29-30	0	0.000375	0.0125	600
I	39-42	3	0.006090	0.2030	200	C	40-41	0	0.000375	0.0125	600
I	40-42	1	0.002796	0.0932	270	C	02-03	0	0.000375	0.0125	600
I	38-42	3	0.002721	0.0907	270	C	05-06	0	0.000375	0.0125	600
I	32-43	1	0.000927	0.0309	1400	C	09-10	0	0.000375	0.0125	600
I	44-42	1	0.003618	0.1206	270						

*Transmission lines of the initial network (type = I)
*Candidate new transmission lines (type = C)

Table 6.96 Parameters of the GENCOs and candidate power generation units for the modified 46 buses in the south Brazilian network

Bus no.	Type	Owner	GENCOs' capacity		Generation cost function coefficients of GENCOs		
			ρ_g^{min} (MW)	ρ_g^{max} (MW)	α_g (\$/MW^{2}h)	β_g (\$/MWh)	γ_g(\$/h)
1	–	–	–	–	–	–	–
2	–	–	–	–	–	–	–
3	–	–	–	–	–	–	–
4	–	–	–	–	–	–	–
5	–	–	–	–	–	–	–
6	–	–	–	–	–	–	–
7	–	–	–	–	–	–	–
8	–	–	–	–	–	–	–
9	–	–	–	–	–	–	–
10	–	–	–	–	–	–	–
11	–	–	–	–	–	–	–
12	–	–	–	–	–	–	–
13	–	–	–	–	–	–	–
14	I	GENCO 1	0	1400.00	0.046	11.50	17.640
15	–	–	–	–	–	–	–
16	I	GENCO 2	0	3000.00	0.043	11.00	21.230
	C	Not specified	0	3000.00	0.043	11.00	21.230
17	I	GENCO 3	0	1250.00	0.057	16.00	28.310
18	–	–	–	–	–	–	–
19	I	GENCO 4	0	1800.00	0.031	10.00	16.450
20	–	–	–	–	–	–	–
21	–	–	–	–	–	–	–
22	–	–	–	–	–	–	–
23	–	–	–	–	–	–	–
24	–	–	–	–	–	–	–
25	–	–	–	–	–	–	–
26	–	–	–	–	–	–	–
27	I	GENCO 5	0	220.00	0.074	11.50	43.240
28	I	GENCO 6	0	1000.00	0.064	11.50	34.670
	C	Not specified	0	1000.00	0.064	11.50	34.670
29	–	–	–	–	–	–	–
30	–	–	–	–	–	–	–
31	I	GENCO 7	0	900.00	0.062	16.00	31.710
	C	Not specified	0	900.00	0.062	16.00	31.710

(continued)

Table 6.96 (continued)

Bus no.	Type	Owner	GENCOs' capacity ρ_g^{min} (MW)	ρ_g^{max} (MW)	Generation cost function coefficients of GENCOs α_g ($/MW^{2}h)	β_g ($/MWh)	γ_g($/h)
32	I	GENCO 8	0	700.00	0.067	16.00	36.630
	C	Not specified	0	700.00	0.067	16.00	36.630
33							
34	I	GENCO 9	0	900.00	0.070	17.00	41.110
35	–	–	–	–	–	–	–
36	–	–	–	–	–	–	–
37	I	GENCO 10	0	400.00	0.051	17.00	29.540
38	–	–	–	–	–	–	–
39	I	GENCO 11	0	700.00	0.073	16.00	43.720
40	–	–	–	–	–	–	–
41	–	–	–	–	–	–	–
42	–	–	–	–	–	–	–
43	–	–	–	–	–	–	–
44	–	–	–	–	–	–	–
45	–	–	–	–	–	–	–
46	I	GENCO 12	0	800.00	0.013	22.00	24.140
	C	Not specified	0	800.00	0.013	22.00	24.140

*Generation units of the initial network (type = I)
*Candidate new generation units (type = C)

Table 6.97 Parameters of the DISCOs and demand for the modified 46 buses in the south Brazilian network

Bus No.	DISCO No.	Demand (MW)	DISCOs' capacity ρ_d^{min} (MW)	ρ_d^{max} (MW)	Benefit function coefficients of DISCOs α_d ($/MW^{2}h)	β_d ($/MWh)	γ_d($/h)
1	–	–	–	–	–	–	–
2	DISCO 1	443.10	0	900.00	−0.054	15.00	21.240
3	–	–	–	–	–	–	–
4	DISCO 2	300.70	0	750.00	−0.013	14.00	24.370
5	DISCO 3	238.00	0	650.00	−0.031	15.00	27.250
6	–	–	–	–	–	–	–
7	–	–	–	–	–	–	–
8	DISCO 4	72.20	0	400.00	−0.052	17.00	31.640
9	–	–	–	–	–	–	–
10	–	–	–	–	–	–	–
11	–	–	–	–	–	–	–

(continued)

Table 6.97 (continued)

Bus No.	DISCO No.	Demand (MW)	DISCOs' capacity ρ_d^{min} (MW)	ρ_d^{max} (MW)	Benefit function coefficients of DISCOs α_d ($/MW^{2}h)	β_d ($/MWh)	γ_d($/h)
12	DISCO 5	511.90	0	1450.00	–0.034	20.00	31.870
13	DISCO 6	185.80	0	650.00	–0.037	18.00	22.420
14	–	–	–	–	–	–	–
15	–	–	–	–	–	–	–
16	–	–	–	–	–	–	–
17	–	–	–	–	–	–	–
18	–	–	–	–	–	–	–
19	–	–	–	–	–	–	–
20	DISCO 7	1091.20	0	2400.00	–0.041	19.00	29.350
21	–	–	–	–	–	–	–
22	DISCO 8	81.90	0	350.00	–0.026	17.00	28.740
23	DISCO 9	458.10	0	1100.00	–0.073	16.00	26.620
24	DISCO 10	478.20	0	1250.00	–0.055	20.00	26.120
25	–	–	–	–	–	–	–
26	DISCO 11	231.90	0	550.00	–0.059	21.00	27.350
27	–	–	–	–	–	–	–
28	–	–	–	–	–	–	–
29	–	–	–	–	–	–	–
30	–	–	–	–	–	–	–
31	–	–	–	–	–	–	–
32	–	–	–	–	–	–	–
33	DISCO 12	229.10	0	650.00	–0.015	23.00	37.410
34	–	–	–	–	–	–	–
35	DISCO 13	216.00	0	600.00	–0.061	17.00	23.670
36	DISCO 14	90.10	0	350.00	–0.057	19.00	27.210
37	–	–	–	–	–	–	–
38	DISCO 15	216.00	0	600.00	–0.071	22.00	31.050
39	–	–	–	–	–	–	–
40	DISCO 16	262.10	0	700.00	–0.025	25.00	37.450
41	–	–	–	–	–	–	–
42	DISCO 17	1607.90	0	2950.00	–0.040	25.00	32.015
43	–	–	–	–	–	–	–
44	DISCO 18	79.10	0	350.00	–0.071	23.00	39.010
45	DISCO 19	86.70	0	400.00	–0.059	24.00	41.310
46	–	–	–	–	–	–	–

Table 6.98 Parameters of atmospheric emissions and fuel consumption

GENCO No.	Type of atmospheric emission						Fuel consumption parameters	
	CO_2 atmospheric emission		SO_2 atmospheric emission		NO_x atmospheric emission			
	$E_{g_{eu}}^{CO_2}, E_{g_{iu}}^{CO_2}$ (ton/MWh)	$E_{p,g}^{CO_2,\max}$ (ton)	$E_{g_{eu}}^{SO_2}, E_{g_{iu}}^{SO_2}$ (ton/MWh)	$E_{p,g}^{SO_2,\max}$ (ton)	$E_{g_{eu}}^{NO_x}, E_{g_{iu}}^{NO_x}$ (ton/MWh)	$E_{p,g}^{NO_x,\max}$ (ton)	$F_{g_{eu}}, F_{g_{iu}}$ (m^3/MWh)	$F_{p,g}^{\max}$ (m^3)
GENCO 1	0.983	2780.169	0.0038	10.747	0.0031	8.767	0.383	1083.219
GENCO 2	0.983	4423.500	0.0038	17.100	0.0031	13.949	0.383	1723.350
GENCO 3	0.983	2322.337	0.0038	8.977	0.0031	7.323	0.383	904.837
GENCO 4	0.983	3693.622	0.0038	14.278	0.0031	11.648	0.383	1439.122
GENCO 5	0.983	486.585	0.0038	1.881	0.0031	1.534	0.383	189.585
GENCO 6	0.983	1769.400	0.0038	6.840	0.0031	5.580	0.383	689.400
GENCO 7	0.983	1548.225	0.0038	5.985	0.0031	4.882	0.383	603.225
GENCO 8	0.983	1105.875	0.0038	4.275	0.0031	3.487	0.383	430.875
GENCO 9	0.983	1654.389	0.0038	6.395	0.0031	5.217	0.383	644.589
GENCO 10	0.983	663.525	0.0038	2.565	0.0031	2.092	0.383	258.525
GENCO 11	0.983	1327.050	0.0038	5.130	0.0031	4.185	0.383	517.050
GENCO 12	0.983	1548.225	0.0038	5.985	0.0031	4.882	0.383	603.225

Table 6.99 Parameters of the available financial resources and the costs of the construction of the power generation units

GENCO No.	Parameters			
	$GIC^{f}_{g_{iu}}$ (M$)	$GIC_{g_{iu}}$ (M$/MW)	$GIC^{max}_{p,g}$ (M$)	GIC^{max}_{g} (M$)
GENCO 1	0.175	0.325	187.20	424.45
GENCO 2	0.210	0.390	210.25	538.30
GENCO 3	0.199	0.371	192.70	465.28
GENCO 4	0.199	0.371	201.34	502.65
GENCO 5	0.196	0.364	117.25	287.52
GENCO 6	0.175	0.325	157.56	352.35
GENCO 7	0.196	0.364	172.35	276.30
GENCO 8	0.168	0.312	132.24	260.75
GENCO 9	0.182	0.338	164.24	362.65
GENCO 10	0.175	0.325	124.23	385.32
GENCO 11	0.213	0.397	147.36	351.20
GENCO 12	0.203	0.377	137.56	315.30

Table 6.100 Data of the bidding strategy parameters for the GENCOs and the DISCOs

GENCO	GENCO bidding strategy parameters				DISCO	DISCO bidding strategy parameters			
	$\xi^{min}_{1,g}$	$\xi^{max}_{1,g}$	$\xi^{min}_{2,g}$	$\xi^{max}_{2,g}$		$\xi^{min}_{1,d}$	$\xi^{max}_{1,d}$	$\xi^{min}_{2,d}$	$\xi^{max}_{2,d}$
1	1	2.5	1	2.5	1	1	2.5	0	1
2	1	2.5	1	2.5	2	1	2.5	0	1
3	1	2.5	1	2.5	3	1	2.5	0	1
4	1	2.5	1	2.5	4	1	2.5	0	1
5	1	2.5	1	2.5	5	1	2.5	0	1
6	1	2.5	1	2.5	6	1	2.5	0	1
7	1	2.5	1	2.5	7	1	2.5	0	1
8	1	2.5	1	2.5	8	1	2.5	0	1
9	1	2.5	1	2.5	9	1	2.5	0	1
10	1	2.5	1	2.5	10	1	2.5	0	1
11	1	2.5	1	2.5	11	1	2.5	0	1
12	1	2.5	1	2.5	12	1	2.5	0	1
–	–	–	–	–	13	1	2.5	0	1
–	–	–	–	–	14	1	2.5	0	1
–	–	–	–	–	15	1	2.5	0	1
–	–	–	–	–	16	1	2.5	0	1
–	–	–	–	–	17	1	2.5	0	1
–	–	–	–	–	18	1	2.5	0	1
–	–	–	–	–	19	1	2.5	0	1

Table 6.101 Data of the weighting coefficients

No.	Weighting coefficient	Value
1	$W_{OF_1,g}^{PD\text{-}GEP}$	1
2	$W_{OF_2,g}^{PD\text{-}GEP}$	1
3	$W_{OF_3,g}^{PD\text{-}GEP}$	1

Table 6.102 Other required information

No.	Parameter	Value
1	P	10 (year)
2	K	22
3	Ir	5 (%)
4	ECOC	250 ($/MWh)
5	$IC_{p,g}^{max}$	$50\ (\%) \cdot \sum_{eu \in \Psi_g^{EU}} \rho_{(p-1),g_{eu}}$
6	RM_p^{min}	10 (%)
7	RM_p^{max}	40 (%)
8	$PERC_{p,g}^{max}$	70 (%)
9	ECO^{max}	3 (%)
10	Annual electrical power demand growth	4 (%)

Table 6.103 Parameter adjustments of the multi-objective SOSA

No.	Multi-objective SOSA parameters	Value	No.	Multi-objective SOSA parameters	Value
1	$BW_{g,p}^{min}$	$0.4; \forall\{g \in [1,3], p \in [1]\}$	17	$PAR_{g,p}^{min}$	$0.35; \forall\{g \in [2], p \in [3]\}$
2	$BW_{g,p}^{min}$	$0.35; \forall\{g \in [1,2,3], p \in [2]\}$	18	$PAR_{g,p}^{max}$	$2, \forall\{g \in [1,3], p \in [1]\}$
3	$BW_{g,p}^{min}$	$0.30; \forall\{g \in [2], p \in [1]\}$	19	$PAR_{g,p}^{max}$	$1.8; \forall\{g \in [1,2,3], p \in [2]\}$
4	$BW_{g,p}^{min}$	$0.25; \forall\{g \in [2], p \in [3]\}$	20	$PAR_{g,p}^{max}$	$1.6; \forall\{g \in [2], p \in [1]\}$
5	$BW_{g,p}^{max}$	$0.95; \forall\{g \in [1,3], p \in [1]\}$	21	$PAR_{g,p}^{max}$	$1.75; \forall\{g \in [2], p \in [3]\}$
6	$BW_{g,p}^{max}$	$0.9; \forall\{g \in [1,2,3], p \in [2]\}$	22	$PMCR_{g,p}$	$0.85; \forall\{g \in [1,3], p \in [1]\}$
7	$BW_{g,p}^{max}$	$0.85; \forall\{g \in [2], p \in [1]\}$	23	$PMCR_{g,p}$	$0.8; \forall\{g \in [1,2,3], p \in [2]\}$
8	$BW_{g,p}^{max}$	$0.8; \forall\{g \in [2], p \in [3]\}$	24	$PMCR_{g,p}$	$0.9; \forall\{g \in [2], p \in [1]\}$

(continued)

Table 6.103 (continued)

No.	Multi-objective SOSA parameters	Value	No.	Multi-objective SOSA parameters	Value
9	MNI-E	100	25	$PMCR_{g,p}$	0.95; $\forall\{g \in [2], p \in [3]\}$
10	MNI-SISS	200	26	$PMS_{g,p}$	220; $\forall\{g \in [1,3], p \in [1]\}$
11	MNI-GISSHMG	200	27	$PMS_{g,p}$	160; $\forall\{g \in [1,2,3], p \in [2]\}$
12	MNI-GISSIME	300	28	$PMS_{g,p}$	180; $\forall\{g \in [2], p \in [1]\}$
13	NHMG	3	29	$PMS_{g,p}$	200; $\forall\{g \in [2], p \in [3]\}$
14	$PAR_{g,p}^{\min}$	0.2; $\forall\{g \in [1,3], p \in [1]\}$	30	PN_g	2; $\forall\{g \in [1,3]\}$
15	$PAR_{g,p}^{\min}$	0.3; $\forall\{g \in [1,2,3], p \in [2]\}$	31	PN_g	3; $\forall\{g \in [2]\}$
16	$PAR_{g,p}^{\min}$	0.25; $\forall\{g \in [2], p \in [1]\}$	–	–	–

Table 6.104 Parameter adjustments of the multi-objective TMS-EMSA

No.	Multi-objective TMS-EMSA parameters	Value	No.	Multi-objective TMS-EMSA parameters	Value
1	$BW_p^{\min}$	0.4; $\forall\{p \in [1,2]\}$	12	$PAR_p^{\min}$	0.30; $\forall\{p \in [5,6,7]\}$
2	$BW_p^{\min}$	0.35; $\forall\{p \in [3,4]\}$	13	$PAR_p^{\max}$	2; $\forall\{p \in [1,2]\}$
3	$BW_p^{\min}$	0.30; $\forall\{p \in [5,6,7]\}$	14	$PAR_p^{\max}$	1.75; $\forall\{p \in [3,4]\}$
4	$BW_p^{\max}$	0.9; $\forall\{p \in [1,2]\}$	15	$PAR_p^{\max}$	1.9; $\forall\{p \in [5,6,7]\}$
5	$BW_p^{\max}$	0.85; $\forall\{p \in [3,4]\}$	16	$PMCR_p$	0.85; $\forall\{p \in [1,2]\}$
6	$BW_p^{\max}$	0.8; $\forall\{p \in [5,6,7]\}$	17	$PMCR_p$	0.8; $\forall\{p \in [3,4]\}$
7	MNI-E	100	18	$PMCR_p$	075; $\forall\{p \in [5,6,7]\}$
8	MNI-SISS	300	19	PMS_p	220; $\forall\{p \in [1,2]\}$
9	MNI-GISS	400	20	PMS_p	180; $\forall\{p \in [3,4]\}$
10	$PAR_p^{\min}$	0.2; $\forall\{p \in [1,2]\}$	21	PMS_p	160; $\forall\{p \in [5,6,7]\}$
11	$PAR_p^{\min}$	0.35; $\forall\{p \in [3,4]\}$	22	PN	7

Table 6.105 Parameter adjustments of the multi-objective continuous/discrete TMS-MSA

No.	Multi-objective continuous/discrete TMS-MSA parameters	Value
1	BW^{min}	0.4
2	BW^{max}	0.9
3	MNI-E	100
4	MNI-SISS	300
5	MNI-PGISS	400
6	PAR^{min}	0.2
7	PAR^{max}	2
8	PMCR	0.85
9	PMS	200
10	PN	7

Table 6.106 Parameter adjustments of the multi-objective SS-HSA

No.	Multi-objective SS-HSA parameters	Value
1	BW	0.75
2	HMCR	0.85
3	HMS	300
4	MNI-E	100
5	MNI-I	700
6	PAR	0.9

Table 6.107 Parameter adjustments of the multi-objective SS-IHSA

No.	Multi-objective SS-IHSA parameters	Value
1	BW^{min}	0.4
2	BW^{max}	0.9
3	HMCR	0.85
4	HMS	300
5	MNI-E	100
5	MNI-I	700
6	PAR^{min}	0.2
7	PAR^{max}	2

Table 6.108 Parameter adjustments of the NSGA-II

No.	NSGA-II parameters	Value
1	Crossover probability	0.95
2	Mutation probability	0.01
3	Population	400
4	MNI-E	100
4	MNI-I	700

Table 6.109 Line data for the modified IEEE 30-bus test system

No.	Type[a]	Corridor		Technical features of transmission lines				Data related to the outage events	
		From bus	To bus	R (p.u.)	X (p.u.)	Length (km)	Limit (MW)	Failure rate (occurrence/ year)	Repair rate (h)
1	I	1	2	0.004762	0.057500	490	130	0.22	60
2	I	1	3	0.016576	0.185200	490	130	0.18	60
3	I	2	4	0.009197	0.173700	320	65	0.24	16
4	I	3	4	0.003661	0.037900	480	130	0.18	60
5	I	2	5	0.016188	0.198300	510	130	0.14	60
6	I	2	6	0.006717	0.176300	490	65	0.30	16
7	I	4	6	0.003454	0.041400	367	90	0.20	35
8	I	5	7	0.008246	0.116000	422	70	0.22	35
9	I	6	7	0.003323	0.082000	508	130	0.12	60
10	I	6	8	0.001513	0.042000	166	32	0.54	11
11	I	6	9	0.014803	0.208000	290	65	0.32	16
12	I	6	10	0.021080	0.556000	438	32	0.52	11
13	I	9	11	0.009661	0.208000	376	65	0.28	16
14	I	9	10	0.008729	0.110000	510	65	0.28	16
15	I	4	12	0.013219	0.256000	148	65	0.26	16
16	I	12	13	0.005688	0.140000	330	65	0.28	16
17	I	12	14	0.007048	0.123100	266	32	0.52	11
18	I	12	15	0.011670	0.130400	320	32	0.52	11
19	I	12	16	0.014911	0.198700	143	32	0.46	11
20	I	14	15	0.017129	0.199700	440	16	0.33	10
21	I	16	17	0.006337	0.193200	155	16	0.33	10
22	I	15	18	0.021310	0.218500	482	16	0.36	10
23	I	18	19	0.012832	0.129200	361	16	0.34	10
24	I	19	20	0.005228	0.068000	289	32	0.46	11
25	I	10	20	0.008147	0.209000	156	32	0.38	11
26	I	10	17	0.004301	0.084500	270	32	0.56	11
27	I	10	21	0.002483	0.074900	310	32	0.56	11
28	I	10	22	0.010087	0.149900	202	32	0.56	11
29	I	21	22	0.001688	0.023600	412	32	0.52	11
30	I	15	23	0.016714	0.202000	148	16	0.34	10
31	I	22	24	0.008408	0.179000	227	16	0.32	10
32	I	23	24	0.012802	0.270000	401	16	0.42	10
33	I	24	25	0.028791	0.329200	389	16	0.42	10
34	I	25	26	0.027521	0.380000	402	16	0.38	10
35	I	25	27	0.011604	0.208700	670	16	0.38	10
36	I	28	27	0.037598	0.396000	383	65	0.26	16
37	I	27	29	0.039548	0.415300	510	16	0.48	10

(continued)

Table 6.109 (continued)

No.	Type[a]	Corridor		Technical features of transmission lines				Data related to the outage events	
		From bus	To bus	R (p.u.)	X (p.u.)	Length (km)	Limit (MW)	Failure rate (occurrence/ year)	Repair rate (h)
38	I	27	30	0.024698	0.602700	396	16	0.48	10
39	I	29	30	0.039905	0.453000	388	16	0.48	10
40	I	8	28	0.009973	0.200000	615	32	0.39	11
41	I	6	28	0.003581	0.059900	317	32	0.47	11

[a]Transmission lines of the initial network (type = I)

Table 6.110 Candidate line data for the modified IEEE 30-bus test system

No.	Type[a]	Corridor		Technical features of transmission lines				Data related to the outage events	
		From bus	To bus	R (p.u.)	X (p.u.)	Length (km)	Limit (MW)	Failure rate (occurrence/ year)	Repair rate (h)
1	I	1	2	0.004762	0.057500	490	130	0.22	60
2	I	1	3	0.016576	0.185200	490	130	0.18	60
3	I	2	6	0.006717	0.176300	490	65	0.30	16
4	I	4	12	0.013219	0.256000	148	65	0.26	16
5	I	6	8	0.001513	0.042000	166	32	0.54	11
6	I	12	16	0.014911	0.198700	143	32	0.46	11
7	I	28	27	0.037598	0.396000	383	65	0.26	16
8	I	10	22	0.010087	0.149900	202	32	0.56	11
9	I	27	30	0.024698	0.602700	396	16	0.48	10
10	C	2	8	0.030690	0.846000	230	98	0.41	40
11	C	9	12	0.014948	0.346000	140	55	0.32	20
12	C	10	11	0.014980	0.212000	120	130	0.22	35
13	C	10	12	0.038582	0.744000	202	145	0.48	10
14	C	13	14	0.010374	0.103800	210	59	0.54	11
15	C	15	16	0.013990	0.144400	380	68	0.18	60
16	C	19	24	0.021941	0.512000	140	200	0.22	60
17	C	23	25	0.067057	0.671000	180	340	0.25	35
18	C	18	16	0.005213	0.139000	380	88	0.47	11

[a]Transmission lines of the initial network (type = I) *Candidate new transmission lines (type = C)

Table 6.111 Parameters associated with GENCOs of the modified IEEE 30-bus test system

		GENCOs' capacity		Generation cost function coefficients of GENCOs			Data related to the outage events	
Bus No.	GENCO No.	ρ_g^{min} (MW)	ρ_g^{max} (MW)	α_g (\$/MW^{2}h)	β_g (\$/MWh)	γ_g(\$/h)	Failure rate (occurrence/ year)	Repair rate (h)
1	GENCO 1	0	200	0.046	11.5	17.640	9.13	40
2	GENCO 2	0	200	0.043	11.0	21.230	9.13	40
3	GENCO 3	0	50	0.031	10.0	28.310	4.42	20
4	–	–	–	–	–	–	–	–
5	–	–	–	–	–	–	–	–
6	–	–	–	–	–	–	–	–
7	–	–	–	–	–	–	–	–
8	GENCO 4	0	130	0.07	11.5	16.450	2.98	60
9	–	–	–	–	–	–	–	–
10	–	–	–	–	–	–	–	–
11	GENCO 5	0	120	0.064	11.5	23.240	9.13	40
12	–	–	–	–	–	–	–	–
13	GENCO 6	0	120	0.062	16.0	32.430	7.30	50
14	GENCO 7	0	160	0.067	16.0	37.630	7.62	100
15	GENCO 8		100	0.070	17.0	41.540	4.42	20
16	–	–	–	–	–	–	–	–
17	–	–	–	–	–	–	–	–
18	GENCO 9	0	100	0.051	17.0	38.240	7.30	50
19	–	–	–	–	–	–	–	–
20	–	–	–	–	–	–	–	–
21	–	–	–	–	–	–	–	–
22	GENCO 10	0	150	0.073	16.0	27.140	7.30	50
23	GENCO 11	0	150	0.057	16.0	29.370	7.30	50
24	–	–	–	–	–	–	–	–
25	–	–	–	–	–	–	–	–
26	–	–	–	–	–	–	–	–
27	GENCO 12	0	120	0.013	22.0	33.670	2.98	60
28	–	–	–	–	–	–	–	–
29	–	–	–	–	–	–	–	–
30	–	–	–	–	–	–	–	–

Table 6.112 Parameters associated with the electrical power demand and DISCOs of the modified IEEE 30-bus test system

Bus No.	DISCO No.	Electrical power demand (MW)	DISCOs' capacity		Benefit function coefficients of DISCOs		
			ρ_d^{min} (MW)	ρ_d^{max} (MW)	α_d (\$/MW^{2}h)	β_d (\$/MWh)	γ_d(\$/h)
1	–	–	–	–	–	–	–
2	–	–	–	–	–	–	–
3	DISCO 1	160	0	350	–0.054	15.00	21.240
4	–	–	–	–	–	–	–
5	DISCO 2	85	0	200	–0.013	14.00	24.370
6	–	–	–	–	–	–	–
7	DISCO 3	50	0	120	–0.031	15.00	27.250
8	DISCO 4	25	0	90	–0.052	17.00	31.640
9	–	–	–	–	–	–	–
10	DISCO 5	190	0	420	–0.034	20.00	31.870
11	–	–	–	–	–	–	–
12	DISCO 6	50	0	130	–0.037	18.00	22.420
13	DISCO 7	20	0	80	–0.041	19.00	29.350
14	DISCO 8	50	0	140	–0.026	17.00	28.740
15	DISCO 9	50	0	120	–0.073	16.00	26.620
16	DISCO 10	9	0	50	–0.055	20.00	26.120
17	–	–	–	–	–	–	–
18	DISCO 11	35	0	100	–0.059	21.00	27.350
19	DISCO 12	17	0	70	–0.015	23.00	37.410
20	DISCO 13	60	0	150	–0.061	17.00	23.670
21	DISCO 14	34	0	100	–0.057	19.00	27.210
22	–	–	–	–	–	–	–
23	DISCO 15	85	0	180	–0.071	22.00	31.050
24	DISCO 16	9	0	50	–0.025	25.00	37.450
25	DISCO 17	9	0	50	–0.040	25.00	32.015
26	–	–	–	–	–	–	–
27	DISCO 18	27	0	80	–0.071	23.00	39.010
28	–	–	–	–	–	–	–
29	DISCO 19	35	0	90	–0.059	24.00	41.310
30	–	–	–	–	–	–	–

Table 6.113 Data of the bidding strategy parameters of GENCOs and DISCOs in the modified IEEE 30-bus test system

GENCO	GENCOs' bidding strategy parameters				DISCO	DISCOs' bidding strategy parameters			
	$\xi_{1,g}^{\min}$	$\xi_{1,g}^{\max}$	$\xi_{2,g}^{\min}$	$\xi_{2,g}^{\max}$		$\xi_{1,d}^{\min}$	$\xi_{1,d}^{\max}$	$\xi_{2,d}^{\min}$	$\xi_{2,d}^{\max}$
1	1	2.5	1	2.5	1	1	2.5	0	1
2	1	2.5	1	2.5	2	1	2.5	0	1
3	1	2.5	1	2.5	3	1	2.5	0	1
4	1	2.5	1	2.5	4	1	2.5	0	1
5	1	2.5	1	2.5	5	1	2.5	0	1
6	1	2.5	1	2.5	6	1	2.5	0	1
7	1	2.5	1	2.5	7	1	2.5	0	1
8	1	2.5	1	2.5	8	1	2.5	0	1
9	1	2.5	1	2.5	9	1	2.5	0	1
10	1	2.5	1	2.5	10	1	2.5	0	1
11	1	2.5	1	2.5	11	1	2.5	0	1
12	1	2.5	1	2.5	12	1	2.5	0	1
–	–	–	–	–	13	1	2.5	0	1
–	–	–	–	–	14	1	2.5	0	1
–	–	–	–	–	15	1	2.5	0	1
–	–	–	–	–	16	1	2.5	0	1
–	–	–	–	–	17	1	2.5	0	1
–	–	–	–	–	18	1	2.5	0	1
–	–	–	–	–	19	1	2.5	0	1

Table 6.114 Data of the weighting coefficients related to the proposed PD-TEP problem in the modified IEEE 30-bus test system

No.	Weighting coefficient	Value
1	$W_{OF_1}^{PD\text{-}TEP}$	1
2	$W_{OF_2}^{PD\text{-}TEP}$	1
3	$W_{OF_3}^{PD\text{-}TEP}$	1

Table 6.115 SCDF data for different sectors of each electrical load in each bus of the modified IEEE 30-bus test system

	Costumer sector								
	Residential [outage duration (h) and outage costs ($/kW)]			Commercial [outage duration (h) and outage costs ($/kW)]			Industrial [outage duration (h) and outage costs ($/kW)]		
Bus no.	1 h	4 h	8 h	1 h	4 h	8 h	1 h	4 h	8 h
1	–	–	–	–	–	–	–	–	–
2	–	–	–	–	–	–	–	–	–
3	0.16562	1.6121	5.8103	25.7958	91.4771	153.9140	10.1739	33.2731	52.0299
4	–	–	–	–	–	–	–	–	–
5	0.18381	1.7120	4.9835	31.1583	104.0825	181.9411	10.1918	31.0844	51.8122
6	–	–	–	–	–	–	–	–	–
7	0.21685	2.4750	4.9761	28.8876	85.7297	176.4380	8.7408	31.2133	59.4031
8	0.22608	1.8749	3.4565	30.1829	114.1715	182.2230	10.8322	28.0048	54.8090
9	–	–	–	–	–	–	–	–	–
10	0.21767	2.6128	4.1306	36.2295	117.9442	203.7876	8.1669	31.3648	52.3087
11	–	–	–	–	–	–	–	–	–
12	0.13863	2.1811	4.3052	26.6466	126.8157	178.0384	10.7630	27.4764	53.3631
13	0.24038	1.6775	5.5037	29.4872	126.2858	217.2180	9.4946	25.4032	49.0371
14	0.19760	2.5162	4.057	33.4842	104.0042	194.1219	9.3753	31.0279	58.0569
15	0.19177	2.4409	3.9722	31.3608	101.6126	195.9111	10.4432	27.2429	50.5290
16	0.14884	1.9581	4.8066	30.1149	99.6405	213.8384	10.2044	31.9861	51.0511
17	–	–	–	–	–	–	–	–	–
18	0.20272	1.8543	3.9321	30.1986	118.5372	184.7880	8.2156	30.3447	58.7322
19	0.20016	1.8125	3.5249	28.9849	120.9763	195.8289	10.6758	26.4715	47.8279
20	0.16559	1.9681	5.386	28.8974	103.5784	214.1976	9.8731	26.9016	46.9196
21	0.15827	2.7110	4.7237	37.6805	107.9055	220.6991	9.0878	29.6376	44.8317
22	–	–	–	–	–	–	–	–	–

(continued)

Table 6.115 (continued)

	Costumer sector								
	Residential [outage duration (h) and outage costs ($/kW)]			Commercial [outage duration (h) and outage costs ($/kW)]			Industrial [outage duration (h) and outage costs ($/kW)]		
Bus no.	1 h	4 h	8 h	1 h	4 h	8 h	1 h	4 h	8 h
23	0.21936	1.5729	4.6725	27.3460	92.1261	209.4245	8.6024	29.1662	45.4151
24	0.23230	2.6712	5.5677	35.6705	119.0799	194.2461	9.2774	27.1946	53.5116
25	0.22858	2.2974	5.6302	36.9250	101.0121	163.9637	8.2630	31.8740	59.2444
26	–	–	–	–	–	–	–	–	–
27	0.15451	1.7938	4.2401	36.4621	93.3547	152.4347	9.0189	25.4327	56.8353
28	–	–	–	–	–	–	–	–	–
29	0.21487	2.3851	4.0186	26.6355	106.7916	212.5862	8.2323	27.3755	51.6634
30	–	–	–	–	–	–	–	–	–

Table 6.116 Other required information related to the proposed PD-TEP problem in the modified IEEE 30-bus test system

No.	Parameter	Value
1	P	10 (year)
2	K	22
3	Ir	10 (%)
4	CCL^{b}	1200$/MW km
5	$\iota_{(s,r)}^{\max}$	4
6	$c\rho_{p}^{\max}$	30 (%)
7	$c\rho_{p,b}^{\max}$	30 (%)
8	$c\rho_{p,b,s}^{\max}$	30 (%)
9	$TIC_{p}^{\max}$	$90 M
10	$TIC^{\max}$	$750 M
11	$w_{p,\ b,\ s}$	1
12	Annual electrical power demand growth	5 (%)
13	Annual price growth	4 (%)

Table 6.117 Line data for the modified Iranian 400 kV transmission network

	Type*		Corridor		Technical features			Data related to the outage events	
Line No.	Condition	Number of bundles	From bus	To bus	R (p.u.)	X (p.u.)	Length (km)	Failure rate (occurrence/ year)	Repair rate (h)
1	I	3	1	2	0.001182	0.021	118	0.24	16
2	C	3	1	49	0.001912	0.032	165	0.26	15
3	I	3	2	3	0.000739	0.024	72	0.34	10
4	I	2	2	5	0.000133	0.002	215	0.12	40
5	C	3	2	49	0.002506	0.028	60	0.34	10
6	I	3	3	4	0.000691	0.015	90	0.33	11
7	C	2	3	52	0.003489	0.048	250	0.27	35
8	I	3	4	5	0.002375	0.037	110	0.24	16
9	C	2	4	50	0.002848	0.041	220	0.27	35
10	C	2	4	52	0.000168	0.002	200	0.36	22
11	I	3	5	6	0.000871	0.028	145	0.32	18
12	I	2	5	7	0.001693	0.035	175	0.42	25
13	I	2	5	28	0.001029	0.021	185	0.42	25
14	C	2	5	25	0.003251	0.042	200	0.36	22
15	I	3	6	7	0.001908	0.021	130	0.38	20
16	C	2	6	50	0.001516	0.027	170	0.42	25
17	C	3	6	8	0.001996	0.029	100	0.24	16
18	I	3	7	8	0.002586	0.032	140	0.38	20
19	I	2	7	10	0.004074	0.048	170	0.42	25

(continued)

Table 6.117 (continued)

Line No.	Type* Condition	Number of bundles	Corridor From bus	To bus	Technical features R (p.u.)	X (p.u.)	Length (km)	Data related to the outage events Failure rate (occurrence/ year)	Repair rate (h)
20	C	2	7	12	0.001248	0.025	250	0.12	40
21	I	2	7	23	0.003504	0.038	330	0.18	32
22	I	2	7	24	0.002801	0.036	210	0.27	35
23	I	2	7	25	0.001237	0.038	220	0.27	35
24	I	3	8	9	0.004373	0.061	144	0.32	18
25	C	3	9	10	0.001701	0.032	100	0.24	16
26	I	3	9	11	0.006914	0.072	91	0.33	11
27	I	2	10	11	0.002942	0.069	221	0.27	35
28	I	3	10	12	0.003132	0.051	130	0.38	20
29	I	3	11	14	0.006344	0.082	160	0.26	15
30	I	3	12	13	0.002277	0.069	156	0.26	15
31	C	3	13	14	0.003267	0.036	80	0.33	11
32	I	2	13	17	0.007533	0.092	275	0.18	32
33	I	2	14	15	0.006223	0.078	260	0.12	40
34	I	2	15	16	0.003375	0.034	174	0.42	25
35	I	3	15	17	0.003201	0.051	150	0.32	18
36	I	3	16	17	0.000568	0.014	160	0.26	15
37	I	2	16	18	0.003237	0.045	195	0.36	22
38	C	2	17	18	0.005182	0.071	200	0.36	22
39	I	2	17	19	0.003183	0.041	315	0.12	40
40	I	2	18	19	0.001698	0.032	342	0.18	32
41	I	2	19	20	0.001714	0.048	231	0.27	35
42	I	2	19	47	0.001070	0.012	510	0.10	50
43	I	2	20	21	0.002470	0.038	270	0.27	35
44	C	2	20	45	0.000984	0.025	320	0.12	40
45	I	2	21	22	0.003255	0.038	312	0.42	25
46	I	2	21	34	0.001940	0.057	270	0.27	35
47	C	2	21	35	0.002321	0.061	200	0.36	22
48	I	2	21	44	0.002199	0.070	296	0.25	28
49	I	2	21	45	0.001854	0.048	300	0.27	35
50	I	3	22	23	0.002224	0.054	134	0.38	20
51	I	3	22	34	0.001769	0.026	85	0.33	11
52	I	3	23	24	0.003695	0.050	90	0.33	11
53	I	2	23	29	0.002087	0.051	280	0.25	28
54	I	2	23	32	0.002490	0.034	272	0.25	28
55	I	3	23	34	0.000683	0.012	142	0.32	18
56	I	3	24	25	0.002903	0.046	100	0.24	16
57	I	2	24	26	0.001151	0.025	210	0.36	22

(continued)

Table 6.117 (continued)

	Type*		Corridor		Technical features			Data related to the outage events	
Line No.	Condition	Number of bundles	From bus	To bus	R (p.u.)	X (p.u.)	Length (km)	Failure rate (occurrence/ year)	Repair rate (h)
58	C	2	24	29	0.001156	0.038	180	0.42	25
59	I	2	24	32	0.003968	0.047	246	0.27	35
60	C	2	25	28	0.002208	0.028	200	0.36	22
61	I	3	26	27	0.002269	0.041	135	0.38	20
62	I	3	26	29	0.000965	0.021	160	0.26	15
63	I	3	27	28	0.002236	0.035	165	0.26	15
64	C	3	27	52	0.001753	0.031	150	0.32	18
65	I	3	29	30	0.005165	0.052	85	0.33	11
66	I	3	29	32	0.001513	0.021	75	0.34	10
67	I	3	30	31	0.003073	0.037	100	0.24	16
68	I	3	30	32	0.002076	0.029	135	0.38	20
69	I	3	31	32	0.001110	0.032	147	0.32	18
70	I	2	31	36	0.001592	0.038	230	0.12	40
71	I	2	31	37	0.002247	0.047	275	0.27	35
72	I	2	32	33	0.002199	0.038	250	0.12	40
73	I	3	32	34	0.005788	0.059	120	0.24	16
74	C	3	33	34	0.002576	0.047	180	0.42	25
75	C	3	33	35	0.002262	0.058	150	0.26	15
76	I	3	33	36	0.002389	0.047	110	0.24	16
77	I	2	33	37	0.002159	0.063	220	0.12	40
78	I	2	34	35	0.001768	0.048	200	0.36	22
79	I	2	35	40	0.002827	0.083	300	0.16	45
80	C	2	35	36	0.002449	0.072	200	0.36	22
81	I	3	36	37	0.003423	0.055	125	0.38	20
82	I	3	36	38	0.004660	0.057	100	0.24	16
83	I	3	37	38	0.000709	0.012	145	0.32	18
84	I	3	37	39	0.004949	0.068	175	0.42	25
85	I	3	38	40	0.001836	0.045	130	0.38	20
86	C	2	39	40	0.004007	0.061	250	0.12	40
87	I	3	39	41	0.001340	0.036	67	0.34	10
88	C	2	40	41	0.003241	0.054	200	0.36	22
89	I	2	40	44	0.002767	0.065	235	0.25	28
90	I	3	41	42	0.007039	0.073	170	0.42	25
91	I	2	42	43	0.002544	0.042	215	0.27	35
92	I	2	43	44	0.004612	0.058	312	0.16	45
93	I	2	43	48	0.001755	0.043	230	0.25	28
94	C	2	43	51	0.003969	0.056	390	0.16	45
95	I	3	44	45	0.004303	0.054	160	0.26	15

(continued)

Table 6.117 (continued)

Line No.	Type*		Corridor		Technical features			Data related to the outage events	
	Condition	Number of bundles	From bus	To bus	R (p.u.)	X (p.u.)	Length (km)	Failure rate (occurrence/ year)	Repair rate (h)
96	I	2	45	46	0.004777	0.056	200	0.36	22
97	I	2	46	47	0.004944	0.067	210	0.36	22
98	I	3	46	48	0.004406	0.073	150	0.32	18
99	C	2	47	48	0.003485	0.059	540	0.14	55
100	C	2	47	51	0.004656	0.058	600	0.14	55
101	C	2	48	51	0.004578	0.062	400	0.12	40
102	C	3	49	50	0.001805	0.051	107	0.24	16

*Transmission lines of the initial network (type = I) *Candidate new transmission lines (type = C)

Table 6.118 Parameters associated with GENCOs of the modified Iranian 400 kV transmission network

Bus No.	GENCO No.	GENCOs' capacity		Generation cost function coefficients of GENCOs			Data related to the outage events	
		$\rho_g^{\min}$ (MW)	$\rho_g^{\max}$ (MW)	α_g (\$/MW^{2}h)	β_g (\$/MWh)	γ_g. (\$/h)	Failure rate (occurrence/ year)	Repair rate (h)
1	GENCO 1	0	300	0.037	12.4	24.270	4.42	20
2	GENCO 2	0	400	0.051	17.5	32.260	4.42	20
3	GENCO 3	0	350	0.047	14.6	31.430	4.42	20
4	GENCO 4	0	160	0.042	22.5	22.840	2.98	60
5	–	–	–	–	–	–	–	–
6	GENCO 5	0	1500	0.052	18.5	24.740	7.30	50
7	GENCO 6	0	7100	0.031	10.0	28.310	7.96	150
8	–	–	–	–	–	–	–	–
9	–	–	–	–	–	–	–	–
10	–	–	–	–	–	–	–	–
11	GENCO 7	0	2100	0.067	26.5	37.270	9.22	50
12	–	–	–	–	–	–	–	–
13	–	–	–	–	–	–	–	–
14	–	–	–	–	–	–	–	–
15	GENCO 8	0	700	0.047	14.5	22.150	4.47	40
16	GENCO 9	0	800	0.054	28.2	27.280	4.47	40
17	GENCO 10	0	650	0.066	18.5	31.750	4.47	40

(continued)

Table 6.118 (continued)

Bus No.	GENCO No.	GENCOs' capacity ρ_g^{min} (MW)	ρ_g^{max} (MW)	Generation cost function coefficients of GENCOs α_g ($/MW^{2}h)	β_g ($/MWh)	γ_g. ($/h)	Data related to the outage events Failure rate (occurrence/ year)	Repair rate (h)
18	–	–	–	–	–	–	–	–
19	GENCO 11	0	600	0.072	21.5	33.240	4.47	40
20	–	–	–	–	–	–	–	–
21	GENCO 12	0	1000	0.058	14.5	27.350	7.30	50
22	GENCO 13	0	1800	0.070	17.0	41.540	9.13	40
23	–	–	–	–	–	–	–	–
24	GENCO 14	0	1200	0.047	21.2	29.540	7.30	50
25	–	–	–	–	–	–	–	–
26	–	–	–	–	–	–	–	–
27	GENCO 15	0	600	0.013	22.0	33.670	4.47	40
28	GENCO 16	0	1000	0.066	14.5	23.250	7.30	50
29	GENCO 17	0	1000	0.070	11.5	16.450	7.30	50
30	GENCO 18	0	4700	0.051	17.0	38.240	7.62	100
31	–	–	–	–	–	–	–	–
32	GENCO 19	0	2000	0.062	16.0	32.430	9.22	50
33	–	–	–	–	–	–	–	–
34	GENCO 20	0	1200	0.057	16.0	29.370	7.30	50
35	–	–	–	–	–	–	–	–
36	–	–	–	–	–	–	–	–
37	GENCO 21	0	100	0.067	26.5	41.720	2.98	60
38	GENCO 22	0	1000	0.046	11.5	17.640	7.30	50
39	–	–	–	–	–	–	–	–
40	GENCO 23	0	500	0.054	19.4	27.320	4.47	40
41	GENCO 24	0	500	0.052	22.6	31.460	4.47	40
42	–	–	–	–	–	–	–	–
43	GENCO 25	0	1900	0.043	11.0	21.230	9.22	50
44	–	–	–	–	–	–	–	–
45	GENCO 26	0	820	0.064	11.5	23.240	4.47	40
46	–	–	–	–	–	–	–	–
47	GENCO 27	0	250	0.067	16.0	37.630	2.98	60
48	GENCO 28	0	100	0.073	16.0	27.140	2.98	60
49	–	–	–	–	–	–	–	–
50	–	–	–	–	–	–	–	–
51	–	–	–	–	–	–	–	–
52	–	–	–	–	–	–	–	–

Table 6.119 Parameters associated with the electrical power demand and DISCOs of the modified Iranian 400 kV transmission network

Bus No.	DISCO No.	Power demand (MW)	DISCOs' capacity		Benefit function coefficients of DISCOs		
			ρ_d^{min} (MW)	ρ_d^{max} (MW)	α_d ($/MW^{2}h)	β_d ($/MWh)	γ_d($/h)
1	DISCO 1	670	0	1450	−0.042	21.00	28.320
2	DISCO 2	400	0	950	−0.038	16.00	23.270
3	DISCO 3	345	0	650	−0.036	32.00	39.640
4	DISCO 4	180	0	420	−0.034	27.00	32.380
5	DISCO 5	990	0	2750	−0.013	14.00	24.370
6	DISCO 6	900	0	1700	−0.046	13.00	20.120
7	DISCO 7	6000	0	8700	−0.041	19.00	29.350
8	DISCO 8	275	0	520	−0.064	23.00	34.150
9	DISCO 9	690	0	1750	−0.052	17.00	31.640
10	DISCO 10	500	0	1150	−0.072	24.00	41.830
11	DISCO 11	550	0	1240	−0.068	21.00	34.540
12	DISCO 12	140	0	310	−0.048	22.00	29.370
13	DISCO 13	200	0	460	−0.027	16.00	27.360
14	DISCO 14	435	0	910	−0.033	19.00	26.270
15	DISCO 15	415	0	840	−0.037	24.00	31.870
16	DISCO 16	850	0	1950	−0.023	14.00	24.370
17	DISCO 17	570	0	1350	−0.026	13.00	27.430
18	DISCO 18	440	0	1050	−0.031	17.00	29.740
19	DISCO 19	375	0	860	−0.015	23.00	37.410
20	DISCO 20	100	0	230	−0.038	24.00	32.680
21	DISCO 21	460	0	950	−0.047	28.00	41.710
22	DISCO 22	1000	0	2700	−0.057	19.00	27.210
23	DISCO 23	500	0	1350	−0.034	21.00	40.120
24	DISCO 24	700	0	1840	−0.037	18.00	22.420
25	DISCO 25	400	0	940	−0.031	17.00	28.560
26	DISCO 26	500	0	1240	−0.040	25.00	32.015
27	DISCO 27	800	0	2300	−0.054	15.00	21.240
28	DISCO 28	600	0	1750	−0.061	17.00	23.670
29	DISCO 29	1050	0	2350	−0.038	22.00	31.350
30	DISCO 30	1700	0	3800	−0.055	20.00	26.120
31	DISCO 31	1000	0	2650	−0.026	17.00	28.740
32	DISCO 32	800	0	2100	−0.071	22.00	31.050
33	DISCO 33	560	0	1400	−0.034	20.00	31.870
34	DISCO 34	1700	0	3400	−0.059	21.00	27.350
35	DISCO 35	340	0	940	−0.025	25.00	37.450
36	DISCO 36	190	0	430	−0.054	21.00	34.110
37	DISCO 37	285	0	520	−0.051	19.00	37.680
38	DISCO 38	760	0	1650	−0.033	16.00	27.610
39	DISCO 39	465	0	1150	−0.071	23.00	39.010

(continued)

Table 6.119 (continued)

Bus No.	DISCO No.	Power demand (MW)	DISCOs' capacity		Benefit function coefficients of DISCOs		
			ρ_d^{min} (MW)	ρ_d^{max} (MW)	α_d (\$/MW^{2}h)	β_d (\$/MWh)	γ_d(\$/h)
40	DISCO 40	435	0	920	−0.042	21.00	30.150
41	DISCO 41	380	0	750	−0.059	24.00	41.310
42	DISCO 42	500	0	1250	−0.063	26.00	40.390
43	DISCO 43	1000	0	2450	−0.031	15.00	27.250
44	DISCO 44	400	0	1350	−0.073	16.00	26.620
45	DISCO 45	580	0	1240	−0.042	27.00	36.140
46	DISCO 46	200	0	430	−0.031	14.00	24.620
47	DISCO 47	400	0	1050	−0.071	23.00	39.010
48	DISCO 48	140	0	320	−0.046	24.00	33.470
49	–	–	–	–	–	–	–
50	–	–	–	–	–	–	–
51	–	–	–	–	–	–	–
52	–	–	–	–	–	–	–

Table 6.120 Data of the bidding strategy parameters of GENCOs and DISCOs in the modified Iranian 400 kV transmission network

GENCO	GENCOs' bidding strategy parameters				DISCO	DISCOs' bidding strategy parameters			
	$\xi_{1,g}^{min}$	$\xi_{1,g}^{max}$	$\xi_{2,g}^{min}$	$\xi_{2,g}^{max}$		$\xi_{1,d}^{min}$	$\xi_{1,d}^{max}$	$\xi_{2,d}^{min}$	$\xi_{2,d}^{max}$
1	1	2.5	1	2.5	1	1	2.5	0	1
2	1	2.5	1	2.5	2	1	2.5	0	1
3	1	2.5	1	2.5	3	1	2.5	0	1
4	1	2.5	1	2.5	4	1	2.5	0	1
5	1	2.5	1	2.5	5	1	2.5	0	1
6	1	2.5	1	2.5	6	1	2.5	0	1
7	1	2.5	1	2.5	7	1	2.5	0	1
8	1	2.5	1	2.5	8	1	2.5	0	1
9	1	2.5	1	2.5	9	1	2.5	0	1
10	1	2.5	1	2.5	10	1	2.5	0	1
11	1	2.5	1	2.5	11	1	2.5	0	1
12	1	2.5	1	2.5	12	1	2.5	0	1
13	1	2.5	1	2.5	13	1	2.5	0	1
14	1	2.5	1	2.5	14	1	2.5	0	1
15	1	2.5	1	2.5	15	1	2.5	0	1
16	1	2.5	1	2.5	16	1	2.5	0	1
17	1	2.5	1	2.5	17	1	2.5	0	1
18	1	2.5	1	2.5	18	1	2.5	0	1

(continued)

Table 6.120 (continued)

GENCO	GENCOs' bidding strategy parameters				DISCO	DISCOs' bidding strategy parameters			
	$\xi_{1,g}^{\min}$	$\xi_{1,g}^{\max}$	$\xi_{2,g}^{\min}$	$\xi_{2,g}^{\max}$		$\xi_{1,d}^{\min}$	$\xi_{1,d}^{\max}$	$\xi_{2,d}^{\min}$	$\xi_{2,d}^{\max}$
19	1	2.5	1	2.5	19	1	2.5	0	1
20	1	2.5	1	2.5	20	1	2.5	0	1
21	1	2.5	1	2.5	21	1	2.5	0	1
22	1	2.5	1	2.5	22	1	2.5	0	1
23	1	2.5	1	2.5	23	1	2.5	0	1
24	1	2.5	1	2.5	24	1	2.5	0	1
25	1	2.5	1	2.5	25	1	2.5	0	1
26	1	2.5	1	2.5	26	1	2.5	0	1
27	1	2.5	1	2.5	27	1	2.5	0	1
28	1	2.5	1	2.5	28	1	2.5	0	1
–	–	–	–	–	29	1	2.5	0	1
–	–	–	–	–	30	1	2.5	0	1
–	–	–	–	–	31	1	2.5	0	1
–	–	–	–	–	32	1	2.5	0	1
–	–	–	–	–	33	1	2.5	0	1
–	–	–	–	–	34	1	2.5	0	1
–	–	–	–	–	35	1	2.5	0	1
–	–	–	–	–	36	1	2.5	0	1
–	–	–	–	–	37	1	2.5	0	1
–	–	–	–	–	38	1	2.5	0	1
–	–	–	–	–	39	1	2.5	0	1
–	–	–	–	–	40	1	2.5	0	1
–	–	–	–	–	41	1	2.5	0	1
–	–	–	–	–	42	1	2.5	0	1
–	–	–	–	–	43	1	2.5	0	1
–	–	–	–	–	44	1	2.5	0	1
–	–	–	–	–	45	1	2.5	0	1
–	–	–	–	–	46	1	2.5	0	1
–	–	–	–	–	47	1	2.5	0	1
–	–	–	–	–	48	1	2.5	0	1

Table 6.121 Data of the weighting coefficients related to the proposed PD-TEP problem in the modified Iranian 400 kV transmission network

No.	Weighting coefficient	Value
1	$W_{OF_1}^{PD-TEP}$	1
2	$W_{OF_2}^{PD-TEP}$	1
3	$W_{OF_3}^{PD-TEP}$	1

Table 6.122 SCDF data for different sectors of each electrical load on each bus of the modified Iranian 400 kV transmission network

Bus no.	Costumer sector								
	Residential [outage duration (h) and outage costs ($/kW)]			Commercial [outage duration (h) and outage costs ($/kW)]			Industrial [outage duration (h) and outage costs ($/kW)]		
	1 h	4 h	8 h	1 h	4 h	8 h	1 h	4 h	8 h
1	0.1550	1.5781	4.2774	30.6659	98.2935	178.6162	9.0330	25.0354	46.8411
2	0.1411	1.7670	3.8827	30.2986	96.8092	209.8338	9.7914	28.8680	49.2853
3	0.1670	1.6027	3.6440	33.8104	113.8792	185.1510	8.9895	32.1880	55.1201
4	0.1526	2.0843	4.3073	36.3907	115.0748	205.2094	8.2984	27.9202	56.4596
5	0.1817	1.5883	4.1024	33.6494	119.3292	192.6278	10.6668	30.2421	55.4916
6	0.1508	2.0014	3.7092	32.8816	104.1142	186.2789	10.8522	28.8845	53.6196
7	0.1834	1.8756	3.7811	33.9649	117.3304	174.7121	9.6251	28.7617	54.3801
8	0.1730	2.0632	4.3653	34.7031	95.1197	187.7122	8.9827	28.3176	51.5599
9	0.1768	1.8158	4.0375	27.8839	92.5071	167.1831	8.3501	26.0696	49.1589
10	0.1422	1.8316	3.5294	33.5127	101.1079	163.6462	10.9545	25.4279	45.0873
11	0.1574	1.6517	4.2084	32.9011	116.8279	198.7600	10.0205	29.5867	57.3388
12	0.1539	1.8939	4.4301	30.7827	112.5084	190.0236	10.5463	30.1294	49.1955
13	0.1782	1.7566	3.6459	30.2596	93.4596	195.9770	8.9715	26.0165	46.4157
14	0.1783	1.9761	3.5972	30.7095	93.4911	185.8439	9.6067	30.9926	56.3044
15	0.1427	1.8985	4.1163	33.2443	100.9212	210.7314	9.6442	30.6474	57.6927
16	0.1737	1.7802	3.5922	35.2005	93.3625	198.7229	8.8494	30.8813	51.2097
17	0.1417	1.7275	4.1703	32.4387	99.8045	197.6056	8.9533	29.9880	47.7639
18	0.1446	1.6682	3.4781	32.3136	96.0984	172.2697	10.4411	28.2866	52.7018
19	0.1711	2.0353	4.2743	36.5309	100.6489	159.8728	9.4062	29.9189	56.6520
20	0.1841	1.6928	4.0385	35.5104	101.2872	162.5587	9.7545	29.0558	57.1883
21	0.1792	2.0524	4.1378	35.5662	102.1772	191.3278	9.8365	27.6672	57.7536
22	0.1718	1.5738	4.2433	35.5536	110.4344	163.8667	8.7185	25.1278	51.2517
23	0.1557	1.9606	4.0291	36.6659	118.1362	169.6987	8.3583	30.5637	52.3042
24	0.1754	1.7135	3.6849	28.4467	118.0099	185.5347	10.8866	30.7502	58.9687
25	0.1569	1.6312	3.8966	34.1559	109.8845	163.6433	8.7906	29.2149	53.1346
26	0.1500	1.7381	4.3229	32.9808	121.8663	195.6830	10.8313	26.4995	47.0755
27	0.1841	1.6306	3.4352	30.5503	91.6470	205.4742	10.0252	27.0959	47.1485
28	0.1640	1.7768	4.3111	30.3727	107.4535	176.9374	10.6311	32.6219	47.0658
29	0.1625	2.0471	3.6196	27.5968	116.6091	182.2068	8.2795	28.6018	45.4223
30	0.1769	1.9129	3.6811	30.9113	105.3619	190.0355	9.3317	30.5204	55.1386
31	0.1678	1.8771	4.3675	36.1940	109.9761	176.0715	9.3369	30.1214	46.3107
32	0.1585	1.9294	4.2331	32.0409	107.8057	165.6977	10.0211	26.3742	49.8267
33	0.1777	2.0618	3.7248	34.0195	93.2361	185.7325	9.4465	31.1736	56.9265
34	0.1504	1.5615	3.5793	33.2233	100.0361	213.0896	8.5115	30.6429	56.3287
35	0.1843	2.0114	4.0250	36.1193	95.0097	208.5219	9.9917	27.1140	49.5140
36	0.1616	1.6970	4.4308	32.1294	111.6811	162.3043	10.1519	30.8626	50.0452
37	0.1523	1.9735	3.4951	33.2508	98.0518	194.2300	9.6374	30.6876	46.4686
38	0.1807	1.8459	3.7871	30.3840	115.6904	184.3292	8.8194	26.6492	54.8997

(continued)

Table 6.122 (continued)

Bus no.	Costumer sector								
	Residential [outage duration (h) and outage costs ($/kW)]			Commercial [outage duration (h) and outage costs ($/kW)]			Industrial [outage duration (h) and outage costs ($/kW)]		
	1 h	4 h	8 h	1 h	4 h	8 h	1 h	4 h	8 h
39	0.1536	1.9471	3.6554	36.6652	92.7547	185.3256	9.8293	30.9843	50.6554
40	0.1653	1.5904	3.6759	33.8377	104.3654	212.6559	9.7717	30.1751	50.9636
41	0.1763	1.6196	4.2791	36.8263	107.1142	175.0184	9.5861	33.1906	46.9540
42	0.1624	2.0717	3.5756	31.1442	113.9737	193.6477	9.1371	30.6492	49.8958
43	0.1821	1.8553	3.8949	32.3011	111.5671	187.8619	9.9542	32.6762	49.9119
44	0.1678	1.8752	4.5471	30.7527	98.3799	193.3991	10.0409	28.2995	54.4747
45	0.1507	1.8576	4.2033	31.6476	110.6966	212.3220	9.5786	30.2727	50.4667
46	0.1522	1.8536	4.5258	36.2138	93.4250	164.0043	8.8890	25.4924	54.3049
47	0.1676	1.9370	3.4111	35.7622	103.7882	188.1816	9.3558	28.8184	45.8933
48	0.1564	1.9957	3.5243	35.0004	113.9396	181.3844	9.3754	31.1923	50.2822
49	–	–	–	–	–	–	–	–	–
50	–	–	–	–	–	–	–	–	–
51	–	–	–	–	–	–	–	–	–
52	–	–	–	–	–	–	–	–	–

Table 6.123 Other required information related to the proposed PD-TEP problem in the modified Iranian 400 kV transmission network

No.	Parameter	Value
1	P	10 (year)
2	K	22
3	Ir	10 (%)
4	CCL^b	350 $/MW-km (three bundles) 200 $/MW-km (two bundles)
5	$\iota_{(s,r)}^{\max}$	3
6	$cp_p^{\max}$	30 (%)
7	$cp_{p,b}^{\max}$	30 (%)
8	$cp_{p,b,s}^{\max}$	30 (%)
9	$TIC_p^{\max}$	60 m$
10	$TIC^{\max}$	450 m$
11	Annual electrical power demand growth	5 (%)
12	Annual price growth	4 (%)

Table 6.124 Data of the weighting coefficients for objective functions of the PD-TEP problem in the modified 46-bus south Brazilian system

No.	Weighting coefficient	Value
1	$W_{OF_1}^{PD\text{-}TEP}$	1
2	$W_{OF_2}^{PD\text{-}TEP}$	1
3	$W_{OF_3}^{PD\text{-}TEP}$.	1

Table 6.125 SCDF data for different sectors of each load in each bus of the modified 46-bus south Brazilian system

	Costumer sector								
	Residential [outage duration (h) and outage costs ($/kW)]			Commercial [outage duration (h) and outage costs ($/kW)]			Industrial [outage duration (h) and outage costs ($/kW)]		
Bus no.	1 h	4 h	8 h	1 h	4 h	8 h	1 h	4 h	8 h
1	–	–	–	–	–	–	–	–	–
2	0.179835	2.085652	4.138239	31.047679	103.140230	187.044358	8.635216	26.332949	54.792670
3	–	–	–	–	–	–	–	–	–
4	0.189331	2.007984	3.847017	28.367840	110.728322	187.059288	9.485870	29.699639	53.535442
5	0.161321	2.166189	4.208584	27.636994	102.822072	188.187245	9.482508	27.470208	54.739740
6	–	–	–	–	–	–	–	–	–
7	–	–	–	–	–	–	–	–	–
8	0.187873	1.941476	4.682850	33.308597	109.922116	176.851316	9.984609	31.499958	48.201184
9	–	–	–	–	–	–	–	–	–
10	–	–	–	–	–	–	–	–	–
11	–	–	–	–	–	–	–	–	–
12	0.186381	2.053447	4.163610	33.285289	109.433497	159.900977	8.668633	27.155426	51.408511
13	0.180252	1.847057	4.055120	31.043898	111.531093	184.563933	10.505970	29.676203	51.812020
14	–	–	–	–	–	–	–	–	–
15	–	–	–	–	–	–	–	–	–
16	–	–	–	–	–	–	–	–	–
17	–	–	–	–	–	–	–	–	–
18	–	–	–	–	–	–	–	–	–
19	–	–	–	–	–	–	–	–	–
20	0.193492	1.892728	4.183917	33.116147	110.374291	194.443991	9.897000	27.783911	51.421069
21	–	–	–	–	–	–	–	–	–
22	0.157196	2.101601	3.912436	33.220884	105.873563	193.692210	8.989189	28.377374	48.613454

(continued)

Table 6.125 (continued)

	Costumer sector								
	Residential [outage duration (h) and outage costs ($/kW)]			Commercial [outage duration (h) and outage costs ($/kW)]			Industrial [outage duration (h) and outage costs ($/kW)]		
Bus no.	1 h	4 h	8 h	1 h	4 h	8 h	1 h	4 h	8 h
23	0.186174	1.987935	4.170319	32.971390	93.630996	162.368334	10.203410	27.688675	52.175998
24	0.183035	2.095012	4.026103	31.146759	107.306144	177.281982	9.707482	30.927296	48.932153
25	–	–	–	–	–	–	–	–	–
26	0.184105	1.897544	3.956217	31.664418	97.750273	159.593041	9.570976	30.498172	49.272482
27	–	–	–	–	–	–	–	–	–
28	–	–	–	–	–	–	–	–	–
29	–	–	–	–	–	–	–	–	–
30	–	–	–	–	–	–	–	–	–
31	–	–	–	–	–	–	–	–	–
32	–	–	–	–	–	–	–	–	–
33	0.182489	1.765202	4.554936	31.343334	106.008299	160.625993	9.643568	29.387244	49.729126
34	–	–	–	–	–	–	–	–	–
35	0.171478	1.782793	4.137203	32.381025	94.770921	171.040945	8.932263	29.562715	55.300659
36	0.182591	1.898506	4.360735	29.893283	94.204153	176.622257	9.810454	29.736142	55.633489
37	–	–	–	–	–	–	–	–	–
38	0.176553	1.988066	4.646792	32.364391	99.452169	176.209260	10.428181	29.849270	47.627630
39	–	–	–	–	–	–	–	–	–
40	0.160829	1.857337	4.037371	30.372781	106.848692	162.411840	8.949742	31.372893	47.436526
41	–	–	–	–	–	–	–	–	–
42	0.175381	1.960529	4.359542	27.551647	90.600424	195.043125	9.359928	31.549581	49.132985
43	–	–	–	–	–	–	–	–	–
44	0.157655	1.894546	4.562781	32.628187	105.605852	162.611876	9.572249	31.203334	48.257245
45	0.156512	1.897518	4.297082	30.716812	108.045223	165.587202	8.833844	27.631408	51.655021
46	–	–	–	–	–	–	–	–	–

Table 6.126 Other required information related to the proposed PD-G&TEP problem in the modified 46-bus south Brazilian system

No.	Parameter	Value
1	P	10 (year)
2	K	22
3	Ir	10 (%)
4	ECOC	250 ($/MWh)
5	$IC_{p,g}^{\max}$	$50\ (\%) * \sum_{eu \in \Psi_g^{EU}} \rho_{(p-1),g_{eu}}$
6	$RM_p^{\min}$	10 (%)
7	$RM_p^{\max}$	40 (%)
8	$PERC_{p,g}^{\max}$	70 (%)
9	$ECO^{\max}$	3 (%)
10	CCL^{b}	1200 ($/MW-km)
11	$\iota_{(s,r)}^{\max}$	4
12	$c\rho_p^{\max}$	30 (%)
13	$c\rho_{p,b}^{\max}$	30 (%)
14	$c\rho_{p,b,s}^{\max}$	30 (%)
15	$TIC_p^{\max}$	$90M
16	$TIC^{\max}$	$750M
17	$w_{p,\ b,\ s}$	1
18	Annual electrical power demand growth	5 (%)
19	Annual price growth	4 (%)

Table 6.127 Predicted power demand in each load level of each period for the modified three-phase medium-voltage open-loop distribution test network

	Power demand (MVA)								
	Period 1			Period 2			Period 3		
Bus no.	Load level 1	Load level 2	Load level 3	Load level 1	Load level 2	Load level 3	Load level 1	Load level 2	Load level 3
1	1.2	0.72	0.24	1.2	0.72	0.24	1.2	0.72	0.24
2	0	0	0	1.2	0.72	0.24	1.2	0.72	0.24
3	0	0	0	0	0	0	1.2	0.72	0.24
4	1.2	0.72	0.24	1.2	0.72	0.24	1.2	0.72	0.24
5	1.2	0.72	0.24	1.2	0.72	0.24	1.2	0.72	0.24
6	1.2	0.72	0.24	1.2	0.72	0.24	1.2	0.72	0.24
7	0	0	0	1.2	0.72	0.24	1.2	0.72	0.24
8	1.2	0.72	0.24	1.2	0.72	0.24	1.2	0.72	0.24
9	0	0	0	1.2	0.72	0.24	1.2	0.72	0.24
10	0	0	0	0	0	0	2.4	3.6	1.2
11	1.2	0.72	0.24	1.2	2.4	0.48	2.4	3.6	1.2
12	0	0	0	1.2	0.72	0.24	1.2	0.72	0.24
13	1.2	0.72	0.24	1.2	2.4	0.48	2.4	3.6	1.2
14	0	0	0	0	0	0	2.4	1.2	0.48

(continued)

Table 6.127 (continued)

Bus no.	Power demand (MVA)								
	Period 1			Period 2			Period 3		
	Load level 1	Load level 2	Load level 3	Load level 1	Load level 2	Load level 3	Load level 1	Load level 2	Load level 3
15	0	0	0	0	0	0	2.4	1.2	0.48
16	1.2	0.72	0.24	1.2	0.72	0.24	1.2	0.72	0.24
17	0	0	0	1.2	0.72	0.24	2.4	1.2	0.48
18	0	0	0	0	0	0	1.2	2.4	0.48
19	1.2	0.72	0.24	1.2	2.4	0.48	1.2	2.4	0.48
20	0	0	0	1.2	0.72	0.24	1.2	0.72	0.24
21	1.2	0.72	0.24	1.2	2.4	0.48	1.2	2.4	0.48
22	0	0	0	0	0	0	1.2	0.72	0.24
23	0	0	0	0	0	0	1.2	0.72	0.24
24	1.2	0.72	0.24	1.2	0.72	0.24	1.2	0.72	0.24

Table 6.128 Branch length data for the modified three-phase medium-voltage open-loop distribution test network

Branch length data					
Branch			Branch		
From bus	To bus	L_{eb}, L_{rb}, L_{ib} (km)	From bus	To bus	L_{eb}, L_{rb}, L_{ib} (km)
1	2	2.2	10	22	1.6
1	5	2.0	11	22	1.4
1	17	1.8	11	26	1.2
2	3	2.2	12	16	2.8
3	4	2.2	12	24	1.8
3	19	1.2	12	26	2.0
4	8	1.6	13	14	2.2
5	6	2.4	13	20	1.8
5	10	2.9	14	15	2.4
5	25	2.0	14	21	1.1
6	7	2.2	15	16	1.5
7	8	2.4	15	23	0.9
7	19	1.1	17	18	1.6
7	26	1.4	18	25	0.8
8	12	2.4	20	25	1.8
9	10	1.6	21	27	1.3
9	13	2.6	22	23	0.9
9	25	1.8	23	24	1.0
10	11	1.8	23	27	1.1
10	21	2.2			

Table 6.129 Load level data for the modified three-phase medium-voltage open-loop distribution test network

Load level	Load level data	
	Δt_m (h/day)	$p\lambda_{p,\,m}$ (\$/MWh)
1	3	70
2	13	49
3	8	35

Table 6.130 Number of customers for each bus and participation of load sector for the modified three-phase medium-voltage open-loop distribution test network

Bus no.	Customer data											
	Period 1				Period 2				Period 3			
	No.	Res.	Com.	Ind.	No.	Res.	Com.	Ind.	No.	Res.	Com.	Ind.
1	100	0.5	0.2	0.3	100	0.5	0.2	0.3	100	0.5	0.2	0.3
2	–	–	–	–	10	0	0.1	0.9	10	0	0.1	0.9
3	–	–	–	–	–	–	–	–	50	0.2	0.2	0.6
4	10	0	0.2	0.8	10	0	0.2	0.8	10	0	0.2	0.8
5	50	0.2	0.2	0.6	50	0.2	0.2	0.6	50	0.2	0.2	0.6
6	30	0.15	0.2	0.65	30	0.15	0.2	0.65	30	0.15	0.2	0.65
7	–	–	–	–	10	0	0	1	10	0	0	1
8	50	0.1	0.2	0.7	50	0.1	0.2	0.7	50	0.1	0.2	0.7
9	–	–	–	–	50	0.1	0.2	0.7	100	0.1	0.2	0.7
10	–	–	–	–	–	–	–	–	10	0	0	1
11	50	0.5	0.4	0.1	50	0.5	0.4	0.1	50	0.5	0.4	0.1
12	–	–	–	–	50	0.4	0.4	0.2	50	0.4	0.4	0.2
13	10	0	0.15	0.85	10	0	0.15	0.85	10	0	0.15	0.85
14	–	–	–	–	–	–	–	–	15	1	0	0
15	–	–	–	–	–	–	–	–	15	1	0	0
16	50	0.3	0.2	0.5	50	0.3	0.2	0.5	50	0.3	0.2	0.5
17	–	–	–	–	10	0	0.25	0.75	15	0	0.25	0.75
18	–	–	–	–	–	–	–	–	20	0.8	0.2	0
19	10	0	0	1	10	0	0	1	10	0	0	1
20	–	–	–	–	50	0	0	1	50	0	0	1
21	100	0.6	0.25	0.15	100	0.6	0.25	0.15	100	0.6	0.25	0.15
22	–	–	–	–	–	–	–	–	10	0	1	0
23	–	–	–	–	–	–	–	–	10	1	0	0
24	50	0.5	0.2	0.3	50	0.5	0.2	0.3	50	0.5	0.2	0.3

Table 6.131 SCDF data for the modified three-phase medium-voltage open-loop distribution test network

Customer sector data					
	[Duration (min) and outage costs ($/kW)]				
Customer sector	1 min	20 min	60 min	240 min	480 min
Residential	0.001	0.093	0.482	4.914	15.69
Commercial	0.381	2.969	8.552	31.32	83.01
Industrial	1.625	3.868	9.085	25.16	55.81

Table 6.132 Economic lifetime data for the modified three-phase medium-voltage open-loop distribution test network

	Economic lifetime data	
Equipment	Parameter	Value (year)
Transformer	δ_s	15
Branch	δ_b	25
DGR	δ_{dg}	25

Table 6.133 Technical and economic features of the substations for the modified three-phase medium-voltage open-loop distribution test network

	Substations																		
	Existing substations in the initial configuration of the open-loop distribution test network			Candidate new substations															
				Substation option 1								Substation option 2							
Bus no.	R (Ω)	$MC_{p,\,es}$ ($/MW)	$I_{es}^{\max}$ (p.u.)	R (Ω)	$IC_{p,is}^{f}$ (M$)	$IC_{p,rs}^{v}$ (M$)	$IC_{p,is}^{v}$ (M$)	$MC_{p,\,rs}$ ($/MW)	$MC_{p,\,is}$ ($/MW)	$I_{rs}^{\max}$ (p.u.)	$I_{is}^{\max}$ (p.u.)	R (Ω)	$IC_{p,is}^{f}$ (M$)	$IC_{p,rs}^{v}$ (M$)	$IC_{p,is}^{v}$ (M$)	$MC_{p,\,rs}$ ($/MW)	$MC_{p,\,is}$ ($/MW)	$I_{rs}^{\max}$ (p.u.)	$I_{is}^{\max}$ (p.u.)
25	0.13	5200	15	0.25	0.5	4.875	4.875	4100	1000	12	12	0.16	0.5	5.100	5.100	4500	1100	15	15
26	0.13	5200	15	0.25	0.5	4.875	4.875	4100	1000	12	12	0.16	0.5	7.100	7.100	4500	1100	15	15
27	–	–	–	0.25	0.5	4.875	4.875	4100	1000	12	12	0.16	0.5	7.100	7.100	4500	1100	15	15

Table 6.134 Technical and economic features of the branches for the modified three-phase medium-voltage open-loop distribution test network

	Branches																			
	Existing branches in the initial configuration of the open-loop distribution test network				Candidate new branches															
					Branch option 1								Branch option 2							
Branch status	R (Ω/km)	Z (Ω/km)	$MC_{p,eb}$ ($/km)	I_{eb}^{max} (p.u.)	R (Ω/km)	Z (Ω/km)	$IC_{p,rb}$ (M $/km)	$IC_{p,ib}$ (M $/km)	$MC_{p,rb}$ ($/km)	$MC_{p,ib}$ ($/km)	I_{rb}^{max} (p.u.)	I_{ib}^{max} (p.u.)	R (Ω/km)	Z (Ω/km)	$IC_{p,rb}$ (M $/km)	$IC_{p,ib}$ (M $/km)	$MC_{p,rb}$ ($/km)	$MC_{p,ib}$ ($/km)	I_{rb}^{max} (p.u.)	I_{ib}^{max} (p.u.)
Existing	0.76	1	500	12	–	–	–	–	–	–	–	–	–	–	–	–	–	–	–	–
Reinforcement	–	–	–	–	0.44	0.7	0.568	–	500	–	10	–	0.25	0.5	1.079	–	500	–	12	–
Installation	–	–	–	–	0.76	1	–	1.136	–	500	–	11	0.44	0.7	–	1.420	–	500	–	15

Table 6.135 Data for the weighting coefficients related to the proposed PD-DEP problem in the modified three-phase medium-voltage open-loop distribution test network

No.	Weighting coefficient	Value
1	$W_{OF_1}^{PD\text{-}DEP}$	1
2	$W_{OF_2}^{PD\text{-}DEP}$	1
3	$W_{OF_3}^{PD\text{-}DEP}$	1

Table 6.136 Other required information related to the proposed PD-DEP problem in the modified three-phase medium-voltage open-loop distribution test network

No.	Parameter	Value
1	P	3 (year)
2	M	3
3	Ir	10 (%)
4	Failure rate	0.4 (failure per year)
5	Repair rate	10 (h)
6	$V_{b_{ge}}^{min}$	0.95 (p.u.)
7	$V_{b_{el}}^{min}$	0.95 (p.u.)
8	$V_{b_{vc}}^{min}$	1.05 (p.u.)
9	$V_{b_{ge}}^{max}$	1.05 (p.u.)
10	$V_{b_{el}}^{max}$	1.05 (p.u.)
11	$V_{b_{vc}}^{max}$	1.05 (p.u.)
12	$IC_{p,\ id}$	0.89 (M$/MW)
13	$OC_{p,\ m,\ id}$	50 ($/MW h)
14	ρ_{id}^{max}	5 MW (%)
15	$\rho_{p}^{max\text{-}dg}$	10 MW (%)
16	$\rho^{max\text{-}dg}$	25 MW (%)
17	DIC_{p}^{max}	$25 M
18	DIC^{max}	$70 M
19	Annual electrical power demand growth	5 (%)
20	Annual price growth	4 (%)

References

1. N. Alguacil, A.L. Motto, A.J. Conejo, Transmission expansion planning: a mixed-integer LP approach. IEEE Trans. Power Syst. **18**(3), 1070–1077 (2003)
2. M.J. Rider, A.V. Garcia, R. Romero, Power system transmission network expansion planning using AC model. IET Gener. Transm. Dis. **1**(5), 731–742 (2007)
3. P. Murugan, S. Kannan, S. Baskar, Application of NSGA-II algorithm to single-objective transmission constrained generation expansion planning. IEEE Trans. Power Syst. **24**(4), 1790–1797 (2009)
4. R.C. Leou, A multi-year transmission planning under a deregulated market. Int. J. Electr. Power Energy Syst. **33**(3), 708–714 (2011)
5. S.J. Kazempour, A.J. Conejo, C. Ruiz, Strategic generation investment using a complementarity approach. IEEE Trans. Power Syst. **26**(2), 940–948 (2011)

6. C.H. Antunes, A.G. Martins, I.S. Brito, A multiple objective mixed integer linear programming model for power generation expansion planning. Energy **29**(4), 613–627 (2004)
7. C. Unsihuay-Vila, J.W. Marangon-Lima, A.C. Zambroni de Souza, I.J. Perez-Arriaga, P. Balestrassi, A model to long-term, multiarea, multistage, and integrated expansion planning of electricity and natural gas systems. IEEE Trans. Power Syst. **25**(2), 1154–1168 (2010)
8. I. Sharan, R. Balasubramanian, Integrated generation and transmission expansion planning including power and fuel transportation constraints. Energy Policy **43**, 275–284 (2012)
9. G.A. Bakirtzis, P.N. Biskas, V. Chatziathanasiou, Generation expansion planning by MILP considering mid-term scheduling decisions. Electr. Power Syst. Res. **86**(1), 98–112 (2012)
10. F. Chen, G. Huang, Y. Fan, A linearization and parameterization approach to tri-objective linear programming problems for power generation expansion planning. Energy **87**, 240–250 (2015)
11. A. Sheffrin, California power crisis: failure of market design or regulation? IEEE Power Eng. Rev. **22**(8), 8–11 (2002)
12. H.K. Alfares, M. Nazeeruddin, Electric load forecasting: Literature survey and classification of methods. Int. J. Syst. Sci. **33**(1), 23–34 (2002)
13. A.K. Singh, I. Ibraheem, S. Khatoon, M. Muazzam, D. Chaturvedi, Load forecasting techniques and methodologies: a review, in *2012 2nd International Conference on Power, Control and Embedded Systems (ICPCES)*, vol. 33, No. 1, (2012), pp. 1–10
14. L.L. Lai, *Power System Restructuring and Deregulation, Trading Performance and Information Technology* (Wiley, Chichester, 2001)
15. A. Mazer, *Electric Power Planning for Regulated and Deregulated Markets* (IEEE Press, Piscataway, 2007)
16. H. Seifi, M.S. Sepasian, *Electric Power Systems Planning: Issues, Algorithms and Solutions* (Springer, Berlin, 2011)
17. W.E. Walker, P. Harremoës, J. Rotmans, J.P. van der Sluijs, M.B.A. van Asselt, P. Janssen, M.P. Krayer von Krauss, Defining uncertainty: a conceptual basis for uncertainty management in model-based decision support. Integr. Assess. **4**(1), 5–17 (2003)
18. J. Wang, M. Shahidehpour, Z. Li, A. Botterud, Strategic generation capacity expansion planning with incomplete information. IEEE Trans. Power Syst. **24**(2), 1002–1010 (2009)
19. Q. Chen, C. Kang, Q. Xia, J. Zhong, Power generation expansion planning model towards low-carbon economy and its application in China. IEEE Trans. Power Syst. **25**(2), 1117–1125 (2010)
20. H. Tekiner, D.W. Coit, F.A. Felder, Multi-period multi-objective electricity generation expansion planning problem with Monte-Carlo simulation. Electr. Power Syst. Res. **80**, 1394–1405 (2010)
21. S. Jin, S.M. Ryan, Capacity expansion in the integrated supply network for an electricity market. IEEE Trans. Power Syst. **26**(4), 2275–2284 (2011)
22. S. Wogrin, E. Centeno, J. Barquín, Generation capacity expansion in liberalized electricity markets: a stochastic MPEC approach. IEEE Trans. Power Syst. **26**(4), 2526–2532 (2011)
23. Y. Wu, S. Lou, S. Lu, A model for power system interconnection planning under low-carbon economy with CO_2 emission constraints. IEEE Trans. Sustain. Energy **2**(3), 205–214 (2011)
24. F. Careri, C. Genesi, P. Marannino, M. Montagan, S. Rossi, I. Siviero, Generation expansion planning in the age of green economy. IEEE Trans. Power Syst. **26**(4), 2214–2223 (2011)
25. S.J. Kazempour, A.J. Conejo, Strategic generation investment under uncertainty via benders decomposition. IEEE Trans. Power Syst. **27**(1), 424–432 (2012)
26. A.J. Pereira, J.T. Saraiva, A long term generation expansion planning model using system dynamics—case study using data from the Portuguese/Spanish generation system. Electr. Power Syst. Res. **97**, 41–50 (2013)
27. S. Wogrin, J. Barquín, E. Centeno, Capacity expansion equilibria in liberalized electricity markets: an EPEC approach. IEEE Trans. Power Syst. **28**(2), 1531–1539 (2013)
28. Y. Feng, S.M. Ryan, Scenario construction and reduction applied to stochastic power generation expansion planning. Comput. Oper. Res. **40**, 9–23 (2013)

29. H. Sadeghi, M. Rashidinejad, A. Abdollahi, A comprehensive sequential review study through the generation expansion planning. Renew. Sustain. Energy Rev. **67**, 1369–1394 (2017)
30. R. Billinton, R.N. Allan, *Reliability evaluation of power systems*, 2nd edn. (Plenum Press, New York, 1994)
31. M. Aien, A. Hajebrahimi, M. Fotuhi-Firuzabad, A comprehensive review on uncertainty modeling techniques in power system studies. Renew. Sustain. Energy Rev. **57**, 1077–1089 (2016)
32. Y. Ben-Haim, *Info-gap decision theory: decisions under severe uncertainty*, 2nd edn. (Academic Press, Amsterdam, 2006)
33. I.J. Silva, M.J. Rider, R. Romero, A.V. Garcia, C.A. Murari, Transmission network expansion planning with security constraints. IEE Proc. Generat. Transm. Distrib. **152**(6), 828–836 (2005)
34. S. Haffner, A. Monticelli, A. Garcia, R. Romero, Specialised branch-and-bound algorithm for transmission network expansion planning. IEE Proc. Generat. Transm. Distrib. **148**(5), 482–488 (2001)
35. S. Haffner, A. Monticelli, A. Garcia, J. Mantovani, R. Romero, Branch and bound algorithm for transmission system expansion planning using a transportation model. IEE Proc. Generat. Transm. Distrib. **147**(3), 149–156 (2000)
36. M. Shivaie, M.T. Ameli, An implementation of improved harmony search algorithm for scenario-based transmission expansion planning. Soft Comput. **18**(8), 1615–1630 (2014)
37. H. Chen, X. Wang, X. Zhao, Generation planning using Lagrangian relaxation and probabilistic production simulation. Electr. Power Energy Syst. **26**, 597–605 (2004)
38. Y. Zhou, Y.P. Li, G.H. Huang, Y. Zhou, A robust approach for planning electric power systems associated with environmental policy analysis. Electr. Power Syst. Res. **95**, 99–111 (2013)
39. S. Kaplan, Power plants: characteristics and costs, in *CRS Report for Congress* (Congressional Research Service, Washington, DC, 2008), pp. 1–108
40. P. Maghouli, S.H. Hosseini, M.O. Buygi, M. Shahidehpour, A scenario-based multi-objective model for multi-stage transmission expansion planning. IEEE Trans. Power Syst. **26**(1), 470–478 (2011)
41. A. Ben-Tal, L.E. Ghaoui, A. Nemirovski, *Robust Optimization* (Princeton University Press, Princeton, 2009)
42. K. Deb, A. Pratap, S. Agarwal, T. Meyarivan, A fast and elitist multiobjective genetic algorithm: NSGA-II. IEEE Trans. Evol. Comput. **6**(2), 182–197 (2002)
43. Y.-X. Jin, H.-Z. Cheng, J.-Y. Yan, L. Zhang, New discrete method for particle swarm optimization and its application in transmission network expansion planning. Electr. Power Syst. Res. **77**(3–4), 227–233 (2007)
44. E. Bustamante-Cedeño, A. Arora, Stochastic and minimum regret formulations for transmission network expansion planning under uncertainties. J. Oper. Res. Soc. **59**(11), 1547–1556 (2008)
45. L.P. Garces, A.J. Conejo, R. Garcia-Bertrand, R. Romero, A bilevel approach to transmission expansion planning within a market environment. IEEE Trans. Power Syst. **24**(3), 1513–1522 (2009)
46. A.M. Leite da Silva, L.S. Rezende, L.A.F. Manso, L.C. de Resende, Reliability worth applied to transmission expansion planning based on ant colony system. Int. J. Electr. Power Energy Syst. **32**(10), 1077–1084 (2010)
47. H. Fan, H.Z. Cheng, Multistage transmission network expansion planning in competitive electricity market based on bi-level programming method. Int. Trans. Electr. Energy Syst. **21** (5), 1719–1730 (2011)
48. H. Zhang, V. Vittal, G.T. Heydt, A mixed-integer linear programming approach for multi-stage security-constrained transmission expansion planning. IEEE Trans. Power Syst. **27**(2), 1125–1133 (2012)
49. M.C. Rocha, J.T. Saraiva, A discrete evolutionary PSO based approach to the multiyear transmission expansion planning problem considering demand uncertainties. Int. J. Electr. Power Energy Syst. **45**(1), 427–442 (2013)

50. C.A.C. Florez, R.A.B. Ocampo, A.H.E. Zuluaga, Multi-objective transmission expansion planning considering multiple generation scenarios. Int. J. Electr. Power Energy Syst. **62**, 398–409 (2014)
51. A.S. Sousa, E.N. Asada, Long-term transmission system expansion planning with multi-objective evolutionary algorithm. Electr. Power Syst. Res. **119**, 149–156 (2015)
52. O. Alizadeh-Mousavi, M. Zima-Bockarjova, Efficient Benders cuts for transmission expansion planning. Electr. Power Syst. Res. **131**, 275–284 (2016)
53. S. Lumbreras, A. Ramos, The new challenges to transmission expansion planning. Survey of recent practice and literature review. Electr. Power Syst. Res. **134**, 19–29 (2016)
54. R. Billinton, R.N. Allan, *Reliability Evaluation of Engineering Systems: Concepts and Techniques* (Plenum Press, New York, 1992)
55. M. Shivaie, M.S. Sepasian, M.K. Sheikh-El-Eslami, Multi-objective transmission expansion planning based on reliability and market considering phase shifter transformers by fuzzy-genetic algorithm. Int. Trans. Electr. Energy Syst. **23**(8), 1468–1489 (2013)
56. M.O. Buygi, Transmission Expansion Planning in Deregulated Power Systems, Ph.D. dissertation, Department of Electrical Power Systems Institute, Darmstadt University of Technology, Darmstadt, Germany, (2004)
57. M. Moeini-Aghtaie, A. Abbaspour, M. Fotuhi-Firuzabad, Incorporating large-scale distant wind farms in probabilistic transmission expansion planning—Part II: Case studies. IEEE Trans. Power Syst. **27**(3), 1594–1601 (2012)
58. TAVANIR. Annual Generation and Transmission Reports. [Online]. http://www.tavanir.org.ir
59. W. Wangdee, Bulk electric system reliability simulation and application, Ph.D. dissertation, University of Saskatchewan, Saskatoon, Canada, (2005)
60. J.H. Roh, M. Shahidehpour, Y. Fu, Market-based coordination of transmission and generation capacity planning. IEEE Trans. Power Syst. **22**(4), 1406–1419 (2007)
61. J.H. Roh, M. Shahidehpour, L. Wu, Market-based generation and transmission planning with uncertainties. IEEE Trans. Power Syst. **24**(3), 1587–1598 (2009)
62. A. Motamedi, H. Zareipour, M.O. Buygi, W.D. Rosehart, A transmission planning framework considering future generation expansions in electricity markets. IEEE Trans. Power Syst. **25**(4), 1987–1995 (2010)
63. A. Motamedi, H. Zareipour, M.O. Buygi, W.D. Rosehart, Capacity expansion in the integrated supply network for an electricity market. IEEE Trans. Power Syst. **26**(4), 2275–2284 (2011)
64. M.R. Hesamzadeh, D.R. Biggar, N. Hosseinzadeh, P.J. Wolfs, Transmission augmentation with mathematical modeling of market power and strategic generation expansion—Part I. IEEE Trans. Power Syst. **26**(4), 2040–2048 (2011)
65. M.R. Hesamzadeh, D.R. Biggar, N. Hosseinzadeh, P.J. Wolfs, Transmission augmentation with mathematical modeling of market power and strategic generation expansion—Part II. IEEE Trans. Power Syst. **26**(4), 2049–2057 (2011)
66. B. Alizadeh, S. Jadid, Reliability constrained coordination of generation and transmission expansion planning in power systems using mixed integer programming. IET Gener. Transm. Distrib. **5**(9), 948–960 (2011)
67. D. Pozo, E.E. Sauma, J. Contreras, A three-level static MILP model for generation and transmission expansion planning. IEEE Trans. Power Syst. **28**(1), 202–210 (2013)
68. M. Mirhosseini Moghaddam, M.H. Javidi, M. Parsa Moghaddam, M.O. Buygi, Coordinated decisions for transmission and generation expansion planning in electricity markets. Int. Trans. Electr. Energy Syst. **23**(8), 1452–1467 (2013)
69. M. Jenabi, S.M.T. Fatemi Ghomi, Y. Smeers, Bi-level game approaches for coordination of generation and transmission expansion planning within a market environment. IEEE Trans. Power Syst. **28**(3), 2639–2650 (2013)
70. S. Jin, S.M. Ryan, A tri-level model of centralized transmission and decentralized generation expansion planning for an electricity market—Part I. IEEE Trans. Power Syst. **29**(1), 132–141 (2014)

71. S. Jin, S.M. Ryan, A tri-level model of centralized transmission and decentralized generation expansion planning for an electricity market—Part II. IEEE Trans. Power Syst. **29**(1), 142–148 (2014)
72. J. Aghaei, N. Amjady, A. Baharvandi, M.A. Akbari, Generation and transmission expansion planning: MILP-based probabilistic model. IEEE Trans. Power Syst. **29**(4), 1592–1601 (2014)
73. B. Alizadeh, S. Jadid, A dynamic model for coordination of generation and transmission expansion planning in power systems. Int. J. Electr. Power Energy Syst. **65**, 408–418 (2015)
74. M. Vaziri, K. Tomsovic, A. Bose, A directed graph formulation of the multistage distribution expansion problem. IEEE Trans. Power Deliv. **19**(3), 1335–1341 (2004)
75. H. Zhao, Z. Wang, D.C. Yu, X. Chen, New formulations and hybrid algorithms for distribution system planning. Electr. Power Compon. Syst. **35**(4), 445–460 (2007)
76. S. Haffner, L.F.A. Pereira, L.A. Pereira, L.S. Barreto, Multistage model for distribution expansion planning with distributed generation—Part I: problem formulation. IEEE Trans. Power Deliv. **23**(2), 915–923 (2008)
77. S. Haffner, L.F.A. Pereira, L.A. Pereira, L.S. Barreto, Multistage model for distribution expansion planning with distributed generation—Part II: numerical results. IEEE Trans. Power Deliv. **23**(2), 924–929 (2008)
78. Ž.N. Popović, D.S. Popović, Graph theory based formulation of multi-period distribution expansion problems. Electr. Power Syst. Res. **80**(10), 1256–1266 (2010)
79. A. Soroudi, M. Ehsan, A distribution network expansion planning model considering distributed generation options and techno-economical issues. Energy **35**(8), 3364–3374 (2010)
80. R.C. Lotero, J. Contreras, Distribution system planning with reliability. IEEE Trans. Power Deliv. **26**(4), 2552–2562 (2011)
81. H. Falaghi, C. Singh, M.R. Haghifam, M. Ramezani, DG integrated multistage distribution system expansion planning. Int. J. Electr. Power Energy Syst. **33**(8), 1489–1497 (2011)
82. I. Ziari, G. Ledwich, A. Ghosh, G. Platt, Integrated distribution systems planning to improve reliability under load growth. IEEE Trans. Power Deliv. **27**(2), 757–765 (2012)
83. A.S. Bin Humayd, K. Bhattacharya, Comprehensive multi-year distribution system planning using back-propagation approach. IET Gener. Transm. Distrib. **7**(12), 1415–1425 (2013)
84. J.F. Franco, M.J. Rider, R. Romero, A mixed-integer quadratically-constrained programming model for the distribution system expansion planning. Int. J. Electr. Power Energy Syst. **62**, 265–272 (2014)
85. G. Muñoz-Delgado, J. Contreras, J.M. Arroyo, Joint expansion planning of distributed generation and distribution networks. IEEE Trans. Power Syst. **30**(5), 2579–2590 (2015)
86. S. Heidari, M. Fotuhi-Firuzabad, S. Kazemi, Power distribution network expansion planning considering distribution automation. IEEE Trans. Power Syst. **30**(3), 1261–1269 (2015)
87. S. Ganguly, N.C. Sahoo, D. Das, Recent advances on power distribution system planning: a state-of-the-art survey. Energy Syst. **4**(2), 165–193 (2013)
88. S.K. Khator, L.C. Leung, Power distribution planning: a review of models and issues. IEEE Trans. Power Syst. **12**(3), 1151–1159 (1997)
89. H.L. Willis, *Power Distribution Planning Reference Book*, 2nd edn. (Marcel Dekker, New York, 2004)
90. G.W. Ault, C.E.T. Foote, J.R. McDonald, Distribution system planning in focus. IEEE Power Eng. Rev. **22**(1), 60–62 (2002)
91. R. Billinton, P. Wang, Distribution system reliability cost/worth analysis using analytical and sequential simulation Techniques. IEEE Trans. Power Syst. **13**(4), 1245–1250 (1998)
92. N. Deo, *Graph Theory with Applications to Engineering and Computers Science* (Prentice-Hall, Englewood Cliffs, 1974)

Chapter 7
Power Filters Planning

7.1 Introduction

In modern distribution networks, increasing the number and variety of disturbing nonlinear loads and the susceptibility of these loads to voltage/current/frequency perturbations has given rise to serious power quality problems. Strictly speaking, power quality problems can effectively escalate the techno-economic losses for customers in different areas, such as residential, industrial, commercial, and agricultural. In addition, irreversible economic effects are imposed on distribution companies or service providers when such power quality problems arise. This is due to the fact that customers are freely authorized to select their service providers in modern deregulated distribution networks. As a result, power quality has undergone major changes, especially from a practical standpoint. Hence, power quality problems and related improvement strategies have been heavily addressed by distribution network activists in these deregulated environments. Despite myriad technical publications on power quality studies, there is still no sound description for power quality that is recognized by researchers in this field. Nevertheless, a practical explanation for power quality must take into account the various aspects of the behavior of distribution networks; for example, achieving an acceptable level of power availability—continuity—and quality for techno-economic provisions of customers. Availability refers to an uninterrupted power supply service, while quality encompasses different types of disturbances associated with the distribution networks that affect characteristics of voltage/current waveforms. Accordingly, the definition used for power quality in this chapter has been adapted from the work by Caramia et al. [1]. Power quality can be simultaneously expressed from two different perspectives: (1) the capability of a distribution network to supply power to customers without any interruptions or damage to their equipment and (2) the capability of customers to operate without any perturbations or reduction in productivity of their distribution networks. The first

M. Kiani-Moghaddam et al., *Modern Music-Inspired Optimization Algorithms for Electric Power Systems*, Power Systems,
https://doi.org/10.1007/978-3-030-12044-3_7

characteristic is mainly associated with voltage quality at the point of common coupling (PCC) between customers and distribution utilities. The second characteristic, however, is mostly, but not exclusively, related to the current quality of customer load waveforms. Any deficiency in monitoring this ability by the distribution networks or customer loads could lead to power quality problems. Because the systematic categorization of the power quality problems depends on different indices, this process is a serious and challenging task. In the literature, there are various classifications for power quality problems. The taxonomy reported by the Institute of Electrical and Electronic Engineers (IEEE) is one of the classifications addressed for power quality problems that have been widely dealt with [2]. Power quality disturbances in accordance with this IEEE taxonomy could be classified into seven types: (1) transients; (2) short-term fluctuations; (3) long-term fluctuations; (4) voltage imbalances; (5) waveform distortions; (6) voltage fluctuations; and, (7) frequency fluctuations. For a detailed discussion about these features of power quality disturbances, interested readers can refer again to the study by Caramia et al. [1].

In power quality disturbances, waveform distortions are more concerning because of their destructive effects on the distribution networks and customer equipment. The waveform distortions usually refer to the harmonic and interharmonic components. Core saturation of transformers, static power converters, and nonlinear and time variant loads (i.e., rectifiers, inverters, drivers of speed adjust) are some instances of waveform distortion sources. Harmonic and interharmonic distortions have malicious effects on the distribution network equipment. Some of the most significant destructive effects of the harmonics can be observed in torque and heat fluctuations in induction motors, increased degradation of the insulation in rotating machines and transformers, increased casualties, and a reduction in the useful life of equipment.

Traditionally, the solution to reducing and eliminating problems associated with the harmonics is to use passive harmonic power filters (PHPFs) [3]. However, large size, series and parallel resonances, and applications for specific harmonics are the major disadvantages of the PHPFs. As a result, active harmonic power filters (AHPFs) were developed to cope with such problems [4]. In general, AHPFs could play a significant role in the elimination of current and voltage harmonics, correction of power factor, reduction of unbalancing, etc. But they are a costly choice for improving power quality. Consequently, a hybrid harmonic power filter (HHPF) design consisting of multiple passive and/or active power filters is increasingly used for improving power quality [5]. In HHPFs, the size and cost of the harmonic power filters are greatly diminished. In HHPF structure, the passive and active parts are utilized to compensate for the low- and high-order harmonics, respectively. As a result, the use of the hybrid configuration for harmonic power filters could give rise to increased flexibility in controlling and inhibiting the harmonics.

Most of the frameworks reported for harmonic power filter planning and harmonic analysis in distribution networks cannot accurately model and scrutinize the real behavior—stochastic behavior—of nonlinear loads. This is due to the fact that these frameworks employ nondeterministic mathematical models. Therefore, dealing with stochastic models with the aim of investigating the malicious effects of the harmonics in distribution networks is recommended in different harmonic standards [6, 7]. For the reason identified above, incorporating the stochastic, or time-varying, behavior of harmonic currents generated by nonlinear loads is considered a remarkable issue in harmonic power filter planning studies.

With that in mind, the authors will focus on five targets in the field of harmonic power filter planning in distribution networks, as follows:

- Target 1: Provide a techno-economic multi-objective framework for the HHPF planning problem in distribution networks.
- Target 2: Incorporate the practical features and limitations of the PHPFs and AHPFs in the proposed techno-economic multi-objective framework.
- Target 3: Enhance the flexibility of the proposed techno-economic multi-objective framework in the harmonic filtering process in distribution networks by using a well-designed parallel-connected combination of three types of PHPFs and four different types of AHPFs.
- Target 4: Consider a well-designed strategy to handle uncertainty in demand and harmonic currents injected by nonlinear loads in the HHPF planning problem.
- Target 5: Solve the proposed framework by using the offered modern multi-objective music-inspired optimization algorithms, which were addressed in Chap. 4, and compare the obtained results with powerful state-of-the-art multi-objective optimization algorithm—non-dominated sorting genetic algorithm-II (NSGA-II).

The authors feel that addressing the basic concepts of harmonic power filter planning studies is beyond the scope of this chapter and, thus, will assume that readers have a working knowledge of the preliminary details of these studies. However, if needed, readers may wish to refer to relevant studies that cover these preliminary details about harmonic power filter planning studies. The rest of this chapter is organized as follows: First, an overview of harmonic power filter planning studies is briefly described in Sect. 7.2. In Sect. 7.3, the proposed techno-economic multi-objective framework for the HHPF planning problem is discussed in detail, including a mathematical model of the framework, solution method, simulation results, case studies, and discussion of the results. Finally, a brief summary and some concluding remarks are summarized in Sect. 7.4.

7.2 A Brief Review of Harmonic Power Filter Planning Studies

In this section, the authors will focus on the different aspects of harmonic power filter planning studies in distribution networks, more specifically, perspectives that are involved in the proposed techno-economic multi-objective framework.

7.2.1 Nonlinear Loads and Their Malicious Effects

Due to the existence of different types of nonlinear loads in various configurations, the interactions of these nonlinear loads with the alternating current (AC) system impedance, as well as the difference in the harmonic spectrum produced by these nonlinear loads, the harmonic analysis process of nonlinear loads as the main step of harmonic power filter planning studies is an arduous process. Therefore, accurate recognition of the features of nonlinear loads can be helpful in reducing the complexity of this process. The existing nonlinear loads in distribution networks can be classified into three main types: (1) current-stiff nonlinear loads; (2) voltage-stiff nonlinear loads; and, (3) a combination of current- and voltage-stiff nonlinear loads. The main characteristic of the current-stiff nonlinear loads, which include most of the existing nonlinear loads in distribution networks, is that their current waveforms are distorted on the AC side. A thyristor rectifier for direct current (DC) drives is one of the most common current-stiff nonlinear loads. The special feature of voltage-stiff loads is that they lead to non-sinusoidal and discrete currents in which case high harmonic distortion of current, low power factor, and harmonic distortion of the AC terminal voltage are created in the PCC of the distribution networks. Diode rectifiers for AC drives, electronic-based devices, etc. are most renowned voltage-stiff nonlinear loads. The third type of nonlinear loads is a combination of current- and voltage-stiff nonlinear loads that simultaneously possess their characteristics. The variable-speed drive is a well-known example of this type of nonlinear load. In the variable-speed drive structure, variable-frequency voltage-source inverter-fed AC motor drive and current-source inverter-fed AC motor drive sections act as the voltage- and current-stiff nonlinear loads, respectively. The authors refer interested readers to the study by Fuchs and Masoum [8] for details on other features of all types of nonlinear loads.

The first consequence of the existence of these nonlinear loads in distribution networks is the occurrence of the phenomenon of harmonics. The harmonics phenomenon is formed when the current waveform and, consequently, the voltage waveform are taken out of their sinusoidal states and oscillated by coefficients of fundamental frequency. The harmonics have numerous destructive effects that affect the performance of different parts of the power system, especially the distribution sector. Some of the most important destructive effects of harmonics that have been recognized thus far are (1) additional losses, mechanical fluctuations, and high heat in synchronous and induction machines; (2) disturbance in the safe operation of protection relays, due to current waveform deformation; (3) reduction of the useful life of equipment, especially high-voltage insulation; (4) increased losses in power transformers and transmission lines; (5) decreased power factor; (6) insulation failure in transformers, cables, rotating machines, etc., due to the harmonic overvoltage in the distribution network; (7) interference with telecommunication systems; (8) damage to the power factor correction capacitors; and, (9) disturbance in the performance of the microprocessor-based systems. It should be noted that these destructive effects surely depend on the types of harmonic sources and their locations in the distribution network, electrical characteristics of the distribution network, etc.

The authors acknowledge that providing a complete discussion of the harmonics and their role in distribution network studies is a challenging process that is beyond the scope of this chapter. Hence, the authors refer interested readers to studies by Fuchs and Masoum [8] and Arrillaga and Smith [9].

7.2.2 *Harmonic Power Filters*

In the past few decades, the number and variety of nonlinear loads have sharply increased, owing to a significant growth in power demand and emerging equipment technology in distribution networks. These nonlinear loads lead to an intensification of known problems originating from the harmonics as well as creating new difficulties. Much effort has been put forth by researchers and engineers from around the world to overcome these problems. An inexpensive yet simple solution to eliminate or reduce the harmful effects of the harmonics that can be employed by customers and distribution network operators (DNOs) is the use of harmonic power filters. The earliest type of harmonic power filter is the PHPF. These filters are built by using a specific arrangement of passive elements—resistors (R), inductors (L), and capacitors (C). These elements are adjusted in order to compensate for a specific or a set of the harmonic frequencies of the currents and voltages. Conceptually, the PHPFs act as a trap to absorb the frequency of the current and/or voltage harmonics that must be attenuated. The PHPFs can be operated in different configurations: parallel-connected, series-connected, and a combination of parallel- and series-connected for single-phase, three-phase three-wire, and three-phase four-wire networks. Parallel-connected PHPFs absorb the harmonic currents by all sources connected to the distribution network by providing slight impedance paths only at their tuned harmonic frequencies. In contrast, the series-connected PHPFs create high-impedance paths only at their tuned harmonic frequencies so that these paths avoid the flow of harmonic currents. Parallel-connected PHPFs also pass only a small portion of the load current, while their series counterparts carry the total current of the load. Accordingly, parallel-connected PHPFs, in comparison with their series counterparts, have several advantages, such as (1) fewer power ratings; (2) more suitable performance in supply reactive power compensation at the principal frequency; and, (3) cost-effective prices. The design of an optimal configuration for the PHPFs depends on many criteria, including the harmonic spectrum and the nature of the distortion. However, some of the most well-known PHPFs designed to improve power quality in industrial, commercial, and residential zones are (1) first-order damped and undamped high-pass PHPFs; (2) single-tuned low-pass PHPFs, known as band-pass or second-order series resonant band-pass PHPFs; (3) second-order damped high-pass PHPFs; (4) fourth-order double band-pass PHPFs; and, (5) composite PHPFs consisting of two second-order series resonant band-pass PHPFs and one second-order damped high-pass PHPF. For a comprehensive description of these configurations of the PHPFs, please refer to the work by Fuchs and Masoum [8] and Das [10].

Basically, the choice of suitable PHPFs must be made by considering different aspects, such as type of nonlinear load, type of compensation (i.e., harmonic current or voltage, reactive power), and required compensation levels. In general, PHPFs have many advantages that allow these filters to be consistently considered as an efficient option for improving power quality in distribution networks. Some of the most important ones are (1) low maintenance costs; (2) quick response time; (3) non-interference in short-circuit currents; (4) usability in large systems in terms of megavolt-ampere reactive (MVAr); (5) affordability and low cost; and, (6) powerful performance. With all of the strengths of these filters, however, weaknesses are also observed in these types of harmonic power filters. Examples of these weaknesses include (1) fixed compensation; (2) large size; (3) impossibility of adjusting the harmonic frequency or changing the filter size after installation; (4) changing the filter performance, due to changes in the operation conditions of the distribution network; (5) compensating for a limited number of harmonic components; and, (6) the possibility of creating resonance with the distribution network impedance and/or other PHPFs at fundamental or other harmonic frequencies.

An appropriate and well-designed solution for coping with the weaknesses associated with the PHPFs is the AHPFs [11]. By changing the harmonic spectrum generated by nonlinear loads or the distribution network configuration, the PHPFs lose their practical efficiency. In this situation, the AHPFs can be employed to provide dynamic compensation. That is, unlike the PHPFs, the AHPFs always adapt their performance to the operating conditions of the distribution networks. The AHPFs should act in such a way that the malicious effects of the nonlinear loads can be neutralized from the distribution network perspective. That is to say that the integration of the AHPFs with nonlinear loads in the PCC should be considered as a linear load from the distribution network perspective. To reach this target, these filters must inject harmonic currents into the PCC of the distribution network to eliminate non-sinusoidal characteristics of the nonlinear loads. The amplitude and phase angles of the injected harmonic currents should be equal and opposite to the harmonic currents generated by nonlinear loads, respectively. The AHPFs can be operated in series-connected or parallel-connected architectures. Selecting a well-suited architecture for the AHPFs depends on different indices, such as types of nonlinear loads and structural effects of the AHPFs on the operating condition of the distribution network. Some of the key advantages of AHPFs over PHPFs include (1) ability to compensate for a wide range of power quality problems; (2) show fine performance, even in the case of changes in the distribution network impedance; (3) ability of a single AHPF to compensate for more than one harmonic; and, (4) exhibit suitable responses to load and harmonic spectrum variations. One major disadvantage, however, is their power rating, which is very close to the load. Due to their high power rating and complex structure, AHPFs are an expensive option for improving power quality in most practical applications, especially for small facilities. AHPFs also do not afford suitable performance when simultaneously dealing with voltage and current power quality disturbances. The authors refer interested readers to the work by Morán et al. [4] and Fuchs and Masoum [8] for a detailed discussion of the role of AHPFs in distribution networks.

A suitable solution for taking full advantage of the PHPFs and AHPFs is the use of HHPFs. In the structure of HHPFs, passive and active sections are responsible for the filtering process in dominant harmonic frequencies (i.e., fifth and seventh) and the higher harmonic orders, respectively [8]. Thus, the size and cost of harmonic power filters installed on the distribution network are effectively reduced.

7.2.3 Harmonic Power Flow

In the literature on power systems, one can find different methodologies for solving the harmonic power flow problem. Classification of these methodologies can be accomplished from different perspectives and classified using three main perspectives: (1) modeling technique; (2) distribution network condition; and, (3) solution approach. From the point of view of the modeling technique, employed to simulate the components of the distribution network and nonlinear loads, these methodologies can further be classified into three main categories: (1) time domain; (2) harmonic domain; and, (3) hybrid time/harmonic domain. Time domain methodologies, which are essentially established under the transient analysis, have several strong points, such as suitable flexibility and high precision. However, one of the biggest weaknesses of these methodologies is that they require a high computational burden. This computational burden increases dramatically with increases in distribution network size, number of nonlinear loads, and harmonic couplings. In contrast to the time domain methodologies, harmonic domain methodologies are developed for computing the frequency response of the distribution network. Although harmonic domain methodologies are simple and require less computational burden than their counterparts in time domain methodologies, obtaining a precise frequency model of distribution network components and nonlinear loads by these methodologies is somewhat difficult. In order to cope with the weaknesses of the time and harmonic domain methodologies, and to benefit from their strengths, the hybrid time/harmonic domain methodologies were developed. In these hybrid methodologies, the time and harmonic domain sections are employed in order to simulate the distribution network components and nonlinear loads, respectively. Hence, the precision of the first category and the simplicity of the second are included in the hybrid time/harmonic domain methodologies. From the perspective of the condition of the distribution network, these methodologies can be divided into four main categories: (1) single phase; (2) three phase; (3) balanced; and, (4) unbalanced. Unbalanced methodologies are complicated and include a large amount of data; thus, their computational burden is not comparable with other methodologies. From the standpoint of the solution approach, these methodologies can be categorized by using two main categories: coupled and decoupled. Naturally, distribution network components and nonlinear loads create couplings among different harmonics. But, if harmonic couplings can be ignored without losing the generality of the destructive effects of the harmonics, the decoupled methodologies can be employed to solve the harmonic power flow problem; otherwise, the coupled methodologies must be applied.

In related literature, many different methodologies have been employed by researchers to solve the harmonic power flow problem; examples include Newton-based coupled harmonic power flow [12, 13], decoupled harmonic power flow [14, 15], fast harmonic power flow [15], modified fast decoupled harmonic power flow [16], fuzzy harmonic power flow [17], probabilistic harmonic power flow [18], modular harmonic power flow [19], and backward/forward sweep-based harmonic power flow [20]. The authors again refer interested readers to the work by Fuchs and Masoum [8] for a more detailed description on this topic.

7.2.4 Harmonic Power Filter Planning Problem

Basically, the placement and sizing problem of HHPFs for power quality improvement in distribution networks are known as the HHPF planning problem. The HHPF planning problem is presented in the form of an optimization problem, which can contain a wide range of objective functions and constraints. Here, the authors divide the objective functions of the harmonic power filter planning problem into two key categories: technical and economical objective functions. Technical objective functions attempt to reduce the destructive effects of the harmonics on distribution network components [21–31]. Total harmonic distortion of voltage (THDV), total harmonic distortion of current (THDI), transmission line loss arising from harmonics (TLLH), motor load loss function (MLLF), and telephone influence factor (TIF) are some of the most well-known and most used technical objective functions in the HHPF problem. Economical objective functions try to minimize the cost of installing the HHPFs in distribution networks [25, 26, 29–31]. It is important to note that the cost of installing PHPFs in distribution networks is directly related to the size of the passive components used, such as tuned capacitors and tuned inductors. Likewise, the cost of installing AHPFs in distribution networks is directly associated with the harmonic currents that they inject into the PCC of the distribution network. The authors also separate the constraints of the HHPF planning problem into four key categories [21–31]: (1) harmonic standard-based constraints; (2) technical constraints; (3) economic constraints; and, (4) logical constraints. The first category refers to the harmonic constraints that can be defined by various standards, such as the IEEE-519 standard. The harmonic standards applied to the individual harmonic distortions of voltage and current, the THDV and THDI in different locations of the distribution network, are among the most known harmonic constraints, as determined by the IEEE-519 standard. The second category specifies the technical constraints associated with the PHPFs, the AHPFs, and the distribution network. This category usually contains the limitations related to the principal reactive power compensation by the PHPFs, limitations associated with the series and parallel resonances, detuning limitations related to the PHPFs and the distribution network, restrictions pertaining to the discrete nature of the components of the AHPFs, limitations relevant to the maximum injected current by the AHPFs, and restrictions associated with the maximum installed size of PHPFs. The third category describes economic constraints. Limitations associated with the required budget for the purchase and installation of the PHPFs and AHPFs in the distribution network are placed in this category. Finally, the restrictions related to the number of allowable PHPFs and AHPFs that can be installed on a given bus of the distribution network are placed in the fourth category.

7.3 Hybrid Harmonic Power Filter Planning: A Techno-economic Framework

In this section, a techno-economic multi-objective framework is developed for HHPF planning problems in distribution networks that have to deal with uncertainty in demand and the amount of harmonic current injected by nonlinear loads. The proposed framework is divided into a harmonic power flow problem and an HHPF planning problem. The harmonic power flow problem acts as the central core of the HHPF planning problem. To solve the harmonic power flow problem, a probabilistic decoupled harmonic power flow (PDHPF) methodology is introduced. This proposed methodology has been practically developed so that the stochastic behavior of demand and nonlinear loads are scrutinized by an efficient two-point estimate method (two-PEM). Implementing the proposed PDHPF methodology requires the development of the deterministic decoupled harmonic power flow (DDHPF) methodology. The DDHPF methodology for the distribution network analysis at harmonic frequencies also requires the acquisition of information obtained from the power flow problem at the principal frequency. Thus, the loadability-based Newton-Raphson power flow (LBNRPF) methodology is employed to solve the power flow problem at the principal frequency. In this methodology, maximization of the loadability of the distribution network is considered as an objective function, while two different types of limitations—including Kirchhoff's point and loop rules and operational limits on the distribution network equipment—are modeled as constraints. In the HHPF planning problem, the DNO simultaneously optimizes the THDV, THDI, TLLH, and total cost of the harmonic power filters (TCHF) as four different objective functions. Concurrently, harmonic standard-based limitations, techno-economic, and logical constraints are taken into account as four different categories of the HHPF planning problem constraints. The individual harmonic distortions of voltage and current, the THDV and THDI in different locations of the distribution network, are constraints that belong to the first category. Limitations that fall into the second category include the principal reactive power compensation by the PHPFs, series and parallel resonances, detuning of the PHPFs and the distribution network, discrete nature of the components of the AHPFs, and maximum injected current by the AHPFs in the distribution network. The limitation of the maximum budget available to the DNO for purchase and installation of the PHPFs and the AHPFs in the distribution network is placed in the third category. The logical constraints comprise the number of allowable PHPFs and AHPFs that can be installed on each bus of the distribution network.

In the proposed techno-economic multi-objective framework, the DNO is authorized to use a parallel-connected combination of three different types of PHPFs and four different types of AHPFs to increase the flexibility in the harmonic filtering process in the distribution network and satisfy the predetermined acceptable harmonic levels by the IEEE-519 standard. These three types of PHPFs, which are configured in a parallel-connected format, are the fifth and seventh second-order series resonant band-pass PHPFs and the second-order damped high-pass PHPF. The difference between these four different types of AHPFs, which are configured in a parallel-connected format, is also in the maximum amount of injected harmonic current into the PCC of the distribution network. The following assumptions are employed in the formulation process of the proposed techno-economic multi-objective framework:

- The harmonic domain methodology is used to compute the frequency response of the distribution network.
- Coupling between different harmonic orders is ignored.
- The LBNRPF methodology is utilized to solve the power flow problem at the principal frequency.
- The PDHPF methodology is employed to solve the power flow problem at harmonic frequencies.
- The condition of the distribution network is considered to be a three-phase balanced condition.
- Uncertainties related to the demand/load and harmonic currents injected by nonlinear loads are considered and applied as a nondeterministic model in the techno-economic multi-objective framework.
- The efficient two-PEM method is widely used to scrutinize these uncertainty parameters in the proposed framework.
- For harmonic and reactive power compensation in the distribution network, the DNO is authorized to choose among various types of PHPFs and different types of AHPFs in an integrated framework.

To recognize, classify, and compare the outstanding characteristics of the previous harmonic power filter planning framework reported in the literature and the techno-economic multi-objective framework proposed in this chapter, an attribute table was designed (see Table 7.1).

7.3.1 Mathematical Model of the Techno-economic Multi-objective Framework

In this section, the mathematical model of the proposed techno-economic multi-objective framework is thoroughly developed and discussed.

Table 7.1 Attributes of previous frameworks reported in the harmonic power filter literature and the techno-economic multi-objective framework proposed in this study

Reference	Type and structure of harmonic power filters	Single or multi-objective framework	Objective function	Nondeterministic framework	VAr compensation	Series and parallel resonances	Detuning constraints of the PHPFs	Maximum permissible size of the AHPFs	Uncertainty consideration	Techno-economic aspects	Solution method
[22]	1 AHPF	Single objective	(1) THDV; (2) MLLF; (3) TLLH; and (4) AHPF size	No	No	No	No	Yes	No	No	Discrete particle swarm optimization
[23]	1 PHPF	Single objective	(1) THDV	No	No	Yes: Series resonance	No	No	No	No	Particle swarm optimization
[24]	1 AHPF	Single objective	(1) THDV and (2) AHPFs size	No	No	No	No	No	No	No	Modified particle swarm optimization
[25]	1 PHPF	Single objective	(1) Total cost	Nondeterministic framework using the Gram–Charlier series expansion	Yes	No	No	No	Yes: (1) demand and (2) harmonic current	No	Adaptive dynamic clone selection algorithm
[26]	1 AHPF	Single objective	(1) THDV; (2) voltage deviation; and (3) total cost	No	No	No	No	Yes	No	No	Improved discrete firefly algorithm
[27]	A series combination of 1 AHPF and 1 PHPF	Single objective	(1) THDV and (2) THDI	No	Yes	No	No	No	No	No	FORTRAN feasible sequential quadratic programming
[28]	1 PHPF	Single objective	(1) THDV and (2) THDI	No	Yes	Yes: Series and parallel resonances	No	No	No	No	FORTRAN feasible sequential quadratic programming

(continued)

Table 7.1 (continued)

Reference	Type and structure of harmonic power filters	Single or multi-objective framework	Objective function	Nondeterministic framework	VAr compensation	Series and parallel resonances	Detuning constraints of the PHPFs	Maximum permissible size of the AHPFs	Uncertainty consideration	Techno-economic aspects	Solution method
[29]	A parallel combination of 4 PHPFs	Multi-objective	(1) THDV; (2) THDI; (3) initial investment cost; and (4) fundamental power losses	No	Yes	Yes: Series resonance	Yes	No	No	Yes	Modified bat optimization algorithms
[30]	1 PHPF	Single objective	(1) TIC plus cost of energy lost	No	Yes	No	No	No	No	No	Bacterial foraging optimization
[31]	1 AHPF	Multi-objective	(1) THDV; (2) installation cost plus cost of losses; and (3) total demand distortion of current	Nondeterministic framework using trade-off/risk method	No	No	No	Yes	Yes: (1) harmonic current	Yes	Iterative approach
Proposed framework	A parallel-connected combination of 3 PHPFs and 4 AHPFs	Multi-objective	(1) THDV; (2) THDI; (3) TLLH; and (4) TCHF	Nondeterministic framework using the efficient 2-PEM	Yes	Yes: Series and parallel resonances	Yes	Yes	Yes: (1) demand and (2) harmonic current	Yes	Five new multi-objective music-inspired optimization algorithms (see Chap. 4): (1) SS-HSA; (2) SS-IHSA; (3) continuous/discrete TMS-MSA; (4) TMS-EMSA; and (5) SOSA

7.3.1.1 Deterministic Decoupled Harmonic Power Flow Methodology

It is well known that the central core of most power system studies is based on power flow problem calculations. Hence, the power flow problem is the most common problem that power system researchers/engineers deal with and are also most familiar with. Similar to most power system problems, the power flow problem is formed from a set of known parameters and unknown quantities that must be computed.

The purpose of the implementation of the power flow problem is to determine the active and reactive power flow in transmission lines and the magnitude and phase angle of the voltage on a certain set of network buses, provided that the network configuration and its primary data are known. The network is nearly linear in nature; however, this feature does not lead to a linear power flow problem. The nonlinearity of the power flow problem arises for two reasons: the presence of various types of nonlinear equipment in the distribution network and the power calculation procedure of multiplying the voltage and current. In the power flow problem, time variations of electrical loads, generation, and network configuration are ignored. The power flow problem is modeled in the sinusoidal steady state; hence, the power flow equations take an algebraic form instead of a differential form.

The network is usually operated under a phase-driven balanced condition. Therefore, a single-phase representation of the distribution network is sufficient for power flow problem calculations. In the power system literature, there are different numerical methodologies in varying versions employed by power researchers and engineers to solve the power flow problem. Each of these methodologies has its strengths and weaknesses. To investigate these methodologies in an orderly manner, please refer to the studies by Frank et al. [32, 33]. Here, though, the authors employ the LBNRPF methodology to solve the power flow problem at the principal frequency.

In general, bus b of the network can be one of the three types: slack bus, voltage-controlled bus—also known as PV bus—and load bus—also known as PQ bus. This bus of the network is also defined using four parameters: voltage amplitude, voltage phase angle, active power, and reactive power. For bus b, two parameters are ordinarily known, depending on the type of bus. The amplitude and phase angle of the voltage are known on the slack bus. Active power and the amplitude of the voltage are known parameters on PV buses. In addition, active and reactive powers are known parameters on PQ buses.

The step-by-step, or iterative, solution procedure of the LBNRPF methodology at the principal frequency can be summarized, as follows:

1. Consider bus 1 and buses 2 through B of the network as slack bus and linear PQ or PV buses.
2. Create the admittance matrix of the network, Y_{bus}, with dimensions $B \times B$, according to Eqs. (7.1) through (7.3):

$$\mathrm{Y}_{\mathrm{bus}}=\begin{bmatrix} Y_{1,1} & \cdots & Y_{1,b} & \cdots & Y_{1,\mathrm{B}} \\ \vdots & & \vdots & & \vdots \\ Y_{b,1} & \cdots & Y_{b,b} & \cdots & Y_{b,\mathrm{B}} \\ \vdots & & \vdots & & \vdots \\ Y_{\mathrm{B},1} & \cdots & Y_{\mathrm{B},2} & \cdots & Y_{\mathrm{B},\mathrm{B}} \end{bmatrix}; \quad \forall\{b\in\Psi^{\mathrm{B}}\} \tag{7.1}$$

$$Y_{b,b}=y_b^{\mathrm{PA}}+\sum_{\substack{\hat{b}\in\Psi_b^{\mathrm{B}}\\ \hat{b}\neq b}} y_{b,\hat{b}}; \quad \forall\{b\in\Psi^{\mathrm{B}},\ \hat{b}\in\Psi_b^{\mathrm{B}},\ \hat{b}\neq b\} \tag{7.2}$$

$$Y_{b,b'}=-y_{b,b'}; \quad \forall\left\{b,b'\in\Psi^{\mathrm{B}},\ b\neq b'\right\} \tag{7.3}$$

In the admittance matrix of the network, the main diagonal element, $Y_{b,\,b}$, is computed as the sum of the admittances that connect to bus b of the network, including parallel-connected admittance—parallel-connected capacitor bank—that is available in the initial configuration of the network [see Eq. (7.2)]. The off-diagonal element, $Y_{b,b'}$, is also the negative of the admittance that exists between buses b and b' of the network [see Eq. (7.3)].

3. Consider an initial guess for the network buses voltage vector, D, with dimensions $2(\mathrm{B}-1)\times 1$, according to Eq. (7.4):

$$\mathrm{D}=\left[\phi_2\ \ |V_2|\ \cdots\ \phi_b\ \ |V_b|\ \cdots\ \phi_{\mathrm{B}}\ \ |V_{\mathrm{B}}|\right]^{\mathrm{T}}; \quad \forall\{b\in\Psi^{\mathrm{B}}\} \tag{7.4}$$

In Eq. (7.4), the sign "T" means transpose. It should be noted that initial guesses for all network buses are considered as 1.0 per-unit (p.u.) volt and zero phase angle.

4. Set the iteration counter ($u=1$).
5. Set $D^{(u)}=\mathrm{D}$.
6. Calculate the mismatch power vector, $\Delta W^{(u)}(D^{(u)})$, with dimensions $2(\mathrm{B}-1)\times 1$, in accordance with Eq. (7.5):

$$\Delta W^{(u)}\left(D^{(u)}\right) = \Bigg[\overbrace{\underbrace{P_2+F_{r,2},}_{\Delta P_2}\ \ \underbrace{Q_2+F_{i,2},}_{\Delta Q_2}}^{\Delta W_2}\ \cdots\ \overbrace{\underbrace{P_b+F_{r,b},}_{\Delta P_b}\ \ \underbrace{Q_b+F_{i,b},}_{\Delta Q_b}}^{\Delta W_b}\ \cdots\ \overbrace{\underbrace{P_{\mathrm{B}}+F_{r,\mathrm{B}},}_{\Delta P_{\mathrm{B}}}\ \ \underbrace{Q_{\mathrm{B}}+F_{i,\mathrm{B}}}_{\Delta Q_{\mathrm{B}}}}^{\Delta W_{\mathrm{B}}}\Bigg]^{\mathrm{T}}$$
$$\forall\{b\in\Psi^{\mathrm{B}}\} \tag{7.5}$$

The mismatch power on bus b of the network, ΔW_b, can be broken down into two parts: the real part, ΔP_b, and the imaginary part, ΔQ_b [see Eq. (7.5)]. Real mismatch power on bus b of the network, ΔP_b, is also made up of two parts: active power demand on bus P_b and the sum of the active power in power transmission lines connected to bus $F_{r,\,b}$ [see Eq. (7.5)]. In addition, imaginary mismatch power on bus b of the network, ΔQ_b, is made up of two parts: reactive power demand on bus Q_b and the sum of the reactive power in transmission lines connected to bus $F_{i,\,b}$ [see Eq. (7.5)]. Note that $F_{r,\,b}$ and $F_{i,\,b}$ are defined using Eqs. (7.6) and (7.7), respectively:

$$F_{r,b} = \sum_{\substack{\hat{b}\in\Psi_b^{\mathrm{B}} \\ b\neq\hat{b}}} Y_{b,\hat{b}} \cdot V_{\hat{b}} \cdot V_b \cdot \cos\left(-\theta_{b,\hat{b}} - \phi_{\hat{b}} + \phi_b\right); \quad \forall\left\{b \in \Psi^{\mathrm{B}}, \hat{b} \in \Psi_b^{\mathrm{B}}, b \neq \hat{b}\right\} \tag{7.6}$$

$$F_{i,b} = \sum_{\substack{\hat{b}\in\Psi_b^{\mathrm{B}} \\ b\neq\hat{b}}} Y_{b,\hat{b}} \cdot V_{\hat{b}} \cdot V_b \cdot \sin\left(-\theta_{b,\hat{b}} - \phi_{\hat{b}} + \phi_b\right); \quad \forall\left\{b \in \Psi^{\mathrm{B}}, \hat{b} \in \Psi_b^{\mathrm{B}}, b \neq \hat{b}\right\} \tag{7.7}$$

7. Stop if the mismatch power vector is small enough; otherwise, go to the next step.

   It is important to note that convergence tolerance may be different in each study. Here, convergence tolerance is considered to be 0.0001.
8. Create a Jacobian matrix, J, with dimensions 2(B − 1) × 2(B − 1), according to Eq. (7.8):

$$\mathrm{J} = \begin{bmatrix} \frac{\partial \Delta P_2}{\partial \phi_2} & \frac{\partial \Delta P_2}{\partial V_2} & \cdots & \frac{\partial \Delta P_2}{\partial \phi_b} & \frac{\partial \Delta P_2}{\partial V_b} & \cdots & \frac{\partial \Delta P_2}{\partial \phi_{\mathrm{B}}} & \frac{\partial \Delta P_2}{\partial V_{\mathrm{B}}} \\ \frac{\partial \Delta Q_2}{\partial \phi_2} & \frac{\partial \Delta Q_2}{\partial V_2} & \cdots & \frac{\partial \Delta Q_2}{\partial \phi_b} & \frac{\partial \Delta Q_2}{\partial V_b} & \cdots & \frac{\partial \Delta Q_2}{\partial \phi_{\mathrm{B}}} & \frac{\partial \Delta Q_2}{\partial V_{\mathrm{B}}} \\ \vdots & \vdots & & \vdots & \vdots & & \vdots & \vdots \\ \frac{\partial \Delta P_b}{\partial \phi_2} & \frac{\partial \Delta P_b}{\partial V_2} & \cdots & \frac{\partial \Delta P_b}{\partial \phi_b} & \frac{\partial \Delta P_b}{\partial V_b} & \cdots & \frac{\partial \Delta P_b}{\partial \phi_{\mathrm{B}}} & \frac{\partial \Delta P_b}{\partial V_{\mathrm{B}}} \\ \frac{\partial \Delta Q_b}{\partial \phi_2} & \frac{\partial \Delta Q_b}{\partial V_2} & \cdots & \frac{\partial \Delta Q_b}{\partial \phi_b} & \frac{\partial \Delta Q_b}{\partial V_b} & \cdots & \frac{\partial \Delta Q_b}{\partial \phi_{\mathrm{B}}} & \frac{\partial \Delta Q_b}{\partial V_{\mathrm{B}}} \\ \vdots & \vdots & & \vdots & \vdots & & \vdots & \vdots \\ \frac{\partial \Delta P_{\mathrm{B}}}{\partial \phi_2} & \frac{\partial \Delta P_{\mathrm{B}}}{\partial V_2} & \cdots & \frac{\partial \Delta P_{\mathrm{B}}}{\partial \phi_b} & \frac{\partial \Delta P_{\mathrm{B}}}{\partial V_b} & \cdots & \frac{\partial \Delta P_{\mathrm{B}}}{\partial \phi_{\mathrm{B}}} & \frac{\partial \Delta P_{\mathrm{B}}}{\partial V_{\mathrm{B}}} \\ \frac{\partial \Delta Q_{\mathrm{B}}}{\partial \phi_2} & \frac{\partial \Delta Q_{\mathrm{B}}}{\partial V_2} & \cdots & \frac{\partial \Delta Q_{\mathrm{B}}}{\partial \phi_b} & \frac{\partial \Delta Q_{\mathrm{B}}}{\partial V_b} & \cdots & \frac{\partial \Delta Q_{\mathrm{B}}}{\partial \phi_{\mathrm{B}}} & \frac{\partial \Delta Q_{\mathrm{B}}}{\partial V_{\mathrm{B}}} \end{bmatrix}; \quad \forall\left\{b \in \Psi^{\mathrm{B}}\right\} \tag{7.8}$$

In the Jacobian matrix, the main diagonal and the off-diagonal elements are addressed on the basis of Eqs. (7.9) through (7.16):

$$\frac{\partial \Delta P_b}{\partial \phi_b} = -\sum_{\substack{b' \in \Psi^{B} \\ b' \neq b}} Y_{b,b'} \cdot V_b \cdot V_{b'} \cdot \sin\left(\phi_b - \phi_{b'} - \theta_{b,b'}\right); \quad \forall\left\{b, b' \in \Psi^{B}, b \neq b'\right\} \tag{7.9}$$

$$\frac{\partial \Delta P_b}{\partial \phi_{b'}} = Y_{b,b'} \cdot V_b \cdot V_{b'} \cdot \sin\left(\phi_b - \phi_{b'} - \theta_{b,b'}\right); \quad \forall\left\{b, b' \in \Psi^{B}, b \neq b'\right\} \tag{7.10}$$

$$\frac{\partial \Delta P_b}{\partial V_b} = \sum_{\substack{b' \in \Psi^{B} \\ b' \neq b}} Y_{b,b'} \cdot V_{b'} \cdot \cos\left(\phi_b - \phi_{b'} - \theta_{b,b'}\right) + 2 \cdot V_b \cdot Y_{b,b} \cdot \cos\left(-\theta_{b,b}\right); \quad \forall\left\{b, b' \in \Psi^{B}, b \neq b'\right\} \tag{7.11}$$

$$\frac{\partial \Delta P_b}{\partial V_{b'}} = Y_{b,b'} \cdot V_b \cdot \cos\left(\phi_b - \phi_{b'} - \theta_{b,b'}\right); \quad \forall\left\{b, b' \in \Psi^{B}, b \neq b'\right\} \tag{7.12}$$

$$\frac{\partial \Delta Q_b}{\partial \phi_b} = -\sum_{\substack{b' \in \Psi^{B} \\ b' \neq b}} Y_{b,b'} \cdot V_b \cdot V_{b'} \cdot \cos\left(\phi_b - \phi_{b'} - \theta_{b,b'}\right); \quad \forall\left\{b, b' \in \Psi^{B}, b \neq b'\right\} \tag{7.13}$$

$$\frac{\partial \Delta Q_b}{\partial \phi_{b'}} = -Y_{b,b'} \cdot V_b \cdot V_{b'} \cdot \cos\left(\phi_b - \phi_{b'} - \theta_{b,b'}\right); \quad \forall\left\{b, b' \in \Psi^{B}, b \neq b'\right\} \tag{7.14}$$

$$\frac{\partial \Delta Q_b}{\partial V_b} = \sum_{\substack{b' \in \Psi^{B} \\ b' \neq b}} Y_{b,b'} \cdot V_{b'} \cdot \sin\left(\phi_b - \phi_{b'} - \theta_{b,b'}\right) + 2 \cdot V_b \cdot Y_{b,b} \cdot \sin\left(-\theta_{b,b}\right); \quad \forall\left\{b, b' \in \Psi^{B}, b \neq b'\right\} \tag{7.15}$$

$$\frac{\partial \Delta Q_b}{\partial V_{b'}} = Y_{b,b'} \cdot V_b \cdot \sin\left(\phi_b - \phi_{b'} - \theta_{b,b'}\right); \quad \forall\left\{b, b' \in \Psi^{B}, b \neq b'\right\} \tag{7.16}$$

9. Calculate the corrective vector, $\Delta D^{(u)}$, based on Eq. (7.17):

$$\Delta D^{(u)} = J^{-1} \cdot \Delta W^{(u)}\left(D^{(u)}\right) \tag{7.17}$$

10. Update the network buses' voltage vector, $D^{(u)}$, according to Eq. (7.18):

$$D^{(u+1)} = D^{(u)} - \Delta D^{(u)} \tag{7.18}$$

11. Set $u = u + 1$and go to step 6.

The power flow methodologies perform the power flow problem calculations at the principal harmonic order, $h = 1$, which is called the principal frequency: $f^1 = 50$ or 60 Hz. But, when network harmonic analysis is required, these methodologies are inefficient. To meet this need, different methodologies have been developed by power researchers/engineers, as described in Sect. 7.2.3. Among these, the Newton-based harmonic power flow methodology has the highest accuracy. The reason for this high accuracy is that this methodology takes into account harmonic coupling at all frequencies. However, the Newton-based harmonic power flow methodology requires a very large amount of computational burden and involves convergence difficulties, especially in real-world large-scale networks with a large number of nonlinear loads [8]. In this circumstance, a decoupled harmonic power flow methodology is a more appropriate and economical choice [8, 14, 15]. In this methodology, to decrease the computational burden and memory required for storing data, harmonic coupling at all frequencies is ignored. For a more detailed description of this methodology, please refer to the work by Chin [14] and Teng and Chang [15]. Without loss of generality of the harmonic power flow, the current authors employed the DDHPF methodology in order to solve the power flow problem at harmonic frequencies. Before introducing the step-by-step solution procedure of the DDHPF methodology, it is necessary to describe some concepts related to the power in the presence of harmonic distortions.

First, the definitions of voltage and current in the presence of their root mean square (RMS) harmonic components are given by Eqs. (7.19) and (7.20), respectively:

$$v(t) = \mathrm{V}_0 + \sum_{h \in \{1,\ldots,\infty\}} \sqrt{2} \cdot V^h \cdot \sin\left(h \cdot \omega_1 \cdot t + \theta_v^h\right); \quad \forall \{h \in \{1, \ldots, \infty\}\} \tag{7.19}$$

$$i(t) = \mathrm{I}_0 + \sum_{h \in \{1,\ldots,\infty\}} \sqrt{2} \cdot I^h \cdot \sin\left(h \cdot \omega_1 \cdot t + \theta_i^h\right); \quad \forall \{h \in \{1, \ldots, \infty\}\} \tag{7.20}$$

Using these definitions, active, reactive, and apparent power are presented by Eqs. (7.21) through (7.23), respectively:

$$\mathrm{P} = \mathrm{V}_0 \cdot \mathrm{I}_0 + \sum_{h \in \{1,\ldots,\infty\}} V^h \cdot I^h \cdot \cos\left(\theta_v^h - \theta_i^h\right); \quad \forall \{h \in \{1, \ldots, \infty\}\} \tag{7.21}$$

$$\mathrm{Q} = \sum_{h\in\{1,\ldots,\infty\}} V^h \cdot I^h \cdot \sin\left(\theta_v^h - \theta_i^h\right); \quad \forall\{h \in \{1,\ldots,\infty\}\} \tag{7.22}$$

$$\mathrm{S} = \sqrt{\left(\sum_{h\in\{0,\ldots,\infty\}} \left(V^h\right)^2\right) \cdot \left(\sum_{h\in\{0,\ldots,\infty\}} \left(I^h\right)^2\right)}; \quad \forall\{h \in \{0,\ldots,\infty\}\} \tag{7.23}$$

In the absence of harmonic distortions, the relationship between active, reactive, and apparent power is described by Eq. (7.24):

$$\mathrm{S} = \sqrt{\mathrm{P}^2 + \mathrm{Q}^2} \tag{7.24}$$

In the presence of harmonic distortions, Eq. (7.24) is not true; hence, to integrate distortion power, E, this equation must be rewritten as shown in Eq. (7.25):

$$\mathrm{S} = \sqrt{\mathrm{P}^2 + \mathrm{Q}^2 + \mathrm{E}^2} \tag{7.25}$$

The harmonic frequency at harmonic order h can be calculated according to Eq. (7.26):

$$f^h = h \cdot f^1; \quad \forall\left\{h \in \Psi^{\mathrm{H}}\right\} \tag{7.26}$$

Given these descriptions, the step-by-step solution procedure of the DDHPF methodology can be summarized as follows:

1. Run the LBNRPF methodology for the principal frequency to compute the amplitude and phase angle of voltage on all buses of the network, as previously indicated in Eqs. (7.1) through (7.18). In this step, all nonlinear loads available on the network are treated as linear loads.
2. Construct the harmonic order vector, HOV, with the dimensions of one times the number of harmonic orders by using Eq. (7.27):

$$\mathrm{HOV} = \left[5 \quad 7 \quad \cdots \quad h \quad \cdots \quad \mathrm{H}\right]; \quad \forall\left\{h \in \Psi^{\mathrm{H}}\right\} \tag{7.27}$$

3. Set the harmonic order counter ($hc = 1$).
4. Specify the harmonic order as $h = \mathrm{HV}(hc)$.
5. Establish the linear load model that is available on bus b of the network for harmonic order h by using Eq. (7.28):

$$y_b^h = \frac{\left(P_b^1 - \mathrm{j}\frac{Q_b^1}{h}\right)}{\left|V_b^1\right|^2}; \quad \forall\left\{b \in \Psi^{\mathrm{B}}, h \in \Psi^{\mathrm{H}}\right\} \tag{7.28}$$

Note that the letter "j" in Eq. (7.28) illustrates the operator $\sqrt{-1}$. In the principal frequency, the linear loads available on the network are considered as PV or PQ buses. In the harmonic frequencies, however, linear loads are modeled as a parallel-connected admittance [see Eq. (7.28)].

6. Organize the parallel-connected admittance model that is available on bus b of the initial configuration of the network for harmonic order h according to Eq. (7.29):

$$y_b^{h,\mathrm{PA}} = h \cdot y_b^{1,\mathrm{PA}} = h \cdot y_b^{\mathrm{PA}}; \quad \forall \{ b \in \Psi^{\mathrm{B}}, h \in \Psi^{\mathrm{H}} \} \tag{7.29}$$

As previously mentioned, the parallel-connected admittance refers to the parallel-connected capacitor bank that is available in the initial configuration of the network.

7. Arrange the transmission line model that is located between buses b and b' of the network for harmonic order h on the basis of Eq. (7.30):

$$y_{b,b'}^{h} = \frac{1}{R_{b,b'} + \mathrm{j}h \cdot X_{b,b'}}; \quad \forall \left\{ b, b' \in \Psi^{\mathrm{B}}, b \neq b', h \in \Psi^{\mathrm{H}} \right\} \tag{7.30}$$

Again, the letter "j" in Eq. (7.30) represents the operator $\sqrt{-1}$.

8. Create the harmonic admittance matrix of the network for harmonic order h, Y_{bus}^{h}, with dimensions $\mathrm{B} \times \mathrm{B}$, based on Eqs. (7.31) through (7.33):

$$Y_{\mathrm{bus}}^{h} = \begin{bmatrix} Y_{1,1}^{h} & \cdots & Y_{1,b}^{h} & \cdots & Y_{1,\mathrm{B}}^{h} \\ \vdots & & \vdots & & \vdots \\ Y_{b,1}^{h} & \cdots & Y_{b,b}^{h} & \cdots & Y_{b,\mathrm{B}}^{h} \\ \vdots & & \vdots & & \vdots \\ Y_{\mathrm{B},1}^{h} & \cdots & Y_{\mathrm{B},2}^{h} & \cdots & Y_{\mathrm{B},\mathrm{B}}^{h} \end{bmatrix}; \quad \forall \{ b \in \Psi^{\mathrm{B}}, h \in \Psi^{\mathrm{H}} \} \tag{7.31}$$

$$Y_{b,b}^{h} = y_b^{h,\mathrm{PA}} + \sum_{\substack{\hat{b} \in \Psi_b^{\mathrm{B}} \\ \hat{b} \neq b}} y_{b,\hat{b}}^{h}; \quad \forall \{ b \in \Psi^{\mathrm{B}}, \hat{b} \in \Psi_b^{\mathrm{B}}, \hat{b} \neq b, h \in \Psi^{\mathrm{H}} \} \tag{7.32}$$

$$Y_{b,b'}^{h} = -y_{b,b'}^{h}; \quad \forall \left\{ b, b' \in \Psi^{\mathrm{B}}, b \neq b', h \in \Psi^{\mathrm{H}} \right\} \tag{7.33}$$

In the admittance matrix of the network for harmonic order h, the main diagonal element $Y_{b,b}^{h}$ is computed as the sum of the admittances connected to bus b of the network at harmonic order h, including parallel-connected admittance that is available in the initial configuration of the network [see Eq. (7.32)]. The off-diagonal element $Y_{b,b'}^{h}$ is also the negative of the admittance that exists between buses b and b' of the network for harmonic order h [see Eq. (7.33)]. For a principal harmonic order, the admittance matrix of the network, Y^1_{bus}, is quite similar to the Y_{bus} that can be calculated using Eq. (7.1).

9. Specify the nonlinear load model that is available on bus b of the network for harmonic order h in accordance with Eqs. (7.34) and (7.35):

$$I_b^{h,\text{NLL}} = A_b^h \cdot I_b^{1,\text{NLL}}; \quad \forall\{b \in \Psi^{\text{B}}, h \in \Psi^{\text{H}}\} \tag{7.34}$$

$$I_b^{1,\text{NLL}} = \left(\frac{\left(P_b^1 + \text{j}Q_b^1\right)}{\left|V_b^1\right|}\right)^*; \quad \forall\{b \in \Psi^{\text{B}}\} \tag{7.35}$$

In Eq. (7.35), the star symbol represents the complex conjugate. In the proposed DDHPF methodology, nonlinear loads are modeled as decoupled current sources that inject harmonic currents into the PCC of the network. The current injected by the nonlinear load that is located on bus b of the network into the PCC for the harmonic order h is expressed as a ratio of the current injected by it at the principal frequency [see Eq. (7.34)]. The current injected by the nonlinear load that is available on bus b of the network for the principal frequency depends on its active and reactive power at the principal frequency and also on the voltage magnitude on the relevant bus for the principal frequency [see Eq. (7.35)].

10. Solve the decoupled power flow relationship for harmonic order h by using Eq. (7.36):

$$\begin{bmatrix} I_1^h \\ \vdots \\ I_b^h \\ \vdots \\ I_{\text{B}}^h \end{bmatrix} = \begin{bmatrix} Y_{1,1}^h & \cdots & Y_{1,b}^h & \cdots & Y_{1,\text{B}}^h \\ \vdots & & \vdots & & \vdots \\ Y_{b,1}^h & \cdots & Y_{b,b}^h & \cdots & Y_{b,\text{B}}^h \\ \vdots & & \vdots & & \vdots \\ Y_{\text{B},1}^h & \cdots & Y_{\text{B},2}^h & \cdots & Y_{\text{B},\text{B}}^h \end{bmatrix} \cdot \begin{bmatrix} V_1^h \\ \vdots \\ V_b^h \\ \vdots \\ V_{\text{B}}^h \end{bmatrix}; \quad \forall\{b \in \Psi^{\text{B}}, h \in \Psi^{\text{H}}\} \tag{7.36}$$

11. $hc = hc + 1$; if $hc \leq$ size (HOV) go to step (4); otherwise, go to the next step.
12. Stop.

7.3.1.2 Passive and Active Harmonic Power Filters

In this section, the mathematical models of the PHPFs and AHPFs in the proposed techno-economic multi-objective framework are thoroughly developed.

Passive harmonic power filters: As stated earlier, in the proposed techno-economic multi-objective framework, a parallel-connected combination of the fifth and seventh second-order series resonant band-pass PHPFs and a second-order damped high-pass PHPF is used to compensate harmonics and reactive power, as shown in Fig. 7.1. To model the characteristics of the PHPFs, it is assumed that each

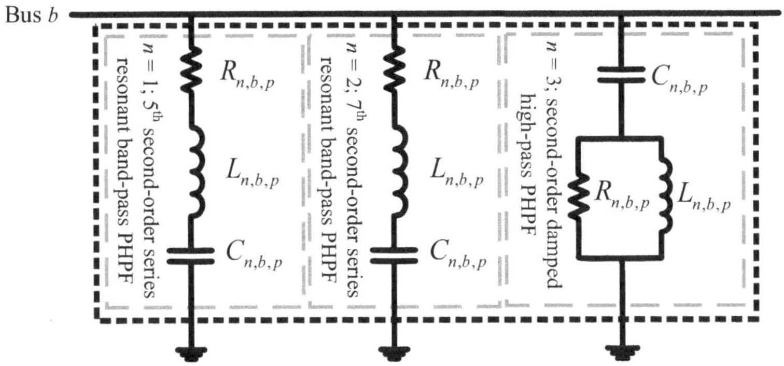

Fig. 7.1 The proposed architecture of the PHPFs on bus b of the distribution network

bus of the distribution network could be a candidate bus for installing "P" candidate branches associated with PHPFs of type n.

In order to more easily control and appropriately adjust the resonance frequency of PHPF p of type n on bus b of the distribution network near its expected worth, two parameters—the tuned frequency and the quality coefficient—are defined and formulated according to Eqs. (7.37) and (7.38), respectively:

$$\chi_{n,b,p} = \begin{cases} \dfrac{1}{\omega_0 \cdot \sqrt{L_{n,b,p} \cdot C_{n,b,p}}}; & \forall \{n \in \Psi^{\mathrm{N}-1}, b \in \Psi^{\mathrm{B}}, p \in \Psi_n^{\mathrm{P}}\} \\ \dfrac{1}{\omega_0 \cdot R_{n,b,p} \cdot C_{n,b,p}}; & \forall \{n = \mathrm{N}, b \in \Psi^{\mathrm{B}}, p \in \Psi_n^{\mathrm{P}}\} \end{cases} \tag{7.37}$$

$$\gamma_{n,b,p} = \begin{cases} \dfrac{\sqrt{\dfrac{L_{n,b,p}}{C_{n,b,p}}}}{R_{n,b,p}}; & \forall \{n \in \Psi^{\mathrm{N}-1}, b \in \Psi^{\mathrm{B}}, p \in \Psi_n^{\mathrm{P}}\} \\ \dfrac{L_{n,b,p}}{\left(R_{n,b,p}\right)^2 . C_{n,b,p}}; & \forall \{n = \mathrm{N}, b \in \Psi^{\mathrm{B}}, p \in \Psi_n^{\mathrm{P}}\} \end{cases} \tag{7.38}$$

In the formulation of PHPFs, the fifth and seventh second-order series resonant band-pass PHPFs are modeled from superscript "1" to superscript "N − 1"; superscript "N" is devoted to the second-order damped high-pass PHPF. Given these equations, the parameters pertaining to PHPF p of type n on bus b of the distribution network can be determined in accordance with Eqs. (7.39) and (7.40):

$$R_{n,b,p} = \begin{cases} \dfrac{1}{\chi_{n,b,p} \cdot \omega_0 \cdot C_{n,b,p} \cdot \gamma_{n,b,p}}; & \forall \{n \in \Psi^{\mathrm{N}-1}, b \in \Psi^{\mathrm{B}}, p \in \Psi_n^{\mathrm{P}}\} \\ \dfrac{1}{\chi_{n,b,p} \cdot \omega_0 \cdot C_{n,b,p}}; & \forall \{n = \mathrm{N}, b \in \Psi^{\mathrm{B}}, p \in \Psi_n^{\mathrm{P}}\} \end{cases} \tag{7.39}$$

$$L_{n,b,p}=\begin{cases}\dfrac{1}{(\chi_{n,b,p})^2\cdot(\omega_0)^2\cdot C_{n,b,p}}; & \forall\{n\in\Psi^{N-1},b\in\Psi^{B},p\in\Psi_n^{P}\}\\ \dfrac{\gamma_{n,b,p}}{(\chi_{n,b,p})^2\cdot(\omega_0)^2\cdot C_{n,b,p}}; & \forall\{n=N,b\in\Psi^{B},p\in\Psi_n^{P}\}\end{cases} \tag{7.40}$$

Since it is possible that some types of PHPFs are not employed by the DNO, a binary variable, $\alpha_{n,\,b,\,p}$, must be defined that addresses the presence ($\alpha_{n,\,b,\,p}=1$) or absence ($\alpha_{n,\,b,\,p}=0$) of PHPF p of type n on bus b of the distribution network. The equivalent admittance related to PHPF p of type n on bus b of the distribution network for the harmonic order h can be derived using Eq. (7.41):

$$Y_{n,b,p}^{h}=\begin{cases}\dfrac{\alpha_{n,b,p}}{\dfrac{1}{\chi_{n,b,p}\cdot\omega_0\cdot C_{n,b,p}\cdot\gamma_{n,b,p}}+\mathrm{j}\left(\dfrac{h\cdot\omega}{(\chi_{n,b,p})^2\cdot(\omega_0)^2\cdot C_{n,b,p}}-\dfrac{1}{h\cdot\omega\cdot C_{n,b,p}}\right)};\\ \forall\{n\in\Psi^{N-1},b\in\Psi^{B},p\in\Psi_n^{P},h\in\Psi^{H}\}\\ \dfrac{\alpha_{n,b,p}}{\dfrac{\dfrac{(\gamma_{n,b,p}\cdot h\cdot\omega)^2}{\chi_{n,b,p}\cdot\omega_0\cdot C_{n,b,p}}}{(\gamma_{n,b,p}\cdot h\cdot\omega)^2+(\chi_{n,b,p}\cdot\omega_0)^2}\cdot\left(1+\mathrm{j}\left[\dfrac{\chi_{n,b,p}\cdot\omega_0}{\gamma_{n,b,p}\cdot h\cdot\omega}-\dfrac{\left[(\gamma_{n,b,p}\cdot h\cdot\omega)^2+(\chi_{n,b,p}\cdot\omega_0)^2\right]\cdot\chi_{n,b,p}}{\dfrac{(\gamma_{n,b,p})^2\cdot(h\cdot\omega)^3}{\omega_0}}\right]\right)};\end{cases}$$

$$\forall\{n=N,b\in\Psi^{B},p\in\Psi_n^{P},h\in\Psi^{H}\} \tag{7.41}$$

An additional admittance matrix resulting from the installation of various types of PHPFs on different buses of the distribution network must be formed according to Eq. (7.42) and then added to the admittance matrix of the distribution network:

$$\Delta Y^{h}=\mathrm{diagonal}\left[\sum_{n\in\Psi^{N}}\sum_{p\in\Psi_n^{P}}\alpha_{n,1,p}\cdot Y_{n,1,p}^{h}\cdots\sum_{n\in\Psi^{N}}\sum_{p\in\Psi_n^{P}}\alpha_{n,b,p}\cdot Y_{n,b,p}^{h}\cdots\sum_{n\in\Psi^{N}}\sum_{p\in\Psi_n^{P}}\alpha_{n,B,p}\cdot Y_{n,B,p}^{h}\right]$$
$$\forall\{n\in\Psi^{N},b\in\Psi^{B},p\in\Psi_n^{P},h\in\Psi^{H}\} \tag{7.42}$$

This additional admittance matrix can be expressed in another way, based on Eqs. (7.43) and (7.44):

$$\Delta Y^{h}=\sum_{n\in\Psi^{N}}\Delta Y_n^{h};\quad\forall\{n\in\Psi^{N},h\in\Psi^{H}\} \tag{7.43}$$

$$\Delta Y_n^{h}=\mathrm{diagonal}\left[\sum_{p\in\Psi_n^{P}}\alpha_{n,1,p}\cdot Y_{n,1,p}^{h}\quad\cdots\quad\sum_{p\in\Psi_n^{P}}\alpha_{n,b,p}\cdot Y_{n,b,p}^{h}\quad\cdots\right.$$
$$\left.\sum_{p\in\Psi_n^{P}}\alpha_{n,B,p}\cdot Y_{n,B,p}^{h}\right];\ \forall\{n\in\Psi^{N},b\in\Psi^{B},p\in\Psi_n^{P},h\in\Psi^{H}\} \tag{7.44}$$

The harmonic current passing through the capacitance relevant to PHPFs p of type n on bus b of the distribution network for harmonic order h and harmonic voltage drop on the capacitance associated with PHPFs p of type n on bus b of the distribution network for harmonic order h are two new relationships that arise according to Eqs. (7.45) and (7.46), respectively:

$$I_{n,b,p}^{h,\text{cap}} = Y_{n,b,p}^{h} . V_b^h; \quad \forall \{n \in \Psi^{\text{N}}, b \in \Psi^{\text{B}}, p \in \Psi_n^{\text{P}}, h \in \Psi^{\text{H}}\} \tag{7.45}$$

$$V_{n,b,p}^{h,\text{cap}} = -\text{j}\frac{I_{n,b,p}^{h,\text{cap}}}{h \cdot \omega \cdot C_{n,b,p}}; \quad \forall \{n \in \Psi^{\text{N}}, b \in \Psi^{\text{B}}, p \in \Psi_n^{\text{P}}, h \in \Psi^{\text{H}}\} \tag{7.46}$$

As before, the letter "j" expresses the operator $\sqrt{-1}$.

Active harmonic power filters: As previously described, in the proposed techno-economic multi-objective framework, a parallel-connected configuration of four types of AHPFs is employed to tackle power quality problems in the distribution networks, as shown in Fig. 7.2. The difference between these AHPFs is in their nominal power or the maximum current that they can inject into the PCC of the distribution network. In general, an AHPF is modeled as a current source, which injects non-sinusoidal harmonic contents of nonlinear loads with equal amplitude and opposite phase angle except principal component into the PCC of the distribution network to clear corresponding harmonic distortions. The phasor model related to AHPF a of type m on bus b of the distribution network for harmonic order h can be expressed by Eq. (7.47):

$$I_{m,b,a}^{h} = I_{m,b,a}^{h,r} + \text{j}I_{m,b,a}^{h,i}; \quad \forall \{m \in \Psi^{\text{M}}, b \in \Psi^{\text{B}}, a \in \Psi_m^{\text{A}}\} \tag{7.47}$$

In addition, the RMS current of this AHPF can be calculated using Eq. (7.48):

$$I_{m,b,a} = \sqrt{\sum_{h \in \Psi^{\text{H}}} \left(\left(I_{m,b,a}^{h,r} \right)^2 + \left(I_{m,b,a}^{h,i} \right)^2 \right)}; \quad \forall \{m = \text{M}, b \in \Psi^{\text{B}}, a \in \Psi_m^{\text{A}}\} \tag{7.48}$$

For the same reason as described for the PHPFs, a binary variable, $\beta_{m,\, b,\, a}$, is also defined in the design process of the AHPFs that describes the presence ($\beta_{m,\, b,\, a} = 1$) or absence ($\beta_{m,\, b,\, a} = 0$) of AHPF a of type m on bus b of the distribution network. Hence, Eqs. (7.47) and (7.48) can be rewritten as shown in Eqs. (7.49) and (7.50), respectively:

$$I_{m,b,a}^{h} = \beta_{m,b,a} \cdot \left(I_{m,b,a}^{h,r} + \text{j}I_{m,b,a}^{h,i} \right); \quad \forall \{m \in \Psi^{\text{M}}, b \in \Psi^{\text{B}}, a \in \Psi_m^{\text{A}}\} \tag{7.49}$$

$$I_{m,b,a} = \beta_{m,b,a} \cdot \left(\sqrt{\sum_{h \in \Psi^{\text{H}}} \left(\left(I_{m,b,a}^{h,r} \right)^2 + \left(I_{m,b,a}^{h,i} \right)^2 \right)} \right); \quad \forall \{m \in \Psi^{\text{M}}, b \in \Psi^{\text{B}}, a \in \Psi_m^{\text{A}}\} \tag{7.50}$$

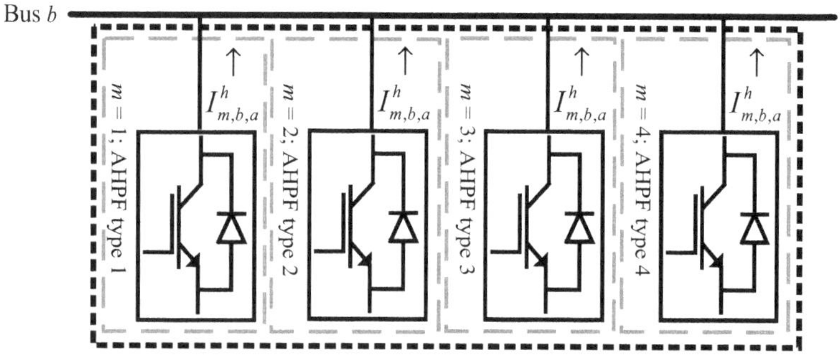

Fig. 7.2 The proposed architecture of the AHPFs on bus b of the distribution network

An additional harmonic current coefficient vector arising from the installation of different types of AHPFs on different buses of the distribution network must be created on the basis of Eq. (7.51) and then added to the harmonic current coefficient vector of the nonlinear loads.

$$\Delta I^h = \left[\sum_{m\in\Psi^{\mathrm{M}}} \sum_{a\in\Psi^{\mathrm{A}}_m} \beta_{m,1,a} \cdot I^h_{m,1,a} \quad \cdots \quad \sum_{m\in\Psi^{\mathrm{M}}} \sum_{a\in\Psi^{\mathrm{A}}_m} \beta_{m,b,a} \cdot I^h_{m,b,a} \quad \cdots \quad \sum_{m\in\Psi^{\mathrm{M}}} \sum_{a\in\Psi^{\mathrm{A}}_m} \beta_{m,\mathrm{B},a} \cdot I^h_{m,\mathrm{B},a} \right]^{\mathrm{T}}; \ \forall\{m\in\Psi^{\mathrm{M}}, b\in\Psi^{\mathrm{B}}, a\in\Psi^{\mathrm{A}}_m, h\in\Psi^{\mathrm{H}}\} \tag{7.51}$$

The letter "T" in Eq. (7.51) means transpose. This additional admittance matrix can alternatively be expressed as shown in Eqs. (7.52) and (7.53):

$$\Delta I^h = \sum_{m\in\Psi^{\mathrm{M}}} \Delta I^h_m; \quad \forall\{m\in\Psi^{\mathrm{M}}, h\in\Psi^{\mathrm{H}}\} \tag{7.52}$$

$$\Delta I^h_m = \mathrm{diagonal}\left[\sum_{a\in\Psi^{\mathrm{A}}_m} \beta_{m,1,a} \cdot I^h_{m,1,a} \quad \cdots \quad \sum_{a\in\Psi^{\mathrm{A}}_m} \beta_{m,b,a} \cdot I^h_{m,b,a} \quad \cdots \quad \sum_{a\in\Psi^{\mathrm{A}}_m} \beta_{m,\mathrm{B},a} \cdot I^h_{m,\mathrm{B},a} \right]; \quad \forall\{m\in\Psi^{\mathrm{M}}, b\in\Psi^{\mathrm{B}}, a\in\Psi^{\mathrm{A}}_m, h\in\Psi^{\mathrm{H}}\} \tag{7.53}$$

7.3.1.3 Hybrid Harmonic Power Filter Planning Problem

In this section, the authors present the mathematical model of the HHPF planning problem in detail. In the proposed techno-economic multi-objective framework, the DNO makes optimal filtering plans in the distribution network by solving the HHPF planning problem. The THDV, THDI, TLHH, and TCHF are considered as objective functions in the HHPF planning problem. At the same time, the HHPF planning problem is subject to the following sets of constraints: harmonic standard-based constraints, technical constraints, economic constraints, and logical constraints.

Investigation into the level of power quality disturbances caused by perturbing loads, especially nonlinear loads, and preservation at this level at predetermined standard limits in order to provide safe and reliable electrical power without any disturbances, is the most important task of the DNO. One of the most important of these disturbances, having the most devastating effects on the distribution network and consumers' equipment, is the harmonic distortion of voltage. These equipment types are designed to operate in the ideal sinusoidal mode; therefore any deviation in the voltage waveform from the ideal sinusoidal mode affects the safe operation of the equipment. As a result, the first target of the proposed HHPF planning problem is formulated with the aim of minimizing the THDV index, as shown in Eqs. (7.54) and (7.55):

$$\mathrm{OF}_1^{\mathrm{HHPFs}} = \mathrm{THDV} = \sum_{b \in \Psi^{\mathrm{B}}} THDV_b; \quad \forall \{ b \in \Psi^{\mathrm{B}} \} \tag{7.54}$$

$$THDV_b = \frac{1}{|V_b^1|} \cdot \sqrt{\sum_{h \in \Psi^{\mathrm{H}}} |V_b^h|^2}; \quad \forall \{ b \in \Psi^{\mathrm{B}}, h \in \Psi^{\mathrm{H}} \} \tag{7.55}$$

Equation (7.54) explains the THDV index which is comprised of the summation of this index on all buses of the distribution network. The THDV index on bus b of the distribution network generally depends on the amplitude of the voltage on this bus for the principal frequency and harmonic order h [see Eq. (7.55)]. Strictly speaking, in a certain operating condition of the distribution network, harmonic distortion of the current is greater than harmonic distortion of the voltage and affects the THDV index. Accordingly, considering the THDI index can offer assurance to the DNO that the filtering process in the distribution network is carried out with more precision and elegance. The DNO can also accurately analyze the suitability and adequacy of the harmonic power filters by using this index. Hence, the second target of the proposed HHPF planning problem is minimization of the total harmonic distortion of the current, namely the THDI index, as given by Eqs. (7.56) and (7.57):

$$\mathrm{OF}_2^{\mathrm{HHPFs}} = \mathrm{THDI} = \sum_{l \in \Psi^{\mathrm{L}}} THDI_l; \quad \forall \{ l \in \Psi^{\mathrm{L}} \} \tag{7.56}$$

$$THDI_l = \frac{1}{|I_l^1|} \cdot \sqrt{\sum_{h \in \Psi^{\mathrm{H}}} |I_l^h|^2}; \quad \forall \{ l \in \Psi^{\mathrm{L}}, h \in \Psi^{\mathrm{H}} \} \tag{7.57}$$

Equation (7.56) represents the THDI index, which is formed from the sum of this index on all lines of the distribution network. The THDI index at line l of the distribution network is affected by the amplitude of the current on this line for the principal frequency and harmonic order h [see Eq. (7.57)].

The harmonics have destructive effects on the distribution network equipment, and identifying these effects requires careful analysis. However, some of these destructive effects—power transmission capacity occupation, additional losses, increasing heat, reducing the useful life, etc.—are fully recognized by the researchers/engineers. In many instances, specialists in the field are relatively indifferent to certain destructive effects of the harmonic distortions on the distribution networks that impose high costs on the DNO for the operation of the distribution networks. Subsequently, considering the TLLH can be a great help for the DNO for reducing the operational costs of the distribution network and improving its efficiency. Hence, the third target of the proposed HHPF planning problem is considered to minimize the TLLH index, as determined by Eqs. (7.58) and (7.59):

$$\mathrm{OF}_3^{\mathrm{HHPFs}} = \mathrm{TLLH} = \sum_{h \in \Psi^{\mathrm{H}}} TLLH_h; \quad \forall \{h \in \Psi^{\mathrm{H}}\} \tag{7.58}$$

$$TLLH_h = \sum_{b \in \Psi^{\mathrm{B}}} \sum_{\substack{b' \in \Psi^{\mathrm{B}} \\ b' \neq b}} \frac{R_{b,b'}}{\left(Z_{b,b'}^h\right)^2} \cdot \left|V_b^h - V_{b'}^h\right|^2; \quad \forall \left\{b, b' \in \Psi^{\mathrm{B}}, b \neq b', h \in \Psi^{\mathrm{H}}\right\} \tag{7.59}$$

Equation (7.58) shows the TLLH index, which is created by the summation of this index at all harmonic orders. The TLLH index at harmonic order h is influenced by the resistance and impedance of the connected line between buses b and b' of the distribution network for harmonic order h and the voltage on these buses for harmonic order h [see Eq. (7.59)]. It should also be noted that there is another approach for calculating the TLLH index. In this approach, the TLLH index is computed via harmonic currents passing through the lines of the distribution network according to Eq. (7.60):

$$TLLH_h = \sum_{l \in \Psi^{\mathrm{L}}} R_l \cdot \left(I_l^h\right)^2; \quad \forall \{l \in \Psi^{\mathrm{L}}, h \in \Psi^{\mathrm{H}}\} \tag{7.60}$$

Essentially, the implementation of the proposed HHPF planning problem is a complicated and costly process that requires a large budget, while the DNO usually faces limitations in providing financial resources. Therefore, a need is felt by the DNO to consider an economic index that can reduce investment costs. The DNO can create significant savings in the required investment costs to perform harmonic power filtering in the distribution network by minimizing this economic index.

The fourth target of the HHPF planning problem is, thus, minimization of the TCHF index—as an economic index—as given by Eqs. (7.61) through (7.63):

$$\mathrm{OF}_{4}^{\mathrm{HHPFs}} = \mathrm{TCHF} = \sum_{n\in\Psi^{\mathrm{N}}}\sum_{b\in\Psi^{\mathrm{B}}}\sum_{p\in\Psi_{n}^{\mathrm{P}}} \alpha_{n,b,p}\cdot \Upsilon\left(\vartheta_{n,b,p}^{\mathrm{PHPF}}\right) + \sum_{m\in\Psi^{\mathrm{M}}}\sum_{b\in\Psi^{\mathrm{B}}}\sum_{a\in\Psi_{m}^{\mathrm{A}}} \beta_{m,b,a}\cdot \Upsilon\left(\vartheta_{m,b,a}^{\mathrm{AHPF}}\right);\forall\left\{n\in\Psi^{\mathrm{N}}, m\in\Psi^{\mathrm{M}}, b\in\Psi^{\mathrm{B}}, p\in\Psi_{n}^{\mathrm{P}}, a\in\Psi_{m}^{\mathrm{A}}\right\} \tag{7.61}$$

$$\Upsilon\left(\varpi_{n,b,p}^{\mathrm{PHPF}}\right) = \tau_{n,b,p}^{\mathrm{PHPF}} + \kappa_{n}^{\mathrm{PHPF}}\cdot \upsilon_{n,b,p}^{\mathrm{PHPF}} + \varpi_{n,b,p}^{\mathrm{PHPF}}\cdot \upsilon_{n,b,p}^{\mathrm{PHPF}};\quad \forall\left\{n\in\Psi^{\mathrm{N}}, b\in\Psi^{\mathrm{B}}, p\in\Psi_{n}^{\mathrm{P}}\right\} \tag{7.62}$$

$$\Upsilon\left(\varpi_{m,b,a}^{\mathrm{AHPF}}\right) = \tau_{m,b,a}^{\mathrm{AHPF}} + \varpi_{m,b,a}^{\mathrm{AHPF}}\cdot \upsilon_{m,b,a}^{\mathrm{AHPF}};\quad \forall\left\{m\in\Psi^{\mathrm{M}}, b\in\Psi^{\mathrm{B}}, a\in\Psi_{m}^{\mathrm{A}}\right\} \tag{7.63}$$

Equation (7.61) illustrates the TCHF index, which is established by the summation of the cost function of various types of PHPFs located on the distribution network plus the sum of the cost function of different types of AHPFs installed on the distribution network. The cost function of a given type of PHPF has three main components: fixed, semifixed, and variable costs [see Eq. (7.62)]. The fixed cost of a given type of PHPF includes the required indirect electric costs to install the filter that are not dependent on the electrical characteristics of the PHPF [see Eq. (7.62)]. The semifixed cost of a given type of PHPF is also related to the cost of the inductance and the resistance, which is considered as a certain proportion of the cost of the capacitance of the PHPF [see Eq. (7.62)]. In addition, the variable cost of a given type of PHPF depends directly on the nominal capacity of the capacitance of the PHPF [see Eq. (7.62)]. Likewise, the cost function of an AHPF has two main components: fixed and variable costs [see Eq. (7.63)]. The fixed cost of a given type of AHPF includes the required indirect electric costs to install the filters that are not dependent on the electrical characteristics of the AHPF [see Eq. (7.63)]. The variable cost of a given type of AHPF is also related to the nominal capacity of the AHPF [see Eq. (7.63)]. The nominal capacity of a given type of AHPF depends mainly on the maximum harmonic current that the AHPF can inject into the PCC of the distribution network [see Eq. (7.63)].

As stated earlier, the constraints of the HHPF planning problem are divided into four distinct categories: (1) harmonic standard-based constraints; (2) technical constraints; (3) economic constraints; and, (4) logical constraints.

Harmonic standard-based constraints: From a practical point of view, the DNO must always monitor and control the harmonic distortions of voltage and current; otherwise, these distortions will have destructive impacts on the safe operation of the distribution network and costumers' equipment. The value of the individual harmonic distortions of voltage and current should, therefore, satisfy the predetermined acceptable levels by the relevant standards, according to Eqs. (7.64) and (7.65):

$$\left|V_b^h\right| \leq V_b^{h,\max}; \quad \forall\left\{b \in \Psi^{\mathrm{B}}, h \in \Psi^{\mathrm{H}}\right\} \tag{7.64}$$

$$\left|I_b^h\right| \leq I_b^{h,\max}; \quad \forall\left\{b \in \Psi^{\mathrm{B}}, h \in \Psi^{\mathrm{H}}\right\} \tag{7.65}$$

Likewise, the value of the THDV and the THDI must satisfy pre-established acceptable levels by the relevant standards, as given in Eqs. (7.66) and (7.67):

$$THDV_b \leq THDV_b^{\max}; \quad \forall\left\{b \in \Psi^{\mathrm{B}}\right\} \tag{7.66}$$

$$THDI_b \leq THDI_b^{\max}; \quad \forall\left\{b \in \Psi^{\mathrm{B}}\right\} \tag{7.67}$$

When the DNO employs the PHPFs with the AHPF cutoff, or the AHPFs with the PHPF cutoff, or the HHPFs, the constraints related to the individual harmonic distortions of voltage and current, the THDV and THDI must be met.

Technical constraints: Practically speaking, the DC source in the AHPF structure is created by using discrete elements such as capacitors and inductors. The AHPF size, hence, must be considered as a discrete parameter. To achieve this goal, some limitations should be made for the AHPF size. It is also evident that the size of an AHPF is similar to the amount of its injected current. First, the injected current of an AHPF on bus b of the distribution network must be equal to or less than the maximum current that the corresponding AHPF is allowed to inject into the PCC of the distribution network, as given in Eq. (7.68):

$$I_{m,b,a} \leq I^{\max}; \quad \forall\left\{b \in \Psi^{\mathrm{B}}\right\} \tag{7.68}$$

Second, the injected current of this AHPF enters into Eq. (7.69) in order to determine the type of AHPF. The type of AHPF is equal to the smallest type, in terms of current value, of set Ψ^{M} for which its current is greater than the injected current by the relevant AHPF:

$$I_{m,b,a} \in \Psi^{\mathrm{M}}; \quad \forall\left\{m \in \Psi^{\mathrm{M}}, b \in \Psi^{\mathrm{B}}, a \in \Psi_m^{\mathrm{A}}\right\} \tag{7.69}$$

In Eq. (7.69), Ψ^{M} is a set of discrete values including zero and admissible values. In other studies in which AHPF size is considered as a continuous parameter, Ψ^{M} consists of a set of real and nonnegative values.

One of the important advantages of the PHPFs is their capability for the principal reactive power compensation to improve the power factor in the distribution networks. However, due to the use of various types of PHPFs by the DNO, some limitations must be made in the principal reactive power compensation process. First, principal reactive power compensation by PHPF p of type n on bus b of the distribution network must be operated under its predetermined limits according to Eq. (7.70):

$$Q_{n,b}^{\min} \leq Q_{n,b,p} \leq Q_{n,b}^{\max}; \quad \forall\left\{n \in \Psi^{\mathrm{N}}, b \in \Psi^{\mathrm{B}}, p \in \Psi_n^{\mathrm{P}}\right\} \tag{7.70}$$

Second, the total principal reactive power compensation by various types of PHPFs on bus b of the distribution network must satisfy Eq. (7.71):

$$Q_b^{\min} \leq \sum_{n \in \Psi^{\mathrm{N}}} \sum_{p \in \Psi_n^{\mathrm{P}}} Q_{n,b,p} \leq Q_b^{\max}; \quad \forall \{n \in \Psi^{\mathrm{N}}, b \in \Psi^{\mathrm{B}}, p \in \Psi_n^{\mathrm{P}}\} \tag{7.71}$$

Third, the total principal reactive power compensation by all PHPFs on all buses of the distribution network must be operated under its predetermined limits according to Eq. (7.72):

$$\mathrm{Q}^{\min} \leq \sum_{b \in \Psi^{\mathrm{B}}} \sum_{n \in \Psi^{\mathrm{N}}} \sum_{p \in \Psi_n^{\mathrm{P}}} Q_{n,b,p} \leq \mathrm{Q}^{\max}; \quad \forall \{n \in \Psi^{\mathrm{N}}, b \in \Psi^{\mathrm{B}}, p \in \Psi_n^{\mathrm{P}}\} \tag{7.72}$$

Series and parallel resonances are two of the most important limitations that must be considered in the HPHF design process. Ignoring these limitations increases the possibility of occurrence of parallel and serial resonances at the principal or harmonic frequencies. Therefore, in the PHPF design process, the inequality limitations associated with the series and parallel resonances are taken into account according to Eqs. (7.73) and (7.74), respectively:

$$\mathrm{Im}\left[Z_{\mathrm{bus}}^h + \left[\Delta Z_1^h \parallel \cdots \parallel \Delta Z_n^h \parallel \cdots \parallel \Delta Z_{\mathrm{N}}^h\right]\right] \neq 0; \quad \forall \{h \in \Psi^{\mathrm{H}}\} \tag{7.73}$$

$$\mathrm{Im}\left[Y_{\mathrm{bus}}^h \parallel \left[\Delta Y_1^h \parallel \cdots \parallel \Delta Y_n^h \parallel \cdots \parallel \Delta Y_{\mathrm{N}}^h\right]\right] \neq 0; \quad \forall \{h \in \Psi^{\mathrm{H}}\} \tag{7.74}$$

Since three types of PHPFs are considered in the proposed techno-economic multi-objective framework, Eqs. (7.73) and (7.74) can be rewritten on the basis of Eqs. (7.75) and (7.76), respectively:

$$\mathrm{Im}\left[Z_{\mathrm{bus}}^h + \Delta Z_1^h \parallel \Delta Z_2^h \parallel \Delta Z_3^h\right] \neq 0; \quad \forall \{h \in \Psi^{\mathrm{H}}\} \tag{7.75}$$

$$\mathrm{Im}\left[Y_{\mathrm{bus}}^h \parallel \Delta Y_1^h \parallel \Delta Y_2^h \parallel \Delta Y_3^h\right] \neq 0; \quad \forall \{h \in \Psi^{\mathrm{H}}\} \tag{7.76}$$

When the PHPFs encounter a fault, some adjunct constraints must be regarded to complete the restrictions related to series and parallel resonances. Since three types of PHPFs are considered in the proposed framework, all single and double faults must be taken into account. As an instance for the single faults, the occurrence of a fault on the fifth second-order series resonant band-pass PHPF results in the cutoff of this filter, while the other PHPFs work normally. In this case, adjunct constraints for the series and parallel resonances are constructed in accordance with Eqs. (7.77) and (7.78), respectively:

$$\mathrm{Im}\left[Z_{\mathrm{bus}}^h + \Delta Z_2^h \parallel \Delta Z_3^h\right] \neq 0; \quad \forall \{h \in \Psi^{\mathrm{H}}\} \tag{7.77}$$

$$\mathrm{Im}\left[Y_{\mathrm{bus}}^h \parallel \Delta Y_2^h \parallel \Delta Y_3^h\right] \neq 0; \quad \forall \{h \in \Psi^{\mathrm{H}}\} \tag{7.78}$$

Other single faults are examined in a similar manner. When the fifth and seventh second-order series resonant band-pass PHPFs are interrupted by a fault, while the second-order damped high-pass PHPF works normally, this is an instance of all dual combinations of faults. Hence, adjunct constraints for the series and parallel resonances are considered according to Eqs. (7.79) and (7.80), respectively:

$$\mathrm{Im}\left[Z_{\mathrm{bus}}^{h} + \Delta Z_{3}^{h}\right] \neq 0; \quad \forall\left\{h \in \Psi^{\mathrm{H}}\right\} \tag{7.79}$$

$$\mathrm{Im}\left[Y_{\mathrm{bus}}^{h} \parallel \Delta Y_{3}^{h}\right] \neq 0; \quad \forall\left\{h \in \Psi^{\mathrm{H}}\right\} \tag{7.80}$$

Other double faults are investigated in the same way. On the one hand, aging and high temperature cause changes in the values of the PHPF elements—capacitance, reactance, and resistance. On the other hand, the actual distribution network parameters of impedance and frequency may be varied under different operating conditions. The occurrence of any of these changes can lead to perturbation problems for the PHPFs. Hence, to achieve a robust design for the PHPFs, perturbation problems should be considered in the form of limitations known as detuning constraints. When the principal frequency of the distribution network varies over the interval 49.5–50.5 Hz according to Eq. (7.81), the performance of the PHPFs should be such that they still are able to satisfy the constraints presented in Eqs. (7.64) through (7.67) and Eqs. (7.70) through (7.80):

$$-1\% \leq \frac{\Delta f^{1}}{f^{1}} \leq 1\% \tag{7.81}$$

By changing the elements of the PHPFs (up to 5% for capacitance, 3% for reactance, and 3% for resistance) due to aging, high temperature, or frequency variations in accordance with Eqs. (7.82) through (7.84), the PHPFs must still be able to meet the requirements of Eqs. (7.64) through (7.67) and Eqs. (7.70) through (7.80):

$$-5\% \leq \frac{\Delta C_{n,b,p}}{C_{n,b,p}} \leq 5\%; \quad \forall\left\{n \in \Psi^{\mathrm{N}}, b \in \Psi^{\mathrm{B}}, p \in \Psi_{n}^{\mathrm{P}}\right\} \tag{7.82}$$

$$-3\% \leq \frac{\Delta L_{n,b,p}}{L_{n,b,p}} \leq 3\%; \quad \forall\left\{n \in \Psi^{\mathrm{N}}, b \in \Psi^{\mathrm{B}}, p \in \Psi_{n}^{\mathrm{P}}\right\} \tag{7.83}$$

$$-3\% \leq \frac{\Delta R_{n,b,p}}{R_{n,b,p}} \leq 3\%; \quad \forall\left\{n \in \Psi^{\mathrm{N}}, b \in \Psi^{\mathrm{B}}, p \in \Psi_{n}^{\mathrm{P}}\right\} \tag{7.84}$$

Additionally, by changing the distribution network impedance up to 20% according to Eq. (7.85), the PHPFs must still be able to meet the requirements of Eqs. (7.64) through (7.67) and Eqs. (7.70) through (7.80):

$$-20\% \leq \frac{\Delta Z_{\text{bus}}}{Z_{\text{bus}}} \leq 20\% \tag{7.85}$$

Consider that perturbation problems related to the PHPFs and the distribution network represent a completely conservative hypothesis in the design process of various types of PHPFs. Because this hypothesis provides a feasibility study of the detuning impacts on these filters, it can lead to a more realistic design of the PHPFs. It is important to note that all possible combinations of perturbation problems associated with the PHPFs and the distribution network should be considered in the design process of various types of PHPFs. The most inappropriate combination is when the distribution network impedance decreases by 20%; the distribution network frequency increases by 1%; and the capacitance, reactance, and resistance of the PHPFs increase by 5%, 3%, and 3%, respectively.

Economic constraint: In practice, the DNO is faced with some limitations in preparing the overall budget for purchasing and installing the PHPFs and AHPFs in the distribution network to overcome the power quality problems. The TCHF must, therefore, meet the condition specified by Eq. (7.86):

$$\text{TCHF} \leq \text{TCHF}^{\max} \tag{7.86}$$

Logical constraints: In order to satisfy practical and rational assumptions, the DNO must consider a collection of logical constraints (i.e., discrete or numerical constraints). Considering these logical constraints makes the plans for installing PHPFs and AHPFs in the distribution network more realistic and practical. Next, some of the most important logical constraints are described.

The number of installed AHPFs on each candidate bus of the distribution network must meet the conditions specified by Eq. (7.87):

$$\sum_{m\in\Psi^{\text{M}}} \sum_{a\in\Psi_m^{\text{A}}} \beta_{m,b,a} \leq 1; \quad \forall\{m \in \Psi^{\text{M}}, b \in \Psi^{\text{B}}, a \in \Psi_m^{\text{A}}\} \tag{7.87}$$

More precisely, the DNO is allowed to install only one AHPF on each bus of the distribution network. The number of installed PHPFs of type n on each candidate bus of the distribution network must satisfy Eq. (7.88):

$$\sum_{p\in\Psi_n^{\text{P}}} \alpha_{n,b,p} \leq 1; \quad \forall\{n \in \Psi^{\text{N}}, b \in \Psi^{\text{B}}, p \in \Psi_n^{\text{P}}\} \tag{7.88}$$

That is, the DNO is allowed to install only one PHPF of each type on each bus of the distribution network.

7.3.1.4 Probabilistic Decoupled Harmonic Power Flow Methodology

Basically, the demand and amount of harmonic currents generated by nonlinear loads in the distribution networks have stochastic, or time-varying, behavior. Ignoring these uncertainty parameters in the HHPF planning problem leads to impractical results. However, by considering these uncertainty parameters, traditional harmonic power flow methodologies are unable to evaluate the effects of these parameters in the HHPF problem. Therefore, it is necessary to develop a new harmonic power flow methodology that can accurately assess the effects of these uncertainty parameters. In the literature related to power systems, there are different techniques for handling uncertainty parameters; these are reported in Sect. 6.3.3 of Chap. 6. Here, uncertainty parameters in the HHPF planning problem arising from the demand and amount of harmonic currents generated by nonlinear loads are introduced in the PDHPF problem by means of the efficient two-PEM. The efficient two-PEM is essentially an extended version of the original point estimate method (PEM) and original two-PEM. The original PEM was first introduced in the mid-1970s by Emilio Rosenblueth as an efficient tool for analyzing the problems that are involved with uncertainties [34]. The original two-PEM is also a particular expression of the original PEM [35]. Inherently, the PEM was developed to compute the statistic moments of an uncertainty parameter. This uncertainty parameter can be a function of one or more uncertainty parameters. As a first case, suppose that Y is a function of one uncertainty parameter, X—$Y = g(X)$. Suppose that this uncertainty parameter also has the probability density function (PDF)—$f(X)$. In this case, the original two-PEM needs two probability concentrations located at two points in order to replace $f(X)$ by matching the first three moments of $f(X)$. For the second case, let Y be a function of K uncertainty parameters. Now suppose that each of these uncertainty parameters also has a PDF. In this scenario, the original two-PEM needs 2^K probability concentrations located at 2^K points in order to substitute the original joint PDF of these uncertainty parameters by aligning the second- and third-order non-crossed moments. Additionally, the moment of Y, $E(Y^i)$; $\forall i \in \Psi^{\{1,2\}}$, can be computed by weighting the value of Y to the power of i at each of the 2^K points. For more details regarding the original PEM and original two-PEM, interested readers should look at the work by Rosenblueth [34, 35]. However, in most real-world engineering problems, the number of uncertainty parameters involved in the problem is high. For example, when the number of uncertainty parameters involved in a particular problem is equal to 10, the use of 2^{10} probability concentrations by the original two-PEM requires a lot of computing and is not economical. Hence, a well-suited version of the original two-PEM, called an efficient two-PEM, was presented by Hong [36] in order to cope with the pitfalls of the original two-PEM. The efficient two-PEM uses only 2K probability concentrations, unlike the original one, which requires 2^K probability concentrations. The number of concentrations for each uncertainty parameter in the original PEM depends on the assumed PDF of the input uncertainty parameters. Certainly, a higher order PEM (i.e., three-PEM or five-PEM) would be required if the distributions of input uncertainty parameters were of a higher order. When one is concerned only with the first three moments (i.e., average, standard deviation, and skewness), only two concentrations for each uncertainty parameter are necessary. The third moment equals zero, if a normal distribution for input uncertainty parameters is assumed. Further information about the efficient two-PEM can be found in the work by Hong [36].

Here, though, because of the importance of the first three moments, the second order of the PEM is sufficient. In the proposed techno-economic multi-objective framework, the demand and amount of injected harmonic currents by nonlinear loads—the real and imaginary parts of the current of the nonlinear loads for each harmonic order—are considered as input uncertainty parameters, as given by Eq. (7.89):

$$\mathrm{X} = \left[P_b, I_b^{h,\mathrm{NLL}}\right]; \quad \forall\left\{b \in \Psi^{\mathrm{B}}, h \in \Psi^{\mathrm{H}}\right\} \tag{7.89}$$

Therefore, the PDHPF methodology is a multivariate nonlinear function, function g, of these input uncertainty parameters and is expressed by Eq. (7.90):

$$\mathrm{Y} = g(\mathrm{X}) \tag{7.90}$$

The output uncertainty parameters, which are directly affected by input uncertainty parameters, can be represented by Eq. (7.91):

$$\begin{aligned} \mathrm{Y} = \; & \left[\mathrm{THDV}, \mathrm{THDV}_b, \mathrm{THDI}, \mathrm{THDI}_l, \mathrm{TLLH}, \mathrm{TLLH}_h, V_b^h\right]; \\ & \forall\left\{b \in \Psi^{\mathrm{B}}, l \in \Psi^{\mathrm{L}}, h \in \Psi^{\mathrm{H}}\right\} \end{aligned} \tag{7.91}$$

Furthermore, the output uncertainty parameters, which are indirectly affected by input uncertainty parameters, can be expressed by Eq. (7.92):

$$\begin{aligned} &\tilde{\mathrm{Y}} = \left[I_{m,b,a}^h, R_{n,b,p}, L_{n,b,p}, C_{n,b,p}^h, \mathrm{TCHF}\right]; \\ &\forall\left\{m \in \Psi^{\mathrm{M}}, b \in \Psi^{\mathrm{B}}, a \in \Psi_m^{\mathrm{A}}, h \in \Psi^{\mathrm{H}}, n \in \Psi^{\mathrm{N}}, p \in \Psi_n^{\mathrm{P}}\right\} \end{aligned} \tag{7.92}$$

In general, the efficient two-PEM and its original version for calculating the moments of output parameters require that the PDF of input uncertainty parameters be known. However, if the PDF of input uncertainty parameters is unknown, these methods can still calculate the moments of output parameters. In this circumstance, by obtaining the three moments of output parameters, the related PDF cannot be reached. Therefore, the PDF associated with output parameters can be of any type with the same first three moments. Nonetheless, it is expected that the PDF of output parameters tends to be similar to the PDF of input parameters, provided that the input parameters are identical and known.

The step-by-step procedure of the probabilistic PDHPF methodology based on the efficient two-PEM can be summarized as follows:

1. Set the number of input uncertainty parameters equal to K.
2. Set E(Y) = 0 and E(Y^2) = 0.
3. Set the input uncertainty parameter counter: $k = 1$.
4. Compute the locations of two concentrations, $\delta_{k,\,1}$ and $\delta_{k,\,2}$, and probabilities of two concentrations, $\rho_{k,\,1}$ and $\rho_{k,\,2}$, according to Eqs. (7.93) through (7.96):

$$\delta_{k,1} = \frac{\lambda_{k,3}}{2} + \sqrt{\mathrm{K} + \left(\frac{\lambda_{k,3}}{2}\right)^2} \tag{7.93}$$

$$\delta_{k,2} = \frac{\lambda_{k,3}}{2} - \sqrt{\mathrm{K} + \left(\frac{\lambda_{k,3}}{2}\right)^2} \tag{7.94}$$

$$\rho_{k,1} = \frac{-\delta_{k,2}}{2\mathrm{K} \cdot \sqrt{\mathrm{K} + \left(\frac{\lambda_{k,3}}{2}\right)^2}} \tag{7.95}$$

$$\rho_{k,2} = \frac{\delta_{k,1}}{2\mathrm{K} \cdot \sqrt{\mathrm{K} + \left(\frac{\lambda_{k,3}}{2}\right)^2}} \tag{7.96}$$

5. Compute two concentration points, $x_{k,\ 1}$ and $x_{k,\ 2}$, using Eqs. (7.97) and (7.98):

$$x_{k,1} = \mu_{\mathrm{X},k} + \delta_{k,1} \cdot \sigma_{\mathrm{X},k} \tag{7.97}$$

$$x_{k,2} = \mu_{\mathrm{X},k} + \delta_{k,2} \cdot \sigma_{\mathrm{X},k} \tag{7.98}$$

6. Run the DDHPF for two concentration points, $x_{k,\ 1}$ and $x_{k,\ 2}$, obtained in the previous step in accordance with vector X. This vector can be determined according to Eq. (7.99):

$$\mathrm{X} = \begin{bmatrix} \mu_{\mathrm{X},1} & \mu_{\mathrm{X},2} & \cdots & x_{k,i} & \cdots & \mu_{\mathrm{X,K}} \end{bmatrix}; \quad \forall \{i \in \Psi^{\mathrm{I}}\} \tag{7.99}$$

7. Update E(Y) and E(Y^2) in accordance with Eqs. (7.100) and (7.101), respectively:

$$\mathrm{E(Y)} \cong \sum_{k \in \Psi^{\mathrm{K}}} \sum_{i \in \Psi^{\mathrm{I}}} \rho_{k,i} \cdot g(\mathrm{X}); \quad \forall \{k \in \Psi^{\mathrm{K}}, i \in \Psi^{\mathrm{I}}\} \tag{7.100}$$

$$\mathrm{E}\left(\mathrm{Y}^2\right) \cong \sum_{k \in \Psi^{\mathrm{K}}} \sum_{i \in \Psi^{\mathrm{I}}} \rho_{k,i} \cdot g(\mathrm{X})^2; \quad \forall \{k \in \Psi^{\mathrm{K}}, i \in \Psi^{\mathrm{I}}\} \tag{7.101}$$

8. Compute the mean and standard deviation of Y according to Eqs. (7.102) and (7.103), respectively:

$$\mu_{\mathrm{Y}} = \mathrm{E(Y)} \tag{7.102}$$

$$\sigma_{\mathrm{Y}} = \sqrt{\mathrm{E}\left(\mathrm{Y}^2\right) - \mu_{\mathrm{Y}}^2} \tag{7.103}$$

9. $k = k + 1$; if $k \leq \mathrm{K}$ go to step (4); otherwise go to the next step.
10. Stop.

7.3.2 Solution Method and Implementation Considerations

In the proposed techno-economic multi-objective framework, there are two decision-making variables of the solution vector of the optimization algorithm: (1) the capacitance related to PHPF p of type n on bus b of the distribution network and (2) the current associated with AHPF a of type m on bus b of the distribution network for harmonic order h. The process of solving the techno-economic multi-objective framework starts with the implementation of the PDHPF methodology for solving the harmonic power flow problem [see Eqs. (7.93) through (7.103)]. As described earlier, implementing the PDHPF methodology requires the development of the DDHPF methodology [see Eqs. (7.27) through (7.36)]. To analyze the distribution network at harmonic frequencies by using the DDHPF methodology, the distribution network must first be analyzed at the principal frequency using the LBNRPF methodology [see Eqs. (7.1) through (7.18)]. After completing the PDHPF methodology, the obtained harmonic data—the current and impedance of transmission line, the voltage magnitude and phase angle, etc.—are transferred to the HHPF planning problem. In the HHPF planning problem, the DNO simultaneously minimizes four objective functions—the THDV, THDI, TLHH, and TCHF, subject to harmonic standard-based constraints, technical constraints, economic constraints, and logical constraints—in order to determine the newly installed PHPFs and AHPFs [see Eqs. (7.54) through (7.88)] through a multi-objective optimization algorithm. More precisely, in the HHPF planning problem, the DNO tries to find the minimum values of the THDV, THDI, TLLH, and TCHF by solving Eqs. (7.54) through (7.88) in a multi-objective manner by using the Pareto optimality concept. Solving the HHPF planning problem by using the Pareto optimality concept leads to the formation of an optimal solution set, called Pareto-optimal solutions, instead of a single optimal solution. After calculation of the Pareto-optimal solutions, the DNO is only allowed to choose one solution from among the Pareto-optimal solutions that illustrates a compromise between different objectives—again, the THDV, THDI, TLLH, and TCHF. However, the main questions are which solution should be chosen by the DNO and how is it selected. The DNO selects the best solution based on its requirements and preferences. In the relevant literature, many methods have been reported on how to choose a trade-off solution among the Pareto-optimal solutions (see Chap. 2). Here, though, the authors employ the fuzzy satisfying method (FSM), based on a conservative methodology—min-max formulation—to choose the final solution. The reason for selecting this method is its simplicity and resemblance to human ratiocination. This method is fully discussed in Chap. 2. Figure 7.3 illustrates the flowchart of the proposed techno-economic multi-objective framework.

7.3.3 Simulation Results and Case Studies

In this section, two test systems, including a modified IEEE 18-bus distorted test network and a modified 34-bus distribution test network, are employed in order to analyze the performance of the proposed techno-economic multi-objective framework.

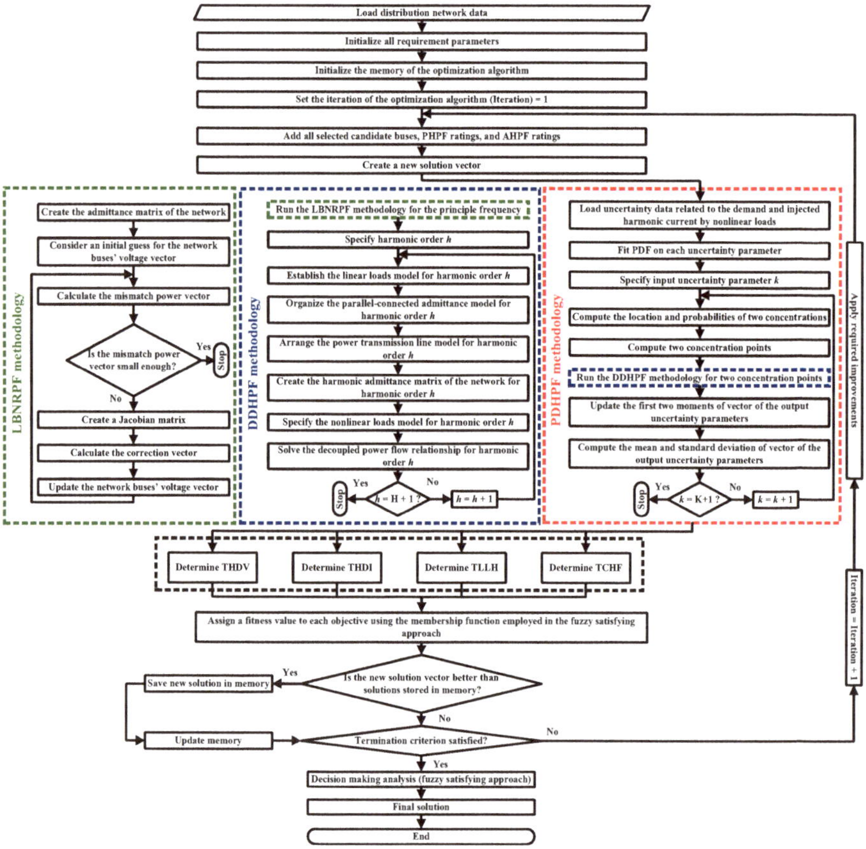

Fig. 7.3 Flowchart of the proposed techno-economic multi-objective framework

7.3.3.1 IEEE 18-Bus Distorted Test Network

Figure 7.4 shows how the authors applied a modified IEEE 18-bus distorted test network to the proposed techno-economic multi-objective framework. This test network is a popular test network that was widely used in several previous studies in the context of the harmonic power filter planning problem with minor variations in the number, location, and type of nonlinear loads [8, 21, 22, 24, 25, 31]. The modified IEEE 18-bus distorted test network consists of two different sides. Two buses of this test network include buses 50 and 51 located on the high-voltage side.

The voltage for this side of the modified IEEE 18-bus distorted test network is equal to 138 kV. Sixteen buses of this test network include buses 2–9 and 20–26, which are placed on low-voltage side. The base apparent power and voltage for this side of the modified IEEE 18-bus distorted test network are equal to 10 MVA and 12.5 kV, respectively. Network data for the modified IEEE 18-bus distorted test

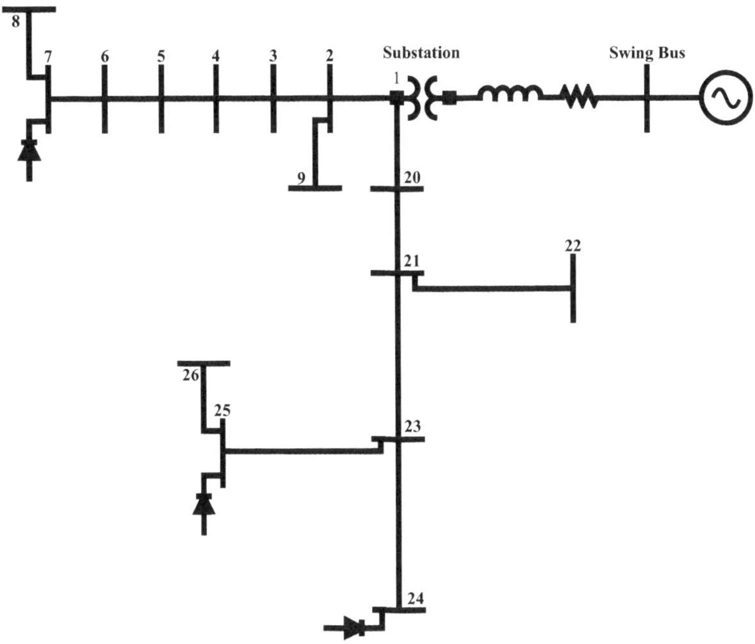

Fig. 7.4 The modified IEEE 18-bus distorted test network

network are given in Appendix 3. Data for the power transmission lines associated with the modified IEEE 18-bus distorted test system are tabulated in Table 7.24. Bus data related to the modified IEEE 18-bus distorted test network are presented in Table 7.25. The active and reactive power demands presented in this table are the predicted values.

The modified IEEE 18-bus distorted test network contains three nonlinear loads with 2.5 MW six-pulse converters that are located on buses 7, 24, and 25. These nonlinear loads generate the 5th, 7th, 11th, 13th, 17th, 19th, 23th, and 25th harmonic orders; harmonic orders above the 25th harmonic order are discarded. The harmonic contents—the predicted injected harmonic currents—of the nonlinear loads are presented in Table 7.26. As stated previously, three types of PHPFs and four types of AHPFs are considered in the techno-economic multi-objective framework. The characteristic coefficients pertaining to the cost of the various types of PHPFs are given in Table 7.27. The difference between these AHPFs is at the maximum harmonic current, where they can inject into the PCC of the modified IEEE 18-bus distorted test network. Table 7.28 shows the data for the maximum harmonic current relevant to the different types of AHPFs. The characteristic coefficients associated with the cost of the different types of AHPFs are tabulated in Table 7.29. Maximum permissible values of the individual harmonic distortions of the voltage and current on bus b for all harmonic orders along with the maximum admissible values of the

THDV and THDI on bus b of the modified IEEE 18-bus distorted test network are presented in Table 7.30. The values of Table 7.30 have been extracted according to the American National Standards Institute (ANSI) and IEEE 519-1992 standards. The minimum and maximum admissible values of the principal reactive power compensation by the fifth second-order series resonant band-pass PHPF p on bus b of the modified IEEE 18-bus distorted test network are set at 0.1 MVAr and 0.4 MVAr, respectively. The minimum and maximum permissible values of the principal reactive power compensation by the seventh second-order series resonant band-pass PHPF p on bus b of the modified IEEE 18-bus distorted test network are considered to be 0.04 MVAr and 0.175 MVAr, respectively. The minimum and maximum admissible values of the principal reactive power compensation by the second-order damped high-pass PHPF p on bus b of the modified IEEE 18-bus distorted test network are 0.06 MVAr and 0.25 MVAr, respectively. The minimum and maximum admissible values of the principal reactive power compensation by all PHPFs on bus b of the modified IEEE 18-bus distorted test network are also set to 0.2 MVAr and 0.825 MVAr, respectively. In addition, the minimum and maximum permissible values of the principal reactive power compensation by all PHPFs on all buses of the modified IEEE 18-bus distorted test network are set to 1 MVAr and 4 MVAr, respectively. Total cost of the inductance and the resistance of the PHPFs of type n is considered as a certain proportion—$\kappa_n^{\mathrm{PHPF}} = 7\%;\ \forall n \in \Psi^{\mathrm{N}}$—of the cost of the capacitance of the corresponding PHPFs. The maximum available budget for the DNO to install harmonic power filters in the modified IEEE 18-bus distorted test network is considered to be equal to \$6M. These assumptions are summarized in Table 7.31. The proposed PDHPF methodology is employed to solve the harmonic power flow problem and calculate the required harmonic data in the modified IEEE 18-bus distorted test network. After completing the harmonic power flow problem, the initial conditions of the voltage distortions for all harmonic orders, the THDV and THDI indices on all buses, and also the TLLH index for all harmonic orders in the modified IEEE 18-bus distorted test network without taking into account the harmonic power filters are presented in Table 7.2. In this study, all calculated numerical results were rounded to six decimal places. For convenience in the comparison process, the presented values for the THDV and THDI indices are considered as the average values, while the presented value for the TLLH index is the original value. The presented values for the THDV, THDI, and TLLH indices are also expressed in terms of percentage. From the analysis of the results reported in Table 7.2, it clearly appears that the modified IEEE 18-bus distorted test network has considerable unacceptable harmonic distortion levels, related to the ANSI and IEEE 519-1992 standards, by considering harmonic current sources (nonlinear loads).

According to this table, the average values of the THDV and THDI indices and the original value of the TLLH index were calculated to be 11.140752%, 19.393611%, and 3.920245%, respectively. It is obvious that these values cannot meet the predetermined acceptable harmonic distortion levels by the relevant standards. Based on the mathematical model of the proposed techno-economic multi-objective framework in this study, three types of PHPFs and four types of AHPFs are

Table 7.2 The initial conditions of the voltage distortions for all harmonic orders, the THDV and THDI indices on all buses, and also the TLLH index for all harmonic orders in the modified IEEE 18-bus distorted test network without taking into account the harmonic power filters

					Voltage distortion for each harmonic order (%)							
No.	Bus No.	Line No.	$THDV_b$ (%)	$THDI_l$ (%)	V_b^5	V_b^7	V_b^{11}	V_b^{13}	V_b^{17}	V_b^{19}	V_b^{23}	V_b^{25}
1	1	1	9.214336	17.896021	6.218979	2.451724	2.640305	5.562730	0.546168	1.953736	0.408554	0.251475
2	2	2	9.453980	18.149912	6.728896	2.285242	2.496109	5.017685	0.806495	2.914945	0.545811	0.358486
3	3	3	8.440937	17.156425	7.184377	1.901224	1.788241	2.780820	1.096228	2.425700	0.218753	0.078117
4	4	4	7.928969	16.880456	7.351016	1.681468	1.278095	1.453090	1.154607	1.717214	0.033885	0.097379
5	5	5	8.430044	17.012921	7.562093	1.130967	0.433678	3.259369	1.098390	1.521733	0.663766	0.529704
6	6	6	8.506295	17.233696	7.384344	0.976694	0.869541	3.821856	0.743791	1.683734	0.365661	0.164796
7	7	7	9.628267	18.326532	6.906645	0.986683	2.105034	5.726243	1.130823	2.592022	0.492195	0.607822
8	8	8	9.573803	18.271339	6.884426	0.976694	2.101757	5.766577	1.120012	2.590942	0.481472	0.607821
9	9	9	9.432195	18.028486	6.695568	2.274143	2.498294	5.006784	0.804333	2.904145	0.543666	0.318892
10	20	10	14.454929	22.543335	8.107559	4.394014	4.464595	10.206515	1.801100	0.717126	0.224115	0.211881
11	21	11	16.165119	23.658249	9.287365	5.559388	4.963817	10.682884	2.157861	1.667534	0.127606	0.097379
12	22	12	16.034405	23.205659	9.175161	5.437301	4.945246	10.816965	2.151374	1.662134	0.140152	0.097382
13	23	13	12.004018	20.368701	10.120562	6.036636	2.214273	1.684189	1.033524	0.916927	0.428928	0.122420
14	24	14	13.801352	21.571925	10.843777	6.635971	1.934621	5.379595	1.357852	0.829447	0.299177	0.155165
15	25	15	12.427753	20.600515	10.286091	5.825759	0.858618	3.321504	3.563087	1.790655	0.547956	0.261106
16	26	–	12.755630	–	10.270538	5.825759	0.859710	3.383640	3.563282	1.791735	0.556534	0.261112
Average in bus or line			11.140752	19.393611	8.187962	3.398729	2.278246	5.241903	1.508058	1.854983	0.379889	0.263808
$TLLH_h$ (%)					0.899841	0.707204	0.623973	0.581963	0.439953	0.354873	0.213904	0.098534

Table 7.3 The situation of the modified IEEE 18-bus distorted test network buses in defined scenarios

No.	Bus No.	Bus situation in each scenario		
		First scenario	Second scenario	Third scenario
1	1	×	√	√
2	2	×	√	√
3	3	×	×	√
4	4	×	×	√
5	5	×	√	√
6	6	×	√	√
7	7	√	√	√
8	8	×	×	√
9	9	×	×	√
10	20	×	√	√
11	21	×	×	√
12	22	×	×	√
13	23	√	√	√
14	24	√	√	√
15	25	√	√	√
16	26	√	√	√

available to the DNO to mitigate harmonic distortions and to compensate for reactive power. To perform an exhaustive analysis of the performance of the techno-economic multi-objective framework in the modified IEEE 18-bus distorted test network, three cases are defined and applied, as follows:

- First case: The techno-economic multi-objective framework is run only in the presence of the PHPFs.
- Second case: The techno-economic multi-objective framework is run only in the presence of the AHPFs.
- Third case: The techno-economic multi-objective framework is run in the presence of both PHPFs and AHPFs, namely the HHPFs.

For an accurate evaluation of each case study, three scenarios are also defined and applied according to Table 7.3. The difference between the defined scenarios is in the number of distribution network buses that can be candidates for installation of PHPFs and/or AHPFs. In Table 7.3, marks "√" and "×" illustrate the availability and unavailability of the PHPFs and AHPFs on the relevant bus of the distribution network, respectively.

In the first scenario, the DNO is only allowed to install PHPFs and/or AHPFs on buses 7, 23, 24, 25, and 26 of the distribution network—a confined search pattern. In the second scenario, more candidate buses of the distribution network for PHPF and AHPF installation are available to the DNO in comparison to the first scenario. In this scenario, the DNO is allowed to install PHPFs and AHPFs on buses 1, 2, 5, 6, 7, 20, 23, 24, 25, and 26 of the distribution network—a semi-confined search pattern. In the third scenario, however, the DNO can install PHPFs and AHPFs on all buses of the distribution network. In a precise sense, buses 1, 2, 3, 4, 5, 6, 7, 8, 9, 20, 21, 22, 23, 24, 25, and 26 of the distribution network are considered as candidate buses for installation of PHPFs and AHPFs—a full search pattern. The proposed techno-economic multi-objective framework is implemented by the DNO and solved by using the proposed multi-objective multi-stage computational multi-dimensional multiple-homogeneous enhanced melody search algorithm (MMM-EMSA) or multi-objective multi-stage computational multi-dimensional single-inhomogeneous enhanced melody search algorithm (MMS-EMSA), named a multi-objective symphony orchestra search algorithm (SOSA), which was addressed in Chap. 4. The parameter adjustments of the newly developed multi-objective SOSA are tabulated in Table 7.32.

7.3.3.1.1 First Case: Simulation Results and Discussion

In the first case, the ability and capability of the proposed techno-economic multi-objective framework are described, while the AHPFs are cut off. Simply put, in the first case, the DNO is only allowed to employ PHPFs. Table 7.4 gives the obtained optimal location and size—optimal plan—for the PHPFs resulting from the implementation of the techno-economic multi-objective framework by the DNO under the first, second, and third scenarios of the first case. From the analysis of the optimal results in Table 7.4, it appears that by increasing the candidate buses of the IEEE 18-bus distorted test network to install the PHPFs, the techno-economic multi-objective framework is allowed to be more flexible. Increasing the flexibility of the framework leads to the installation of more stable PHPFs from a techno-economic perspective. In the first scenario, based on a confined search pattern, the implementation of the framework brings about the installation of five PHPFs of each type on buses 7, 23, 24, 25, and 26 of the IEEE 18-bus distorted test network. In the second scenario, founded on a semi-confined search pattern, the implementation of the framework also leads to the installation of six PHPFs of each type on buses 5, 7, 23, 24, 25, and 26 of the IEEE 18-bus distorted test network.

In the third scenario, dependent on a full search pattern, the implementation of the framework results in the installation of seven PHPFs of each type on buses 5, 7, 8, 21, 23, 24, and 25 of the IEEE 18-bus distorted test network. However, the PHPFs installed in the third scenario are of smaller size than the PHPFs installed in the first and second scenarios. The PHPFs installed in the second scenario have also smaller size than the PHPFs installed in the first scenario. Table 7.5 presents the calculated values for different objective functions of the HHPF planning problem, resulting from the implementation of the techno-economic multi-objective framework by the DNO under the first, second, and third scenarios of the first case. The values for the THDV and THDI indices are the average values, while the value for the TLLH index is the original value. For comparative purposes, the value of the objective functions of the HHPF planning problem without consideration of the harmonic power filters is also presented in this table. To illustrate the optimal results presented in Table 7.5, consider a first scenario, where, for the PHPFs, the calculated average value for the THDV index is 0.896973%.

Table 7.4 Optimal location and size of the PHPFs under the first, second, and third scenarios of the first case

Scenario No.	Type of PHPF	Design parameters	{Bus number of located PHPF}: Value
1	5th second-order series resonant band-pass PHPF	$C_{1,\,b,\,1}$ (μF)	{7}: 97.702084, {23}: 91.087142, {24}: 96.943748, {25}: 94.731524, {26}: 89.458451
		$L_{1,\,b,\,1}$ (mH)	{7}: 9.217162, {23}: 8.712652, {24}: 9.145802, {25}: 9.035346, {26}: 8.632635
	7th second-order series resonant band-pass PHPF	$C_{2,\,b,\,1}$ (μF)	{7}: 28.116674, {23}: 26.628542, {24}: 29.183512, {25}: 29.761644, {26}: 26.356242
		$L_{2,\,b,\,1}$ (mH)	{7}: 16.228645, {23}: 15.018423, {24}: 17.693856, {25}: 17.142522, {26}: 15.154522
	Second-order damped high-pass PHPF	$C_{3,\,b,\,1}$ (μF)	{7}: 51.127024, {23}: 49.235325, {24}: 52.283685, {25}: 51.535361, {26}: 50.568704
		$L_{3,\,b,\,1}$ (mH)	{7}: 4.961765, {23}: 4.721502, {24}: 5.021974, {25}: 4.865784, {26}: 4.712325
2	5th second-order series resonant band-pass PHPF	$C_{1,\,b,\,1}$ (μF)	{5}: 72.504216, {7}: 75.889641, {23}: 74.721185, {24}: 78.129086, {25}: 80.889321, {26}: 72.928635
		$L_{1,\,b,\,1}$ (mH)	{5}: 7.979912, {7}: 8.334474, {23}: 8.555126, {24}: 8.758903, {25}: 8.86585, {26}: 7.942653
	7th second-order series resonant band-pass PHPF	$C_{2,\,b,\,1}$ (μF)	{5}: 21.721436, {7}: 23.869935, {23}: 22.633888, {24}: 24.936116, {25}: 24.528216, {26}: 23.326321
		$L_{2,\,b,\,1}$ (mH)	{5}: 14.501324, {7}: 15.079043, {23}: 14.893217, {24}: 15.925937, {25}: 15.254035, {26}: 14.835631
	Second-order damped high-pass PHPF	$C_{3,\,b,\,1}$ (μF)	{5}: 40.673884, {7}: 43.274512, {23}: 38.761385, {24}: 42.809358, {25}: 44.068396, {26}: 39.521468
		$L_{3,\,b,\,1}$ (mH)	{5}: 4.103198, {7}: 4.276532, {23}: 3.460691, {24}: 4.616198, {25}: 4.628947, {26}: 3.545201

(continued)

Table 7.4 (continued)

Scenario No.	Type of PHPF	Design parameters	{Bus number of located PHPF}: Value
3	5th second-order series resonant band-pass PHPF	$C_{1,b,1}$ (μF)	{5}: 63.063991, {7}: 54.525734, {8}: 56.751339, {21}: 61.273729, {23}: 55.346645, {24}: 53.853204, {25}: 52.924904
		$L_{1,b,1}$ (mH)	{5}: 7.381541, {7}: 5.215563, {8}: 5.234129, {21}: 6.503955, {23}: 5.824113, {24}: 5.555045, {25}: 5.160336
	7th second-order series resonant band-pass PHPF	$C_{2,b,1}$ (μF)	{5}: 17.200952, {7}: 18.256907, {8}: 16.144317, {21}: 16.588459, {23}: 15.636514, {24}: 18.176603, {25}: 17.920026
		$L_{2,b,1}$ (mH)	{5}: 11.442573, {7}: 12.936118, {8}: 11.189767, {21}: 11.030579, {23}: 10.837121, {24}: 12.879218, {25}: 12.697418
	Second-order damped high-pass PHPF	$C_{3,b,1}$ (μF)	{5}: 31.330924, {7}: 33.560033, {8}: 30.722364, {21}: 32.047029, {23}: 31.721842, {24}: 35.442656, {25}: 36.917441
		$L_{3,b,1}$ (mH)	{5}: 2.166111, {7}: 3.220529, {8}: 2.604476, {21}: 2.215619, {23}: 2.297352, {24}: 3.401192, {25}: 3.542718

Table 7.5 Optimal values of different objective functions of the HHPF planning problem under the first, second, and third scenarios of the first case

No.	Objective functions of the HHPF planning problem	Before applying harmonic power filters	After applying PHPFs		
			First scenario	Second scenario	Third scenario
1	Average THDV (%)	11.140752	0.896973	0.746517	0.563600
2	Average THDI (%)	19.393611	1.070351	0.941298	0.812672
3	TLLH (%)	3.920245	0.062254	0.048033	0.033487
4	TCHF (M$)	–	4.517321	4.369404	3.842451

Therefore, the average value of the THDV index has been recovered in comparison with the average value of the THDV index before considering PHPFs, which is 11.140752%. In this scenario, the average value of the THDI index after applying the PHPFs is 1.070351%, compared to an average value of 19.393611% before applying the PHPFs. In addition, the original value of the TLLH index with the PHPFs decreased from 3.920245 to 0.062254% without the PHPFs. In this scenario, the TCHF index is $4.517321M. The results presented in Table 7.5 for the other scenarios are analyzed in the same way. As a general result, after applying the PHPFs in all scenarios, the THDV, THDI, and TLLH indices significantly improved and are placed at acceptable levels in the sense of the relevant standards. Another result from the analysis of the results presented in Table 7.5 is that by increasing the candidate buses of the IEEE 18-bus distorted test network to install the PHPFs—the third scenario compared to the first and second scenarios and also the second scenario compared to the first scenario—the indices of the HHPF planning problem have more favorable values. Table 7.6 gives the voltage distortions for all harmonic orders, the THDV and THDI indices on all buses of the IEEE 18-bus distorted test network, and also the TLLH index for all harmonic orders in the modified IEEE 18-bus distorted test network arising from the implementation of the proposed techno-economic multi-objective framework by the DNO under the third scenario of the first case. Since the third scenario is the completed state of the first and second scenarios, this information is provided only for the third scenario. From the analysis of the results presented in this table, it is evident that the voltage distortions for all harmonic orders, the THDV and THDI indices on all buses, and also the TLLH index for all harmonic orders in the modified IEEE 18-bus distorted test network considering the PHPFs have been enhanced in comparison to the conditions in which the PHPFs are not considered (Table 7.6 vs. Table 7.2). In more precise terms, after applying the PHPFs, the voltage distortions for all harmonic orders, the THDV and THDI indices on all buses, and the TLLH index for all harmonic orders in the modified IEEE 18-bus distorted test network are able to meet the predetermined acceptable levels by the relevant standards.

As a result, after applying the PHPFs, the modified IEEE 18-bus distorted test network is in a stable and acceptable condition in terms of harmonic pollution. As previously mentioned in Sect. 7.2.2, the simplest way to deal with harmonic problems in the distribution networks is the use of the PHPFs. Given the strengths

Table 7.6 Voltage distortions for all harmonic orders, the THDV and THDI indices on all buses, and also the TLLH index for all harmonic orders in the modified IEEE 18-bus distorted test network under the third scenario of the first case

No.	Bus No.	Line No.	$THDV_b$ (%)	$THDI_l$ (%)	Voltage distortion for each harmonic order (%)							
					V_b^5	V_b^7	V_b^{11}	V_b^{13}	V_b^{17}	V_b^{19}	V_b^{23}	V_b^{25}
1	1	1	0.466145	0.749917	0.314612	0.124030	0.133570	0.281413	0.027630	0.098837	0.020668	0.012722
2	2	2	0.478268	0.760556	0.340408	0.115608	0.126275	0.253840	0.040799	0.147464	0.027612	0.018136
3	3	3	0.427019	0.718925	0.363451	0.096181	0.090465	0.140679	0.055457	0.122714	0.011067	0.003952
4	4	4	0.401119	0.707361	0.371881	0.085063	0.064657	0.073510	0.058410	0.086872	0.001714	0.004926
5	5	5	0.426468	0.712912	0.382559	0.057214	0.021939	0.164888	0.055566	0.076983	0.033579	0.026797
6	6	6	0.430325	0.722163	0.373567	0.049410	0.043989	0.193344	0.037627	0.085178	0.018498	0.008337
7	7	7	0.487085	0.767957	0.349400	0.049915	0.106491	0.289685	0.057207	0.131128	0.024900	0.030749
8	8	8	0.484330	0.765644	0.348276	0.049410	0.106326	0.291725	0.056660	0.131073	0.024357	0.030749
9	9	9	0.477166	0.755468	0.338722	0.115046	0.126386	0.253288	0.040690	0.146918	0.027504	0.016132
10	20	10	0.731261	0.944659	0.410154	0.222289	0.225859	0.516338	0.091116	0.036278	0.011338	0.010719
11	21	11	0.817778	0.991378	0.469839	0.281244	0.251115	0.540437	0.109164	0.084359	0.006455	0.004926
12	22	12	0.811166	0.972413	0.464163	0.275068	0.250175	0.547220	0.108836	0.084085	0.007090	0.004926
13	23	13	0.607272	0.853533	0.511990	0.305388	0.112018	0.085201	0.052285	0.046386	0.021699	0.006193
14	24	14	0.698198	0.903953	0.548576	0.335707	0.097870	0.272148	0.068692	0.041961	0.015135	0.007850
15	25	15	0.628708	0.863247	0.520364	0.294719	0.043436	0.168031	0.180253	0.090587	0.027721	0.013209
16	26	–	0.645295	–	0.519577	0.294719	0.043492	0.171175	0.180263	0.090642	0.028155	0.013209
Average in bus or line			0.563600	0.812672	0.414221	0.171938	0.115254	0.265183	0.076291	0.093841	0.019218	0.013345
$TLLH_h$ (%)					0.007695	0.006388	0.005166	0.004791	0.003515	0.002718	0.001948	0.001266

Table 7.7 Optimal location and size of the AHPFs under the first, second, and third scenarios of the second case

Scenario No.	Type of AHPF	Design parameters	{Bus number of located AHPF}: Value
1	AHPF type 1	$I_{1,\ b,\ 1}$ (p.u.)	{7}: 0.069578, {24}: 0.068785
	AHPF type 2	$I_{2,\ b,\ 1}$ (p.u.)	{23}: 0.056706, {25}: 0.059975, {26}: 0.058302
	AHPF type 3	$I_{3,\ b,\ 1}$ (p.u.)	–
	AHPF type 4	$I_{4,\ b,\ 1}$ (p.u.)	–
2	AHPF type 1	$I_{1,\ b,\ 1}$ (p.u.)	{7}: 0.067372, {25}: 0.069067
	AHPF type 2	$I_{2,\ b,\ 1}$ (p.u.)	{23}: 0.059761, {24}: 0.058636
	AHPF type 3	$I_{3,\ b,\ 1}$ (p.u.)	{5}: 0.049958
	AHPF type 4	$I_{4,\ b,\ 1}$ (p.u.)	–
3	AHPF type 1	$I_{1,\ b,\ 1}$ (p.u.)	–
	AHPF type 2	$I_{2,\ b,\ 1}$ (p.u.)	{7}: 0.061698, {24}: 0.057468
	AHPF type 3	$I_{3,\ b,\ 1}$ (p.u.)	{6}: 0.045189, {25}: 0.048561
	AHPF type 4	$I_{4,\ b,\ 1}$ (p.u.)	{23}: 0.031254, {26}: 0.030755

of the PHPFs, there are some disadvantages that affect their use in the distribution networks. The most technical solution for overcoming the shortcomings of the PHPFs is the use of AHPFs, which is discussed in the next section.

7.3.3.1.2 Second Case: Simulation Results and Discussion

In the second case, the ability and capability of the proposed techno-economic multi-objective framework are evaluated, while the PHPFs are cut off. In the second case, the DNO is only allowed to use AHPFs. Table 7.7 gives the location and size—optimal plan—for the AHPFs arising from the implementation of the techno-economic multi-objective framework by the DNO under the first, second, and third scenarios of the second case. From the analysis of the results in Table 7.7, it becomes clear that by increasing the candidate buses of the IEEE 18-bus distorted test network to install the AHPFs, the proposed techno-economic multi-objective framework has a higher freedom rate to decide on the installation of the APHFs. This higher freedom rate can lead to the installation of more stable AHPFs from a techno-economic point of view. In the first scenario that is placed in a confined search pattern, the implementation of the framework gives rise to the installation of five AHPFs: two of type 1 on buses 7 and 24, and three of type 2 on buses 23, 25, and 26 of the IEEE 18-bus distorted test network. In the second scenario, that is formed on a semi-confined search pattern, the implementation of the framework leads to the installation of five AHPFs: two of type 1 on buses 7 and 25; two of type 2 on buses 23 and 24; and one of type 3 on bus 5 of the IEEE 18-bus distorted test network. In addition, in the third scenario, based on a full search pattern, the implementation of the framework affords the installation of six AHPFs; two of type 2 on buses 7 and 24; two of type 3 on buses 6 and 25; and two of type 4 on buses 23 and 26 of the IEEE

Table 7.8 Optimal values of different objective functions of the HHPF planning problem under the first, second, and third scenarios of the second case

No.	Objective functions of the HHPF planning problem	Before applying harmonic power filters	After applying AHPFs		
			First scenario	Second scenario	Third scenario
1	Average THDV (%)	11.140752	0.721621	0.592461	0.481427
2	Average THDI (%)	19.393611	0.949170	0.780038	0.694129
3	TLLH (%)	3.920245	0.053012	0.039215	0.028126
4	TCHF (M$)	–	5.200426	5.063952	4.587297

18-bus distorted test network. Nevertheless, the AHPFs installed in the third scenario have lower injected harmonic currents or are of smaller size than the AHPFs installed in the first and second scenarios. The AHPFs installed in the second scenario have lower injected harmonic currents or are smaller than the AHPFs installed in the first scenario. Table 7.8 shows the computed values for different objective functions of the HHPF planning problem arising from the implementation of the proposed techno-economic multi-objective framework by the DNO under the first, second, and third scenarios of the second case. To describe the optimal results provided in Table 7.8, consider a scenario in which, after applying the AHPFs, the average values of the THDV and the THDI indices are 0.721621% and 0.949170%, respectively. Accordingly, the THDV and THDI indices, by applying AHPFs, improved in comparison with ignoring the AHPFs, which resulted in values of 11.140752% and 19.393611%, respectively. The original value obtained for the TLLH index after considering the AHPFs is 0.053012%. This value represents a significant improvement, compared with conditions before considering the AHPFs (3.920245%). The obtained value for the TCHF index is $5.200426M. The results presented in Table 7.8 for the other scenarios were investigated in the same way. As a consequence, after considering the AHPFs in all scenarios, the THDV, THDI, and TLLH indices improved considerably and are located at acceptable levels with respect to the relevant standards. From the analysis of the results reported in Table 7.8, it can be concluded that by increasing the candidate buses of the IEEE 18-bus distorted test network to install the AHPFs—the third scenario compared to the first and second scenarios, and the second scenario compared to the first scenario—the indices of the HHPF planning problem have more favorable values. Table 7.9 presents the voltage distortions for all harmonic orders, the THDV and THDI indices on all buses, and the TLLH index for all harmonic orders in the modified IEEE 18-bus distorted test network arising from the implementation of the proposed techno-economic multi-objective framework by the DNO under the third scenario of the second case.

For the same reason as described for the first case, this information in the second case is provided only for the third scenario. From the analysis of the optimal results provided in this table, it clearly appears that, by considering the AHPFs, the voltage distortions for all harmonic orders, the THDV and THDI indices on all buses, and the TLLH index for all harmonic orders in the modified IEEE 18-bus distorted test network improved in comparison to the conditions in which the PHPFs are not considered (Table 7.9 vs. Table 7.2). That is to say that, after considering the AHPFs, the voltage distortions

Table 7.9 Voltage distortions for all harmonic orders, the THDV and THDI indices on all buses, and the TLLH index for all harmonic orders in the modified IEEE 18-bus distorted test network under the third scenario of the second case

					Voltage distortion for each harmonic order (%)							
No.	Bus No.	Line No.	$THDV_b$ (%)	$THDI_l$ (%)	V_b^5	V_b^7	V_b^{11}	V_b^{13}	V_b^{17}	V_b^{19}	V_b^{23}	V_b^{25}
1	1	1	0.398180	0.640528	0.268741	0.105946	0.114095	0.240383	0.023601	0.084427	0.017654	0.010867
2	2	2	0.408536	0.649615	0.290776	0.098752	0.107864	0.216830	0.034851	0.125964	0.023586	0.015491
3	3	3	0.364759	0.614057	0.310459	0.082157	0.077275	0.120168	0.047371	0.104822	0.009453	0.003375
4	4	4	0.342635	0.604179	0.317660	0.072661	0.055230	0.062792	0.049894	0.074206	0.001464	0.004208
5	5	5	0.364288	0.608921	0.326781	0.048872	0.018740	0.140847	0.047464	0.065758	0.028683	0.022890
6	6	6	0.367583	0.616822	0.319100	0.042206	0.037575	0.165154	0.032141	0.072759	0.015801	0.007121
7	7	7	0.416067	0.655937	0.298457	0.042637	0.090965	0.247449	0.048866	0.112009	0.021269	0.026265
8	8	8	0.413714	0.653961	0.297497	0.042206	0.090823	0.249192	0.048399	0.111962	0.020805	0.026265
9	9	9	0.407594	0.645269	0.289336	0.098272	0.107959	0.216359	0.034757	0.125497	0.023493	0.013780
10	20	10	0.624643	0.806863	0.350353	0.189879	0.192929	0.441055	0.077831	0.030989	0.009684	0.009156
11	21	11	0.698545	0.846768	0.401336	0.240238	0.214502	0.461641	0.093248	0.072059	0.005514	0.004208
12	22	12	0.692897	0.830569	0.396487	0.234963	0.213699	0.467435	0.092967	0.071826	0.006056	0.004208
13	23	13	0.518731	0.729029	0.437341	0.260862	0.095685	0.072779	0.044661	0.039623	0.018535	0.005290
14	24	14	0.596400	0.772095	0.468593	0.286761	0.083601	0.232469	0.058677	0.035843	0.012928	0.006705
15	25	15	0.537042	0.737326	0.444494	0.251749	0.037103	0.143532	0.153972	0.077379	0.023678	0.011283
16	26	–	0.551211	–	0.443822	0.251749	0.037150	0.146217	0.153980	0.077426	0.024049	0.011283
Average in bus or line			0.481427	0.694129	0.353827	0.146869	0.098450	0.226519	0.065168	0.080159	0.016416	0.011400
$TLLH_h$ (%)					0.006579	0.005761	0.004642	0.003707	0.002872	0.002288	0.001359	0.000918

for all harmonic orders, the THDV and THDI indices on all buses, and the TLLH index for all harmonic orders in the modified IEEE 18-bus distorted test network easily meet the predetermined acceptable levels by the relevant standards. As a result, after considering the AHPFs, the modified IEEE 18-bus distorted test network is in a stable and admissible condition in terms of harmonic pollution.

As previously described in Sect. 7.2.2, the most technical solution for overcoming the disadvantages of the PHPFs is the use of AHPFs. However, one of the biggest disadvantages of the AHPFs is their high cost, which causes these filters not to be considered as an economical solution for dealing with the harmonic problems in the distribution networks. A well-designed solution for overcoming the disadvantages of the PHPFs and AHPFs and also simultaneously acquiring both available benefits is the use of HHPFs. Consequently, the HHPFs are taken into account as a techno-economic solution for dealing with the harmonic problem in the distribution network, which is discussed in the next section.

7.3.3.1.3 Third Case: Simulation Results and Discussion

In the third case, the ability and capability of the proposed techno-economic multi-objective framework are examined, while the PHPFs and AHPFs are simultaneously considered. Put another way, in the third case, the DNO can use different options of the harmonic power filters—namely, three types of PHPFs along with four types of AHPFs. Table 7.10 gives the location and size—optimal plan—for the HHPFs arising from the implementation of the techno-economic multi-objective framework by the DNO under the first, second, and third scenarios of the third case. From the analysis of the results in this table, it clearly appears that, in all three scenarios, the implementation of the proposed framework leads to the installation of a specified number of harmonic power filters consisting of three PHPFs of each type and two AHPFs. In the first scenario, these three PHPFs are installed on buses 7, 24, and 25, while the two AHPFs of type 1 are installed on buses 7 and 25.

In the second scenario, these three PHPFs are also installed on buses 7, 24, and 25, while the two AHPFs of type 2 and type 3 are installed on buses 24 and 7, respectively. In addition, in the third scenario, these three PHPFs are installed on buses 7, 24, and 25, while the two AHPFs of type 3 and type 4 are installed on buses 7 and 25, respectively. However, it is obvious that the AHPFs installed in the third scenario have lower injected harmonic current or are smaller than the AHPFs installed in the first and second scenarios, and also the PHPFs installed in the third scenario are smaller than the PHPFs installed in the first and second scenarios. The AHPFs installed in the second scenario have lower injected harmonic current or are smaller than the AHPFs installed in the first scenario, and also the PHPFs installed in the second scenario are smaller than the PHPFs installed in the first scenario. Table 7.11 gives the calculated values for different objective functions of the HHPF planning problem arising from the implementation of the techno-economic multi-objective framework by the DNO under the first, second, and third scenarios of the third case. To demonstrate the results presented in this table, consider the first scenario.

Table 7.10 Optimal location and size of the HHPFs under the first, second, and third scenarios of the third case

Scenario No.	Type of PHPF and AHPF	Design parameter	{Bus number of located PHPF and/or AHPF}: Value
1	5th second-order series resonant band-pass PHPF	$C_{1,\,b,\,1}$ (μF)	{7}: 59.524311, {24}: 62.851722, {25}: 60.923439
		$L_{1,\,b,\,1}$ (mH)	{7}: 7.615418, {24}: 8.254894, {25}: 7.960187
	7th second-order series resonant band-pass PHPF	$C_{2,\,b,\,1}$ (μF)	{7}: 19.256514, {24}: 17.176194, {25}: 19.919621
		$L_{2,\,b,\,1}$ (mH)	{7}: 15.377008, {24}: 14.571302, {25}: 15.139928
	Second-order damped high-pass PHPF	$C_{3,\,b,\,1}$ (μF)	{7}: 35.559228, {24}: 33.441818, {25}: 36.916612
		$L_{3,\,b,\,1}$ (mH)	{7}: 3.272012, {24}: 3.152671, {25}: 3.502271
	AHPF type 1	$I_{1,\,b,\,1}$ (p.u.)	{7}: 0.060524, {25}: 0.063476
	AHPF type 2	$I_{2,\,b,\,1}$ (p.u.)	–
	AHPF type 3	$I_{3,\,b,\,1}$ (p.u.)	–
	AHPF type 4	$I_{4,\,b,\,1}$ (p.u.)	–
2	5th second-order series resonant band-pass PHPF	$C_{1,\,b,\,1}$ (μF)	{7}: 60.165308, {24}: 56.704342, {25}: 58.702912
		$L_{1,\,b,\,1}$ (mH)	{7}: 7.849819, {24}: 7.341882, {25}: 7.501309
	7th second-order series resonant band-pass PHPF	$C_{2,\,b,\,1}$ (μF)	{7}: 17.136527, {24}: 18.411854, {25}: 20.189012
		$L_{2,\,b,\,1}$ (mH)	{7}: 14.428857, {24}: 14.588706, {25}: 15.852613
	Second-order damped high-pass PHPF	$C_{3,\,b,\,1}$ (μF)	{7}: 33.075676, {24}: 34.924215, {25}: 32.439486
		$L_{3,\,b,\,1}$ (mH)	{7}: 2.904928, {24}: 3.059896, {25}: 2.894894
	AHPF type 1	$I_{1,\,b,\,1}$ (p.u.)	–
	AHPF type 2	$I_{2,\,b,\,1}$ (p.u.)	{24}: 0.055022
	AHPF type 3	$I_{3,\,b,\,1}$ (p.u.)	{7}: 0.042448
	AHPF type 4	$I_{4,\,b,\,1}$ (p.u.)	–

(continued)

Table 7.10 (continued)

Scenario No.	Type of PHPF and AHPF	Design parameter	{Bus number of located PHPF and/or AHPF}: Value
3	5th second-order series resonant band-pass PHPF	$C_{1,\,b,\,1}$ (μF)	{7}: 57.226401, {24}: 51.844735, {25}: 54.830827
		$L_{1,\,b,\,1}$ (mH)	{7}: 7.009371, {24}: 6.454181, {25}: 6.766542
	7th second-order series resonant band-pass PHPF	$C_{2,\,b,\,1}$ (μF)	{7}: 15.609498, {24}: 17.174262, {25}: 18.030086
		$L_{2,\,b,\,1}$ (mH)	{7}: 11.715113, {24}: 13.420358, {25}: 13.888001
	Second-order damped high-pass PHPF	$C_{3,\,b,\,1}$ (μF)	{7}: 31.559673, {24}: 34.437865, {25}: 33.882126
		$L_{3,\,b,\,1}$ (mH)	{7}: 2.658912, {24}: 2.942292, {25}: 2.889904
	AHPF type 1	$I_{1,\,b,\,1}$ (p.u.)	–
	AHPF type 2	$I_{2,\,b,\,1}$ (p.u.)	–
	AHPF type 3	$I_{3,\,b,\,1}$ (p.u.)	{7}: 0.044875
	AHPF type 4	$I_{4,\,b,\,1}$ (p.u.)	{25}: 0.033564

Table 7.11 Optimal values of different objective functions of the HHPF planning problem under the first, second, and third scenarios of the third case

No.	Objective functions of the HHPF planning problem	Before applying harmonic power filters	After applying PHPFs and/or AHPFs		
			First scenario	Second scenario	Third scenario
1	Average THDV (%)	11.140752	0.519219	0.438367	0.311235
2	Average THDI (%)	19.393611	0.723729	0.625779	0.562879
3	TLLH (%)	3.920245	0.041248	0.028305	0.020921
4	TCHF (M$)	–	3.742652	3.258802	2.951422

In this scenario, after considering the HHPFs, the average values of the THDV and THDI indices are 0.519219% and 0.723729%, respectively. The average values obtained for the THDV and THDI indices after considering the HHPFs show a significant improvement compared with the conditions, where the HHPFs are ignored—11.140752% and 19.393611%, respectively. The original value obtained for the TLLH index after applying the HHPFs is 0.041248%. This value represents a remarkable improvement compared with conditions before applying the HHPFs, where this index has a value of 3.920245%. The calculated value for the TCHF index is $3.742652M. The optimal results reported in Table 7.11 for the other scenarios are examined in the same way. As a general result, after applying HHPFs in all scenarios, the THDV, THDI, and TLLH indices have achieved and are placed at acceptable levels, with respect to the relevant standards.

From the analysis of the results presented in Table 7.8, it clearly appears that by increasing the candidate buses of the IEEE 18-bus distorted test network to install HHPFs—the third scenario compared to the first and second scenarios, and also the second scenario compared to the first scenario—the indices of the HHPF planning problem have more desirable values.

Table 7.12 provides the voltage distortions for all harmonic orders, the THDV and THDI indices on all buses, and the TLLH index for all harmonic orders in the modified IEEE 18-bus distorted test network arising from the implementation of the proposed techno-economic multi-objective framework by the DNO under the third scenario of the third case. For the same reason as described for the first and second cases, this information in the third case is also provided only for the third scenario. From the analysis of the results presented in Table 7.12, it can be observed that, by applying HHPFs, the voltage distortions for all harmonic orders, the THDV and THDI indices on all buses, and the TLLH index for all harmonic orders in the modified IEEE 18-bus distorted test network have been improved in comparison to the conditions in which HHPFs are not applied (Table 7.12 vs. Table 7.2).

This means that after applying HHPFs, the voltage distortions for all harmonic orders, the THDV and THDI indices on all buses, and the TLLH index for all harmonic orders in the modified IEEE 18-bus distorted test network easily satisfy the predetermined acceptable levels by the relevant standards. As a consequence, after applying HHPFs, the modified IEEE 18-bus distorted test network is in a stable and

Table 7.12 Voltage distortions for all harmonic orders, the THDV and THDI indices on all buses, and the TLLH index for all harmonic orders in the modified IEEE 18-bus distorted test network under the third scenario of the third case

No.	Bus No.	Line No.	$THDV_b$ (%)	$THDI_l$ (%)	Voltage distortion for each harmonic order (%)							
					V_b^5	V_b^7	V_b^{11}	V_b^{13}	V_b^{17}	V_b^{19}	V_b^{23}	V_b^{25}
1	1	1	0.240125	0.498446	0.229559	0.090500	0.097461	0.205335	0.020161	0.072118	0.015081	0.009283
2	2	2	0.248971	0.505513	0.248381	0.084354	0.092138	0.185216	0.029770	0.107598	0.020147	0.013233
3	3	3	0.211577	0.532603	0.265194	0.070179	0.066009	0.102647	0.040465	0.089539	0.008075	0.002884
4	4	4	0.192679	0.485603	0.271345	0.062067	0.047178	0.053637	0.042620	0.063387	0.001251	0.003595
5	5	5	0.211175	0.490261	0.279137	0.041747	0.016008	0.120312	0.040544	0.056171	0.024501	0.019553
6	6	6	0.213989	0.523995	0.272576	0.036052	0.032097	0.141075	0.027455	0.062151	0.013498	0.006083
7	7	7	0.255404	0.527923	0.254943	0.036421	0.077702	0.211371	0.041742	0.095678	0.018168	0.022436
8	8	8	0.253394	0.525995	0.254122	0.036052	0.077581	0.212860	0.041343	0.095639	0.017772	0.022436
9	9	9	0.248167	0.518680	0.247151	0.083945	0.092219	0.184814	0.029690	0.107200	0.020068	0.011771
10	20	10	0.433569	0.648509	0.299271	0.162195	0.164800	0.376749	0.066483	0.026471	0.008273	0.007821
11	21	11	0.496697	0.680582	0.342821	0.205212	0.183228	0.394333	0.079652	0.061553	0.004710	0.003595
12	22	12	0.491872	0.676972	0.338679	0.200705	0.182542	0.399283	0.079413	0.061354	0.005173	0.003595
13	23	13	0.343099	0.614881	0.373577	0.222828	0.081735	0.062168	0.038150	0.033846	0.015833	0.004519
14	24	14	0.409444	0.620564	0.400272	0.244951	0.071412	0.198575	0.050122	0.030617	0.011043	0.005728
15	25	15	0.358741	0.592647	0.379687	0.215044	0.031694	0.122606	0.131523	0.066098	0.020227	0.009638
16	26	–	0.370844	–	0.379113	0.215044	0.031734	0.124899	0.131530	0.066138	0.020543	0.009638
Average in bus or line			0.311235	0.562879	0.302239	0.125456	0.084096	0.193492	0.055666	0.068472	0.014023	0.009738
$TLLH_h$ (%)					0.005023	0.004111	0.003696	0.002905	0.002135	0.001423	0.000908	0.000720

acceptable condition in terms of harmonic pollution. Providing a techno-economic analysis on three scenarios of three cases helps to better understand the performance of the PHPFs, AHPFs, and HHPFs in dealing with harmonics in the modified IEEE 18-bus distorted test network. For this purpose, consider the first scenario of all three cases. Once again, it is necessary to mention that the first scenario is based on a confined pattern. In the first scenario of the first case, then, the obtained investment cost to install the PHPFs in the modified IEEE 18-bus distorted test network is equal to \$4.517321M. With this investment cost, the DNO is able to reduce the average value of the THDV index from 11.140752 to 0.896973%, decrease the average value of the THDI index from 19.393611 to 1.070351%, and also decrease the original value of the TLLH index from 3.920245 to 0.062254% (see Table 7.5). In the first scenario of the second case, the calculated investment cost to install the AHPFs in the modified IEEE 18-bus distorted test network is \$5.200426M. For this investment cost by the DNO, the average value of the THDV index is reduced from 11.140752 to 0.721621%, the average value of the THDI index is decreased from 19.393611 to 0.949170%, and the original value of the TLLH index is reduced from 3.920245 to 0.053012% (see Table 7.8). Finally, for the first scenario of the third case, the computed investment cost to install the HHPFs in the modified IEEE 18-bus distorted test network is \$3.742652M. For this investment cost, the DNO is able to reduce the average value of the THDV index from 11.140752 to 0.519219%, decrease the average value of the THDI index from 19.393611 to 0.723729%, and decrease the original value of the TLLH index from 3.920245 to 0.041248% (see Table 7.11). As a result, the average values obtained for the THDV and THDI indices, and the original value for the TLLH index in the first scenario of the second case, are more favorable than those in the first scenario of the first case. However, this desirability is not seen in the TCHF index, where the obtained investment cost for installing the AHPFs in the first scenario of the second case is higher than the calculated investment cost for installing the PHPFs in the first scenario of the first case (Table 7.5 vs. Table 7.8). Moreover, the average values calculated for the THDV and THDI indices, the original value for the TLLH index, and the value for the TCHF index in the first scenario of the third case are more desirable than those in the first scenario of the first and second cases. That is, in the first scenario of the third case, the DNO, with much lower investment costs than for the first scenario of the first and second cases, is able to obtain more favorable average values for the THDV and THDI indices and the original value of the TLLH index (Table 7.11 vs. Tables 7.5 and 7.8). The second and third scenarios of all three cases are examined in the same way.

7.3.3.1.4 Investigation of Passive Harmonic Power Filter Performance

In this section, a harmonic attenuation (HA) coefficient is defined and used to evaluate the filtering performance of the allocated PHPFs in the first and third cases. The HA coefficient is generally formulated according to Eq. (7.104) [37]:

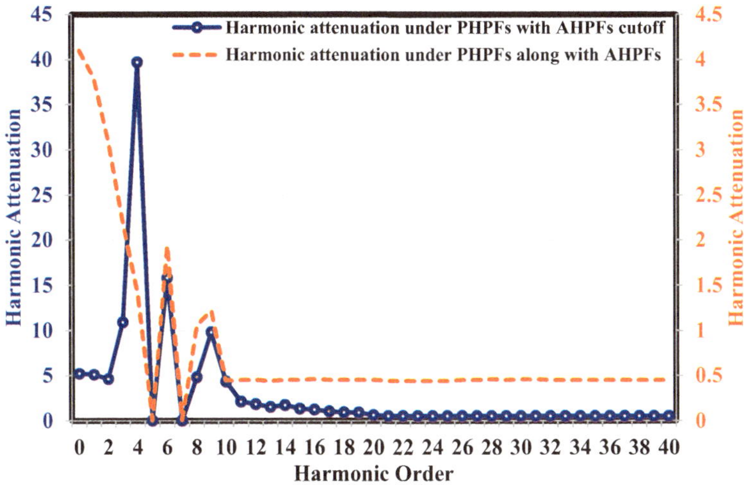

Fig. 7.5 The HA coefficient of the PHPFs under the first and third cases

$$\zeta^h = \frac{I^{h,\mathrm{DN}}}{I^{h,\mathrm{NLL}}} = \left| \frac{\Delta Z^h}{\Delta Z^h + Z^h_{\mathrm{bus}}} \right| = \left| \frac{\sum\limits_{n \in \Psi^{\mathrm{N}}} \Delta Z^h_n}{\sum\limits_{n \in \Psi^{\mathrm{N}}} \Delta Z^h_n + Z^h_{\mathrm{bus}}} \right|; \quad \forall \left\{ h \in \Psi^{\mathrm{H}}, h \in \Psi^{\mathrm{H}} \right\} \quad (7.104)$$

The HA coefficient represents the ratio of the current of the source of the distribution network to the current of nonlinear loads for harmonic order h. In other words, this coefficient illustrates the impedance observed from a nonlinear load point. In general, by decreasing the HA coefficient, the filtering performance of the PHPFs improves. More precisely, the harmonic compensation performance by the PHPFs is improved if the HA coefficient decreases. Figure 7.5 illustrates the HA coefficient of the PHPFs under the first and third cases. The obtained HA coefficient of the PHPFs under the first and third cases relates to the third scenario of these cases. From the analysis of Fig. 7.5, it is possible to observe that in the first and third cases—considering the PHPFs in the first case and considering the HHPFs in the third case—the harmonic currents produced by nonlinear loads are well attenuated. There are also no resonance points in either case due to consideration of the limitations associated with series and parallel resonances. In addition, in the third case, where the HHPFs are considered, the HA coefficient has a much lower value than the first case in which the PHPFs are considered. As a result, the HHPFs have much better performance than the PHPFs in the process of harmonic mitigation.

7.3.3.2 The 34-Bus Distribution Test Network

In this section, the proposed techno-economic multi-objective framework is applied to a modified 34-bus distribution test network [8, 30].

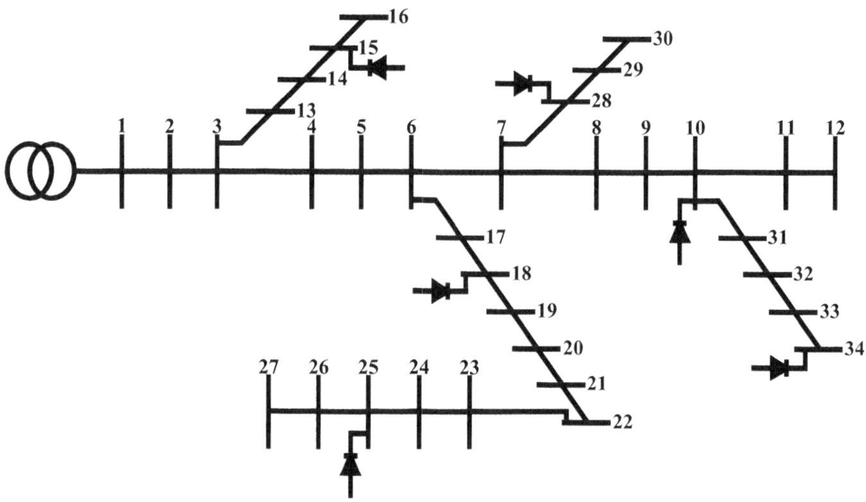

Fig. 7.6 The modified 34-bus distribution test network

The single-line diagram of this test network is illustrated in Fig. 7.6. The modified 34-bus distribution test network is a three-phase network with a voltage of 10 kV. Network data for the modified 34-bus distribution test network are presented in Appendix 3. Data for transmission lines related to the modified 34-bus distribution test network are provided in Table 7.33. Bus data for the modified 34-bus distribution test network are tabulated in Table 7.34. The active and reactive power demands presented in this table are the predicted values. In the modified 34-bus distribution test network, there are six identical nonlinear loads with 2.5 MW six-pulse converters located on buses 10, 15, 18, 25, 28, and 34. It is assumed that all nonlinear loads produce the 5th, 7th, 11th, 13th, 17th, 19th, 23th, and 25th harmonic orders. Harmonic orders above the 25th can be ignored. The harmonic contents of the nonlinear loads are given in Table 7.26. The injected harmonic currents by nonlinear loads presented in this table are the predicted values. The characteristic coefficients relevant to various types of PHPFs are tabulated in Table 7.27. Data pertaining to the different types of AHPFs and their characteristic coefficients of cost are given in Tables 7.28 and 7.29, respectively. In accordance with the ANSI and IEEE 519-1992 standards, the maximum admissible values of the individual harmonic distortions of voltage and current on bus b of the modified 34-bus distribution test network for all harmonic orders, along with maximum THDV and THDI on bus b of the modified 34-bus distribution test network, are tabulated in Table 7.30.

Provided in Table 7.35 is other required information, including data related to the minimum and maximum permissible values of principal reactive power compensation by PHPF p of type n on bus b; the minimum and maximum admissible values of the principal reactive power compensation by all PHPFs on bus b; the minimum and maximum permissible values of the principal reactive power compensation by all

Table 7.13 Average values of the THDV and THDI indices and the original value of the TLLH index in the modified 34-bus distribution test network without taking into account the harmonic power filters

No.	Objective functions of the HHPF planning problem	Value
1	Average THDV (%)	18.147589
2	Average THDI (%)	31.590981
3	TLLH (%)	5.858904

PHPFs on all buses; and the proportion that indicates the cost of the inductance and the resistance for each type of PHPF and the maximum available budget for the DNO to install harmonic power filters in the modified 34-bus distribution test network. To determine the harmonic pollution condition of the modified 34-bus distribution test network in the presence of nonlinear loads, the DNO employs the proposed PDHPF methodology to solve the harmonic power flow problem. After solving this harmonic power flow problem, the required harmonic data in the modified 34-bus distribution test network are identified. The average values of the THDV and THDI indices and the original value of the TLLH index in the presence of nonlinear loads and without taking into account the harmonic power filters are tabulated in Table 7.13. All numerical results are rounded to six decimal places. The results presented in this table indicate that the DNO faces high harmonic pollution in the modified 34-bus distribution test network.

In accordance with this table, the average values of the THDV and THDI indices and the original value of the TLLH index are 18.147589%, 31.590981%, and 5.858904%, respectively. It is clearly evident that these values cannot meet the predetermined acceptable harmonic distortion levels by the ANSI and IEEE 519-1992 standards. According to the mathematical model of the proposed techno-economic multi-objective framework in this study, three types of PHPFs and four types of AHPFs are available to the DNO to mitigate harmonic distortions and to compensate for reactive power. Hence, three cases are employed to investigate and analyze the performance of the techno-economic multi-objective framework in the modified 34-bus distribution test network. The distinction between the first, second, and third cases is in the type of harmonic power filters utilized by the DNO to deal with the power quality problems. The first, second, and third cases are as follows:

- First case: The techno-economic multi-objective framework considers only the PHPFs.
- Second case: The techno-economic multi-objective framework considers only the AHPFs.
- Third case: The techno-economic multi-objective framework simultaneously considers both the PHPFs and AHPFs.

The techno-economic multi-objective framework is implemented by the DNO and solved by using the multi-objective SOSA, which is fully described in Chap. 4. The parameter adjustments of the newly developed multi-objective SOSA are given in Table 7.31.

7.3.3.2.1 First Case: Simulation Results and Discussion

In this section, the ability and capability of the proposed techno-economic multi-objective framework are examined only in the presence of PHPFs. Table 7.14 gives the location and size—optimal plan—for the PHPFs resulting from the implementation of the techno-economic multi-objective framework by the DNO under the first case. The analysis of the results presented in Table 7.14 shows that the implementation of the proposed framework leads to the installation of 13 PHPFs of each type on buses 3, 7, 9, 10, 14, 15, 18, 21, 25, 28, 29, 33, and 34.

The calculated values for different objective functions of the HHPF planning problem resulted from the implementation of the techno-economic multi-objective

Table 7.14 Optimal location and size of the PHPFs under the first case

No.	Type of PHPF	Design parameters	{Bus number of located PHPF}: Value
1	5th second-order series resonant band-pass PHPF	$C_{1,b,1}$ (μF)	{3}: 54.576484, {7}: 59.604181, {9}: 52.677703, {10}: 83.195015, {14}: 56.913886, {15}: 83.155113, {18}: 68.266873, {21}: 61.352269, {25}: 84.913733, {28}: 73.109551, {29}: 51.355208, {33}: 65.815238, {34}: 82.457022
		$L_{1,b,1}$ (mH)	{3}: 6.134591, {7}: 7.123365, {9}: 5.978747, {10}: 8.487894, {14}: 6.611762, {15}: 8.483823, {18}: 8.173759, {21}: 7.431612, {25}: 8.663244, {28}: 7.432681, {29}: 5.641203, {33}: 6.659942, {34}: 8.412604
2	7th second-order series resonant band-pass PHPF	$C_{2,b,1}$ (μF)	{3}: 17.685439, {7}: 19.467511, {9}: 17.305914, {10}: 22.994683, {14}: 18.155433, {15}: 22.983654, {18}: 22.603412, {21}: 20.675415, {25}: 23.469728, {28}: 21.592302, {29}: 17.021265, {33}: 21.082242, {34}: 22.790706
		$L_{2,b,1}$ (mH)	{3}: 11.760556, {7}: 13.748509, {9}: 11.048458, {10}: 16.293115, {14}: 12.340959, {15}: 16.285301, {18}: 15.651151, {21}: 14.716525, {25}: 16.629713, {28}: 14.861708, {29}: 10.885323, {33}: 14.024521, {34}: 16.148585
3	Second-order damped high-pass PHPF	$C_{3,b,1}$ (μF)	{3}: 34.226245, {7}: 35.687543, {9}: 32.704953, {10}: 47.070265, {14}: 37.626544, {15}: 47.047689, {18}: 43.407566, {21}: 37.651061, {25}: 48.042685, {28}: 42.189304, {29}: 30.235785, {33}: 38.959529, {34}: 46.652721
		$L_{3,b,1}$ (mH)	{3}: 2.642829, {7}: 2.879809, {9}: 2.581241, {10}: 4.517016, {14}: 2.797403, {15}: 4.514849, {18}: 3.853966, {21}: 3.161402, {25}: 4.610332, {28}: 3.547889, {29}: 2.245852, {33}: 2.693526, {34}: 4.476947

Table 7.15 Optimal values of different objective functions of the HHPF planning problem under the first case

No.	Objective functions of the HHPF planning problem	Before applying harmonic power filters	After applying PHPFs
1	Average THDV (%)	18.147589	0.658647
2	Average THDI (%)	31.590981	0.838822
3	TLLH (%)	5.858904	0.046909
4	TCHF (M$)	–	7.925440

framework by the DNO under the first case are presented in Table 7.15. From the analysis of these results, it can be seen that, after considering the PHPFs, the average values obtained for the THDV and THDI indices are 0.658647% and 0.838822%, respectively. Therefore, the average values of the THDV and THDI indices are recovered when compared with the average values of the THDV and THDI indices before considering PHPFs, which are 18.147589% and 31.590981%, respectively. The original value of the TLLH index after applying the PHPFs is 0.046909%, compared to an original value of 5.858904% before applying the PHPFs. The obtained value for the TCHF index is $7.925440M. As a consequence, considering the PHPFs has brought about a significant improvement in the THDV, THDI, and TLLH indices; therefore, these indices are located within acceptable levels with respect to the relevant standards. The PHPFs are the most traditional way to deal with harmonic problems in distribution networks. However, the PHPFs suffer from a variety of problems that heavily influence the decision of the DNO to select these filters as a way to deal with the power quality problems arising from harmonics (see Sect. 7.2.2).

The most technical way to solve the deficiency of PHPFs is with the use of AHPFs, which are discussed in the next section.

7.3.3.2.2 Second Case: Simulation Results and Discussion

In this section, the ability and capability of the proposed techno-economic multi-objective framework are evaluated only in the presence of AHPFs. Table 7.16 gives the location and size—optimal plan—for the AHPFs arising from the implementation of the techno-economic multi-objective framework by the DNO under the second case. Based on the analysis of the results presented in Table 7.16, it is apparent that the implementation of the framework gives rise to the installation of 11 AHPFs: 2 of type 1 on buses 18 and 28; 2 of type 2 on buses 15 and 25; 4 of type 3 on buses 7, 10, 31, and 34; and 3 of type 4 on buses 4, 13, and 22 of the modified 34-bus distribution test network. The values for different objective functions of the HHPF planning problem arising from the implementation of the proposed techno-economic multi-objective framework by the DNO under the second case are given in Table 7.17. From the analysis of the results reported in Table 7.17, it becomes clear that after applying the AHPFs, the calculated average value for the THDV index is

Table 7.16 Optimal location and size of the AHPFs under the second case

No.	Type of AHPF	Design parameters	{Bus number of located AHPF}: Value
1	AHPF type 1	$I_{1, b, 1}$ (p.u.)	{18}: 0.067272, {28}: 0.068561
2	AHPF type 2	$I_{2, b, 1}$ (p.u.)	{15}: 0.058773, {25}: 0.059276
3	AHPF type 3	$I_{3, b, 1}$ (p.u.)	{7}: 0.047920, {10}: 0.048993, {31}: 0.047111, {34}: 0.048575
4	AHPF type 4	$I_{4, b, 1}$ (p.u.)	{4}: 0.0353505, {13}: 0.036256, {22}: 0.039122

0.566564%. Therefore, the average value of the THDV index is improved in comparison with the average value of the THDV index before applying the AHPFs, which is 18.147589%. The average value of the THDI index after considering the AHPFs is 0.718039%, compared with an average value of 31.590981% before applying the AHPFs. In addition, the original value of the TLLH index with the AHPFs decreased from 5.858904 to 0.037013% without the AHPFs. The calculated original value of the TCHF index is $9.092039M. As a result, by applying the AHPFs, a remarkable improvement is observed in the THDV, THDI, and TLLH indices, and these indices are acceptable with respect to the relevant standards.

Practically speaking, the use of AHPFs by the DNO to overcome the shortcomings of the PHPFs is a completely logical and technical solution. However, the high cost of the AHPFs and the limited budget available to the DNO restrict the use of these filters in the distribution networks to overcome the power quality problems resulting from harmonics. As a result, the DNO must seek alternative solutions to conquer the disadvantages of the PHPFs and AHPFs, and realize the benefits of both.

A well-designed solution that simultaneously considers techno-economic features is HHPF, which is discussed in the next section.

7.3.3.2.3 Third Case: Simulation Results and Discussion

In this section, the ability and capability of the proposed techno-economic multi-objective framework are described in the simultaneous presence of the PHPFs and AHPFs. Table 7.18 gives the location and size—optimal plan—for the HHPFs arising from the implementation of the techno-economic multi-objective framework by the DNO under the third case. From the analysis of the results reported in this table, it appears that the implementation of the framework results in the installation of six PHPFs of each type along with three AHPFs. These six PHPFs are installed on buses 10, 15, 18, 25, 28, and 34; for the three AHPFs, two of type 3 are installed on buses 10 and 34, and one of type 4 on bus 18. The values obtained for different objective functions of the HHPF planning problem arising from the implementation of the techno-economic multi-objective framework under the third case are tabulated in Table 7.19. From the analysis of these results, it clearly appears that, after considering

Table 7.17 Optimal values of different objective functions of the HHPF planning problem under the second case

No.	Objective functions of the HHPF planning problem	Before applying harmonic power filters	After applying AHPFs
1	Average THDV (%)	18.147589	0.566564
2	Average THDI (%)	31.590981	0.718039
3	TLLH (%)	5.858904	0.037013
4	TCHF (M$)	–	9.092039

Table 7.18 Optimal location and size of the HHPFs under the third case

No.	Type of PHPF and AHPF	Design parameter	{Bus number of located PHPF and/or AHPF}: Value
1	5th second-order series resonant band-pass PHPF	$C_{1,b,1}$ (μF)	{10}: 62.470985, {15}: 81.918506, {18}: 61.428506, {25}: 71.433739, {28}: 65.162443, {34}: 52.220281
		$L_{1,b,1}$ (mH)	{10}: 7.570296, {15}: 8.357659, {18}: 7.354985, {25}: 7.908201, {28}: 7.624738, {34}: 6.388447
2	7th second-order series resonant band-pass PHPF	$C_{2,b,1}$ (μF)	{10}: 20.508791, {15}: 22.641862, {18}: 20.339204, {25}: 21.107875, {28}: 19.245189, {34}: 18.125272
		$L_{2,b,1}$ (mH)	{10}: 14.531717, {15}: 16.043121,{18}: 14.083358, {25}: 15.248183, {28}: 13.246284, {34}: 12.302221
3	Second-order damped high-pass PHPF	$C_{3,b,1}$ (μF)	{10}: 36.660621, {15}: 46.348039, {18}: 35.460067, {25}: 39.073766, {28}: 37.603269, {34}: 32.518778
		$L_{3,b,1}$ (mH)	{10}: 2.877705, {15}: 4.447708, {18}: 2.568081, {25}: 3.421388, {28}: 3.162229, {34}: 2.464093
4	AHPF type 1	$I_{1,b,1}$ (p.u.)	–
5	AHPF type 2	$I_{2,b,1}$ (p.u.)	–
6	AHPF type 3	$I_{3,b,1}$ (p.u.)	{10}: 0.047054, {34}: 0.044075
7	AHPF type 4	$I_{4,b,1}$ (p.u.)	{18}: 0.039794

the HHPFs, the average values of the THDV and THDI indices are calculated as 0.301934% and 0.515084%, respectively. The average values obtained for the THDV and THDI indices after considering the HHPFs demonstrate a considerable improvement compared with the conditions in which the HHPFs are ignored and equaled 18.147589% and 31.590981%, respectively. The original value calculated for the TLLH index after applying the HHPFs is 0.020812%. This value illustrates a notable improvement over conditions before applying the HHPFs, with an index equal to 5.858904%. The computed value for the TCHF index is equal to $5.676107M. As a consequence, and after considering the HHPFs, the THDV,

Table 7.19 Optimal values of different objective functions of the HHPF planning problem under the third case

No.	Objective functions of the HHPF planning problem	Before applying harmonic power filters	After applying PHPFs and/or AHPFs
1	Average THDV (%)	18.147589	0.301934
2	Average THDI (%)	31.590981	0.515084
3	TLLH (%)	5.858904	0.020812
4	TCHF (M$)	–	5.676107

THDI, and TLLH indices were recovered and located at acceptable levels with respect to the relevant standards.

Providing a techno-economic analysis on the performance of harmonic power filters in mitigating the harmonics in the modified 34-bus distribution test network, based on the results obtained under three cases, can greatly help readers to recognize the advantages of these devices. In the first case, the investment cost to install PHPFs in the modified 34-bus distribution test network is $7.925440M. For this investment cost, the DNO is able to reduce the average value of the THDV index from 18.147589 to 0.658647%, decrease the average value of the THDI index from 31.590981 to 0.838822%, and decrease the original value of the TLLH index from 5.858904 to 0.046909% (see Table 7.15). In the second case, the computed investment cost to install the AHPFs in the modified 34-bus distribution test network is $9.092039M. For this investment cost by the DNO, the average value of the THDV index is reduced from 18.147589 to 0.566564%, the average value of the THDI index is decreased from 31.590981 to 0.718039%, and the original value of the TLLH index is reduced from 5.858904 to 0.037013% (see Table 7.17). In the third case, the computed investment cost to install the HHPFs in the modified 34-bus distribution test network is $5.676107M. For this investment cost, the DNO is able to reduce the average value of the THDV index from 18.147589 to 0.301934%, decrease the average value of the THDI index from 31.590981 to 0.515084%, and decrease the original value of the TLLH index from 5.858904 to 0.020812% (see Table 7.19). As a result, the obtained average values for the THDV and THDI indices, and the original value obtained for the TLLH index in the second case, are more desirable than those in the first case. However, the TCHF index lacks this desirability. That is, the investment cost for installing the AHPFs in the second case is higher than the calculated investment cost for installing the PHPFs in the first case (Table 7.17 vs. Table 7.15). Moreover, the calculated average values for the THDV and THDI indices, the computed original value for the TLLH index, and the value obtained for the TCHF index in the third case are more desirable than those in the first and second cases. More precisely, in the third case, the DNO, with much lower investment costs than the noted investment costs in the first and second cases, is able to obtain more favorable values for the average values of the THDV and THDI indices and the original value of the TLLH index (Table 7.19 vs. Tables 7.15 and 7.17).

7.3.3.2.4 Performance Evaluation of the Proposed Optimization Algorithms: Simulation Results and Discussion

In this section, the performance of the proposed modern multi-objective music-inspired optimization algorithms, which was addressed in Chap. 4—namely, the multi-objective single-stage computational single-dimensional harmony search algorithm (SS-HSA), multi-objective single-stage computational single-dimensional improved harmony search algorithm (SS-IHSA), multi-objective continuous/discrete two-stage computational multi-dimensional melody search algorithm (TMS-MSA), multi-objective two-stage computational multi-dimensional enhanced melody search algorithm (TMS-EMSA), and the multi-objective SOSA—is compared with the NSGA-II for the first, second, and third cases. The parameter adjustments of the multi-objective TMS-EMSA, multi-objective continuous/discrete TMS-MSA, multi-objective SS-HSA, multi-objective SS-IHSA, and NSGA-II are tabulated in Tables 7.36, 7.37, 7.38, 7.39, and 7.40, respectively. For more information about the NSGA-II, please refer to the work by Deb et al. [38]. Table 7.20 shows the computed values for different objective functions of the HHPF planning problem, namely the THDV, THDI, TLLH, and

Table 7.20 The calculated values of the objective functions of the HHPF planning problem by the proposed multi-objective optimization algorithms under the first, second, and third cases

Case No.	Multi-objective optimization algorithm	Objective functions of the HHPF planning problem			
		Average THDV (%)	Average THDI (%)	TLLH (%)	TCHF (M $)
1	SOSA	0.658647	0.838822	0.046909	7.925440
	TMS-EMSA	0.709121	0.893204	0.051497	8.488031
	Continuous/discrete TMS-MSA	0.741446	0.931743	0.053006	8.873846
	SS-HSA	0.805802	1.022369	0.058463	9.790201
	SS-IHSA	0.776547	0.990082	0.056715	9.487055
	NSGA-II	0.830124	1.056556	0.060280	10.174024
2	SOSA	0.566564	0.718039	0.037013	9.092039
	TMS-EMSA	0.603136	0.762913	0.040581	9.757206
	Continuous/discrete TMS-MSA	0.642188	0.805055	0.043416	10.185109
	SS-HSA	0.710125	0.882522	0.047163	11.157047
	SS-IHSA	0.681096	0.850906	0.045483	10.749603
	NSGA-II	0.732629	0.914272	0.047963	11.468800
3	SOSA	0.301934	0.515084	0.020812	5.676107
	TMS-EMSA	0.330893	0.551899	0.022519	6.053870
	Continuous/discrete TMS-MSA	0.344578	0.583478	0.023664	6.421810
	SS-HSA	0.381392	0.636683	0.026002	7.040946
	SS-IHSA	0.368013	0.613994	0.024875	6.772870
	NSGA-II	0.390012	0.654455	0.027027	7.370896

TCHF, by the proposed multi-objective optimization algorithms under the first, second, and third cases. Bearing in mind the optimal results reported in Table 7.20, it clearly appears that the proposed techno-economic multi-objective framework under the first, second, and third cases by the SOSA gives rise to more efficient results than other multi-objective optimization algorithms.

The evaluation of the performance of these optimization algorithms relative to each other was performed using an index of cost saving (ICS).

A detailed description of the ICS is provided in Sect. 5.3.5.3 of Chap. 5. The ICS for the proposed multi-objective optimization algorithms under the first, second, and third cases is tabulated in Tables 7.21, 7.22, and 7.23, respectively.

To clarify, the ICS shows 7.117826, 11.167232, 18.261930, 15.182596, and 20.656793% superiority—positive sign—of the multi-objective SOSA performance compared with the performances of the multi-objective TMS-EMSA, multi-objective continuous/discrete TMS-MSA, multi-objective SS-HSA, multi-objective SS-IHSA, and NSGA-II, respectively, from the perspective of the average value of the THDV index by corresponding optimization algorithms. Also, the ICS represents 6.088418, 9.972814, 17.953107, 15.277522, and 20.607899% superiority—positive sign—of the multi-objective SOSA performance compared with the performances of the multi-objective TMS-EMSA, multi-objective continuous/discrete TMS-MSA, multi-objective SS-HSA, multi-objective SS-IHSA, and NSGA-II, respectively, from the standpoint of the average value of the THDI index by corresponding optimization algorithms. In addition, the ICS illustrates 8.909256, 11.502471, 19.762927, 17.289958, and 22.181486% superiority—positive sign—of the multi-objective SOSA performance compared with the performances of the multi-objective TMS-EMSA, multi-objective continuous/discrete TMS-MSA, multi-objective SS-HSA, multi-objective SS-IHSA, and NSGA-II, respectively, from the point of view of the original value of the TLLH index by corresponding optimization algorithms. Moreover, the ICS demonstrates 6.628050, 10.687654, 19.047218, 16.460482, and 22.101225% superiority—positive sign—of the multi-objective SOSA performance compared with the performances of the multi-objective TMS-EMSA, multi-objective continuous/discrete TMS-MSA, multi-objective SS-HSA, multi-objective SS-IHSA, and NSGA-II, respectively, from the perspective of the TCHF index by corresponding optimization algorithms. The ICS illustrates 7.663285% weakness—negative sign—of the multi-objective TMS-EMSA performance compared with the performance of the multi-objective SOSA, from the perspective of the average value of the THDV index by the multi-objective SOSA, and 4.359724, 11.998108, 8.682797, and 14.576497% superiority—positive sign—of the multi-objective TMS-EMSA performance compared with the performances of the multi-objective continuous/discrete TMS-MSA, multi-objective SS-HSA, multi-objective SS-IHSA, and NSGA-II, respectively, from the perspective of the average value of the THDV index by corresponding optimization algorithms. Also, the ICS shows 6.483139% weakness—negative sign—of the multi-objective TMS-EMSA performance compared with the performance of the multi-objective SOSA, from the standpoint of the average value of the THDI index by the multi-objective SOSA, and 4.136226, 12.633892, 9.784846, and 15.460799% superiority—positive sign—of

Table 7.21 The ICS for different objectives of the HHPF planning problem based on various multi-objective optimization algorithms under the first case

Case No.	Perspective	Multi-objective optimization algorithm	Multi-objective optimization algorithm					
			SOSA	TMS-EMSA	Continuous/discrete TMS-MSA	SS-HSA	SS-IHSA	NSGA-II
First case	Average THDV	SOSA	0	7.117826	11.167232	18.261930	15.182596	20.656793
		TMS-EMSA	−7.663285	0	4.359724	11.998108	8.682797	14.576497
		Continuous/discrete TMS-MSA	−12.571074	−4.558460	0	7.986577	4.520138	10.682500
		SS-HSA	−22.342013	−13.633921	−8.679795	0	−3.767318	2.929923
		SS-IHSA	−17.900332	−9.508391	−4.734127	3.630544	0	6.454096
		NSGA-II	−26.034735	−17.063801	−11.960142	−3.018359	−6.899389	0
	Average THDI	SOSA	0	6.088418	9.972814	17.953107	15.277522	20.607899
		TMS-EMSA	−6.483139	0	4.136226	12.633892	9.784846	15.460799
		Continuous/discrete TMS-MSA	−11.077559	−4.314691	0	8.864314	5.892340	11.813193
		SS-HSA	−21.881519	−14.460862	−9.726501	0	−3.261043	3.235701
		SS-IHSA	−18.032431	−10.846122	−6.261275	3.158057	0	6.291573
		NSGA-II	−25.957116	−18.288319	−13.395646	−3.343900	−6.713989	0
	TLLH	SOSA	0	8.909256	11.502471	19.762927	17.289958	22.181486
		TMS-EMSA	−9.780639	0	2.846847	11.915228	9.200387	14.570338
		Continuous/discrete TMS-MSA	−12.997506	−2.930267	0	9.334108	6.539716	12.067020
		SS-HSA	−24.630668	−13.527001	−10.295060	0	−3.082077	3.014266
		SS-IHSA	−20.904304	−10.132629	−6.997321	2.989925	0	5.914067
		NSGA-II	−28.504125	−17.055362	−13.722974	−3.107948	−6.285815	0

(continued)

Table 7.21 (continued)

Case No.	Perspective	Multi-objective optimization algorithm	Multi-objective optimization algorithm					
			SOSA	TMS-EMSA	Continuous/discrete TMS-MSA	SS-HSA	SS-IHSA	NSGA-II
	TCHF	SOSA	0	6.628050	10.687654	19.047218	16.460482	22.101225
		TMS-EMSA	−7.098545	0	4.347776	13.300748	10.530391	16.571545
		Continuous/discrete TMS-MSA	−11.966604	−4.545400	0	9.359920	6.463639	12.779387
		SS-HSA	−23.528801	−15.341249	−10.326469	0	−3.195364	3.772578
		SS-IHSA	−19.703827	−11.769796	−6.910295	3.096422	0	6.752185
		NSGA-II	−28.371724	−19.863181	−14.651798	−3.920481	−7.241119	0

Table 7.22 The ICS for different objectives of the HHPF planning problem based on various multi-objective optimization algorithms under the second case

Case No.	Perspective	Multi-objective optimization algorithm	Multi-objective optimization algorithm					
			SOSA	TMS-EMSA	Continuous/discrete TMS-MSA	SS-HSA	SS-IHSA	NSGA-II
Second case	Average THDV	SOSA	0	6.063640	11.775990	20.216299	16.815838	22.666997
		TMS-EMSA	−6.455051	0	6.081085	15.066220	11.446257	17.675112
		Continuous/discrete TMS-MSA	−13.347830	−6.474824	0	9.566907	5.712557	12.344720
		SS-HSA	−25.338885	−17.738785	−10.578989	0	−4.262101	3.071677
		SS-IHSA	−20.215192	−12.925774	−6.058661	4.087871	0	7.033983
		NSGA-II	−29.310899	−21.469950	−14.083259	−3.169019	−7.566187	0
	Average THDI	SOSA	0	5.881928	10.808702	18.637835	15.614768	21.463306
		TMS-EMSA	−6.249521	0	5.234673	13.553089	10.341095	16.555138
		Continuous/discrete TMS-MSA	−12.118562	−5.523827	0	8.777911	5.388491	11.945788
		SS-HSA	−22.907252	−15.677934	−9.622572	0	−3.715569	3.472708
		SS-IHSA	−18.504148	−11.533818	−5.695387	3.582460	0	6.930760
		NSGA-II	−27.329017	−19.839614	−13.566402	−3.597644	−7.446886	0
	TLLH	SOSA	0	8.792291	14.748019	21.521107	18.622342	22.830098
		TMS-EMSA	−9.639856	0	6.529850	13.955855	10.777653	15.391030
		Continuous/discrete TMS-MSA	−17.299327	−6.986027	0	7.944787	4.544555	9.480224
		SS-HSA	−27.422797	−16.219413	−8.630458	0	−3.693687	1.667952
		SS-IHSA	−22.883852	−12.079544	−4.760917	3.562114	0	5.170652
		NSGA-II	−29.584200	−18.190778	−10.473097	−1.696244	−5.452586	0

(continued)

Table 7.22 (continued)

Case No.	Perspective	Multi-objective optimization algorithm	Multi-objective optimization algorithm: SOSA	TMS-EMSA	Continuous/discrete TMS-MSA	SS-HSA	SS-IHSA	NSGA-II
	TCHF	SOSA	0	6.817187	10.732040	18.508553	15.419769	20.723711
		TMS-EMSA	−7.315927	0	4.201260	12.546698	9.231940	14.923915
		Continuous/discrete TMS-MSA	−12.022276	−4.385507	0	8.711426	5.251300	11.192897
		SS-HSA	−22.712265	−14.346740	−9.542735	0	−3.790316	2.718270
		SS-IHSA	−18.230938	−10.170913	−5.542346	3.651898	0	6.270900
		NSGA-II	−26.141122	−17.541845	−12.603605	−2.794225	−6.690451	0

Table 7.23 The ICS for different objectives of the HHPF planning problem based on various multi-objective optimization algorithms under the third case

Case No.	Perspective	Multi-objective optimization algorithm	Multi-objective optimization algorithm: SOSA	TMS-EMSA	Continuous/discrete TMS-MSA	SS-HSA	SS-IHSA	NSGA-II
Third case	Average THDV	SOSA	0	8.751771	12.375717	20.833682	17.955615	22.583407
		TMS-EMSA	−9.591168	0	3.971524	13.240707	10.086600	15.158251
		Continuous/discrete TMS-MSA	−14.123616	−4.135778	0	9.652535	6.367981	11.649385
		SS-HSA	−26.316347	−15.261428	−10.683792	0	−3.635469	2.210188
		SS-IHSA	−21.885246	−11.218127	−6.801072	3.507939	0	5.640595
		NSGA-II	−29.171276	−17.866500	−13.185403	−2.260141	−5.977778	0
	Average THDI	SOSA	0	6.670604	11.721778	19.098829	16.109277	21.295734
		TMS-EMSA	−7.147377	0	5.412200	13.316517	10.113291	15.670443
		Continuous/discrete TMS-MSA	−13.278223	−5.721880	0	8.356591	4.970081	10.845207
		SS-HSA	−23.607606	−15.362231	−9.118595	0	−3.695313	2.715541
		SS-IHSA	−19.202693	−11.251152	−5.230017	3.563625	0	6.182396
		NSGA-II	−27.057917	−18.582385	−12.164468	−2.791342	−6.589803	0
	TLLH	SOSA	0	7.580265	12.052062	19.960003	16.333668	22.995523
		TMS-EMSA	−8.201998	0	4.838573	13.395123	9.471356	16.679616
		Continuous/discrete TMS-MSA	−13.703633	−5.084595	0	8.991616	4.868341	12.443112
		SS-HSA	−24.937536	−15.466939	−9.879986	0	−4.530653	3.792503
		SS-IHSA	−19.522391	−10.462276	−5.117478	4.334281	0	7.962407
		NSGA-II	−29.862579	−20.018650	−14.211460	−3.942004	−8.651256	0

(continued)

Table 7.23 (continued)

Case No.	Perspective	Multi-objective optimization algorithm	Multi-objective optimization algorithm					
			SOSA	TMS-EMSA	Continuous/discrete TMS-MSA	SS-HSA	SS-IHSA	NSGA-II
	TCHF	SOSA	0	6.240024	11.612037	19.384312	16.193474	22.992984
		TMS-EMSA	−6.655318	0	5.729537	14.019082	10.615883	17.867922
		Continuous/discrete TMS-MSA	−13.137578	−6.077765	0	8.793363	5.183327	12.876127
		SS-HSA	−24.045336	−16.304876	−9.641144	0	−3.958085	4.476389
		SS-IHSA	−19.322451	−11.876700	−5.466683	3.807386	0	8.113341
		NSGA-II	−29.858299	−21.755108	−14.779104	−4.686160	−8.829728	0

the multi-objective TMS-EMSA performance compared with the performances of the multi-objective continuous/discrete TMS-MSA, multi-objective SS-HSA, multi-objective SS-IHSA, and NSGA-II, respectively, from the standpoint of the average value of the THDI index by corresponding optimization algorithms. In addition, the ICS represents 9.780639% weakness—negative sign—of the multi-objective TMS-EMSA performance compared with the performance of the multi-objective SOSA, from the point of view of the original value of the TLLH index by the multi-objective SOSA, and 2.846847, 11.915228, 9.200387, and 14.570338% superiority—positive sign—of the multi-objective TMS-EMSA performance compared with the performances of the multi-objective continuous/discrete TMS-MSA, multi-objective SS-HSA, multi-objective SS-IHSA, and NSGA-II, respectively, from the point of view of the original value of the TLLH index by corresponding optimization algorithms. Moreover, the ICS demonstrates 7.098545% weakness—negative sign—of the multi-objective TMS-EMSA performance compared with the performance of the multi-objective SOSA, from the perspective of the TCHF index by the multi-objective SOSA, and 4.347776, 13.300748, 10.530391, and 16.571545% superiority—positive sign—of the multi-objective TMS-EMSA performance compared with the performances of the multi-objective continuous/discrete TMS-MSA, multi-objective SS-HSA, multi-objective SS-IHSA, and NSGA-II, respectively, from the perspective of the TCHF index by corresponding optimization algorithms.

The results presented in Table 7.21 for the other multi-objective optimization algorithms are analyzed in the same way. The results presented in Tables 7.22 and 7.23 for all multi-objective optimization algorithms are examined in the same way.

7.4 Conclusions

In this chapter, the authors, with a new point of view, developed a techno-economic multi-objective framework for the HHPF planning problem in distribution networks with consideration of uncertainty in demand and harmonic currents injected by nonlinear loads. The proposed framework was broken down into two problems: (1) harmonic power flow and (2) HHPF planning. The harmonic power flow problem acted as the central core of the HHPF planning problem. In the newly proposed framework, not only was the PDHPF methodology addressed in order to solve the harmonic power flow problem, but also the efficient two-PEM was employed in order to handle uncertainty in demand and harmonic currents injected by nonlinear loads. To implement the proposed PDHPF methodology on the basis of the efficient two-PEM, the DDHPF methodology was developed. The DDHPF methodology for the distribution network evaluation at harmonic frequencies was needed for the acquisition of information determined from the power flow problem at the principal frequency. At this point, the LBNRPF methodology was modeled and applied with the aim of solving the power flow problem at the principal frequency. Maximization of the loadability of the distribution network was considered as the objective function in the LBNRPF methodology, subject to Kirchhoff's point and loop rules and operational restrictions on the distribution network equipment. In the HHPF planning problem, however, the THDV, THDI, TLLH, and TCHF were simultaneously optimized as four objective functions from the perspective of the DNO. At the same time, harmonic standard-based limitations and techno-economic and logical constraints were regarded as HHPF planning problem constraints.

To increase the flexibility and filtering capability of the proposed framework against the harmonics and principal reactive power compensation, the DNO was authorized to employ a parallel-connected combination of three types of PHPFs and four types of AHPFs. The fifth and seventh second-order series resonant band-pass PHPFs and the second-order damped high-pass PHPF were considered as three types of PHPFs. The maximum amount of the injected harmonic current into the PCC of the distribution network was also the main distinction of the four types of AHPFs.

The HHPF planning methodology was successfully implemented on the IEEE 18-bus distorted test network, based on three cases to demonstrate the feasibility and effectiveness of the proposed techno-economic multi-objective framework. The DNO was allowed to employ the PHPFs, AHPFs, and HHPFs under the first, second, and third cases, respectively. In addition, each case was run with respect to the confined, semi-confined, and full search patterns as three different scenarios in order to investigate the effects associated with the number of candidate buses on optimal outputs of the proposed framework.

The results from the case studies illustrate that the proposed techno-economic multi-objective framework under the second scenario (i.e., a semi-confined search pattern) in all three cases has a more desirable performance in comparison with that under the first scenario (i.e., a confined search pattern) in all three cases. This functionality was much better for the proposed framework on the basis of the third scenario than that according to the first and second scenarios. More precisely, it was inferred from these results that more flexibility of the proposed framework and more favorable outputs were achieved by increasing the number of candidate buses for the installation of harmonic power filters under the third scenario in all three cases. The implementation of the framework under all three scenarios of the second case (i.e., when only the AHPFs were used by the DNO) brought about a significant improvement in the THDV, THDI, and TLLH indices in comparison with the three scenarios of the first case (i.e., when only the PHPFs were employed by the DNO). However, this desirability was not observed in the TCHF index. Put another way, the investment cost for installing the AHPFs under all three scenarios of the second case was higher than the calculated investment cost for installing the PHPFs under all three scenarios of the first case. In addition, the accomplishment of the proposed framework under all three scenarios of the third case (i.e., when the DNO was authorized to use both PHPFs and AHPFs) resulted in the desired values for the THDV, THDI, TLLH, and TCHF indices in comparison with all three scenarios of the first and second cases (i.e., when the DNO was allowed to utilize only PHPFs or only AHPFs). In other words, the DNO, with much lower investment costs for installing HHPFs in the third case versus the performed investment costs for installing PHPFs and AHPFs in the first and second cases, respectively, was able to obtain more favorable values for the THDV, THDI, and TLLH indices.

To assess the filtering performance of allocated PHPFs in the first and third cases, the HA coefficient was defined according to the ratio of the current of the source of the distribution network to the current of nonlinear loads for harmonic order *h*. From the analysis of results associated with the HA coefficient, it was shown that the HHPFs in the third case have better performance in harmonic mitigation on the distribution network than the PHPFs in the first case.

Next, the HHPF planning methodology was also successfully applied on the modified 34-bus distribution test network according to three cases. The results from the case studies demonstrate that the proposed techno-economic multi-objective framework in the second case (i.e., when only the AHPFs were used by the DNO) gave rise to favorable values for the THDV, THDI, and TLLH indices in comparison with the first case (i.e., when only the PHPFs were used by the DNO). Nevertheless, the use of AHPFs imposed more investment costs on the DNO in comparison with the utilization of PHPFs. Moreover, the values related to the THDV, THDI, TLLH, and TCHF indices were significantly improved in the third case (i.e., when the DNO was allowed to utilize both PHPFs and AHPFs) compared to the values in the first and second cases. That is to say that the DNO was able to achieve better values for the THDV, THDI, and TLLH indices with low investment costs in the third case than for the other cases.

The authors accomplished an exhaustive analysis in such a way that the proposed multi-objective SOSA, multi-objective TMS-EMSA, multi-objective continuous/discrete TMS-MSA, multi-objective SS-HSA, and multi-objective SS-IHSA were compared with current state-of-the-art multi-objective optimization algorithm making up the NSGA-II. The proposed multi-objective SOSA, multi-objective TMS-EMSA, multi-objective continuous/discrete TMS-MSA, multi-objective SS-HSA, and multi-objective SS-IHSA were thoroughly discussed in Chap. 4. The results obtained from the implementation of the proposed multi-objective optimization algorithms, representing the framework in all three cases by the proposed multi-objective SOSA, indicate more efficient outputs than the other multi-objective optimization algorithms.

The efficiency of the proposed optimization algorithms relative to each other was also scrutinized by using the ICS. It was deduced from these results that the efficiency of the proposed multi-objective SOSA was more economical and effective when compared to other multi-objective optimization algorithms. In addition, the multi-objective TMS-EMSA, multi-objective continuous/discrete TMS-MSA, multi-objective SS-IHSA, multi-objective SS-HSA, and NSGA-II were also ranked as the next best multi-objective optimization algorithms. As a consequence, the newly developed multi-objective meta-heuristic music-inspired optimization algorithms discussed in Chap. 4, especially the multi-objective SOSA, may be worthy techniques for large-scale power system optimization problems with big data, in comparison to other multi-objective optimization algorithms, when feasible regions of the solution space and/or dimensions of the problems with complexities—such as mixed-integer decision-making variables, multiple conflicting objective functions, nonlinearity, and discontinuity—increase.

Appendix 1: List of Abbreviations and Acronyms

AC	Alternating current
AHPFs	Active harmonic power filters
ANSI	American National Standards Institute
C	Capacitance
DC	Direct current
DDHPF	Deterministic decoupled harmonic power flow
DNO	Distribution network operator
FSM	Fuzzy satisfying method
HA	Harmonic attenuation
HHPFs	Hybrid harmonic power filters
ICS	Index of cost saving
IEEE	Institute of Electrical and Electronic Engineers
L	Inductance
LBNRPF	Loadability-based Newton-Raphson power flow
MLLF	Motor load loss function
MMM-EMSA	Multi-stage computational multi-dimensional multiple-homogeneous enhanced melody search algorithm
MMS-EMSA	Multi-stage computational multi-dimensional single-inhomogeneous enhanced melody search algorithm
MVAr	Megavolt-ampere reactive power
NSGA-II	Non-dominated sorting genetic algorithm-II
PCC	Point of common coupling
PDHPF	Probabilistic decoupled harmonic power flow
PHPFs	Passive harmonic power filters
R	Resistance
SOSA	Symphony orchestra search algorithm
SS-HSA	Single-stage computational single-dimensional harmony search algorithm
SS-IHSA	Single-stage computational single-dimensional improved harmony search algorithm
THDV	Total harmonic distortion of voltage
THDI	Total harmonic distortion of current
TIF	Telephone influence factor
TLLH	Transmission line loss arising from harmonics
TMS-EMSA	Two-stage computational multi-dimensional single-homogeneous enhanced melody search algorithm
TMS-MSA	Two-stage computational multi-dimensional single-homogeneous melody search algorithm
Two-PEM	Two-point estimate method

Appendix 2: List of Mathematical Symbols

Index	
a	Index for AHPFs of type m running from 1 to A
$b, b', \hat{b}$	Index for buses of the distribution network running from 1 to B
h	Index for harmonic orders running from 2 to H
hc	Index for harmonic order counters running from 1 to HC
i	Index of concentrations for each uncertainty parameter running from 1 to I
k	Index for uncertainty parameters running from 1 to K
m	Index for types of AHPFs running from 1 to M
n	Index for types of PHPFs running from 1 to N
p	Index for PHPFs of type n running from 1 to P
u	Index for iterations of the LBNRPF methodology running from 1 to U
Set	
Ψ_m^{A}	Set of indices of AHPFs of type m
Ψ^{B}	Set of indices of buses of the distribution network
Ψ_b^{B}	Set of indices of buses of the distribution network for buses connected to bus b of the distribution network
Ψ^{H}	Set of indices of harmonic orders
Ψ^{I}	Set of indices of concentrations
Ψ^{K}	Set of indices of uncertainty parameters
Ψ^{M}	Set of indices of types of AHPFs
Ψ^{N}	Set of indices of types of PHPFs
Ψ_n^{P}	Set of indices of PHPFs of type n
Parameters	
A_b^h	Ratio of the current of a nonlinear load on bus b of the distribution network for harmonic order h to its current in the principal frequency [%]
D	Distribution network bus voltage vector at initial guess
f^1	Principal frequency [Hz]
f^h	Harmonic frequency at harmonic order h [Hz]
HOV	Harmonic order vector
$I_b^{1,\mathrm{NLL}}$	Current of a nonlinear load on bus b of the distribution network for the principal frequency [kA or p.u.]
$I_b^{h,\mathrm{NLL}}$	Current of a nonlinear load on bus b of the distribution network for harmonic order h [kA or p.u.]
$I_b^{h,\max}$	Maximum admissible value of the current on bus b of the distribution network for harmonic order h [kA or p.u. or %]
$\lvert I_l^1 \rvert$	Magnitude of the current at line l of the distribution network for the principal frequency [kA or p.u.]
$\mathrm{I}^{\max}$	Maximum admissible value of the injected current by each AHPF [kA or p.u.]
P_b	Active load—demand—on bus b of the distribution network [MW or p.u.]
P_b^1	Active load—demand—on bus b of the distribution network for the principal frequency [MW or p.u.]
Q_b	Reactive load—demand—on bus b of the distribution network [MVAr or p.u.]
Q_b^1	Reactive load—demand—on bus b of the distribution network for the principal frequency [MVAr or p.u.]

(continued)

Q^{max}	Maximum admissible value of the principal reactive power compensation by all PHPFs on all buses of the distribution network [MVAr or p.u.]
Q_b^{max}	Maximum admissible value of the principal reactive power compensation by all PHPFs on bus *b* of the distribution network [MVAr or p.u.]
$Q_{n,b}^{max}$	Maximum admissible value of the principal reactive power compensation by each PHPF of type *n* on bus *b* of the distribution network [MVAr or p.u.]
Q^{min}	Minimum admissible value of the principal reactive power compensation by all PHPFs on all buses of the distribution network [MVAr or p.u.]
Q_b^{min}	Minimum admissible value of the principal reactive power compensation by all PHPFs on bus *b* of the distribution network [MVAr or p.u.]
$Q_{n,b}^{min}$	Minimum admissible value of the principal reactive power compensation by each PHPF of type *n* on bus *b* of the distribution network [MVAr or p.u.]
$R_{b,\ b'}$	Resistance between buses *b* and b' of the distribution network [Ω or p.u.]
R_l	Resistance of line *l* of the distribution network [Ω or p.u.]
$TCHF^{max}$	Maximum allowable value of the TCHF index [M$]
$THDI_b^{max}$	Maximum admissible value of THDI on bus *b* of the distribution network [p.u. or %]
$THDV_b^{max}$	Maximum admissible value of THDV on bus *b* of the distribution network [p.u. or %]
$V_b^{h,\max}$	Maximum admissible value of voltage on bus *b* of the distribution network for harmonic order *h* [p.u. or %]
$X_{b,\ b'}$	Inductance between buses *b* and b' of the distribution network [Ω or p.u.]
y_b^{PA}	Parallel-connected admittance connected to bus *b* of the distribution network that is available in the initial configuration of the distribution network [℧ or p.u.]
$y_b^{1,PA}$	Parallel-connected admittance connected to bus *b* of the distribution network that is available in the initial configuration of the distribution network for the principal frequency [℧or p.u.]
$y_b^{h,PA}$	Parallel-connected admittance connected to bus *b* of the distribution network that is available in the initial configuration of the distribution network for harmonic order *h* [℧or p.u.]
$y_{b,b'}$	Existing admittance between buses *b* and b' of the distribution network [℧or p.u.]
$y_{b,b'}^h$	Existing admittance between buses *b* and b' of the distribution network for harmonic order *h* [℧or p.u.]
$y_{b,\hat{b}}$	Existing admittance between buses *b* and $\hat{b}$ of the distribution network [℧or p.u.].
$y_{b,\hat{b}}^h$	Existing admittance between buses *b* and $\hat{b}$ of the distribution network for harmonic order *h* [℧or p.u.]
y_b^h	Admittance of a linear load on bus *b* of the distribution network for harmonic order *h* [℧or p.u.]
$y_{b,b'}^h$	Existing admittance between buses *b* and b' of the distribution network for harmonic order *h* [℧or p.u.]
$Y_{b,\ b}$	Element (*b*,*b*)—main-diagonal element—of the admittance matrix
$Y_{b,b}^h$	Element (*b*,*b*)—main-diagonal element—of the admittance matrix for harmonic order *h*
$Y_{b,b'}$	Element (b,b')—off-diagonal element—of the admittance matrix
$Y_{b,b'}^h$	Element (b,b')—off-diagonal element—of the admittance matrix for harmonic order *h*
$Y_{b,\hat{b}}$	Element ($b,\hat{b}$)—off-diagonal element—of the admittance matrix

(continued)

Y_{bus}	Admittance matrix of the distribution network
Y_{bus}^{h}	Admittance matrix of the distribution network for harmonic order h
Z_{bus}	Impedance matrix of the distribution network
Z_{bus}^{h}	Impedance matrix of the distribution network for harmonic order h
$Z_{b,b'}^{h}$	Impedance between buses b and b' of the distribution network for harmonic order h [Ω or p.u.]
$\theta_{b,b}$	Phase angle of the element (b,b)—main-diagonal element—of the admittance matrix [rad]
$\theta_{b,b'}$	Phase angle of the element (b,b')—off-diagonal element—of the admittance matrix [rad]
$\theta_{b,\hat{b}}$	Phase angle of the element $(b,\hat{b})$—off-diagonal element—of the admittance matrix [rad]
ω_0	The rated angular frequency of the distribution network [rad/sec]
$\tau_{n,b,p}^{\text{PHPF}}$	Characteristic coefficient corresponding to the fixed cost of PHPF p of type n on bus b of the distribution network [M$]
$\tau_{m,b,a}^{\text{AHPF}}$	Characteristic coefficient corresponding to the fixed cost of AHPF a of type m on bus b of the distribution network [M$]
$\upsilon_{n,b,p}^{\text{PHPF}}$	Characteristic coefficient corresponding to the variable cost of PHPF p of type n on bus b of the distribution network [M$/p.u.]
$\upsilon_{m,b,a}^{\text{AHPF}}$	Characteristic coefficient corresponding to the variable cost of AHPF a of type m on bus b of the distribution network [M$/p.u.]
κ_{n}^{PHPF}	Certain portion that connects the cost of the inductance and resistance of the PHPFs of type n to the cost of capacitance of corresponding type of PHPFs [%]
Variables	
$C_{n,b,p}$	Capacitance of PHPF p of type n on bus b of the distribution network [μF or p.u.]
$D^{(u)}$	Distribution network bus voltage vector at iteration u of the LBNRPF methodology
E(Y)	First moment of Y
$E(Y^2)$	Second moment of Y
$F_{i,b}$	Reactive power in power transmission lines connected to bus b of the distribution network [MVAr or p.u.]
$F_{r,b}$	Active power in power transmission lines connected to bus b of the distribution network [MW or p.u.]
I^{h}	Current for harmonic order h [kA or p.u.]
$I^{h,\text{DN}}$	Current of the distribution network's source for harmonic order h [kA or p.u.]
I_{b}^{h}	Current on bus b of the distribution network for harmonic order h [kA or p.u.]
$\lvert I_{b}^{h}\rvert$	Magnitude of the current on bus b of the distribution network for harmonic order h [kA or p.u.]
I_{l}^{h}	Current at line l of the distribution network for harmonic order h [kA or p.u.]
$I_{m,b,a}$	RMS injected current by AHPF a of type m on bus b of the distribution network [kA or p.u.]
$I_{m,b,a}^{h}$	Injected current by AHPF a of type m on bus b of the distribution network for harmonic order h [kA or p.u.]
$I_{m,b,a}^{h,\text{r}}$	Real part of the injected current by AHPF a of type m on bus b of the distribution network for harmonic order h [kA or p.u.]
$I_{m,b,a}^{h,\text{i}}$	Imaginary part of the injected current by AHPF a of type m on bus b of the distribution network for harmonic order h [kA or p.u.]
$I_{n,b,p}^{h,\text{cap}}$	

(continued)

	Harmonic current passing through the capacitance relevant to PHPF p of type n on bus b of the distribution network for harmonic order h [kA or p.u.]
$\left\|I_l^h\right\|$	Magnitude of the current at line l of the distribution network for harmonic order h [kA or p.u.]
J	Jacobian matrix of the distribution network
$L_{n,\ b,\ p}$	Inductance of PHPF p of type n on bus b of distribution network [mH or p.u.]
$N_{m,b,a}^{\text{AHPFs}}$	Number of installed AHPFs on each candidate bus of the distribution network
$N_{n,b,p}^{\text{PHPFs}}$	Number of installed PHPFs on each candidate bus of the distribution network
$\text{OF}_1^{\text{HHPFs}}$	First objective function of the HHPF planning problem [p.u.]
$\text{OF}_2^{\text{HHPFs}}$	Second objective function of the HHPF planning problem [p.u.]
$\text{OF}_3^{\text{HHPFs}}$	Third objective function of the HHPF planning problem [p.u.]
$\text{OF}_4^{\text{HHPFs}}$	Fourth objective function of the HHPF planning problem [M$]
$Q_{n,\ b,\ p}$	Principal reactive power compensation by PHPF p of type n on bus b of the distribution network [MVAr or p.u.]
$R_{n,\ b,\ p}$	Resistance of PHPF p of type n on bus b of the distribution network [Ω or p.u.]
TCHF	Total cost of harmonic power filters [M$]
THDI	Total harmonic distortion of the current [p.u.]
$THDI_l$	Total harmonic distortion of the current at line l of the distribution network [p.u.]
THDV	Total harmonic distortion of the voltage [p.u.]
$THDV_b$	Total harmonic distortion of the voltage on bus b of the distribution network [p.u.]
TLLH	Total line loss arising from harmonics [p.u.]
$TLLH_h$	Total line loss arising from harmonic order h [p.u.]
V_b	Magnitude of the RMS voltage on bus b of the distribution network [kV or p.u.]
$V_{\hat{b}}$	Magnitude of the RMS voltage on bus $\hat{b}$ of the distribution network [kV or p.u.]
$V_{b'}$	Magnitude of the RMS voltage on bus b' of the distribution network [kV or p.u.]
V_b^h	RMS voltage on bus b of the distribution network for harmonic order h [kV or p.u.]
$V_{b'}^h$	RMS voltage on bus b' of the distribution network for harmonic order h [kV or p.u.]
V^h	Voltage at harmonic order h [kV or p.u.]
$V_{n,b,p}^{h,\text{cap}}$	Harmonic voltage drop on the capacitance associated with PHPF p of type n on bus b of the distribution network for harmonic order h [kV or p.u.]
$\|V_b\|$	Magnitude of the RMS voltage on bus b of the distribution network [kV or p.u.]
$\left\|V_b^1\right\|$	Magnitude of the RMS voltage on bus b of the distribution network for the principal frequency [kV or p.u.]
$\left\|V_b^h\right\|$	Magnitude of the voltage on bus b of the distribution network for harmonic order h [kV or p.u.]
X	Input uncertainty parameter vector
Y	Output uncertainty parameter vector, which is directly affected by input uncertainty parameters
$\tilde{\text{Y}}$	Output uncertainty parameter vector, which is indirectly affected by input uncertainty parameters
$Y_{n,b,p}^h$	Equivalent admittance associated with PHPF p of type n on bus b of the distribution network for harmonic order h [℧or p.u.]
$\alpha_{n,\ b,\ p}$	

(continued)

	Binary variable, which is 0 if there is no PHPF p of type n on bus b of the distribution network; 1 if there is a PHPF p of type n on bus b of the distribution network
$\beta_{m,\ b,\ a}$	Binary variable, which is 0 if there is no AHPF a of type m on bus b of the distribution network; 1 if there is an AHPF a of type m on bus b of the distribution network
$\varpi_{n,b,p}^{\mathrm{PHPF}}$	Nominal capacity of the capacitance of PHPF p of type n on bus b of the distribution network [μF or p.u.]
$\varpi_{m,b,a}^{\mathrm{AHPF}}$	Nominal capacity of AHPF a of type m on bus b of the distribution network [kA or p.u.]
$\delta_{k,\ 1}$	Location of the first concentration of probabilistic parameter k
$\delta_{k,\ 2}$	Location of the second concentration of probabilistic parameter k
$\lambda_{k,\ 3}$	Skewness of probabilistic parameter k
$\rho_{k,\ 1}$	Probability of the first concentration of probabilistic parameter k
$\rho_{k,\ 2}$	Probability of the second concentration of probabilistic parameter k
$x_{k,\ 1}$	First concentration point of probabilistic parameter k
$x_{k,\ 2}$	Second concentration point of probabilistic parameter k
$\mu_{\mathrm{X},\ k}$	Mean value of vector X_k
$\sigma_{\mathrm{X},\ k}$	Standard deviation of vector X_k
$\eta_{b,\ a}$	Correction coefficient related to AHPF a on bus b of the distribution network
ζ^h	Harmonic attenuation coefficient for harmonic order h
ϕ_b	Phase shift of the voltage on bus b of the distribution network with respect to the voltage phase angle of the slack bus [rad]
ϕ_b	Phase angle of the RMS voltage on bus b of the distribution network [rad]
$\phi_{\hat{b}}$	Phase angle of the RMS voltage on bus $\hat{b}$ of the distribution network [rad]
$\phi_{b'}$	Phase angle of the RMS voltage on bus b' of the distribution network [rad]
$\chi_{n,\ b,\ p}$	Tuned frequency of PHPF p of type n on bus b of the distribution network
$\gamma_{n,\ b,\ p}$	Quality coefficient of PHPF p of type n on bus b of the distribution network
ω	The angular frequency [rad/s]
$\Delta C_{n,\ b,\ p}$	Variation of the capacitance of PHPF p of type n on bus b of the distribution network due to perturbation problems [μF or p.u.]
$\Delta D^{(u)}$	Corrective vector at iteration u of the LBNRPF methodology
Δf^1	Variation of the principal frequency of the distribution network due to perturbation problems [Hz]
ΔI^h	Additional harmonic current coefficient vector arising from the installation of different types of AHPFs on different buses of the distribution network for harmonic order h
ΔI_m^h	Additional harmonic current coefficient vector arising from the installation of AHPFs of type m on different buses of the distribution network for harmonic order h
$\Delta L_{n,\ b,\ p}$	Variation of the inductance of PHPF p of type n on bus b of the distribution network due to perturbation problems [mH or p.u.]
$\Delta R_{n,\ b,\ p}$	Variation of the resistance of PHPF p of type n on bus b of the distribution network due to perturbation problems [Ω or p.u.]
ΔP_b	Active mismatch power on bus b of the distribution network [MW]

(continued)

ΔQ_b	Reactive mismatch power on bus b of the distribution network [MVAr]
$\Delta W^{(u)}(D^{(u)})$	Mismatch power vector at iteration u of the LBNRPF methodology
ΔW_b	Element b of the mismatch power vector or mismatch power on bus b of the distribution network
ΔY^h	Additional admittance matrix resulting from the installation of various types of PHPFs on different buses of the distribution network for harmonic order h
ΔY_n^h	Additional admittance matrix resulting from the installation of PHPFs of type n on different buses of the distribution network for harmonic order h
ΔY_1^h	Additional admittance matrix resulting from the installation of the fifth second-order series resonant band-pass PHPFs on different buses of the distribution network for harmonic order h
ΔY_2^h	Additional admittance matrix resulting from the installation of the seventh second-order series resonant band-pass PHPFs on different buses of the distribution network for harmonic order h
ΔY_3^h	Additional admittance matrix resulting from the installation of the second-order damped high-pass PHPFs on different buses of the distribution network for harmonic order h
ΔZ^h	Additional impedance matrix resulting from the installation of various types of PHPFs on different buses of the distribution network for harmonic order h
ΔZ_n^h	Additional impedance matrix resulting from the installation of PHPFs of type n on different buses of the distribution network for harmonic order h
ΔZ_1^h	Additional impedance matrix resulting from the installation of the fifth second-order series resonant band-pass PHPFs on different buses of the distribution network for harmonic order h
ΔZ_2^h	Additional impedance matrix resulting from the installation of the seventh second-order series resonant band-pass PHPFs on different buses of the distribution network for harmonic order h
ΔZ_3^h	Additional impedance matrix resulting from the installation of the second-order damped high-pass PHPFs on different buses of the distribution network for harmonic order h
ΔZ_{bus}	Variation of the impedance of the distribution network due to perturbation problems [Ω]
$\Upsilon\left(\vartheta_{n,b,p}^{\text{PHPF}}\right)$	Cost function of PHPF p of type n on bus b of the distribution network [M$]
$\Upsilon\left(\vartheta_{m,b,a}^{\text{AHPF}}\right)$	Cost function of AHPF a of type m on bus b of the distribution network [M$]

Appendix 3: Input Data

Table 7.24 Data for the transmission lines of the modified IEEE 18-bus distorted test network

Line No.	From bus	To bus	R (%)	X (%)	Line charge (%)	Length (mi)	Base impedance (Ω)	Harm only
1	1	2	0.431	1.204	0.0035	0.318	15.625	0
2	2	3	0.601	1.677	0.0049	0.443	15.625	0
3	3	4	0.316	0.882	0.0026	0.233	15.625	0
4	4	5	0.896	2.502	0.0073	0.661	15.625	0
5	5	6	0.295	0.824	0.0024	0.218	15.625	0
6	6	7	1.720	2.120	0.0046	0.455	15.625	0
7	7	8	4.070	3.053	0.0051	0.568	15.625	0
8	2	9	1.706	2.209	0.0043	0.451	15.625	0
9	1	20	2.910	3.768	0.0074	0.769	15.625	0
10	20	21	2.222	2.877	0.0056	0.587	15.625	0
11	21	22	4.803	6.218	0.0122	1.269	15.625	0
12	21	23	3.985	5.160	0.0101	1.053	15.625	0
13	23	24	2.910	3.768	0.0074	0.769	15.625	0
14	23	25	3.727	4.593	0.0100	0.985	15.625	0
15	25	26	3.208	2.720	0.0059	0.583	15.625	0
16	50	1	0.312	6.753	0.0000	0.000	0.000	0
17	50	51	0.050	0.344	0.0000	0.000	0.000	0
18	51	0	0.000	0.010	0.0000	0.000	0.000	1

Table 7.25 Bus data of the modified IEEE 18-bus distorted test network

Bus No.	Predicted active power of load (%)	Predicted reactive power of load (%)	Parallel-connected load (%)	Bus voltage (%)
1	0	0	0	0
2	2	1.2	−10.5	0
3	4	2.5	−6	0
4	15	9.3	−6	0
5	30	22.6	−18	0
6	8	5	0	0
7	2	1.2	−6	0
8	10	6.2	0	0
9	5	3.1	0	0
20	10	6.2	−6	0
21	3	1.9	−12	0
20	2	1.2	0	0
23	8	5	0	0
24	5	3.1	−15	0
25	10	6.2	−9	0
26	2	1.2	0	0
50	0	0	−12	0
51	0	0	0	105

Table 7.26 Predicted harmonic contents of the nonlinear loads in the modified IEEE 18-bus distorted test network

Harmonic No.	Harmonic order	Predicted current (p.u.)
1	5th	0.0551
2	7th	0.0450
3	11th	0.0388
4	13th	0.0380
5	17th	0.0251
6	19th	0.0210
7	23th	0.0100
8	25th	0.0051

Table 7.27 The characteristic coefficients associated with the cost of the PHPFs in the modified IEEE 18-bus distorted test network

No. of PHPFs	Types of PHPFs	Characteristic coefficients	
		$\tau_{n,b,p}^{\text{PHPF}}$ (M$)	$\upsilon_{n,b,p}^{\text{PHPF}}$ (M$)
1	5th second-order series resonant band-pass PHPF	0.05	1.1
2	7th second-order series resonant band-pass PHPF	0.05	1.2
3	Second-order damped high-pass PHPF	0.05	1.6

Table 7.28 The characteristic coefficients related to the cost of the AHPFs in the modified IEEE 18-bus distorted test network

No. of AHPFs	Types of AHPFs	Maximum current of each type of AHPF (%)
1	AHPF type 1	7
2	AHPF type 2	6
3	AHPF type 3	5
4	AHPF type 4	4

Table 7.29 The characteristic coefficients related to the cost of the AHPFs in the modified IEEE 18-bus distorted test network

No. of AHPFs	Types of AHPFs	Characteristic coefficients	
		$\tau_{m,b,a}^{\text{AHPF}}$ (M$)	$\upsilon_{m,b,a}^{\text{AHPF}}$ (M$)
1	AHPF type 1	0.09	7.5
2	AHPF type 2	0.09	7.1
3	AHPF type 3	0.09	6.7
4	AHPF type 4	0.09	6.3

Table 7.30 Maximum permissible values of the individual harmonic distortion of voltage and current along with the maximum admissible values of the THDV and THDI in accordance with the ANSI and IEEE 519-1992 standards

No.	Parameter	Value (%)
1	$V_b^{h,\max};\ \forall\{b \in \Psi^{\mathrm{B}}, h \in \Psi^{\mathrm{H}}\}$	3
2	$I_b^{h,\max};\ \forall\{b \in \Psi^{\mathrm{B}}, h \in \{5,7\}\}$	4
3	$I_b^{h,\max};\ \forall\{b \in \Psi^{\mathrm{B}}, h \in \{11,13\}\}$	2
4	$I_b^{h,\max};\ \forall\{b \in \Psi^{\mathrm{B}}, h \in \{17,19\}\}$	1.5
5	$I_b^{h,\max};\ \forall\{b \in \Psi^{\mathrm{B}}, h \in \{23,25\}\}$	0.6
6	$THDV_b^{\max};\ \forall b \in \Psi^{\mathrm{B}}$	5
7	$THDI_b^{\max};\ \forall b \in \Psi^{\mathrm{B}}$	6

Table 7.31 Other required information in the modified IEEE 18-bus distorted test network

No.	Parameter	Value
5	$Q_{n,b}^{\min};\ n = 1$ (5th second-order series resonant band-pass PHPF), $\forall b \in \Psi^{\mathrm{B}}$	0.1 (MVAr)
6	$Q_{n,b}^{\max};\ n = 1$ (5th second-order series resonant band-pass PHPF), $\forall b \in \Psi^{\mathrm{B}}$	0.4 (MVAr)
7	$Q_{n,b}^{\min};\ n = 2$ (7th second-order series resonant band-pass PHPF), $\forall b \in \Psi^{\mathrm{B}}$	0.04 (MVAr)
8	$Q_{n,b}^{\max};\ n = 2$ (7th second-order series resonant band-pass PHPF), $\forall b \in \Psi^{\mathrm{B}}$	0.175 (MVAr)
9	$Q_{n,b}^{\min};\ n = 3$ (second-order damped high-pass PHPF), $\forall b \in \Psi^{\mathrm{B}}$	0.06 (MVAr)
10	$Q_{n,b}^{\min};\ n = 3$ (second-order damped high-pass PHPF), $\forall b \in \Psi^{\mathrm{B}}$	0.250 (MVAr)
11	$Q_b^{\min};\ \forall b \in \Psi^{\mathrm{B}}$	0.2 (MVAr)
12	$Q_b^{\max};\ \forall b \in \Psi^{\mathrm{B}}$	0.825 (MVAr)
13	$Q^{\min}$	1 (MVAr)
14	$Q^{\max}$	4 (MVAr)
15	$\kappa_n^{\mathrm{PHPF}};\ \forall n \in \Psi^{\mathrm{N}}$	7 (%)
16	$\mathrm{TCHF}^{\max}$	6 (M$)

Table 7.32 Parameter adjustments of the multi-objective SOSA

No.	Multi-objective SOSA parameters	Value
1	$BW_{g,p}^{\min}$	$0.4; \forall\{g \in [1,3], p \in [1]\}$
2	$BW_{g,p}^{\min}$	$0.35; \forall\{g \in [1,2,3], p \in [2]\}$
3	$BW_{g,p}^{\min}$	$0.30; \forall\{g \in [2], p \in [1]\}$
4	$BW_{g,p}^{\min}$	$0.25; \forall\{g \in [2], p \in [3]\}$
5	$BW_{g,p}^{\max}$	$0.95; \forall\{g \in [1,3], p \in [1]\}$
6	$BW_{g,p}^{\max}$	$0.9; \forall\{g \in [1,2,3], p \in [2]\}$
7	$BW_{g,p}^{\max}$	$0.85; \forall\{g \in [2], p \in [1]\}$
8	$BW_{g,p}^{\max}$	$0.8; \forall\{g \in [2], p \in [3]\}$
9	MNI-E	100
10	MNI-SISS	200
11	MNI-GISSHMG	200
12	MNI-GISSIME	300
13	NHMG	3
14	$PAR_{g,p}^{\min}$	$0.2; \forall\{g \in [1,3], p \in [1]\}$
15	$PAR_{g,p}^{\min}$	$0.3; \forall\{g \in [1,2,3], p \in [2]\}$
16	$PAR_{g,p}^{\min}$	$0.25; \forall\{g \in [2], p \in [1]\}$
17	$PAR_{g,p}^{\min}$	$0.35; \forall\{g \in [2], p \in [3]\}$
18	$PAR_{g,p}^{\max}$	$2; \forall\{g \in [1,3], p \in [1]\}$
19	$PAR_{g,p}^{\max}$	$1.8; \forall\{g \in [1,2,3], p \in [2]\}$
20	$PAR_{g,p}^{\max}$	$1.6; \forall\{g \in [2], p \in [1]\}$
21	$PAR_{g,p}^{\max}$	$1.75; \forall\{g \in [2], p \in [3]\}$
22	$PMCR_{g,\ p}$	$0.85; \forall\{g \in [1,3], p \in [1]\}$
23	$PMCR_{g,\ p}$	$0.8; \forall\{g \in [1,2,3], p \in [2]\}$
24	$PMCR_{g,\ p}$	$0.9; \forall\{g \in [2], p \in [1]\}$
25	$PMCR_{g,\ p}$	$0.95; \forall\{g \in [2], p \in [3]\}$
26	$PMS_{g,\ p}$	$220; \forall\{g \in [1,3], p \in [1]\}$
27	$PMS_{g,\ p}$	$160; \forall\{g \in [1,2,3], p \in [2]\}$
28	$PMS_{g,\ p}$	$180; \forall\{g \in [2], p \in [1]\}$
29	$PMS_{g,\ p}$	$200; \forall\{g \in [2], p \in [3]\}$
30	PN_g	$2; \forall\{g \in [1,3]\}$
31	PN_g	$3; \forall\{g \in [2]\}$

Table 7.33 Data for the transmission lines of the modified 34-bus distribution test network

Line No.	From bus	To bus	Length (km)	R (p.u.)	X (p.u.)	Tap setting
1	1	2	0.60	0.0967	0.0397	1
2	2	3	0.55	0.0886	0.0364	1
3	3	4	0.55	0.1359	0.0377	1
4	4	5	0.50	0.1236	0.0343	1
5	5	6	0.50	0.1236	0.0343	1
6	6	7	0.60	0.2598	0.0446	1
7	7	8	0.40	0.1732	0.0298	1
8	8	9	0.60	0.2598	0.0446	1
9	9	10	0.40	0.1732	0.0298	1
10	10	11	0.25	0.1083	0.0186	1
11	11	12	0.20	0.0866	0.1488	1
12	3	13	0.30	0.1299	0.0223	1
13	13	14	0.40	0.1732	0.0298	1
14	14	15	0.20	0.0866	0.1488	1
15	15	16	0.10	0.0433	0.0074	1
16	6	17	0.60	0.1483	0.0412	1
17	17	18	0.55	0.1359	0.0377	1
18	18	19	0.55	0.1718	0.0391	1
19	19	20	0.50	0.1562	0.0355	1
20	20	21	0.50	0.1562	0.0355	1
21	21	22	0.50	0.2165	0.0372	1
22	22	23	0.50	0.2165	0.0372	1
23	23	24	0.60	0.2598	0.0446	1
24	24	25	0.40	0.1732	0.0298	1
25	25	26	0.25	0.1083	0.0186	1
26	26	27	0.20	0.0866	0.1488	1
27	7	28	0.30	0.1299	0.0223	1
28	28	29	0.30	0.1299	0.0223	1
29	29	30	0.30	0.1299	0.0223	1
30	10	31	0.30	0.1299	0.0223	1
31	31	32	0.40	0.1732	0.0298	1
32	32	33	0.30	0.1299	0.0223	1
33	33	34	0.20	0.0866	0.1488	1

Table 7.34 Bus data for the modified 34-bus distribution test network

Bus No.	Predicted active power of load (MW)	Predicted reactive power of load (MVAr)	Injected MVAr	Bus voltage (%)
1	0	0	0	105
2	0.2300	0.1425	0	0
3	0	0	0	0
4	0.2300	0.1425	0	0
5	0.2300	0.1425	0	0
6	0	0	0	0

(continued)

Table 7.34 (continued)

Bus No.	Predicted active power of load (MW)	Predicted reactive power of load (MVAr)	Injected MVAr	Bus voltage (%)
7	0	0	0	0
8	0.2300	0.1425	0	0
9	0.2300	0.1425	0	0
10	0	0	0	0
11	0.2300	0.1425	0	0
12	0.1370	0.0840	0	0
13	0.0720	0.0450	0	0
14	0.0720	0.0450	0	0
15	0.0720	0.0450	0	0
16	0.0135	0.0075	0	0
17	0.2300	0.1425	0	0
18	0.2300	0.1425	0	0
19	0.2300	0.1425	0	0
20	0.2300	0.1425	0	0
21	0.2300	0.1425	0	0
22	0.2300	0.1425	0	0
23	0.2300	0.1425	0	0
24	0.2300	0.1425	0	0
25	0.2300	0.1425	0	0
26	0.2300	0.1425	0	0
27	0.1370	0.0850	0	0
28	0.0750	0.0480	0	0
29	0.0750	0.0480	0	0
30	0.0750	0.0480	0	0
31	0.0570	0.0345	0	0
32	0.0570	0.0345	0	0
33	0.0570	0.0345	0	0
34	0.0570	0.0345	0	0

Table 7.35 Other required information in the modified 34-bus distribution test network

No.	Parameter	Value
5	$Q_{n,b}^{\min}$; $n = 1$ (5th second-order series resonant band-pass PHPF), $\forall b \in \Psi^{\mathrm{B}}$	0.15 (MVAr)
6	$Q_{n,b}^{\max}$; $n = 1$ (5th second-order series resonant band-pass PHPF), $\forall b \in \Psi^{\mathrm{B}}$	0.6 (MVAr)
7	$Q_{n,b}^{\min}$; $n = 2$ (7th second-order series resonant band-pass PHPF), $\forall b \in \Psi^{\mathrm{B}}$	0.06 (MVAr)
8	$Q_{n,b}^{\max}$; $n = 2$ (7th second-order series resonant band-pass PHPF), $\forall b \in \Psi^{\mathrm{B}}$	0.27 (MVAr)
9	$Q_{n,b}^{\min}$; $n = 3$ (second-order damped high-pass PHPF), $\forall b \in \Psi^{\mathrm{B}}$	0.09 (MVAr)
10	$Q_{n,b}^{\min}$; $n = 3$ (second-order damped high-pass PHPF), $\forall b \in \Psi^{\mathrm{B}}$	0.38 (MVAr)
11	$Q_{b}^{\min}$; $\forall b \in \Psi^{\mathrm{B}}$	0.3 (MVAr)

(continued)

Table 7.35 (continued)

No.	Parameter	Value
12	$Q_b^{\max};\ \forall b \in \Psi^{\mathrm{B}}$	1.25 (MVAr)
13	$Q^{\min}$	2 (MVAr)
14	$Q^{\max}$	6 (MVAr)
15	$\kappa_n^{\mathrm{PHPF}};\ \forall n \in \Psi^{\mathrm{N}}$	7 (%)
16	$\mathrm{TCHF}^{\max}$	10 (M$)

Table 7.36 Parameter adjustments of the multi-objective TMS-EMSA

No.	Multi-objective TMS-EMSA parameters	Value
1	$BW_p^{\min}$	$0.4; \forall\{p \in [1,2]\}$
2	$BW_p^{\min}$	$0.35; \forall\{p \in [3,4]\}$
3	$BW_p^{\min}$	$0.30; \forall\{p \in [5,6,7]\}$
4	$BW_p^{\max}$	$0.9; \forall\{p \in [1,2]\}$
5	$BW_p^{\max}$	$0.85; \forall\{p \in [3,4]\}$
6	$BW_p^{\max}$	$0.8; \forall\{p \in [5,6,7]\}$
7	MNI-E	100
8	MNI-SISS	300
9	MNI-GISS	400
10	$PAR_p^{\min}$	$0.2; \forall\{p \in [1,2]\}$
11	$PAR_p^{\min}$	$0.35; \forall\{p \in [3,4]\}$
12	$PAR_p^{\min}$	$0.30; \forall\{p \in [5,6,7]\}$
13	$PAR_p^{\max}$	$2; \forall\{p \in [1,2]\}$
14	$PAR_p^{\max}$	$1.75; \forall\{p \in [3,4]\}$
15	$PAR_p^{\max}$	$1.9; \forall\{p \in [5,6,7]\}$
16	$PMCR_p$	$0.85; \forall\{p \in [1,2]\}$
17	$PMCR_p$	$0.8; \forall\{p \in [3,4]\}$
18	$PMCR_p$	$075; \forall\{p \in [5,6,7]\}$
19	PMS_p	$220; \forall\{p \in [1,2]\}$
20	PMS_p	$180; \forall\{p \in [3,4]\}$
21	PMS_p	$160; \forall\{p \in [5,6,7]\}$
22	PN	7

Table 7.37 Parameter adjustments of the multi-objective continuous/discrete TMS-MSA

No.	Multi-objective continuous/discrete TMS-MSA parameters	Value
1	$\mathrm{BW}^{\min}$	0.4
2	$\mathrm{BW}^{\max}$	0.9
3	MNI-E	100
4	MNI-SISS	300
5	MNI-PGISS	400

(continued)

Table 7.37 (continued)

No.	Multi-objective continuous/discrete TMS-MSA parameters	Value
6	PAR^{min}	0.2
7	PAR^{max}	2
8	PMCR	0.85
9	PMS	200
10	PN	7

Table 7.38 Parameter adjustments of the multi-objective SS-HSA

No.	Multi-objective SS-HSA parameters	Value
1	BW	0.75
2	HMCR	0.85
3	HMS	300
4	MNI-E	100
5	MNI-I	700
6	PAR	0.9

Table 7.39 Parameter adjustments of the multi-objective SS-IHSA

No.	Multi-objective SS-IHSA parameters	Value
1	BW^{min}	0.4
2	BW^{max}	0.9
3	HMCR	0.85
4	HMS	300
5	MNI-E	100
6	MNI-I	700
7	PAR^{min}	0.2
8	PAR^{max}	2

Table 7.40 Parameter adjustments of the NSGA-II

No.	NSGA-II parameters	Value
1	Crossover probability	0.95
2	Mutation probability	0.01
3	Population	400
4	MNI-E	100
5	MNI-I	700

References

1. P. Caramia, G. Carpinelli, P. Verde, *Power Quality Indices in Liberalized Markets* (Wiley, Oxford, 2009)
2. IEEE 1159 (1995) recommended practice for monitoring electric power quality, November
3. T.D.C. Busarello, J.A. Pomilio, M.G. Simões, Passive filter aided by shunt compensators based on the conservative power theory. IEEE Trans. Ind. Appl. **52**(4), 3340–3347 (2016)
4. L. Morán, D. Dixon, M. Torres, Active power filters, in *Power Electronics Handbook*, 4th edn., (2018), pp. 1341–1379
5. S. Ostroznik, P. Bajec, P. Zajec, A study of a hybrid filter. IEEE Trans. Ind. Electron. **57**(3), 935–942 (2010)
6. Y. Baghzouz et al., Time-varying harmonics: Part I–Characterizing measured data. IEEE Trans. Power Deliv. **13**(3), 938–944 (1998)
7. Y. Baghzouz et al., Time-varying harmonics: Part II–Harmonic summation and propagation. IEEE Trans. Power Deliv. **17**(1), 279–285 (2002)
8. E.F. Fuchs, M.A.S. Masoum, *Power Quality in Power Systems and Electrical Machines* (Elsevier Academic Press, Burlington, 2008)
9. J. Arrillaga, B.C. Smith, *Power System Harmonic Analysis* (John Wiley and Sons, New York, 1997)
10. J.C. Das, Passive filters—potentialities and limitations. IEEE Trans. Ind. Appl. **40**(1), 232–241 (2004)
11. L.B.G. Campanhol, S.A.O. da Silva, A. Goedtel, Application of shunt active power filter for harmonic reduction and reactive power compensation in three-phase four-wire systems. IET Power Electron. **7**(11), 2825–2836 (2014)
12. D. Xia, G.T. Heydt, Harmonic power flow studies Part I—Formulation and solution. IEEE Trans. Power App. Syst. **101**(6), 1257–1265 (1982)
13. D. Xia, G.T. Heydt, Harmonic power flow studies—Part II implementation and practical application. IEEE Trans. Power App. Syst. **101**(6), 1266–1270 (1982)
14. H.C. Chin, Optimal shunt capacitor allocation by fuzzy dynamic programming. Electr. Power Syst. Res. **35**(2), 133–139 (1995)
15. J.H. Teng, C.Y. Chang, A fast harmonic load flow method for industrial distribution systems, in *International Conference on Power System Technology*, vol. 3 (2000), pp. 1149–1154
16. I.M. Elamin, Fast decoupled harmonic load flow method, in *Conference Record of the 1990 IEEE Industry Applications Society Annual Meeting* (1990), pp. 1749–1756
17. Y.Y. Hong, J.S. Lin, C.H. Liu, Fuzzy harmonic power flow analyses, in *International Conference on Power System Technology*, vol. 1 (2000), pp. 121–125
18. T. Esposito, G. Carpinelli, P. Varilone, P. Verde, Probabilistic harmonic power flow for percentile evaluation, in *Canadian Conference on Electrical and Computer Engineering*, vol. 2 (2001), pp. 831–838
19. C.N. Bathurst, B.C. Smith, N.R. Watson, J. Arrillaga, A modular approach to the solution of the three-phase harmonic power-flow. IEEE Tran. Power Deliv. **15**(3), 984–989 (2000)
20. J.H. Teng, C.Y. Teng, Backward/forward sweep-based harmonic analysis method for distribution systems. IEEE Trans. Power Deliv. **22**(3), 1665–1672 (2007)
21. M. Kiani-Moghaddam, M. Shivaie, A. Salemnia, M.T. Ameli, Probabilistic multi-objective framework for multiple active power filters planning. Electr. Power Compon. Syst **45**(18), 2062–2077 (2017)
22. I. Ziari, A. Jalilian, A new approach for allocation and sizing of multiple active power line conditioners. IEEE Trans. Power Deliv. **25**(2), 1026–1035 (2010)
23. H.H. Zeineldin, A.F. Zobaa, Particle swarm optimization of passive filters for industrial plants in distribution networks. Electr. Power Compon. Syst. **39**(16), 1795–1808 (2011)
24. I. Ziari, A. Jalilian, Optimal allocation and sizing of active power line conditioners using a new particle swarm optimization-based approach. Electr. Power Compon. Syst. **40**(12), 273–291 (2012)

25. Y.Y. Hong, W.J. Liao, Optimal passive filter planning considering probabilistic parameters using cumulant and adaptive dynamic clone selection algorithm. Int. J. Electr. Power Energy Syst. **45**(1), 159–166 (2013)
26. M. Farhoodnea, A. Mohamed, H. Shareef, H. Zayandehroodi, Optimal placement of active power conditioner in distribution systems using improved discrete firefly algorithm for power quality enhancement. Appl. Soft Comput. **23**, 249–258 (2014)
27. A.F. Zobaa, Optimal multiobjective design of hybrid active power filters considering a distorted environment. IEEE Trans. Ind. Electron. **61**(1), 107.114 (2014)
28. A.F. Zobaa, S.H.E. Abdel Aleem, A new approach for harmonic distortion minimization in power systems supplying nonlinear loads. IEEE Trans. Ind. Inf. **10**(2), 1401–1412 (2014)
29. N.C. Yang, M.D. Le, Optimal design of passive power filters based on multi-objective bat algorithm and Pareto front. Appl. Soft Comput. **35**, 257–266 (2015)
30. M. Mohammadi, Bacterial foraging optimization and adaptive version for economically optimum sitting, sizing and harmonic tuning orders setting of LC harmonic passive power filters in radial distribution systems with linear and nonlinear loads. Appl. Soft Comput. **29**, 345–356 (2015)
31. G. Carpinelli, D. Proto, A. Russo, Optimal planning of active power filters in a distribution system using trade-off/risk method. IEEE Trans. Power Deliv. **32**(2), 841–851 (2017)
32. S. Frank, I. Steponavice, S. Rebennack, Optimal power flow: a bibliographic survey I. Energy Syst. **3**(3), 221–258 (2012)
33. S. Frank, I. Steponavice, S. Rebennack, Optimal power flow: a bibliographic survey II. Energy Syst. **3**(3), 259–289 (2012)
34. E. Rosenblueth, Point estimation for probability moments. Proc. Natl. Acad. Sci. U S A **72**(10), 3812–3814 (1975)
35. E. Rosenblueth, Two-point estimates in probability. Appl. Math. Model. **5**, 329–335 (1981)
36. H.P. Hong, An efficient point estimate method for probabilistic analysis. Reliab. Eng. Syst. Saf. **59**, 261–267 (1998)
37. N. He, D. Xu, L. Huang, The application of particle swarm optimization to passive and hybrid active power filter design. IEEE Trans. Ind. Electron. **56**(8), 2841–2851 (2009)
38. K. Deb, A. Pratap, S. Agarwal, T. Meyarivan, A fast and elitist multiobjective genetic algorithm: NSGA-II. IEEE Trans. Evol. Comput. **6**(2), 182–197 (2002)

Index

M. Kiani-Moghaddam et al., *Modern Music-Inspired Optimization Algorithms for Electric Power Systems*, Power Systems,
https://doi.org/10.1007/978-3-030-12044-3

R

S

T

Printed by Printforce, the Netherlands